# *Drosophila* Protocols

http://www.cshprotocols.org

# *Drosophila*
# Protocols

Edited by

**William Sullivan**
*University of California, Santa Cruz*

**Michael Ashburner**
*University of Cambridge*

**R. Scott Hawley**
*University of California, Davis*

**CSH PRESS**

**COLD SPRING HARBOR LABORATORY PRESS**
**Cold Spring Harbor, New York • www.cshlpress.org**

# *Drosophila* Protocols

*Developmental Editor:* Marilu Hoeppner
*Project Coordinator:* Mary Cozza
*Production Editor:* Dorothy Brown
*Desktop Editor:* Susan Schaefer
*Book Designer:* Denise Weiss
*Cover design:* Tony Urgo

*Front cover (printed hardcover):* In situ hybridization to the histone gene cluster provides a means of identifying compound chromosome-2-bearing sperm during the process of individualization in the testes of *C(2)EN/O* males. (Courtesy of Abby Dernburg from Dernburg et al., *Genetics 143:* 1629 [1996].)

*Back cover (printed hardcover):* Individualized sperm bearing a compound chromosome 2 are identified through in situ hybridization to the histone gene cluster in the testes of *C(2)EN/O* males. (Courtesy of Abby Dernburg from Dernburg et al., *Genetics 143:* 1629 [1996].)

Library of Congress Cataloging-in-Publication Data

Drosophila protocols / [edited by] William Sullivan, Michael Ashburner, R. Scott Hawley.
    p. cm.
    Includes bibliographical references.
    ISBN 978-0-87969-586-6 (cloth) -- ISBN 978-0-87969-827-0 (pbk.)
    I. Drosophila--Laboratory manuals. I. Sullivan, William, 1957- II. Ashburner, M. III.
Hawley, R. Scott.

QL537.D76 D765 2000
595.77'4--dc21

                                    00-020100

# Contents

**v**

## MOLECULAR BIOLOGY

## APPENDICES

# Preface

LARRY SANDLER WOULD OFTEN TELL HIS STUDENTS "I can make a living pushing flies, but you can't." He was referring to the fact that during his days as a graduate student and post-doc, formal genetic analysis yielded many of the most exciting findings in *Drosophila*. However, since that time, *Drosophila* has become an organism of choice for molecular, biochemical, and cellular studies, and many key insights rely on a combination of approaches. For students just beginning to work with *Drosophila*, mastering techniques and approaches to be used in combination with "fly pushing," Larry correctly concluded, would be an essential element of their training.

With new techniques being generated at an ever increasing rate, Larry's advice is even more appropriate. Although the variety of approaches currently available to the fly community makes *Drosophila* research particularly exciting, it is also daunting. It requires all of us to step outside the comfort zone of tried-and-true procedures we learned as graduate students and post-docs. Establishing a new technique in the lab is often a leap of faith and requires a large commitment of time and resources. Therefore, reducing this activation barrier was foremost in our minds when we developed a strategy for revising the Laboratory Manual that accompanied Michael Ashburner's Drosophila: *A Laboratory Handbook* (1989), known to the community as the Grey Book. Unlike the original edition, this book makes no attempt to be comprehensive. We felt liberated from this responsibility because there exists a number of other excellent *Drosophila* protocol manuals, including Ashburner's original more comprehensive work (Drosophila: *A Laboratory Manual* 1989). Instead, we chose to provide in-depth descriptions of a select set of protocols that are most likely to be used by the *Drosophila* community in the next decade. Each protocol includes enough basic information to be useful to the uninitiated, in sufficient detail to serve as a useful reference for the more experienced.

We arrived at a "Top 37" list by soliciting the advice of the *Drosophila* community and surveying the literature for the most frequently used protocols. Some of the protocols, such as RNA interference, are currently used by only a few laboratories, but are likely soon to become more routine. Some widely used protocols, such as transformation, are not included because there already exist a number of excellent published descriptions of this technique.

The protocols are grouped into six sections: Chromosomes, Cell Biology, Molecular Biology, Genomics, Biochemistry, and The Organism. Several of the protocols assume a knowledge of basic molecular techniques and may require users to refer to one of the many excellent sources of such methods. Very little genetics is included in this volume, because much of this material is covered in the Grey Book. New advances in *Drosophila* genetic techniques will be included in a planned second edition of the Grey Book. In addition to an introduction describing the purpose of each procedure, the chapters include many figures, tables, illustrations, and examples of the kinds of data that can be produced. We believe that these features make for a more useful and interesting reference. It is our hope that it will be read between time-points for ideas and new approaches.

The manual also includes six appendices. Reprinted in Appendix 1 is the Table of Contents of the original edition of the protocol manual. This has been included because the revision required the sacrifice of many useful protocols from that book. (Although it was published more than a decade ago, most labs still have a copy and the taped covers attest to its continued value as a reference.) Appendix 2 consists of line drawings of many key aspects of *Drosophila* biology and development. These have been plucked from a number of sources and slightly modified to serve as templates for figures and transparencies. Appendix 3 includes a number of commonly used solutions, buffers, and recipes used by the *Drosophila* community. Appendix 4 lists the toxic and dangerous reagents used in these protocols. Also included are Suppliers and Trademarks in Appendices 5 and 6, respectively.

This book is in print only because of the enthusiastic support of both the *Drosophila* community and the good folk at Cold Spring Harbor Laboratory. To all of the contributing authors, we owe a hearty thanks. Without exception the authors put a great deal of time, effort, and thought into the project. The majority of the chapters required only minor changes, and the authors diligently responded to any of our suggestions. This made our job relatively painless and sometimes even pleasant. (And to those authors who were late returning manuscripts and wondering if they were the very last, rest easy, the person concerned atoned with an excellent chapter and is forgiven.)

The publishing group at Cold Spring Harbor was a pleasure to work with. Throughout the entire project, it was clear that quality was their top priority. John Inglis and his team had an extraordinary ability to focus our rather vague ideas about the book and marshal the resources and personnel to carry them out. We thank Dotty Brown for the wonderful job she did with the copy editing, drawings, figures, appendices, and indexing. We also want to thank Marilu Hoeppner for cheerfully carrying out that most difficult task of ensuring the accuracy and consistency of the book. Finally, we wish to give a special thanks to Mary Cozza for her exceptional organizational skills and equally important her firm, but gentle, prodding.

**W. Sullivan**
**M. Ashburner**
**R. Scott Hawley**

# Abbreviations

| | |
|---|---|
| Acetyl CoA | acetyl coenzyme A |
| ACh | acetylcholine |
| ADH | alcohol dehydrogenase |
| ADP | adenosine diphosphate |
| AEL | after egg laying |
| AL | antennal lobe |
| ALH | after larval hatching |
| AP | alkaline phosphatase |
| BAC | bacterial artificial chromosome |
| BCIP | 5-bromo-4-chloro-3-indolyl phosphate (see X-phosphate) |
| BDGP | Berkeley *Drosophila* Genome Project |
| BES | $N,N$-bis(2-hydroxyethyl)-2-aminoethanesulfonic acid |
| BFD | Berkeley Fly Database |
| BLAST | basic local alignment search tool |
| BrdU | 5-bromo-2′-deoxyuridine |
| BSA | bovine serum albumin |
| BSS | balanced salt solution |
| b/w | black/white |
| BWSV | black widow spider venom |
| CALI | chromophore-assisted laser inactivation |
| CAT | chloramphenicol acetyltransferase |
| CCD | charge-coupled device |
| CDS | coding sequence |
| CF | chemical fixation |
| CHEF | contour-clamped homogeneous electric field |
| CLSM | confocal laser-scanning microscopy |
| CMFDG | 5-chloromethylfluorescein di-β-D-galactopyranoside |
| CNS | central nervous system |
| ConA | concanavalin A |
| CPD | criticial point dried |
| CPITC | coumarin phenylisothiocyanate |
| CPRG | chlorophenol red β-D-galactopyranoside |
| DAB | 3,3′-diaminobenzidine |
| DABCO | 1,4-diazabicyclo-(2,2,2)-octane |
| DAPI | 4′,6-diamidino-2-phenylindole |

| | |
|---|---|
| DCV | *Drosophila* C virus |
| DDBJ | DNA Database of Japan |
| DEPC | diethyl pyrocarbonate |
| DHFR | dihydrofolate reductase |
| DIC | differential interference contrast |
| DIG | digoxigenin |
| DLM | dorsal longitudinal flight muscle |
| DMF | *N,N*-dimethylformamide |
| DNA | deoxyribonucleic acid |
| dNTPs | deoxynucleoside triphosphates |
| DOP-PCR | degenerate oligonucleotide-primed PCR |
| DRES | *Drosophila*-related EST sequences |
| DSHB | The Developmental Studies Hybridoma Bank (University of Iowa) |
| dsRNA | double-stranded RNA |
| DT-A | diphtheria toxin A subunit |
| DTT | dithiothreitol |
| ECL | enhanced chemiluminescence |
| EDAC | 1-ethyl-3-(3-dimethylaminopropyl)-carbodiimide |
| EDGP | European *Drosophila* Genome Project |
| EGFP | "enhanced" GFP |
| EJC | excitatory junctional current |
| EJP | excitatory junctional potential |
| EM | electron microscopy |
| EMBL | European Molecular Biology Laboratory |
| EMS | ethyl methane sulfonate |
| ENU | ethyl nitrosourea |
| EP | enhancer promoter |
| ERG | electroretinogram |
| EST | expressed sequence tag |
| FAQs | frequently asked questions |
| FCS | fetal calf serum |
| FETi | fast extensor tibiae motor neuron |
| FIGE | field inversion gel electrophoresis |
| FISH | fluorescent in situ hybridization |
| FITC | fluorescein isothiocyanate |
| FLP-FRT | site-specific FLP recombinase–FLP recombination target |
| FRT | FLPase recombination target |
| FLUOS | 5(6)-carboxyfluorescein-*N*-hydroxysuccinimide ester |
| FTP | file transfer protocol |
| GF | giant fiber |
| GFP | green fluorescent protein |
| GIFTS | Gene Interaction in the Fly Transworld Server |
| GST | glutathione-*S*-transferase |
| GTPase | guanosine triphosphatase |
| HAME | *p*-hydroxyl benzoic acid methyl ester |
| HEPES | *N*-(2-hydroxyethyl)piperazine-*N'*-(2-ethanesulfonic acid) |
| HGT | high gelling temperature |
| HHMI | Howard Hughes Medical Institute |
| HL | hemolymph-like |

| | |
|---|---|
| HMDS | hexamethyldisilazane |
| HMW | high molecular weight |
| HPF | high-pressure freezing or high-pressure freezer |
| HPF-FS | high-pressure freezing/freeze-substitution |
| HPLC | high-performance liquid chromatography |
| HRP | horseradish peroxidase |
| hsp | heat shock promoter |
| HU | hydroxyurea |
| iACT | inner antenno-cerebral tract |
| IdU | 5-iodo-2′-deoxyuridine |
| iEM | immunoEM |
| ip | intraperitoneal |
| IPTG | isopropyl-$\beta$-D-thiogalactopyranoside |
| IR | infrared |
| kb | kilobase |
| KC | Kenyon cell |
| LB medium | Luria-Bertani medium |
| LM | light microscopy |
| LMW | low molecular weight |
| LocI | local interneuron |
| MAb | monoclonal antibody |
| Mb | megabase |
| MB | mushroom body |
| mEJCs | miniature EJCs |
| $[Mg^{++}]$ | magnesium ion concentration |
| MM3 medium | modified M3 medium |
| MRP | mechanoreceptor potential |
| MT | microtubule |
| Mt | metallothionein |
| $[Na^+]$ | sodium ion concentration |
| NA | numerical aperature |
| ND | neutral density |
| NBT | nitroblue tetrazolium |
| NIDA | National Institute on Drug Abuse |
| NGS | normal goat serum |
| NMJ | neuromuscular junction |
| NPG | $n$-propyl gallate |
| NOR | nucleolus organizer region |
| n-syb | neuronal synaptobrevin |
| NVOC-C1 | 6-nitroveratryl chloroformate |
| OD | optical density |
| PBS | phosphate-buffered saline |
| PCR | polymerase chain reaction |
| PDMN | posterior dorsal mesothoracic nerve |
| PEG | polyethylene glycol |
| PFGE | pulsed-field gel electrophoresis |
| PI | propidium iodide |
| PIPES | piperazine-$N,N'$-bis(2-ethanesulfonic acid) |
| PLT | progressive lowering of temperature |

| | |
|---|---|
| PMSF | phenylmethylsulfonyl fluoride |
| PNS | peripheral nervous system |
| PPF | paired-pulse facilitation |
| ppm | peripodial membrane |
| PPD | $p$-phenylenediamine |
| PTP | post-tetanic potentiation |
| Rac | ricin A chain |
| RH | relative humidity |
| RiI | relay interneurons |
| RNA | ribonucleic acid |
| RNA-i | dsRNA genetic interference |
| SDS | sodium dodecyl sulfate |
| SDS-PAGE | SDS-polyacrylamide gel electrophoresis |
| SEM | scanning electron microscopy |
| SETi | slow extensor tibiae motor neuron |
| SIT | silicon intensifier target |
| SNF | soluble nuclear fraction |
| STF | short-term facilitation |
| STS | sequence tagged site |
| syb | synaptobrevin |
| TBS | Tris-buffered saline |
| TdT | terminal deoxynucleotidyl transferase |
| TEM | transmission electron microscopy |
| TEMED | $N,N,N',N'$,-tetramethylethylenediamine |
| TEP | transepithelial potential |
| TES | $N$-Tris(hydroxymethyl)methyl-2-aminoethanesulphonic acid |
| TeTxLC | tetanus toxin light chain |
| TEVC | two-electrode voltage-clamp |
| TLC | thin-layer chromatography |
| TRITC | tetramethylrhodamine isothiocyanate |
| TSA | tyramide signal amplification system |
| TTM | tergotrochanteral jump muscle |
| TTMn | tergotrochanteral motor neuron |
| TUNEL | terminal deoxynucleotidyl transferase [TdT]-mediated dUTP nick end-labeling |
| UV | ultraviolet |
| UAS | upstream activation sequence |
| URL | uniform resource locator |
| UTR | untranslated regions |
| VNC | ventral nerve cord |
| WGA | wheat germ agglutinin |
| WPP | white prepupal stage |
| WWW | World Wide Web |
| X-gal | 5-bromo-4-chloro-3-indolyl β-D-galactopyranoside |
| X-phosphate | 5-bromo-4-chloro-3-indolyl phosphate |
| YAC | yeast artifical chromsome |

# Chromosomes

# CONTENTS

# 1

# Preparation and Analysis of *Drosophila* Mitotic Chromosomes

Sergio Pimpinelli,* Silvia Bonaccorsi, Laura Fanti,*† and Maurizio Gatti*

*Istituto Pasteur, Fondazione Cenci Bolognetti
Dipartimento di Genetica e Biologia Molecolare, Universita' di Roma
"La Sapienza"
P.A. Moro 5, 00185 Roma, Italy

†Istituto di Genetica, Universita' di Bari
V. Amendola 165/A, 70126 Bari, Italy

Mitotic chromosome cytology has an important role in many areas of *Drosophila* research. It is routinely needed for characterization of the mitotic phenotypes elicited by mutations affecting chromosome structure and/or behavior (for review, see Gatti and Goldberg 1991). In addition, mitotic cytology has proven to be essential for cytogenetic analysis of heterochromatin (for review, see Gatti and Pimpinelli 1992). Heterochromatin cannot be cytologically dissected by polytene chromosome analysis, because the bulk of this material is included in the chromocenter. However, heterochromatin breakpoints can be precisely determined in mitotic chromosomes processed with high-resolution banding techniques, such as quinacrine, Hoechst, and N-banding (for review, see Gatti and Pimpinelli 1992). Finally, good mitotic preparations are essential for fine mapping of repetitive DNA sequences along heterochromatin by in situ hybridization (see, e.g., Palumbo et al. 1994; Pimpinelli et al. 1995; Dernburg et al. 1996) and for immunolocalization of chromosomal proteins (see, e.g., Pak et al. 1997; Fanti et al. 1998; Platero et al. 1998).

In this chapter, we present the protocols routinely used in our laboratories for mitotic chromosome preparation, chromosome banding, and fluorescent in situ hybridization (FISH). In addition, we describe our fixation and immunostaining procedures for protein localization along mitotic chromosomes.

## PREPARATION AND STAINING OF LARVAL BRAIN MITOTIC CHROMOSOMES

Although mitotic chromosome preparations can be obtained from embryonic and gonial cells of both sexes, the tissue that provides the best mitotic figures is the larval brain. This tissue contains two major types of dividing cells: the neuroblasts and the ganglion mother cells (Hofbauer and Campos-Ortega 1990). The neuroblasts divide either symmetrically, producing two neuroblast stem cells, or asymmetrically, producing another neuroblast and a smaller cell called the ganglion mother cell. The ganglion mother cell divides only once, producing two daughter cells that differentiate into neurons. Several squashing techniques have been developed for preparation of larval brain mitotic chromosomes (Ashburner 1989; Gonzalez and Glover 1993). Below, we present a series of squashing protocols that are routinely used in our laboratories for a variety of experimental purposes. These procedures are minor modifications of a basic technique developed 25 years ago (Gatti et al. 1974) and can be successfully used for preparing mitotic chromosomes of various *Drosophila* and mosquito species (Gatti et al. 1976; Pimpinelli et al. 1976; Bonaccorsi et al. 1980, 1981). To characterize various aspects of mitotic chromosome morphology and behavior, larval brains can be squashed either in aceto-orcein to obtain orcein-stained chromosomes (Protocol 1.1) or in 45% acetic acid to obtain unstained preparations (Protocol 1.2). Unstained material can be then stained with Giemsa to obtain permanent preparations, processed with a variety of banding techniques, or used for in situ hybridization.

## Preparation of Orcein-stained Chromosomes

Depending on the experimental purpose, aceto-orcein squashes can be prepared by three different experimental regimes, which are summarized in Protocol 1.1. First, dissected brains can be squashed in aceto-orcein without colchicine treatment and hypotonic shock (i.e., Protocol 1.1, with the omission of steps 3 and 4). This procedure allows observation of all phases of mitosis and permits evaluation of the mitotic index and the frequency of anaphases (Gatti and Baker 1989). However, chromosome morphology is poorly defined in these preparations.

In the second regime, colchicine treatment is omitted, but brains are incubated in hypotonic solution (i.e., Protocol 1.1, with the omission of step 3). Hypotonic treatment improves metaphase chromosome spreading and causes sister chromatid separation, allowing examination of chromosome condensation and detection of hyperploid and polyploid metaphases. However, hypotonic shock disrupts anaphase (Brinkley et al. 1980), and anaphase figures are almost absent in hypotonically treated brains.

In the third regime (Protocol 1.1), brains are incubated in vitro with colchicine, treated with hypotonic solution, fixed, and squashed. This procedure provides a large number of well-spread metaphase figures (200–400 per brain) that can be analyzed for chromosome morphology, the presence of chromosome aberrations, and the degree of ploidy. However, because colchicine disrupts spindle microtubules, inducing metaphase arrest followed by chromosome overcontraction, colchicine treatment must be omitted if the degree of chromosome condensation has to be evaluated.

## Preparation of Unstained Chromosomes

Unstained brain chromosomes can be prepared according to the same experimental regimes described above for aceto-orcein squashes. The procedure for this type of preparation is given in Protocol 1.2.

## Giemsa Staining

Although well-sealed aceto-orcein squashes remain in good condition for 1–2 months, there are cases in which permanent chromosome preparations are needed. This can be done by staining preparations obtained according to Protocol 1.2 with 2% Giemsa in a phosphate buffer at pH 7.0. We routinely use Giemsa from Merck, but other Giemsa brands work just as well. The timing of Giemsa staining varies with the Giemsa brand and should be adjusted to obtain the desired staining. After Giemsa staining, chromosome preparations are differentiated by washing the slides in tap water. Giemsa stain is additive, and thus if chromosomes are not sufficiently stained, the slides can be stained again in 2% Giemsa, until the desired stain is obtained. After washing in tap water, the slides are air-dried and then mounted in Euparal (Carolina Biological Supply) or similar medium.

## PROTOCOL 1.1*

## Aceto-Orcein Squashes of Larval Brains

### Materials

### *Supplies and Equipment*

Siliconized slides, used only as a support for drops of either saline or hypotonic solution (for preparation, see Protocol 12.6 [Dernberg] or Protocol 6.1 [Kennison] in this volume.)

Dissecting tools
- dissecting microscope
- forceps (2 pairs; e.g., Dumont #5 Biologie)

Petri dish (35 x 10 mm) with cover

Nonsiliconized slides and coverslips (20 x 20 mm or 22 x 22 mm)

Blotting paper

Depilatory wax (found in most cosmetic shops) or nail polish

### *Solutions and Reagents*

Saline (0.7% NaCl in $H_2O$)

(*Optional*) **Colchicine** ($10^{-3}$ M) in $H_2O$

(*Optional*) Hypotonic solution (0.5% sodium citrate • $2H_2O$ in $H_2O$)

#### *Fixative*

Acetic acid/**methanol**/$H_2O$ (ratio 11:11:2)
Use freshly prepared fixative.

#### *Preparation of 2% Aceto-orcein*

Boil synthetic orcein powder (Gurr, BDH Laboratory Supplies, Poole, Dorset BH15 1TD England; Phone: +44 1202 660444; Fax: +44 1202 666856; Web site: http://www.bdh.com) in 45% **acetic acid** for 45 minutes in a reflux condenser. We usually prepare 5% orcein, which is subsequently diluted to 2% with 45% acetic acid.

Before use, remove particulate matter from aceto-orcein either by filtration through blotting paper or by centrifugation in a microcentrifuge.

**CAUTION: acetic acid, colchicine, methanol** (see Appendix 4)

*In all protocols, $H_2O$ indicates glass distilled and deionized.

## Method

1. Dissect larval brains as follows:

   a. Transfer drops of saline (~50 µl) onto a siliconized slide. Place one to three actively crawling third-instar larvae in each drop.

   b. Use two pairs of thin forceps (e.g., Dumont #5 Biologie) to grab the larval mouth parts and the larval body at a position approximately two thirds of its length in the posterior part, and then pull apart.

   *Note:* The brain usually remains attached to the head together with several imaginal disks and the salivary glands.

   c. Use the forceps to remove the brain from the remainder of the head.

   *Note:* There is no need to completely remove the imaginal discs, which will not interfere with squashing, but the more rigid mouth parts should be completely removed.

2. Wash the dissected brains in saline at room temperature for 1–5 minutes.

3. Transfer the brains to a 35 × 10-mm petri dish containing 2 ml of saline supplemented with a drop (~100 µl) of $10^{-3}$ M colchicine. Cover the petri dish and incubate at 25°C for 1.5 hours.

4. Transfer the brains into a drop (~50 µl) of hypotonic solution placed on a siliconized slide and incubate at room temperature for 10 minutes.

   *Note:* The incubation time is critical because treatments that exceed 10 minutes may induce sister chromatid separation in metaphase chromosomes.

5. Fix the brains at room temperature in a freshly prepared mixture of acetic acid/methanol/$H_2O$ (11:11:2). Check fixation under the dissecting microscope, and keep the brains in the fixative until they begin to become transparent (10–20 seconds).

   *Note:* We usually fix the brains in 2 ml of fixative placed in a 35 × 10-mm petri dish.

6. Transfer fixed brains individually into small drops (2 µl) of 2% aceto-orcein placed on a very clean, dust-free *nonsiliconized* coverslip (20 × 20 or 22 × 22 mm). Between one and four brains may be transferred to a corresponding number of drops placed on the same coverslip.

7. Leave the brains in the aceto-orcein drops for 1–2 minutes and then lower a very clean, dust-free nonsiliconized slide onto the coverslip; surface tension will cause the coverslip to adhere to the slide. Invert the sandwich and squash between two or three sheets of blotting paper.

   *Note:* Squashing should be performed in two steps: First exert a gentle pressure to remove the excess of aceto-orcein and then squash very hard. The gentle presquashing prevents coverslip sliding and the consequent damage to the preparation.

8. Seal the edges of the coverslip either with melted depilatory wax or with nail polish.

   *Note:* Well-sealed slides can be stored at 4°C for 1–2 months without substantial deterioration.

# Unstained Squashes of Larval Brains

## Materials

### Supplies and Equipment

Sliconized slides, used only as a support for drops of either saline or hypotonic solution
(for preparation, see Appendix 3)
Dissecting tools
- dissecting microscope
- forceps (2 pairs; e.g., Dumont #5 Biologie)
Petri dish (35 x 10 mm) with cover
Nonsiliconized slides and siliconized coverslips (20 x 20 mm or 22 x 22 mm)
Blotting paper
Razor blade

### Solutions and Reagents

Saline (0.7% NaCl in $H_2O$)
(*Optional*) **Colchicine** ($10^{-3}$ M) in $H_2O$
(*Optional*) Hypotonic solution (0.5% sodium citrate • $2H_2O$ in $H_2O$)
**Acetic acid** (45%)
Absolute **ethanol**, chilled at –20ºC
**Liquid nitrogen** or **dry ice**

**CAUTION: acetic acid, colchicine, dry ice, ethanol, liquid nitrogen** (see Appendix 4)

## Method

1. Perform steps 1–5 of Protocol 1.1 above.

   *Note:* Incubation in $10^{-3}$ M colchicine (step 3) and hypotonic treatment (step 4) may be omitted depending on the desired type of preparation.

2. Transfer fixed brains individually into small drops (2 µl) of 45% acetic acid placed on a very clean, dust-free *siliconized* coverslip (20 x 20 or 22 x 22 mm). Between one and four brains may be transferred to a corresponding number of drops placed on the same coverslip.

   *Note*: Siliconized coverslips are used instead of nonsiliconized coverslips (as in step 6, Protocol 1.1).

3. Leave the brains in the 45% acetic acid drops for 1–2 minutes and then lower a very clean, dust-free nonsiliconized slide onto the coverslip; surface tension will cause the coverslip to adhere to the slide. Invert the sandwich and squash between two or three sheets of blotting paper.

   *Note:* Squashing should be performed in two steps: First exert a gentle pressure to remove the excess acetic acid and then squash very hard. The gentle presquashing prevents coverslip sliding and the consequent damage to the preparation.

4. Freeze the slides either on dry ice or in liquid nitrogen.

5. Flip off the coverslip with a razor blade and immediately immerse the slides in absolute ethanol at –20°C, and leave for 15 minutes. Air-dry the slides and store them at 4°C until further processing.

## CHROMOSOME-BANDING TECHNIQUES

The classical chromosome-banding techniques developed for mammalian chromosomes do not differentiate the euchromatic arms of *Drosophila* mitotic chromosomes. However, some of these techniques produce a sharp and highly reproducible banding of *Drosophila* heterochromatin (see, e.g., Holmquist 1975; Gatti et al. 1976). For example, the use of quinacrine-, Hoechst- and N-banding differentiates *Drosophila* heterochromatin into 61 cytological entities (Figure 1.1C), allowing precise localization of heterochromatic breakpoints (see, e.g., Gatti and Pimpinelli 1983, 1992; Pimpinelli and Dimitri 1989; Dimitri

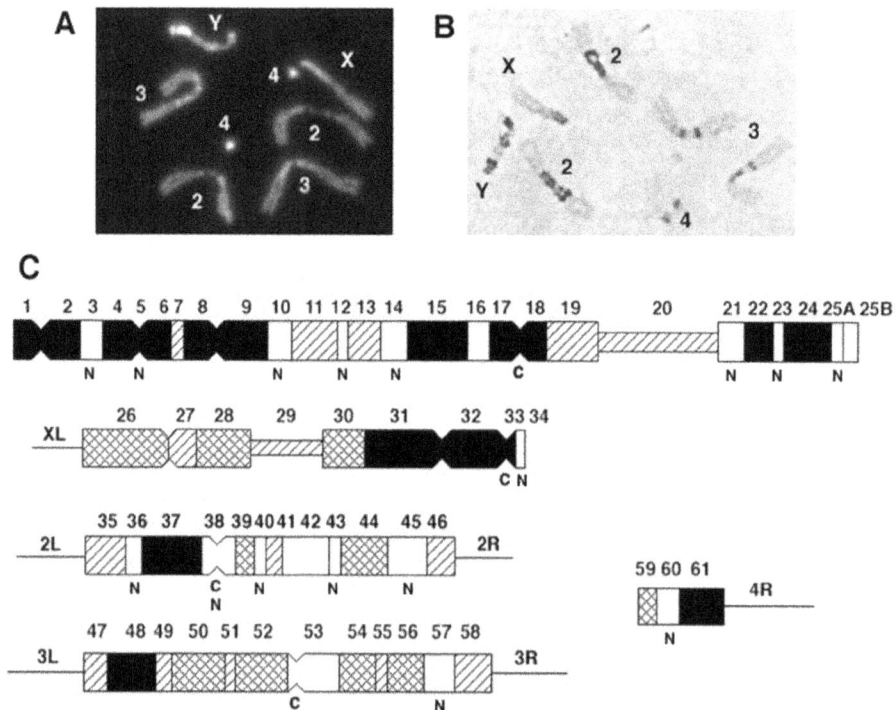

**Figure 1.1.** Examples of Hoechst-stained (*A*) and N-banded (*B*) metaphases. (*C*) A cytological map of *D. melanogaster* heterochromatin. The diagrams have been constructed by examining prometaphase chromosomes sequentially processed for quinacrine-, Hoechst- and N-banding. The entirely heterochromatic *Y* chromosome, the *X* chromosome, and the second, third, and fourth chromosome heterochromatin are schematically represented left to right and from top to bottom. Only the heterochromatic portions of chromosomes are shown; euchromatin is depicted as a straight line. The narrow crosshatched regions of the *X* and *Y* chromosomes represent the nucleolus organizer regions. C indicates the position of the centromere; the location of the fourth chromosome centromere has not been precisely determined. (*Solid segments*) Bright fluorescence: (*crosshatched segments*) moderate fluorescence; (*open segments*) no fluorescence. The N letters below each diagram indicate the regions that are consistently stained by the N-banding procedure (*B* Reprinted, with permission, from Pimpinelli et al. 1976.) For example, in *B*, the third chromosome heterochromatin exhibits an intensely stained block and two less heavily stained regions. The latter are not consistently observed and were not considered as N-bands.

1991). These banding techniques can also be successfully used to differentiate mitotic heterochromatin of various *Drosophila* and mosquito species (Gatti et al. 1976; Pimpinelli et al. 1976; Bonaccorsi et al. 1980, 1981).

Hoechst 33258 and 4′,6-diamidino-2-phenylindole (DAPI) stainings produce identical banding patterns, differentiating mitotic heterochromatin into blocks that differ for the degree of fluorescence (Figure 1.1A). Both compounds are general indicators of AT-richness along the chromosomes (Verma and Babu 1989; Sumner 1990), and the Hoechst- and DAPI-bright regions have been shown to contain large numbers of AT-rich satellite repeats (Bonaccorsi and Lohe 1991; Lohe et al. 1993).

Quinacrine staining produces a banding pattern similar but not identical to the Hoechst banding. The major difference between these two banding patterns is in the bulk of *X*-chromosome heterochromatin (region h31–h32) which is Hoechst bright and quinacrine dull. This region, which accommodates the relatively AT-rich 359-bp satellite DNA (Hilliker and Appels 1982; Lohe et al. 1993), can therefore be cytologically identified by sequential staining with Hoechst and quinacrine (see below).

The N-banding procedure specifically stains the nucleolus organizer regions (NORs) of several animal and plant species (Matsui and Sasaki 1973; Funaki et al. 1975). In *Drosophila melanogaster*, N-banding stains 16 major heterochromatic blocks (see Figure 1.1B) that are not fluorescent after Hoechst, DAPI, or quinacrine staining and do not correspond to NORs. N-banded regions are all enriched in the AAGAG repeats cloned from the 1.705 g/cm$^3$ satellite DNA (Bonaccorsi and Lohe 1991; Lohe et al. 1993). In addition, N-banded heterochromatic blocks are specifically immunostained with antibodies directed to the GAGA factor (see Figure 1.2C).

As shown in Figure 1.1, some heterochromatic regions that are poorly differentiated by Hoechst, DAPI, or quinacrine banding are sharply defined by the N-banding procedure. The cytological map of *D. melanogaster* heterochromatin (Figure 1.1) indicates which banding procedure should be applied for the best differentiation of each heterochromatic region. The most efficient method in our experience for defining a heterochromatic breakpoint is sequential application of two banding procedures such as Hoechst- (or DAPI-) and N-banding.

The degree of resolution of the banding techniques depends on the degree of elongation of heterochromatic regions. Thus, for all of the banding procedures described below, chromosomes should be prepared according to Protocol 1.2, with the omission of colchicine treatment.

PROTOCOL 1.3

# Hoechst Staining

The following procedure (Gatti et al. 1976; Gatti and Pimpinelli 1983) is a modification of a protocol developed by Latt (1973) for mammalian chromosomes. For other Hoechst-staining protocols, see Holmquist (1975) and Hazelrigg et al. (1982).

## Materials
### Supplies and Equipment

Slides, prepared according to Protocol 1.2, with the omission of colchicine treatment
Rubber cement

### Solutions and Reagents

**Hoechst 33258** (0.5 µg/ml) dissolved in HB

*Hoechst Buffer (HB)*
> 150 mM NaCl
> 30 mM **KCl**
> 10 mM **Na$_2$HPO$_4$**
> Adjust to pH 7.0.

*Mounting Buffer*
> 160 mM Na$_2$HPO$_4$
> 40 mM sodium citrate
> Adjust to pH 7.0.

**CAUTION: Hoechst 33258, KCl, Na$_2$HPO$_4$** (see Appendix 4)

## Method

1. Rehydrate air-dried slides in HB at room temperature for 5 minutes.

2. Stain slides in Hoechst 33258 (0.5 µg/ml in HB) at room temperature for 10 minutes.

3. Wash briefly (~5 seconds) in HB.

4. Place slides in a vertical position and allow to air-dry.

5. Mount in either HB or Mounting Buffer.

6. Seal around the edges of the coverslip with rubber cement.

> *Notes:* The slides may be stored in the dark at 4ºC for 1–2 days before observation; we have found that this treatment reduces fluorescence fading.
> 
> Fluorescence fading depends on both the lamp and the set of filters used for epifluorescence. If the degree of fading is too high, the slides may be mounted in Vectashield (H-1000; Vector Laboratories) or in similar media containing glycerol and antifading compounds (see Appendix 3).

PROTOCOL 1.4

# DAPI Staining

## Materials
### Supplies and Equipment

Slides, prepared according to Protocol 1.2, with the omission of colchicine treatment

### Solutions and Reagents

2x SSC (see Appendix 3)
**DAPI** (0.2 µg/ml), dissolved in 2x SSC
Vectashield or other mounting media (H-1000; Vector Laboratories)

**CAUTION: DAPI** (see Appendix 4)

### Method

1. Rehydrate air-dried slides in 2x SSC at room temperature for 5 minutes.

2. Stain in DAPI (0.2 µg/ml in 2x SSC) at room temperature for 5 minutes.

3. Wash briefly (~5 seconds) in 2x SSC.

4. Place slides in a vertical position and allow to air-dry.

5. Mount in Vectashield or in similar media containing glycerol and antifading compounds.

PROTOCOL 1.5

# Quinacrine Staining

The following quinacrine staining protocol was developed by Gatti et al. (1976). For other quinacrine-banding techniques, see Vosa (1970), Ellison and Barr (1971), and Faccio Dolfini (1974).

### Materials

### *Supplies and Equipment*

Slides, prepared according to Protocol 1.2, with the omission of colchicine treatment
Rubber cement

### *Solutions and Reagents*

**Quinacrine dihydrocloride** (0.5%; Gurr), dissolved in absolute **ethanol**
Absolute ethanol

**CAUTION: ethanol, quinacrine dihydrocloride** (see Appendix 4)

### Method

1. Immerse slides in absolute ethanol for 5 minutes.

2. Stain slides in 0.5% quinacrine dihydrochloride (in absolute ethanol) for 10 minutes.

3. Wash slides twice (5 seconds each) in absolute ethanol.

4. Place slides vertically and allow to air-dry.

5. Mount in $H_2O$.

6. Seal around the edges of the coverslip with rubber cement. Store the slides in the dark at 4°C before observation.

> *Note:* This treatment improves the degree of differentiation and reduces fluorescence fading. Here again (see Protocol 1.3, step 6) if fluorescence fading is too high, mount the slides in Vectashield (H-1000; Vector Laboratories) or similar media.

# N-Banding

The N-banding procedure presented below is essentially that of Funaki et al. (1975) with minor modifications (Pimpinelli et al. 1976). Best results are obtained when preparations obtained according to Protocol 1.2 are aged for 2–5 days at 4°C before processing; these preparations give a sharper N-banding compared to either fresh or older slides.

## Materials
### Supplies and Equipment

Slides, prepared according to Protocol 1.2, with the omission of colchicine treatment

### Solutions and Reagents

$NaH_2PO_4$ (1 M, pH 4.2)
**Giemsa** (4%; Merck), dissolved in phosphate buffer
Euparal (Carolina Biological Supply) or similar medium for mounting

> *Phosphate Buffer*
> 8 mM **KH$_2$PO$_4$**
> 6 mM **Na$_2$HPO$_4$**
> Adjust to pH 7.0.

**CAUTION: Giemsa, KH$_2$PO$_4$, Na$_2$HPO$_4$** (see Appendix 4)

## Method

1. Immerse slides in 1 M $NaH_2PO_4$ at 85°C and incubate for 15 minutes.

2. Rinse slides in $H_2O$ at room temperature.

3. Stain slides in 4% Giemsa (in phosphate buffer at pH 7.0) at room temperature for 20 minutes.

4. Rinse slides in tap water.

   *Note:* Remember that Giemsa staining is additive and that insufficiently stained preparations can be restained (see above).

5. Let the slides air-dry and mount in Euparal or similar medium.

6. Examine slides using phase-contrast optics.

# Sequential Quinacrine-, Hoechst-, and N-banding

The following protocol describes sequential quinacrine-, Hoechst- and N-banding. However, sequential application of Hoechst- and N-banding is sufficient for precise determination of most heterochromatic breakpoints. Quinacrine staining may be used only for breakpoints falling into regions that are Hoechst-bright and quinacrine-dull (Gatti and Pimpinelli 1983).

## Materials
### Supplies and Equipment

Slides, prepared according to Protocol 1.2, with the omission of colchicine treatment
Coplin jar
Rubber cement

### Solutions and Reagents

**Quinacrine dihydrocloride** (0.5%; Gurr), dissolved in absolute **ethanol**
Ethanol (70%, 95%, and absolute)
Destain (**methanol/acetic acid** [3:1])
HB (see p. 10)
Hoechst 33258 (0.5 µg/ml), dissolved in HB
Giemsa (4%; Merck), dissolved in phosphate buffer (see p. 12) at pH 7.0
Mounting media

**CAUTION: acetic acid, ethanol, methanol, quinacrine dihydrocloride** (see Appendix 4)

## Method

1. Perform quinacrine staining according to Protocol 1.5.

2. After analysis of quinacrine banding, peel off the rubber cement from the coverslip edges and place the slides into a Coplin jar containing $H_2O$. Keep the slides in $H_2O$ until the coverslip slides off. To facilitate coverslip sliding, shake the jar and/or gently tap against the bench.

3. Place slides vertically and allow to air-dry.

4. Wash slides in 70% ethanol for 5 minutes.

5. Destain slides by washing three times (30 minutes each) in a 3:1 solution of methanol/acetic acid.

6. Place slides vertically and allow to air-dry.

7. Stain slides with Hoechst 33258 according to Protocol 1.3.

8. After analysis of Hoechst banding, dismount the slides as described in step 2.

9. Wash slides in 95% ethanol for 5 minutes.

10. Air-dry slides and store them at 4ºC for 2–5 days.

11. Process slides with the N-banding technique, according to Protocol 1.6.

## IN SITU HYBRIDIZATION _____

The fluorescent in situ hybridization (FISH) technique permits fine mapping of both middle and highly repetitive DNA sequences along *D. melanogaster* heterochromatin (see, e.g., Abad et al. 1992; Palumbo et al. 1994; Pimpinelli et al. 1995; Berghella and Dimitri

1996; Dernburg et al. 1996). Best results are obtained when this technique is coupled with DAPI staining and digital recording of fluorescent signals. For example, if digital images of the FISH signals and DAPI fluorescence are detected separately using a charge coupled device (CCD camera), they can be then pseudocolored and merged using suitable computer programs. This allows precise overlapping of the DAPI banding (which is identical to the Hoechst 33258 banding) and the FISH signals, facilitating the assignment of the repetitive sequence under study to specific regions of the cytological map of *D. melanogaster* heterochromatin (Figure 1.1). We describe here the FISH procedures that are routinely used in our laboratories. These procedures have been adapted to *Drosophila* chromosomes by modifying the FISH technique of D'Aiuto et al. (1993). For other FISH procedures for *Drosophila* chromosomes, see Gonzalez and Glover (1993).

PROTOCOL 1.8

# Fluorescent In Situ Hybridization

## Materials

### *Supplies and Equipment*

Slides (2–3 days old) prepared according to Protocol 1.2. The most suitable chromosome preparations for in situ hybridization are those prepared according to Protocol 1.2 without colchicine pretreatment, but with hypotonic shock.

Coplin jars
Coverslips (22 x 22 mm)
Moist chamber at 37ºC
Rubber cement

### *Solutions and Reagents*

Biotin-11-dUTP (Enzo) *or* **digoxigenin**-11-dUTP (Boehringer Mannheim)
Nick translation kit (Boehringer Mannheim)
Sonicated salmon sperm DNA (3 µg per slide)
Sodium acetate (3 M, pH 4.5)
**Ethanol** (70%, 90%, and absolute) at room temperature
Ethanol (70%) chilled at –20ºC
Ethanol (90%) chilled at 4ºC or on ice
2x SSC (see Appendix 3)
0.1x SSC
Detection reagents
   *Labeling with a single probe.* Depending on the label, use either:
   • Fluorescein isothiocyanate (FITC)-conjugated avidin (DCS grade; Vector Laboratories; 3.3 µg/ml, diluted in SBT)
   • Rhodamine-conjugated antidigoxigenin sheep IgG, Fab fragments (Boehringer Mannheim; 2 µg/ml, diluted in SBT)
   *Double labeling.* Use a mixture of the following two reagents:
   • 3.3 µg/ml FITC-conjugated avidin (Vector Laboratories)
   • 2 µg/ml rhodamine-conjugated antidigoxigenin sheep IgG, Fab fragment (Boehringer Mannheim), diluted in SBT

DAPI (0.2 μg/ml), dissolved in 2x SSC
Vectashield H-1000 (Vector Laboratories) or other mounting media (see Appendix 3)
Rubber cement

### Hybridization Mixture (10 μl per slide)

| | |
|---|---|
| ultrapure **formamide** (J.T. Baker) | 5 μl |
| 50% dextran sulfate | 2 μl |
| 20x SSC | 1 μl |
| $H_2O$ | 2 μl |

### Denaturation Solution

| | |
|---|---|
| ultrapure formamide (J.T. Baker) | 35 ml |
| 20x SSC | 5 ml |
| $H_2O$ | 10 ml |

Mix all components and pour into a Coplin jar.

### Wash Buffer 1

50% formamide
2x SSC

### Blocking Solution

3% bovine serum albumin (BSA)
0.1% Tween-20
4x SSC

### SBT Solution

1% BSA
0.1% Tween-20
4x SSC

### SSC/Tween-20 Solution

0.1% Tween-20
4x SSC

**CAUTION: ethanol, DAPI, digoxigenin, formamide** (see Appendix 4)

## Method

### Probe Preparation and Denaturation

1. Label 1 μg of probe by nick translation using either biotin-11-dUTP or digoxigenin-11-dUTP, following the manufacturer's instructions.

2. Mix the labeled DNA (40–80 ng per slide) with sonicated salmon sperm DNA (3 μg per slide). Add 0.1 volume of 3 M sodium acetate (pH 4.5) and 2 volumes of cold absolute ethanol (chilled at –20ºC). Place at –80ºC for 15 minutes and centrifuge at 13,600*g* for 15 minutes to pellet the DNA. Aspirate the ethanol. Dry the pellet in a Savant speed vac.

3. Resuspend the DNA pellet in the hybridization mixture by vortexing.

4. Denature the probe mixture at 80°C for 8 minutes. Place the probe on ice until used.

## Slide Denaturation

1. Place the Coplin jar containing the Denaturation Solution in a water bath at 70°C.

2. Dehydrate 2–3-day-old slides (prepared according to Protocol 1.2) by immersion in 70%, 90%, and absolute ethanol at room temperature (3 minutes each). Air-dry slides.

3. Transfer the slides into the Denaturation Solution at 70°C and leave for exactly 2 minutes. To avoid a temperature drop of this solution, do not immerse more than three to four slides at any given time.

4. Quickly transfer the slides to cold 70% ethanol at –20°C and incubate for 3 minutes. Then dehydrate sequentially through cold 90% and absolute ethanol at 4°C (3 minutes each). Air-dry slides.

## Hybridization

Do not allow slides to dry during the following passages.

1. Apply 10 µl of Hybridization Mixture to each of the denatured slides. Cover each slide with a 22 x 22-mm coverslip; avoid forming air bubbles. Seal the edges of the coverslip with rubber cement.

2. Incubate slides in a moist chamber at 37°C overnight.

   *Note:* The hybridization temperature we routinely use for middle repetitive probes is 37°C. However, for simple sequence satellite DNAs, other hybridization temperatures should be used, depending on their nucleotide composition and sequence. For each satellite probe, the hybridization temperature is maintained at 8–12°C below its melting temperature value (for details, see Bonaccorsi and Lohe [1991] and Lohe et al. [1993]).

3. Peel off the rubber cement, gently remove the coverslip, and wash slides three times in Wash Buffer 1 at 42°C for 5 minutes (each wash).

4. Wash slides three times in 0.1x SSC at 60°C for 5 minutes (each wash).

5. Remove excess liquid from around the specimen and apply 100 µl of Blocking Solution to each slide. Cover with a 22 x 22-mm coverslip and incubate at 37°C for 30 minutes.

## Biotin-labeled DNA Detection

1. Gently remove the coverslip and blot excess liquid from around the specimen.

2. To each slide, apply 80 µl of 3.3 µg/ml FITC-conjugated avidin, diluted in SBT. Cover with a 22 x 22-mm coverslip and incubate at 37°C for 30 minutes in a dark, humid chamber.

3. Remove the coverslips. Wash slides three times in SSC/Tween-20 at 42°C for 5 minutes (each wash).

4. Stain with 0.2 µg/ml DAPI (in 2x SSC) at room temperature for 5 minutes.

5. Rinse in 2x SSC at room temperature.

6. Mount slides in Vectashield H-1000 (Vector Laboratories) or similar antifading mountants. Seal the coverslip edges with either rubber cement or nail polish and store in the dark at 4°C.

   *Note:* Slides can be stored for weeks without substantial deterioration.

### Digoxigenin-labeled DNA Detection

1. Gently remove the coverslip and blot excess liquid from around the specimen.

2. To each slide, apply 80 µl of 2 µg/ml rhodamine-conjugated antidigoxigenin sheep IgG, Fab fragments, diluted in SBT. Cover with a 22 x 22-mm coverslip and incubate at 37°C for 30 minutes in a dark, humid chamber.

3. Proceed as described in steps 3–6 above in Biotin-labeled DNA Detection.

### Double Labeling

For simultaneous in situ hybridizations with biotin- and digoxigenin-labeled probes, perform the following steps:

1. Mix together 40–80 ng (per slide) of each labeled probe.

2. Follow the procedure for single probes:

   a. Probe Preparation and Denaturation (steps 2–4, omit step 1)

   b. Slide Denaturation (steps 1–4)

   c. Hybridization (steps 1–5)

3. (*Signal detection*) Add 80 µl of 3.3 µg/ml FITC-conjugated avidin, 2 µg/ml rhodamine-conjugated antidigoxigenin sheep IgG, Fab fragment, diluted in SBT. Cover with a 22 x 22-mm coverslip and incubate at 37°C for 30 minutes in a dark, humid chamber.

4. Proceed as described in steps 3–6 above in Biotin-labeled DNA Detection.

## IMMUNOSTAINING PROCEDURES

The methanol/acetic acid fixation techniques described above preserve chromosome morphology very well, but they remove a substantial fraction of chromosomal proteins. Thus, preparations obtained by these techniques are not suitable for immunolocalization of proteinaceous components of metaphase chromosomes. We have therefore developed a series of fixation procedures that result in good chromosomal quality with minimal removal of proteins. We stress here that different fixation protocols result in differential extraction of chromosomal proteins. Thus, for each protein under study, develop a fixation procedure that minimizes its removal from the chromosomes. In addition, we have found that the timing of hypotonic treatment is also critical. For example, certain proteins cannot be visualized by immunostaining hypotonically treated preparations (e.g., the Modulo protein, Method 4 [Protocol 1.9]), suggesting that the hypotonic shock removes these proteins from metaphase chromosomes. Despite these problems, by playing with the above parameters, we have empirically determined the conditions that allow immunolocalization of several proteins along mitotic chromosomes. These experiments showed that certain fixation procedures result in artifacts. One of the most common artifacts is the absence of immunostaining of condensed chromosomes, which appear embedded into a fluorescent halo (Figure 1.2a). For example, this immunofluorescence pattern is usually observed when chromosomes fixed according to Method 2 (Protocol 1.9) are stained with antibodies directed to HP1. However, if chromosomes are fixed according to Method 1 (Protocol 1.9), the HP1 protein remains associated with the chromosomes and exhibits a preferential localization in the heterochromatic regions (Figure 1.2b). An obvious interpretation of these results is that the lack of immunostaining of mitotic chromosomes reflects the fixative-induced extraction of HP1. Thus, before concluding that a particular protein is not associated with chromosomes, it is highly advisable to examine its behavior using a variety of fixation techniques.

Below are four fixation/immunostaining protocols. Each of these protocols proved to be suitable for the immunolocalization of one or more proteins along mitotic chromosomes.

## PROTOCOL 1.9

## Fixation and Immunostaining

### Materials

### *Supplies and Equipment*

Silconized slides, used only as a support for drops of either saline or hypotonic solution (for preparation, see Appendix 3)

Dissecting tools
- dissecting microscope
- forceps (2 pairs; Dumont #5 Biologie)

Syringe or tungsten needle

Nonsiliconized slides and siliconized coverslips (18 x 18 mm)

Blotting paper

Razor blade

### *Solutions and Reagents*

Saline (0.7% NaCl in $H_2O$)

Hypotonic solution (0.5% sodium citrate • $2H_2O$ in $H_2O$)

**Liquid nitrogen**

1x PBS (phosphate-buffered saline)

Primary and secondary antibodies, diluted in PBS/BSA (see below)

**DAPI** (0.05 µg/ml), dissolved in 2x SSC

Vectashield H-1000 (Vector Laboratories)

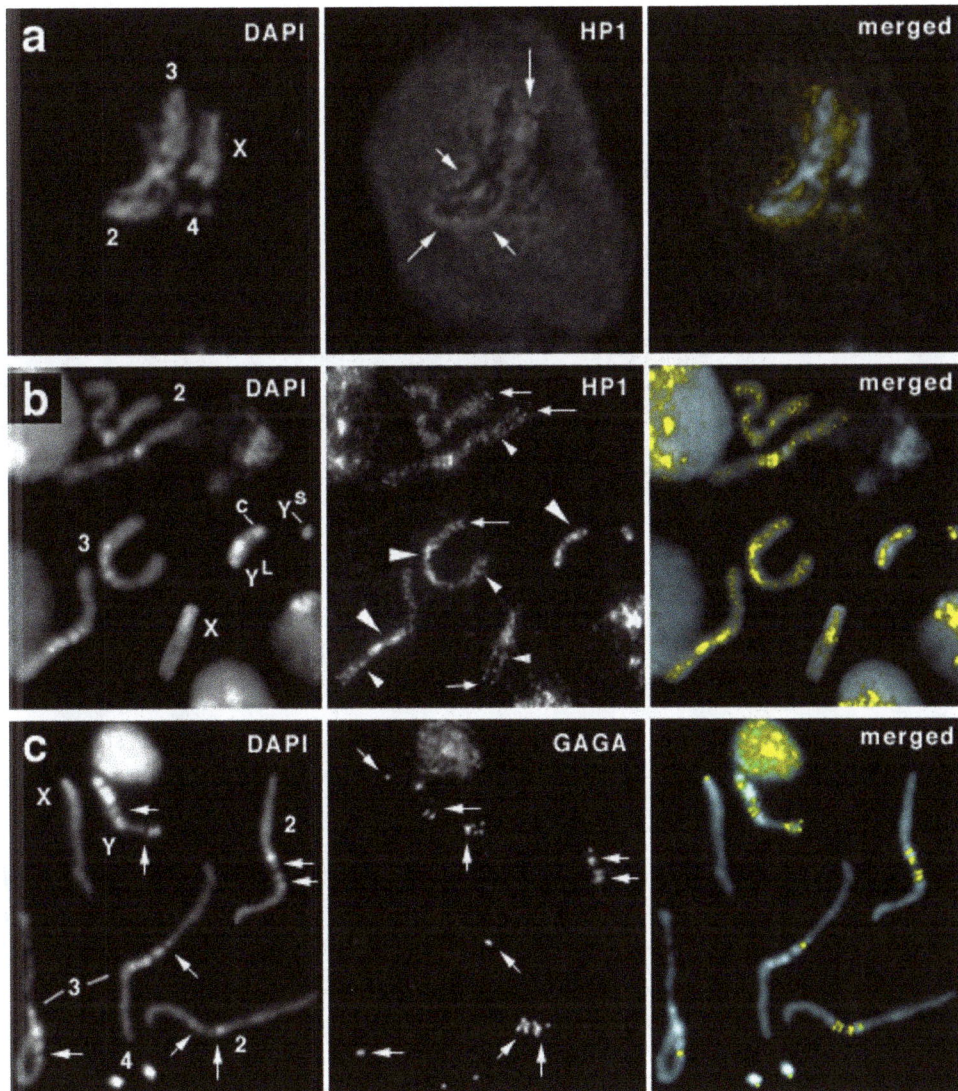

**Figure 1.2.** Immunostaining of larval brain cells with antibodies directed to the HP1 and GAGA proteins. (*a*) A prometaphase figure fixed and immunostained for HP1 according to Method 2 (Protocol 1.9). Note that the chromosomes are unstained and surrounded by a fluorescent halo (*arrows*). (*b*) A prometaphase figure fixed according to Method 1 (Protocol 1.9) and immunostained for HP1. HP1 is clearly associated with the chromosomes and enriched at the heterochromatic (*large arrowheads*), telomeric (*arrows*), and several euchromatic (*small arrowheads*) regions. (Reprinted, with permission, from Fanti et al. 1998.) (*c*) A prometaphase figure fixed according to Method 2 (Protocol 1.9) and immunostained for the GAGA factor. Note the accumulation of this protein in discrete heterochromatic regions, which correspond to the N bands (*arrows*).

*Fixative 1*

> **Methanol/acetic acid**/$H_2O$ (ratio 10:2:1)
> Use freshly prepared solution.

*Fixative 2*

> Methanol/acetic acid/$H_2O$ (ratio 5:2:3)
> Use freshly prepared solution.

*Fixative 3*

> 45% acetic acid containing 2% **formaldehyde**
> Use freshly prepared solution.

*PTX Solution*

> 1x PBS
> 1% Triton X-100

*PBS/BSA*

> 1x PBS
> 1% BSA
> Alternatively, substitute 1% nonfat dry milk for BSA.

**CAUTION: acetic acid, DAPI, formaldehyde, liquid nitrogen, methanol** (see Appendix 4)

## Method 1

The following procedure has been successfully applied for immunostaining with anti-HP1 antibodies (Fanti et al. 1998; the antibodies used were produced by James et al. 1989). For another protocol for HP1 immunostaining, see Pak et al. (1997).

1.  Dissect brains in saline as described in Protocol 1.1, step 1.

2.  Transfer brains into a drop of hypotonic solution (~50 μl) placed on a siliconized slide and leave at room temperature for 2 minutes.

3.  Transfer brains into a 5-μl drop of freshly prepared Fixative 1 placed on a clean, dust-free 18 × 18-mm siliconized coverslip. During fixation, which should not exceed 2 minutes, use either a syringe or a tungsten needle to fragment the brains.

    *Note:* Brains fixed in solutions containing a low proportion of acetic acid are rather hard and need to be fragmented before squashing.

4.  Lower a clean, dust-free nonsiliconized slide on the coverslip, invert the sandwich, and squash between two to three sheets of blotting paper.

5.  Freeze the slides in liquid nitrogen and flip off the coverslip with a razor blade.

6.  Immediately immerse the slides in PBS at room temperature and leave for 5 minutes.

7.  Transfer the slides into PTX and leave at room temperature for 10 minutes.

8.  Transfer the slides into PBS/BSA and incubate at room temperature for 30 minutes.

9.  Remove excess of PBS/BSA from around the specimen and cover it with 15 μl of primary antibody diluted in PBS/BSA. Place a clean 18 × 18-mm nonsiliconized coverslip over the preparation and incubate at 4°C overnight in a humid chamber.

10. Gently remove the coverslip and wash the slides three times in PBS at room temperature (5 minutes each wash).

11. Remove excess of PBS/BSA from around the specimen and cover it with 15 μl of secondary antibody diluted in PBS/BSA. Place a clean 18 × 18-mm nonsiliconized coverslip over the preparation and incubate at room temperature for 1 hour in a dark, humid chamber.

12. Gently remove the coverslip and wash the slides three times in PBS at room temperature (5 minutes each wash).

13. Stain the slides with 0.05 µg/ml DAPI (in 2x SSC) at room temperature for 5 minutes.

14. Wash slides in PBS at room temperature for 5 minutes.

15. Mount in Vectashield H-1000 (Vector Laboratories) or similar antifading media; seal the edges of the coverslip with either rubber cement or nail polish.

## Method 2

Method 2 has been used for immunolocalization of the GAGA (Huang et al. 1998), E(var)3-93D, and Psc proteins (L. Fanti and S. Pimpinelli, unpubl.); the antibodies used were produced by Tsukiyama et al. (1994), Dorn et al. (1993), and Rastelli et al. (1993), respectively. Method 2 is identical to Method 1, with the exception of steps 2 and 3.
For Method 2, note the following changes to the procedure given for Method 1:

- In step 2, brains are hypotonically swollen for 10 minutes.
- In step 3, brains are immersed in Fixative 2 for approximately 3 minutes and then fragmented using either a syringe or a tungsten needle.

## Method 3

This procedure has proven to be suitable for immunolocalization of the Ph protein (the antibody used was produced by DeCamillis et al. [1992]; the immunolocalization experiments are unpublished results of L. Fanti and S. Pimpinelli). Method 3 differs from Method 1 only in steps 2 and 3.
For Method 3, note the following changes to the procedure given for Method 1:

- In step 2, brains are hypotonically treated for 10 minutes.
- In step 3, brains are immersed in Fixative 3 for 8 minutes and then transferred to a 5-µl drop of the same fixative placed on a 18 x 18-mm siliconized coverslip (in this method, brains do not need to be fragmented).

## Method 4

This procedure is identical to Method 3 with the omission of the hypotonic treatment (step 2). It has been used by Perrin et al. (1998) to immunolocalize the Modulo protein using antibodies generated by Garzino et al. (1987).

## REFERENCES _____

Abad J.P., Carmena M., Baars S., Saunders R.D., Glover D.M., Ludena P., Sentis C., Tyler-Smith C., and Villasante A. 1992. Dodeca satellite: A conserved G+C-rich satellite from the centromeric heterochromatin of *Drosophila melanogaster*. *Proc. Natl. Acad. Sci.* **89:** 4663–4667.

Ashburner M. 1989. Drosophila: *A laboratory handbook.* Cold Spring Harbor Laboratory Press, Cold Spring Harbor, New York.

Berghella L. and Dimitri P. 1996. The heterochromatic *rolled* gene of *Drosophila melanogaster* is extensively polytenized and transcriptionally active in the salivary gland chromocenter. *Genetics* **144:** 117–125.

Bonaccorsi S. and Lohe A. 1991. Fine mapping of satellite DNA sequences along the Y chromosome of *Drosophila melanogaster*: Relationships between satellite sequences and fertility factors. *Genetics* **129:** 177–189.

Bonaccorsi S., Pimpinelli S., and Gatti M. 1981. Cytological dissection of sex chromosome heterochromatin of *Drosophila hydei*. *Chromosoma* **84:** 391–403.

Bonaccorsi S., Santini G., Gatti M., Pimpinelli S., and Coluzzi M. 1980. Intraspecific polymorphism of sex chromosome heterochromatin in two species of the *Anopheles gambiae* complex. *Chromosoma* **76:** 57–64.

Brinkley B., Cox S.M., and Pepper D.A. 1980. Structure of the mitotic apparatus and chromosomes after hypotonic treatment of mammalian cells in vitro. *Cytogenet. Cell Genet.* **26:** 165–176.

D'Aiuto L., Antonacci R., Marzella R., Archidiacono N., and Rocchi M. 1993. Cloning and comparative mapping of a human chromosome 4-specific alpha satellite DNA sequence. *Genomics* **18:** 230–235.

DeCamillis M., Cheng N.S., Pierre D., and Brock H.W. 1992. The *polyhomeotic* gene of *Drosophila* encodes a chromatin protein that shares polytene chromosome-binding sites with Polycomb. *Genes Dev.* **6:** 223–232.

Dernburg A.F., Sedat J.W., and Hawley S.R. 1996. Direct evidence of a role for heterochromatin in meiotic chromosome segregation. *Cell* **86:** 135–146.

Dimitri P. 1991. Cytogenetic analysis of the second chromosome heterochromatin of *Drosophila melanogaster*. *Genetics* **127:** 553–564.

Dorn R., Krauss V., Reuter G. and Saumweber H. 1993. The enhancer of position-effect variegation of *Drosophila*, *E(var)3-93D*, codes for a chromatin protein containing a conserved domain common to several transcriptional regulators. *Proc. Natl. Acad. Sci.* **90:** 11376–11380.

Ellison J.R. and Barr H.J. 1971. Differences in the quinacrine staining of the chromosomes of a pair of sibling species: *Drosophila melanogaster* and *Drosophila simulans*. *Chromosoma* **44:** 424–435.

Faccio Dolfini S. 1974. The distribution of repetitive DNA in the chromosomes of cultured cells of *Drosophila melanogaster*. *Chromosoma* **44:** 383–391.

Fanti L., Giovinazzo G., Berloco M., and Pimpinelli S. 1998. The heterochromatin protein 1 prevents telomere fusions in *Drosophila*. *Mol. Cell* **2:** 527–538.

Funaki K., Matsui S., and Sasaki M. 1975. Location of nucleolar organizers in animal and plant chromosomes by means of an improved N-banding technique. *Chromosoma* **49:** 357–370.

Garzino V., Moretti C., and Pradel J. 1987. Nuclear antigens differentially expressed during early development of *Drosophila melanogaster*. *Biol. Cell* **61:** 5–13.

Gatti M. and Baker B.S. 1989. Genes controlling essential cell cycle functions in *Drosophila melanogaster*. *Genes Dev.* **3:** 438–453.

Gatti M. and Goldberg M.L. 1991. Mutations affecting cell division in *Drosophila*. *Methods Cell Biol.* **35:** 543–585.

Gatti M. and Pimpinelli S. 1983. Cytological and genetic analysis of the *Y* chromosome of *Drosophila melanogaster*. I. Organization of the fertility factors. *Chromosoma* **88:** 349–373.

———. 1992. Functional elements in *Drosophila melanogaster* heterochromatin. *Annu. Rev. Genet.* **26:** 239–275.

Gatti M., Pimpinelli S., and Santini G. 1976. Characterization of *Drosophila* heterochromatin. I. Staining and decondensation with Hoechst 33258 and quinacrine. *Chromosoma* **57:** 351–375.

Gatti M., Tanzarella C., and Olivieri G. 1974. Analysis of the chromosome aberrations induced by X-rays in somatic cells of *Drosophila melanogaster*. *Genetics* **77:** 701–719.

Gonzalez C. and Glover D.M. 1993. Techniques for studying mitosis in *Drosophila*. In *The cell cycle: A practical approach* (ed. P. Fantes and R. Brooks), pp. 143–175. IRL Press, Oxford.

Hazelrigg T., Fornili P., and Kaufman T.C. 1982. A cytogenetic analysis of X-ray induced male steriles on the Y chromosome of *Drosophila melanogaster*. *Chromosoma* **87:** 535–559.

Hilliker A.J. and Appels R. 1982. Pleiotropic effects associated with the deletion of heterochromatin surrounding rDNA on the X chromosome of *Drosophila*. *Chromosoma* **86:** 469–490.

Hofbauer A. and Campos-Ortega J.A. 1990. Proliferation pattern and early differentiation of the optic lobes in *Drosophila melanogaster*. *Roux's Arch. Dev. Biol.* **198:** 264–274.

Holmquist G. 1975. Hoechst 33258 fluorescent staining of *Drosophila* chromosomes. *Chromosoma* **49:** 333–356.

Huang D.W., Fanti L., Pak D.T., Botchan M.R., Pimpinelli S., and Kellum R. 1998. Distinct cytoplasmic and nuclear fractions of *Drosophila* heterochromatin protein 1: Their phosphorylation levels and associations with origin recognition complex proteins. *J. Cell Biol.* **142:** 307–318.

James T.C., Eissenberg J.C., Craig C., Dietrich V., Hobson A., and Elgin S.C.R. 1989. Distribution patterns of HP1, a heterochromatin-associated nonhistone chromosomal protein of *Drosophila*. *Eur. J. Cell Biol.* **50:** 170–180.

Latt S.A. 1973. Microfluorimetric detection of deoxyribonucleic acid replication in human metaphase chromosomes. *Proc. Natl. Acad. Sci.* **70:** 3395–3399.

Lohe A.R., Hilliker A.J., and Roberts P.A. 1993. Mapping simple repeated DNA sequences in heterochromatin of *Drosophila melanogaster*. *Genetics* **134:** 1149–1174.

Matsui S. and Sasaki M. 1973. Differential staining of nucleolus organizers in mammalian chromosomes. *Nature* **246:** 148–150.

Pak D.T.S., Pflumm M., Chesnokov I., Huang D.W., Kellum R., Marr J., Romanowski P., and Botchan M.R. 1997. Association of the origin recognition complex with heterochromatin and HP1 in higher eukaryotes. *Cell* **91:** 311–323.

Palumbo G., Bonaccorsi S., Robbins L., and Pimpinelli S. 1994. Genetic analysis of *Stellate* elements of *Drosophila melanogaster*. *Genetics* **138:** 1181–1197.

Perrin L., Demakova O., Fanti L., Kallenbach S., Saingery S., Mal'ceva N.I., Pimpinelli S., Zhimulev I., and Pradel J. 1998. Dynamics of the sub-nuclear distribution of Modulo and the regulation of position-effect variegation by nucleolus in *Drosophila*. *J. Cell Sci.* **111:** 2753–2761.

Pimpinelli S. and Dimitri P. 1989. Cytogenetic organization of the *Rsp (Responder)* locus in *Drosophila melanogaster*. *Genetics* **121:** 765–772.

Pimpinelli S., Santini G., and Gatti M. 1976. Characterization of *Drosophila* heterochromatin. II. C- and N-banding. *Chromosoma* **57:** 377–386.

Pimpinelli S., Berloco M., Fanti L., Dimitri P., Bonaccorsi S., Marchetti E., Caizzi R., Caggese C., and Gatti M. 1995. Transposable elements are stable structural components of *Drosophila melanogaster* heterochromatin. *Proc. Natl. Acad. Sci.* **92:** 3804–3808.

Platero J.S., Csink A.K., Quintanilla A., and Henikoff S. 1998. Changes in chromosomal localization of heterochromatin-binding proteins during the cell cycle in *Drosophila*. *J. Cell Biol.* **140:** 1297–1306.

Rastelli L., Chan C.S., and Pirrotta V. 1993. Related chromosome binding sites for *zeste*, suppressors of *zeste* and *Polycomb* group proteins in *Drosophila* and their dependence on *Enhancer of zeste function*. *EMBO J.* **12:** 1513–1522.

Sumner A.T. 1990. *Chromosome banding*. Unwin Hyman Ltd., London.

Tsukiyama T., Becker P.B., and Wu C. 1994. ATP-dependent nucleosome disruption at a heat-shock promoter mediated by binding of GAGA transcription factor. *Nature* **367:** 525–532.

Verma R.S. and Babu A. 1989. *Human chromosomes. Manual of basic techniques*. Pergamon Press, New York.

Vosa C.G. 1970. The discriminating fluorescence patterns of the chromosomes of *Drosophila melanogaster*. *Chromosoma* **31:** 446–451.

# CONTENTS

# In Situ Hybridization to Somatic Chromosomes

**Abby F. Dernburg**
*Department of Developmental Biology*
*Stanford University School of Medicine*
*Stanford, California 94305*

Pardue and Gall (1969; Gall and Purdue 1969) pioneered the technique of in situ hybridization as a means to visualize and physically map specific sequences on *Drosophila* polytene chromosomes. Hybridization techniques can also be used to localize sequences on smaller, diploid chromosomes, such as condensed mitotic chromosomes (see Pimpinelli et al., this volume). Variations of the method also allow the hybridization of probes to chromosomes within intact cells and tissues, rather than to chromosomes isolated from their cellular context and flattened on slides. This chapter presents methods for hybridizing fluorescent probes to chromosomes in whole-mount *Drosophila* tissues. These methods allow the investigation of nuclear organization even at stages where chromosomes are decondensed (as in interphase) or, for other reasons, cannot be discriminated in the light microscope. Consequently, they are useful for addressing a variety of cell biological questions.

In addition to enhancing our understanding of somatic chromosome organization, this experimental approach has also revealed interactions among meiotic chromosomes in *Drosophila* females, which spend much of meiosis in a compact ball called the karyosome (Dernburg et al. 1996a). Fluorescent in situ hybridization (FISH) methods can also be used to karyotype individual nuclei using chromosome-specific markers (Dernburg et al. 1996b). With appropriate fixation conditions, hybridization to chromosomal DNA can be performed in conjunction with immunostaining, allowing the colocalization of cellular or chromosomal proteins (see, e.g., Franke et al. 1996; Marshall et al. 1996, 1997).

## CHOICE OF TISSUE FOR INVESTIGATION OF CHROMOSOMES

The *Drosophila* embryo has provided an excellent model system for a variety of questions in chromosome biology, including investigations of somatic chromosome architecture and chromosome-associated proteins. Its advantages include the ease of obtaining populations of staged specimens (see Sisson, this volume), as well as a variety of techniques for experimental manipulation, fixation, and staining of specific cellular components. The organization of the late blastoderm embryo, with a monolayer of nearly synchronous nuclei at the surface, greatly facilitates microscopic imaging and analysis. For many reasons, embryos are often the tissue of choice for cell biological studies in *Drosophila*.

Significant differences, however, have been documented when chromosome architecture in embryos has been compared to that of more "typical" somatic nuclei. In the rapidly dividing blastoderm embryo, chromosomes never have enough time between mitoses to relax from the polarized configuration induced by anaphase chromosome movement. Consequently, they retain a "Rabl" configuration throughout interphase (Rabl 1885): In early embryonic nuclei, centromeric regions are clustered at the nuclear pole closest to the embryo surface, with the chromosome arms extending down along the long axis of the nucleus, resulting in a constrained position along this axis for any given locus (Hiraoka et al. 1993; Dernburg et al. 1996c; Marshall et al. 1996). Conversely, nuclei that experience longer interphases, such as those in larval neuroblasts or imaginal tissues, reveal a marked reduction in this polarity, probably as a consequence of chromosome diffusion (Dernburg et al. 1996c; Cenci et al. 1997; Marshall et al. 1997). Somatic pairing between homologous chromosomes only becomes extensive at the end of blastoderm development (Hiraoka et al. 1993; Fung et al. 1998), probably as another consequence of the rapid pace of early cell cycles. Even in nuclei that divide much less frequently, the extent of somatic pairing can be increased by lengthening the cell cycle (Golic and Golic 1996), implying that chromosomes retain the capacity to diffuse within the nucleus and that homologous regions can associate throughout interphase.

These established differences between nuclei at different developmental stages, as well as other stage- or tissue-specific variations yet to be discovered, warrant careful consideration of the experimental approach. Embryo tissues are the most easily isolated, fixed, hybridized, and imaged. They also tend to produce superior ratios of signal to background in hybridization experiments compared to other whole-mount somatic tissues. However, if answers to the questions being addressed may be influenced by developmental stage, then chromosome organization must be examined within the appropriate context.

## METHODS FOR PROBE SYNTHESIS AND LABELING

Either single-copy or repetitive DNA elements can be targeted in FISH experiments, but the more abundant the target, the greater the likelihood of a detectable signal. Nevertheless, good detection equipment (see p. 54 [Microscopy and Image Analysis]) and some of the better fluorescent labeling reagents have made it possible to detect probes covering as little as a few kilobases of single-copy genomic DNA. Further improvements in labeling technology may push the sensitivity still further.

### Why Use Fluorescence-based Detection?

Compared with other histochemical methods, fluorescent probe detection provides several major advantages for in situ localization of DNA sequences. Fluorescent probes remain physically bound to the site of hybridization, providing optimal spatial localization. The signal-to-background provided is also superior, because the tissue itself does not usually fluoresce strongly. Multiple fluorescent probes can be discriminated in the same sample through the use of specific excitation and emission filters. Fluorescent reagents are stable for long periods if stored frozen and protected from light, and detection is simple and rapid. Microscopy techniques that reduce out-of-focus information allow fluorescent signals to also be localized with great precision within a three-dimensional volume.

FISH probes may be labeled directly with fluorophores, usually by incorporation of fluorescently conjugated nucleotides, or with haptens such as biotin or digoxigenin, which are then detected with fluorescent-binding reagents. Direct labeling is very convenient and tends to produce optimal signal/background staining. In particular cases, fluorescent probes (especially short oligos) will penetrate tissues better than secondary detection reagents such as antibodies or streptavidin.

Hapten-labeled probes coupled with secondary detection, on the other hand, are more economical, given the costliness of some of the better fluorescent dNTPs (deoxynucleoside triphosphates). Hapten labeling also provides versatility, because the same probe may be detected with different secondary reagents in different experiments. Another small advantage is that the fluorescent reagents are only added toward the end of a hybridization procedure, minimizing photodamage during sample handling. Perhaps most importantly, secondary detection allows for signal amplification, because an antibody or avidin molecule that binds to a single hapten molecule can itself carry multiple fluorophores. This may be particularly helpful when laser illumination is used to excite the sample, as in confocal microscopy, or when the signal will be detected by eye. However, the use of multiple layers of antibodies and/or biotin-avidin reagents to enhance the signal also tends to produce high fluorescent background in whole-mount tissues. For this reason, a single application of secondary detection reagent is usually optimal.

## Choice of Fluorophores

A wide spectrum of probe labeling and detection reagents is available through several commercial vendors. These reagents can be very expensive, and the abundance of options makes the choice confusing. Outlined in this section are criteria for selecting reagents, and specific recommendations are presented.

FISH probes may be labeled and detected using a variety of fluorophores. The cyanine dyes Cy2 and Cy3 are more photostable and brighter than their spectral counterparts, fluorescein and rhodamine. Indodicarbocyanine, or Cy5, absorbs red light and emits in the far red, providing an additional region of the spectrum for probe detection. With either laser or mercury arc lamp illumination, Cy3 is the best readily available fluorophore, in terms of brightness and photostability (Wessendorf and Brelje 1992; Wiegant et al. 1996). However, it is not useful for FISH if a red-emitting dye such as propidium iodide is used as a counterstain.

With conventional fluorescence technology, up to three FISH probes plus a counterstain can be distinguished in the same sample, using blue (4′,6-diamidino-2-phenylindole [DAPI]), green (fluorescein/Cy2), red (rhodamine/Cy3/Texas Red), and infrared (Cy5) emission. Combinatorial labeling schemes or other tricks can be used to increase the number of probes that can be used together. For experiments involving multiple probes, some consideration should be given to the relative sensitivities of different reagents, fluors, and detection strategies (see below), so that the smallest target is probed in the most easily detectable way.

Fluorescent nucleotides are available from several commercial suppliers. Fluorescein- and other green-labeled dNTPs have not shown dramatic differences in performance. Cy2 is not yet available directly conjugated to nucleotides. Cy3 is available conjugated to 2′-deoxyuridine-5′-triphosphate (UTP) and 2′-deoxycytidine-5′-triphosphate (dCTP) (Amersham). Cy5 is currently the sole option for far red emission, and it is also available as nucleotide conjugates (Amersham). At this time, no blue-emitting fluorophore works reliably for FISH probes. Accordingly, this wavelength range is conveniently reserved for a counterstain such as DAPI or Hoechst dyes 33258 or 33342. With laser illumination, propidium iodide is frequently used as a counterstain.

Digoxigenin-labeled probes may be detected using fluorescein- and rhodamine-conjugated antidigoxigenin F(ab) fragments (Boehringer Mannheim). Commercially available antibodies that recognize digoxin (Jackson ImmunoResearch), although not marketed for FISH, can also detect digoxigenin-labeled probes and yield excellent signal/background. The latter are available conjugated with Cy2, Cy3, and Cy5, which may provide superior performance in terms of brightness, photostability, and versatility.

Biotin can be detected using fluorescent avidin, modified avidins such as UltraAvidin (Leinco Technologies) or NeutraLite Avidin (Molecular Probes), streptavidin, or antibiotin antibodies. Different secondary reagents may work better in different tissues; unmodified avidin tends to result in higher background than the other choices.

Table 2.1 lists some of the many choices available for labeling and detecting probes in the four most useful wavelength ranges. This is not meant to be a comprehensive list, but to point to some basic reagents and commercial suppliers. It is impossible to provide a recommendation of labeling and detection reagents that are optimal for every situation, because success depends on many parameters, including the labeling method used, autofluorescence spectra of different samples, and microscope illumination sources and fluorescence filters. Moreover, new reagents are constantly becoming available.

Fluorescent nucleotides are the most costly component of in situ hybridization experiments. Unfortunately, only a fraction of the dNTP added to the terminal transferase-labeling reaction is incorporated into the probe; the unincorporated nucleotides are generally discarded after ethanol precipitation. For laboratories carrying out FISH on a large scale, it may be worth the effort to reclaim unincorporated nucleotides by high-performance liquid chromatography (HPLC) purification. Protocols for repurifying Cy3- and Cy5-labeled dNTPs have been developed in Pat Brown's lab at Stanford and are available on their Web Site at: http://cmgm.stanford.edu/pbrown/mguide/hplc.html

## Probe Synthesis

Good probes for whole-mount FISH must meet two criteria: The DNA fragments must be very small, and they must be highly labeled. A labeling scheme (see Protocols 2.1 and 2.2) that has proven reliable involves fragmenting the probe DNA and then adding a mixture of labeled and unlabeled nucleotides to the 3′ ends using the enzyme terminal deoxynucleotidyl transferase (TdT). This method can be used to label a variety of DNA probes, regardless of their initial size (e.g., plasmid, cosmid, or P1 clones, PCR products, or total genomic DNA). Short oligonucleotides may also be labeled in this way without digestion, because their small size allows them to diffuse through thick tissues. A potential advantage of end-labeling is that the modified nucleotides are not incorporated into the complementary probe sequence itself and may thus interfere less with hybridization. The free 3′ tail may also make haptens more accessible to detection reagents.

## Oligonucleotide Probes to Repetitive Sequences from *Drosophila*

The *Drosophila* genome contains a large number of satellite sequences comprising short, regular, or nearly regular repeats, 5–12 bases in length. Many of these satellites have been mapped by hybridization to mitotic chromosomes (Dimitri 1991; Abad et al. 1992; Lohe et al. 1993). Some of these sequences are extremely useful for FISH experiments, e.g., several are specific to particular chromosomes. They also provide excellent FISH targets because of their great abundance and simplicity. Probes to such sequences can be generated as synthetic oligonucleotides, 25–35 bases in length, and labeled according to the

**Table 2.1.** A Useful Subset of Available Labeling and Detection Reagents

| Type of label or probe | Emission | | | |
| --- | --- | --- | --- | --- |
| | blue[a] | green | red | infrared |
| Direct label | Cascade Blue-dUTP (M) | FITC-dUTP (A,B,M,N) | Cy3-dCTP/dUTP (A) | Cy5-dCTP/dUTP (A) |
| Biotinylated probe (B,G,N) | AMCA-antibiotin (J)<br>AMCA-streptavidin (J)<br>AMCA-NeutraLite avidin (M) | Cy2-antibiotin (J)<br>Cy2-streptavidin (J,A)<br>FITC-UltraAvidin (L) | Cy3-antibiotin (J)<br>Cy3-streptavidin (J,A) | Cy5-antibiotin (J)<br>Cy5-streptavidin (J,A) |
| Digoxigenin-labeled probe (B) | AMCA-antidigoxigenin (B)<br>AMCA-antidigoxin (J) | FITC-antidigoxigenin (B)<br>Cy2-antidigoxin (J) | Rhod-antidigoxigenin (B)<br>Cy3-antidigoxin (J) | Cy5-antidigoxin (J) |
| Counterstain | DAPI (S)<br>Hoechst 33242/58 (S) | OliGreen (M)[b] | propidium iodide (S) | |

(A) Amersham; (B) Boehringer Mannheim; (G) GIBCO-BRL/Life Technologies; (J) Jackson ImmunoResearch; (L) Leinco Technologies; (M) Molecular Probes; (N) New Life Science Products; (S) Sigma.

[a]UV-excitable, blue-emitting dyes are less easily detected than others and are thus recommended only for detection of very abundant sequences in experiments involving multiple probes.

[b]There is not currently a well-characterized green fluorescent DNA counterstain that is as generally useful as DAPI, Hoechst, or propidium iodide. OliGreen is marketed as a single-stranded nucleic acid detection reagent, but it has been useful as a green DNA counterstain in Drosophila tissues. It should be tested at a dilution of 1:1000–1:10,000 from the solution provided. YOYO-1 is marketed as a green DNA counterstain, but it may bind to other cellular components in whole-mount tissues. Reprinted, with permission, from Dernburg (1999 [© Oxford University Press].)

**Table 2.2.** Oligonucleotide Probes to Useful Satellite Sequences

| Oligonucleotide sequence | Chromosome(s) arm | Total amount of target | Potential uses | Aliases |
|---|---|---|---|---|
| (AATAC)$_6$ | YL | 3.5 Mb | sexing/karyotyping | |
| (AATAG)$_6$ | 2L (and Y) | 0.2 Mb | 2-specific if no Y is present | |
| (AACAC)$_6$ | 2R | unknown | 2-specific (also C(2)EN) | |
| (AATAACATAG)$_3$ | 2L and 3L | 1.5–2 Mb per chromosome | autosomal centromeres | 1.686 g/cm$^3$ satellite |
| 5′-<u>CCCGTACTGGT</u> <u>CCCGTACTGGT</u> <u>CCCGTACTCGGT</u> <u>CCCGTACTCGGT</u>-3′ | 3R | unknown | 3-specific | dodeca satellite |
| $\{$(AAGAG)$_6$ $($AAGAGAG$)_5$ | all chromosomes predominantly 2, 3, Y | genome contains >10 Mb | bulk heterochromatin, $bw^D$ insertion | 1.705 g/cm$^3$ satellite |
| (AATAT)$_6$ | 4 (and Y) | 3.5 Mb on 4 (dispersed) | major signal on 4 if no Y is present | 1.672 g/cm$^3$ satellite |

method described in Protocol 2.2. Some of the most useful sequences are listed in Table 2.2 and are shown in graphical form in Figure 2.1.

### The 359-bp Repeat on the *X* Chromosome

The proximal heterochromatin of the *Drosophila melanogaster X* chromosome contains a huge number of copies of a 359-base sequence originally known as the 1.688 g/cm$^3$ satellite. This sequence provides an excellent FISH target that is specific to the *X* chromosome (see Figure 2.4). The repeated sequence can be amplified by polymerase chain reaction (PCR) using genomic DNA as a template with the following primers, which were based on the published sequence (Hsieh and Brutlag 1979): 5′-CGGTCATCAAATAATCATT-TATTTTGC-3′ (27-mer) and 5′-CGAAATTTGGAAAAACAGACTCTGC-3′ (25-mer). Using *Taq* polymerase and standard PCR buffer containing 2 mM Mg$^{++}$, 25 cycles with an annealing temperature of 55°C, and a 30-second extension time at 72°C should produce a single product of 361 bases. Higher-molecular-weight multimers may also be detected. This PCR product can be effectively digested into fragments suitable for labeling using a single enzyme, either *Alu*I or (slightly better) *Tsp*509I.

### Probe Template Amplification by DOP-PCR

It may be useful in some cases to amplify DNA before labeling it. DOP-PCR (degenerate oligonucleotide primed-PCR; Telenius et al. 1992) allows amplification of any complex DNA, regardless of sequence. One example of the technique's value for chromosome hybridization is to generate a probe covering hundreds of kilobases of single-copy sequence. A pool of P1 or YAC (yeast artificial chromosome) clones would be difficult to prepare on a large enough scale, but DNA minipreparations of several P1s or gel slices of one or more YACs could each be amplified and then pooled to make a probe. A more extreme version of this technique is chromosome painting, in which a large regional probe, or even a probe covering an entire chromosome arm, is generated from a minute quantity of DNA microdissected from a polytene chromosome squash. A procedure for microdissection of polytene chromosomes is given in Protocol 2.3.

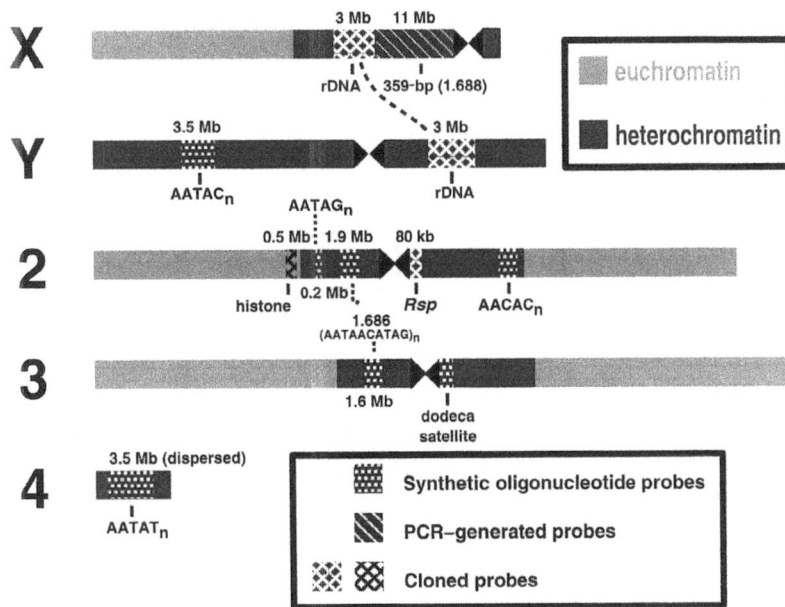

**Figure 2.1.** Probes to chromosome-specific repetitive sequences in the *Drosophila melanogaster* genome. The five chromosomes are shown schematically, with heterochromatic regions shaded *dark gray* and euchromatin in *light gray*. Probes that localize to only one or two chromosomes are useful tools for cytological studies. For many short, simple satellite repeats within the pericentric heterochromatin, oligonucleotide probes of 30–50 bases in length can be used to mark specific chromosomal targets (see Table 2.2). For the 359-bp repeat on the *X* chromosome (Hsieh and Brutlag 1979), a single repeat unit can be synthesized by PCR and labeled to make a probe, as described in the text. For some complex sequences such as the 4.8-kb histone gene repeat (Lifton et al. 1978), the 240-bp *Rsp* repeat (Wu et al. 1988), and the ribosomal DNA (28S and 18S subunits), cloned sequences have been used. The positional data in this diagram are adapted primarily from Lohe et al. (1993). The dodeca satellite was discovered and mapped by Villasante and coworkers (Abad et al. 1992; Carmena et al. 1993). The *Rsp* locus was mapped genetically by Brittnacher and Ganetzky (1989) and cytologically by Pimpinelli and Dimitri (1989). The AACAC repeat in *2R* heterochromatin was discovered by Zhimulev and coworkers (Makunin et al. 1995). Estimates of the amount of each repeat on the particular chromosome are indicated if available from published work. Certain satellite repeats, including the AATAT and AATAG repeats, are abundant on the *Y* chromosome, in addition to their unique autosomal positions (Lohe et al. 1993), and are thus primarily useful for studying female tissues. Their distribution on the *Y* chromosome is not shown here. (Reprinted, with permission, from Dernburg and Sedat 1998.)

DOP-PCR is performed using a single primer that has a fixed sequence at its 5′ end, and several degenerate bases near its 3′ end that allow it to anneal at low stringency to many sites in complex DNA. Following several PCR cycles with an annealing temperature of only 30°C, the annealing temperature is raised so that only products with the primer sequence at each end are amplified in the reaction. An oligonucleotide that has worked well as a primer for *Drosophila* DNA has a recognition site for the enzyme *Bsm*I, which cuts 3′ to its recognition sequence, so that it can be removed following amplification: 5′-CCCAACGATGCGAATGCNNNNNNCAGG-3′ (*Bsm*I recognition sequence is underlined; N indicates an equal mixture of all four bases). This primer has resulted in better probes than the most widely used primer for DOP-PCR (5′-CCCGACTCG-GNNNNNNATGTGG-3′) in direct comparisons.

For DOP-PCR amplification of microdissected templates, extreme caution must be taken to avoid any DNA contamination of reagents or tubes, because only a few picograms of template will be available. Gloves and barrier tips, along with UV irradiation of the solutions (other than those containing template or nucleotides), may help to ensure specific amplification of the desired template. For primary and secondary DOP-PCR procedures, see Protocols 2.4 and 2.5.

PROTOCOL 2.1*

# Restriction Enzyme Fragmentation of Probe DNA

## Materials

### *Supplies and Equipment*

Polypropylene microcentrifuge tubes (1.5 ml)
Equipment for agarose gel electrophoresis

### *Solutions and Reagents*

Template DNA

BSA (bovine serum albumin) (10 mg/ml; molecular biology grade; New England Biolabs or other supplier)

**DTT** (dithiothreitol) (100 mM) aqueous solution, stored frozen in aliquots

Restriction enzymes: *AluI*, *HaeIII*, *MseI*, *MspI*, *RsaI*, and *Sau3AI* (New England Biolabs; order at high concentration where available)

Sodium acetate (3 M, pH 5.2)

**Ethanol** (cold 70% and absolute)

Buffers and reagents for agarose gel electrophoresis

---

*5× 4BC buffer*

40 mM Tris-Cl (pH 7.5)
250 mM NaCl
40 mM **MgCl₂**

*TE′*

10 mM Tris-Cl (pH 7.5)
0.1 mM EDTA

---

**CAUTION: DTT, ethanol, MgCl₂** (see Appendix 4)

## Method

1. Set up the following reaction in a 1.5-ml polypropylene microcentrifuge tube:

   | | |
   |---|---|
   | Probe DNA | 25 µg |
   | 5× 4BC buffer | 50 µl |
   | 10 mg/ml BSA | 2.5 µl |
   | 100 mM DTT | 2.5 µl |
   | H₂O (see Note below) | to ~225 µl |

   *Note:* The final reaction volume should be 250 µl. Adjust volume of H₂O depending on the volume of enzymes required (step 2, below). This reaction is for 25 µg of DNA and may be scaled up or down, as desired.

*For all protocols, H₂O indicates glass distilled and deionized.

2. Add 25 units of each restriction enzyme (add all six enzymes as in a multiple digest). Mix and incubate at 37°C from 2 hours to overnight.

3. Precipitate the digested products as follows:

   a. Add 1/10 volume of 3 M sodium acetate and then 2.5 volumes of cold absolute ethanol. *Optional:* Chill briefly at –20°C.

   b. Centrifuge in a microcentrifuge at 12,000*g* for 15 minutes to pellet the DNA.

   c. Wash DNA pellet with cold 70% ethanol, allow to dry briefly, and then resuspend DNA in TE′ to a final concentration of 1 μg/μl.

4. Analyze 0.5 μg (2%) of the products by electrophoresis through a 2% agarose gel.

   *Note:* Depending on the complexity of the DNA, either a discrete set or a smear of fragments with an average size of 100–150 bp should be detected.

5. Measure the concentration accurately by UV absorption.

PROTOCOL 2.2

## Labeling with Terminal Deoxynucleotidyl Transferase

This procedure labels 10 μg of probe DNA, enough for 20–500 FISH samples, depending on the length and complexity of the target sequence. It can easily be scaled down and carried out in a smaller volume.

## Materials

### *Supplies and Equipment*

Microcentrifuge tubes (1.5 ml) (amber microcentrifuge tubes are useful for labeling and storage of fluorescent probes)
(*Optional*) Sephadex G-25 spin column
Water bath at 95°C

### *Solutions and Reagents*

5x TdT reaction buffer (Promega)
Labeled and unlabeled dNTPs
TdT (Promega)
Glycogen (20 mg/ml) (molecular biology grade; Boehringer Mannheim or other supplier)
**Ammonium acetate** (4 M)
**Ethanol** (cold 75% and absolute)
TE (pH 7.5; see Appendix 3)

**CAUTION: ammonium acetate, ethanol** (see Appendix 4)

## Method

1. Place a 1.5-ml microcentrifuge tube on ice. Add 10 μg of DNA fragments in TE′ (from step 3c in Protocol 2.1). Heat the tube in a 95°C water bath for 2 minutes to denature the DNA. Chill rapidly on ice and spin briefly to collect condensate.

2. Add the following to the denatured DNA:

| | |
|---|---|
| 5× TdT labeling buffer | 20 µl |
| unlabeled dNTP (whichever the labeled nucleotide analog is derived from, i.e., use dTTP if incorporating a modified dUTP) | 13.5 nmoles |
| labeled dNTP (usually supplied as a 1 mM stock solution) | 6.75 nmoles (6.75 µl of a 1 mM stock solution) |
| $H_2O$ | adjust volume to 97 µl |

3. Add 60 units of TdT and incubate at 37°C for 1 hour.

4. Add 2 µl of 20 mg/ml glycogen to act as a carrier.

5. (*Optional*) Transfer the probe to a 1-ml Sephadex G-25 spin column (or a commercial version) and centrifuge to remove unincorporated nucleotides.

   *Note:* This may help to reduce background staining (reports vary) but will also result in loss of some precious probe. Spin column purification should not be used when labeling oligonucleotide probes.

6. Precipitate the probe as follows:

   a. Add ammonium acetate to 2 M followed by 2.5 volumes of absolute ethanol. Chill at 4°C.

   *Note:* Sodium acetate is not used here to minimize the coprecipitation of unincorporated nucleotides.

   b. Centrifuge in a microcentrifuge at 12,000*g* for 15 minutes to pellet the DNA.

   *Note:* Probes labeled with fluorescent nucleotides should give a visibly colored pellet.

   c. Aspirate the ethanol carefully, wash the pellet with cold 75% ethanol, and dry briefly.

7. Resuspend in TE. Store at –20°C.

   *Note:* It is convenient to store probes at a concentration of 50–200 ng/µl. Probes are stored in TE, but made up during the hybridization procedure in hybridization buffer (see p. 47).

PROTOCOL 2.3

# Microdissection of Polytene Chromosomes for DOP–PCR

## Materials

### Supplies and Equipment

Razor blade

Dissection needles (1-mm glass capillary tubes; one needle will be required per chromosome) (World Precision Instruments 1B100F-4)

Needle puller

Plastic dish containing a strip of double-sided tape or modeling clay for holding needles

Stratalinker or other short-wave **UV** source

PCR tubes (0.5 ml)

Specialized equipment for microdissection:

- an inverted microscope (preferably equipped with a freely rotating stage)
- microinjection/micromanipulation apparatus
- good-quality phase lens (10–20×) with a long working-distance, to observe chromosomes through the glass slide

Water bath at 95°C

### *Solutions and Reagents*

Template DNA
**Liquid nitrogen**
TE (see Appendix 3)
**Ethanol** (95%)

**CAUTION: ethanol, liquid nitrogen, UV radiation** (see Appendix 4)

## Method

1. Prepare squashes of salivary glands from female third-instar larvae (raised on a rich diet under uncrowded conditions). Use lactic-acetic acid without orcein or other DNA dyes (see Pimpinelli et al., this volume [Protocol 1.2]).

   *Note:* The ideal squash will have chromosome arms extended with long straight regions that are well separated from each other. This can be achieved by using a slightly greater volume of liquid and/or more pressure than normal in the spreading step of the squash. One good slide can provide the material needed for a large number of dissections.

2. Freeze the slides in liquid nitrogen. Immediately crack the coverslips using a new razor blade as a wedge at one corner and transfer to 95% ethanol. After at least 10 minutes, remove slides and allow to air dry.

   *Notes:* This step minimizes exposure to acid, which can depurinate the template.
   Slides dipped in liquid nitrogen can crack or splinter. Wear appropriate eye protection.

3. Prepare dissection needles. Wear gloves to avoid contaminating the glass needles. Pull each capillary tube into two needles using any program for microinjection needles on a needle puller (the ends are broken off later, so the exact profile of the needle is not critical).

4. Break off the tip of each needle under a dissecting microscope. Make a clean, even break by drawing the tip of one needle along a broader portion of another needle held perpendicularly, with light and even pressure. With practice, most needles break cleanly to leave a smooth circular edge. *Discard* any needles with jagged edges. Place the needles in the plastic dish containing a strip of double-sided tape or modeling clay. Irradiate needles by placing the dish in a Stratalinker (or other short-wave UV source) for 10 minutes at full power.

5. Pipette 20 μl of TE into each of several clean PCR tubes. Irradiate the tubes for 10 minutes with UV light.

6. Get ready to dissect! Turn on the microdissection equipment.

7. Place a clean needle into the needle holder of the microdissection apparatus and mount over the stage. Adjust the apparatus to give maximum range of motion with-

in the visual field of the microscope. Turn the back pressure on the microinjector to a positive but low value (4–6 psi). Mount a slide on the stage, chromosome-side up, and use clips or tape to hold it to the stage securely.

8. Focus the objective and move the stage until a target chromosome is within view (Figure 2.2A). Choosing a specific target requires the ability to recognize the banded pattern of the polytene chromosomes under phase optics without stains such as orcein. To test the procedure, an arbitrary chromosome arm may be chosen.

9. Focus on a plane above the slide, and slowly lower the needle and move it laterally until its tip is in focus. Lower the focus again to see the chromosomes (Figure 2.2B).

10. At a low-speed setting on the micromanipulator, lower the tip of the needle carefully onto the chromosome. Then scrape along the chromosome arm, essentially using the edge of the needle as a bulldozer, moving it laterally while holding it against the slide (Figure 2.2C). As the needle pushes against the chromosome, it will lift off the slide and bunch up into a ball of material. When the end of the chromosome (or the desired region) is reached, continue to move the needle laterally very slowly while lifting it with the micromanipulator; the microdissected material will usually stick to it. Lift the needle well away from the stage.

11. Very carefully, move a tube containing 20 µl of UV-irradiated TE (from step 5, above) over the needle so that the tip of the needle is immersed in the liquid. Raise the back-pressure on the injector to about 30–50 psi. Still holding the plastic tube over the needle with one hand, remove the needle holder from its mount with the other hand. Break off the tip of the needle against the wall of the tube while still holding it immersed in the solution. Gas should start to bubble through the liquid. Withdraw the rest of the needle, trying to avoid removing any liquid from the tube. Set the needle down, cap the tube, label it appropriately, and discard the needle. Turn the injector back pressure down again when this is completed. Repeat the above procedure (steps 7–11) for as many chromosomes as desired. Transfer only one dissected chromosome/region per tube; successfully amplified samples may be pooled *after* the PCR.

12. Heat the tubes at 95ºC in a water bath for 5 minutes.

## PROTOCOL 2.3 NOTES

If the air in the microdissection room is dry, dissected chromosome pieces may fly away instead of sticking to the glass dissection needles. The environment may be humidified by heating an open water bath in the room to 80–90ºC.

PROTOCOL 2.4

# Primary DOP-PCR Amplification

The following amplification procedure may be performed on microdissected material, small volumes of melted gel slice (1–5 µl), or small quantities of other DNA templates. Always perform a negative control with no added template.

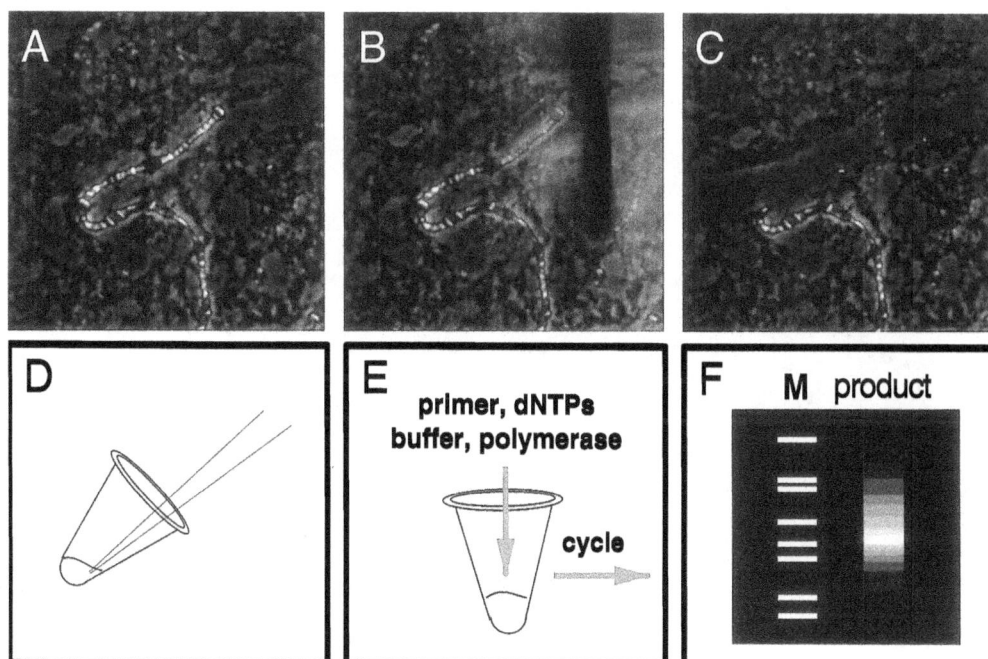

**Figure 2.2.** Synthesis of chromosome painting probes by microdissection and DOP-PCR. This figure illustrates the steps described in the text for polytene chromosome microdissection and DOP-PCR amplification to generate large regional probes. (*A*) Portion of a squashed, unstained polytene chromosome arm (*2*R), visualized with phase optics. (*B*) The broad tip of a microdissection needle is brought down onto the slide at a position near the end of the chromosome arm. (*C*) Microdissection has been carried out by dragging the needle from right to left across the chromosome, then lifting it off the slide; the missing distal region of the chromosome arm is now stuck to the tip of the needle. This tiny quantity of DNA (on the order of 1–10 pg) is then transferred to a small volume of liquid in a 0.5-ml microcentrifuge tube (*D*). Other components required for DOP-PCR are added (*E*), and the amplification is carried out as described in the text. Following a successful amplification, the products can be visualized by gel electrophoresis (*F*), and will normally comprise a distribution of fragment sizes.

## Materials

### *Supplies and Equipment*

PCR tubes (0.5 ml)
Thermal cycler
Equipment for agarose-gel electrophoresis

### *Solutions and Reagents*

Template DNA
**MgCl$_2$** (50 mM)
*Bsm*I primer (10 μM stock): 5′-CCCAACGATGCGAATGCNNNNNCAGG-3′
dNTPs (2 mM each dATP, dCTP, dGTP, and dTTP)
*Taq* DNA polymerase (10 units/μl)
(*Optional*) Mineral oil
DNase 1 (1 μg/ml)
Buffers and reagents for agarose-gel electrophoresis

*10x DOP-PCR Buffer (Meltzer et al. 1992)*

100 mM Tris-Cl (pH 8.3)
500 mM **KCl**
1 mg/ml gelatin (molecular biology grade)

*10x DNase Buffer*

0.5 M Tris-Cl (pH 7.5)
0.1 M **MgSO₄**
1 mM dithiothreitol (**DTT**)
500 μg/ml bovine serum albumin (BSA)
Store in aliquots at –20°C.

**CAUTION: DTT, KCl, MgCl₂, MgSO₄** (see Appendix 4)

## Method

1. To each tube containing microdissected material (from step 12, Protocol 2.3) or 10 pg to 5 ng DNA template, or no template, add:

   | | |
   |---|---|
   | 10x DOP-PCR buffer | 5 μl |
   | 50 mM MgCl₂ | 2.5 μl |
   | *Bsm*I primer (10 μM stock) | 10 μl |
   | 2 mM dNTPs | 5 μl |
   | H₂O | to 48 μl |
   | 2.5 units *Taq* DNA polymerase (DNase treated, see Notes below) | 2 μl |

   *Notes:* Overlay PCRs with mineral oil or use a thermal cycler with a heated lid.

   DNase treatment is essential to degrade DNA that may contaminate commercial preparations of *Taq* DNA polymerase. It is convenient to perform this reaction in a PCR tube and the incubations in a thermal cycler. This recipe provides enough enzyme for five PCR samples; scale up or down as needed.

   Place a 0.5-ml PCR tube on ice. Add the following to the tube:

   | | |
   |---|---|
   | H₂O | 6.6 μl |
   | *Taq* DNA polymerase (5 units/μl) | 3 μl |
   | 10x DNase buffer | 1.2 μl |
   | 1 μg/ml DNase | 1.2 μl |

   Incubate at 37°C for 15 minutes. Heat at 90°C for 10 minutes to inactivate the DNase.

2. Perform thermal cycling to amplify the DNA using the following program:

   93°C for 4 minutes

   3 cycles:
       94°C for 30 seconds
       30°C for 1 minute

   Ramp to 72°C over 5 minutes (ramp = 7.2 seconds/°C or 84°C/minute) and hold 1 second.

   3 cycles:
       94°C for 30 seconds
       30°C for 1 minute
       72°C (no ramp) for 2 minutes

36 cycles:
    94°C for 20 seconds
    56°C for 1 minute
    72°C for 2 minutes

Extension cycle at 72°C for 10 minutes.

3.  Electrophorese 2 μl of each reaction in DNA sample buffer through a 1.4% agarose/TBE gel. Compare samples with pBR322 digested with *Msp*I or other appropriate size markers.

    *Notes:* There should be a smear of fragments in each lane from reactions including template DNA. Do not be alarmed if there is also a significant background smear in the control lane (minus template), particularly if the lanes with template look brighter or the fragment size distribution is different (usually longer). If the template is relatively small (e.g., a P1 or a YAC), a discrete banding pattern may be seen, but from microdissected chromosome templates, the mixture of products is usually too complex to see discrete products. In fact, a banding pattern could be an indication of contamination, particularly if the bands are consistent among different lanes.

    No amplification at all indicates a problem. If the negative control lane looks identical to the samples with template, pick just one or two of the samples to test as probes, because this may indicate contamination with extraneous DNA in a solution or in the handling procedure.

4.  Store the reaction mixture at –20°C.

    *Notes:* It is not necessary to remove oil.

    A successful primary amplification provides a nearly inexhaustible source of probe. All of the reaction products now have the primer sequence at each end, so they are simply amplified by a normal, high-stringency PCR using the same primer. In this procedure, contamination is less of a concern, but it is always a good idea to be scrupulously clean when performing the PCR. To generate a probe, simply reamplify a tiny amount (0.5 μl) using Protocol 2.5.

## PROTOCOL 2.5

# Secondary DOP-PCR Amplification

### Materials

### *Supplies and Equipment*

PCR tubes (0.5 ml)
Thermal cycler
Microcon-30 columns (500-μl capacity; Amicon)

### *Solutions and Reagents*

10x DOP-PCR buffer (see p. 38)
**MgCl$_2$** (50 mM)
dNTPs (1.25 mM each dATP, dCTP, dGTP, and dTTP)
*Bsm*I primer (10 μM stock)
*Taq* DNA polymerase (10 units/μl)
(*Optional*) mineral oil
(*Optional*) NEB Buffer 2
(*Optional*) *Bsm*I enzyme (New England Biolabs, 5 units/μl)
(*Optional*) BSA (bovine serum albumin; 10 mg/ml; molecular biology grade; New England Biolabs or other supplier)

(*Optional*) EDTA (500 mM)

TE (see Appendix 3)

**CAUTION: MgCl₂** (see Appendix 4)

## Method

1. To a PCR tube, add the following:

| | |
|---|---:|
| H₂O | 22.5 μl |
| 10x DOP-PCR buffer | 5 μl |
| 50 mM MgCl₂ | 2 μl |
| 1.25 mM dNTPs | 8 μl |
| *Bsm*I primer | 11.5 μl |
| template (primary amplification mixture from step 4, Protocol 4) | 0.5 μl |
| *Taq* DNA polymerase (DNase treatment not necessary) | 0.5 μl |

2. Perform thermal cycling using the following program:

   93°C for 4 minutes

   16 cycles:
   94°C for 30 seconds
   56°C for 1 minute
   72°C for 2 minutes

   Extension cycle at 72°C for 10 minutes.

3. (*Optional*) Perform enzymatic digestion with *Bsm*I enzyme to cleave the ends as follows:

   *Note:* If the *Bsm*I primer was used for DOP-PCR, the primer sequences can be removed by digestion of the products with *Bsm*I restriction enzyme. This may be useful because it removes the nonspecific sequences that do not contribute to probe specificity, but no experimental comparison has been performed.

   a. Without removing the oil, add:

   | | |
   |---|---:|
   | NEB Buffer 2 | 8 μl |
   | 10 mg/ml BSA | 1 μl |
   | H₂O | 40 μl |
   | *Bsm*I enzyme (5 units/μl; New England Biolabs) | 3 μl |

   b. If oil was used in the PCR, centrifuge the tube briefly to mix the aqueous phases. Incubate at 65°C for 90 minutes.

   c. Add 1 μl of 500 mM EDTA to stop the reaction.

4. Purify products as follows:

   a. If an oil overlay was used, remove the oil.

   b. Remove primers, cleaved ends, and other small molecules by centrifuging the sample through a Microcon-30 filter: Pipette the oil-free PCR into the upper chamber of a Microcon-30 filter. Add TE to 450 μl. Centrifuge according to the manufacturer's instructions (14,000*g* for ~12 minutes). Add 450 μl of TE and centrifuge again. Collect retentate. If residual volume is very small (<10 μl), pipette 10–20 μl of TE onto the filter, mix by tapping, invert, and place into collection tube. Centrifuge again to pool this rinse with the retentate.

5. Determine the DNA concentration by measuring UV absorbance before proceeding with probe synthesis.

6. Cut and label the PCR products as described in Protocols 2.1 and 2.2 (above) to generate a probe.

## FIXATION METHODS

The fixation step is absolutely critical for success with whole-mount in situ hybridization, as defined by two criteria: (1) reliable detection of a specific signal and (2) preservation of the nuclear and tissue structure to the degree required by the experiment. Probe molecules must be able to diffuse through the tissue, so the tissue must be somewhat permeable. At the same time, the sample must be able to withstand the high salt, heat, and formamide required for hybridization without undergoing large-scale structural changes. How this is accomplished is the least generalizable facet of FISH experiments, because various tissues respond differently to identical fixation conditions.

In general, fixation of tissues with moderate concentrations (1.5–5%) of formaldehyde provides the best combination of structural preservation and accessibility to hybridization. Ideally, fixation is carried out in an aqueous medium, preferably in an isotonic buffer (e.g., Ringer's solution or other balanced salts). However, the signal/background can often be increased at the expense of morphology by fixing with a mixture of acetic acid and alcohol, or alcohol alone. In special cases, formaldehyde fixation may not be compatible with other experimental goals (e.g., preservation of cytoskeletal elements) and other fixatives may be substituted. Postfixation with formaldehyde may help to preserve a sample fixed initially with alcohol or acid/alcohol. Glutaraldehyde is an unsuitable fixative as it increases the background fluorescence and prohibits hybridization.

With certain tissues, additional steps will be required for permeability. Such treatments might include incubation with nonionic detergents or dissection to remove physical barriers. Modifications that have proven to be useful with particular tissues include (1) warming the fixative solution to 30–37°C before addition to the sample; (2) fixing at low temperature, typically 0–4°C for 1 hour or more; (3) reducing the formaldehyde concentration to 1–2%; and (4) including both 0.1% deoxycholate and 0.1% Triton X-100 in the dissection buffer. Any of these may affect chromosome preservation, however.

Fixation conditions that are compatible with immunodetection are often a good starting point for FISH. Both procedures may even be performed sequentially on the same sample. By combining immunofluorescence and FISH, interactions between chromosomal regions and other components of nuclear architecture may be examined. Immunostaining with antibodies to the nuclear lamina or to nuclear pore proteins is also a good way to delineate the nuclear periphery (Dernburg et al. 1996c; Gotta et al. 1996; Marshall et al. 1996). In general, it is simplest to perform immunostaining steps following FISH. Surprisingly, however, antibodies bound prior to the hybridization can also be detected. Most antibodies show good reactivity after hybridization, although the epitopes recognized by certain monoclonal antibodies may be altered. Soluble proteins may also be extracted to some degree during the hybridization procedure.

### Embryos

For *Drosophila* embryos, the best method to use is also the most common fixation procedure for immunostaining, as described by Rothwell and Sullivan (this volume; see Protocol 9.3). Briefly, after bleach dechorionation, transfer embryos to a test tube or vial containing 3.7% formaldehyde (freshly prepared from paraformaldehyde) in buffer under an equal volume of heptane. Agitate for 10 minutes to keep the phases mixed, then transfer embryos to methanol/EGTA (ethyleneglycol-bis-[β-aminoethylether]-$N,N,N',N'$-tetraacetic acid) equilibrated on dry ice. After several minutes on dry ice, heat-shock to remove the vitelline membrane. Once the embryos settle, immediately begin to rehydrate the embryos through a methanol:$H_2O$ series (9:1, 7:3, 5:5, 3:7, 1:9, 5–10 minutes each) followed by 100% buffer. (Embryos left for an extended time in methanol become overextracted and float instead of sinking during the hybridization procedure.)

There are at least two reasons that other embryo fixation procedures might be used, however: (1) Cytoskeletal elements, particularly microtubules, are not preserved well by the method above and (2) mutations affecting the chemistry of the vitelline membrane may preclude successful devitellinization. In either case, a one-step fixation/devitellinization in methanol/EGTA solution may be acceptable, although the nuclear morphology is not as well stabilized by this procedure.

## Egg Chambers

*Drosophila* egg chambers are more challenging to fix than embryos. They are easily overfixed such that probes will not penetrate well into the oocyte nucleus, even if some hybridization is detected in the follicle and nurse cells closer to the surface. If late-stage oocytes are to be examined, the tissue must also be protected against osmotic shock before and during fixation, because this will activate metaphase-arrested nuclei to undergo the metaphase-anaphase transition. The large-scale preparatory method described by Matthies et al. (this volume) for immunostaining may be used, but reduce the formaldehyde solution to 3.7% for younger egg chambers and 5–8% to optimize hybridization to older (stage 10 and above) chambers. Warming the buffered formaldehyde to 30°C immediately prior to fixation tends to enhance probe penetration. In contrast to reported results with immunostaining, FISH is equally successful whether fixation is performed with freshly prepared formaldehyde or commercial formaldehyde solutions.

Manual dissection of fixed ovarioles gives a much higher yield per adult than the blender technique (facilitating analysis of mutants), and younger egg chambers in particular will be enriched, although the procedure itself is more laborious. A procedure for manual dissection and fixation of egg chambers is given in Protocol 2.6.

## Other Tissues

Protocol 2.7 should serve as a starting point for other tissues and is a general approach to fixing tissue or cells onto microscope slides in a manner compatible with whole-mount FISH analysis. The procedure includes a simple formaldehyde fixation in buffer, followed by postfixation in cold ethanol. Postfixation of the tissue in methanol or ethanol can markedly improve permeability. In some instances, it may be preferable to fix tissue onto coverslips, because microscope optics are usually designed to optimize imaging immediately adjacent to the coverslip. On the other hand, coverslips are much more fragile and harder to manipulate.

To evaluate the success of a particular fixation procedure, the hybridized sample should be compared to identically fixed tissue that has not undergone FISH. In particular, the appearance of condensed mitotic or meiotic chromosomes with and without hybridization can provide a useful criterion for deciding whether preservation is adequate (Figure 2.3). Underfixation will result in deterioration of morphology; overfixation will lead to weak or undetectable FISH signals and high fluorescent background. With some tissues, the middle ground may be a challenge to find. The ultimate goal is to maintain the organization found in the living state. However, this requirement can be relaxed somewhat if the question being asked does not address chromosome organization, e.g., if only the presence or absence of particular sequences is being assayed. As with any cytological procedure, consideration should be given to ensuring that results do not represent fixation artifacts. When different fixation procedures converge to yield the same answer, confidence in the experimental results is enhanced.

PROTOCOL 2.6

# Manual Dissection and Fixation of Egg Chambers

## Materials

### Supplies and Equipment

Dissecting tools

- dissecting surface; e.g., a silicone rubber dissecting pad made with Sylgard silicone elastomer (Dow Corning) cured in a large petri plate, or a siliconized multiwell slide (see below)
- fine dissecting forceps, such as DuMont et Fils #5 (Roboz)
- stereo dissecting microscope

Pasteur pipette

Polypropylene tube(s) (0.5 ml), such as those used for PCR

### Preparation of Siliconized Glass Slides and Coverslips

Immerse slides or coverslips briefly in a 1–2% solution of Surfasil (Pierce) in **chloroform**. Rinse once with chloroform and allow to air dry. Store at room temperature.

### Solutions and Reagents

**Ether** or **$CO_2$**

1x Modified Robb's saline (MRS) (see Matthies et al. [Protocol 4.1])

Cacodylate Fixative Buffer (see Matthies et al. [Protocol 4.1])

### 2x SSCT

0.3 M NaCl
0.03 M sodium citrate
0.1% Tween-20

### Preparation of 37% Formaldehyde

To a 100-ml Pyrex screw-cap tube, add 1.85 g of EM grade **paraformaldehyde** (Polysciences) and 3.5 ml of $H_2O$. Cap loosely and place in a boiling water bath.

Add 90 µl of 1 N **NaOH** and agitate for approximately 1 minute. The solution should become nearly clear. Cool by running the bottom of the tube under a stream of tap water. Pass the solution through a nonsterile 0.22-µm syringe filter into a clean screw-cap airtight vial. Total volume is ~5 ml. Use the same day.

CAUTION: **cacodylate, chloroform, $CO_2$, ethanol, ether, NaOH, paraformaldehyde**
(see Appendix 4)

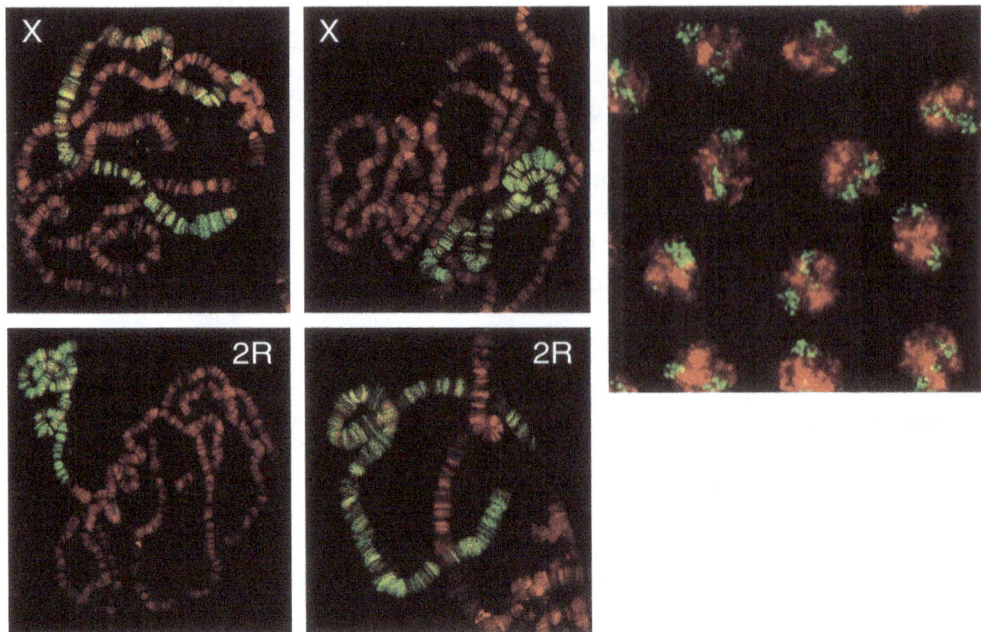

**Figure 2.3.** Chromosome painting probes on polytene and embryonic chromosomes. These images show different regional probes generated by microdissection and DOP-PCR. (*Four panels at left*) Two different probes, one to the X chromosome and one to 2R, are hybridized to polytene chromosome squashes to assess their coverage and specificity. (*Green*) Hybridization signal; (*red*) DNA counterstain (DAPI). These successful probes give very good coverage of complete or nearly complete euchromatic arms. Very limited cross-hybridization is detected on other chromosome arms, mostly in telomeric regions. (*Right*) Whole-arm probe (here, 3L) hybridized in situ to a whole-mount blastoderm embryo. In most of the nuclei, two separate large signals are detected, representing the separate domains occupied by the two homologous arms. In one nucleus, the homologous signals are either immediately adjacent to each other or interspersed with each other. At later nuclear cycles, homologous pairing is much more extensive (Hiraoka et al. 1993), and the two homologous signals are rarely separate.

## Method

1. Anesthetize four to eight females with carbon dioxide or ether. Transfer to a 100-μl drop of MRS on a silicone rubber dissecting plate or siliconized slide.

2. Rupture the abdomens with forceps, and transfer the intact ovaries to a fresh 100-μl drop of MRS.

3. When all the ovaries have been dissected, transfer them to a 100-μl drop of fixative (5% formaldehyde made by diluting freshly prepared 37% solution in cacodylate buffer, prewarmed to 37°C) on the dissecting surface. Start a timer set for 4 minutes.

4. During the fixation (4 minutes), begin to separate individual ovarioles by teasing them apart with forceps.

5. At the end of the fixation period, transfer the ovaries individually to a large drop of 2x SSCT. Continue separating the ovaries into individual ovarioles. This process will take several minutes.

6. Use a pasteur pipette or forceps to transfer the ovarioles into a 0.5-ml tube containing 2x SSCT. Wash with two changes of 2x SSCT.

7. Repeat steps 1–6 until sufficient material for hybridization has been obtained. Proceed with hybridization as described in Protocol 2.8.

## PROTOCOL 2.7

## Formaldehyde Fixation onto Slides for Whole-mount FISH

### Materials

### *Supplies and Equipment*

Dissecting tools

- dissecting surface; e.g., a surface made with Sylgard silicone elastomer (Dow Corning), a siliconized multiwell slide, or siliconized coverslips (see p. 43)
- fine dissecting forceps, such as DuMont et Fils #5 (Roboz)
- scalpel or razor blade
- stereo dissecting microscope

Siliconized coverslips (18 × 18 mm; see p. 43)

Glass slides. For some tissues, it is helpful to first treat slides with agents that improve adhesion, including polylysine solutions, aminoalkylsilane, and gelatin "subbing" solutions. Commercially available treated slides such as Superfrost Plus (Fisher) may also be useful.

Whatman #3 filter paper

Hemostats, the tips covered with short pieces of silicone rubber tubing, for handling slides

### *Solutions and Reagents*

Dissecting buffer (typically a buffered saline solution such as Ringer's or phosphate-buffered saline [PBS]; see Appendix 3)

**Formaldehyde** (37%) prepared fresh from paraformaldehyde

**Liquid nitrogen** in a Dewar flask, or an aluminum block on **dry ice**, to freeze slides

Ethanol (95%), chilled to –20ºC in a Coplin jar or other slide-staining container

2× SSCT (see p. 43)

**CAUTION: aminoalkylsilane, dry ice, ethanol, formaldehyde, liquid nitrogen, silane** (see Appendix 4)

### Method

1. On a dissecting surface or siliconized glass coverslip, dissect or otherwise isolate the desired tissue in a suitable buffer, e.g., Ringer's solution or PBS. Transfer the tissue to a 20–30-µl drop of buffer on a siliconized coverslip and remove large contaminants with forceps.

2. Add an equal volume of 2× formaldehyde in dissecting buffer (typically 3–8% formaldehyde made by diluting freshly prepared 37% solution into buffer).

3. Remove all but 10–20 µl of the fixative by careful pipetting.

4. Touch the center of a glass slide to the drop of fixative solution and invert carefully to sandwich the tissue between the slide and coverslip.

5. Wick away excess buffer with pieces of absorbent paper, e.g., Whatman #3 filter paper.

   *Note:* The aim of this step is to press the tissue gently against the slide, but not to flatten it.

6. Fix for a total period of 5–10 minutes.

7. Immerse the slide and coverslip into liquid nitrogen, or place on a flat aluminium block on dry ice, to freeze the sample. Once frozen, crack the coverslip using a new single-edged razor blade, and quickly transfer the slide to a staining jar containing 95% ethanol at –20°C. Store in ethanol at –20°C if hybridization is not performed immediately.

8. Repeat steps 1–7 until sufficient material has been obtained. When sufficient samples have been prepared, allow the ethanol to rise to above 0°C (i.e., until frost on outside of container melts). Transfer samples to a container filled with 2x SSCT. Proceed with hybridization as described in Protocol 2.9.

## HYBRIDIZATION METHODS

The general strategy for FISH is to equilibrate the tissue in buffered formamide, to add the probe(s), and to denature both the chromosomal DNA and probe together by heat treatment. The probe is then allowed to anneal for several hours at an appropriate temperature, unbound probe is washed away, and secondary detection (if required) is performed. The nuclear DNA is then usually counterstained, and the sample is mounted for microscopy. For experiments in which immunolocalization of other cellular components is desired, these staining steps can conveniently be performed after hybridization.

The key strategic decision that must be made is whether the tissue can be handled in suspension or whether it would be better fixed to microscope slides. Procedures for hybridizing tissues in suspension and on microscope slides are given.

### Hybridization to Tissues in Suspension

Samples such as *Drosophila* embryos or egg chambers are easily manipulated in suspension in 0.5-ml PCR tubes (Protocol 2.8), in which solutions are changed by allowing the tissue to settle to the bottom of the tube, aspirating the liquid carefully, and replacing with 400–500 μl of the new solution. This is by far the most convenient way to perform the hybridization and requires a minimum volume for all washing steps. The denaturation step is also easy to perform consistently, because the sample can be heated and annealed in a thermal cycler or conventional water bath.

### Hybridization to Tissues on Microscope Slides

Other tissues may be too large, delicate, transparent, buoyant, or have other physical properties that make fixation on slides a more suitable approach. If the tissue is not amenable to handling in suspension, it can be fixed directly onto a microscope slide or coverslip, which is then carried through the procedure in Coplin jars, larger staining jars, or coverslip-staining jars (Thomas Scientific). This procedure has been useful for spermatocytes, spermatids, and imaginal disks (see Protocol 2.9). Coplin jars require 50–60 ml of solution to cover the sample, larger staining jars require 200 ml, and coverslip-staining jars hold about 7–10 ml. Following hybridization, staining with antibodies or other detection reagents can be performed in a humid chamber. Pipette 25–50 μl of staining solution directly on to the tissue and cover with a cut piece of Parafilm. Wash in Coplin or staining jars, allowing the Parafilm to float away from the glass.

PROTOCOL 2.8

# Hybridization to Tissue in Liquid Suspension

## Materials

### Supplies and Equipment

Thermal cycler or other means of heating tubes to 91ºC and incubating at 37ºC
Aspirator with fine tip and liquid trap
Glass slides
Coverslips (22 x 22 mm)

### Solutions and Reagents

Fixed material in standard or thin-walled 0.5-ml polypropylene tubes (Protocol 2.6)
2x SSCT (see p. 43)
20x SSC
Tween-20 (a 10% v/v stock)
2x SSCT with increasing **formamide** concentrations is made from 20x SSC, 10% Tween-20, and formamide in H$_2$O
(*Optional*) RNase A (10 mg/ml stock in H$_2$O; used at a final concentration of 10–100 μg/ml)
Formamide (Fluka 47670)
Labeled probe(s)
Reagents for immunostaining and secondary detection
- primary antibody
- 0.5% protein blocking solution
- fluorescent detection reagents, e.g., fluorescently labeled streptavidin, antibiotin, or antidigoxigenin antibodies (see Table 2.1)

Mounting medium. Mounting solutions for fluorescence microscopy are typically buffered 90% glycerol solutions containing an antifading agent such as **DABCO** (1,4-diazabicyclo-[2,2,2]-octane), PPD (*p*-phenylenediamine), or NPG (*n*-propyl gallate). A commercially available mounting medium such as VectaShield (Vector) could be used. For preparation of glycerol/NPG, see below.
Nail polish

### Preparation of 1.1x Hybridization Solution

Add the following components to a 15-ml conical plastic tube with volume markings:

| | |
|---|---|
| dextran sulfate | 1 g |
| 20x SSC (see Appendix 3) | 1.5 ml |
| formamide | 5 ml |

Make up the final volume to 9.0 ml with H$_2$O. Dextran sulfate will take several hours to dissolve at room temperature; the hybridization solution may be stored for months at 4ºC.

### Preparation of Glycerol/NPG

To make an inexpensive all-purpose mounting medium, dissolve 4% NPG (w/v) in high-quality glycerol. Agitate for several hours. The solution may be stored at room temperature indefinitely.

Adjust the pH by adding 70 µl of 2 M Tris base (no HCl) and 30 µl of H$_2$O to 900 µl of glycerol/NPG. Mix thoroughly. This pH-adjusted mountant should be used within a day or two.

**CAUTION: DABCO, DAPI, DNA counterstains (e.g., DAPI, propidium iodide), formamide, phenylenediamine, propidium iodide, n-propylgallate** (see Appendix 4)

## Method

1. Wash the tissue three times with 2× SSCT. Incubate for at least 10 minutes per wash.

2. (*Optional*) Treat the sample with 10 µg/ml boiled RNase A in 2× SSCT for 30 minutes.

   *Note:* This step is usually dispensable, but probes to highly transcribed sequences, e.g., rDNA, may also bind to RNA. Therefore, this step may serve as a useful control. This treatment is also useful if the nuclear DNA will be counterstained with propidium iodide or another dye that binds to RNA (see step 13, below).

3. Step the tissue gradually into 2× SSCT containing 50% formamide, by adding sequentially and incubating for 10 minutes in each of the following solutions:

   2× SSCT containing 20% formamide
   2× SSCT containing 40% formamide
   2× SSCT containing 50% formamide

4. Add fresh 2× SSCT/50% formamide, place samples at 37°C, and incubate for at least 30 minutes.

   *Note:* This prehybridization incubation can have a marked impact on the permeability of some tissues and should be increased to as long as several hours if inconsistent probe penetration or high background is observed.

5. Aspirate as much of the prehybridization solution as possible. Take care not to aspirate the tissue. Add the Probe Solution. Mix gently by flicking the tube.

### Probe Solution

Add 25–500 ng of each probe (depending on complexity) to 36 µl of 1.1× hybridization buffer. Adjust the final volume of the probe mixture to 40 µl with H$_2$O and mix thoroughly by repeated pipetting. The final buffer composition should be 3× SSC, 50% formamide, 10% dextran sulfate.

*Note:* The probe concentration must be optimized empirically, but it is usually 0.5–10 ng/µl. Beyond an optimal level, increasing the concentration of double-stranded probe molecules may preferentially enhance their ability to self-anneal. This is because the kinetics for self-annealing follow the square of probe concentration, whereas hybridization to the chromosomes should increase only linearly with probe concentration.

6. Denature the probe and chromosomal DNA by heating the sample to 91°C for 2 minutes.

   *Note:* The denaturation temperature has been optimized by examining signal/noise and morphological preservation in formaldehyde-fixed samples (see Figure 2.4). With fixatives other than formaldehyde, denaturation at 70–80°C may be sufficient.

7. Reduce the sample temperature to 37ºC and allow the probe to anneal for several hours (typically overnight).

    *Note:* This stringency will work for most probes, but for very AT-rich sequences, if hybridization fails, annealing temperatures of 30–37ºC should be tested.

8. Add 500 μl of 2x SSCT/50% formamide to the sample, mix, and allow to settle.

9. Wash the sample with three changes of 2x SSCT/50% formamide at the annealing temperature (see step 7 and Note, above) over a total period of at least 1 hour.

10. Step the sample back out of formamide by washing for 10 minutes each with:

    2x SSCT/40% formamide
    2x SSCT/20% formamide

11. Wash with three changes of 2x SSCT (without formamide).

12. (*For secondary detection or immunostaining*) For avidin and many antibodies (including anti-digoxigenin F[ab] fragments from Boehringer Mannheim), perform staining in 2x SSCT, but if another buffer (e.g., PBS, TBS [Tris-buffered saline]) is preferred, it is simple to switch at this point. Perform the following steps:

    a. Block the sample with a 0.5% (w/v) protein solution (e.g., BSA or normal serum proteins). Incubate the sample in blocking solution for 30 minutes.

    b. Add diluted antibody (typical dilutions, 1:200–1:500) and/or avidin (1:2000 dilution). Incubate the sample with antibody/avidin solution for 2 hours, protected from light, with agitation on a nutator or similar device.

    c. Wash tissue with three changes of buffer for at least 10 minutes each wash to remove unbound detection reagents.

13. Counterstain DNA with DAPI or another fluorescent counterstain in the second wash of step 12c.

    *Note:* For nucleic acid dyes that also bind to RNA (e.g., propidium iodide), an RNase step before counterstaining is crucial (see step 2, above).

14. Mount the samples as follows:

    a. Exchange the tissue into mounting medium and then transfer to a slide or coverslip. Alternatively, pipette the tissue in buffer onto a slide or coverslip (poly-L-lysine-treated glass may facilitate this), gently remove most of the buffer, and overlay with mounting medium.

    b. If the sample is on a coverslip, hold a glass slide horizontally over the drop of mounting medium and then touch the center of the slide to the drop. If the sample is on a slide, invert the slide and touch the drop to a coverslip lying flat on a benchtop.

    c. Allow the liquid to spread out between the two layers of glass, then invert carefully. If necessary, remove excess mounting medium from the edge of the coverslip by blotting or gentle aspiration. Use clear nail polish to seal the edges of the coverslip to the slide.

        *Notes:* A minimal volume of mounting medium should be used, so that the samples are as close to the coverslips as possible and are immobilized by sandwiching between the two layers of glass.

        Samples tend to deteriorate over time in glycerol and should therefore be stored in buffer that has been used to perform the staining and washing steps (i.e., PBS, TBS, etc., made to 1 mM EDTA) at 4ºC until shortly before imaging.

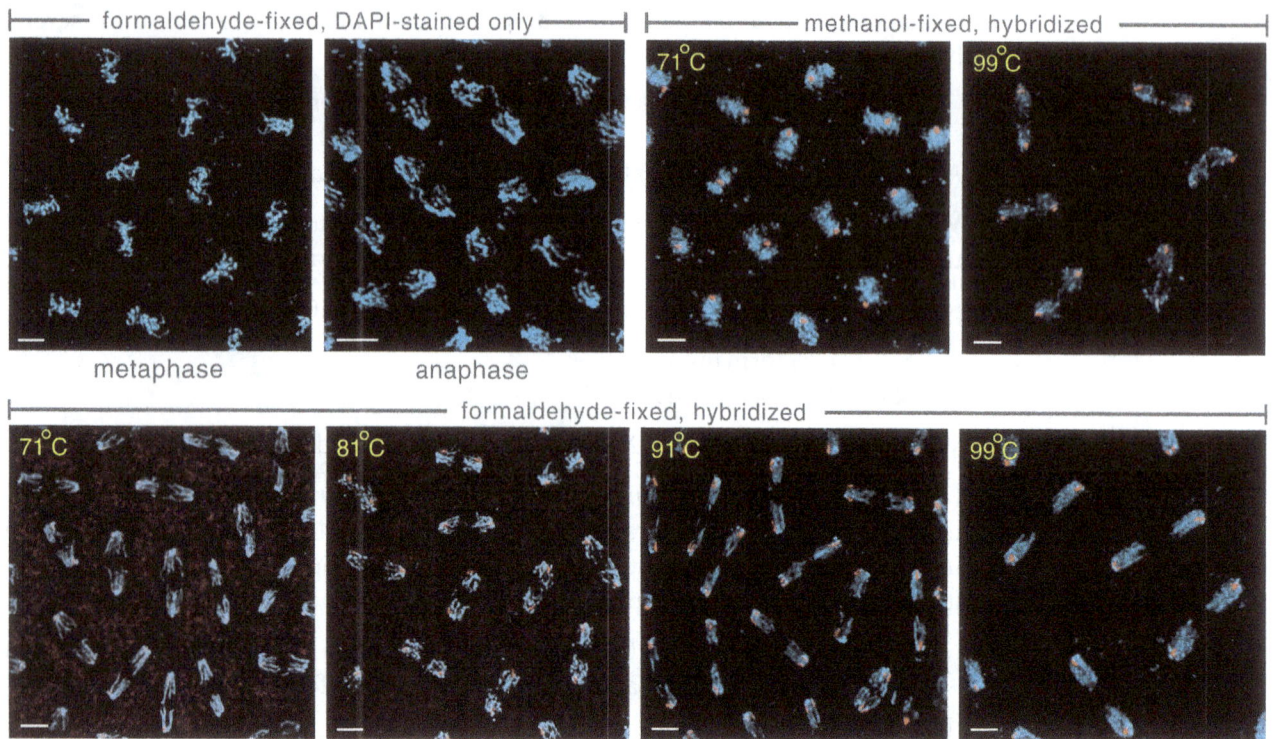

**Figure 2.4.** Comparison of the effects of different fixation methods and hybridization temperatures on chromosome morphology following FISH. Two panels (*top left*) show mitotic chromosomes that have been fixed with formaldehyde according to standard procedures, and not subjected to hybridization. Their appearance defines the standard to which chromosome morphology should be compared following hybridization. It is easier to compare mitotic chromosomes to interphase nuclei, which have diffuse DNA staining even under optimal conditions. The remaining six panels show chromosomes hybridized with a probe to the 359-bp repeat on the *X* chromosome. This is a very abundant sequence, and the probe can easily be detected even under suboptimal conditions. When embryos are fixed with methanol alone (two panels, *top right*), hybridization is successful and signal/noise is very good, even with relatively low denaturation temperatures. However, chromosome morphology is better preserved by formaldehyde fixation (four panels, *bottom*). Following fixation with 3.7% formaldehyde, the denaturation step is crucial for achieving good signal/background. In 50% formamide, as the denaturation temperature is increased from about 70ºC to 91ºC, the signal gradually improves, and the cytoplasmic background fluorescence is reduced, to an optimal point at about 91ºC. Beyond this, higher denaturation temperatures do not improve the signal/noise and will begin to degrade the chromosome morphology perceptibly. (Reprinted, with permission, from Spector et al. 1998.)

PROTOCOL 2.9

# Hybridization to Tissue on Slides or Coverslips

## Materials

### *Supplies and Equipment*

Coplin or slide-staining jars (VWR or other supplier) or coverslip-staining jars (Thomas Scientific)

Aspirator with fine tip and liquid trap

Coverslips (22 x 22 mm)

Kimwipes

Humidified denaturation chamber. A simple homemade version may be constructed by setting an aluminium heat block to 94ºC and covering the block with a plastic lid containing a layer of Kimwipes moistened with $H_2O$ around the inner rim (Figure 2.5).

Humidified incubation chamber, consisting of a watertight plastic container (e.g., Tupperware) containing a few layers of paper towels moistened with 2x SSC, overlaid with a sheet of Parafilm

Glass slides

Hemostats, the tips covered with short pieces of silicone rubber tubing, for handling slides

### *Solutions and Reagents*

Fixed tissue on slides or coverslips (from Protocol 2.7)

2x SSCT (see p. 43)

(*Optional*) RNase A (10 mg/ml in $H_2O$)

Labeled probe(s)

1.1x Hybridization Solution (see p. 47)

**Formamide** (Fluka 47670)

Reagents for immunostaining and secondary detection

- primary antibody
- 0.5% protein blocking solution
- fluorescent detection reagents, e.g., fluorescently labeled streptavidin, antibiotin, or antidigoxigenin antibodies (see Table 2.1)

Mounting medium (see p. 48)

DNA counterstains (e.g., **DAPI, propidium iodide**; see Appendix 3)

**CAUTION: DAPI, formamide, propidium iodide** (see Appendix 4)

### Method

1.  Place the slide(s) in a Coplin jar containing 2x SSCT. Incubate for at least 10 minutes. Repeat wash with two changes of 2x SSCT.

2.  (*Optional*) Treat the sample with RNase A (final concentration 10–100 µg/ml).

    *Note:* RNase may be added to the buffer in the Coplin jar; or smaller volumes can be used by transferring the slide(s) to a humid chamber, pipetting the RNase solution directly onto the sample, and covering with a piece of Parafilm or a coverslip.

3.  Step the tissue gradually into 2x SSCT containing 50% formamide, by transferring sequentially and incubating in each of the following solutions for 10 minutes:

    2x SSCT containing 25% formamide
    2x SSCT containing 50% formamide

4.  Transfer samples to fresh 2x SSCT/50% formamide, place the staining jar in a 37°C water bath, allow the temperature to equilibrate, and incubate for at least 30 minutes before addition of the probe.

    *Note:* This prehybridization incubation can have a marked impact on the permeability of some tissues and should be increased to as long as several hours if inconsistent probe penetration or high background is observed.

5.  Carefully remove the slides/coverslips from the prehybridization solution and drain on paper towels. (Save the prehybridization solution for the first posthybridization wash.) Wearing gloves, use the tip of an aspirator and/or Kimwipes to remove as much of the prehybridization solution as possible without damaging the tissue. Wipe the back of the glass dry with Kimwipes.

6.  Pipette the Probe Solution onto a clean 22 x 22-mm coverslip or, if the sample is on a coverslip, pipette it onto a slide. Touch the drop of solution to the sample and invert the slide.

### Probe Solution

Add 10–200 ng of each probe (depending on complexity) to 1.1x Hybridization Solution and adjust volume to 15 µl with $H_2O$. The final buffer composition should be 3x SSC, 50% formamide, 10% dextran sulfate.

> *Note:* The probe concentration must be optimized empirically, but it is usually 0.5–10 ng/µl. Beyond an optimal level, increasing the concentration of double-stranded probe molecules may preferentially enhance their ability to self-anneal. This is because the kinetics for self-annealing follow the square of probe concentration, whereas hybridization to the chromosomes should increase only linearly with probe concentration.

7.  Denature probe and chromosomal DNA by placing the slide onto a hot block pre-equilibrated to 94°C. Cover with a humidifying lid (Figure 2.5) and incubate for 2 minutes.

> *Note:* As an alternative to constructing a humidified denaturation chamber, seal the coverslip to the slide with a gasket of rubber cement, most easily applied using a syringe with a wide-bore needle. Allow the rubber cement to dry, and then denature the slide directly on a heat block at 94°C.

8.  Transfer the slides to a humidified chamber at 37°C and incubate for several hours to overnight.

9.  Transfer slides to a Coplin or staining jar containing 2x SSCT/50% formamide at the annealing temperature.

> *Note:* If the samples are on slides, allow the coverslips to fall off and discard them. If rubber cement was used, this should be carefully peeled off using forceps while the slide is immersed in washing solution. If the samples are on coverslips, retrieve them carefully using flat-tipped forceps (Millipore) and transfer to coverslip-staining jars containing 2x SSCT/50% formamide, at the annealing temperature.

10. Wash the sample with one further change of 2x SSCT/50% formamide at the annealing temperature over a total period of at least 1 hour.

11. Transfer the sample into 2x SSCT/25% formamide at room temperature and incubate for at least 10 minutes.

12. Wash with three changes of 2x SSCT (without formamide).

13. (*For secondary detection or immunostaining*) For avidin and many antibodies (including antidigoxigenin and antidigoxin antibodies), perform staining in 2x SSCT, but if another buffer (e.g., PBS and TBS) is preferred, it is simple to switch at this point. For staining steps (below), transfer the slides into a humid chamber, pipette the blocking or staining solution onto the sample, and overlay with a piece of Parafilm cut to approximately 25 x 25 mm (or as needed to cover the specimen).

    a.  Block the sample with a 0.5% (w/v) protein solution (e.g., BSA and normal serum proteins). Incubate the sample in blocking solution for 30 minutes.

    b.  Add diluted antibody (typical dilutions, 1:200 to 1:500) and/or avidin (1:2000 dilution). Incubate the sample with antibody/avidin solution for 2 hours, protected from light.

    c.  Wash tissue with three changes of 2x SSCT for at least 10 minutes each wash to remove unbound detection reagents.

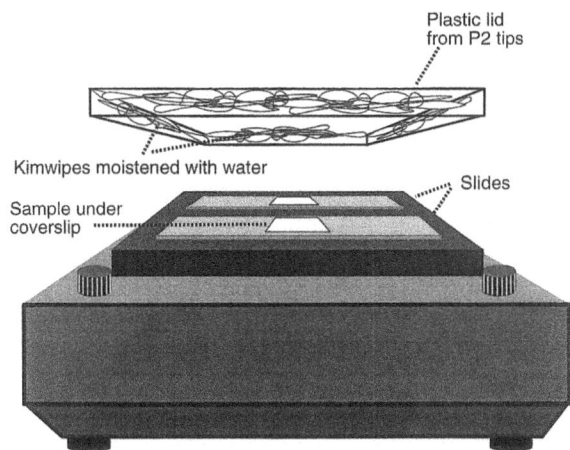

**Figure 2.5.** Denaturation of FISH samples on slides using a humidified heat block. This diagram illustrates a simple way to heat-denature samples on glass slides using a conventional heating device (VWR or other supplier) with a removable aluminum block. The heat block is first equilibrated at 94°C using a thermometer placed in one of the wells, after which, the aluminum block is then inverted so that its flat side faces up. A humidifying chamber can then be assembled using a shallow plastic box; the lids from racks of P2 Pipetman tips are perfectly sized for this. Kimwipes are placed around the interior rim of the lid and moistened with water (salt would damage the anodized aluminum surface). The lid is placed over the slides during the 2-minute denaturation period. This type of block accommodates two slides at a time. (Reprinted, with permission, from Dernburg 1999 [© Oxford University Press].)

16. Counterstain with DAPI or another fluorescent dye in the second wash.

17. Mount samples for microscopy as follows:

    a. Pipette 12–15 µl of mounting medium onto a clean 22 X 22-mm coverslip.

    b. Transfer the slide briefly to a staining jar containing 50 mM Tris-Cl (pH 7.5) to remove most of the salt.

    c. Remove slide and drain briefly on paper towels. Use Kimwipes and/or an aspirator to remove as much buffer as possible from the slide without damaging the sample or allowing it to dry out.

    d. Touch the sample to the drop of mounting medium, invert the slide, and seal with clear nail polish.

## TROUBLESHOOTING

The most common problem in hybridization experiments is the absence of a detectable signal. This is also the most difficult situation upon which to improve, because the source of the problem is hard to diagnose and may lie with the fixation method, probe synthesis, or other experimental factors. It is extremely helpful to embark on any attempt at hybridization using the type of probe most likely to work: an oligonucleotide probe to a simple, abundant, repetitive sequence. The *Drosophila* genome contains a wide choice of such sequences, including some that are restricted to specific chromosomal loci (Table 2.2). After labeling, test the probe on an acid-fixed mitotic chromosome preparation using methods such as those described by Pimpinelli et al. (this volume). Chances are high that such a probe, if labeled successfully, will produce at least a weak signal in embryos or other whole-mount tissues, allowing the experimental conditions to be refined and optimized.

Single-copy probes can also be tested on squash preparations of polytene chromosomes (see Pardue, this volume).

More often than not, disappointing or inconsistent results are due to over- or under-fixation of the tissue. Potential problems attributable to fixation include failure to detect the probe, nonspecific binding of probe or detection reagents, and unacceptable morphological preservation. If the probe can be readily detected on flattened or acid-fixed preparations, then effort should be focused on improving the whole-mount fixation.

## MICROSCOPY AND IMAGE ANALYSIS

FISH experiments frequently require the detection of faint signals at several different wavelengths within thick biological samples. High-quality oil immersion lenses, well-designed fluorescence filters, and sensitive detection devices such as cooled CCD cameras all help to extend the sensitivity and versatility of the technique.

Diploid *Drosophila* nuclei are typically about 3–5 μm in diameter, only one order of magnitude greater than the resolution limit of traditional optical microscopes. Questions regarding the subnuclear distribution of specific sequences are thus close to the edge of what can be accomplished with fluorescent detection. To achieve high resolution, it is valuable to apply techniques that remove out-of-focus information from fluorescence images. Two widely used methods are confocal laser-scanning microscopy (CLSM) and wide-field deconvolution microscopy, in which the out-of-focus light is collected but later restored to its proper position within the object using computational algorithms.

## ACKNOWLEDGMENT

This work was supported in part by a postdoctoral fellowship (DRG-1392) from the Cancer Research Fund of the Damon Runyon–Walter Winchell Foundation.

## REFERENCES

Abad J.P., Carmena M., Baars S., Saunders R.D., Glover D.M., Ludena P., Sentis C., Tyler S.C., and Villasante A. 1992. Dodeca satellite: A conserved G+C-rich satellite from the centromeric heterochromatin of *Drosophila melanogaster. Proc. Natl. Acad. Sci.* **89:** 4663–4667.

Brittnacher J.G. and Ganetzky B. 1989. On the components of segregation distortion in *Drosophila melanogaster.* IV. Construction and analysis of free duplications for the *Responder* locus. *Genetics* **121:** 739–750.

Carmena M., Abad J.P., Villasante A., and Gonzalez C. 1993. The *Drosophila melanogaster* dodecasetallite sequence is closely linked to the centromere and can form connections between sister chromatids during mitosis. *J. Cell Sci.* **105:** 41–50.

Cenci G., Rawson R.B., Belloni G., Castrillon D.H., Tudor M., Petrucci R., Goldberg M.L., Wasserman S.A., and Gatti M. 1997. UbcD1, a *Drosophila* ubiquitin-conjugating enzyme required for proper telomere behavior. *Genes Dev.* **11:** 863–875.

Dernburg A.F. 1999. Fluorescence *in situ* hybridization in whole-mount tissues. In *Chromosome structural analysis: A practical approach* (ed. W.A. Bickmore). Oxford University Press, United Kingdom. (In press.)

Dernburg A.F. and Sedat J.W. 1998. Mapping three-dimensional chromosome architecture in situ. *Methods Cell Biol.* **53:** 187–233.

Dernburg A.F., Sedat J.W., and Hawley R.S. 1996a. Direct evidence of a role for heterochromatin in meiotic chromosome segregation. *Cell* **86:** 135–146.

Dernburg A.F., Daily D.R., Yook K.J., Corbin J.A., Sedat J.W., and Sullivan W. 1996b. Selective loss of sperm bearing a compound chromosome in the *Drosophila* female. *Genetics* **143:** 1629–1642.

Dernburg A.F., Broman K.W., Fung J.C., Marshall W.F., Philips J., Agard D.A., and Sedat J.W. 1996c. Perturbation of nuclear architecture by long-distance chromosome interactions. *Cell* **85:** 745–759.

Dimitri P. 1991. Cytogenetic analysis of the second chromosome heterochromatin of *Drosophila melanogaster. Genetics* **127:** 553–564.

Franke A., Dernburg A., and Baker G. 1996. Evidence that MSL-mediated dosage compensation in *Drosophila* begins at blastoderm. *Development* **122:** 2751–2760.

Fung J.C., Marshall W.F., Dernburg A., Agard D.A., and Sedat J.W. 1998. Homologous chromosome pairing in *Drosophila melanogaster* proceeds through multiple independent initiations. *J. Cell. Biol.* **141:** 5–20.

Gall J.G. and Pardue M.L. 1969. Formation and detection of RNA-DNA hybrid molecules in cytological preparations. *Proc. Natl. Acad. Sci.* **63:** 378–383.

Golic M.M. and Golic K.G. 1996. A quantitative measure of the mitotic pairing of alleles in *Drosophila melanogaster* and the influence of structural heterozygosity. *Genetics* **143:** 385–400.

Gotta M., Laroche T., Formenton A., Maillet L., Scherthan H., and Gasser S.M. 1996. The clustering of telomeres and colocalization with Rap1, Sir3, and Sir4 proteins in wild-type *Saccharomyces cerevisiae. J. Cell. Biol.* **134:** 1349–1363.

Hiraoka Y., Dernburg A.F., Parmelee S.J., Rykowski M.C., Agard D.A., and Sedat J.W. 1993. The onset of homologous chromosome pairing during *Drosophila melanogaster* embryogenesis. *J. Cell. Biol.* **120:** 591–600.

Hsieh T. and Brutlag D. 1979. Sequence and sequence variation within the 1.688 g/cm3 satellite DNA of *Drosophila melanogaster. J. Mol. Biol.* **135:** 465–481.

Lifton R.P., Goldberg M.L., Karp R.W., and Hogness D.S. 1978. The organization of histone genes in *Drosophila melanogaster:* Functional and evolutionary implications. *Cold Spring Harbor Symp. Quant. Biol.* **42:** 1047–1051.

Lohe A.R., Hilliker A.J., and Roberts P.A. 1993. Mapping simple repeated DNA sequences in heterochromatin of *Drosophila melanogaster. Genetics* **134:** 1149–1174.

Makunin I.V., Pokholkova G.V., Zakharkin S.O., Kholodilov N.G., and Zhimulev. I.F. 1995. Isolation and characterization of the repeated DNA sequences from the centromeric heterochromatin of the second chromosome of *Drosophila malnogaster. Doklady Academii Nauk* **344:** 266–269.

Marshall W.F., Dernburg A.F., Harmon B., Agard D.A., and Sedat J.W. 1996. Specific interactions of chromatin with the nuclear envelope: Positional determination within the nucleus in *Drosophila melanogaster. Mol. Biol. Cell.* **7:** 825–842.

Marshall W.F., Straight A., Marko J.F., Swedlow J., Dernburg A., Belmont A., Murray A.W., Agard D.A., and Sedat J.W. 1997. Interphase chromosomes undergo constrained diffusional motion in living cells. *Curr. Biol.* **7:** 930–939.

Meltzer P.S., Guan X.Y., Burgess A., and Trent J.M. 1992. Rapid generation of region specific probes by chromosome microdissection and their application. *Nat. Genet.* **1:** 24–28.

Pardue M.L. and Gall J.G. 1969. Molecular hybridization of radioactive DNA to the DNA of cytological preparations. *Proc. Natl. Acad. Sci.* **64:** 600–604.

Pimpinelli S. and Dimitri P. 1989. Cytogenetic analysis of segregation distortion in *Drosophila melanogaster:* The cytological organization of the *Responder (Rsp)* locus. *Genetics* **121:** 765–772.

Rabl C. 1885. Über Zelltheilung. *Morpholog. Jahrbuch* **10:** 214–330.

Spector D.L., Goldman R., and Leinwand L.A. 1988. *Cells: A laboratory manual.* Cold Spring Harbor Laboratory Press, Cold Spring Harbor, New York.

Telenius H., Carter N.P., Bebb C.E., Nordenskjold M., Ponder B.A., and Tunnacliffe A. 1992. Degenerate oligonucleotide-primed PCR: General amplification of target DNA by a single degenerate primer. *Genomics* **13:** 718–725.

Wessendorf M.W. and Brelje T.C. 1992. Which fluorophore is brightest? A comparison of the staining obtained using fluorescein, tetramethylrhodamine, lissamine rhodamine, Texas red, and cyanine 3.18. *Histochemistry* **98:** 81–85.

Wiegant J., Verwoerd N., Mascheretti S., Bolk M., Tanke H.J., and Raap A.K. 1996. An evaluation of a new series of fluorescent dUTPs for fluorescence in situ hybridization. *J. Histochem. Cytochem.* **44:** 525–529.

Wu C.I., Lyttle T.W., Wu M.L., and Lin G.F. 1988. Association between satellite DNA sequence and the Responder of Segregation Distorter in *D. melanogaster. Cell* **54:** 179–189.

# CONTENTS

# 3

# BrdU Labeling of Chromosomes

**Antony W. Shermoen**
*Department of Biochemistry/Biophysics*
*University of California at San Francisco*
*San Francisco, California 94143-0448*

T HE RAPID ADVANCES IN UNDERSTANDING the regulation of the cell cycle in organisms from yeast to humans have included the need to determine accurately what part of the cell cycle is being experimentally regulated. Presented in this chapter is an easy method for marking and detecting the S phase in *Drosophila* embryos. This technique involves the incorporation of the thymidine analog, BrdU (5-bromo-2′-deoxyuridine), into DNA and its detection by immunofluorescence or histochemistry (see Figure 3.1) (Edgar and O'Farrell 1990; Duronio and O'Farrell 1994). Also presented is a description of labeling larval, pupal, and adult tissues (Truman and Bate 1988; Awad and Truman 1997; Britton and Edgar 1998). These methods can be expanded with a double label of different parts of S phase using a combination of BrdU and a second analog, IdU (5-iodo-2′-deoxyuridine) (Aten et al. 1994). In addition, combining BrdU labeling with in situ hybridization enables determination of what sequences are being replicated (Calvi et al. 1998; A.W. Shermoen and P.H. O'Farrell, in prep.).

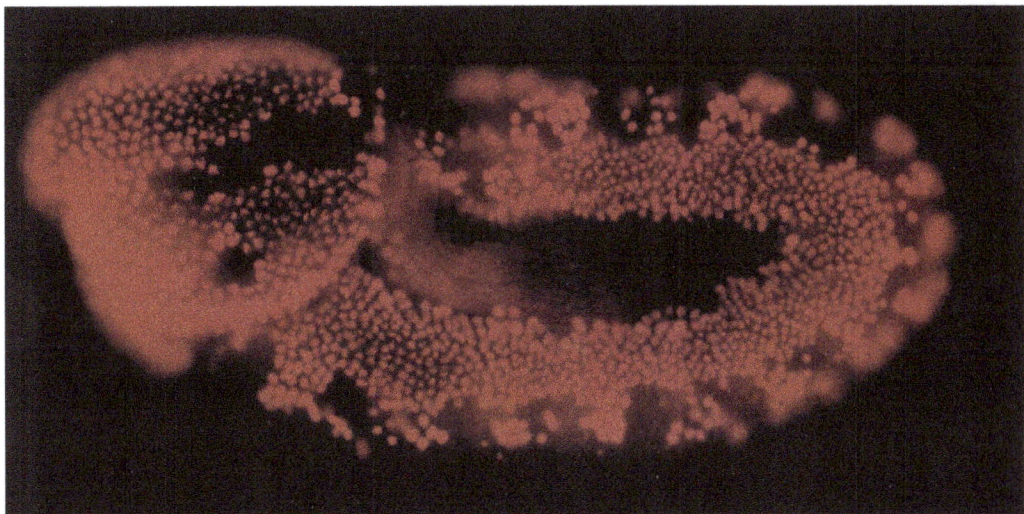

**Figure 3.1.** Stage-11 embryo labeled for 16 minutes with BrdU and stained with monoclonal antibody anti-BrdU and a goat anti-mouse-Rho secondary antibody. The image was taken using a 20x objective on a Leica DMRD microscope.

---

────── FLOW CHART 1 ──────

**BrdU LABELING**

**Embryos**

Collect embryos ⟹ 50% bleach × 2 minutes ⟹ Rinse with H$_2$O × ~1 minute ⟹ Blot × 15–30 seconds ⟹ Octane × 4 minutes ⟹ Blot × 30–45 seconds ⟹ BrdU labeling medium × ≥5 minutes ⟹ Fix in 5% formaldehyde/1× PBS × 20 minutes ⟹ methanol × 20 seconds ⟹ Store at –20°C in methanol.

**Tissues (Imaginal Discs, Salivary Glands, CNS)**

Clean larvae with H$_2$O ⟹ Dissect tissue in Schneider's medium (or PBS) ⟹ Transfer the tissue to a basket with fine-mesh Nitex (not more than 45-μm hole; it is useful to leave discs attached to other tissue if using a larger-mesh Nitex) in Schneider's medium ⟹ BrdU labeling medium × ≥10 minutes ⟹ 5% formaldehyde × 30 minutes ⟹ PTX (3 × 10 minutes) ⟹ May be stored in PTX at 4°C overnight.

---

PROTOCOL 3.1*

# BrdU Labeling

A detailed procedure for BrdU labeling of embryos is given below. Flowchart 1 presents a summary of this procedure for embryos and other tissues.

## Materials

### Media

Grape agar plates (for preparation, see Appendix 3)
Schneider's medium (GIBCO BRL 11720-034)

### Supplies and Equipment

Culture dishes (60 × 15 mm; Fisher 8-757-13A)
A black surface under the embryos to allow easier visibility
Squirt bottle containing H$_2$O
Paint brush (#2)
Kimwipes
Pipetman (1 ml)
Watch glass (2) (Fisher 15-355; these beveled-edge watch glasses are deeper)
Dissecting microscope
Glass scintillation vials (20 ml) or Eppendorf tubes for fixation and devitellinization
Pasteur pipette
Dissecting tools for BrdU labeling of larvae (see Flowchart 1)
- Dumont tweezers (#5; Ted Pella 5622)
- micro slides (75 × 50 mm, Corning 2947-75x50); Sigmacote (Sigma SL-2) or RAIN-X (UNELKO Corp.; commonly sold in auto supply stores) is used to coat the slide to ensure that the dissecting medium beads into drops and traps the larvae to be dissected

*For all protocols, H$_2$O indicates glass distilled and deionized.

### Preparation of Baskets for Egg Collections, Treatments, and Incubations

The key consideration of what size basket to use is that the embryos must be approximately one monolayer or more. Described below are two baskets used by the author for large numbers of embryos. For descriptions of other baskets, see Rothwell and Sullivan ( this volume).

The first basket is made from a 50-ml plastic screw-cap centrifuge tube, 2.5 cm in height. Cut a hole in the cap as large as possible, but still maintaining enough of the structure to be able to screw the cap on firmly. Cut an 8-cm square piece of Nitex (~70-μm mesh, Sefar America, Inc. [4221 NE 34th St., Kansas City, Missouri 64117 [Phone: 800-283-8182]). Screw the top on with the mesh between the top and body as taut and flat as possible. Use somewhat greater volumes than used for a commercial basket. For preparation of baskets for larval tissues, use fine-mesh Nitex (not more than 45-μm hole).

Commercial embryo baskets may be obtained from Labworks, Inc. (721 Bolero Ct., Novato, California 94945 [Phone: 707-484-7998; Web site: www.labworksinc.com], LW-401; customer must specify size). Dimensions: 3.2 cm (inner diameter), 3.8 cm (outer diameter), 2 cm (height).

## Solutions and Reagents

**Bleach** (50%)
**Octane** (Sigma O 2001)
1x Phosphate-buffered saline (PBS) (pH 7.2; for preparation, see Appendix 3)
**Formaldehyde** (37%; Sigma F 1635) or dilute to 5% formaldehyde in 1x PBS
**Heptane**
**Methanol**

### BrdU-labeling Medium

Prepare 1 mg/ml (BrdU; Sigma B 5002) in Schneider's medium. Alternatively, use 1x PBS, which is just as effective.

CAUTION: **bleach, formaldehyde, heptane, methanol, octane** (see Appendix 4)

## Method

1. (*Embryo collection*) Collect the embryos as follows:

    a. Squeeze $H_2O$ from a squirt bottle onto the grape agar plate containing eggs. Use a paintbrush to loosen the embryos from the surface of the agar.

    b. Pour the $H_2O$ containing the embryos into a basket. Rinse the grape agar plate with more $H_2O$. Transfer the rinse to the basket.

    c. Use Kimwipes to blot the embryos dry from the bottom of the basket.

    > *Note:* A cage of approximately 5 g of adult flies will produce approximately 40–60 embryos/minute (or more than double that amount obtained on the first day with fresh flies, properly fed).

2. (*Dechorionation*) Immerse the embryos in 50% bleach and agitate occasionally for 2 minutes. Wash these embryos in running $H_2O$ (deionized or distilled) until there is no longer any bleach odor (~1 minute).

3. Use Kimwipes to blot the $H_2O$ from the inner and outer sides. Then blot the underside of the basket to remove $H_2O$ from around the embryos.

*Note:* The time for blotting depends on the number of embryos obtained—a monolayer should be blotted for approximately 15 seconds, whereas more than a monolayer of embryos must be blotted for approximately 30 seconds. This is a critical step because too much blotting will collapse the vitelline membrane against the embryo. These desiccated embryos do not devitellinize efficiently. In addition, too little blotting slows down the uptake of BrdU and makes the start of labeling more heterogeneous among the embryos.

4. (*Permeabilization*) Immerse the embryos in octane (in a watch glass) to permeabilize them. Use a 1-ml Pipetman to spurt 1 ml of octane on the embryos to agitate them. Continue this agitation occasionally for the duration of this step (4 minutes). During this period, add approximately 2 ml of the BrdU-labeling Medium to a 60 x 15-mm culture dish (see step 6, Notes, below).

   *Note:* The state of the embryos may be monitored under a dissecting microscope during the agitation. Ensure that there is no sheen around the embryos. The presence of a sheen would be indicative of an $H_2O$ envelope preventing the octane from permeabilizing the embryo.

5. After 4 minutes, lift the basket from the octane, blot the underside with Kimwipes, and blow gently on the embryos until there is no octane odor (~30–45 seconds).

6. (*BrdU labeling*) Perform BrdU labeling as follows:

   a. Place the basket into the 60 x 15-mm culture dish containing BrdU-labeling Medium (see step 4, above).

   b. Pour about 6–8 ml of the BrdU-labeling Medium over the embryos (the embryos should be ~1–4 mm below the surface of the liquid). Again monitor the embryos with a dissecting microscope while spritzing the labeling medium onto the embryos with a 1-ml Pipetman. Take up the medium from the culture dish outside of the basket and use a strong spurt to dislodge the embryos from the bottom of the basket (the embryos should at least cycle through a monolayer condition at the surface).

      *Notes:* Ensure that there is no sheen over the embryos, which would be indicative of a coating of octane that would act as a barrier to the entry of BrdU. During the labeling period, occasionally spurt medium over the embryos to agitate them and keep them off the sides, and rotating in the middle of the basket. For optimal labeling times, see Troubleshooting below.

      If precise timing of labeling is important, start step 6a with Schneider's medium alone or 1x PBS. After the embryos are on the surface and there is no octane "sheen" around them, transfer the basket to BrdU-labeling Medium.

      If different periods of labeling from the same sample are desired, collect these samples by using a small square of Nitex to scoop up the quantity of embryos needed and transfer these to a vessel containing heptane and formaldehyde (proceed to step 7, below).

7. Fix and devitellinize the embryos as follows:

   a. Transfer the embryos to a container containing equal volumes of formaldehyde and heptane. The container should be sufficiently large so that the embryos are at the interface when rocking. Fix the embryos while rocking (3–4 minutes for 37% formaldehyde or 20 minutes for 5% formaldehyde in 1x PBS). At the end of this period, use a pasteur pipette or Pipetman to withdraw the formaldehyde from below the heptane layer (the embryos will form a layer at the interface between the two solutions).

   b. Now add about 2–2.5 volumes of methanol and shake vigorously for 20 seconds. The devitellinized embryos will fall to the bottom of the container.

   c. Aspirate the embryos at the interface and the upper layer of heptane, and then add as much of the methanol as possible without disturbing the embryos. Transfer the embryos to a new Eppendorf tube and rinse twice with fresh methanol.

      *Note:* These embryos may be stored indefinitely at –20°C in methanol.

# BrdU Detection

Presented here is an in-depth description of the BrdU detection procedure for embryos. Flowchart 2 provides a summary of this procedure for embryos as well as for larval, pupal, and adult tissues with the use of fluor-conjugated secondary antibody.

The choice of secondary antibody is between a fluorescently labeled antibody or an antibody that leaves a histochemical deposition for the detection of the BrdU, such as alkaline phosphatase (AP) deposition of nitroblue tetrazolium (NBT) or horseradish peroxidase (HRP) deposition of diaminobenzidine (DAB). These histochemical detection methods are about ten times more sensitive than that of normal fluorescently labeled antibodies. However, one consequence of the chemical deposit is to quench any costaining with a fluorescently labeled antibody in the region of BrdU label. Both the AP reaction, which leaves a blue precipitate, and the HRP reaction, which leaves a brown precipitate, for a double-label could be used. One shortcoming of this approach is the difficulty in detecting subcellular localization, or the colocalization of two targets. Both of these methods are outlined in this protocol.

## Materials

### *Supplies and Equipment*

Eppendorf tubes (0.5 ml and 1 ml)
Sequencing tips
Aluminum foil
Multiwell culture dish (24-well; Fisher 08-772-4G)
Nutator
P-200 Pipetman
Coverslips (22 x 22 mm)
Micro slides (25 x 75 mm, Fisher 12-518-100B) for mounting embryos
Nail polish

### *Solutions and Reagents*

**Sodium tetraborate** (100 mM)
Normal goat serum (NGS; Jackson ImmunoResearch Laboratories 005-000-121)
Hoechst 33258 (Molecular Probes H-1398)
Mountant (e.g., **Fluoromount-G**, Southern Biotechnology Associates 100-01)
Primary antibody. Monoclonal antibody (Mab) anti-**BrdU** (Becton Dickinson 347580); for a table of primary antibodies, see Dolbeare and Shelden (1994)

---
**FLOW CHART 2**

### BrdU DETECTION

### Embryos and Tissues (Imaginal discs, Salivary Glands, CNS)

HCl treat x 30 minutes ⇒ Borax (2 x 2 minutes) ⇒ PTX (2 x 3 minutes) ⇒ 10%NGS/PTX x 45 minutes ⇒ Mab anti-BrdU x 1.5 hours ⇒ PTX x 2 minutes ⇒ 10% NGS/PTX (3 x 20 minutes) ⇒ Goat anti-mouse • Rho x 1 hour ⇒ PTX (1 x 2 minutes, 2 x 20 minutes) ⇒ Hoechst x 2 minutes ⇒ PTX (2 x 5 minutes) ⇒ Mount.

Secondary antibodies (Jackson ImmunoResearch Laboratories, Inc.); use either:
*Anti-mouse • Fluor, Goat anti-mouse • Rho (115-026-072)*
*Anti-mouse secondary antibodies for histochemical detection*
    *(1) anti-mouse • AP (115-056-072) or*
    *(2) anti-mouse • HRP (115-036-006)*
**Sodium azide** (5%)

## HCl/Triton X-100 Solution

2.2 N **HCl**
0.1% Triton X-100
Use freshly prepared solution.

## PTX Solution

1x PBS
0.1% Triton X-100

## 10% NGS/PTX

Prepare a solution of 10% NGS in PTX. Alternatively, use 5% milk (dried) in PTX instead of 10% NGS/PTX. These solutions are used for blocking and for dilution of antibodies.

## Preadsorbed Antibodies

Dilute the commercial concentration of the *secondary* antibody 30–50 times with 10% NGS/PTX (to make it 10x concentration). Mix the antibody solution with an equal volume of embryos at room temperature for 1–2 hours.

The embryos are usually from an overnight collection, which is dechorionated, fixed, devitellinized, and stored at –20°C in **methanol** for such a purpose. The embryos are equilibrated with PTX before use. These stock antibodies can be used for months if stored with the embryos at 4°C.

## AP Detection Solution 1

100 mM NaCl
50 mM **MgCl$_2$**
100 mM Tris-Cl (pH 9.5)
1 mM Levamisole
0.1% Triton X-100

## AP Detection Solution 2

This solution comprises AP Detection Solution 1 (above) containing 4.5 μl/ml **NBT** (Boehringer Mannheim 1 383 213) and 3.5 μl/ml **BCIP** (5-bromo-4-chloro-3-indolyl phosphate; Boehringer Mannheim 1 383 221).

## Staining Solution for HRP Detection

1x PBS
0.5 mg/ml **DAB**
0.003% **H$_2$O$_2$**
0.05% **NiCl$_2$**
Use freshly prepared staining solution.

**CAUTION: BCIP, BrdU, DAB, Fluoromount-G, HCl, H$_2$O$_2$, methanol, MgCl$_2$, NBT, NiCl$_2$, sodium azide, sodium tetraborate** (see Appendix 4)

## Method

1. Aspirate the methanol from the embryos (use a Pipetman with a sequencing tip to withdraw the last of the methanol). Start with approximately 5–200 µl of labeled embryos (200–8000 embryos).

2. Denature the DNA as follows:

   a. Add at least 10 volumes of freshly prepared HCl/Triton X-100 Solution to the embryos.

   b. Rock on a Nutator for 15 minutes. Allow the embryos to settle to the bottom of the Eppendorf tube.

   c. Withdraw most of the HCl/Triton X-100 Solution. Repeat steps 2a and 2b (above). Withdraw all of the HCl/Triton X-100 Solution.

3. Neutralize the embryos as follows:

   a. Add approximately 10 volumes of 100 mM sodium tetraborate (borax) to the embryos.

   b. Rock the embryos in the solution for approximately 2 minutes. Allow the embryos to settle to the bottom of the Eppendorf tube.

   c. Aspirate most of the borax solution. Repeat steps 3a and 3b (above). Withdraw all of the borax solution.

4. Rinse the embryos twice with PTX for 3 minutes. In preparation for antibody detection of BrdU, incubate the embryos in 10% NGS/PTX or 5% milk/PTX for 45 minutes.

5. Dilute the stock monoclonal antibody anti-BrdU 1/20 with 10% NGS/PTX. Add 100 µl of antibody for 20–40 µl of embryos (the minimum volume of antibody should be 100 µl). Rock the embryos on a Nutator for 1.5 hours. Withdraw the monoclonal antibody anti-BrdU, which may be saved at 4ºC and used a second time.

6. Rinse the embryos with PTX briefly (<2 minutes). Then rinse again with 10% NGS/PTX or 5% milk/PTX (3x 20 minutes). Aspirate off the blocking solution. Withdraw the final volume of solution using a P-200 and a sequencing tip, to avoid losing embryos.

7. Add 100–200 µl of the preadsorbed secondary antibody at a final dilution of approximately 1/300.

8. There is a choice of secondary antibody detection using either of the following methods (for a discussion of these methods, see above):

   a. Fluorescent detection using Goat anti-mouse • fluor (Rho, FITC, Cy5; Jackson ImmunoResearch Laboratories). These are usually used at about 1/300–1/500 in 10%NGS/PTX or 5% milk/PTX at a final volume of 100–200 µl.

      i. Wrap the tubes containing embryos and fluorescent antibodies in aluminum foil to prevent any bleaching of signal and rock at room temperature for 1 hour. Allow the embryos to settle and withdraw the secondary antibody. (This secondary antibody may also be stored at 4ºC and used again.)

      ii. Fill the Eppendorf tube with PTX and rock the embryos several times and allow them settle. Aspirate off most of the PTX using a sequencing pipette tip over the tip of a pasteur tip for greater control.

   iii. Refill the tube with PTX and rock on the Nutator for 20 minutes. Repeat this wash a second time.

   iv. Stain these embryos with 1 μg/ml Hoechst 33258 for 2 minutes. Aspirate the Hoechst solution and rinse with PTX twice for 5 minutes each.

> *Notes:* The Hoechst stain is for quality control of the DNA denaturation step. The euchromatin of interphase chromosomes produces a fuzzy Hoechst stain if properly denatured. Heterochromatin and mitotic chromosomes stain fairly well, despite the denaturation.
>
> Additional stains that may be used are antibodies to the nuclear pore such as wheat germ agglutinin (WGA) conjugated with a fluor (WGA-fluorescein; Molecular Probes W-834); antibodies to chromosomes in mitosis from prophase to telophase (rabbit anti-phosphohistone 3; Upstate Biotechnology 06-570; this antibody is used 1/1000 and is *not* preadsorbed and leaves no background); or antibodies to in situ hybridization probes (see Dernberg, this volume; Calvi et al. 1998; A.W. Shermoen and P.H. O'Farrell, in prep.). Another option for a fluorescent signal is to use the tyramide signal amplification system (TSA) developed by New England Nuclear. This method uses HRP activity to deposit a tyramide-fluor at and in the immediate region of the target antibody with easily a fivefold or greater amplification (Wilkie and Davis 1998; Speel et al. 1999).

  b. Histochemical detection of BrdU using either goat anti-mouse • AP or goat anti-mouse • HRP:

*Goat anti-mouse • AP*

   i. Add 100–200 μl (final diluted volume) of preadsorbed antibody diluted to 1x with 10% NGS/PTX. Incubate on a Nutator for 1 hour and then withdraw the secondary antibody.

   ii. Rinse the embryos briefly with PTX (several rocks in the hand). Aspirate off most of the PTX and wash with three changes, 20 minutes each, of PTX.

   iii. Rinse the embryos twice for 5 minutes each with AP Detection Solution 1.

   iv. Stain with AP Detection Solution 2.

> *Note:* Staining is usually performed in a multiwell culture dish covered with aluminum foil; check the staining reaction periodically (after 10 minutes initially) under a dissecting microscope. There is less background if the staining is done wrapped in aluminum foil.

*Goat anti-mouse • HRP*

   i. Perform steps i and ii, above, as for goat anti-mouse • AP.

   ii. Withdraw the PTX from the embryos. Add 300 μl to 1 ml of the Staining Solution for HRP detection. Check the staining every few minutes. When the stain has developed sufficiently, stop the reaction by adding approximately 25 μl of 5% sodium azide.

9. Mount embryos as follows:

  a. Withdraw the embryos using a Pipetman (P-200) with a cut-off pipette tip and deposit them on a coverslip (22 x 22 mm). Withdraw the PTX using a sequencing tip on a P-200.

  b. Cover the embryos with 35–40 μl of mountant (the volume will vary with the number of embryos). Spread the embryos with a sequencing tip. Gently touch the middle of a slide to the mountant/embryos. Let the mountant spread to the sides of the coverslip (if it does not spread, add a bit more mountant at the side of the coverslip).

  c. Seal the edges of the coverslip with nail polish (this will prevent mountant from mixing with the oil used with oil-immersion lens, an optically fatal condition).

## TROUBLESHOOTING

- *Poor yield at the devitellinization step* (i.e., step 7 in Protocol 3.1). Experience has shown that this is invariably due to too much drying at the step between the dechorionation and the octane permeabilization (i.e., step 3 in Protocol 3.1). The second most common problem is too much drying at the step between octane permeabilization and the immersion into BrdU labeling medium (i.e., step 5 in Protocol 3.1).

- *Gradient of BrdU labeling.* This problem usually occurs from inadequate clearing of octane from part of the surface of the embryo (see step 4 in Protocol 3.1). A gradient of labeling may also be seen if the vitelline membrane is only partially removed. The vitelline membrane, which is left, will adhere closely to the embryo and may be seen as a diffraction pattern around the embryo using differential interference contrast (DIC) optics.

- *Differences in labeling efficiency between embryos of different ages.* Older embryos (nuclear cycles 15 and 16; stages 8–11) label much more efficiently (≥5 minutes), whereas the younger embryos take longer to label (≥8 minutes).

## ACKNOWLEDGMENTS

I thank Justine Melo, Smruti Vidwans, and Tin Tin Su for comments on this procedure description. The work during which the previously published methods were modified was supported by National Institutes of Health grant RO1-GM-37193 to Patrick H. O'Farrell.

## REFERENCES

Aten J.A., Stap J., Hoebe R., and Bakker P.J.M. 1994. Application and detection of IdUrd and CldUrd as two independent cell-cycle markers. *Methods Cell Biol.* **41:** 317–326.

Awad T.A. and Truman J.W. 1997. Postembryonic development of the midline glia in the CNS of *Drosophila:* Proliferation, programmed cell death, and endocrine regulation. *Dev. Biol.* **187:** 283–297.

Britton J.S. and Edgar B.A. 1998. Environmental control of the cell cycle in *Drosophila:* Nutrition activates mitotic and endoreplicative cells by distinct mechanisms. *Development* **125:** 2149–2158.

Calvi B.R., Lilly M.A., and Spradling A.C. 1998. Cell cycle control of chorion gene amplification. *Genes Dev.* **12:** 734–744.

Dolbeare F. and Shelden J.R. 1994. Immunochemical quantitation of bromodeoxyuridine application to cell-cycle kinetics. *Methods Cell Biol.* **41:** 297–316.

Duronio R.J. and O'Farrell P.H. 1994. Developmental control of a G1-S transcriptional program in *Drosophila. Development* **120:** 1503–1515.

Edgar B.A. and O'Farrell P.H. 1990. The three postblastoderm cell cycles of *Drosophila* embryogenesis are regulated in G2 by *string. Cell* **62:** 469–480.

Speel E.J.A., Hopman A.H.N., and Komminoth P. 1999. Amplification methods to increase the sensitivity of *in situ* hybridization: Play CARD(S). *J. Histochem. Cytochem.* **47:** 281–288.

Truman J.W. and Bate M. 1988. Spatial and temporal patterns of neurogenesis in the central nervous system of *Drosophila melanogaster. Dev. Biol.* **125:** 145–157.

Wilkie G. and Davis I. 1998. Visualizing mRNA *in situ* hybridization using "high resolution" and sensitive tyramide signal amplification. *Technical Tips Online.* 22/7/98 at http://www.biomednet.com/db/tto.

# CONTENTS

# 4

# Analysis of Meiosis in Fixed and Live Oocytes by Light Microscopy

Heinrich J.G. Matthies,* Michael Clarkson,† Robert B. Saint,†
Ruria Namba,* and R. Scott Hawley*
*Section of Molecular and Cellular Biology, University of California, Davis
†Department of Genetics, University of Adelaide, Adelaide, Australia

*D*ROSOPHILA FEMALE MEIOSIS HAS SERVED NOT ONLY AS THE CLASSIC SYSTEM for understanding the fundamental basis of the inheritance of genetic material (Bridges 1916), but also as an excellent model system for studying the mechanics of female meiosis (for reviews, see Hawley et al. 1993; Orr-Weaver 1995). The success of the meiotic analysis of *Drosophila* females reflects both the wealth of existing meiotic mutants (Hawley 1993) and the fact that chromosome behavior can easily be monitored using marked chromosomes. Characteristics of many of these mutants have been analyzed by molecular (Zhang et al. 1990; Kerrebrock et al. 1992; Rasooly et al. 1994; Hari et al. 1995; Sekelsky et al. 1995; McKim et al. 1996; Bickel et al. 1997; McKim et al. 1998) and biochemical (McDonald et al. 1990; Walker et al. 1990; Afshar et al. 1995) studies, leading to a mechanistic understanding of certain aspects of meiosis. To directly visualize the effects of these mutations on the meiotic machinery, procedures have been developed that allow study of *Drosophila* female meiosis by classical immunocytochemical (Theurkauf and Hawley 1992), FISH (fluorescent in situ hybridization; Dernburg et al. 1996; see Dernburg, this volume), and GFP (green fluorescent protein; Wang and Hazelrigg 1994; Endow and Komma 1996, 1997; Moore et al. 1998; also see Hazelrigg, this volume) methods.

Much has been learned about *Drosophila* meiosis by immunocytochemical studies of whole-mount oocytes lacking functional meiotic proteins and by localization of some of the relevant proteins (McKim et al. 1993; White-Cooper et al. 1993; Afshar et al. 1995; Endow and Komma 1996; Matthies et al. 1996; Moore et al. 1998). However, none of these classic methods easily point out kinetic problems, nor are the earliest points of deviation always readily obvious by static methods. For these reasons, and because the behavior of the highly dynamic components of the meiotic spindle may be perturbed in meiotic mutants, methods have been developed to directly observe the cytoskeleton and chromosomes in living egg chambers (Theurkauf 1994a; Endow and Komma 1996; Matthies et al. 1996).

These cytochemical and live imaging methods are discussed in this chapter. For additional protocols on immunocytochemical analysis of oogenesis, see Theurkauf (1994b) and Verheyen and Cooley (1994), and for a protocol on in vitro activation of oocytes, see Page and Orr-Weaver (1997). Finally, oogenesis has been thoroughly studied by electron microscopy (for reviews, see Mahowald and Kambysellis 1980; Spradling 1994).

## ANATOMY AND CELL BIOLOGY OF *DROSOPHILA* OOGENESIS: A BRIEF SUMMARY

Several reviews written about oogenesis should be referred to for more details and references (see, e.g., Mahowald and Kambysellis 1980; Spradling 1994). The purpose of this cursory summary is to aid in early attempts at applying the methods described below and identifying the morphological features of the egg chamber stages. These stages are arbitrary, but they are useful for discussions and evaluations of oogenesis.

Figure 4.1 presents a schematic representation of the female reproductive system. Egg chambers develop in a pair of ovaries (*Ov*) consisting of parallel ovarioles (*Ovl*) in anterior (germarium, *Grm*) and posterior (vitellarium, *Vtl*) compartments. Egg chamber development begins with the division of the stem cells in the germarium, and all of the subsequent stages are, in a more or less linear and chronological array, in the vitellarium, with the latest stage closest to the lateral oviduct (*Odl*). The first asymmetric stem-cell division leads to a new stem cell and a developing cystoblast. The cystoblast goes through four mitotic divisions with incomplete cytokinesis, generating a 16-cell cyst in which the cells are connected by cytoplasmic bridges called ring canals. These ring canals can be nicely visualized by the actin stain, phalloidin.

The 16-cell cyst is surrounded by somatically derived epithelial cells (the follicle cells), which eventually divide mitotically four to five times. In a single view through the midsection of an egg chamber, the follicle cells stained with DAPI (4′,6-diamidino-2-phenylindole) are seen as brightly staining dots at the perimeter (Figure 4.2).

The entire cyst moves into and eventually down the vitellarium. Two of the cells in the cyst begin to differentiate into oocytes, and the synaptonemal complex begins to be laid down in either of these stage-2 cells (Carpenter 1975). Eventually, 1 cell becomes an oocyte, with the remaining 15 cells becoming metabolically active polyploid nurse cells.

Division of the follicle cells ceases in stage-6 egg chambers, and they become polyploid during the next stage. Since both the nurse and follicle cells become increasingly polyploid during various stages, whereas the oocyte nucleus remains diploid, it is easy to understand why the oocyte nucleus, in comparison, becomes relatively weakly marked by DNA indicators (dyes or GFP-histone).

At stage 8, the oocyte is readily distinguished as it begins the endocytic uptake of the yolk (stage 8, Figure 4.2, top). Stage-9 egg chambers can be recognized by the position of the oocyte nucleus, which at this time is adjacent to the follicle cell/nurse cell interface (see weakly stained dot in DAPI image of stage 9, Figure 4.2, bottom). During stage 11, the nurse cells dump their contents into the oocyte. The volume of the oocyte dramatically increases at the expense of the nurse cell's cytoplasm (see stage 11, Figure 4.2, bottom). By stage 12, the nurse cells begin to degenerate and at the beginning of stage 13, only 15–11 nurse cell nuclei are remaining. The oocyte chromatin of stage-13 oocytes is much less intensely stained by DNA indicators than the nurse cells. To clearly denote the position of the karyosome in stage-13 oocytes, the image intensities of the karyosome in the example of this stage were exaggerated (Figure 4.2, bottom, stage 13). The karyosome is readily

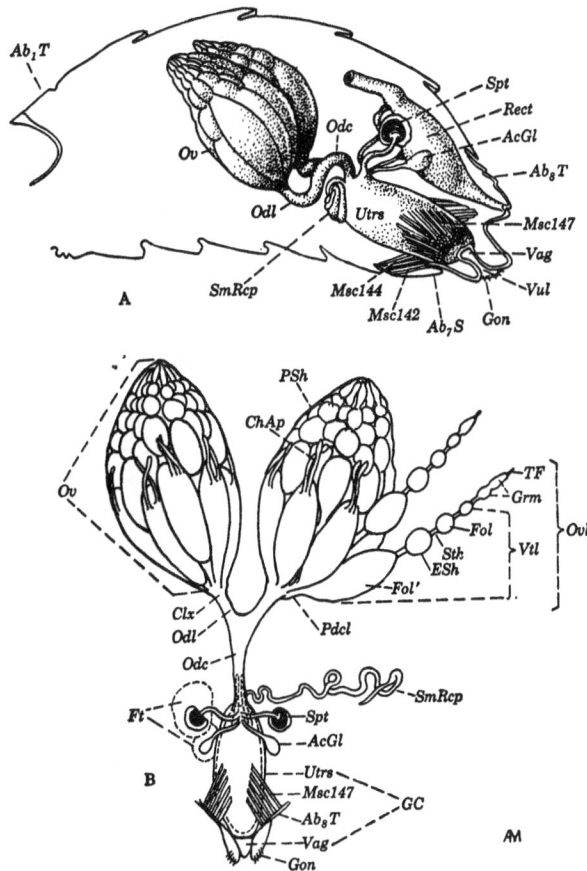

**Figure 4.1.** Female reproductive system. (*Top*) Side view of the abdomen with the left wall removed. (*Ov*) Ovary; (*Odl*) lateral oviduct; (*Odc*) common oviduct; (*Utrs*) uterus; (*Vag*) vagina; (*Vul*) vulva; (*Gon*) gonopod; (*Rect*) rectum; (*Ab$_1$T* and *Ab$_8$T*) first and eighth abdominal tergites. (*Bottom*) Dorsal view of the reproductive organs in isolation. The abbreviations for the bottom image are the same with the addition of the following: (*PSh*) Peritoneal sheath; (*ChAp*) chorionic appendage or dorsal appendage; (*Ovl*) ovariole; (*TF*) terminal filament; (*Grm*) germarium; (*Vtl*) vitellarium; (*Fol*) follicle; (*Stk*) interfollicular stalk; (*ESh*) epithelial sheath; (*Fol'*) basal follicle; (*Pdcl*) pedicle; (*SmRcp*) seminal receptacle; (*Spt*) spermatheca; (*AcGl*) accessory gland. (Reprinted from Miller 1950.)

obvious at the dorsal-anterior part of the oocyte, adjacent to one of the dorsal appendages or filaments (compare Nomarski and DAPI images).

Around stage 13, the oocyte nuclear envelope breaks down and the meiotic spindle assembles around the chromatin. The meiotic spindle arrests at metaphase I until fertilization. The metaphase I spindle is highly tapered and lacks astral microtubules. The chromatin is stretched on the spindle but is held together by chiasmata. The always nonexchange fourth chromosomes are found adjacent to the main mass that contains the exchange chromosomes.

In fixed cells, the chorion and follicle cells are removed, so the investigator must be able to recognize the shape of the stage-14 oocyte. In intact oocytes, the dorsal appendages are generated by a particular class of follicle cells during stage 13 and are completed along with the chorion during stage 14. Study of these appendages will allow the investigator to recognize the overall morphology of the oocyte (~500 μm in length), because this is the region where the metaphase I spindle (12–15 μm in length) is located.

| Stage | Length | Identification | Nomarski | DAPI |
|---|---|---|---|---|
| 2 | 8 hr | average size: 25 × 25 μ; karyosome has not yet formed (arrow); oocyte and nurse cell nuclei similar in size | | |
| 3 | 8 hr | average size: 35 × 35 μ; karyosome (k) and endobody are visible in germinal vesicle | | |
| 4 | 6 hr | average size: 40 × 50 μ; nurse cell nuclei contain similar amounts of DNA and appear polytene (arrow) | | |
| 5 | 5 hr | average size: 55 × 75 μ; nurse cell nuclei no longer polytene (arrow); posterior nurse cells have more DNA than anterior ones | | |
| 6 | 3 hr | average size: 60 × 85 μ; follicle cell mitoses cease; nurse cell ploidy equal | | |
| 7 | 6 hr | average size: 70 × 115 μ; more elongated in shape than previous oval-shaped stages; nurse cells have higher ploidy at posterior; no yolk visible in oocyte | | |
| 8 | 6 hr | yolk visible in oocyte (arrow); follicle cell layer still uniform | | |

| Stage | Length | Identification | Nomarski | DAPI |
|---|---|---|---|---|
| 8 | 6 hr | yolk visible in oocyte; follicle cell layer uniform | | |
| 9 | 6 hr | follicle cells in process of migration over oocyte, resulting in anterior-posterior gradient of cell thickness (arrows); oocyte is ~1/3 of egg chamber | | |
| 10A | 6 hr | follicle cells form columner epithelium over oocyte; centripetal migration not yet visible; oocyte is 1/2 of egg chamber | | |
| 10B | 4 hr | centripetal migration in progress (arrows); dorsal follicle cells thicker than ventral (triangle); vitelline membrane extends into opercular region | | |
| 11 | 0.5 hr | oocyte larger than nurse chamber due to onset of nurse cell dumping, but dumping still incomplete | | |
| 12 | 2 hr | dumping complete; 15 nurse cell nuclei remain at anterior; dorsal filaments not visible in nurse chamber | | |
| 13 | 1 hr | some nurse cell nuclei still remain; dorsal filaments visible at anterior end | | |
| 14 | >2 hr | no nurse cell nuclei remain; dorsal filaments complete their elongation | | |

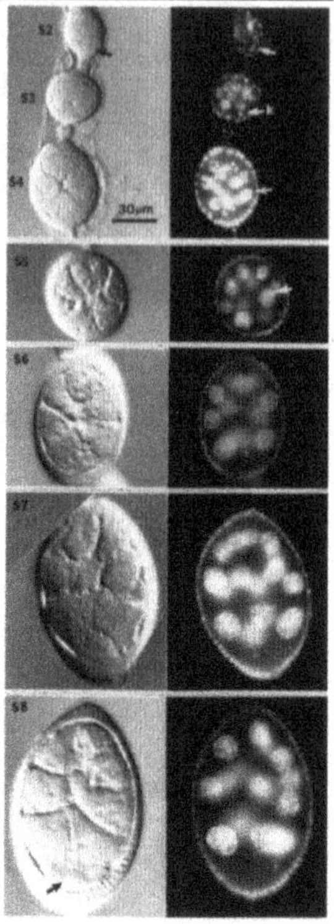

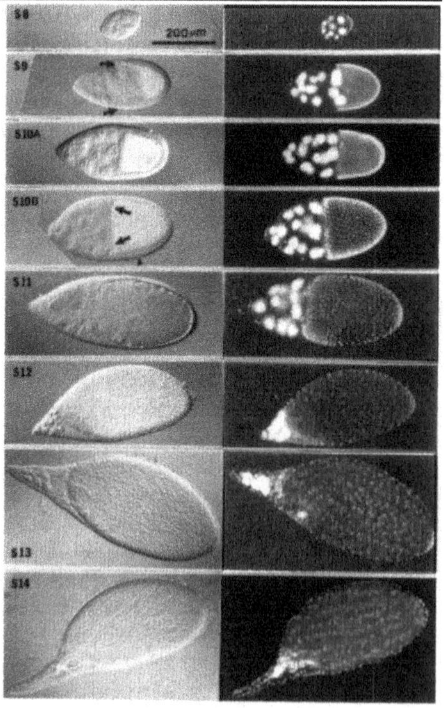

**Figure 4.2.** Stages of oogenesis. (*Top*) Stage 2 to stage 8. The various stages are shown viewed by Nomarski and fluorescence (DAPI) optics. Stage durations are from transplantation studies into host females. (*Bottom*) Stage 8 to stage 14. (Reprinted, with permission, from Spradling 1994.)

## METHODS TO VISUALIZE OOGENESIS AND FEMALE MEIOSIS

Described in this chapter are protocols and reagents used for the analysis of fixed or live egg chambers using light microscopy. Methods for fixing egg chambers are the same as standard fixation methods, but with additional steps to remove vitelline and chorion layers. Depending on the ease of obtaining a reasonable number of mothers, egg chambers can be isolated either by a mass method (see Protocol 4.1) or, if necessary, by hand dissection (see Protocol 4.2). Protocol 4.3 describes procedures for removal of the chorion and vitelline membranes from late-stage egg chambers and permeabilization of the vitelline membrane of early-stage egg chambers. The protocols for fixation can also be used for enhancer-trap experiments utilizing the *lacZ* gene, which is part of the *P-lac W* element used in many mutant screens. This allows the identification of the expression pattern of a P-element-disrupted gene required for oogenesis by a standard in situ β-galactosidase assay (see Protocol 4.5).

Large numbers of mutants can be screened by using dissected, live egg chambers observed by differential interference optics (DIC) (Figure 4.2) (see Verheyen and Cooley 1994), fluorescence microscopy, or time-lapse laser confocal microscopy. The overall structure and development of the egg chambers can be quickly monitored in unfixed egg chambers, by taking advantage of the autofluorescence of chorion and by the use of DNA-binding dyes. These dyes can penetrate live, early-stage egg chambers. The overall state of chromatin, tubulin, and actin can be rapidly determined using flies expressing GFP-histone2AvD, GFP-ncd (Endow and Komma 1997), or GFP-moesin (Edwards et al. 1997). However, the GFP-labeled moesin and ncd do not label all structures equally, so labeled actin, tubulin, and/or DNA-binding dyes can be injected into living egg chambers to directly observe dynamic events. Protocol 4.4 describes procedures for labeling egg chambers with antibodies or DNA-binding dyes.

## IMAGING WHOLE-MOUNT FIXED EGG CHAMBERS

The methods described in this section were developed by Bill Theurkauf in Bruce Alberts' laboratory (Theurkauf and Hawley 1992), and further refined in the Hawley (McKim et al. 1993) and Goldstein laboratories (Matthies et al. 1996).

## PROTOCOL 4.1*

# Mass Isolation and Fixation of Egg Chambers

Egg chambers are obtained essentially according to the procedure of Theurkauf and Hawley (1992). Two alternative methods for mass isolation can be used and are described below. These protocols were developed to allow antibodies to penetrate the oocyte and to avoid premature activation of the oocyte, as hypotonic buffers lead to activation (Mahowald et al. 1983). Method 1 describes the isolation and subsequent fixation of egg chambers. Method 2 is essentially similar to Method 1, but involves isolating egg chambers in the presence of fixative. When oocytes are isolated in fixative, there is not much difference in the overall morphology except that the fourth chromosomes are more consistently found attached to the chromosomal mass. An early indicator of activation is fourth chromosome movement to the poles. In both cases, but especially for the second method, speed is the most critical component.

*For all protocols, H$_2$O indicates glass distilled and deionized.

## Materials

### *Media*

> **Bottles Containing Yeast Paste**
>
> Prepare standard corn molasses fly food (for preparation, see Appendix 3) in half-pint jars. Combine 2.7 g of yeast in 6 ml of $H_2O$. Mix to form a paste. Apply generously to one side of the bottle. This amount of yeast should be sufficient for approximately 10 bottles.

### *Supplies and Equipment*

Modified containers; use either:
- 50-ml Falcon tubes, each with the bottom removed and a hole cut into the cap
- 100-ml plastic beakers with the bottoms removed

Nitex screen (350 μm, 250 μm, 125 μm, 86 μm, and 64 μm, each 2 sq. inches)

(*Optional*) Rubber bands

Ring stands (2)

Funnels (2)

Beakers (2) (500 ml)

**CO₂** station

Blender. Use either an 8-ounce or a 5-cup container. An 8-ounce blender container minimizes the surface area, but the standard, 5-cup blender containers are satisfactory.

Dissecting microscope

Falcon tubes (15 ml)

### *Solutions and Reagents*

(*Optional*) **Ether**

1x PBS (see Appendix 3)

**Formaldehyde** (16%, EM grade, methanol-free; Ted Pella)

> **5x Modified Robb's Medium**
>
> 275 mM sodium acetate
> 40 mM potassium acetate
> 100 mM sucrose
> 10 mM glucose
> 2.2 mM **MgCl₂**
> 0.05 mM **CaCl₂**
> 100 mM HEPES (pH 7.4)
>
> **1x Robb's Medium Containing 1% Bovine Serum Albumin (BSA)**
>
> Prepare 400 ml of 1x Robb's Medium from the 5x stock containing 1% BSA. Use as a freshly prepared solution.
> This solution is required for performing Method 1.
>
> **4% Paraformaldehyde in 1x Robb's Medium**
>
> Prepare 400 ml of 1x Robb's Medium from the 5x stock containing 4% **paraformaldehyde**. Warm to 30ºC. Use as a freshly prepared solution.
> This solution is required for performing Method 2.

*EBR (1% BSA)*

130 mM NaCl
4.7 mM **KCl**
2.2 mM $MgCl_2$
1.9 mM $CaCl_2$
10 mM HEPES (pH 7.1)

*2× Fix Buffer (pH 7.2)*

200 mM **potassium cacodylate**
200 mM sucrose
80 mM sodium acetate
20 mM EGTA

*1× Fix Buffer*

Combine 1 part of 2× Fix Buffer with 1 part of 16% formaldehyde (i.e., 1:1 ratio). The final concentration of formaldehyde is 8%.

*PBST*

1× PBS
0.1% Triton X-100

CAUTION: $CaCl_2$, $CO_2$, ether, formaldehyde, KCl, $MgCl_2$, paraformaldehyde, potassium cacodylate (see Appendix 4)

## Method 1

### Isolation and Subsequent Fixation of Egg Chambers

1.  Hold several hundred 2–3-day-old female flies with males for 2–4 days in well-yeasted bottles.

    *Note:* This procedure causes the flies to maximize egg production and laying. When fewer flies are kept in the bottle, all stages of oogenesis are found in proportion to the duration of each stage (see Figure 4.2). In more crowded bottles, a much higher percentage of stage-14 oocytes is observed.

2.  Set up filters to separate ground flies from egg chambers and to collect purified egg chambers at the desired stage. Attach the Nitex filters of the appropriate mesh size to the top of a modified 50-ml Falcon tube by screwing on the cap or by using a rubber band at the bottom of a modified 100-ml beaker.

    a.  To separate ground flies from egg chambers, attach a 350-μm filter to either a modified Falcon tube or a beaker on a ring stand. Set up a funnel above this unit and place a 500-ml beaker below to collect the filtrate.

    b.  To collect purified egg chambers at the desired stage, set up another funnel above two more filters in series (a 250-μm filter followed by a 125-μm filter) on a ring stand and place a 500-ml beaker below.

        *Notes:* This assembly allows the investigator to quickly pour the ground flies onto the filters and then regrind the flies retained on the 350-μm filter at least once.

        The 125-μm filter retains mostly stage-11–14 egg chambers. If earlier stages are desired, an 86-μm filter will trap stages 9–14, allowing stages 1–8 to flow through; these can be trapped by a 64-μm net.

3. Pour 150 ml of freshly prepared 1x Robb's Medium containing 1% BSA into the blender.

   *Notes:* The BSA helps minimize egg chamber losses and helps in the purification step, as many fly parts float in this buffer, whereas egg chambers sink.

   If the main focus of the study is not meiosis, it may be better to use EBR (containing 1% BSA), which is iso-osmotic, rather than Robb's Medium, which is hypertonic.

4. Anesthetize the flies with either $CO_2$ or ether. Transfer flies to the blender.

5. Pulse the flies three to five times at the lowest setting (2 seconds each time).

6. Pour all of the mix from the blender onto the 350-μm filter. The egg chambers flow through and are collected in the beaker.

7. Remove the 350-μm Nitex filter and rinse the flies into the blender container with another 150 ml of 1x Robb's Medium containing 1% BSA. Blend again as in step 5.

   *Note:* It is tempting to blend for too long, but the result is that more legs are isolated and the eyes can become disrupted. Verify the efficiency of releasing the egg chambers and optimize the blending by looking at the filtrate using a dissecting microscope. Use the least blending possible to get a reasonable quantity of egg chambers. Blending too vigorously produces more legs with their sticky bristles, making egg chamber rolling difficult (see Protocol 4.3).

8. Pour the filtrate through the 350-μm filter. Collect the filtrate in the beaker containing the filtrate from step 6.

9. Quickly transfer the filtrate through the funnel above the 250-μm and 125-μm filters. Rinse the eggs on the filters with another 50 ml of 1x Robb's Medium containing BSA. Then rinse the contents from the 125-μm filter into a 15-ml tube with a minimum of PBS. Allow the egg chambers to settle for approximately 5 minutes (there is no need to wait longer because the remnants that settle slower are not egg chambers).

10. Aspirate the buffer and add 1x Fix Buffer (warmed to 25°C). Fix with gentle rocking for no more than 4–5 minutes.

11. Allow the eggs to settle for 1 minute and then quickly remove the fix and add 10 ml of PBST. Allow the eggs to settle and repeat this three times.

    *Notes:* If the aim is to study germaria through stage-10b egg chambers, skip the next step (i.e., Protocol 4.3 [Removal of the Chorion and Vitelline Membranes]). Proceed to washing egg chambers (Protocol 4.3 [Washing and Permeabilizing]).

    A good approach to monitor fixation is by looking at the extensive microtubules found in well-fixed follicle cells. In overfixed egg chambers, histone antibodies do not label nurse cell nuclei very well.

## Method 2

### *Simultaneous Isolation and Fixation of Egg Chambers*

1. Follow steps 1–6 and step 9 as described above in Method 1 using 4% paraformaldehyde in 1x Robbs Medium.

   *Note:* In Method 2, flies are blended only once, poured through the 350-μm filter, and subsequently over the 250-μm and 125-μm filters (see step 2, below). This should take less than 1 minute.

2. Perform steps 10–11 as described in Method 1.

PROTOCOL 4.2

# Hand Dissection of Egg Chambers

When it is difficult to generate large numbers of females, hand dissection is an option. Egg chambers from 16 flies are dissected in Modified Robb's Medium. If the fly is healthy, the ovaries will take up a substantial portion of the abdomen.

## Materials

### *Media*

Bottles containing yeast paste (see p. 72)

### *Supplies and Equipment*

Dissecting surface, either slides or spot plates (Pyrex)
Black surface for placing slides or spot plates
Fiber-optic light source
Dissecting tools
- dissecting microscope
- Dumont #5 tweezers (2 pairs)

Kimwipes

### *Solutions and Reagents*

1x Robb's Medium containing 1% BSA (see p. 72)

### *Fixative*
1x Fix Buffer (from 2x stock, see p. 73)
4% fresh **paraformaldehyde**

CAUTION: **paraformaldehyde** (see Appendix 4)

## Method

1. Raise flies as described in Protocol 4.1, step 1.

2. Add a drop of Modified Robb's Medium either onto a slide or onto spot plates (Pyrex), placed on a black surface and illuminated from the two sides with a fiber-optic light source.

3. Anesthetize two flies at a time and collect the ovaries from each fly as follows:

   a. Use a pair of Dumont #5 tweezers to grab the constriction at the junction of the thorax and abdomen (see Figure 4.1, around Ab$_1$T). Use another pair of tweezers to pull at the tip of the posterior abdomen (region that appears as a lip around the vulva in Figure 4.1) while the fly is on its dorsal side. The uterus and any oocyte about to emerge will be attached to a piece of the abdomen and some internal organs will also come along.

   b. Pull until the organs rupture, wipe the tweezers on a Kimwipe, and remove the remainder of any exposed organs.

c. While still maintaining the grip on the most proximal abdomen ($Ab_1T$ region), remove a portion of the cuticle. At this point, the paired ovaries will come out of the fly, but sometimes a little pressure behind the ovaries will facilitate this. Dissect the other fly.

4. Put both pairs of ovaries in a well with Fixative for 4 minutes. During this time, remove the muscle-containing sheath surrounding the ovarioles and tease them apart into separate egg chambers.

5. Proceed with the next appropriate step (see Protocol 4.3) depending on the stage of interest, i.e., removing the outer layers (rolling to remove both chorion and vitelline membranes for late-stage egg chambers) or washing and permeabilizing the egg chambers (detergent extraction for stages where the vitelline membrane is still extractable up to stage 13).

PROTOCOL 4.3

# Outer Membrane Removal of Egg Chambers

The follicle cells, chorion, and vitelline layers are removed from late-stage egg chambers by a process called "rolling." This allows antibodies to penetrate the eggs. For early-stage egg chambers, the vitelline membrane can be permeabilized with a nonionic detergent.

## Materials

### Supplies and Equipment

Slides, end-frosted
Dissecting microscope

### Solutions and Reagents

PBST (see p. 73)

## Method

### Removal of the Chorion and Vitelline Membranes

1. Transfer the fixed egg chambers to the frosted portion of a slide. Remove most of the buffer.

2. Place the frosted portion of another slide on top and roll the egg chambers between the slides. Alternatively, place a 50 × 22-mm coverslip on top of the embryos and roll the egg chambers between the coverslip and slide.

> Notes: Roll the eggs hard enough to remove both the chorion and the vitelline layers, to allow antibodies to penetrate. The eggs are fairly tough, but the trick is to have just the right amount of eggs and to remove most of the buffer. Too few eggs tend to get damaged; on the other hand, if there are too many eggs, it is difficult to remove the vitelline layer. At the point when the egg chambers have been rolled enough, some will appear to be breaking open.
>
> In the beginning, learn to recognize the vitelline membrane in a dissecting microscope. If necessary, look at the eggs at 488 nm under a fluorescent microscope; under 488-nm excitation, the vitelline membrane is obvious and appears as an autofluorescent green sheet. Vitelline membrane that tends to come off in small pieces indicates that the formaldehyde is getting old.

### *Washing and Permeabilization*

1. Rinse the egg chambers into a 15-ml Falcon tube with PBST and allow them to settle by gravity.

2. Remove the PBST. Add 5 ml of PBST and gently rock the egg chambers for 30 minutes.

3. Allow the egg chambers to settle. Remove PBST and rinse three times with 10 ml of PBST.

PROTOCOL 4.4

## Labeling the Egg Chambers

### Materials

### *Supplies and Equipment*

Microcentrifuge tubes (0.5 ml; Eppendorf)
Coverslips (#1.5)
Disposable transfer pipettes
Nail polish

### *Solutions and Reagents*

Primary antibodies
  *For labeling of α- and β-tubulin:*
  - DM1A, anti-α-tubulin, mouse monoclonal (Sigma T 9026)
  - DM1B, anti-β-tubulin, mouse monoclonal (Amersham N357)
  - YL1/2, anti-α-tubulin, rat monoclonal (Accurate)
  - YOL1/34, anti-α-tubulin, rat monoclonal (Accurate)
  *For direct labeling of α- and β-tubulin:*
  - Cy3-labeled TUB2.1, anti-β-tubulin, mouse monoclonal (Sigma)
  - FITC-labeled DM1A, anti-α-tubulin, mouse monoclonal (Sigma)
  *For histone labeling:*
  - anti-core histone, monoclonal IgG2 (Chemicon MAB052; monoclonal antibody to histone H1 + core proteins)
Secondary antibodies
1x PBS (see Appendix 3)
(*Optional*) DNA-binding dyes (e.g., OliGreen, SYBR Green I, or PicoGreen; Molecular Probes)
(*Optional*) Rhodamine-phalloidin (Molecular Probes R-415), 10-μl aliquot at –20ºC

### *Blocking Solution*

PBST (see p. 73)
Blocking reagent
Some examples of blocking reagents include BSA (0.1–5%); normal serum from the same host species as that in which the secondary antibodies were raised (serum from various species is available from Jackson ImmunoResearch Laboratories); Blotto (prepared by dissolving powdered milk [0.1–5%] in PBS and briefly centrifuged).

> *Mounting Medium*
> 1x PBS
> 90% glycerol
> 1 mg/ml *p*-phenylenediamine

**CAUTION:** *p*-phenylenediamine (see Appendix 4)

## Method

1. Transfer the egg chambers to a 0.5-ml microcentrifuge tube. Incubate the egg chambers in Blocking Solution for 1 hour. Then rinse the egg chambers five times with Blocking Solution.

   *Note:* If only microtubules, chromosomes, or actin are being labeled, this blocking step can be skipped with little degradation of the signal. However, for other antibodies, the minimum amount and best blocking reagent must be determined empirically. Too much blocking can sometimes obscure the signal, but careful examination is necessary to determine background versus genuine labeling. Preimmune sera for affinity-purified antibodies, preadsorbing antibody with antigen, and secondary antibody alone are good controls, but can be misleading. The best control and method is to use a null mutant; it is simple to determine the best blocking conditions and antibody concentration using null mutants.

2. Add the appropriate primary antibodies at empirically determined concentrations (usually 0.1–10 µg/ml) in 0.2–0.5 ml of Blocking Solution. Incubate either at room temperature for 4 hours or at 4°C overnight.

   *Notes:* When performing double-labeling of microtubule-associated proteins and tubulin, it is sometimes better to add the nontubulin antibody first, incubate, and then add the tubulin antibody; however, adding both simultaneously also works.

   Chromosomes can be visualized using DNA-binding dyes or histone antibodies. For most purposes, dyes are sufficient and the relevant methods are discussed below (see Protocol 4.4 Notes), but antibody labeling gives a crisper outline of the edges of the chromatin. A mouse monoclonal antibody against core histones sold by Chemicon yields excellent results. To label microtubules, several monoclonal antitubulin (α- or β-subunit) antibodies have been used, including mouse monoclonal antibodies DM1A and DM1B and rat monoclonal antibodies YL1/2 and YOL1/34. The rat monoclonal antibodies label meiosis I spindles slightly better, but Sigma sells an FITC-conjugated anti-α-tubulin antibody that works very well and eliminates the need for secondary antibodies. Alternatively, directly label both the core histone antibody and one of the tubulin antibodies by standard methods (Harlow and Lane 1988) or buy a kit sold by several companies (e.g., Sigma and Amersham). The secondary antibodies are thus avoided, which makes the procedure faster and gives less background. Generally, direct immunofluorescence is successful for antigens that are abundant or highly concentrated at one location.

3. Wash the egg chambers six times with Blocking Solution for 15 minutes (each wash).

4. Add the appropriate secondary antibodies in Blocking Solution. Incubate at room temperature for 4 hours. Wash egg chambers as described in step 3 (above).

5. Rinse three times with 500 µl of PBS.

6. (*Optional*) To label actin, incubate with fluorescently conjugated phalloidin as follows:

   a. Place a 10-µl previously frozen aliquot of rhodamine-phalloidin in a SpeedVac and dry for approximately 5 minutes. Resuspend in 100 µl of PBS (final concentration 1 µg/ml).

    b. Incubate at room temperature for 20 minutes.

    c. Rinse three times with 500 µl of PBS.

7. After the final wash, allow egg chambers to settle and then remove as much as possible of the PBS. Add 22–40 µl of Mounting Medium. Make sure that the Mounting Medium and whatever PBS is left are mixed by gently stirring with the tip of a disposable transfer pipette, and then transfer to the surface of a clean #1.5 coverslip.

> *Notes:* It is important to remove as much of the PBS as possible because the refractive index of the mounting media (1.515) is designed to match the refractive index of glass (PBS has a much lower refractive index).
>
>     Most lenses are designed to image through 170-µm coverslips; therefore, the common practice of using #1 coverslips (the thickness of #1.5 coverslips is ~150–180 µm) can be considered to be major tradeoff for the increase in working distance compared to the substantial degradation of the thick egg chamber image. Put a few pieces of broken coverslip into the mounting media on the coverslip to avoid compressing the egg chambers. This also makes it easy to mount the coverslip and slide in a parallel manner. Put the slide over the coverslip and seal with nail polish.

## PROTOCOL 4.4 NOTES

If histone antibodies are not used, DNA can be detected by staining with DAPI (0.1 µg/ml) or with some of the DNA-binding dyes sold by Molecular Probes. DAPI is reliable and resistant to bleaching, but it requires detection with UV, and the oocyte (but not the earlier egg chambers) is autofluorescent in this range. Molecular Probes sells many DNA-binding dyes that work well. Several dyes including TOTO-3 have been used as recommended by Molecular Probes, but SYBR Green or PicoGreen are much less expensive and are very bright even when used at a dilution of 1:10,000. Many *Drosophila* laboratories use OliGreen at 1:5000 to 1:10,000 dilution to label egg chambers treated with 20 µg/ml of RNase A for 30 minutes. Label with DNA-binding dyes after the next to final wash.

## PROTOCOL 4.5

# β-Galactosidase Activity in *Drosophila* Ovaries

### Materials

### *Supplies and Equipment*

Water bath at 37ºC
Depression slide
Dissecting microscope
Slides
Coverslips (#1.5)

### *Solutions and Reagents*

1x PBS (see Appendix 3)
**X-gal** (8%) in **DMSO**, freshly prepared
Mounting Medium without *p*-phenylenediamine (see p. 78)

> **β-*Galactosidase Staining Solution***
>
> 10 mM **sodium phosphate**
> 150 mM NaCl
> 1 mM **MgCl$_2$**
> 3.3 mM **K$_4$[Fe$^{II}$(CN)$_6$]**
> 3.3 mM **K$_3$[Fe$^{III}$(CN)$_6$]**
> 0.3% Triton X-100
> Immediately before use, incubate the staining solution and 8% X-gal solution at 37°C. To 1 ml of warm staining solution, add 25 µl of warm X-gal solution. This is done to increase the solubility of X-gal.

**CAUTION: DMSO, K$_3$[Fe$^{III}$(CN)$_6$], K$_4$[Fe$^{II}$(CN)$_6$], MgCl$_2$, *p*-phenylenediamine, sodium phosphate, X-gal** (see Appendix 4)

## Method

1. Obtain egg chambers by hand dissection and fix as described in Protocol 4.2. Permeabilize the egg chambers as described in Protocol 4.3 (Washing and Permeabilization).

2. Incubate in β-galactosidase Staining Solution at 37°C. Observe the egg chambers in a depression slide with a dissecting microscope. After egg chambers have been stained sufficiently, rinse with PBS.

   *Note:* The time for staining should be determined empirically and will range from 5 minutes to overnight. Some lines that express high levels of activity may need to be rinsed with PBS after several minutes. Others take much longer.

3. Mount the egg chambers in Mounting Solution without the *p*-phenylenediamine.

4. Photograph with a conventional light microscope.

## DATA COLLECTION

The best images are obtained when a confocal microscope is used to avoid out-of-focus light, and images should be obtained by whatever averaging methods are available (at least 9 averages). Images with the least amount of aberrations can be obtained by looking at structures that are close to the surface of the egg chamber on the side next to the coverslip. Lenses should be used with the highest available numerical aperture (NA). To observe structures deep in the egg chambers, use lenses that have a coverslip correction collar. Careful examination of deep structures and a reiterative adjustment of this collar will improve the quality of these images. Data collection should be optimized by taking advantage of the full dynamic range of the data-collection equipment, i.e., set the black level just below the lowest signal and adjust the gain to avoid saturation of the part of the image that will be recorded. Most digital image-collection devices have look-up tables that use displays in pseudo-coloring, making this part of data collection trivial. The same effect can be achieved by postcollection data-processing methods; however, the data lose resolution because fewer data points exist between each intensity value.

## LIVE ANALYSIS OF MEIOSIS

The tradition of observing living *Drosophila* cells injected with fluorescent cellular proteins was initiated in the Alberts laboratory at the University of California, San Francisco. Doug Kellogg injected fluorescent cytoskeletal elements into early embryos (Kellogg et al. 1988), Jon Minden injected labeled histones (Minden et al. 1989), and Bill Theurkauf began injecting egg chambers with labeled tubulin (Theurkauf et al. 1994a). Later, in collaboration with the Theurkauf and Goldstein laboratory, these methods were used to study meiosis in wild-type and *ncd*-null mutants (Matthies et al. 1996).

Tubulin injection and subsequent imaging of live, early egg chambers have been reported by Theurkauf (1994a), but will not be discussed here. No one has reported injecting actin into egg chambers, but this has been done in early embryos, and the tubulin method discussed below should work for fluorochrome-conjugated actin. The overall gross morphology of live egg chambers dissected as described below can be observed with any microscope equipped with DIC optics. OliGreen works well for observing the chromatin by fluorescence. The early stages quickly pick up the dye and label the DNA; however, later stages need to be injected. OliGreen can be diluted 1:100, and a small aliquot can be mixed in the endogenous fluid around the dissected egg chambers on the coverslip. Rhodamine-conjugated tubulin labeled as described by Kellogg et al. (1988) or commercially available conjugated tubulin from either Molecular Probes or Cytoskeleton has been used. All work well, and purchasing the labeled tubulin saves a lot of time. The tubulin from Molecular Probes is brighter, but the tubulin from Cytoskeleton is less expensive.

Chromatin can be imaged by injecting various DNA-binding dyes, but the effect of the dyes on chromatin in vivo is not always easy to determine. Therefore, to facilitate the study of chromosome behavior, we modified a 4.0-kb *Drosophila* His2AvD genomic fragment to encode a fusion protein containing nearly all of His2AvD (an H2AZ class variant histone) fused to the GFP of the jellyfish *Aequorea victoria* (Clarkson and Saint 1999). Since this construct is under the control of endogenous His2AvD regulatory sequences within the genomic fragment, it is expressed in a pattern indistinguishable from that of His2AvD and complements an His2AvD null allele. Preliminary studies with flies bearing this GFP-histone fusion in the presence of wild-type levels of His2AvD show some slight female meiotic chromosome misbehavior.

PROTOCOL 4.6

## Analysis of Meiosis in Live Egg Chambers

### Materials

#### Supplies and Equipment

Injection apparatus
Syringe (10 or 20 ml)
Plastic tubing to connect micropipette holder to syringe

Glass needles (filament: 4 inches long, 1.0 mm diameter)
Coverslips (#1.5; 50 × 22 mm)
Inverted microscope linked to a confocal
**CO$_2$** station
Dissecting tools
- Dumont #5 forceps
- dissecting microscope

Microcentrifuge tubes
Microcentrifuge or Beckman T100 tabletop ultracentrifuge

### Solutions and Reagents

Halocarbon oil
Rhodamine tubulin (10-µl aliquot at –20°C; Molecular Probes or Cytoskeleton)
(*Optional*) Fluorescein or rhodamine actin (Cytoskeleton)
OliGreen (10-µl aliquot at –20°C; Molecular Probes)

**CAUTION: CO$_2$** (see Appendix 4)

## Method

### Pipette Pulling and Setting Up the Injection Apparatus

1. Set up the injection apparatus on the microscope.

   *Notes:* Any standard, simple injection apparatus will suffice; use plastic tubing between the needle holder and a simple 10- or 20-ml syringe.

   Injection needles are pulled similar to the needles pulled for P-element transformation, but the tips are stubbier.

2. Break open the tip of a sealed, pulled needle as follows:

   a. Place a coverslip on the stage of the inverted microscope linked to the confocal. Focus on the middle of the edge of a coverslip.

   b. Move the stage with the mounted coverslip to the side, and bring the needle into focus (the injection holder approaches the stage at a fairly shallow angle). Gently tap the needle against the side of the coverslip and visually verify that the needle tip breaks. Lift the needle holder straight up; this allows the needle to come back into focus quickly.

### Hand Dissection of Egg Chambers

1. Maintain well-fed, 2–3-day-old flies as described for the fixed-oocyte preparations (see step 1 in Protocol 4.1).

2. Anesthetize several females with CO$_2$ and immerse in halocarbon oil.

   *Note:* The halocarbon oil allows for high oxygen levels and helps keep the oocytes bathed in their native extracellular fluid.

3. Remove ovaries as described in step 3 of Protocol 4.2 and tease apart with two forceps on the surface of a clean #1.5 coverslip (50 × 22 mm).

4. Remove whatever loose particles are visible and rotate the oocytes in the appropriate position in a line of 5–10 oocytes.

*Note:* In the next step, the oocytes will be observed through an inverted microscope linked to a confocal; thus, the goal is to orient the late stage-12 to early stage-13 oocyte in such a way as to position the intact nucleus as close as possible to the coverslip. At this stage, the nurse cells have begun to degenerate, but 15–13 nurse cell nuclei can usually still be seen. The nurse cells can be seen as spheres at the most anterior portion of the oocyte. The dorsal appendages are usually not visible, but will extend during the experiment.

5. Observe the oocytes on the inverted microscope linked to the confocal using either DIC or phase-contrast optics.

*Note:* If using GFP-histone-expressing flies, the oocyte karyosome in the nucleus is seen as a dim green spot relative to intensely bright, nurse cell nuclei or even the follicle cells. With these optics, the nucleus (~35 μm) is seen as a clearing of the oocyte cytoplasm. It lies on the posterior and most anterior side of the oocyte. If the oocyte is not positioned right, the nucleus is not as clearly visible because it is not as close as possible to the coverslip. Go back to the dissecting microscope and reorient the oocyte.

## Injection

1. To prepare the dye mixture for injection, quickly thaw out 10-μl aliquots of rhodamine-labeled tubulin and OliGreen. Dilute the OliGreen 1:10 in $H_2O$. Add 1 μl of 1:10 dilution of OliGreen to 1 μl of rhodamine-tubulin. Centrifuge in a microcentrifuge at 12,000*g* at 4°C for 10 minutes or in a Beckman T100 tabletop ultracentrifuge at 20,000*g*.

2. Mount the coverslip on the stage, and pipette 1 μl of the rhodamine-labeled tubulin onto the oil. It is important to perform these steps quickly to avoid tubulin polymerization in the needle. Use a 10X or 20X lens to lower the needle into the tubulin and draw some into the needle.

3. Lift up the needle and bring the first oocyte of the column into focus. Lower the needle and push it into the oocyte until the tip of the needle lies just next to the nucleus and push on the syringe. A small clearing will be seen in the cytoplasm. Pull the needle out and inject the remaining oocytes.

## DATA COLLECTION

Set the confocal to collect every 30 seconds, and choose an oocyte that looks the best using either a 40X or 60X lens. Look for a nice clear circular nuclear outline and the best orientation gauged by the sharpness of the best focus. If the oocyte is lying on the dorsal side parallel to the coverslip, the nucleus will appear perfectly round. If the oocyte is on its side with the nucleus closest to the coverslip, the nucleus can be seen to lie next to the membrane and will appear as a half-circle. The second orientation is more difficult to image, but different aspects can be detected. When nuclear envelope breakdown approaches, the nuclear envelope begins to pucker. Quickly begin collecting data every 4–10 seconds. Take either a single scan or two scans and average. Averaged scans look crisper, but time-resolution is obviously lost. Most confocals are capable of doing time-lapse Z-series, and this makes it easier to follow the karyosome. Soon after the nuclear envelope breaks down, the karyosome moves toward the middle of the nucleoplasm, and thus it is necessary to refocus every 10 seconds until the position of the karyosome has stabilized. If able to observe in four dimensions as many confocals can, this is less of a problem. Eventually, the tubulin signal can be seen around the chromatin. Data are best collected on the hard drive, and data presentation of single images can be carried out using Adobe Photoshop. Videos from the digital images can also be generated with commercially available software packages sold by Adobe Systems.

## REFERENCES

Afshar K., Barton N.R., Hawley R.S., and Goldstein L.S.B. 1995. DNA binding and meiotic chromosomal localization of the *Drosophila* nod kinesin-like protein. *Cell* **81:** 129–138.

Bickel S.E., Wyman D.W., and Orr-Weaver T.L. 1997. Mutational analysis of the *Drosophila* sister-chromatid cohesion protein ORD and its role in the maintenance of centromeric cohesion. *Genetics* **146:** 1319–1331.

Bridges C.B. 1916. Non-disjunction as proof of the chromosome theory of heredity. *Genetics* **1:** 1–52.

Carpenter A.T.C. 1975. Electron microscopy of meiosis in *Drosophila melanogaster* females. I. Structure, arrangement, and temporal change of the synaptonemal complex in wild-type. *Chromosoma* **51:** 157–182.

Clarkson M. and Saint R. 1999. A *His2AvDGFP* fusion gene complements a lethal *His2AvD* mutant allele and provides an in vivo marker for *Drosophila* chromosome behavior. *DNA Cell Biol.* **18:** 457–462.

Dernburg A.F., Sedat J.W., and Hawley R.S. 1996. Direct evidence of a role for heterochromatin in meiotic chromosome segregation. *Cell* **86:** 135–146.

Edwards K.A., Demsky M., Montaque R.A., Weymouth N., and Kiehart D.P. 1997. GFP-moesin illuminates actin cytoskeleton dynamics in living tissue and demonstrates cell shape changes during morphogenesis in *Drosophila*. *Dev. Biol.* **191:** 103–117.

Endow S.A. and Komma D.J. 1996. Centrosome and spindle function of the *Drosophila* Ncd microtubule motor visualized in live embryos using Ncd-GFP fusion proteins. *J. Cell Sci.* **109:** 2429–2442.

———. 1997. Spindle dynamics during meiosis in *Drosophila* oocytes. *J. Cell Biol.* **137:** 1321–1336.

Harlow E. and Lane D. 1988. *Antibodies: A laboratory manual.* Cold Spring Harbor Laboratory, Cold Spring Harbor, New York.

Hawley R.S. 1993. Meiosis as an "M" thing: Twenty-five years of meiotic mutants in *Drosophila*. *Genetics* **135:** 613–618.

Hawley R.S., Irick H., Zitron A.E., Haddox D.A., Lohe A., New C., Whitley M.D., Arbel T., Jang J.K., McKim K., and Childs G. 1993. There are two mechanisms of achiasmate segregation in *Drosophila*, one of which requires heterochromatic homology. *Dev. Genet.* **13:** 440–467.

Hari K.L., Santerre A., Sekelsky J.J., McKim K.S., Boyd J.B., and Hawley R.S. 1995. The mei-41 gene of *D. melanogaster* is a structural and functional homolog of the human ataxia telangiectasia gene. *Cell* **82:** 815–821.

Kellogg D.R., Mitichison T.J., and Alberts B.M. 1988. Behavior of microtubules and actin filaments in living *Drosophila* embryos. *Development* **103:** 676–686.

Kerrebrock A.W., Miyazaki W.Y., Birnby D., and Orr-Weaver T.L. 1992. The *Drosophila* mei-S332 gene promotes sister-chromatid cohesion in meiosis following kinetochore differentiation. *Genetics* **130:** 827–841.

Mahowald A.P. and Kambysellis M.P. 1980. Oogenesis. In *Genetics and biology of* Drosophila (ed. M. Ashburner and T.R.F. Wright), pp. 141–224. Academic Press, London.

Mahowald A.P., Goralski T.J., and Caulton J.H. 1983. In vitro activation of *Drosophila* eggs. *Dev. Biol.* **98:** 437–445.

Matthies H.J., McDonald H.B., Goldstein L.S., and Theurkauf W.E. 1996. Anastral meiotic spindle morphogenesis: Role of the non-claret disjunctional kinesin-like protein. *J. Cell Biol.* **134:** 455–464.

McDonald H.B., Stewart R.J., and Goldstein L.S.B. 1990. The kinesin-like ncd protein of *Drosophila* is a minus end-directed microtubule motor. *Cell* **63:** 1159–1165.

McKim K.S., Dahmus J.B., and Hawley R.S. 1996. Cloning of the *Drosophila melanogaster* meiotic recombination gene mei-218: A genetic and molecular analysis of interval 15E. *Genetics* **144:** 215–228.

McKim K.S., Jang J.K., Theurkauf W.E., and Hawley R.S. 1993. Mechanical basis of meiotic metaphase arrest. *Nature* **362:** 364–366.

McKim K.S., Green-Marroquin B.L., Sekelsky J.J., Chin G., Steinberg C., Khodosh R., and Hawley R.S. 1998. Meiotic synapsis in the absence of recombination. *Science* **279:** 876–878.

Miller A. 1950. The internal anatomy and histology of the image of *Drosophila melanogaster*. In *Biology of* Drosophila (ed. M. DeDemerec), pp. 420–534. Facsimile edition 1994 by Cold Spring Harbor Laboratory Press, Cold Spring Harbor, New York.

Minden J.S., Agard D.A., Sedat J.W., and Alberts B.M. 1989. Direct cell lineage analysis in *Drosophila melanogaster* by time-lapse, three-dimensional optical microscopy of living embryos. *J. Cell Biol.* **109:** 505–516.

Moore D.P., Page A.W., Tang T.T., Kerrebrock A.W., and Orr-Weaver T.L. 1998. The cohesion protein MEI-S332 localizes to condensed meiotic and mitotic centromeres until sister chromatids separate. *J. Cell Biol.* **140:** 1003–1012.

Orr-Weaver T.L. 1995. Meiosis in *Drosophila:* Seeing is believing. *Proc. Natl. Acad. Sci.* **92:** 10443–10449.

Page A.W. and Orr-Weaver T.L. 1997. Activation of the meiotic divisions in *Drosophila* oocytes. *Dev. Biol.* **183:** 195–207.

Rasooly R.S., Zhang P., Tibolla A.K., and Hawley R.S. 1994. A structure-function analysis of NOD, a kinesin-like protein from *Drosophila melanogaster. Mol. Gen. Genet.* **242:** 145–151.

Sekelsky J.J., McKim K.S., Chin G.M., and Hawley R.S. 1995. The *Drosophila* meiotic recombination gene mei-9 encodes a homologue of the yeast excision repair protein Rad1. *Genetics* **141:** 619–627.

Spradling A.C. 1994. Developmental genetics of oogenesis. In *The development of* Drosophila melanogaster (ed. M. Bate and A.M. Arias), pp. 1–71, Cold Spring Harbor Laboratory Press, Cold Spring Harbor, New York.

Theurkauf W.E. 1994a. Premature microtubule-dependent cytoplasmic streaming in cappuccino and spire mutant oocytes. *Science* **265:** 2093–2096.

———. 1994b. Immunofluorescence analysis of the cytoskeleton during oogenesis and early embryogenesis. In Drosophila melanogaster: *Practical uses in cell and molecular biology* (ed. L.S.B. Goldstein and E.A. Fryberg), pp. 489–506. Academic Press, San Diego.

Theurkauf W.E. and Hawley R.S. 1992. Meiotic spindle assembly in *Drosophila* females: Behavior of nonexchange chromosomes and the effects of mutations in the nod kinesin-like protein. *J. Cell Biol.* **116:** 1167–1180.

Verheyen E. and Cooley L. 1994. Looking at oogenesis. In Drosophila melanogaster: *Practical uses in cell and molecular biology* (ed. L.S.B. Goldstein and E.A. Fryberg), pp. 489–506. Academic Press, San Diego.

Walker R.A., Salmon E.D., and Endow S.A. 1990. The *Drosophila* claret segregation protein is a minus-end directed motor molecule. *Nature* **347:** 780–782.

Wang S. and Hazelrigg T. 1994. Implications for bcd mRNA localization from spatial distribution of exu protein in *Drosophila* oogenesis. *Nature* **369:** 400–403.

White-Cooper H., Alphey L., and Glover D.M. 1993. The cdc25 homologue twine is required for only some aspects of the entry into meiosis in *Drosophila. J. Cell Sci.* **106:** 1035–1044.

Zhang P., Knowles B.A., Goldstein L.S., and Hawley R.S. 1990. A kinesin-like protein required for distributive chromosome segregation in *Drosophila. Cell* **62:** 1053–1062.

# CONTENTS

# 5

# Cytological Analysis of Spermatocyte Growth and Male Meiosis in *Drosophila melanogaster*

Silvia Bonaccorsi, Maria Grazia Giansanti,
Giovanni Cenci, and Maurizio Gatti
*Istituto Pasteur, Fondazione Cenci Bolognetti*
*Dipartimento di Genetica e Biologia Molecolare, Università di Roma "La Sapienza"*
*P. A. Moro 5, 00185 Roma, Italy*

*D*ROSOPHILA MALE MEIOSIS IS A HIGHLY SUITABLE SYSTEM for the genetic and molecular dissection of the meiotic process. In the past few years, many genes involved in the regulation and execution of male meiosis have been identified and characterized (for review, see Maines and Wasserman 1998). Due to their large size, meiotic cells of *Drosophila* males have proved to be excellent for the cytological examination of mutant phenotypes and immunolocalization of meiotic gene products (Gunsalus et al. 1995; Williams et al. 1995, 1996; Hime et al. 1996; Molina et al. 1997; Basu et al. 1998; Bonaccorsi et al. 1998; Carmena et al. 1998; Giansanti et al. 1998). It is also of interest that meiotic division can be viewed in testes of third-instar larvae. This allows examination of meiosis in animals homozygous for mitotic mutations that die at the larval/pupal boundary (Ripoll et al. 1985; Gonzales et al. 1988; Sunkel and Glover 1988; Casal et al. 1990; Gunsalus et al. 1995; Molina et al. 1997). In this chapter, we describe the main procedures for the cytological analysis of premeiotic and meiotic stages of *Drosophila* males. To facilitate the understanding of these procedures, an overview of *Drosophila* male meiosis is given below.

## OVERVIEW

Male meiosis occurs in the context of a complex developmental process, called spermatogenesis, that leads to the formation of 64 spermatozoa from each primary gonial cell (for reviews, see Lindsley and Tokuyasu 1980; Fuller 1993; Maines and Wasserman 1998). *Drosophila melanogaster* spermatogenesis begins in the germinal proliferation center, an area located at the apex of the testis which comprises 5–9 germ-line stem cells and 9–17 somatically derived, cyst progenitor cells (Hardy et al. 1979; for review, see Lindsley and Tokuyasu 1980). The asymmetric division of a stem cell generates a primary spermatogonium and another stem cell, which remains anchored at the testis apex. Concomitantly with the formation of this primary gonial cell, the asymmetric division of two neighboring cyst progenitor cells generates two cyst cells, which encyst the primary spermatogonium and its progeny until completion of spermatogenesis.

The primary gonial cell undergoes four rounds of mitotic divisions, giving rise to 16 primary spermatocytes. These cells, after the premeiotic DNA synthesis, enter a growth phase that lasts approximately 90 hours and results in a 25-fold increase in nuclear volume (for review, see Lindsley and Tokuyasu 1980; Fuller 1993; Cenci et al. 1994). Nuclear growth is accompanied by the formation of the Y-chromosome lampbrush-like loops generated by the *kl-5, kl-3,* and *ks-1* fertility factors. At the end of the growth period, the Y loops degenerate, and primary spermatocytes divide meiotically producing 64 spermatids (Bonaccorsi et al. 1988; Cenci et al. 1994).

In both gonial and meiotic divisions, the execution of cytokinesis does not lead to complete separation of the daughter cells, which remain connected by cytoplasmic bridges called the ring canals (Lindsley and Tokuyasu 1980; Hime et al. 1996). Moreover, in cysts containing spermatogonia or immature primary spermatocytes, all of the germ-line cells are linked together by the fusome, a prominent, branched cytoplasmic structure enriched in both F actin and α-spectrin, which passes through the ring canals. The fusome drastically regresses in mature spermatocytes and remains almost absent during meiotic divisions but reforms after meiosis in spermatid cysts (Gunsalus et al. 1995; Hime et al. 1996; Giansanti et al. 1999). A fusome has also been observed in *Drosophila* females, and several genes have been identified that specify fusome components (Robinson and Cooley 1996). Yet, the precise function of this structure is still poorly defined.

Degeneration of the Y-chromosome loops marks the beginning of meiosis and is concomitant with the formation of a very prominent spindle, particularly amenable to cytological analysis (see, e.g., Cenci et al. 1994; Gunsalus et al. 1995; Bonaccorsi et al. 1998; Carmena et al. 1998; Giansanti et al. 1998). Both meiotic divisions occur within a double nuclear membrane that is invaginated and possibly fenestrated at positions underlying the asters but does not otherwise appear to disassemble (Tates 1971; Church and Lin 1982). In addition, during most phases of meioses I and II, the nuclei are surrounded by two additional systems of membranes. Three to five layers of so-called parafusorial membranes lie parallel to the spindle axis, whereas several layers of umbrella-shaped astral membranes cup the nuclear poles associated with the asters (Tates 1971). The actual biological role of these two membrane systems is not currently understood.

During meioses I and II, mitochondria line up along the parafusorial membranes and are equally partitioned between the two daughter cells at each division. If no meiotic errors occur, each spermatid receives the same haploid chromosome complement and the same amount of mitochondria (Tates 1971; Fuller 1993). After completion of the second meiotic division, the mitochondria inherited by each cell fuse to form a complex structure, called the Nebenkern, containing multiple layers of mitochondrial membranes. Thus, immediately following meiosis, at the so-called onion stage, a cyst is composed of 64 spermatids, each containing a single nucleus associated with a single Nebenkern. Onion-stage spermatids can be easily analyzed in living preparations viewed by phase-contrast optics, where nuclei and Nebenkern are phase-light and phase-dark, respectively. In wild type, onion-stage nuclei and Nebenkern have a highly regular appearance and consistently exhibit similar sizes. Deviations from this regularity are diagnostic of failures in the fidelity of meiotic divisions. For example, because the volume of an onion-stage nucleus is proportional to its chromatin content (Gonzalez et al. 1989), the presence of nuclei of different dimensions is indicative of errors in chromosome segregation. On the other hand, spermatids composed of an abnormally large Nebenkern associated with two or four regular-sized nuclei are diagnostic of failures in meiotic cytokinesis (Fuller 1993; Castrillon and Wasserman 1994; Gunsalus et al. 1995; Williams et al. 1995; Basu et al. 1998; Carmena et al. 1998; Giansanti et al. 1998).

In our description of the procedures for cytological analysis of meiosis, we first delineate the methods for preparation of living testes. We then provide a detailed account of the fixation procedures for indirect immunofluorescence with various types of antibodies. Next, we outline the squashing and staining techniques for meiotic chromosome preparations. Finally, we present brief descriptions of both Y-loop development and male meiosis, providing a photographic documentation of these processes. We believe that these pictures facilitate recognition of the various premeiotic and meiotic stages and help in defining departures from normality observed in meiotic mutants.

## PREPARATION OF LIVING TESTES

The easiest and fastest method for the analysis of spermatocyte growth and male meiosis is the examination of living testis squashes in aqueous solutions. These preparations allow unambiguous recognition of most cell types of *Drosophila* spermatogenesis, as well as most phases of male meiosis (see, e.g., Fuller 1993; Cenci et al. 1994; Maines and Wasserman 1998). Living testes squashes can be successfully carried out using a variety of aqueous solutions: Testis Buffer (see below; Hennig 1967; Pisano et al. 1993; Cenci et al. 1994); insect Ringer's (Lifschytz and Hareven 1977; Kemphues et al. 1980; Castrillon et al. 1993); TB1 (7 mM $K_2HPO_4$, 80 mM KCl, 16 mM NaCl, 5 mM $MgCl_2$, 1% PEG 6000; Regan and Fuller 1990); PBS (Castrillon et al. 1993); and 0.7% NaCl (Sunkel and Glover 1988; Gonzalez and Glover 1993). Although these solutions give very similar results, in our laboratory, we routinely use testis buffer (TB). Protocol 5.1 is a standard procedure for preparation of living testes squashes.

PROTOCOL 5.1*

# Preparation of Live Testis Squashes

## Materials

### Supplies and Equipment

Dissecting tools
- fine forceps or tungsten needles
- dissecting microscope

Nonsiliconized coverslips (20 x 20 mm)
Slides
Blotting paper

### Solutions and Reagents

#### Testis Buffer (TB)

| Component and final concentration | Amount to add to make 100 ml |
| --- | --- |
| 183 mM **KCl** | 1.364 g |
| 47 mM NaCl | 0.274 g |
| 10 mM Tris-HCl | 0.121 g |
| 1 mM EDTA | 0.037 g |
| 1 mM **PMSF** | 0.017 g |

Dissolve in approximately 80 ml of $H_2O$. Adjust to pH 6.8 and bring to final volume of 100 ml with $H_2O$. Sterilize by filtration.

**CAUTION: KCl, PMSF** (see Appendix 4)

*For all protocols, $H_2O$ indicates glass distilled and deionized.

## Method

1. Dissect testes in TB from either third-instar larvae, pupae, or young adults.

2. Transfer testes into a 2-μl drop of TB on a clean, nonsiliconized 20 x 20-mm coverslip. Tear open the adult testes using either very thin forceps or tungsten needles.

3. Place a clean slide over the coverslip without pressing and invert the sandwich. If desired, a very mild squashing can be obtained by removing excessive buffer from the edges of the coverslip with a piece of blotting paper.

   *Note:* Steps 1–3 are usually performed at room temperature.

4. Analyze preparations with a 40x objective, using phase-contrast optics.

## PREPARATION AND STAINING OF FIXED TESTES

### Fixation Procedures

Several fixation procedures have been developed to make testis preparations for the purpose of immunostaining. These protocols and their relative advantages and disadvantages are described below.

Protocol 5.2 describes a methanol-acetone fixation technique (Pisano et al. 1993; Cenci et al. 1994) which results in very good preservation of cell morphology. Fixed cells viewed by phase-contrast optics exhibit most of the structural details that can be seen in live material. This allows analysis of unstained fixed preparations and selection of the most suitable ones for immunostaining. Remarkably, the Y loops, which are usually faint and labile in living preparations, become clearly apparent after this type of fixation. Moreover, this fixation protocol results in excellent microtubule preservation for immunostaining with antitubulin antibodies. The main disadvantage of this technique is a poor preservation of chromosome structure. In most instances, the chromosomes do not exhibit a distinct morphology and tend to coalesce into one or more masses of chromatin.

Protocol 5.3 has proved to be particularly suitable for $\gamma$-tubulin and centrosomin immunostaining (Bonaccorsi et al. 1998). It results in preparations having the same characteristics as those obtained by using Protocol 5.2.

Protocol 5.4 describes a fixation procedure (Gunsalus et al. 1995; Hime et al. 1996; Giansanti et al. 1998) that preserves F-actin-containing structures because it does not involve the use of methanol. Thus, it is particularly suitable for contractile ring and fusome visualization. However, it does not preserve cell morphology and microtubules as efficiently as Protocol 5.2 and does not permit a clear detection of the Y loops.

Protocol 5.5 (Williams 1993; Giansanti et al. 1999) was originally developed by Cayetano Gonzalez (EMBL, Heidelberg). It preserves chromosome morphology very well, allowing a clear visualization of the chromosome condensation-decondensation cycle. It also results in excellent preservation of microtubules for immunostaining. The main disadvantage of this fixation technique is its poor preservation of cell morphology, as viewed by phase-contrast optics. In addition, because it involves a rather hard squashing of formaldehyde-fixed testes, meiotic figures result flatter and larger than those obtained using the above fixation protocols.

## Staining Procedures

### Immunostaining with a Single Antibody

Testis preparations fixed according to Protocols 5.2 through 5.5 can be immunostained with a variety of antibodies. The concentration of the primary antibody will vary with both the type of antibody and the type of incubation and should be determined empirically each time. Protocol 5.6 describes a general procedure for immunostaining.

### Double Immunostaining

Double immunostaining can be carried out using two different primary antibodies raised in different species, such as mouse or rabbit. These primary antibodies can be detected separately using species-specific secondary antibodies conjugated with different fluorochromes. One of these secondary antibodies should be conjugated with dyes such as rhodamine or Texas Red, which emit a red fluorescence, and the other with fluorescein, which produces a green fluorescence. In our laboratory, double immunostaining is carried out by repeating twice the steps described in Protocol 5.6 (see Notes).

Detection of double immunofluorescence requires the absence of "bleed through" through the filters used for red and green fluorescence. In addition, double immunofluorescence is greatly facilitated by the use of a cooled charged-coupled device (CCD) that permits separate collection of the fluorescein and rhodamine signals as gray-scale digital images. These images can be then converted into Photoshop format, pseudocolored, and merged (see, e.g., Gunsalus et al. 1995; Bonaccorsi et al. 1998; Giansanti et al. 1998).

### F-actin Staining

Preparations fixed according to Protocol 5.4 can be stained for F actin according to Protocol 5.7. This staining procedure proved to be highly suitable for staining the male fusome and the cytokinetic contractile ring (Gunsalus et al. 1995; Hime et al. 1996; Giansanti et al. 1998).

### Chromatin Staining

To visualize DNA, preparations obtained using Protocols 5.2 through 5.5 can be stained with several DNA-binding dyes (see Protocol 5.8). If the slides are to be examined by a fluorescence microscope equipped with filters that permit UV (ultraviolet) excitation, the most suitable dyes for DNA staining are Hoechst 33258 or DAPI (see, e.g., Bonaccorsi et al. 1988; Cenci et al. 1994; Hime et al. 1996; Carmena et al. 1998). If the slides are to be analyzed with a confocal microscope not equipped with a UV laser, DNA can be stained with either propidium iodide or TOTO-3 iodide (Gonzalez and Glover 1993; Carmena et al. 1998).

### Multiple Stainings

Testis squashes can be stained with antibodies directed to specific proteins, with rhodamine-phalloidin to visualize F actin, and with DNA-binding dyes. Best results are obtained when these stainings are carried out in the following order: immunostaining, F-actin staining, and DNA staining. This implies that multiply stained preparations are all mounted in the media described for DNA staining (see Protocol 5.8, step 2).

# Methanol–Acetone Fixation Technique I

## Materials

### *Supplies and Equipment*

Dissecting tools
- fine forceps or tungsten needles
- dissecting microscope

Nonsiliconized coverslips (20 × 20 mm)
Slides
Blotting paper
Razor blade

### *Solutions and Reagents*

TB (for preparation, see p. 89)
**Liquid nitrogen**
**Methanol**, at –20ºC
**Acetone**, at –20ºC

### *1× PBS*

| Component and final concentration | Amount to add to make 1 liter |
|---|---|
| 8.0 mM $Na_2HPO_4$ | 1.14 g |
| 1.5 mM $KH_2PO_4$ | 0.20 g |
| 137.0 mM NaCl | 8.00 g |
| 2.7 mM KCl | 0.20 g |

Combine all components in approximately 800 ml of $H_2O$ and stir to dissolve; adjust pH to 7.4; bring to final volume of 1 liter; sterilize by autoclaving. We usually prepare a 10× stock solution of PBS and dilute to 1×.

### *PBS/Acetic Acid/Triton X-100 Solution*

1× PBS
0.5% **acetic acid**
1% Triton X-100

CAUTION: acetic acid, acetone, KCl, $KH_2PO_4$, liquid nitrogen, methanol, $Na_2HPO_4$ (see Appendix 4)

## Method

1. Perform steps 1–3 of Protocol 5.1 at room temperature.

2. Freeze the slides in liquid nitrogen.

3. Remove the coverslips with a razor blade, immediately immerse the slides into methanol at –20ºC, and leave for 5 minutes.

4. Transfer the slides into acetone at –20ºC and leave for 1–2 minutes.

5. Transfer the slides into PBS/Acetic Acid/Triton X-100 Solution and leave at room temperature for 10 minutes.

6. Wash the slides twice in PBS (5 minutes each wash).

PROTOCOL 5.3

# Methanol-Acetone Fixation Technique II

## Materials

### Supplies and Equipment

Dissecting tools
- fine forceps or tungsten needles
- dissecting microscope

Nonsiliconized coverslips (20 x 20 mm)
Slides
Blotting paper
Razor blade

### Solutions and Reagents

TB (for preparation, see p. 89)
**Liquid nitrogen**
**Methanol**, at –20ºC
**Acetone**, at –20ºC
1x PBS (see p. 92)

*PBS/Acetic Acid/Tween-20 Solution*
1x PBS
0.1% **acetic acid**
0.1% Tween-20

*PBS/Tween-20 Solution*
1x PBS
0.1% Tween-20

**CAUTION: acetic acid, acetone, liquid nitrogen, methanol** (see Appendix 4)

## Method

1. Perform steps 1–3 of Protocol 5.1 at room temperature.

2. Freeze the slides in liquid nitrogen.

3. Remove the coverslips with a razor blade, immediately immerse the slides into methanol at –20ºC, and leave for 15 minutes.

4. Transfer the slides into acetone at –20ºC and leave for 30 seconds.

5. Transfer the slides into PBS/Acetic Acid/Tween-20 Solution and leave for 10 minutes.

6. Wash the slides three times in PBS/Tween-20 Solution (10 minutes each wash).

7. Wash in PBS for 5 minutes.

PROTOCOL 5.4

# Fixation with Paraformaldehyde

## Materials

### Supplies and Equipment

Dissecting tools
   • fine forceps or tungsten needles
   • dissecting microscope
Nonsiliconized coverslips (20 × 20 mm)
Slides
Blotting paper
Razor blade

### Solutions and Reagents

TB (for preparation, see p. 89)
**Liquid nitrogen**
**Ethanol**, at −20ºC
1× PBS (see p. 92)
**Paraformaldehyde** (4%) dissolved in 1× PBS; use as a freshly prepared solution
PBT (1× PBS containing 0.1% Triton X-100)

**CAUTION: ethanol, liquid nitrogen, paraformaldehyde** (see Appendix 4)

## Method

1. Perform steps 1–3 of Protocol 5.1 at room temperature.

2. Freeze the slides in liquid nitrogen.

3. Remove the coverslips with a razor blade, immediately immerse the slides into ethanol at −20ºC, and leave for 10 minutes.

4. Fix slides in 4% paraformaldehyde in PBS at room temperature for 7 minutes.

5. Wash the slides twice in PBS at room temperature (5 minutes each wash).

6. Immerse the slides for 30 minutes in PBT (PBS containing 0.1% Triton-X) at room temperature.

7. Wash the slides twice in PBS (5 minutes each wash).

PROTOCOL 5.5

# Formaldehyde Fixation

## Materials

### Supplies and Equipment

Dissecting tools
- fine forceps or tungsten needles
- dissecting microscope

Nonsiliconized coverslips (20 x 20 mm)
Slides
Blotting paper
Razor blade

### Solutions and Reagents

Saline (0.7% NaCl in $H_2O$)
**Acetic acid** (45% and 60%)
**Liquid nitrogen**
**Ethanol**, at –20ºC
PBT (1x PBS containing 0.1% Triton X-100)
1x PBS (see p. 92)
**Formaldehyde** (3.7%) dissolved in 1x PBS

**CAUTION: acetic acid, ethanol, formaldehyde, liquid nitrogen** (see Appendix 4)

## Method

1. Dissect larval or adult testes in saline.

2. Fix testes in 3.7% formaldehyde in PBS at room temperature for 30 minutes.

3. Transfer testes into 45% acetic acid and leave for 30 seconds.

4. Transfer testes into a drop (4 μl) of 60% acetic acid placed on a 20 x 20-mm coverslip and keep them at room temperature for 2–3 minutes.

5. Invert the sandwich and squash testes applying a moderate pressure.

6. Freeze the slides in liquid nitrogen.

7. Remove the coverslips with a razor blade, immediately immerse the slides into ethanol at –20ºC, and leave for 15 minutes.

8. Transfer the slides into PBT and leave at room temperature for 10 minutes.

9. Wash the slides twice in PBS at room temperature (5 minutes each wash).

## Immunostaining of Fixed Testes Preparations

### Materials

#### Supplies and Equipment

Slides, prepared as described in Protocols 5.2–5.5

Humid chamber (e.g., a slide box containing two layers of blotting paper soaked with $H_2O$)

Rubber cement

#### Solutions and Reagents

1x PBS (see p. 92)

PBT (1x PBS containing 0.1% Triton X-100)

Vectashield mounting medium H-1000 (Vector Laboratories)

**PBS/BSA**
    1x PBS
    1% BSA

**PBT/BSA**
    1x PBS
    0.1% Triton X-100
    1% BSA

#### Antibodies

Primary antibodies, diluted in PBS/BSA or PBT/BSA

Secondary antibodies, diluted in PBS

#### Method

1. After the final PBS washes described in Protocols 5.2 through 5.5, immerse the slides in PBS/BSA and leave at room temperature for 45 minutes.

2. Remove excess of PBS from around the specimen. (The squashed testes appear as an opaque area on the slide.) Cover with a 20-μl drop of primary antibody diluted in PBS/BSA. Alternatively, dilute the primary antibody in PBT/BSA.

3. Incubate the slides in a humid chamber for the desired time (e.g., either at room temperature for 2 hours or at 4°C overnight).

4. Wash the slides twice in PBS (5 minutes each wash). If the primary antibody was dissolved in PBT/BSA, perform two washes in PBT (5 minutes each) and one wash in PBS for 5 minutes.

5. Remove excess of PBS from around the specimen (see step 2) and cover the preparation with 20 μl of the secondary antibody diluted in PBS.

   *Note:* We routinely use the following secondary antibodies: FLUOS-conjugated anti-mouse sheep IgG F(ab′)2 fragment (FLUOS: 5[6]-carboxyfluorescein-N-hydroxysuccinimide ester), and rhodamine-conjugated anti-mouse sheep IgG, F(ab′)2 fragment, both from Boehringer Mannheim, and both diluted 1:10; FITC-conjugated anti-rabbit goat IgG and TRITC-conjugated anti-rabbit goat IgG, both from Cappel (Durham), and both diluted 1:50.

6. Incubate the slides in the dark in a humid chamber at room temperature for 1 hour.

7. Wash the slides twice in PBS (5 minutes each wash).

8. If preparations are not to be stained with other reagents, mount either in PBS or in Vectashield. However, mounting in Vectashield results in poor cell morphology when slides are examined by phase-contrast optics.

9. Seal with rubber cement. Omit this step if slides are mounted in Vectashield.

## PROTOCOL 5.6 NOTES

In our laboratory, double immunostaining is carried out as follows: Immunostain the slides with one of the primary antibodies and detect it with either a fluorescein- or a rhodamine-conjugated secondary antibody, following steps 1–7 (above). Then immunostain the slides with the second primary antibody and detect it with the appropriate secondary antibody (steps 2–8, above).

PROTOCOL 5.7

# F–Actin Staining

## Materials

### Supplies and Equipment

Slides, prepared as described in Protocol 5.4
Humid chamber (e.g., a slide box containing two layers of blotting paper soaked with $H_2O$)
Rubber cement

### Solutions and Reagents

1x PBS (see p. 92)
Vectashield H-1000 mounting medium (Vector Laboratories)

#### Preparation of Rhodamine-labeled Phalloidin

Dissolve 300 units of rhodamine-labeled phalloidin (Molecular Probes) in 1.5 ml of **methanol**. This stock solution can be stored at –20ºC.
Dry 100 µl of this stock solution under vacuum and resuspend in 200 µl of PBS.

CAUTION: **methanol** (see Appendix 4)

## Method

1. Remove the excess of PBS from around the specimen and cover it with 20 µl of Rhodamine-labeled Phalloidin.

2. Incubate the slides in the dark in a humid chamber at 37ºC for 1.5 hours.

3. Rinse the slides in PBS for 1–2 minutes.

4. Mount the slides either in PBS or in Vectashield H-1000 medium.

5. Seal with rubber cement. Omit this step if slides are mounted in Vectashield.

PROTOCOL 5.8

# Chromatin Staining

## Materials

### Supplies and Equipment

Slides, prepared as described in Protocols 5.2–5.5
Rubber cement

### Solutions and Reagents

1x PBS
**Hoechst 33258** (0.5 µg/ml), dissolved in Hoechst Buffer
**Hoechst 33258** (5 µg/ml), dissolved in Hoechst Buffer
**DAPI** (0.2 µg/ml), dissolved in 1x PBS

#### Hoechst Buffer

| Component and final concentration | Amount to add to make 1 liter |
|---|---|
| 10 mM **$Na_2HPO_4$** | 1.42 g |
| 150 mM NaCl | 8.76 g |
| 30 mM **KCl** | 2.23 g |

Dissolve in approximately 800 ml of $H_2O$; Adjust to pH 7.0 and bring to final volume of 1 liter with $H_2O$. Sterilize by autoclaving.

**CAUTION: DAPI, Hoechst 33258, KCl, $Na_2HPO_4$** (see Appendix 4)

## Method

### Hoechst or DAPI Staining

1.  After final PBS washes, air-dry slides.

2.  Stain with Hoechst or DAPI as follows:

    a.  *For Hoechst staining:* Stain with 0.5 µg/ml Hoechst 33258 dissolved in Hoechst Buffer for 10 minutes. Rinse for 1 minute and mount in Hoechst Buffer. Alternatively, directly mount slides in 5 µg/ml of Hoechst 33258 dissolved in Hoechst Buffer.

    b.  *For DAPI staining:* Stain with 0.2 µg/ml DAPI dissolved in PBS for 10 minutes. Rinse for 1 minute and mount in PBS. Alternatively, directly mount slides in 0.2 µg/ml of DAPI dissolved in PBS.

        Note: Hoechst 33258- and DAPI-stained slides can also be mounted in Vectashield H-1000. However, as mentioned earlier, Vectashield affects the appearance of cells viewed by phase-contrast optics.

3.  Seal with rubber cement.

### Staining with Propidium Iodide or TOTO-iodide

To stain preparations with propidium iodide or TOTO-iodide, add these dyes to 90% glycerol, 2.5% *n*-propyl gallate mounting medium, to a final concentration of 1 µg/ml and 1/200, respectively (Gonzalez and Glover 1993; Carmena et al. 1998).

## PREPARATION OF MEIOTIC CHROMOSOMES

Several techniques have been developed for preparation of male meiotic chromosomes (see, e.g., Cooper 1965; Ashburner 1989; Miyazaki and Orr-Weaver 1992). All of these techniques allow a clear visualization of chromosomes, but not of other cellular structures such as the cytoskeleton and the spindle. We describe here three of these techniques that in our hands give good reproducible results. The first technique, developed by Lifschytz and Hareven (1977) and described below in Protocol 5.9, is particularly useful for the analysis of adult testes. The other two techniques (Ripoll et al. 1985; Cenci et al. 1997) are routinely used in our laboratory for larval and pupal testis preparations. The technique described in Protocol 5.10 is a very simple method for preparing aceto-orcein-stained chromosomes. The other technique outlined in Protocol 5.11 allows preparation of unstained chromosomes that can be subsequently stained with various dyes such as Giemsa, Hoechst 33258, and DAPI, or processed for in situ hybridization.

PROTOCOL 5.9

# Preparation of Meiotic Chromosomes from Adult Testes

## Materials

### Supplies and Equipment

Dissecting tools
- fine forceps or tungsten needle
- dissecting microscope

Slides
Coverslips (20 x 20 mm)
Blotting paper
Nail polish

### Solutions and Reagents

**Acetic acid** (45% and 60%)
Orcein (3%), dissolved in 60% acetic acid (for preparation of 2% aceto-orcein, see p. 100)

#### Ringer's Solution

| Component and final concentration | Amount to add to make 1 liter |
| --- | --- |
| 182 mM **KCl** | 13.6 g |
| 46 mM NaCl | 2.7 g |
| 3 mM **CaCl$_2$** | 0.33 g CaCl$_2$·2H$_2$O |
| 10 mM Tris-HCl | 1.21 g |

Combine all components in approximately 800 ml of H$_2$O and stir to dissolve. Adjust to pH 7.2 and bring to a final volume of 1 liter with H$_2$O. Sterilize by autoclaving.

#### Lacto–Aceto-orcein

**Lactic acid:** 3% orcein in 60% acetic acid (1:1 ratio)

**CAUTION: acetic acid, CaCl$_2$, KCl, lactic acid** (see Appendix 4)

## Method

1. Dissect testes in Ringer's Solution.

2. Fix testes in 45% acetic acid for 1 minute.

3. Stain in 3% orcein dissolved in 60% acetic acid for 5 minutes.

4. Quickly rinse the testes in 60% acetic acid and transfer them into a small drop of 60% acetic acid placed on a slide.

5. Cut testes under dissecting scope to release testis content. Add a drop of Lacto–Aceto-Orcein (1:1 lactic acid to 3% orcein in 60% acetic acid).

6. Gently lower a coverslip over the specimen; remove excess liquid from the edges of the coverslip with a piece of blotting paper.

7. Seal with nail polish.

PROTOCOL 5.10

# Aceto-Orcein Preparation of Meiotic Chromosomes from Larvae or Pupae

## Materials

### Supplies and Equipment

Dissecting tools
- fine forceps or tungsten needle
- dissecting microscope

Coverslips (20 x 20 mm)
Slides
Blotting paper
Nail polish

### Solutions and Reagents

Saline (0.7% NaCl in $H_2O$)

#### 2% Aceto-Orcein

Boil orcein powder (Gurr, BDH Laboratory Supplies, Poole, Dorset BH15 1TD England; Phone: +44 1202 660444; Fax: +44 1202 666856; Web site: http://www.bdh.com) in 45% **acetic acid** for 30–45 minutes in a reflux condenser. We usually make 5% orcein and subsequently dilute it to 2% with 45% acetic acid. Before use, remove particulate matter by either filtration or centrifugation in a microcentrifuge.

**CAUTION: acetic acid** (see Appendix 4)

## Method

1. Dissect larval or pupal testes in saline.

2. Transfer testes into a drop of 2% Aceto-Orcein placed on a coverslip and leave for 2 minutes.

3. Lower a slide over the coverslip and invert the sandwich; squash gently between two sheets of blotting paper.

4. Seal with nail polish.

PROTOCOL 5.11

# Preparation of Unstained Meiotic Chromosomes from Larvae or Pupae

## Materials

### Supplies and Equipment

Dissecting tools
  - fine forceps or tungsten needle
  - dissecting microscope
Siliconized coverslips (20 x 20 mm)
Slides
Blotting paper
Razor blade

### Solutions and Reagents

**Acetic acid** (45%)
**Dry ice** or **liquid nitrogen**

**CAUTION: acetic acid, dry ice, liquid nitrogen** (see Appendix 4)

## Method

1. Perform steps 1–3 of Protocol 5.10 but use 45% acetic acid instead of 2% aceto-orcein, and use siliconized coverslips.

2. Freeze the slides either on dry ice or in liquid nitrogen.

3. Remove the coverslip with a razor blade and let the slides air-dry.

   *Note:* Air-dried slides can be stained with a variety of dyes, processed for chromosome banding, or used for in situ hybridization. For a detailed account of these techniques, see Pimpinelli et al. (this volume).

## SPERMATOCYTE GROWTH AND Y-LOOP DEVELOPMENT _____

Spermatocyte growth and Y-loop unfolding are illustrated in Figure 5.1, which shows cells fixed by the methanol/acetone procedure described in Protocol 5.2 and sequentially stained for tubulin and chromatin. Stage subdivision has been proposed by Cenci et al. (1994) and is based only on morphological criteria. Each stage corresponds to a phase of spermatocyte development that is unambiguously distinguishable from both the preceding and the following one.

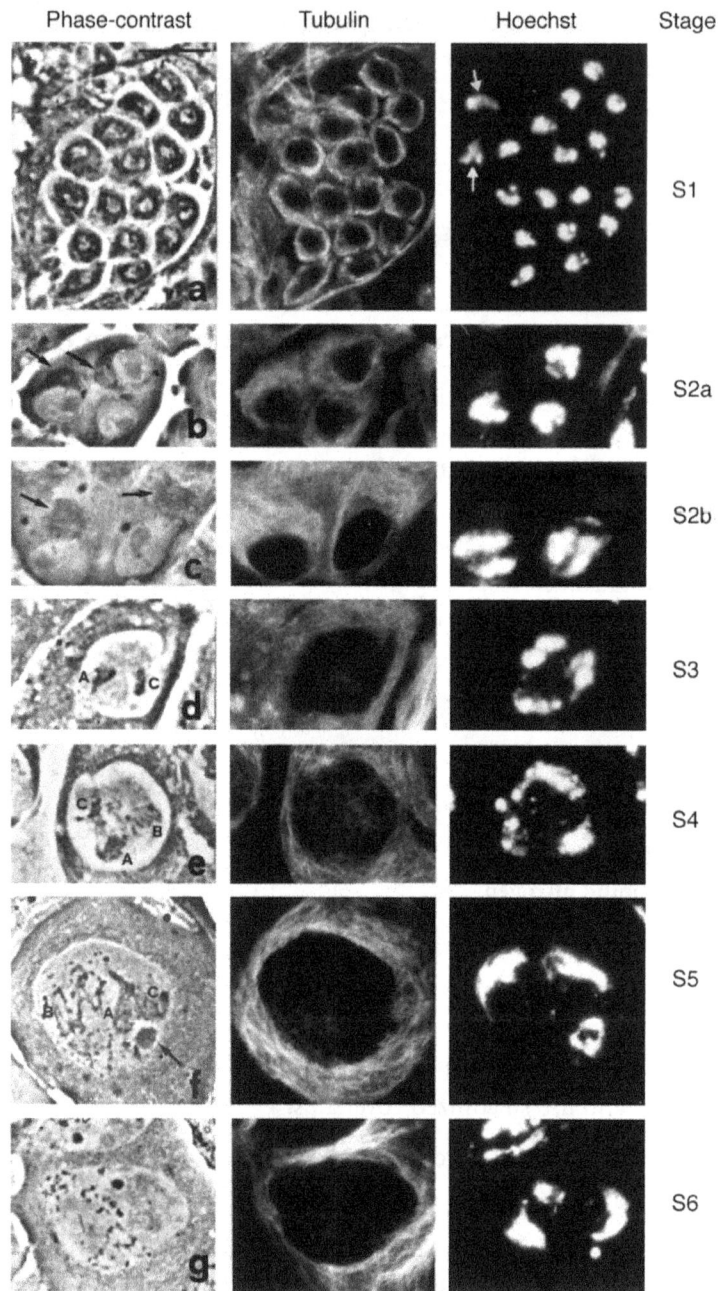

**Figure 5.1.** Primary spermatocyte growth and Y-loop development. (*a*) A complete cyst of 16 young primary spermatocytes; the arrows point to the cyst cells. (*b,c*) Polar spermatocytes; the arrows point to the clusters of mitochondria. (*d*) Young apolar spermatocyte showing the primordia of the *kl-5* (A) and *ks-1* (C) Y-chromosome loops. (*e*) Primary spermatocyte in the S4 stage in which some filaments of the *kl-3* loop (B) have become apparent. (*f*) Mature primary spermatocyte in the S5 stage showing fully developed Y-chromosome loops. (*g*) Primary spermatocyte in the S6 stage with disintegrating Y loops. Bar, 10 μm. (Reprinted, with permission, from Cenci et al. 1994.)

Primary spermatocytes that have just completed premeiotic DNA synthesis (stage S1) are similar in size and morphology to $G_2$ spermatogonia. As the S1 spermatocyte enters the growth period, its nucleus assumes an eccentric position within the cytoplasm, and the mitochondria cluster at the cell pole opposite to the nucleus (stages S2a and S2b). During the S2 stage (also called the polar spermatocyte stage; Tates 1971; Fuller 1993), the nucleus enlarges, and the chromatin that initially occupies the entire nuclear space begins to subdivide into three masses or clumps (stage S2). Two of these masses correspond to the paired large metacentric autosomes and the other corresponds to the sex chromosomes. The paired fourth chromosomes cannot be readily distinguished and are usually associated with the sex chromosome clump (Cooper 1965; Cenci et al. 1994).

As spermatocyte growth continues, the nucleus resumes a central position within the cell and the mitochondria disperse; the chromatin clumps expand moderately, while remaining apposed to the inner nuclear envelope, so that the nuclear space not occupied by the chromatin progressively increases. This space becomes filled by the Y-chromosome loops elaborated by the *kl-5* (loop A), *kl-3* (loop B), and *ks-1* (loop C) fertility factors. These structures develop asynchronously with a characteristic sequence, providing useful markers for staging spermatocyte growth (Bonaccorsi et al. 1988; Cenci et al. 1994). The first appearance of the Y loops marks the S3 spermatocytes, which exhibit two phase-dark masses that correspond to the primordia of the *kl-5* and *ks-1* loops. The *kl-3* loop begins to unfold later during the S4 stage. Then, the three loops enlarge steadily and reach their maximum size in mature spermatocytes (stage S5), where they occupy most of the nuclear space. Although the *kl-5* and *ks-1* loops are composed of a thick thread, and the *kl-3* loop of a thinner and less folded filament, in many cases, the three Y-chromosome loops cannot be distinguished from each other. However, the Y loops can be identified by both cytological and genetic methods.

Cytologically, the Y loops can be differentiated by either Giemsa staining or immunostaining with loop-specific antibodies. Preparations obtained following Protocol 5.2 can be stained with 2% Giemsa (Merck) dissolved in borate buffer (Merck), rinsed in tap water, air-dried, and then mounted in Euparal. In these Giemsa-stained preparations, the *kl-5* and *ks-1* loops are selectively stained, whereas the *kl-3* loop and the chromatin clumps remain unstained (Bonaccorsi et al. 1988). Alternatively, the Y loops can be differentiated by immunostaining with (1) the S5 monoclonal antibody (Saumweber et al. 1980), which reacts strongly with the *kl-5* loop and weakly with the *ks-1* loop, but fails to decorate the *kl-3* loop (Melzer and Glatzer 1985; Bonaccorsi et al. 1988); (2) the T53-1 antibody, which specifically recognizes the *kl-3* loop (Pisano et al. 1993); and (3) either the anti-RB97 antibody (Heatwole and Haynes 1996) or the anti-Boule antibody (Eberhart et al. 1996; Cheng et al. 1998), which specifically bind the *ks-1* loop.

The Y loops can also be distinguished by genetic methods. For example, the use of fertile X-Y translocations with one breakpoint within the X heterochromatin and the other between two adjacent Y-chromosome fertility factors (Kennison 1981) permits construction of males carrying complementary sets of Y loops (Hardy et al. 1981; Bonaccorsi et al. 1988). Because the two elements of these $T(X;Y)$s segregate regularly, $X/Y^D X^P$ and $X^D Y^P/0$ males can be obtained by crossing $T(X;Y)$ males to regular and to *attached-X/0* females, respectively. Thus, if the $T(X;Y)$ used is broken between *kl-5* and *kl-3*, the spermatocytes of $X/Y^D X^P$ males will only exhibit the *kl-5* loop, whereas those of $X^D Y^P/0$ males will carry both the *kl-3* and *ks-1* loops. Males missing one or two specific Y loops can also be constructed using Y chromosomes that are deleted for the corresponding loop-forming fertility factors. These Y deficiencies are usually kept in combination with either $X \cdot YL$ or $X \cdot YS$. Thus, if $X \cdot YL/Df(YL)^*$ or $X \cdot YS(Df(YS)^*$ males [$Df(YL)^*$ and $Df(YS)^*$ denote genetically and cytologically defined Y deficiencies] are crossed to regular females, the male progeny will only exhibit the Y loops carried by $Df(YL)^*$ and $Df(YS)^*$, respectively.

The disintegration of the Y loops marks the end of the growth phase of primary spermatocytes and the beginning of meiosis. During stage S6, the Y loops fall apart into pieces, the nucleolus disappears, and the chromatin clumps begin to condense while remaining apposed to the nuclear envelope.

## THE MEIOTIC DIVISION

The meiotic division, as well as its subdivision into stages (Cenci et al. 1994), is illustrated in Figures 5.2 and 5.3, which show meiotic cells fixed by methanol/acetone (Protocol 5.2), immunostained for tubulin, and then stained with Hoechst 33258. We believe that these figures can provide a useful guide for recognition of meiotic stages and definition of the departures from normality observed in meiotic mutants. The sequence of meiotic phases presented in Figures 5.2 and 5.3 is, in most instances, self-explanatory. We thus focus only on some aspects of the meiotic process that are not immediately obvious from these figures. For a detailed description of the events illustrated in Figures 5.2 and 5.3, see Cenci et al. (1994).

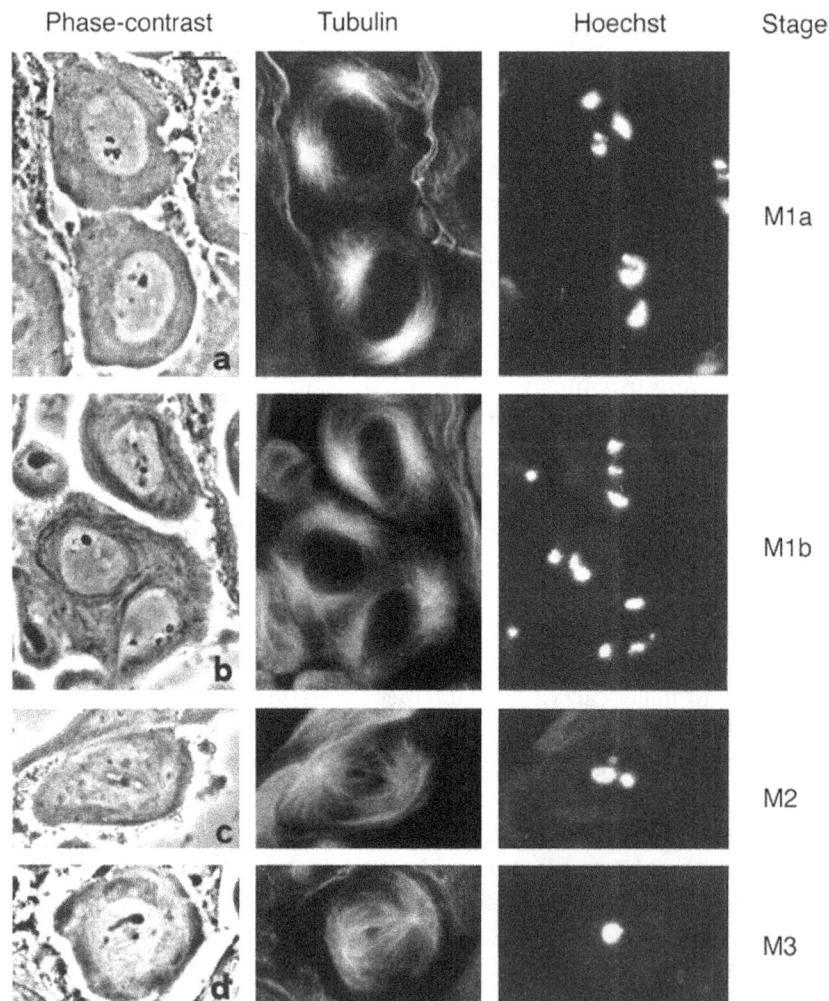

**Figure 5.2.** First meiotic division in *D. melanogaster* males. (*a*) Late prophase nuclei in the M1a stage showing phase-dense granules resulting from disintegration of the Y loops; (*b*) prometaphase nuclei; (*c*) late prometaphase nucleus in which the nuclear-cytoplasmic demarcation is no longer visible; (*d*) metaphase I with congregated bivalents. (*Continued on facing page.*)

The disintegration of the Y loops and the onset of the meiotic division are accompanied by a substantial shrinking of the spermatocyte nucleus. Late prophase cells in the M1 stage have a nuclear diameter 20% shorter than that of mature spermatocytes in the S5 stage. The M1 stage is also characterized by a high degree of chromatin condensation and by the presence of two prominent asters. Interestingly, the process of chromatin condensation, which occurs between the S6 and M1 stages, is rather rapid and no leptotene-pachytene stages can be seen in *Drosophila* primary spermatocytes (Cooper 1965; Cenci et al. 1994). Although the M1a spermatocytes are in late prophase, the M1b cells are probably in early prometaphase; in M1b cells, a few spindle fibers irradiating from the asters appear to reach the bivalents that occupy variable positions within the nucleoplasm. During the transition between early (M1b stage) and late (M2 stage) prometaphase, the nuclear-cytoplasmic demarcation evident in the M1 stages progressively disappears. This phenomenon is not related to the breakdown of the nuclear membrane, but rather to the dispersion of the nuclear lamins (White-Cooper et al. 1993; Eberhart and Wasserman 1995).

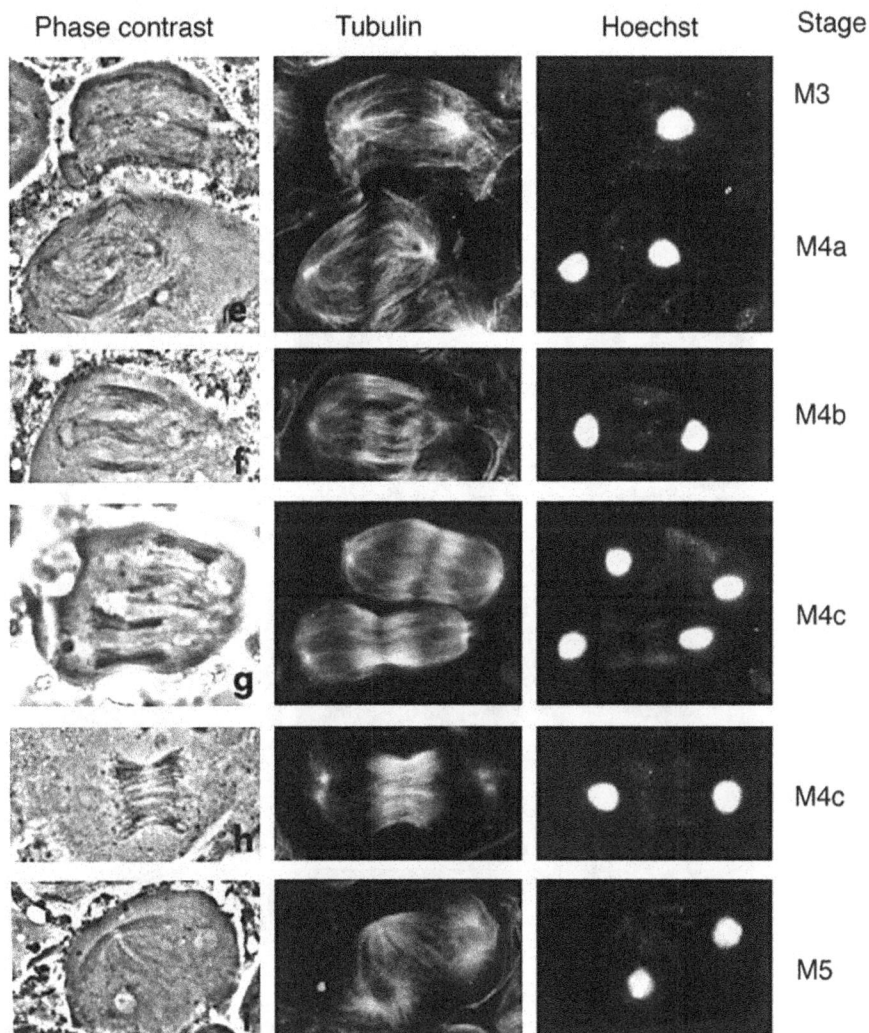

**Figure 5.2.** (*Continued.*) (*e*) Metaphase I (*top*) plus early anaphase (*bottom*); (*f*) mid-anaphase. (*g,h*) late anaphases showing prominent central spindles; (*i*) telophase. Bar, 10 μm. (Reprinted, with permission, from Cenci et al. 1994.)

From late prometaphase I (stage M2) to telophase I (stage M5), the spindle undergoes dramatic transformations. In prometaphase (stage M2), metaphase (stage M3), and anaphase A (stage M4a), the spindle consists of two closely apposed, umbrella-shaped half-spindles. During anaphase B (stages M4b and M4c), the spindle poles move apart and spindle microtubules reorganize: The astral microtubules shorten and the density of

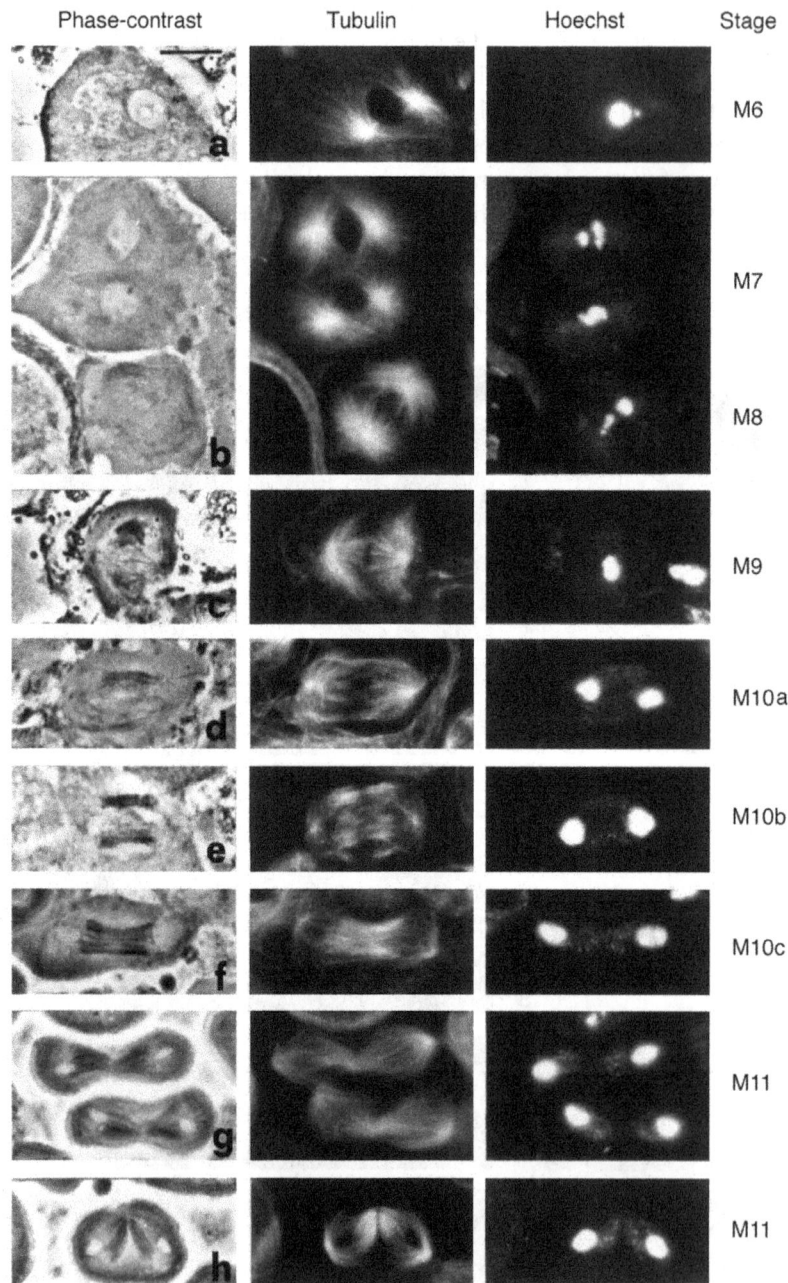

**Figure 5.3.** Second meiotic division in *D. melanogaster* males. (*a*) Prophase secondary spermatocyte nucleus. (*b*) Two prometaphase secondary spermatocyte nuclei (*top*) plus a prometaphase cell (*bottom*) without nuclear-cytoplasmic demarcation. (*c*) Metaphase; (*d*) early anaphase; (*e*) mid-anaphase; (*f*) late anaphase; (*g,h*) telophases. Bar, 10 μm. (Reprinted, with permission, from Cenci et al. 1994.)

microtubules between the segregating sets of chromosomes progressively increases. The latter microtubules form a dense bundle, called the central spindle, which is pinched in the middle during telophase when cytokinesis occurs (stage M5).

After completion of the first meiotic division and a very short prophase (stage M6), secondary spermatocytes enter prometaphase (stages M7 and M8) and then metaphase (stage M9). Here again, nuclear-cytoplasmic demarcation disappears between the M7 and M8 stages, due to the dispersion of nuclear lamins. Subsequent spindle transformations are very similar to those that occur during the first meiotic division. After anaphase A (stage M10a), aster microtubules shorten and a prominent central spindle assembles (stages M10b and M10c), which is pinched in the middle during telophase II (stage M11).

During ana-telophase I and II, the mitochondria appear to line up along the central spindle. The occurrence of cytokinesis divides these organelles into two equal groups at each meiotic division. As a consequence, all of the spermatids receive the same amount of mitochondria and develop Nebenkern of identical sizes.

## REFERENCES

Ashburner M. 1989. Drosophila: *A laboratory handbook.* Cold Spring Harbor Laboratory Press, Cold Spring Harbor, New York.

Basu J., Williams B.C., Li Z., Williams E.V., and Goldberg M. L. 1998. Depletion of a *Drosophila* homolog of yeast Sup35p disrupts spindle assembly, chromosome segregation, and cytokinesis during male meiosis. *Cell Motil. Cytoskelet.* **39:** 286–302.

Bonaccorsi S., Giansanti M.G., and Gatti M. 1998. Spindle self-organization and cytokinesis during male meiosis in *asterless* mutants of *Drosophila melanogaster. J. Cell Biol.* **142:** 751–761.

Bonaccorsi S., Pisano C., Puoti F., and Gatti M. 1988. Y chromosome loops in *Drosophila melanogaster. Genetics* **120:** 1015–1034.

Carmena M., Riparbelli M.G., Minestrini G., Tavares A.M., Adams R., Callaini G., and Glover D.M. 1998. *Drosophila* polo kinase is required for cytokinesis. *J. Cell Biol.* **143:** 659–671.

Casal J., Gonzalez C., Wandosell F., Avila J., and Ripoll P. 1990. Abnormal meiotic spindles cause a cascade of defects during spermatogenesis in *asp* males of *Drosophila. Development* **108:** 251-260.

Castrillon D.H. and Wasserman S.A. 1994. *diaphanous* is required for cytokinesis in *Drosophila* and shares domains of similarity with the products of the *limb deformity* gene. *Development* **120:** 3367–3377.

Castrillon D.H., Gonczy P., Alexander S., Rawson R., Eberhart C.G., Viswanathan S., Di Nardo S., and Wasserman S.A. 1993. Toward a molecular genetic analysis of spermatogenesis in *Drosophila melanogaster:* Characterization of male-sterile mutants generated by single P element mutagenesis. *Genetics* **135:** 489–505.

Cenci G., Bonaccorsi S., Pisano C., Verni F., and Gatti M. 1994. Chromatin and microtubule organization during premeiotic, meiotic, and early postmeiotic stages of *Drosophila melanogaster* spermatogenesis. *J. Cell Sci.* **107:** 3521–3534.

Cenci G., Rawson R.B., Belloni G., Castrillon D.H., Tudor M., Petrucci R., Goldberg M.L., Wasserman S.A., and Gatti M. 1997. UbcD1, a *Drosophila* ubiquitin-conjugating enzyme required for proper telomere behavior. *Genes Dev.* **11:** 863–875.

Cheng M.H., Maines J.Z., and Wasserman S.A. 1998. Biphasic subcellular localization of the DAZ-related protein boule in *Drosophila. Dev. Biol.* **204:** 567–576.

Church K. and Lin H.P.P. 1982. Meiosis in *Drosophila melanogaster.* The prometaphase-I kinetochore microtubule bundle and kinetochore orientation in males. *J. Cell Biol.* **93:** 365–373.

Cooper K.W. 1965. Normal spermatogenesis in *Drosophila.* In *Biology of* Drosophila (ed. M. Demerec), pp. 1–61. Hafner Publing, New York.

Eberhart C.G. and Wasserman S.A. 1995. The *pelota* locus encodes a protein required for meiotic cell division: An analysis of G2/M arrest in *Drosophila* spermatogenesis. *Development* **121:** 3477–3486.

Eberhart C.G., Maines J.Z., and Wasserman S.A. 1996. Meiotic cell cycle requirement for a fly homolog of human Deleted in Azoospermia. *Nature* **381:** 783–785.

Fuller M.T. 1993. Spermatogenesis. In *The development of* Drosophila melanogaster (ed. M. Bate and A.M. Arias), vol. 1, pp. 71–147. Cold Spring Harbor Laboratory Press, Cold Spring Harbor, New York.

Giansanti M.G., Bonaccorsi S., and Gatti M. 1999. The role of anillin during meiotic cytokinesis of *Drosophila* males. *J. Cell Sci.* **112:** 2323–2334.

Giansanti M.G., Bonaccorsi S., Williams B., Williams E.V., Santolamazza C., Goldberg M.L., and Gatti M. 1998. Mutations in the profilin-encoding gene *chickadee* reveal interactions between the central spindle and the contractile ring during meiotic cytokinesis in *Drosophila melanogaster* males. *Genes Dev.* **12:** 396–410.

Gonzalez C. and Glover D.M. 1993. Techniques for studying mitosis in *Drosophila*. In *The cell cycle. A practical approach* (ed. P. Fantes and R. Brooks), pp. 143–175. IRL Press, Oxford.

Gonzalez C., Casal J., and Ripoll P. 1988. Functional monopolar spindles caused by mutation in *mgr*, a cell division gene of *Drosophila melanogaster. J Cell Sci.* **89:** 39–47.

———. 1989. Relationship between chromosome content and nuclear diameter in early spermatids of *Drosophila melanogaster. Genet. Res. Camb.* **54:** 205–212.

Gunsalus K.C., Bonaccorsi S., Williams E., Verni F., Gatti M., and Goldberg M.L. 1995. Mutations in *twinstar*, a *Drosophila* gene encoding a cofilin/ADF homolog, result in defects in centrosome migration and cytokinesis. *J. Cell Biol.* **131:** 1243–1259.

Hardy R.W., Tokuyasu K.T., and Lindsley D.L. 1981. Analysis of spermatogenesis in *Drosophila melanogaster* bearing deletions for *Y*-chromosome fertility genes. *Chromosoma* **83:** 593–617.

Hardy R.W., Tokuyasu K.T., Lindsley D.L., and Garavito M. 1979. The germinal proliferation center in the testis of *Drosophila melanogaster. J. Ultrastruct. Res.* **69:** 180–190.

Heatwole V.M. and Haynes S. 1996. Association of RB97D, an RRM protein required for male fertility, with a Y chromosome lampbrush loop in *Drosophila* spermatocytes. *Chromosoma* **105:** 285–292.

Hennig W. 1967. Untersuchungen zur Struktur und Funktion des Lampenbursten-Y-Chromosoms in der Spermatogenese von *Drosophila. Chromosoma* **22:** 294–357.

Hime G.R., Brill J.A., and Fuller M.T. 1996. Assembly of ring canals in the male germ line from structural components of the contractile ring. *J. Cell Sci.* **109:** 2779–2788.

Kemphues K.J., Raff E.C., Raff R., and Kaufman T.C. 1980. Mutation in a testis-specific β-tubulin in *Drosophila:* Analysis of its effects on meiosis and map location of the gene. *Cell* **21:** 445–451.

Kennison J.A. 1981. The genetical and cytological organization of the *Y* chromosome of *Drosophila melanogaster. Genetics* **98:** 259–548.

Lifschytz E. and Hareven D. 1977. Gene expressions and the control of spermatid morphogenesis in *Drosophila melanogaster. Dev. Biol.* **58:** 276–294.

Lindsley D.L. and Tokuyasu K.T. 1980. Spermatogenesis. In *The genetics and biology of* Drosophila (ed. M. Ashburner and T.R.F. Wright), vol. 2b, pp. 225-294. Academic Press, London.

Maines J. and Wasserman S. 1998. Regulation and execution of meiosis in *Drosophila* males. *Curr. Top. Dev. Biol.* **37:** 301–332.

Melzer S. and Glatzer K.H. 1985. Localization of RNP antigens in primary spermatocytes of *Drosophila melanogaster* by indirect immunofluorescence and their correlation to fertility factors. *Drosophila Info. Services* **61:** 121.

Miyazaki W.Y. and Orr-Weaver T. 1992. Sister-chromatid misbehavior in *Drosophila ord* mutants. *Genetics* **132:** 1047–1061.

Molina I., Baars S., Brill J., Hales K.G., Fuller M.T., and Ripoll P. 1997. A chromatin-associated kinesin-related protein required for normal mitotic chromosome segregation in *Drosophila. J. Cell Biol.* **139:** 1361–1371.

Pisano C., Bonaccorsi S., and Gatti M. 1993. The *kl-3* loop of the Y chromosome of *Drosophila melanogaster* binds a tektin-like protein. *Genetics* **133:** 569–579.

Regan C.L. and Fuller M.T. 1990. Interacting genes that affect microtubule function in *Drosophila melanogaster:* Two classes of mutation revert the failure to complement between *hay^{ncd2}* and mutations in tubulin genes. *Genetics* **125:** 77–90.

Ripoll P., Pimpinelli S., Valdivia M.M., and Avila J. 1985. A cell division mutant of *Drosophila* with a functionally abnormal spindle. *Cell* **41:** 907–912.

Robinson D.N. and Cooley L. 1996. Stable intercellular bridges in development: The cytoskeleton lining the tunnel. *Trends Cell Biol.* **6:** 474–479.

Saumweber H., Symmons P. , Kabish R., Will H., and Bonhoeffer F. 1980. Monoclonal antibodies against chromosomal proteins of *Drosophila melanogaster. Chromosoma* **80:** 253–275.

Sunkel C.E. and Glover D.M. 1988. *polo,* a mitotic mutant of *Drosophila* displaying abnormal spindle poles. *J Cell Sci.* **89:** 25–38.

Tates A.D. 1971. "Cytodifferentiation during spermatogenesis in *Drosophila melanogaster:* An electron microscope study." Ph. D. thesis. Rijksuniversiteit, Leiden.

White-Cooper H., Alphey L., and Glover D.M. 1993. The *cdc25* homologue *twine* is required for only some aspects of the entry into meiosis in *Drosophila. J. Cell Sci.* **106:** 1035–1044.

Williams B.C. 1993. "Studies on the *Drosophila l(1)zw10* gene and its product." Ph. D. thesis. Cornell University, Ithaca, New York.

Williams B.C., Gatti M., and Goldberg M.L. 1996. Bipolar spindle attachments affect redistribution of ZW10, a *Drosophila* centromere/kinetochore component required for accurate chromosome segregation. *J. Cell Biol.* **34:** 1127–1140

Williams B.C., Riedy M.F., Williams E.V., Gatti M., and Goldberg M.L. 1995. The *Drosophila* kinesin-like protein KLP3A is a midbody component required for central spindle assembly and initiation of cytokinesis. *J. Cell Biol.* **129:** 709–723

# CONTENTS

# Preparation and Analysis of Polytene Chromosomes

**James A. Kennison**
*Laboratory of Molecular Genetics*
*National Institute of Child Health and Human Development*
*National Institutes of Health*
*Bethesda, Maryland 20892-2785*

Although the large polytene chromosomes of diptera were originally described in the late 1880s, it was not until the early 1930s that their significance to the study of the genome of *Drosophila* was realized. Theophilus Painter (1933, 1934) first characterized the polytene chromosomes of *Drosophila* and emphasized their importance in studying the structure of chromosomes and the localization of genes. Calvin Bridges (1935, 1937) quickly took up the task of making extremely detailed maps of *Drosophila* salivary gland polytene chromosomes and correlating these maps with the known genetic maps. Bridges (1938) began revising the polytene chromosome maps to include more detail, but died after completing only the *X* chromosome. The remaining revised maps were finished by his son, Philip (Bridges and Bridges 1939; Bridges 1941a,b, 1942). Both the original and revised maps have been reprinted (Lindsley and Grell 1968; Lindsley and Zimm 1992) and continue to be the standards for the field.

Polytene chromosomes are found in several larval and adult tissues, but preparations are usually made of the chromosomes in the larval salivary glands, because the glands are easily dissected and the polytene chromosomes are large.

PROTOCOL 6.1*

## Dissection of Larval Salivary Glands and Polytene Chromosome Preparation

### Materials

#### Supplies and Equipment

Deep depression slide or shallow dish for dissection. Hanging drop slides (Fisher) or depression culture microscope slides (VWR) work well. Alternatively, use a siliconized microscope slide (the siliconized surface keeps the liquid from spreading too much; for preparation of siliconized slides, see below).

*For all protocols, H$_2$O indicates glass distilled and deionized.

Very sharp forceps (2 pairs). Forceps with bent or broken tips make the dissections much more difficult. Fine-point forceps can be obtained from VWR (25607-856) or from Roboz (Web site: www.roboz.com).

Dissecting microscope

Glass coverslips (22 × 22 mm; also called cover glasses) siliconized if the coverslip is to be removed for a permanent preparation or if the chromosomes are to be used for another procedure, such as in situ hybridization (for preparation of siliconized coverslips, see below).

Glass microscope slides need not be treated, but must be brushed clear of any lint or dust just before use. Both microscope slides and coverslips can be obtained from almost any scientific supply company, such as Fisher, Thomas Scientific, or VWR.

Dissecting needle (VWR 25778-000) or other pointed object for tapping on the preparation; the object should not be too sharp or it will break the coverslip when tapped.

Bibulous paper (VWR 28511-007) for blotting out the excess stain

(*Optional*) Razor blade, if the coverslip is to be removed.

### Preparation of Siliconized Slides and Coverslips

Any siliconizing solution, such as Sigmacote (Sigma SL-2), will work. Dip the slides or coverslips in the siliconizing solution in a chemical fume hood, allow to air-dry, and clean by rubbing with a soft cloth or tissue.

Siliconized slides can be stored indefinitely. Make sure to remove all dust and lint just before use. Use a soft paintbrush to remove lint while blowing across the coverslip.

### Solutions and Reagents

Saline (0.7% NaCl)

**Acetic acid** (45%)

(*Optional*) **Liquid nitrogen**, a block of **dry ice**, or an **ethanol**/dry-ice bath, if the coverslip is to be removed

(*Optional*) Ethanol, 95% and absolute, if the slides are to be made permanent

(*Optional*) Mounting medium, e.g., Permount (Fisher), if the slides are to be made permanent

### Staining Solution of Lactic Acid, Acetic Acid, and Orcein

3 parts glacial acetic acid (Sigma A 6283)

2 parts **lactic acid** (Sigma L 1250)

1 part $H_2O$

2% powdered orcein (1 g per 50 ml of solution; Sigma O 7380)

The final concentrations of acetic acid and orcein are 50% and 2%, respectively. Although differing concentrations of lactic acid (between 16% and 30%) are used, the lactic acid concentration suggested above by Lefevre (1976) is preferred.

Heat the staining solution (but do not boil) and filter. If stored tightly capped, it will last for many years. The same solution, but without the orcein stain, can be used to prepare polytene chromosomes for in situ hybridization.

**CAUTION: acetic acid, dry ice, ethanol, lactic acid, liquid nitrogen** (see Appendix 4)

## Method

### *Dissecting Larval Salivary Glands*

Salivary glands are dissected from wandering third-instar larvae. The cultures should be well fed (adding live yeast paste every couple of days helps) and not very crowded. Although larvae raised at 18°C are preferred, larvae raised at higher temperatures are fine if large and healthy. Select large larvae that have crawled out of the medium, but have not yet stopped and everted their spiracles.

1. Place the larvae in saline in a deep depression slide or small dish. At this stage, determine the sex of the larvae if chromosomes from a particular sex are desired. Alternatively, select larvae of the desired genotype.

   *Notes:* The testes develop faster than the ovaries and are visible in living larvae as two clear spheres on either side of the body cavity, approximately two thirds of the way from the anterior to posterior. If the gonads are not clearly visible, the larva is a female.

   Marker mutations can also be used to pick larvae of particular genotypes. For example, *Tubby* is a dominant mutation that makes the larvae short and fat. It is carried by some balancer chromosomes. The homozygous, nonbalancer-carrying larvae can be selected as non-*Tubby* in phenotype. Loss-of-function mutations in *yellow* cause the mouth hooks to be pigmented brown instead of black. Homozygous *red* mutants have dark red granules deposited in the Malpighian tubules (wild-type Malpighian tubules are yellow in color). There are also balancer chromosomes available from the Bloomington *Drosophila* stock center (http://flystocks.bio.indiana.edu/) with transposons carrying GFP (green fluorescent protein) (Brand 1995) that can be used to select larvae of desired genotypes.

2. Perform the dissection in either 45% acetic acid or a saline solution.

   *Note:* Salivary glands can be dissected directly in 45% acetic acid if done quickly (this solution is used for fixing the chromosomes; see step 1). It is usually easier to dissect the salivary glands in a saline solution first, so that there is less need to rush. A 0.7% saline solution is good for dissection, because it is hypotonic and causes the cells to swell.

3. Grasp the mouth hooks firmly with one pair of forceps. Then grasp the body of the larva approximately midway down with the other pair of forceps. Figure 6.1A shows the positions of the forceps. Slowly pull the two pairs of forceps apart. The mouth hooks should separate from the body, with the salivary glands and little else attached (Figure 6.1B). The paired salivary glands are large, clear sausage-shaped cylinders connected at one end by a common duct. A pair of dissected salivary glands is shown in Figure 6.1C.

4. Try to keep the glands attached to each other by the duct, to make it easier to transfer them together. Remove the mouth hooks (they prevent the glands from flattening completely) and any other unwanted tissue by gently teasing with the forceps.

### *Fixation and Staining*

1. Transfer the dissected salivary glands to a drop of 45% acetic acid for approximately 30 seconds.

   *Note:* If the glands are left too long in the acetic acid solution, they fall apart before transfer (step 2, below). If they are not left long enough, the nuclei will not break open and the chromosomes will not spread well. The exact timing of fixation should be determined by trial and error.

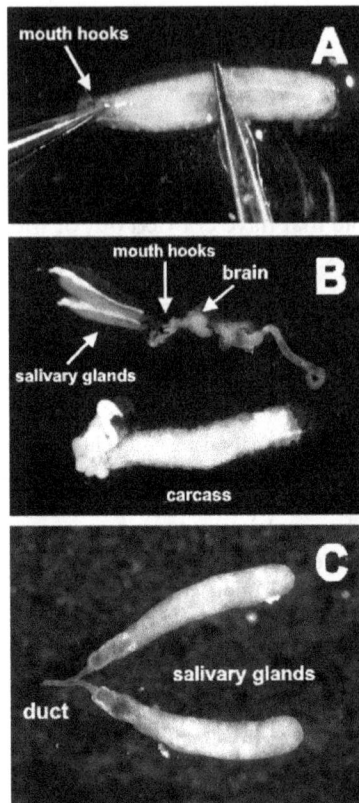

**Figure 6.1.** Dissection of larval salivary glands. (*A*) Position of the forceps for removing mouth hooks and salivary glands from larva. Anterior end of larva with mouth hooks is to the left. (*B*) Dissected mouth hooks with associated tissues, including salivary glands (*top*), and the remaining carcass of the larva (*below*). (*C*) Completely dissected salivary glands still joined at the common duct.

2. Place a drop (~25 μl) of staining solution containing lactic acid, acetic acid, and orcein on a siliconized coverslip. After the glands are fixed sufficiently, transfer them to the drop of staining solution. Allow the glands to sit in the staining solution for several minutes.

   *Note:* I allow the salivary glands to sit in staining solution for only approximately 2 minutes; however, the time should be determined by trial and error. If not stained sufficiently, the chromosomes appear very light. If stained for too long, the polytene bands are be very dark and refractory. Understained chromosomes are usually easier to analyze than overstained chromosomes.

3. Hold a clean microscope slide horizontally above the coverslip and slowly lower it until the slide touches the surface of the drop of stain. Capillary action should lift the coverslip up to the slide. Quickly flip the slide and coverslip over, so that the coverslip is now on top.

### Spreading the Chromosomes

1. Use a pointed object, such as a dissecting needle, to tap on the middle of the coverslip, being very careful not to move the coverslip with respect to the slide. Begin tap-

ping in the middle of the coverslip and slowly work out from the center in a spiral. The preparation should be tapped sufficiently to disperse the cells and spread the chromosomes.

> *Note:* It is important that the dissecting needle be held vertical while tapping. When the dissecting needle hits the coverslip, the tap compresses the stain and forces it away from the center of the preparation. As the dissecting needle moves away, the coverslip moves back up to its original position. The bouncing of the coverslip on the slide caused by the tapping sets up waves of stain that spread the chromosome arms away from each other.

2. Place the slide between two layers of bibulous paper and press firmly on the coverslip to remove the excess stain. Be careful not to move the coverslip across the microscope slide at this stage. The preparation should be squashed as flat as possible. If enough force cannot be applied with the thumb to completely flatten the preparation, place it overnight at 4°C. The stain will evaporate at the edges and flatten the preparation. Do not let it evaporate too long, or the preparation will begin to dry out and air will appear under the edges of the coverslip.

### Making the Preparation Permanent

1. To freeze the preparation, place the slide either in liquid nitrogen, in an ethanol/dry ice bath, or on top of a flat block of dry ice.

2. After several minutes, when the preparation is completely frozen, insert the edge of a razor blade between the edge of the coverslip and the slide, and flip the coverslip off quickly.

3. Dehydrate in 95% ethanol for 10 minutes and then in absolute ethanol for 10 minutes. Air-dry.

4. Place a drop of mounting medium on the slide and add a coverslip. Place a weight on top to flatten the coverslip and allow to dry before examination.

## PROTOCOL 6.1 NOTES

- If large dry areas appear between the coverslip and slide around the edges of the preparation immediately after squashing (i.e., following step 2 in Spreading the chromosomes above), there is probably lint (or debris such as mouth hooks) in the preparation. Such preparations cannot be flattened sufficiently and should be discarded. Make sure that all dust and lint are removed from the microscope slides and coverslips before use. Also make sure that the mouth hooks and any other chitinous larval debris have been completely removed from the dissected salivary glands (see step 4 in Dissecting Larval Salivary Glands above).
- If the polytene chromosomes are not distinct, i.e., they are puffy with indistinct bands, the glands have not been fixed sufficiently before staining. In addition to the poor morphology, underfixed salivary gland nuclei will also not break open very well and the chromosomes will appear enclosed in small spheres. Add 5 to 10 seconds to the fixation time in the 45% acetic acid before staining (see step 1 in Fixation and Staining above). If the chromosomes are brittle and break into a large number of small frag-

ments upon squashing, the glands have been fixed for too long. This is seen less often, as the glands tend to fall apart and cannot be transferred to the staining solution if left too long in the 45% acetic acid.

## EXAMINATION AND ANALYSIS OF POLYTENE CHROMOSOMES

Preparations are best examined on a compound microscope with phase-contrast optics. I prefer to start at a lower magnification (400x final magnification), but switch to a higher magnification (630x or 1000x final magnification) for detailed analyses. Bridges (1935) divided the euchromatic portions of the genome into 102 numbered divisions. Divisions 1 through 20 are on the left arm of the *X* chromosome. Divisions 21 through 60 and 61 through 100 comprise the second and third chromosomes, respectively, with the centromeres between divisions 40 and 41 for the second chromosome and between divisions 80 and 81 for the third chromosome. The small fourth chromosome includes divisions 101 and 102. Each division is subdivided into six lettered subdivisions (letters A through F). In the revised maps, each polytene chromosome band is numbered. Thus, polytene chromosome band 21C2 refers to the second band in the third subdivision of the first numbered division of the left arm of the second chromosome. Lefevre (1976) published photographic interpretations of Bridges maps that are extremely useful. These photographs have been reprinted in Lindsley and Zimm (1992) and are currently available together with copies of Bridges original polytene chromosome maps from *Academic Press* (ISBN: 0-12-450991-6). Photographs of portions of the polytene chromosomes can also be found on the images location of FlyBase at: http://flybase.bio.indiana.edu:82/images/.

It is difficult at first to recognize the chromosome arms, but with practice, it becomes much easier. One good method for learning to recognize the various chromosome arms without instruction is to obtain stocks with simple, described chromosome rearrangements. These serve as good landmarks and allow the identification of those regions with certainty.

## REFERENCES

Brand A.H. 1995. GFP in *Drosophila. Trends Genet.* **11:** 324–325.

Bridges C.B. 1935. Salivary chromosome maps with a key to the banding of the chromosomes of *Drosophila melanogaster. J. Hered.* **26:** 60–64.

———. 1937. Correspondences between linkage maps and salivary chromosome structure, as illustrated in the tip of chromosome 2R of *Drosophila melanogaster. Cytologia (Fujii Jubilee Volume)* 745–755.

———. 1938. A revised map of the salivary gland X-chromosome of *Drosophila melanogaster. J. Hered.* **29:** 11–13.

Bridges C.B. and Bridges P.N. 1939. A new map of the second chromosome. A revised map of the right limb of the second chromosome of *Drosophila melanogaster. J. Hered.* **30:** 475–476.

Bridges P.N. 1941a. A revised map of the left limb of the third chromosome of *Drosophila melanogaster. J. Hered.* **32:** 64–65.

———. 1941b. A revision of the salivary gland 3R-chromosome map of *Drosophila melanogaster. J. Hered.* **32:** 299–300.

———. 1942. A new map of the salivary gland 2L-chromosome of *Drosophila melanogaster. J. Hered.* **33:** 403–408.

Lefevre G. Jr. 1976. A photographic representation and interpretation of the polytene chromosomes of *Drosophila melanogaster* salivary glands. In *The genetics and biology of* Drosophila (ed. M. Ashburner and E. Novitski), vol. 1a, pp. 31–66. Academic Press, New York.

Lindsley D.L. and Grell E.H. 1968. *Genetic variations of* Drosophila melanogaster. Carnegie Institution of Washington Publication No. 627, Washington, D.C.

Lindsley D.L. and Zimm G.G. 1992. *The genome of* Drosophila melanogaster. Academic Press, San Diego.

Painter T.S. 1933. New method for the study of chromosome rearrangements and the plotting of chromosome maps. *Science* **78:** 585–586.

———. 1934. Salivary chromosomes and the attack on the gene. *J. Hered.* **25:** 465–476.

# CONTENTS

# 7

# In Situ Hybridization to Polytene Chromosomes

**Mary-Lou Pardue**
*Department of Biology*
*Massachusetts Institute of Technology*
*Cambridge, Massachusetts 02139*

IN SITU HYBRIDIZATION TO POLYTENE CHROMOSOMES can be used to answer several kinds of questions. For instance, the technique is frequently used to find the chromosomal site of a cloned DNA sequence, as well as to investigate the distribution of families of repeated sequences in the chromosome set and to measure the amount of a repeated sequence at different sites. These different experimental goals are best accomplished by variations of the technique.

Simple mapping of genes and even gene fragments of a few hundred base pairs is easily done using nonradioactive probes and by omitting several steps in slide preparation (these steps are marked "optional" in the protocol). On the other hand, quantitation of sequences at different chromosomal sites is best done using radioactive probes where the signal is measured directly from the bound probe, rather than after the one or two subsequent steps (antibody or streptavidin binding, enzymatic color production) that are required to detect nonradioactive probes. In addition, tritium probes can reliably detect sequences significantly shorter than those detected by nonradioactive probes on polytene chromosomes. For quantitative studies, results are improved by following the complete protocol for slide preparation.

## PREPARATION OF LABELED NUCLEIC ACID PROBES

The simplest and most successful probes are made by random primer labeling (Feinberg and Vogelstein 1983, 1984) of gel-purified fragments of cloned DNA. The DNA can be labeled while still in the gel slice or it can be purified with an Elutip-d minicolumn (Schleicher and Schuell), using the manufacturer's directions.

- *Nonradioactive Probes.* We prefer digoxigenin for nonradioactive labeling. However, biotin-labeled probes have been used by a number of laboratories. Biotin probes can be used in double-label experiments with digoxigenin (DIG) (Hartmann and Jackle 1995). A procedure for labeling gel-purified DNA fragments with digoxigenin is given below (see Protocol 7.1).

- *Radioactive Probes.* Tritium ($^3H$) is by far the best radioisotope for in situ hybridization because of the very low energy of the beta particle that is emitted. These beta particles travel less than 1 μm through autoradiographic emulsion, giving a precise localization for the probe. A procedure for $^3H$ labeling of DNA in agarose is given in Protocol 7.2.

## PROTOCOL 7.1*

# Random Primer Labeling of Gel-purified DNA Fragments with Digoxigenin

### Materials

### Supplies and Equipment

Elutip-d minicolumn (Schleicher and Schuell)
DIG DNA-labeling Kit (Boehringer Mannheim 1175 033)

### *Solutions and Reagents*

Hybridization buffer (see step 5, Protocol 7.4)

### Method

1. Isolate the DNA fragment of interest from the low-melt agarose gel using an Elutip-d minicolumn.

2. Denature 0.5 μg of the DNA by boiling for 2 minutes. Chill on ice.

3. Label the DNA with the DIG DNA-labeling Kit (Boehringer Mannheim) as directed by the manufacturer. Allow the labeling reaction to proceed at room temperature overnight.

4. Stop the reaction by boiling for 10 minutes. There is no need to remove unincorporated nucleotides. Adjust the volume of solution to that desired for hybridization by adding $H_2O$ and salts (and formamide, if desired) (see step 5, Protocol 7.4).

   *Note:* The reaction will yield enough probe for at least 20 preparations.

5. Immediately before applying the hybridization mix to the slides, boil the mixture for 3 minutes and quickly chill.

6. The mix can be stored at –20°C, but should be reboiled each time prior to use.

---

*For all protocols, $H_2O$ indicates glass distilled and deionized.

# Random Primer Labeling of DNA in Agarose with Tritium

## Materials

### Supplies and Equipment

Elutip-d minicolumns (Schleicher & Schuell)

### Solutions and Reagents

Unlabeled dNTP stocks (dATP, dCTP, and dGTP [each 100 μM in Buffer 1, see below]; alternatively, use 100 μM dNTP stocks from Boehringer Mannheim)
$^3$H-labeled TTP
Bovine serum albumin (BSA; 10 mg/ml)
DNA polymerase I, large (Klenow) fragment
Salmon sperm DNA
Absolute **ethanol**
Hybridization buffer (see step 5, Protocol 7.4)

### Buffer 1

200 μM Tris-Cl (pH 8.0)
25 μM $MgCl_2$
50 μM β-**mercaptoethanol**
If commercially prepared 100 μM dNTP stocks are used, this buffer will not be required.

### Buffer 2

1 mM Tris-Cl (pH 7.5)
1 mM EDTA

### Solution X

1.25 M Tris-Cl (pH 8.0)
0.125 M $MgCl_2$

### Solution A

1 ml of Solution X
18 μl of β-mercaptoethanol
5 μl each dATP, dCTP, dGTP (100 μM stocks, see above)

### Solution B

2 M HEPES (pH 6.6)
Adjust pH with 4 M **NaOH**.

### Solution C

Hexadeoxyribonucleotides (90 OD units/ml in Buffer 2)

### OLB

Mixture of Solutions A:B:C (ratio 100:250:150)

CAUTION: β-**mercaptoethanol, ethanol, NaOH, radioactive substances**
(see Appendix 4)

### Method

1. Excise DNA from the gel in as narrow a band as possible (~100–500 ng of DNA per gel slice). Transfer to a preweighed Eppendorf microcentrifuge tube. Determine weight of the gel slice and add 3 ml of $H_2O$ per gram of gel to the tube.

2. Boil for 7 minutes. Then incubate at 37ºC for 10 minutes.

   *Note:* The fragment can be stored at –20ºC after boiling, but it should be reboiled for 3 minutes prior to use.

3. Add 5 μCi of $^3$H-labeled TTP to an Eppendorf microcentrifuge tube. Allow to dry.

4. Add the following (in the order shown) to the Eppendorf tube (step 3, above):

   | | |
   |---|---|
   | $H_2O$ | to give a final volume of 50 μl |
   | OLB | 10 μl |
   | 10 mg/ml BSA | 2 μl |
   | DNA in agarose | up to 32.5 μl (25–100 ng) |
   | DNA polymerase I, large (Klenow) fragment | 5 units |

   *Note:* For random primer labeling of gel-purified DNA, see Protocol 7.2 Notes, below.

5. Incubate at room temperature from 3 hours to overnight.

   *Note:* Overnight labeling is preferred because it gives a higher yield of labeled product to unlabeled template.

6. Purify the product with an Elutip-d minicolumn, as directed by the manufacturer. Count 2 μl of the eluate in a scintillation counter to determine yield.

7. Add salmon sperm DNA to 50 μg/ml and precipitate with 2.5 volumes of ethanol. Centrifuge in a microcentrifuge to pellet the DNA. Dry the pellet. Dissolve in hybridization buffer to give about 100,000 cpm/15 μl of mix (this is the amount applied to each slide). Immediately before use, boil for 3 minutes and quickly chill. This probe can be stored at –20ºC, but it does not have a long storage life.

## PROTOCOL 7.2 NOTES

For purified DNA, the standard random primer reaction involves the following steps:

- Heat DNA to 100ºC for 2 minutes and quick cool.
- Follow Protocol 7.2 for gel slices, but add 50–100 ng of denatured DNA in step 4.

## POLYTENE CHROMOSOME PREPARATIONS

The most important step for in situ hybridization is making the salivary gland squashes. The quality of the chromosome preparations determines both the ease of identifying the chromosome landmarks and the efficiency of binding the probe. Unfortunately, criteria that promote good cytology are generally opposed to those that promote high levels of hybridization, so the most useful chromosomes are a compromise. Protocol 7.3 describes the preparation of polytene chromosome squashes from salivary glands. Details of this technique can also be found in Kennison (this volume).

PROTOCOL 7.3

# Preparation of Polytene Chromosome Squashes

## Materials

### Supplies and Equipment

Dissecting tools
Plastic test tube tops
Coverslips (18 mm², siliconized)
Razor blade
Sturdy dissecting needle (VWR 25 778-000)
Phase-contrast microscope
Dry, dust-free box for holding slides

### Preparation of Subbed Slides

Wash slides, dip in subbing solution, and allow to dry in a dust-free place. These slides may be stored indefinitely in a dry, dust-free container. Subbed slides give a better retention of cytological preparations.

### Siliconized Coverslips

Dip coverslips (18 mm²) in Sigmacote (Sigma SL-2), rinse well, and dry. Siliconization prevents adherence of cytological material to coverslips.

### Solutions and Reagents

**Acetic acid** (45%)
**Liquid nitrogen**
**Ethanol** (95%)

### Subbing Solution

This is an aqueous solution of 0.1% gelatin and 0.01% chrome alum (chromium potassium sulfate: $CrK[SO_4]_2 \cdot 12H_2O$). Dissolve the gelatin in hot water. Allow the solution to cool, and then add chrome alum. Subbing solution can be stored at 4°C for long periods.

### DPBS

130 mM NaCl
7 mM $Na_2HPO_4$
3 mM $KH_2PO_4$

**CAUTION: acetic acid, ethanol, $KH_2PO_4$, liquid nitrogen, $Na_2HPO_4$** (see Appendix 4)

## Method

1. Grow larvae in moderately crowed vials with added yeast. Choose larvae that are climbing up the sides of the tube.

   *Note:* Female larvae have better polytene chromosomes than males and also have twice the amount of *X* chromosome DNA.

2. Dissect the larvae in DPBS. Remove the salivary glands and gently clean off as much fat body as possible.

   *Note:* Take care to avoid injuring the salivary glands. It is better to leave fat body than to injure the glands, because chromosomes from injured glands spread poorly.

3. Place a small drop (~15 μl) of 45% acetic acid on a dust-free siliconized 18-mm² coverslip. Transfer the glands to this drop, being careful not to carry any DPBS to dilute the acid. Cover the drop with a plastic test tube top to prevent evaporation and let sit for 2–5 minutes. (The optimal time may vary from stock to stock.)

4. Invert a subbed slide over the coverslip and press it down lightly.

   *Note:* Place the glands at approximately the same position on each slide, so that they can be found easily as they are carried through the procedure.

5. Turn the slide over so that the coverslip is on top and tap lightly over the glands with a sturdy dissecting needle. The aim is to break the nuclei and to spread the chromosomes. Check spreading with a phase-contrast microscope.

   *Notes:* The coverslip can move around slightly during this step. The amount of liquid should permit gentle movement without shearing chromosomes.
   Nuclei from overlapping chromosomes will not have good morphology after the hybridization.

6. When chromosomes are well spread, place the slide on a paper towel with the coverslip side down and press firmly to flatten the chromosomes. Do not move the coverslip sideways at this point or chromosomes will be stretched.

7. Submerge the preparation in liquid nitrogen at least till bubbling ceases. Remove, flip coverslip off with a razor blade, and immediately plunge slide in 95% ethanol. Leave for 10 minutes. Repeat washes in 95% ethanol twice (10 minutes each) and air-dry the slides.

8. If planning to use ³H-labeled probe, make several (5–10) slides and develop after different autoradiographic exposures to obtain optimal labeling of the preparation. This is especially important if the aim is to quantify label over different sites comparatively. Exposures that give good grain counts at different sites can be chosen and then counts corrected to the same exposure time.

## PROTOCOL 7.3 NOTES

Preparations can be checked dry with a phase microscope. The best chromosomes are flat and gray with no refractivity. Chromosomes with a clearly recognizable banding pattern at this stage will have good morphology at the end of the experiment. Slides may be kept for weeks in a dry, dust-free box.

PROTOCOL 7.4

## Pretreatment and Hybridization

### Materials

#### Supplies and Equipment

Moist chamber. During several steps, preparations are incubated with small amounts of liquid under coverslips. To avoid having to seal the coverslips, place the slides in a moist chamber. A plastic sandwich box makes a good chamber. Place paper towels on the bottom, soaked with a solution. The solution in the chamber should have the same salt concentration as the solution under the coverslip, to prevent distillation and a change in concentration of hybridization buffer. (If the hybridization buffer contains formamide, use 20x SSC in the chambers. It is less expansive and balances the hybridization buffer well.) Slides are supported above the moist paper towels on two parallel glass rods.

Dish to hold slides

Magnetic stirrer

Magnetic stirring bar

#### Solutions and Reagents

20x SSC (see Appendix 3)

2x SSC

**Triethanolamine** (0.1 M, pH 8.0)

**Acetic anhydride**

RNase A (100 μg/ml in 2x SSC)

**Ethanol** (70% and 95%)

**NaOH** (0.07 N; prepared fresh immediately prior to use)

Hybridization buffer (usually 2x or 4x TNS, see below)

*1x TNS*

  0.15 M NaCl

  10 mM Tris-Cl (pH 6.8)

**CAUTION: acetic anhydride, ethanol, NaOH, triethanolamine** (see Appendix 4)

### Method

1. Treat the slides as follows:

   a. Heat the preparations in 2x SSC at 70°C for 30 minutes.

   b. Rinse in 2x SSC at room temperature.

   c. Dehydrate in 70% ethanol for 10 minutes (twice) and then in 95% ethanol for 5 minutes. Air-dry.

      *Note:* This step helps chromosomes adhere to the slide.

2. (*Optional*) Remove endogenous RNA as follows:

   a. Insert slides in a moist chamber and place approximately 50 μl of RNase A (100 μg/ml in 2x SSC) on top of the squash. Cover preparation with a coverslip and incubate at room temperature for 2 hours.

  b. Remove the coverslips by dipping in a beaker of 2x SSC. Perform three washes in 2x SSC (5 minutes each wash).

  c. Dehydrate through ethanol as in step 1c (above).

   *Note:* This step improves hybridization if there are significant endogenous transcripts in the vicinity of the target sequence (a situation that holds for the ribosomal genes and many puffed regions). Endogenous transcripts are seldom troublesome in simple mapping experiments, but their removal may be important for rigorous quantitative analysis of sequences.

3. (*Optional, to reduce background from radioactive probes*) Acetylate the preparation with acetic anhydride (5 ml/liter in 0.1 M triethanolamine [pH 8.0]). Place the slides in a rack and suspend in a dish over a magnetic stirring bar. Cover the slides with tri-ethanolmine and then add acetic anhydride while stirring. When acetic anhydride is completely dissolved, turn off the stirrer and let sit for 10 minutes. Dehydrate through ethanol as in step 1c.

4. Denature DNA of preparation by placing slides in freshly made 0.07 N NaOH for 3 minutes. Wash through three changes of 70% ethanol (10 minutes each) and then two changes of 95% ethanol (5 minutes each). Air-dry.

5. Set up the hybridization reaction. Place 10 µl of probe directly over each preparation and cover with an 18-mm$^2$ coverslip. Place in a moist chamber made with the hybridization buffer (or 20x SSC, if hybridization buffer contains formamide).

  *Note:* Choose the same hybridization buffers and temperatures as those for filter hybridization. We usually use 2x TNS and hybridize overnight at 68ºC. For cross-species hybridization, we increase the salt concentration to 4x or 6x. For the high probe concentrations that we use for nonradioactive experiments, an incubation for 4–6 hours is all that is needed. Much lower probe concentrations are used for radioactive probes and hybridizations are incubated for 12–16 hours.

6. Remove the coverslips by dipping into 2x SSC. Perform 3 washes (10 minutes each) to remove unhybridized probe.

  *Note:* For radioactive probes, these washes are best done at 60ºC; but for nonradioactive probes, room temperature washes seem to work.

7. Slides with nonradioactive probes are now ready for hybrid detection (see Protocol 7.5). Slides with radioactive probes should be dehydrated through ethanol, as in step 1c, and air-dried for autoradiography (see Protocol 7.6).

---

Protocol 7.5

## Detection of Digoxigenin Probes

Digoxigenin probes are detected with an antibody, which is coupled to an enzyme or to a fluorescent dye. This protocol describes use of the antibody coupled to alkaline phosphatase.

### Materials

### *Supplies and Equipment*

Moist chamber (see p. 125)
Phase-contrast microscope

### Solutions and Reagents

**NBT/BCIP (nitroblue tetrazolium/5-bromo-4-chloro-3-indolyl phosphate)** stock solution (Boehringer Mannheim 1681451)
Anti-DIG–alkaline phosphatase conjugate (Boehringer Mannheim 1093274)
Giemsa stain
Phosphate buffer (0.01 M, pH 6.8)

### PBT

DPBS
0.1% Tween-20

### pH 9 Buffer

100 mM NaCl
50 mM **MgCl$_2$**
100 mM Tris-Cl (pH 9.0)
0.1% Tween-20

**CAUTION: MgCl$_2$, NBT/BCIP** (see Appendix 4)

## Method

1. After washing off nonhybridized probe (see step 6 in Protocol 7.4), wash slides three times in PBT (10 minutes each).

2. Drain each slide and wipe around the chromosomes, but do not allow them to dry. Apply 20 μl of antibody solution (1:200 dilution in PBT) over the squash, cover with a coverslip, and place in moist chamber. Incubate at room temperature for 1 hour.

3. Remove coverslips by dipping in PBT. Wash slides as follows:

   a. Wash three times in PBT (3 minutes each).
   b. Wash three times in pH 9 buffer (3 minutes each).

4. Make alkaline phosphatase reaction solution by adding 20 μl of stock NBT/BCIP solution to 1 ml of pH 9 Buffer. Put 50 μl over the preparation and cover with a large coverslip. Allow color to develop at room temperature. Check color development under the microscope. When signal is adequate, stop the reaction by washing in H$_2$O.

   *Note:* The time necessary is usually 15 minutes to 2 hours.

5. Stain the chromosomes with Giemsa stain diluted 1:20 with 0.01 M phosphate buffer (pH 6.8) or in H$_2$O. Stain for 2–3 minutes, rinse in H$_2$O, and check color under the microscope. If desired, return to staining solution to increase color.

   *Note:* Alternatively, slides can be viewed unstained with phase. Place a drop of H$_2$O over the chromosomes and add a coverslip to view. Remove coverslip and store slides dry.

# Autoradiography for Detection of Radioactive Probes

## Materials

### Supplies and Equipment

Dark room, completely light-tight with a double door to allow exit and entry without exposing slides that are drying

Safelight (Wratten series II; Kodak)

Plastic or glass dipping chamber, to fit a microscope slide with minimal waste of emulsion

Rack for drying slides

Water bath at 45°C

Light-tight black microscope slide boxes to hold exposing slides

Black electrical tape

Silica gel

### Solutions and Reagents

Developer (Kodak D-19)

Fixer (Kodak Rapid Fix)

Kodak NTB-2 autoradiographic emulsion diluted 1:1 with $H_2O$ and stored in 10-ml aliquots in a light-tight box. When diluting emulsion, mix gently to avoid making bubbles.

Giemsa stain

Phosphate buffer (0.01 M, pH 6.8)

## Method

1. In the dark, with one safelight, melt an aliquot of the emulsion for 15 minutes in the 45°C water bath. Pour into a dipping chamber, prewarmed at 45°C, taking care to avoid forming bubbles.

2. Dip each slide slowly into the emulsion, drain, and place on the rack to dry for approximately 2 hours.

3. Place the dried slides in the light-tight boxes containing silica gel. Wrap with black electrical tape and store in a refrigerator. (The refrigerator should not contain either radioactive materials or chemicals that might affect the emulsion.)

   *Note:* Time of exposure depends on the probe and the target. For single-copy λ clones on polytene chromosomes, a 24-hour exposure is usually enough. Prepare several slides so that different exposures can be checked.

4. Develop slides in Kodak D-19 for 2 minutes, rocking gently for 6-second periods interspersed by 6-second rests. Stop the reaction by immersing slides in $H_2O$ for 30 seconds and then fix in Rapid Fix for 5 minutes. The room lights may be turned on after slides have been fixed.

5. Wash five times in $H_2O$ (5 minutes each).

6. Stain slides with Giemsa diluted 1:20 with phosphate buffer for 2–3 minutes. Wash with $H_2O$ and air-dry.

## REFERENCES

Feinberg A.P. and. Vogelstein B. 1983. A technique for radiolabelling DNA restriction endonuclease fragments to high specific activity. *Anal. Biochem.* **132:** 6–13.

———. 1984. A technique for radiolabelling DNA restriction endonuclease fragments to high specific activity. (Addendum) *Anal. Biochem.* **137:** 266–267.

Hartmann C. and Jackle H. 1995. Spatiotemporal relationships between a novel *Drosophila* stripe expressing gene and known segmentation genes by simultaneous visualization of transcript patterns. *Chromosoma* **104:** 84–91.

# CONTENTS

# 8

# Mapping Protein Distributions on Polytene Chromosomes by Immunostaining

**Renato Paro**
*ZMBH*
*University of Heidelberg*
*Heidelberg, Germany*

THE FORMIDABLE SIZE AND STRUCTURE OF POLYTENE CHROMOSOMES allow mapping of chromosomal protein distributions at very high resolution. Immunological methods were already used in the early 1970s to identify the chromosomal association of abundant nuclear proteins, such as histones or RNA polymerase II. However, an absolute requirement for a successful and refined mapping of protein distributions is the reproducibility of the polytene-banding patterns throughout the procedure. An important improvement in this respect was the introduction of the modifications described by Silver and Elgin (1976), which resulted in highly reproducible banding patterns comparable to acid-treated chromosome preparations. A further technical advance was the introduction of enzyme-coupled secondary antibodies. Unlike the fading fluorescent signals, this alternative means of detection yielded stable preparations that allowed an extensive and thus more precise cytological mapping of the binding sites of many different regulatory proteins (Zink and Paro 1989). Currently, immunostaining polytene chromosomes is most often used for achieving the following goals:

- Mapping distributions of chromosome-associated proteins. For map positions of some regulatory factors, see FlyBase at: http://flybase.bio.indiana.edu:82/allied-data/lk/cytofeatures.
- Identifying and characterizing *cis*-regulatory DNA elements bound by particular proteins. Transgene constructs containing the element will create a new protein-binding site at the integration site (Saumweber et al. 1990; Zink et al. 1991).
- Mapping functional protein domains necessary for chromosomal binding or other activities, such as the interaction with other partner proteins (Kuroda et al. 1991; Messmer et al. 1992; Rastelli et al. 1993; Platero et al. 1995).

PROTOCOL 8.1*

# Preparation of Chromosome Squashes from Third-instar Larvae

## Materials

### *Media*

#### *Nutrient-rich Fly Medium*

| | |
|---|---|
| agar | 8 g |
| dried yeast | 18 g |
| soybean meal | 10 g |
| molasses | 7 g |
| malt extract | 80 g |
| cornmeal | 80 g |

Adjust volume to 1 liter with $H_2O$ and boil extensively on a hot plate until all ingredients are well dissolved. Add 6.3 ml of propionic acid as a mold inhibitor. Pour into 175-ml bottles and allow to solidify. To each bottle, add a large drop of live baker's yeast on the surface of the medium.

### *Supplies and Equipment*

Dissecting tools

- tweezers
- dissecting microscope

Siliconized coverslips (Corning or equivalent quality; 22 × 22 mm)

Pencil with eraser end

Filter paper (Whatman 3MM)

(*Optional*) Squashing apparatus; for extended chromosome-spreading sessions, we use the custom-made squashing apparatus illustrated in Figure 8.1.

Microscope equipped for phase contrast

Diamond pen

Razor blade

#### *Preparation of Poly-L-lysine-coated Slides*

Place 100–200 slides in racks.

In general, high-quality slides do not require any pretreatment with detergent. However, in some cases, pretreatment with a strong detergent may be necessary, to ensure a homogeneous wetting of the surface by the poly-L-lysine solution. Wash slides for 2 hours in a strong detergent, and then wash for 2 hours in running tap water and rinse twice in $H_2O$. Dip slides twice in 95% ethanol and air-dry.

Dip slides into a poly-L-lysine solution (slide adhesive solution, 0.1% w/v in $H_2O$; Sigma P 8920). After withdrawal of the rack, the solution should wet the glass surface uniformly and stay on slides. Air-dry slides and store at 4ºC.

### *Solutions and Reagents*

**Liquid nitrogen**
(*Optional*) **Methanol**
(*Optional*) **Ammonium sulfate** (50% w/v)

*For all protocols, $H_2O$ indicates glass distilled and deionized.

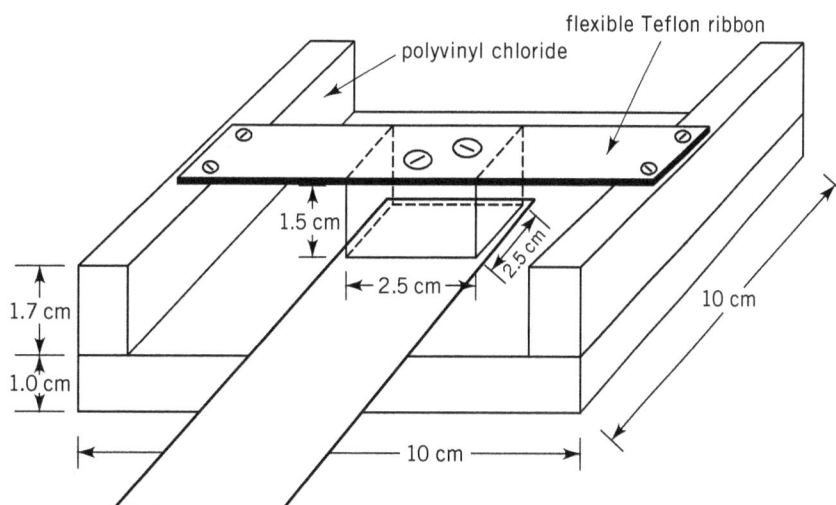

**Figure 8.1.** Schematic diagram of the squashing apparatus. We use this apparatus to ease the effort when squashing chromosomes over an extended period, and to produce more homogeneous pressure over the entire coverslip. A small block made of polyvinyl chloride (2.5 x 2.5 x 1.5 cm) is attached to a flexible Teflon ribbon as shown. The block should hang approximately 2–3 mm above the plane of the holding block (also made of polyvinyl chloride). A filter paper (i.e., Whatmann 3MM) and the slide with the coverslip are then positioned on the holding block and held in place by the suspended block.

### 1× Phosphate-buffered Saline (PBS; pH 7.5)

137 mM NaCl
2.7 mM **KCl**
4.3 mM **Na$_2$HPO$_4$** • 7H$_2$O
1.4 mM **KH$_2$PO$_4$**

### Fixing Solution

| | |
|---|---|
| glacial **acetic acid** | 2.25 ml |
| 37% **formaldehyde** stock solution | 0.25 ml |
| H$_2$O | 2.50 ml |

Important: Prepare Fixing Solution fresh every 2–3 hours!

### 37% Formaldehyde Stock Solution

To 1.85 g of **paraformaldehyde**, add H$_2$O to a final volume of 5 ml. Add 70 µl of 1 N **KOH**. Boil to dissolve the paraformaldehyde, filter, and store in aliquots at –20°C.

**CAUTION: acetic acid, ammonium sulfate, formaldehyde, KCl, KH$_2$PO$_4$, KOH, liquid nitrogen, methanol, Na$_2$HPO$_4$, paraformaldehyde** (see Appendix 4)

## Method

### Raising Third-instar Larvae

1. Raise flies in 175-ml bottles containing Nutrient-rich Fly Medium.

   *Note:* The quality of the chromosomes depends critically on the state of nutrition and well-being of the larvae. This is particularly important to consider in cases where the genetic background (e.g., unhealthy mutants or certain transgenic lines) places a heavy toll on the development of the larvae.

2. Allow the flies to lay eggs just to the point where larvae will hatch under uncrowded conditions (i.e., <100 larvae in a 175-ml bottle). Let the larvae develop at 18°C.

3. For salivary gland preparations, use third-instar larvae that are still crawling and have not yet started to pupate.

### *Dissecting Larval Salivary Glands and Fixing Chromosomes*

1. To prepare one slide, remove two larvae from the bottle and wash in PBS. With tweezers, dissect out the two pairs of salivary glands in PBS (for details on dissection of larval salivary glands, see Kennison, this volume). Carefully try to get rid of most of the attached white fat-body cells, without damaging or separating the two glands.

2. Hold the salivary gland pairs at the common duct with tweezers and transfer the glands to a drop of fixing solution on a siliconized coverslip. Incubate the glands for 10–20 minutes, occasionally stirring with the tip of the tweezers to ensure homogeneous fixation.

   *Note:* The fixation time is an important parameter and needs to be adjusted for every antigen tested. Aim for the shortest time necessary to visualize the signal, as extended fixation times will result in difficulties later in spreading the chromosomes appropriately.

### *Preparing Chromosome Squashes*

1. Take up the coverslip with a poly-L-lysine–treated slide.

   *Note:* At this point, the spreading of the chromosomes begins. Unlike the sole treatment with acids (see Kennison, this volume), the inclusion of formaldehyde keeps the chromosome sticky, making it more difficult to obtain preparations with well-spread chromosome arms.

2. Tap the coverslip with the eraser end of a pencil until the cells are broken up.

   *Note:* This can be monitored best when done against a black background.

3. Move the eraser end of the pencil over the coverslip to spread the chromosomes. To avoid extensive movement of the coverslip, hold down the coverslip with the tip of a finger (use latex gloves to prevent acid burns). Do not continue to spread the specimen if the coverslip sticks to the slide, as this will result in shearing of the chromosomes.

   *Note:* The chromosomes are very brittle at this point of the process. The extent of spreading depends on the constitution and size of the chromosomes, as well as on the time of fixation.

4. Invert the slides and place over one layer of blotting paper. Apply firm pressure with the thumb; the chromosomes become flattened and pressed to the slide.

   *Note:* The squashing apparatus (see Figure 8.1) can substantially ease this step, because it allows a more firm and homogeneous pressure on the chromosomes without the risk of moving the coverslip.

5. Examine the preparation under phase contrast. Use only preparations with well-spread chromosome arms showing high-contrast banding patterns. Mark the position of the coverslip with a diamond pen.

6. Freeze slides in liquid nitrogen and remove coverslip with a razor blade (wear appropriate eye protection).

7. Dip slides immediately in PBS and wash for 15 minutes, slowly shaking the rack. Repeat wash in PBS.

8. Proceed with the immunostaining as described in Protocol 8.2. Alternatively, keep the slides (up to 1 week) in 100% methanol or in 50% (w/v) ammonium sulfate at 4°C. Be aware, however, that certain antigens might be affected by exposure to methanol.

PROTOCOL 8.2

# Immunostaining

Immunostaining can be performed by either of the following three methods. Method 1 describes a procedure for immunostaining using enzyme-coupled secondary antibodies. Where only weak signals are observed with Method 1, amplification with the biotin-avidin system might be helpful (Method 2). Secondary antibodies tagged with a fluorochrome can be used to speed up the detection procedure, as well as for double-labeling experiments (Method 3).

## Materials

### *Supplies and Equipment*

Rack for holding slides
Slide jars
Aluminium foil

### *Solutions and Reagents*

1x PBS (see p. 133)

Primary antibodies. Affinity-purified primary antibodies are diluted in PBS/BSA (bovine serum albumin). The dilutions must be adjusted for each individual primary antibody. Typical dilutions for rabbit polyclonal antibodies range between 1:50 and 1:500. For double-labeling experiments, use primary antibodies raised in two different species.

Secondary antibodies. Secondary antibodies are diluted in PBS containing 2% normal serum, which is obtained from the same species as the secondary antibody. Depending on the immunostaining method used, the following antibodies are used:

> *Method 1:* Anti-rabbit IgG (Fc) horseradish peroxidase (HRP) conjugate; 1:100 dilution.
>
> *Method 2:* Biotin-conjugated secondary antibody and biotin-avidin amplification system (Vectastain Elite ABC Kit; Vector Laboratories PK 6100).
>
> *Method 3:* Fluorescence-conjugated secondary antibody; for double-labeling experiments, use fluorescence-conjugated secondary antibodies raised to the corresponding type of primary antibody; the dilution of each antibody needs to be determined individually.

Solution containing 0.5 mg/ml **DAB** (3,3′-diaminobenzidine tetrahydrochloride; Sigma D 5637) and 0.01% **H$_2$O$_2$** (Merck 107210)

(*Optional*) DAB enhancement kit (metal amplification system for peroxidases; Sigma *FAST* DAB with metal enhancer D 0426)

### *Blocking solution*

Add a small spoon of nonfat dry milk to 40 ml of PBS.

### *Wash Solution 1*

1x PBS
300 mM NaCl
0.2% NP-40 (Nonidet P-40)
0.2% Tween-20–80

*Wash Solution 2*

1x PBS
400 mM NaCl
0.2% NP-40
0.2% Tween-20–80

*PBS/Triton X-100*

1x PBS
1% Triton X-100

*PBS/BSA*

1x PBS
1% BSA

*PBT*

1x PBS
0.1 % BSA

**CAUTION: DAB, $H_2O_2$** (see Appendix 4)

## Method

### Method 1: Use of Enzyme-coupled Secondary Antibodies

1. Wash stored slides twice in PBS (10 minutes each wash) and once in PBS/Triton X-100 for 10 minutes. Incubate in Blocking Solution at room temperature for 1 hour or at 4ºC overnight, slowly shaking the jar containing the slides. Drain off excess solution.

2. Add 40 µl of diluted affinity-purified primary antibody to each slide. Cover with a coverslip and incubate at room temperature for 1 hour and also at 4ºC overnight in a humid chamber.

3. Rinse slides in PBS to remove the coverslips and place in rack. Wash three times in Blocking Solution (5 minutes each wash). Shake rack with slides thoroughly during the washing procedure.

4. Rinse slides in PBS and remove excess solution. Add 40 µl of secondary antibody (i.e., anti-rabbit IgG [Fc] HRP conjugate, 1:100 dilution) to each slide. Cover with a coverslip and incubate at room temperature for 1 hour in a humid chamber.

   *Note:* Often, commercially available secondary antibody preparations show a high affinity for *Drosophila* nuclear proteins, resulting in some excellent signals on the chromosomes. As such, a corresponding control experiment should always be included before a new batch is utilized. Cross-reactivity of secondary antibody preparations can often be substantially reduced by preincubating a lower dilution of the serum overnight with formaldehyde-fixed embryos (for preparation of preadsorbed antibodies, see Rothwell and Sullivan [this volume], Protocol 9.3).

5. Rinse slides in PBS and place in rack. Wash 15 minutes in Wash Solution 1 and 15 minutes in Wash Solution 2. Shake rack thoroughly during the washing procedure.

   *Note:* If background problems persist, the NaCl concentration in Wash Solution 2 can be raised to 500 mM.

6. Rinse slides in PBS. Add 100 µl of a solution containing 0.5 mg/ml DAB and 0.01% $H_2O_2$. Follow the appearance of the signals under the microscope using bright-field

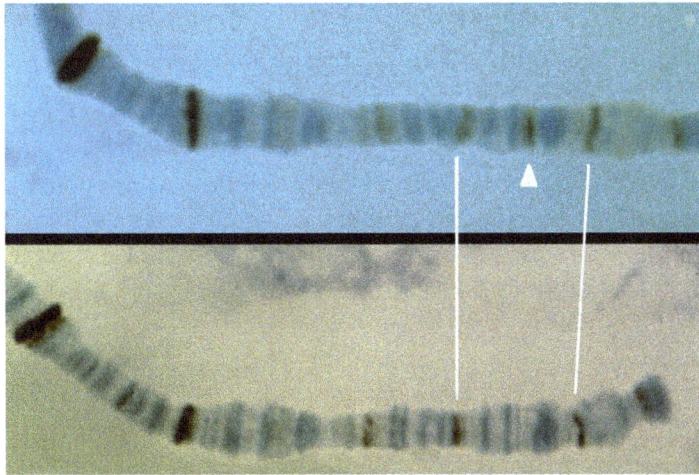

**Figure 8.2.** Distribution of the Polycomb protein at section 43–49 of the second chromosome visualized with enzyme-conjugated (HRP) secondary antibodies. (*Top*) Chromosome with a transgene construct containing a Polycomb-binding site (PRE) at 44E (*arrowhead*); (*bottom*) wild-type chromosome.

optics to prevent overstaining. Stop the reaction by dipping slides in PBS. Wash in PBS for 10 minutes.

> *Note:* Silver treatment of the preparation turns the brown color of a DAB signal to black, thus substantially improving the contrast, which is very convenient when black and white photography is used for documentation. We use the DAB enhancement kit from Sigma and follow the instructions given by the supplier, except that the metal amplification step is shortened to approximately 1 minute.

7. Counterstain with Giemsa as described in Protocol 8.3.

## Method 2: Use of Biotin-conjugated Secondary Antibodies and Enzymatic Detection with the Avidin/Biotin System

1. Perform steps 1–3 as described in Method 1 (above).

2. Perform step 4 of Method 1 using a secondary antibody conjugated to biotin.

3. During secondary antibody incubation, mix the biotin solution and the avidin-HRP solution to allow the formation of complexes. We use reagents from the Vectastain Elite ABC kit. Mix 40 μl of solution A and 40 μl of solution B in 1 ml of PBT at room temperature for 10 minutes.

4. After incubation with secondary antibodies, rinse slides twice in PBT (10 minutes each). Remove excess liquid and place 50 μl of the biotin-avidin-HRP mix on the preparation and cover with a coverslip. Incubate the slides at room temperature for 40 minutes in a humid chamber.

5. Wash slides in Wash Solution 1 for 15 minutes and then in Wash Solution 2 for 15 minutes.

> *Note:* If background problems persist, the NaCl concentration in Wash Solution 2 can be raised to 500 mM.

6. Perform the color reaction as described in Method 1, step 6.

### Method 3: Use of Fluorescence-coupled Secondary Antibodies

1. Perform steps 1–3 described in Method 1, above.

    *Note:* For double-labeling experiments, both primary antibodies (raised in two different species) can be applied simultaneously at the appropriate concentrations.

2. Rinse slides in PBS and remove excess solution. At this point, perform all steps under dimmed-light conditions. During incubations, wrap the slide jars in aluminum foil to minimize bleaching of fluorochromes. Add 40 μl of diluted fluorescence-conjugated secondary antibody solution. Cover with a coverslip and incubate at room temperature for 1 hour in humid chamber.

    *Note:* Also during this step, both secondary antibodies of the appropriate type can be applied simultaneously for double-labeling experiments.

3. Rinse slides in PBS and place in a rack. Wash slides in Wash Solution 1 for 15 minutes and then in Wash Solution 2 for 15 minutes.

    *Note:* If background problems persist, the NaCl concentration in Wash Solution 2 can be raised to 500 mM.

4. Store slides in PBS at 4ºC in the dark.

5. Counterstain with Hoechst 33258 as described in Protocol 8.4.

## PROTOCOL 8.3

# Giemsa Staining

Chromosomes with HRP signals should be counterstained with Giemsa as described below.

### Materials

### *Supplies and Equipment*

Slide jars

### *Solutions and Reagents*

**Giemsa** (Merck 109204)
**Sodium phosphate** buffer (10 mM, pH 6.8)
Glycerol (99.5%)
(*Optional*) Permanent mounting solutions, e.g., Entelan (EM Science)

**CAUTION: Giemsa, sodium phosphate** (see Appendix 4)

### Method

1. Prepare a 1:130 dilution of Giemsa in 10 mM sodium phosphate buffer (pH 6.8). Stain slides for 30 seconds to 1 minute. Rinse slides by dipping several times into $H_2O$.

    *Note:* If chromosome bands appear too weak in a microscope with bright-field optics, repeat staining.

2. Mount in 99.5% glycerol with a coverslip and immediately examine the slides under the microscope.

    *Note:* Giemsa stain will fade within a few hours. However, chromosomes can be washed in PBS and restained. For storage, slides can be frozen at –20ºC. Entelan (EM Science) can be used as a permanent mounting solution.

PROTOCOL 8.4

## Staining with Hoechst 33258

Chromosomes with fluorescent signals can be counterstained with Hoechst 33258 as described below. However, banding patterns produced by the fluorescent DNA stains are often difficult to compare with the published polytene chromosome-banding patterns. We find it more convenient to simply use phase contrast to correlate the fluorescent signal with the corresponding chromosome bands.

### Materials

#### Solutions and Reagents

##### Hoechst Staining Solution

10 μg/ml Hoechst 33258
1x PBS

##### Preparation of the Stock Solution of Mowiol–DABCO 2.5%

Combine 2.4 g of Mowiol (Hoechst), 6 g of glycerol, and 6 ml of $H_2O$ and mix for 3 hours. Add 12 ml of 0.2 M Tris-Cl (pH 8.5) and incubate with mixing at 50ºC for 10 minutes. Centrifuge at 5000*g* for 15 minutes to pellet insoluble material. Add **DABCO** (1,4-diazabicyclo-[2,2,2]-octane; Merck 803456) to final 2.5% to the solution as antibleaching agent. Store in 500-μl aliquots at −20ºC.

**CAUTION: DABCO** (see Appendix 4)

### Method

1. Stain slides in Hoechst staining solution for 5 minutes.

2. Mount preparations with fluorescent signals in Mowiol–DABCO 2.5%.

## REFERENCES

Kuroda M.I., Kernan M., Kreber R., Ganetzky B., and Baker B.S. 1991. The maleless protein associates with the X chromosome to regulate dosage compensation in *Drosophila*. *Cell* **66:** 935–947.

Messmer S., Franke A., and Paro R. 1992. Analysis of the functional role of the Polycomb chromodomain in *Drosophila melanogaster*. *Genes Dev.* **6:** 1241–1254.

Platero J.S., Hartnett T., and Eissenberg J.C. 1995. Functional analysis of the chromo domain of HP1. *EMBO J.* **14:** 3977–3986.

Rastelli L., Chan C.S., and Pirrotta V. 1993. Related chromosome binding sites for zeste, suppressors of zeste and Polycomb group proteins in *Drosophila* and their dependence on Enhancer of zeste function. *EMBO J.* **12:** 1513–1522.

Saumweber H., Frasch M., and Korge G. 1990. Two puff-specific proteins bind within the 2.5 kb upstream region of the *Drosophila melanogaster* Sgs-4 gene. *Chromosoma* **99:** 52–60.

Silver L.M. and Elgin S.C.R. 1976. A method for determination of the in situ distribution of chromosomal proteins. *Proc. Natl. Acad. Sci.* **73:** 423–427.

Zink B. and Paro R. 1989. In vivo binding pattern of a trans-regulator of homoeotic genes in *Drosophila melanogaster*. *Nature* **337:** 468–471.

Zink B., Engstrom Y., Gehring W.J., and Paro R. 1991. Direct interaction of the Polycomb protein with Antennapedia regulatory sequences in polytene chromosomes of *Drosophila melanogaster*. *EMBO J.* **10:** 153–162.

# Cell Biology

# CONTENTS

# 9

# Fluorescent Analysis of *Drosophila* Embryos

**Wendy F. Rothwell and William Sullivan**
*Sinsheimer Laboratories*
*Department of Biology*
*University of California*
*Santa Cruz, California 95064*

Cellular analysis in *Drosophila* is frequently being applied to problems that were once addressed only through genetic, molecular, and biochemical approaches. A number of factors have contributed to this renaissance in *Drosophila* cell biology: Many of the procedures for cellular analysis are efficient, inexpensive, and relatively easy to perform; the generation of highly specific antibodies for immunofluorescent analysis is now routine (see Rebay and Fehon, this volume); most *Drosophila* researchers have ready access to conventional, confocal, or deconvolution fluorescent microscope systems; and finally, a variety of high-quality fluorescent probes are now commercially available (Table 9.1). More recently, the availability of green fluorescent protein (GFP)-tagged proteins has made live fluorescent analysis possible (see Hazelrigg, this volume).

The early *Drosophila* embryo is particularly amenable to fluorescent analysis. Large numbers of specifically staged embryos are easily collected from normal and mutant stocks. There is a wealth of molecular and genetic reagents that make the cellular analysis particularly powerful. The morphological and cellular events of embryogenesis have been extensively characterized (Foe and Alberts 1983; Foe et al. 1993). For example, directly after fertilization, the embryo proceeds through a series of rapid nuclear divisions that rely on the highly coordinated dynamics of the microtubules, microfilaments, and other cytoskeletal components (Warn et al. 1984; Karr and Alberts 1986; Kellogg et al. 1988). During this time, critical events that establish the axis and patterning in the embryo are occurring (for review, see St Johnston and Nüsslein-Volhard 1992). These events have been thoroughly described through fluorescent analysis and provide an excellent resource in which to analyze the primary cellular defect in newly isolated mutations.

As many of these early events are under maternal control, much of this analysis has been performed with maternal-effect mutations (Sullivan et al. 1993). With the development of the FLP-FRT (site-specific FLP recombinase–FLP recombination target) system, it is now possible to efficiently generate germ-line clones of zygotic lethal mutations and analyze the maternal effect of these mutations (Golic 1991; Dang and Perrimon 1992). Approximately 70–80% of the genes that mutate to zygotic lethality are also required in the early embryo. Consequently, this provides a general alternative approach toward examining the cellular basis of some of the more intractable zygotic lethal mutations

**141**

**Table 9.1.** Commercially Available Reagents

| Structure | Probe[a] | Vendor/Part no.[b] |
|---|---|---|
| Nucleus/DNA | DAPI | Sigma/D 9542 |
| | PI | Sigma/P 4170 |
| | anti-histone | Chemicon/MAB052 |
| | Hoechst 33258 | Molecular Probes |
| | OliGreen™ | Molecular Probes |
| Cytoskeleton | | |
| F-actin | FITC-phalloidin | Sigma/P 5282 |
| | TRITC-phalloidin | Sigma/P 1951 |
| actin | anti-actin | ICN Biomedicals/clone C4 |
| tubulin | β-tubulin | Chemicon/MAB380 |
| | α-tubulin | Sigma/T 9026 |
| Membrane | FITC-ConA | Molecular Probes |
| | anti-py99 | Santa Cruz Biotechnology |
| | α-spectrin | Sigma/S 1390 |
| Motors | dynein | Chemicon/MAB1618 |
| | kinesin | DSHB/SUK4,5 |
| Other | adducin-related protein | DSHB/1B1 |
| | even-skipped | DSHB/2B8 |
| | Wingless | SHB/4D4 |
| | engrailed | DSHB/4D9 |
| | syntaxin | DSHB/8C3 |
| | cyclin B | DSHB/F2F4 |
| Secondaries | FITC goat anti-mouse | Chemicon/AP124F |
| | FITC goat anti-rabbit | Chemicon/AP132F |
| | TRITC goat anti-mouse | Chemicon/AP124R |
| | TRITC goat anti-rabbit | Chemicon/AP132R |
| | Cy5 goat anti-mouse | Chemicon/AP124S |
| | Cy5 goat anti-rabbit | Chemicon/AP132S |

[a]Abbreviations: (ConA) concanavalin A; (DAPI) 4,6-diamidino-2-phenylindole; (FITC) fluorescein isothiocyanate; (PI) propidium iodide; (py99) phosphotyrosine; (TRITC) tetramethylrhodamine isothiocyanate.
[b]DSHB is the Developmental Studies Hybridoma Bank (University of Iowa).

(Garcia-Bellido and Robbins 1983; Perrimon et al. 1984, 1989; Perrimon and Mahowald 1986). Transformants bearing GFP-tagged proteins are now frequently used for live cellular analysis (see Hazelrigg, this volume). Fluorescent analysis of fixed samples complements this approach. Fixed analysis allows many more embryos to be examined in a single session on the microscope, and the fixed preparations are stable for long periods. Antiquenching reagents in the mounting media allow extensive documentation of the samples without deterioration of the fluorescent signal (see p. 156). Double- and triple-labeling for colocalization studies are easily performed (see Figure 9.4B,C). Fixed samples also enable one to record the images over multiple planes for three-dimensional reconstructions.

This chapter describes the most common and generally applicable procedures for fixed fluorescent analysis of the *Drosophila* embryo. Pioneers in this field such as Rabinowitz (1941), Zalokar and Erk (1977), Foe and Alberts (1983), and Mitchison and Sedat (1983) laid the groundwork for many of the techniques described here. *Drosophila* embryos are protected by an outer chorion layer as well as an impermeable and opaque vitelline membrane (Figure 9.1). Consequently, preparation of *Drosophila* embryos proceeds as follows: chorion removal, vitelline membrane permeabilization, fixation, and finally, vitelline membrane removal. These procedures as well as procedures for collecting, staining, and mounting the embryos are described below.

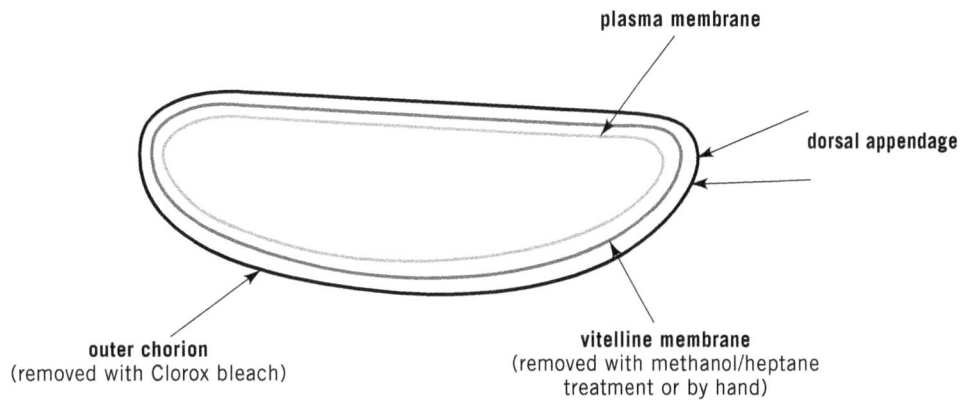

**Figure 9.1.** Membranes of the *Drosophila* embryo. Each embryo contains a plasma membrane, vitelline membrane, and outer chorion.

## PROTOCOL 9.1*

## Embryo Collection

### Materials

### *Supplies and Equipment*

Plastic fly "collection" bottle (6 ounce; Applied Scientific AS-355N)

Lids from 35 × 10-mm disposable plastic tissue-culture dishes

Screw-cap microcentrifuge tubes (1.5 ml)

Razor blades

Synthetic mesh (120-μm, 206 821 7345; Research Nets Inc., 14207 100th Ave Ne, Bothell, Washington 98011-5126. Telephone: 425-821-7345)

Small fine-haired paint brush

Squirt bottle containing $H_2O$

---

#### *Preparation of Grape or Apple Juice Egg Plates*

Plastic-ware and ingredients:

    10 × 35-mm plastic petri dish lids

    700 ml of $H_2O$ (deionized)

    300 ml of juice concentrate (grape or apple)

    0.5 g of methylparaben dissolved in 10 ml of **ethanol**

    30 g of agar (American Bioorganics 00-58010)

Add agar to the $H_2O$ and autoclave for 40 minutes. While autoclaving, spread the tops of the petri dishes face up on a series of trays. Add methylparaben solution to juice concentrate. Quickly and thoroughly mix the concentrate into the autoclaved solution. Immediately begin dispensing media into dishes. Once the juice plates have cooled and hardened, store at 4ºC. This procedure makes 1 liter of solution and yields approximately 170 juice plates.

#### *Yeast Paste*

Mix 1 g of dry yeast and 1.3 ml of $H_2O$ to yield a paste.

---

**CAUTION: ethanol** (see Appendix 4)

---

*For all protocols, $H_2O$ indicates glass distilled and deionized.

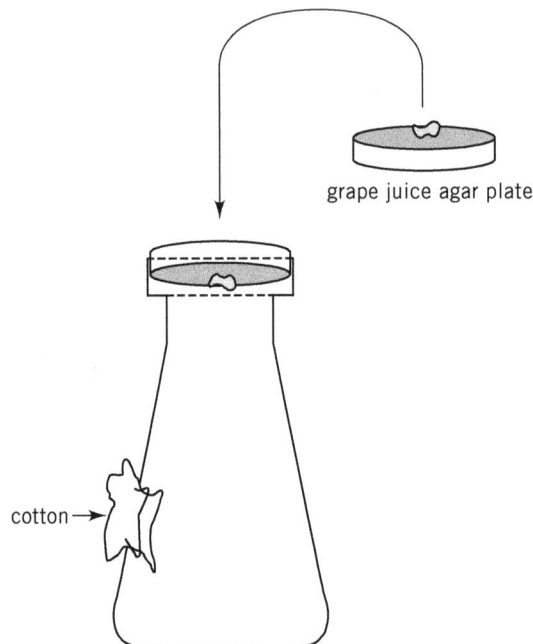

**Figure 9.2.** Collecting bottle and egg plate (see text for details).

## Method

1. Place 200–400 flies in a 6-ounce plastic collection bottle fitted with a small cotton-filled breathing hole. Cover the bottle with a 35 × 10-mm disposable, plastic petri dish lid containing grape juice agar and a small dollop of thick yeast paste (see Figure 9.2). The eggs are deposited on the grape juice agar. Avoid opening the incubator during the egg-laying period, as the flies prefer quiet.

   *Note:* Typically, a single bottle is set up for each line being examined. However, if embryos of short time points (<1 hour) are being collected and/or the line does not lay many embryos, several bottles can be used.

2. While the flies are laying, prepare egg baskets as follows:

   a. Cut 1.5-ml screw-cap microcentrifuge tubes in half (see Figure 9.3). Also cut off the top of each screw cap, so that it forms an open ring that can be screwed back onto the microcentrifuge tube.

   b. Place a piece of 120-μm synthetic mesh around the threaded end of each microcentrifuge tube.

   c. Use the ring made from the screw cap to hold the mesh in place, thus creating a basket.

   *Notes:* If larger quantities of eggs are being fixed, larger egg baskets must be constructed. A good rule of thumb is that enough surface area should be provided so that the embryos can lie in a monolayer on the mesh.

   In general, a single basket is used for each timed embryo collection. Between collections, the egg basket is thoroughly washed with $H_2O$, and the mesh is replaced.

3. Obtain embryo collections at appropriate time points. Transfer the embryos from the agar plates into the egg basket using $H_2O$ and a small paintbrush. If there are relatively few embryos, rapidly pick the embryos from the plate with forceps and transfer them in clusters to the egg basket. In either case, wash all the embryos into the mesh at the

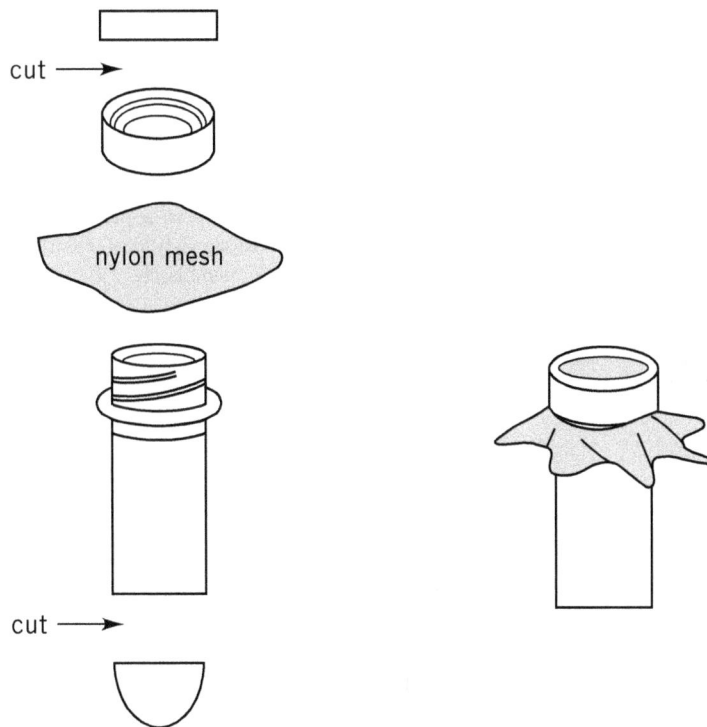

**Figure 9.3.** Constructing an egg basket (see text for details).

bottom of the basket using copious amounts of $H_2O$. Proceed rapidly with the washing to avoid anoxia.

*Note:* For embryo collections from various time points, dechorionation should follow immediately for each collection (see Protocol 9.2).

## PROTOCOL 9.2

## Embryo Dechorionation

### Materials

### Supplies and Equipment

Dissecting microscope
Glass petri dish
Pasteur pipette

### Solutions and Reagents

Squirt bottle containing 1x Embryo Wash Solution
**Bleach** (Clorox)

> ### 10x Embryo Wash Solution
> 7% NaCl
> 0.5% Triton X-100
> Dissolve 70 g of NaCl and 5 ml of Triton X-100 in sufficient $H_2O$. Adjust to 1 liter with $H_2O$.

**CAUTION: bleach** (see Appendix 4)

## Method

1. Place the egg baskets in a glass petri dish partially filled with a 50% bleach (Clorox) solution (the level should be just below the rim of the basket). Use a pasteur pipette to continually rinse the embryos with the bleach solution.

   *Note:* Because the potency of the bleach varies, monitor the dechorionation process by observing the embryos under a dissecting microscope. When the dorsal appendages have dissolved in 80% of the embryos (~1–3 minutes), immediately proceed to step 2, below (extensive washing). Monitoring prevents damaging the embryos through overexposure to bleach.

2. Immediately wash the basket of embryos using a squirt bottle containing Embryo Wash Solution. Wash the inner edges of the basket, so that all of the embryos lie on the mesh.

## EMBRYO FIXATION

Protocol 9.3 presents the following fixation procedures:

- *Formaldehyde-based Fixation Techniques.* The slow formaldehyde fix method (see Protocol 9.3, Method 1) is a modification of the Mitchison and Sedat protocol (1983) and is excellent for preserving many structures deep within the embryo because it allows time for the fix to permeate the embryo. In general, use reagent-grade formaldehyde. However, preservation of some cellular components may require higher-quality (EM-grade) formaldehyde.

  The fast formaldehyde fix method (see Protocol 9.3, Method 2) designed by Theurkauf (1992) is a modification of the slow formaldehyde fixation procedure. It preserves cortical structures and is excellent for fixing dynamic cytoskeletal structures such as microtubules.

- *Fixation with Methanol.* This is a relatively harsh procedure that destroys membranes (see Protocol 9.3, Method 3). In addition, in early embryos, the pole cells are often lost. However, it adequately preserves many structures, such as microtubules. The methanol fix has the advantage that virtually all of the embryos are devitellinized. This is important when analyzing embryos that are limited in quantity.

- *Fixation by Boiling.* The boiling fix method (see Protocol 9.3, Method 4) is often tried when more standard fixations fail. For example, immunofluorescent analysis of a number of centrosome proteins requires this fixation method. This procedure also results in devitellinization of virtually all of the embryos.

- *Fixation without Methanol (Hand Devitellinization of Embryos).* Often, the highest quality preparations are obtained by not exposing the embryos to methanol, which destroys membrane and other structures. In addition, some reagents such as fluorescently labeled phalloidin will not work if the embryos have been exposed to methanol. Foregoing the methanol, however, necessitates removing the vitelline membrane by hand. We find this procedure to be most satisfactory (see Protocol 9.3, Method 5). For examples of embryos prepared by this method, see Figure 9.4B,C.

The appropriate technique depends on a number of variables, including the preservation qualities of the cellular components being examined, the position of the components within the embryos, and the fluorescent probe used. Some general guidelines are presented, but for a new antibody or reagent, it may be necessary to try all of these techniques to determine which one is the most appropriate.

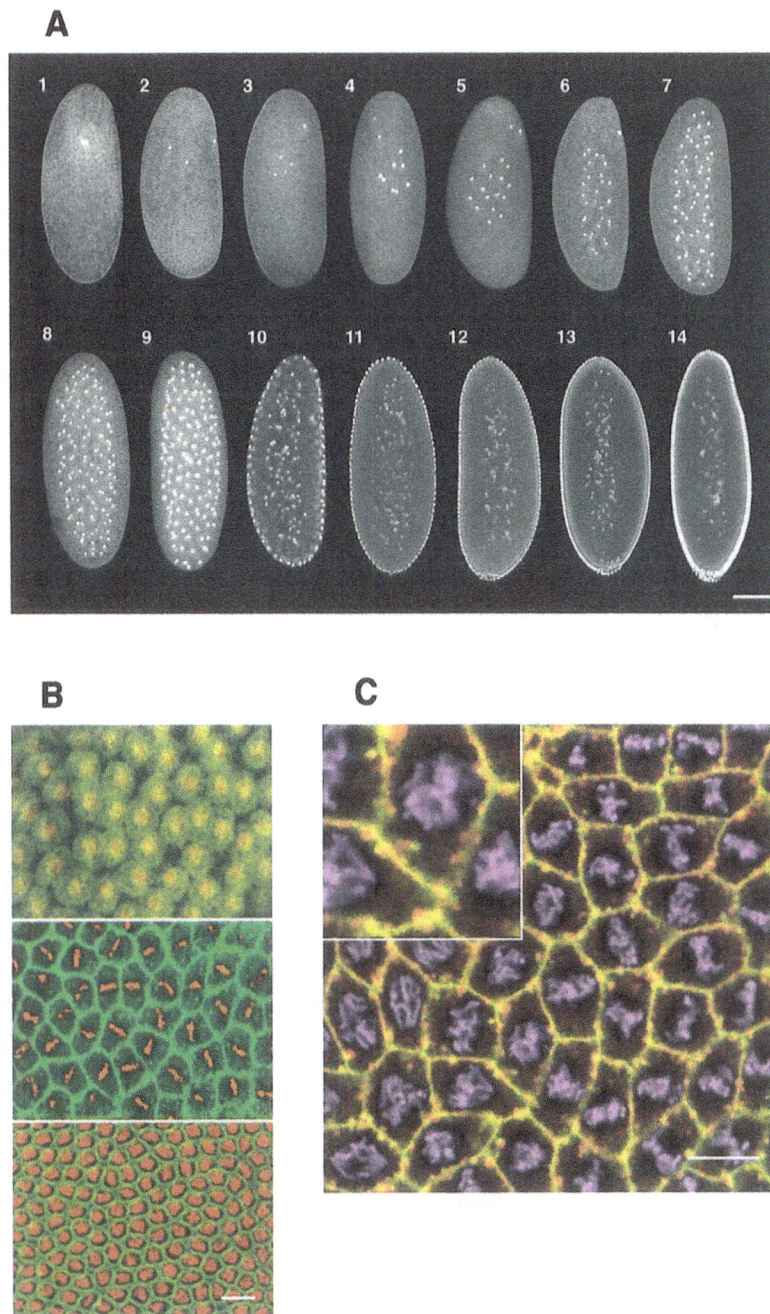

**Figure 9.4.** Confocal micrographs of wild-type *Drosophila* embryos. (*A*) Whole-mount *Drosophila* embryos stained with propidium iodide to visualize the nuclear division cycle. Shown are syncytial embryos in nuclear division cycles 1–13 and interphase of nuclear cycle 14 (cellularization). Bar, 100 μm. (*B*) Surface views of *Drosophila* embryos prepared by hand devitellinization and double-stained for actin (fluorescein phalloidin, *green*) and DNA (propidium iodide, *red*). Interphase actin caps (*top panel*) and actin-based metaphase furrows (*middle panel*) are shown for embryos in nuclear division cycle 13. (*Bottom panel*) Embryo at cellularization. Actin-based cellularization furrows surround each of the nuclei. Bar, 10 μm. (*C*) Surface view of a *Drosophila* embryo in late metaphase/early anaphase of nuclear division cycle 13. The embryo was prepared by hand devitellinization and triple-stained for actin (fluorescein phalloidin, *green*), DNA (propidium iodide, *purple*), and the furrow component, Dah (anti-Dah, Cy5-labeled secondary, *red*). Both actin and Dah localize to the furrows and appear to colocalize in some regions (*yellow* staining, inset). Bar, 10 μm. (*Inset*) 2× magnification.

PROTOCOL 9.3

# Embryo Fixation

## Materials

### Supplies and Equipment

Scissors
Glass vials (5 ml)
Pasteur pipette
Conical tube (50 ml)
3MM Whatman paper
Double-stick tape
Needle (23 gauge)
Syringe (3 ml)
Petri dish lid (35 x 10 mm)

### Solutions and Reagents

**Heptane**
**Formaldehyde** (reagent and EM grade)
**Methanol**
Embryo Wash Solution (for preparation, see p. 145)

### PEM Buffer (Karr and Alberts 1986)

0.1 M PIPES
1 mM **MgCl$_2$**
1 mM EGTA
Adjust pH to 6.9 with **KOH.**

### 10x PBS Solution (Sambrook et al. 1989)

| | |
|---|---|
| NaCl | 80 g |
| **KCl** | 2 g |
| **Na$_2$HPO$_4$** | 14.4 g |
| **KH$_2$PO$_4$** | 2.4 g |

Dissolve all components in 800 ml of H$_2$O. Adjust the pH to 7.4 with **HCl**. Store at room temperature. It is not necessary to sterilize this solution for work with *Drosophila* embryos.

### PBTA Solution (1x PBS, 1% BSA, 0.05% Triton X-100, 0.02% Sodium Azide)

Mix the following components:

| | |
|---|---|
| 10x PBS | 50 ml |
| BSA | 5 g |
| Triton X-100 | 250 µl |
| **Sodium azide** | 0.1 g |

Adjust volume to 500 ml with H$_2$O.

### Preparation of Heptane Saturated with 37% Formaldehyde

1. Combine equal volumes of heptane and 37% formaldehyde in a scintillation vial (for small volumes of 2–10 ml total) or in a 100-ml glass bottle with plastic screw-top (for larger volumes of 25–80 ml total).

2. Secure the lid and shake the mixture vigorously for 15 seconds. Let the solution settle into the 2 phases.

3. Repeat this mixing procedure several (at least three) times before using the solution to ensure that the heptane becomes saturated with the formaldehyde. The saturated heptane is the upper phase in the 1:1 heptane:formaldehyde stock.

It is best to prepare the solution the day before it is to be used, shaking the vial or bottle periodically throughout the day. Because formaldehyde is sensitive to light, wrap the bottle in aluminum foil. The mixture is stored at room temperature and remains active for several months. If crystals form in the mixture upon addition of the formaldehyde, the solution will not work properly and should be discarded. Use formaldehyde from another source to prepare fresh solution.

**CAUTION: formaldehyde, HCl, heptane, KCl, $KH_2PO_4$, KOH, methanol, $MgCl_2$, $Na_2HPO_4$, sodium azide** (see Appendix 4)

## Method 1

### Slow Formaldehyde Fix

1. Remove the mesh containing the dechorionated embryos from the egg basket. The embryos should be in the center of the mesh. Cut away the excess mesh and extract excess liquid by gently blotting with a paper towel.

2. Place the mesh (embryos facing out) on the inner edge of a 5-ml glass vial. Use a pasteur pipette filled with approximately 1 ml of heptane to wash the embryos off the mesh into the vial (contaminating bits of agar will preferentially stick to the mesh).

3. Remove the mesh, and immediately add an equal volume of 3.7% formaldehyde in PEM buffer. Screw the lid on tightly and shake vigorously for 15 seconds.

4. Let the vial stand at room temperature for 20 minutes.

   *Note:* At the end of this period, the embryos will lie at the interface between the lower formaldehyde and the upper heptane layer. Be aware of the time, as longer fixation times result in cross-linking of the plasma membrane and the vitelline membrane and thus significantly decrease the efficiency of devitellinization.

5. Carefully remove the bottom formaldehyde phase of fluid. Use a yellow pipette tip and angle the vial to remove as much formaldehyde as possible without removing the embryos.

6. Add 1.0 ml of methanol. Cap the vial, shake vigorously for 15 seconds, and let stand for 1 minute after shaking.

   *Note:* Usually about half of the embryos are devitellinized and these sink to the bottom of the vial.

7. Remove the upper heptane layer, along with the embryos that did not sink. Add methanol to the embryos remaining until the vial is approximately two-thirds full. Store these embryos in methanol at 4°C.

   *Note:* Storing the embryos overnight at 4°C before staining allows time for clearing and often results in higher-quality images.

## Method 2

### Fast Formaldehyde Fix

Perform steps 1–7 as described for the Slow Formaldehyde Fix (Method 1, above). However, at step 3, add 37% formaldehyde and fix for 5 minutes at room temperature (step 4). As with the Slow Formaldehyde Fix procedure, longer fix times decrease the yield because of poor devitellinization.

## Method 3

### *Methanol Fix*

1.  Perform steps 1 and 2 of the Slow Formaldehyde Fix (Method 1, above).

2.  Remove the mesh and add 1.0 ml of methanol. Cap the vial, shake vigorously for 15 seconds, and let stand for 1 minute after shaking.

    *Note:* Almost all of the embryos are devitellinized and sink to the bottom of the vial.

3.  Remove the upper heptane layer and most of the methanol, leaving the embryos at the bottom of the vial. Add fresh methanol until the vial is approximately two-thirds full. Store these embryos in methanol at 4°C.

## Method 4

### *Boiling Fix*

1.  Heat 5 ml of Embryo Wash Solution to 90–100°C in a 50-ml conical tube. Add dechorionated embryos to the heated Embryo Wash Solution.

    *Note:* This step causes the rapid fixation of the embryos.

2.  Immediately add 40 ml of ice-cold Embryo Wash Solution and place on ice. Allow the embryos to sink to the bottom of the tube (~1 minute).

3.  Use a pasteur pipette to transfer the fixed embryos in approximately 1 ml of Embryo Wash Solution to a 5-ml glass vial. Allow the embryos to sink to the bottom of the vial (~30 seconds).

4.  Remove as much of the Embryo Wash Solution as possible, leaving the embryos in the bottom of the vial. Proceed with the devitellinization by adding 1 ml of heptane followed by 1 ml of methanol to the embryos. Cap the tube, shake it vigorously for 15 seconds, and let stand for 1 minute after shaking.

    *Note:* Almost all of the embryos are devitellinized and sink to the bottom of the vial.

5.  Remove the upper heptane layer and most of the methanol, leaving the embryos at the bottom of the vial. Add fresh methanol until the vial is approximately two-thirds full. Store these embyros in methanol at 4°C.

## Method 5

### *Embryo Fixation without Methanol (Hand Devitellinization of Embryos)*

1.  Wash dechorionated embryos into a 5-ml glass vial with approximately 1 ml of heptane saturated with 37% formaldehyde. Incubate the embryos in this solution at room temperature for 40 minutes.

2.  Use a pasteur pipette to transfer the embryos to a small piece of 3MM Whatman paper and allow the heptane to evaporate (~30 seconds).

    *Note:* Ensure that the embryos stay in the neck of the pipette when transferring them. Embryos entering the body of the pipette tend to stick to the sides and are lost. Also, the embryos tend to sink quickly toward the tip of the pipette. It is best to allow them to sink and to transfer them in 1–2 drops onto the Whatman paper.

3. Transfer the embryos from the Whatman paper to double-stick tape positioned on the bottom of the lid of a small petri dish (35 × 10 mm) by gently placing the Whatman paper embryo-side down onto the tape and very gently tapping the paper until the embryos stick to the tape. Remove the Whatman paper and immediately cover the embryos with PBTA Solution.

4. Remove the vitelline membrane by hand under a dissecting scope using a 23-gauge needle mounted on a 3-ml syringe. This is best achieved by poking a small hole in the vitelline membrane at one end of the embryo and gently "pushing" the embryo out through the hole by applying pressure from the opposite end. The membrane will remain stuck to the tape and the devitellinized embryo will float up into the PBTA Solution.

5. Transfer the devitellinized embryos in PBTA Solution to a 5-ml glass vial. Store embryos in PBTA at 4ºC or proceed immediately with staining of embryos (see pp. 153–156).

## STORAGE OF FIXED EMBRYOS

Hand-devitellinized embryos are stored in PBTA Solution (see p. 148) at 4ºC. For all other fixes, store the embryos in methanol at 4ºC. Overnight storage in methanol clears the embryo and often improves image quality. However, loss of image quality does occur if the embryos are stored in methanol for extended periods of time (weeks). The extent of the deterioration depends on the probe and the antigen being examined.

## REHYDRATION OF EMBRYOS STORED IN METHANOL

Embryos stored in methanol must be rehydrated before staining. Rehydration methods differ slightly for embryos fixed in methanol and embryos fixed in formaldehyde (Methods 1 and 2, below). If a new probe or antibody is being used, a gentle rehydration method is recommended (Method 3, below).

PROTOCOL 9.4

# Rehydration of Embryos

### Materials

### *Supplies and Equipment*

Microcentrifuge tube (1.5 ml)
Pasteur pipette

### *Solutions and Reagents*

**Methanol**
PBTA Solution (for preparation, see p. 148)
1x PBS (for preparation of 10x PBS, see p. 148)

**CAUTION: methanol** (see Appendix 4)

## Method 1

### Rehydration of Embryos Fixed in Methanol

1. Transfer the embryos, in methanol, to a 1.5-ml microcentrifuge tube. Remove as much of the methanol as possible.

2. Add 250 μl of methanol to the embryos. Gently add 250 μl of PBTA Solution, taking care not to shake the tube.

   Note: Do not shake the PBTA:methanol mixture, because bubbles that form interfere with the ability of the embryos to sink to the bottom of the tube.

3. Add PBTA Solution until the tube is two-thirds full. Invert the tube gently two to three times and let the embryos sink to the bottom of the tube.

4. Remove the solution, leaving the embryos in the bottom of the tube, and add 500 μl of PBTA Solution.

5. Allow the embryos to rehydrate in the PBTA Solution at room temperature for 15 minutes on a rotator.

## Method 2

### Rehydration of Embryos Fixed in Formaldehyde

1. Transfer the embryos, in methanol, to a 1.5-ml microcentrifuge tube. Remove as much of the methanol as possible.

2. Add 500 μl of PBTA Solution. Allow embryos to rehydrate in this solution at room temperature for 15 minutes on a rotator.

## Method 3

### Alternative Rehydration Procedure

1. Transfer the embryos, in methanol, to a 1.5-ml microcentrifuge tube. Remove as much of the methanol as possible.

2. Perform 5-minute washes in the following series of methanol:PBS solutions: 80%:20%, 60%:40%, and 20%:80%.

3. Immerse the embryos in 100% PBS, and then 100% PBTA.

4. Incubate the embryos in PBTA at room temperature for 30 minutes on a rotator.

## EMBRYO STAINING

Embryos can be stained with specific fluorescent probes or antibodies through either direct or indirect immunofluorescence. Extensive washing with frequent solution changes reduces background and is key to high-quality preparations. Consequently, any extended break in the preparation (such as leaving overnight) should be made during a washing step. Washes more than 2 hours should be performed at 4°C.

## Nuclear Stains

Several effective probes exist for visualizing DNA (see Table 9.1). Embryo staining procedures are described below for the following:

- Staining with propidium iodide (see Protocol 9.5)
- Staining with 4,6-diamidino-2-phenylindole (DAPI; see Protocol 9.6)

Other fluorescent DNA probes are processed as described for DAPI. Table 9.1 provides a list of commonly used probes and compatible fixations.

## Antibody Staining

Unlabeled primary antibodies require the use of a labeled secondary antibody. As mentioned above, extensive rinsing and washing is the key to good preparations. A procedure for staining using primary unlabeled antibodies is given in Protocol 9.7. Variations of this procedure are discussed in Protocol 9.7 Notes.

PROTOCOL 9.5

# Propidium Iodide Staining

Propidium iodide is a nucleic acid stain that is added to the mounting medium (for preparation, see p. 156; see also Figure 9.4A–C). Before staining with propidium iodide, the embryos must be treated with RNase to remove the RNA.

### Materials

#### *Solutions and Reagents*

RNase (10 mg/ml)
PBTA Solution (for preparation, see p. 148)

*PBS-Azide*
> 1x PBS (for preparation of 10x PBS, see p. 148)
> 0.02% **sodium azide**

CAUTION: **sodium azide** (see Appendix 4)

### Method

1. Allow the rehydrated embryos to settle to the bottom of the tube and remove as much of the PBTA Solution as possible.

2. Add enough 10 mg/ml RNase to cover the embryos and incubate at 37ºC for 2 hours.

3. Remove the RNase (store this RNase at 4ºC; it can be reused many times). Wash the embryos several times in PBTA.

4. Wash embryos with PBS-Azide and mount as described in Protocol 9.8. Alternatively, treat embryos with another probe or antibody.

PROTOCOL 9.6

# DAPI Staining

DAPI is a commonly used DNA-binding dye. Because it is specific for double-stranded DNA, no prior RNase treatment is required. When working with DAPI, wear gloves because it is a potential mutagen.

## Materials

### Solutions and Reagents

PBTA Solution (for preparation, see p. 148)

### 100× DAPI

Dissolve 10 mg of **DAPI** in 100 ml of **methanol**. Store in the dark at –20°C.

**CAUTION: DAPI, methanol** (see Appendix 4)

## Method

1. Allow the rehydrated embryos to settle to the bottom of the tube and remove as much of the PBTA Solution as possible.

2. Add 495 µl of PBTA and 5 µl of 100× DAPI to the embryos. Incubate on a rotator for 5 minutes. To avoid quenching, the tube must be protected from light during and after this staining step.

3. Remove the DAPI solution and discard it as hazardous waste.

4. DAPI-stained embryos require extensive rinsing. Quickly rinse the embryos three times in PBTA Solution, allowing the embryos to settle between rinses. Wash the embryos in PBTA for 1 hour. Repeat this rinse/wash cycle at least once. It is beneficial to include an overnight wash at 4°C.

PROTOCOL 9.7

# Unlabeled Primary Antibody staining

## Materials

### Supplies and Equipment

Pasteur pipette
Microcentrifuge tubes (0.5 ml; Eppendorf)

### Solutions and Reagents

PBS-Azide (see p. 153)
PBTA Solution (for preparation, see p. 148)
Primary and secondary antibodies

## Method

1. Use a pasteur pipette to transfer the embryos (in PBTA) to a 0.5-ml microcentrifuge tube (Eppendorf). Allow embryos to settle to the bottom of the tube. Remove the PBTA Solution and add primary antibody (diluted in PBTA) to the embryos. (If using a new antibody, try concentrations of 0.1 µg/ml, 1 µg/ml, and 10 µg/ml.) Incubate the embryos in primary antibody at room temperature for 1 hour on a rotator (for best results, incubate the embryos at 4°C overnight).

2. Remove the primary antibody and rinse the embryos three times with PBTA, allowing the embryos to settle between rinses. Wash the embryos for at least 1 hour at room temperature. Longer washes and more rinses usually produce cleaner images. Overnight washes should be performed at 4°C.

> *Note:* The primary antibody usually can be reused several times and should be stored at 4°C.

3. Add fluorescently labeled secondary antibody (usually at 1 μg/ml) and incubate at room temperature for 1 hour. Make sure it is directed against the same species in which the primary antibody was generated.

> *Note:* Once the fluorescent secondary antibody is added, the embryos should be kept out of the light as much as possible.

4. Remove the secondary antibody. Proceed through three rinse/wash cycles as described in step 2, above.

5. At this point, either prepare the embryos for mounting (steps 6–7) or counterstain the embryos for another probe (see Protocol 9.7 Notes, Double Labeling).

6. Rinse the embryos four times in PBS-Azide to remove the detergent (Triton X-100). Proceed to Protocol 9.8.

## PROTOCOL 9.7 NOTES

- *Double Labeling.* When performing a double-immunofluorescent analysis, it is possible to save time by mixing the primary antibodies together in step 1 and mixing the secondary antibodies in step 3 of the unlabeled primary antibody staining procedure (above). However, by performing the staining in sequence, the primary antibodies are not mixed and can be reused many times. To perform the staining in sequence, simply repeat the primary and secondary staining (steps 1–4). Make sure to wash the embryos thoroughly between stains to avoid contamination of the primary and secondary antibodies. The following rules apply for double labeling:

  1. When examining whether two structures colocalize, the clearest results are obtained when the separate components are stained with fluorescein and Cy5 (Figure 9.4C). Bleed-through between these channels is avoided because their emission maxima lie far apart (519 nm and 670 nm, respectively).

  2. When double staining with two primary antibodies of the same species, one of the antibodies must be directly labeled (see below).

- *Directly Labeled–Antibody Staining.* Coupling a fluorophore directly to a primary antibody is a simple procedure (Francis-Lang et al. 1999) and produces excellent images. Directly labeled antibodies produced in this way are often used when staining embryos with two antibodies generated in the same species. When counterstaining with an unlabeled primary antibody, take care to stain with the directly labeled antibody last. This ensures that the secondary used to visualize the unlabeled primary does not bind to the directly labeled antibody. To stain embryos with a directly labeled antibody, perform the following steps:

  1. Use a pasteur pipette to transfer the embryos (in PBTA) to a 0.5-ml microcentrifuge tube (Eppendorf) and let the embryos settle at the bottom of the tube.

2. Remove as much PBTA as possible and add 300–500 μl of the labeled antibody (diluted in PBTA) to the embryos.

3. Incubate the embryos in the labeled antibody at room temperature for 1 hour on a rotator in the dark.

- *Using Dyes and Fluorescently Labeled Molecules for Multichannel Labeling.* Dyes, fluorescently labeled molecules, and antibodies can be used together to generate effective double- and triple-labeled images (Figure 9.4B,C). When performing a triple-label stain, take care to remove as much bleed-through from neighboring channels as possible. If two components under investigation are closely positioned and need to be resolved, be sure to stain them with fluorophores with very different emission maxima (e.g., fluorescein and Cy5, see Figure 9.4C).

PROTOCOL 9.8

# Mounting and Storage of Embryos

## Materials

### Supplies and Equipment

Coverslips (22 x 22 mm or 22 x 30 mm)
Glass slides
Nail polish

### Solutions and Reagents

#### Glycerol-based Mounting Medium

For the stock, prepare 10 mg/ml 1,4-**phenylenediamine** (Aldrich P2,396-2) in 10x PBS (see p. 148). Combine 10 ml of the stock with 90 ml of glycerol. Store in 1-ml aliquots at –20°C.

#### Mounting Medium with Propidium Iodide

Prepare mounting medium as described above. Add **propidium iodide** to a final concentration of 1 μg/ml. Store in the dark at –20°C.

CAUTION: **phenylenediamine, propidium iodide** (see Appendix 4)

## Method

Before proceeding with mounting the embryos, be sure that steps 6 and 7 of the procedure for unlabeled primary antibodies (see Protocol 9.7) have been completed.

1. Remove as much of the PBS-Azide solution as possible and add 40 μl of glycerol-based mounting medium to the embryos.

    *Note:* The mounting medium should be kept on ice, and the Mounting Medium with Propidium Iodide should be kept in the dark. Mounting medium with or without propidium iodide can be used depending on the choice of stain.

2. Transfer the embryos (in mounting medium) onto a glass microscope slide using a P-200 Pipetman with the yellow tip cut at an angle to allow pipetting of the viscous solution. Place a coverslip over the embryos and seal with nail polish.

*Note:* It is important that the correct amount of mounting solution (containing embryos) is placed on the slide. For example, a 22 x 22-mm coverslip requires 40 µl of mounting media. Less than this volume results in bubbles and more than this volume results in a floating coverslip that cannot be sealed with nail polish.

3. Store the slides flat at –20ºC in the dark.

*Note:* When stored at –20ºC, the embryos are stable in this medium for many days with little loss of image quality.

# REFERENCES

Foe V.E. and Alberts B.M. 1983. Studies of nuclear and cytoplasmic behavior during the five mitotic cycles that precede gastrulation in *Drosophila* embryogenesis. *J. Cell Sci.* **61:** 31–70.

Foe V.E., Odell G., and Edgar B.A. 1993. Mitosis and morphogenesis in the *Drosophila* embryo: Point and counterpoint. In *The development of* Drosophila melanogaster (ed. M. Bate and A. Martinez Arias), pp. 149–300. Cold Spring Harbor Laboratory Press, Cold Spring Harbor, New York.

Francis-Lang H., Minden J., Sullivan W., and Oegema K. 1999. Live confocal analysis with fluorescently labeled proteins. *Methods Mol. Biol.* **122:** 223–239.

Garcia-Bellido A. and Robbins L. 1983. Viability of female germ-line cells homozygous for zygotic lethals in *Drosophila melanogaster*. *Genetics* **103:** 235–247.

Golic K.G. 1991. Site-specific recombination between homologous chromosomes in *Drosophila*. *Science* **252:** 958–961.

Karr T.L. and Alberts B.M. 1986. Organization of the cytoskeleton in early *Drosophila* embryos. *J. Cell Biol.* **102:** 1494–509.

Kellogg D.R., Mitchison T.J., and Alberts B.M. 1988. Behavior of microtubules and actin filaments in living *Drosophila* embryos. *Development* **103:** 675–686.

Mitchison T.J. and Sedat J. 1983. Localization of antigenic determinants in whole *Drosophila* embryos. *Dev. Biol.* **99:** 261–264.

Perrimon N. and Mahowald A.P. 1986. The maternal role of zygotic lethals during early embryogenesis in *Drosophila*. In *Gametogenesis and the early embryo* (ed. J.G. Gall), pp. 221–237. Alan Liss, New York.

Perrimon N., Engstrom L., and Mahowald A.P. 1984. The effects of zygotic lethal mutations on female germ-line functions in *Drosophila*. *Dev. Biol.* **105:** 404–414.

———. 1989. Zygotic lethals with specific maternal effect phenotypes in *Drosophila melanogaster*. I. Loci on the X chromosome. *Genetics* **121:** 333–352.

Rabinowitz M. 1941. Studies on the cytology and early embryology of the egg of *Drosophila melanogaster*. *J. Morphol.* **69:** 1–49.

Sambrook J., Fritsch E.F., and Maniatis T. 1989. Preparation of reagents and buffers used in molecular cloning. In *Molecular cloning: A laboratory manual*, pp. B.1–B.28. Cold Spring Harbor Laboratory Press, Cold Spring Harbor, New York.

Sullivan W., Fogarty P., and Theurkauf W. 1993. Mutations affecting the cytoskeletal organization of syncytial *Drosophila* embryos. *Development* **118:** 1245–1254.

St Johnston D. and Nüsslein-Volhard C. 1992. The origin of pattern and polarity in the *Drosophila* embryo. *Cell* **68:** 201–219.

Theurkauf W.E. 1992. Behavior of structurally divergent alpha-tubulin isotypes during *Drosophila* embryogenesis: Evidence for post-translational regulation of isotype abundance. *Dev. Biol.* **154:** 205–217.

Warn R.M., Magrath R., and Webb S. 1984. Distribution of F-actin during cleavage of the *Drosophila* syncytial blastoderm. *J. Cell Biol.* **98:** 156–162.

Zalokar M. and Erk I. 1977. Phase-partition fixation and staining of *Drosophila* eggs. *Stain Technol.* **52:** 89–95.

# CONTENTS

# 10

# Imaginal Discs

**Seth S. Blair**
*Department of Zoology*
*University of Wisconsin, Madison*
*Madison, Wisconsin 53706*

THE IMAGINAL DISCS ARE SET ASIDE DURING EMBRYONIC DEVELOPMENT and do not participate appreciably in larval life. During metamorphosis, however, the larval epidermis is
converted into the pupal case, and the imaginal discs (along with the histoblast nests)
form the outer covering of the developing adult. Many different approaches have been
used to analyze the roles of specific gene products in the patterning of imaginal tissues.
This chapter describes methods for generating genetic mosaics, by which genetic changes
can be limited to small groups of imaginal cells, and presents immunohistological techniques for monitoring gene expression.

Imaginal disc primordia appear in embryos as clusters of 20–40 cells, first identifiable
by the expression of various marker genes, that invaginate from the embryonic epithelium. Each disc is a single-layered epithelial sheet or sac that stays connected to the embryonic and larval epithelium by a thin stalk. Eventually, each sac flattens and the two sides
of the sac take on different characteristics, forming on one side the thicker, highly folded disc epithelium and on the other side, the thinner, unfolded peripodial membrane.
Most of the ectodermal adult structures are derived from the disc epithelium. The discs
undergo extensive proliferation during the three larval instars, and by late in the third
instar, just before metamorphosis, each disc contains tens of thousands of cells. Figure
10.1 schematically shows the locations and shapes of the imaginal discs in the third instar
larva.

Just prior to metamorphosis, at approximately 108 hours after egg laying (AEL) at
25ºC, the third-instar larva "wanders" out of the food and up the side of the vial or bottle; the length of this stage is approximately 12 hours, but varies considerably. Toward the
end of this stage, the anterior spiracles evert and, at approximately 120 hours AEL, the
larva pupariates: The larva stops wandering and the larval epithelium forms the barrel-
shaped body wall of the puparium (the white prepupal stage, WPP). After an hour or so,
the body wall tans to a dark brown color. For the first 5–6 hours after pupariation (AP),
the fly is still a prepupa, and the imaginal cells have yet to secrete the pupal cuticle. During
the prepupal stages, the imaginal discs of the appendages (leg, wing, haltere, etc.) go
through the dramatic morphological changes of eversion. For instance, the leg discs
unfold and lengthen into a tube, and the wing discs lengthen and flatten to form the wing

**159**

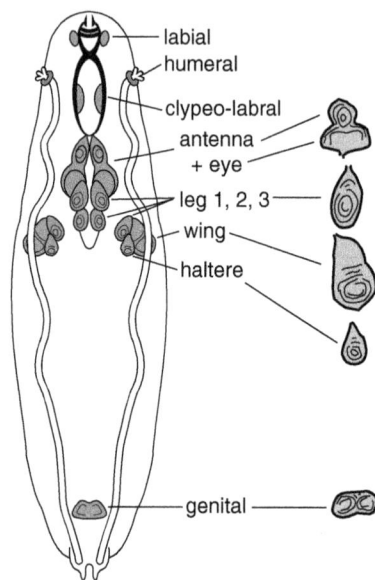

**Figure 10.1.** Schematic representation of late third-instar larvae, showing the approximate locations and shapes of the imaginal discs. (Reprinted, with permission, from Bodenstein 1950.)

blade. During eversion, the disc epithelia break through the peripodial membrane. At approximately 5–6 hours AP, the disc epithelia begin to secrete the pupal cuticle. Later, at about 16 hours AP, the pupal cuticle lifts off the surface of the disc epithelium (apolysis) in preparation for secretion of the adult cuticle, which begins at approximately 40 hours AP. Cell division, epithelial patterning, and morphological changes occur at all stages up to the time of secretion of the adult cuticle and differ from disc to disc. Space limits prevent a detailed description of those changes here, but for several useful reviews, see Bodenstein (1950), Cohen (1993), Fristrom and Fristrom (1993), and Blair (1995, 1999).

Each stage described above presents different advantages and disadvantages for observations. A great deal of patterning takes place during the early stages of disc development, from embryogenesis through mid-third instar. However, at early stages, the imaginal discs are small and difficult to identify. At wandering third instar, discs are easy to isolate. However, the extensive folding of the disc epithelia at wandering third instar obscures some events, and several patterning processes are not yet complete. At prepupal and pupal stages, eversion unfolds some structures, and by 24–40 hours AP, many (although not all) developmental processes are complete. However, once secreted, the pupal and adult cuticle can block the penetration of most probes (although X-gal staining for β-galactosidase activity can penetrate cuticle). Therefore, this chapter concentrates on three easily isolated stages, the wandering third-instar disc, prepupal stage (0–6 hours AP), and pupal stage from 24–40 hours AP, when the pupal cuticle can be removed.

The techniques described below are our standard laboratory protocols for generating mosaic clones and visualizing those clones in wandering larval stage imaginal discs and pupal tissues. The immunohistological techniques are also generally applicable to the description of protein expression. To visualize mitotic recombinant clones in adults, see Lawrence et al. (1986) for appropriate genetic markers.

## GENETIC MOSAICS

### Techniques for Generating Clones

Means exist for both removing and adding back gene funtion to imaginal cells (for review, see Blair 1995, 1999). Particularly useful has been the generation of genetic mosaics, using either mitotic recombination or the FLPout technique (Figure 10.2). In mitotic recombination, homozygotic cells are generated in heterozygotic flies by inducing recombination between homologous chromosomes. This is done either by irradiating heterozygotic larvae with X-rays or γ-rays or by using heat-shock-induced expression of FLPase to induce recombination between FLPase recombination targets (FRTs) that were inserted into

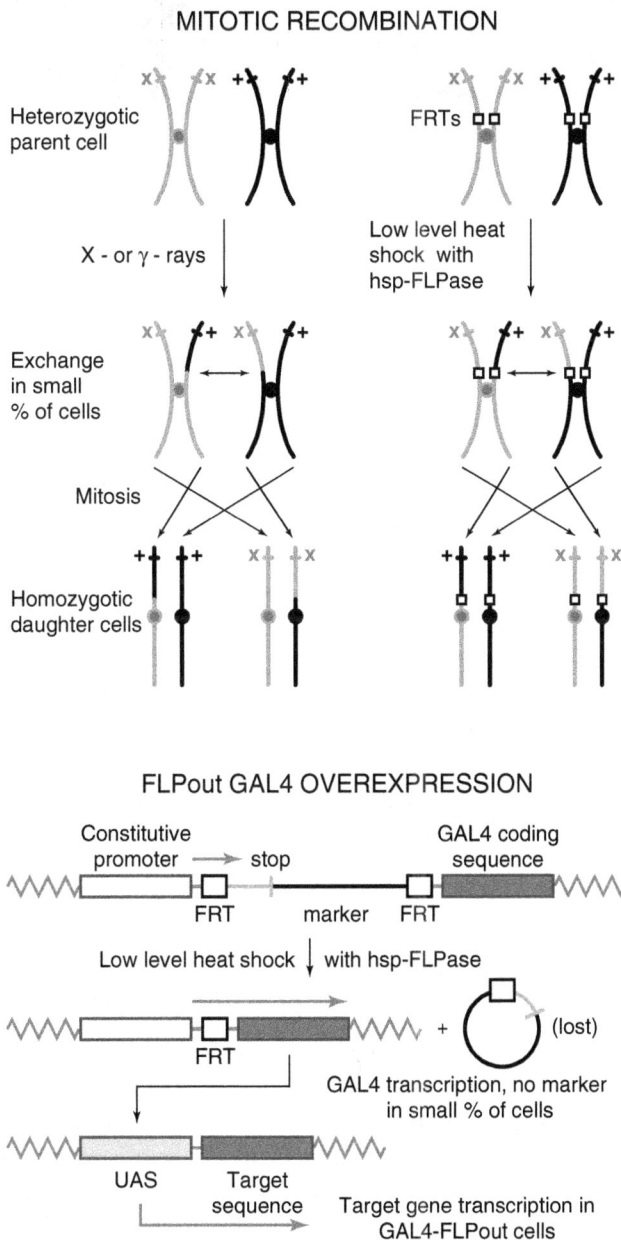

MITOTIC RECOMBINATION

FLPout GAL4 OVEREXPRESSION

**Figure 10.2.** Schematic representation of mitotic recombination and GAL4 FLPouts. See text for details. (Reprinted, with permission, from Blair 1995.)

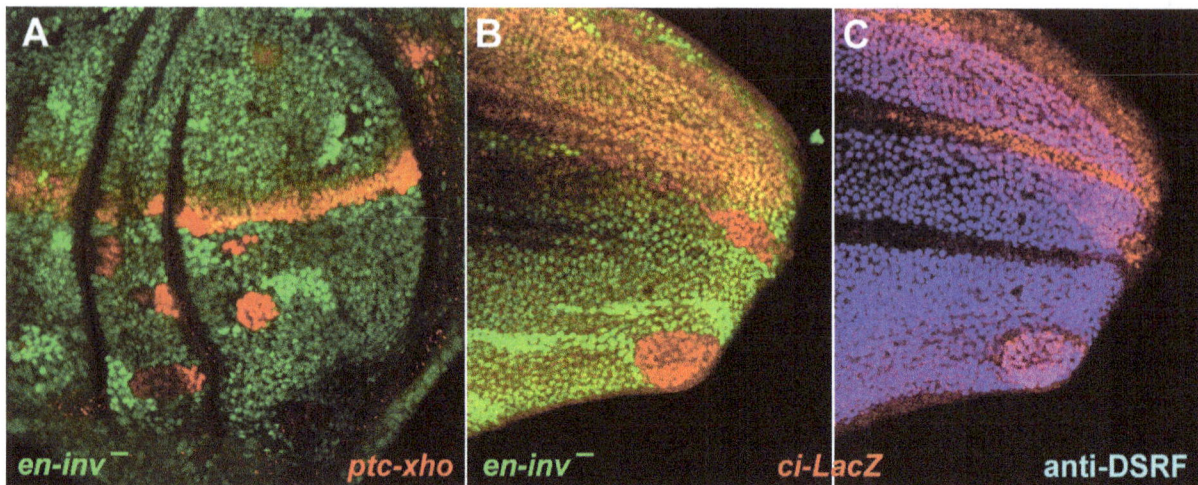

**Figure 10.3.** Mitotic recombinant clones lacking the posterior "selector" transcription factors *engrailed* and *invected* (*en-inv⁻*) in late third-instar wing discs (*A*) or approximately 30-hr AP pupal wings (*B,C*). Clones are identified by the absence of the πM Myc epitope marker construct, as followed using anti-Myc (*green*). The disc in *A* also contains the *patched-xho LacZ* (*ptc-xho*) promoter construct, whose expression was followed using anti-β-galactosidase (*red*). The *ptc-xho* construct is normally expressed in a stripe just anterior to the anterior-posterior compartment boundary, but is ectopically expressed in most posterior *en-inv⁻* clones. The pupal wing in *B* and *C* contains the *cubitus interruptus-LacZ* enhancer trap (*ci-LacZ*; *red*) that is normally repressed in the posterior, but is ectopically expressed in a posterior *en-inv⁻* clone. The wing has also been stained with anti-DSRF (*blue*), which is not expressed in vein nuclei (*C*). Note the ectopic unstained region surrounding the *en-inv⁻* clone, indicating the induction of ectopic vein-like development in neighboring wild-type cells.

selected chromosome arms (Xu and Rubin 1993). In the FLPout technique, heat-shock-induced expression of FLPase joins a ubiquitous promoter to a selected coding sequence by removing blocking DNA flanked by FRTs (Struhl and Basler 1993). A useful variant on the FLPout technique couples it with the upstream activation sequence (UAS)-GAL4 technique of Brand and Perrimon (1993). A GAL4 FLPout clone expresses GAL4, which in turn drives the expression of any gene coupled to the UAS promoter (de Celis and Bray 1997; Pignoni and Zipursky 1997).

With appropriate levels of irradiation or heat shock, mitotic recombination or FLPouts can be generated in a small percentage of disc cells at any stage; each cell then divides and normally forms a coherent patch or "clone" of genetically altered tissue (see Protocol 10.1). However, the location of such clones is random; thus, clones must be identified using genetic or histological markers (see Figure 10.3)

## Identifying Sexes and Genotypes

As most crosses for generating mosaic clones generate several different genotypes, it is often more efficient to screen out the inappropriate larvae before dissection and staining.

- *Sexing Larvae.* From mid-third instar to WPP stages, the genital discs are visible through the body wall as clear spaces on each side of the abdomen, approximately one quarter of the way from the posterior end. Place larvae in H₂O in a dish on a dark surface, illuminate with light from the side, and examine under a dissecting scope. The male genital discs are large and obvious, but the female genital discs are small and often difficult to see (see Figure 10.4).
- *Balancers.* Some balancer chromosomes contain useful larval markers. The dominant mutation *Tubby* (*Tb*) is present on some versions of the third-chromosome balancers *TM6* and *TM6B* and on the second-third–double balancer *SM5a-TM6B*.

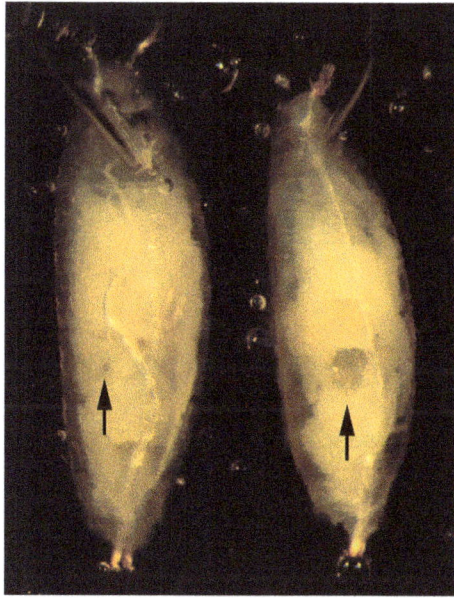

**Figure 10.4.** Distinguishing sexes at the white prepupal (wpp) stage. Two wpp have been pinned on their sides and lit from the side; anterior is up and dorsal right. (*Left*) A female, as identified by the very small genital disc (*arrow*); (*right*) a male, as identified by the large genital disc (*arrow*). Late third-instar larvae can be sexed in the same manner.

*Tb* larvae and pupae are short and squat, and shorten even further if disturbed. Recently constructed versions of the *FM7*, *CyO*, and *TM3, Ser* balancers contain a ubiquitously expressed green fluorescent protein (GFP) construct, easily identified at low power under a fluorescence microscope using the filter set for fluorescein or using a fluorescent dissecting microscope.

- *Other Markers. yellow* (1-0.0) larvae can be identified by placing the larvae on a light background and examining the dark mouth parts through a dissecting microscope; wild-type mouth parts are jet black, whereas *yellow* mouth parts are more tan. *white* (1-1.5) or the combination of *cinnabar* (2-57.5) and *brown* (2-104.5) gives colorless Malpighian tubules, whereas *red Malpighian tubules* (3-53.6) colors them red. *Black cells* (2-80.6) induce small dark specks on the surface of the larvae, a phenotype easily scored under a dissecting microscope using a light background.

PROTOCOL 10.1*

## Generating Clones

### Materials

### *Media*

#### *G Selection Food (Xu and Rubin 1993)*

Take vials (~10 ml) containing standard *Drosophila* food (see Appendix 3) and poke holes in the food. Add approximately 0.25 ml of 25 mg/ml G418 (a.k.a. Geneticin, Sigma G 5013; store at –20ºC) to each vial. Air-dry food. We store G food vials in a "cool room" at approximately 10ºC.

*For all protocols, H$_2$O indicates glass distilled and deionized.

### Stocks

For stock requirements, see Method below.

### Supplies and Equipment

Standard fly-rearing tools

Water bath at 37°C, for expressing genes linked to a heat shock promoter (*hsp*; e.g., *hspFLPase*, or the π*M*, *NM,* or *hsp-CD2* clonal markers)

Plastic petri dishes (35 mm); useful for rearing and heat-shocking larvae and pupae

Fine paintbrush for transferring larvae

Electrical tape for sealing petri dishes

(*For irradiation*) Source of **X-rays** or γ**-rays**

**CAUTION:** γ**-rays, X-rays** (see Appendix 4)

### Method

### FLP-FRT Mitotic Recombination

1. If stocks are not readily available, use meiotic recombination to generate a stock containing the selected mutation coupled to a more proximal FRT insertion on the same chromosome arm. Establish several stocks from individual recombinants, and check them to make sure the selected mutation is present.

   > *Note:* We usually use the FRT insertions of Xu and Rubin (1993), which contain the FRT insertion coupled to a neomycin resistance gene that allows survival of larvae on G Selection Food. Stocks exist with autosomal FRTs linked to more distal *white+* or *yellow+* insertions, or with the *X* chromosome *FRT18A* coupled to the more distal mutation *Bar*. These can be used to follow recombination distal to the FRT. Assuming that the mutation is lethal, mate *y w; autosomal mutant/FRT y+* or *y w; autosomal mutant/FRT w+* females to *autosomal balancer* males, or *mutant/FRT18A Bar* females to *X balancer* males. Grow these on G Selection Food, and look for adults that have lost *y+* or *w+* (*y w; FRT autosomal mutant?/balancer* males) or lost *Bar* (*FRT18A X mutant?/balancer* females). Other FRT insertions are also available, containing different markers.

2. Obtain a stock containing the same *FRT* insertion as the selected *FRT mutation* stock uses, linked to a histologically identifiable *marker*, as well as an *hsp-FLPase*. Mate this stock to the selected *FRT mutation* or *FRT mutation/balancer* stock.

   > *Notes:* Several *marker,FRT,FLPase* stocks are available. The histological markers include a ubiquitously expressed *LacZ* enhancer trap on the *X* chromosome (*WG1296*; Blair 1992), heat-shock promoters linked to proteins carrying a Myc epitope (the π*M* nuclear marker or the *NM* membrane-bound marker; Xu and Rubin 1993), a ubiquitously expressed *armadillo* promoter linked to *LacZ* (Vincent and Lawrence 1994), or a heat shock promoter linked to *CD2* (Jiang and Struhl 1995). There are *hsp-FLPase* insertions on *y* and *y w X* chromosomes and on the *MKRS* third-chromosome balancer, in most stocks balanced over *TM6, Tb*.
   >
   > For an autosomal mutation, use female virgin *y* or *yw FLPase*; autosomal *FRT marker* flies so that both male and female larvae in the next generation will contain the FLPase.
   >
   > For sex-linked mutations, use *X marker FRT*; *MKRS, FLPase/TM6, Tb* flies, and ignore the *Tb* larvae in the next generation.

3. To obtain clones generated at a selected stage of larval development, use either of the following methods:

   a. For the most accurate timing, perform the following steps:

      i. Transfer the mated adults (10–20 females, 1–10 males, depending on the fertility of the stocks) to a fresh food bottle for approximately 2 hours to clear the females of any eggs they may have retained.

ii.   Then transfer the adults to a second bottle for a short period (typically 4 hours), remove the adults, and rear the bottle to the selected stage, typically 2 or 3 days AEL.

iii.   Heat-shock the bottle by immersing in a 37°C water bath for 1.5 hours.

iv.   Rear the bottle until wandering third instar to WPP stage, approximately 5 days AEL.

b.   For less accurate timing, perform the following steps:

i.   Transfer adults (5–10 females, 1–10 males, depending on the fertility of the stocks) to a fresh bottle for 3 days.

ii.   Remove the adults and heat shock the bottle by immersing in a 37°C water bath for 1.5 hours.

iii.   Pick wandering larvae or WPP at selected periods after the heat shock, typically 2 or 3 days.

> *Note:* Waiting for shorter times after the heat shock selects for more frequent and smaller clones, whereas waiting longer times selects for less frequent and larger clones.

4.   Remove the wandering larvae or WPP from the side of the bottle using a fine, moistened paintbrush. Care must be taken in crowded bottles, as immature larvae will wander out of overcrowded food.

> *Note:* If using a *MRKS FLPase/TM6, Tb* stock, select only the non-*Tb* larvae. If the mutation is on the *X* chromosome, select female larvae as outlined on p. 162 (Identifying Sexes and Genotypes).

5.   (*Optional*) Pupal stages are most precisely staged by timing from the WPP stage; age these in a petri dish with a moistened tissue. There is also a useful time window from 24 to 40 hours AP, the period between the apolysis of the pupal cuticle and the secretion of the adult cuticle. It is not always critical to precisely time pupae within this time window, and it is easier to get large quantities of this stage by picking wandering third-instar larvae instead of WPP. Place the wandering larvae in a petri dish with a moistened tissue, and age at 25°C for 1.5 days or at 21°C for 2 days. Most of the resultant pupae will be at 24–40 hours AP.

6.   (*Optional*) If using the *πM* or *NM* marker stocks, heat shock larvae, WPP, or pupae just before dissection to express the Myc epitope. Place larvae, WPP, or pupae in a petri dish with a moist tissue, wrap the edges of the petri dish twice with electrical tape, and float the dish in the 37°C water bath. In our hands, late third-instar to 6-hour AP tissues need 1.5 hours of heat shock, whereas 24–40-hour AP tissues require 3 hours.

> *Note:* The second heat shock is needed because *hsp-marker* expression induced by the first heat shock (in step 3a[iii] or step 3b[ii]) fades fairly quickly. Although this second heat shock will result in expression of both the *marker* genes and the *hsp-FLPase*, the discs or wings will be fixed before the FLPase has time to induce a second round of mitotic recombination.

### *Irradiation-induced Mitotic Recombination*

Irradiation-induced mitotic recombination is useful if no *FRT mutant* stock is available and only a low clone rate is needed. It is also necessary for those mutations that are proximal to any of the available FRT insertions. Clone rates are higher for the *X* chromosome than for the autosomes.

1.   Mate the *mutation* stock to a *marker* stock (see step 2, above).

> *Note:* We use the same *FRT marker* chromosomes as used above for FRT-FLP recombination.

2. Perform steps 3–6 as described for FLP-FRT mitotic recombination, except substitute irradiation of vials or bottles for the first heat shock in step 3.

   *Note:* We use 1–4 Krad of γ-rays from a cesium source irradiator; higher doses induce more tissue damage and produce poorly viable adults.

### GAL4 FLPout Clones

1. Mate virgin female *Actin>CD2>GAL4; MKRS, FLPase/TM6, Tb* flies to *UAS-gene* males.

   *Note:* We use the *X*-chromosome GAL4 FLPout construct of Pignoni and Zipursky (1997).

2. Generate clones as outlined above for FLP-FRT recombination (steps 3–5). Select only non-*Tb* larvae or WPP. The FLPout clones are independently marked by the absence of *CD2* expression; no heat shock is required to express *CD2*.

---

PROTOCOL 10.2

## Dissection

We perform dissections in a simple Ringer's solution; other commonly used salines are PBS and B&E. Some researchers dissect directly in fixative (see Protocol 10.3). For some protocols, such as in situ hybridization, where there are many washes and hybridization steps in covered tubes, it is helpful to leave the discs attached to the body walls so that they are not lost.

### Materials

### *Supplies and Equipment*

Dumont #5 or #55 forceps with "Biologie" tips (Fine Science Tools); superfine tips are, however, easily bent or broken

Petri dishes. Perform dissections in petri dishes; if pinning is necessary (as for prepupae or pupae), use plastic petri dishes in which the bottom has been lined with Dow Corning Sylgard 184 elastomer (Fisher). A dark background is preferable for visualizing the light larval structures, so either use clear Sylgard on a black surface or darken the Sylgard; in our experience, adding copy machine toner to clear Sylgard is superior to the commercially available black Sylgard.

Stainless-steel minutien pins (Fine Science Tools 26002-10)

Glass pasteur pipettes. After dissection, we move discs or body walls with glass pasteur pipettes. To prevent sticking, coat with a commercial siliconizing compound, with fat body from larval carcasses (pipette these up and down a few times), or with spit.

Dissecting microscope

### *Solutions and Reagents*

> **_Drosophila Ringers (Eugene et al. 1979)_**
> 130 mM NaCl
> 5 mM **KCl**
> 1.5 mM **MgCl₂**

*B&E Saline (Ephrussi and Beadle 1936)*

NaCl 7.5 g/liter
KCl 0.35 g/liter
**CaCl$_2$** 0.21 g/liter (0.267 g CaCl$_2$·H$_2$O, 0.278 g CaCl$_2$·2H$_2$O)
Tricine 1.793 g/liter
Adjust pH to 6.9.

*PBS-Triton*

1x PBS (see Appendix 3)
0.3% Triton X-100 (Sigma)

**CAUTION: CaCl$_2$, KCl, MgCl$_2$** (see Appendix 4)

## Method

### Wandering Third-instar Larvae

For dissecting the discs free from the body wall, tear the larva in half and invert the body wall. For dissecting the cluster of dorsal discs (wing, haltere, and second leg disc), pinch the larvae about a third of the way posterior and rip the dorsal anterior side of the larvae a little further anterior. The dorsal cluster is located about a quarter of the way posterior in a dorsolateral position, and often pops out of the resultant tear; it must then be teased loose from the trachea. The other leg discs are found on the ventral side of the ventral brain. The eye-antennal discs are stretched between the surface of the optic lobes of the brain and mouth parts. The discs appear slightly more transparent than the white-yellow fat body and can be easily identified with a little practice.

### Prepupal Stages (0–6 Hours AP)

1. In a Sylgard-lined petri dish containing saline, pin the prepupa down dorsal-side up.

2. The body wall at these stages starts to harden, and it is easiest to peel it away a bit at a time. Beginning with the flat dorsoanterior operculum, tear away sections of the body wall until the discs are exposed. Then tease the structures loose.

   *Note:* Remember that the shapes and positions of the imaginal tissues change during these stages. If tissue is shiny, then the pupal cuticle has been secreted and antibody staining will be difficult.

### Later Pupal Stages (16–40 Hours AP)

Although secreted, at these stages, the pupal cuticle can be removed because it is no longer directly attached to the imaginal epithelium. In unfixed tissue, the epithelia are fragile and sticky. However, if the tissue is strongly fixed, the imaginal epithelium becomes quite tough and the dissection is straightforward. The only disadvantage of this is that some antigens do not survive the long fixation.

1. Take pupae from this stage, place in saline in a petri dish, and pin dorsal-side up through the abdomen.

2. Peel away the cuticle, starting at the anterior. Stop when the anterior half of the pupa is exposed, transfer to a well, and fix at 4°C overnight (see Protocol 10.3).

3. The next day, place the pupa in a petri dish containing PBS-Triton wash solution, and pull completely loose from the pupal case. Tease the distal ends of the appendages off the ventral and lateral surfaces of the pupa. Then find some region of the appendage (usually the base) where there is an obvious space between the appendage and the pupal cuticle, and tear at the cuticle. If it is easier, first remove the appendage from the pupa. With minimal tearing, it is usually possible to pull the appendage from the cuticle, much like pulling lobster meat from a claw.

> *Note:* Care must be taken not to spill too much fat body, as this will adhere to the appendages and obscure them. Performing the dissection in larger volumes can help, and to some extent, the fat body can be washed off dissected appendages. As noted above, for some protocols such as in situ hybridization, it is useful to leave the appendage attached to the body wall so that it is not easily lost. However, if the abdomen still contains fat body it may continue to spill throughout the protocol. If this is a problem, clean out the pupal abdomen before the postfix dissection by inserting a pipette into the abdomen and blowing.

## Protocol 10.3

## Fixation

In this fixation procedure, we use the PIPES-EGTA-NP-40 fixative of Brower (1986) with 2% EM-grade formaldehyde. This is the only fixative that we have been able to get to work for the Myc epitope markers. Other standard fixatives include PIPES-EGTA (lacking the NP-40) with 2% formaldehyde and PBS with 4% formaldehyde.

### Materials

#### Supplies and Equipment

Glass plates (3- or 9-well). Fixation and antibody staining (pp. 170–173) are carried out in 3- or 9-well glass plates (Corning 7223 and 7220), placed in a large petri dish with a moist tissue to prevent evaporation. All fixations, washes, and incubations are done in the same well. The sloped sides of the wells make it easy to add and remove solutions, and incubation volumes as small as 30 μl can be used.

Large petri dish, with a moist tissue

Micropipettes or pasteur pipettes. Solutions and tissue are transferred with either micropipettes or pasteur pipettes. Tissue can stick to glass pipettes (see p. 166); to remove solutions, we use pasteur pipettes while watching through a dissecting microscope to make sure no tissue is removed. Pasteur pipettes pulled out to a fine tip are also helpful.

#### Solutions and Reagents

**Formaldehyde** (8%), at 4°C (diluted from 16% EM-grade formaldehyde; Polysciences)

> #### Brower Fix Buffer (Brower 1986)
> 0.15 M PIPES (1,4-piperazinediethane sulfonic acid)
> 3 mM **MgSO$_4$**
> 1.5 mM EGTA (ethyleneglycol-bis-(β-aminoethylether)-*N,N,N′,N′*-tetraacetic acid)
> 1.5% NP-40 (Nonidet P-40)
> Adjust pH to 6.9. Store at 4°C.

**CAUTION: formaldehyde, MgSO$_4$** (see Appendix 4)

## Method

1. Mix three parts of cold Brower Fix Buffer with one part of cold 8% formaldehyde just before fixation.

2. Following dissection, transfer the tissue to 3- or 9-well glass plates.

3. Remove as much saline as possible and replace with fixative (~100–200 µl per well).

4. For larval through 6-hour AP stages, fix at 4°C for approximately 2 hours. For 16–40-hour AP pupal stages, fix at 4°C overnight.

   *Note:* After fixation, wash the tissue in PBS or PBS-Triton (see step 1 [Protocol 10.4 or 10.5]).

## IMMUNOSTAINING

Protocol 10.4 describes a triple-label, fluorescent-antibody staining procedure that assumes having three different primary antisera generated in three different species. Biotin and rhodamine-streptavidin are used on the most critical of these antibodies, as they have the best signal-to-noise ratio. Fluorescein is used on the second. Cy5 is used on the third; this is only poorly visible in a standard fluorescence microscope, but is used by the third channel on many laser-scanning confocal microscopes. Obviously, other fluorescent tags, as well as double and single fluorescent stains, are possible. For labeling with a single antiserum, the nickel-intensified ABC-HRP staining procedure (Protocol 10.5) is more sensitive, with a lower signal-to-noise ratio, compared to fluorescent labeling.

## General Considerations

The following general procedures apply to Protocols 10.4 and 10.5:

- Unless otherwise noted, all steps for antibody staining are performed in a 3- or 9-well glass plate in PBS-T at 4°C.
- In all steps described as washes, as much of the prior solution is removed from the well as possible and is replaced with the wash solution. This is repeated immediately one to three times, and then the tissue is incubated in the solution for the stated time. Some primary antisera benefit from longer washes.
- Incubation times for all steps are approximate and precision is not critical.

## Steps to Reduce Background Staining

To reduce background staining, the following notes apply to Protocols 10.4 and 10.5:

- *Preadsorbed Primary Antisera.* For polyclonal antisera (containing several species of antibodies), it is often possible to preadsorb out those antibodies that bind general *Drosophila* antigens rather than the desired signal. The best tissues to use for preadsorption are those not expressing the desired signal antigen. If the antigen is not maternally expressed, use embryos before the initiation of most zygotic transcription (0–3 hours AEL). However, even preadsorption with antigen-expressing tissue sometimes takes out more background antigens than signal. The concentrations used must be determined by trial and error. As a starting point, dilute primary antiserum 1/10 in PBS containing 0.1% sodium azide to prevent infection. Incubate three volumes of this solution with one volume of fixed, 0–3-hour AEL embryos at 4ºC overnight or longer.

• *Blocking Solutions.* In some cases, background staining for specific antisera can be lowered using various blocking reagents, such as bovine serum albumin (BSA) or normal goat serum.

PROTOCOL 10.4

# Triple-label Fluorescent Antibody Staining

## Materials

### Supplies and Equipment

Diamond pencil
Slides
Coverslips
Nail polish
(*Optional*) 00 insect pins for arranging tissues (Fine Science Tools 26001-30)

### Solutions and Reagents

1x PBS (see Appendix 3)
PBS-Triton (see p. 167)

#### Blocking Solution

PBS-Triton
5% normal goat serum or 1% BSA
Use as a freshly prepared solution or store at –20ºC as the solution goes bad quite quickly.

#### Mounting Solution

1x PBS
80% glycerol
4% *n*-propyl gallate
Stir overnight to dissolve *n*-propyl gallate. It is added to inhibit photobleaching.

### Antibodies and Detection Reagents

Primary antibodies. The protocol assumes the use of three different primary antisera generated in three different species (A, B, C); the most critical antiserum is generated in species A. For marking clones, use 1/5 of mouse monoclonal anti-Myc supernatant; 1/1000 of rabbit anti-c-Myc (Santa Cruz Biotechnology sc-789; preadsorbing as described above is essential for this antiserum); 1/200 monoclonal mouse anti-β-galactosidase (Promega Z3781); 1/3000 rabbit anti–β-galactosidase (ICN Biomedicals/Cappell 55976); or 1/400 mouse anti-CD2 (Serotec clone MRC Ox-34).

Secondary antibodies. When choosing secondary antisera, make sure to match the secondary antiserum to the species of the primary antisera. Three antisera are required, against the three primary antibody species:

• Biotin anti-species A
• FITC anti-species B
• Cy5 anti-species C

Cross-reactivity can be a problem, especially between anti-mouse and anti-rat antisera; we use labeled antiserum from Jackson that has been preadsorbed against other species. Appropriate dilutions vary depending on the manufacturer. Secondary antisera from Jackson ImmunoResearch Laboratories and Vector Laboratories work well at 1/200.

Streptavidin-rhodamine (Vector Laboratories)

**CAUTION: *n*-propyl gallate** (see Appendix 4)

## Method

1. After fixation, wash the tissue in PBS or PBS-Triton for 1 hour or longer.

   *Note:* If storage longer than 1 day is anticipated, PBS is used, as the detergent in PBS-Triton appears to degrade the tissue, affecting the staining of some antigens.

2. (*Optional*) Preincubate tissue in Blocking Solution at 4°C for 1–3 hours.

3. Remove the wash or Blocking Solution, replace with a mixture of primary antisera (concentrations depend on the antisera used), and incubate overnight.

4. Wash for 1 hour.

5. Remove the wash solution and incubate tissue in secondary antiserum mixture for 1.5 hours (biotin anti-species A, FITC anti-species B, Cy5 anti-species C).

6. Wash for 1 hour.

7. Remove the wash solution and incubate in tertiary mixture (1/800 streptavidin-rhodamine, FITC anti-species B, Cy5 anti-species C) for 1.5 hours.

8. Wash for 1 hour.

9. Wash into PBS and pipette tissue onto a slide. Arrange the tissue using insect pins.

   *Note:* Remember that imaginal discs have two surfaces; at wandering and WPP stages, the prospective wing blade and distal leg structures are best viewed with the peripodial (convex) side up.

10. Sponge off the remaining solution with a tissue and replace it with a small drop of Mounting Solution.

11. Lower a coverslip onto the tissue, and seal the corners of the coverslip to the slide with nail polish to prevent shearing with immersion lenses.

12. Observe under a standard compound fluorescence microscope. Mark discs for confocal microscopy by scoring the underside of the slide with a diamond pencil. Store slides at –20°C to slow fading.

PROTOCOL 10.5

# Nickel-intensified ABC-HRP Staining

## Materials

### Supplies and Equipment

Slides
Coverslips
Nail polish

### Solutions and Reagents

PBS-Triton (see p. 167)
(*Optional*) Blocking Solution (see p. 170)
Mounting Solution (see p. 170)

### Antibodies and Detection Reagents

Primary antibody (species A)
Secondary antibody (biotin anti-species A)
Vectastain Elite ABC kit (standard kit; Vector Laboratories PK-6100)
Tris-Cl (0.05 M, pH 7.6)
**DAB** (2 mg/ml); store in aliquots at –20ºC. The horseradish peroxidase (HRP) substrate diaminobenzidine (DAB) is a potential carcinogen. Use gloves and a chemical fume hood, and dispose of DAB solutions and washes either in bleach or in a permanganate inactivation solution (see below); the permanganate solution is thought to be safer than bleach (Lunn and Sansone 1992). Consult local safety office for guidelines.
**NiCl$_2$** (8%)
**H$_2$O$_2$** (3%)

> ### Permanganate Inactivation Solution for DAB (Lunn and Sansone 1992)
> Make the following separate solutions:
>   0.2 M **KMnO$_4$**
>   2 M **H$_2$SO$_4$**
> Mix 1:1 before use. Let waste DAB sit in this solution in a chemical fume hood cabinet for 1 day before disposal.

**CAUTION: DAB, H$_2$O$_2$, H$_2$SO$_4$, KMnO$_4$, NiCl$_2$** (see Appendix 4)

### Method

1. Wash for 1 hour or more after fixation.

2. (*Optional*) Preincubate tissue in Blocking Solution at 4ºC for 1–3 hours.

3. Dilute the primary antiserum 10–50X more than the dilution used for fluorescent labeling (see Protocol 10.4), and incubate overnight.

4. Wash for 1 hour.

5. Incubate in biotin anti-species A for 1.5 hours (use the same concentration as in Protocol 10.4).

6. Wash for 1 hour. During this washing step, mix together the A and B solutions from the Vectastain Elite ABC Kit in PBS-Triton (each at a 1/100 dilution), and let sit at room temperature for 0.5–1 hour.

7. Replace the wash solution with the diluted A/B solution and incubate for 1.5 hours.

8. Wash for 1 hour.

9. Replace the wash solution with 0.05 M Tris-Cl (pH 7.6).

10. Prepare the following reaction mixture (~200 µl for 2–4 wells):

| | |
|---|---|
| 0.05 M Tris-Cl (pH 7.6) | 150 µl |
| 2 mg/ml DAB | 50 µl |
| 8% NiCl$_2$ | 1 µl |
| 3% H$_2$O$_2$ | 1 µl |

11. Incubate tissue in reaction mixture for 2–30 minutes. To obtain optimal staining, it is necessary to observe the progress of the staining reaction.

12. Wash tissue in 0.05 M Tris-Cl (pH 7.6).

13. Mount on slides as in steps 10–11 (Protocol 10.4).

# REFERENCES

Blair S.S. 1992. *engrailed* expression in the anterior lineage compartment of the developing wing blade of *Drosophila*. *Development* **115:** 21–34.

———. 1995. Compartments and appendage development in *Drosophila*. *BioEssays* **17:** 299–309.

———. 1999. *Drosophila* imaginal disc development: Patterning the adult fly. In *Development—Genetics, epigenetics and environmental regulation* (ed. V.E.A. Russo et al.), pp. 347–370. Springer, Heidelberg.

Bodenstein D. 1950. The postembryonic development of *Drosophila*. In *The biology of* Drosophila (ed. M. Demerec), pp. 275–367. Wiley, New York.

Brand A.H. and Perrimon N. 1993. Targeted gene expression as a means of altering cell fates and generating dominant phenotypes. *Development* **118:** 401–415.

Brower D.L. 1986. *engrailed* gene expression in *Drosophila* imaginal discs. *EMBO J.* **5:** 2649–2656.

Cohen S.M. 1993. Imaginal disc development. In *The development of* Drosophila melanogaster (ed. M. Bate and A. Martinez Arias), pp. 747–841. Cold Spring Harbor Laboratory Press, Cold Spring Harbor, New York.

de Celis J.F. and Bray S. 1997. Feed-back mechanisms affecting Notch activation at the dorsoventral boundary in the *Drosophila* wing. *Development* **124:** 3241–3251.

Ephrussi B. and Beadle G.W. 1936. A technique of transplantation for *Drosophila*. *Am. Nat.* **70:** 218–225.

Eugene O.M., Yund M.A., and Fristrom J.W. 1979. Preparative isolation and short-term organ culture of imaginal discs of *Drosophila melanogaster*. *TCA Man.* **5:** 1055–1062.

Fristrom D.K. and Fristrom J.W. 1993. The metamorphic development of the adult epidermis. In *The development of* Drosophila melanogaster (ed. M. Bate and A. Martinez Arias), pp. 843–897. Cold Spring Harbor Laboratory Press, Cold Spring Harbor, New York.

Jiang J. and Struhl G. 1995. Protein kinase A and hedgehog signaling in *Drosophila* limb development. *Cell* **80:** 563–572.

Lawrence P.A., Johnston P., and Morata G. 1986. Methods of marking cells. In Drosophila: *A practical approach* (ed. D.B. Roberts), pp. 229–242. IRL Press, Oxford.

Lunn G. and Sansone E.B. 1992. Safe disposal of diaminobenzidine. *Trends Genet.* **8:** 7.

Pignoni F. and Zipursky S.L. 1997. Induction of *Drosophila* eye development by *decapentaplegic*. *Development* **124:** 271–278.

Struhl G. and Basler K. 1993. Organizing activity of Wingless protein in *Drosophila*. *Cell* **72:** 527–540.

Vincent J.P. and Lawrence P.A. 1994 *Drosophila wingless* sustains engrailed expression only in adjoining cells: Evidence from mosaic embryos. *Cell* **77:** 909–915.

Xu T. and Rubin G.M. 1993. Analysis of genetic mosaics in developing and adult *Drosophila* tissues. *Development* **117:** 1223–1237.

# CONTENTS

# 11

# The Neuromuscular Junction

**Hugo J. Bellen**
*Howard Hughes Medical Institute*
*Department of Molecular and Human Genetics*
*Program in Developmental Biology*
*Baylor College of Medicine*
*Houston, Texas 77030*

**Vivian Budnik**
*Department of Biology*
*University of Massachusetts*
*Amherst, Massachusetts 01003*

THE PAST 10 YEARS HAVE SEEN A VAST EXPANSION OF OUR KNOWLEDGE of the molecular mechanisms that underlie synaptogenesis, synaptic vesicle exocytosis and endocytosis, and cross-talk between pre- and postsynaptic cells. The *Drosophila* neuromuscular junction (NMJ), unlike most synapses, is amenable to detailed in vivo analysis of development and function (for reviews, see Broadie and Bate 1993; Goodman and Shatz 1993; Keshishian et al. 1993; Littleton and Bellen 1995; Budnik et al. 1996; Wu and Bellen 1997; Davis and Goodman 1998). At present, the NMJs are the only synapses where sophisticated genetic approaches can be used in combination with electrophysiological manipulations (K. Broadie, this volume), dye injection and uptake experiments (Keshishian et al. 1993; Ramaswami et al. 1994; Kuromi and Kidokoro 1998), immunohistochemistry (Littleton et al. 1993), and electron microscopy (Atwood et al. 1993; Jia et al. 1993; Broadie et al. 1995; Thomas et al. 1997). Hence, many investigators have embarked on studying the molecular machinery underlying developmental and functional aspects of NMJs, and these studies should greatly help our basic understanding of synaptogenesis and synaptic transmission.

## DEVELOPMENT AND STRUCTURE OF THE NMJ

NMJs develop late in embryogenesis after the mature muscle pattern has been established, typically 12.5 hours after egg laying (AEL) at 22–25°C. Between 12.5 and 14.5 hours AEL, many of the initial contacts are established between the filopodia of the axonal growth cones and the muscles. Neuronal processes then explore the muscle surfaces and send out branches which finally retract to their terminal positions by approximately 14.5 hours AEL. During the next 2–3 hours (14.5–17 hours), swellings appear along the processes that have contacted the muscles. These swellings or boutons become more numerous and

**175**

larger, as the nerve terminals take the shape of the mature embryonic NMJs by approximately 17 hours of development (Johansen et al. 1989b; Broadie and Bate 1993; Keshishian et al. 1993; Yoshihara et al. 1997). Much of the NMJ maturation process occurs after secretion of the embryonic cuticular sheath is initiated (~15 hours AEL). Hence, staining of mature embryonic NMJs is difficult to achieve with standard immunocytochemical protocols, but rather requires either a sonication procedure (which was first developed for brine shrimp embryos; Manzanares et al. 1993; Guan et al. 1996) or dissection prior to immunohistochemical staining (see Figure 11.1 and Protocols 11.2 and 11.3).

Although many of the anatomical features seen in third larval instar NMJs are present in mature embryos, there is extensive process elongation and development of new synap-

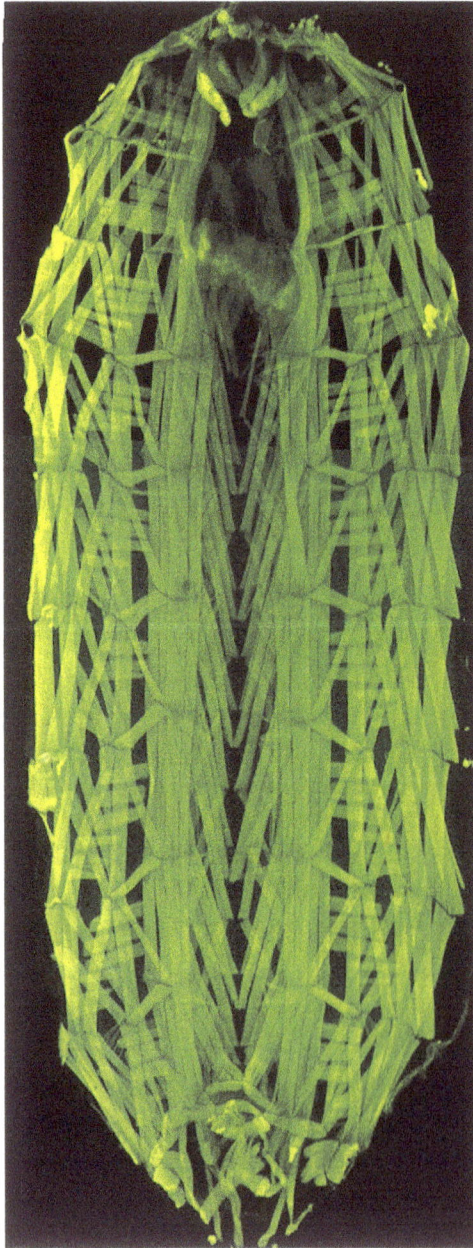

**Figure 11.1.** Body wall muscle fillet of a third instar larva stained with fluorescein isothiocyanate (FITC)-conjugated phalloidin, showing the segmentally repeated pattern of muscles.

tic boutons in the 3-day period after embryogenesis (see Figure 11.2; Gorczyca et al. 1993; see also Figure 6 in Keshishian et al. 1993). Similarly, the postsynaptic apparatus increases in size and complexity throughout the postembryonic period (Guan et al. 1996). During the first-instar larval stage, two classes of motor endings become apparent (Johansen et al. 1989a; Monastirioti et al. 1995). Type I processes are located near the nerve entry point and consist of large boutons which can range from 1 to 5 μm in diameter in third-instar larvae. All type I boutons contain glutamate (Johansen et al. 1989b), but they may also contain proctolin (Keshishian et al. 1993) and/or PACAP-like peptide (Zhong and Pena 1995). These peptides may be contained within dense core vesicles, whereas glutamate is found in small clear vesicles (Atwood et al. 1993; Jia et al. 1993). Type II processes have a more variable arborization pattern, extend over larger surfaces of muscle, and have smaller boutons (<1 μm) when compared to type I processes. They utilize glutamate and octopamine as neurotransmitters (Monastirioti et al. 1995), and at the electron microscopy (EM) level are characterized by the presence of small, clear glutamate-containing vesicles, as well as large, dense elliptical vesicles (Jia et al. 1993). Similar elliptical vesicles are also found in octopaminergic terminals of other arthropod species (Lee and Wyse 1991).

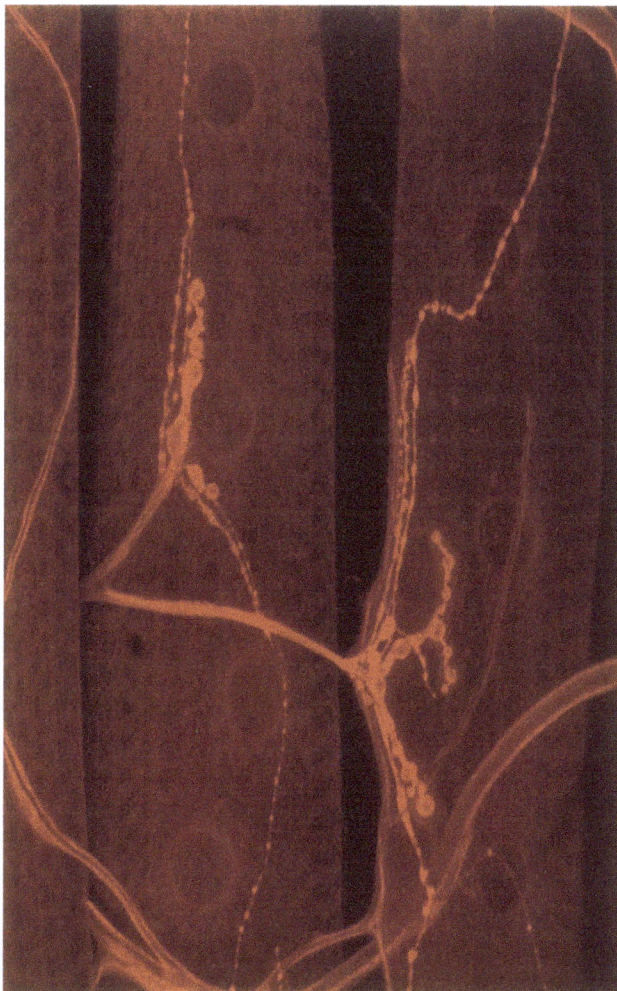

**Figure 11.2.** Neuromuscular junctions at muscles 12 and 13 in the third-instar larva stained with presynaptic antibody anti-horseradish peroxidase (HRP).

In addition to type I and II endings, larval muscles are also innervated by peptidergic endings filled with round, dense core vesicles and are almost, if not completely, devoid of glutamatergic clear vesicles. Examples of these terminals are type III endings that innervate muscle 12 (Jia et al. 1993). Type III endings appear during late first larval instar; they have boutons with an ellipsoid shape and contain an insulin-like peptide (Gorczyca et al. 1993).

## METHODS TO LABEL AND STUDY NMJs

Five methods are available to label synaptic terminals. Here, we describe the protocols and reagents associated with the two most commonly used methods, discussed below: (1) immunocytochemical staining to label synapses of embryos, first-instar larvae, and third-instar larvae and (2) FM1-43 dye uptake at the third-instar larval NMJ.

The following are three other methods: (1) staining embryos or larvae that express a kinesin β-galactosidase or tau–β-galactosidase fusion via an enhancer detector construct; these proteins target the β-galactosidase to the axons and synapses (Giniger et al. 1993; Callahan and Thomas 1994); (2) injecting lipophilic dyes such as Lucifer yellow or Neurobiotin to label cell bodies, axons, and synapses (Halpern et al. 1991; Broadie and Bate 1993); and (3) using confocal or fluorescence microscopy to study NMJs of larvae transformed with constructs that encode green fluorescent protein (GFP) targeted to the synapse (T. Hazelrigg, this volume). The last method is very promising for many applications such as screening for mutants that affect synaptogenesis (G. Pennetta, pers. comm.), following the fate of proteins in vesicles and membranes (M. Estes and M. Ramaswami, pers. comm.), following vesicular transport in vivo, and determining the kinetics of release. The protocol involves crossing flies, which carry an improved version of GFP under the control of an upstream activating sequence (UAS) (Brand and Perrimon 1993) or a GFP fused to a pre- or postsynaptic protein under the control of UAS to flies that express the *GAL4* gene in neurons or muscles. Live preparations are then viewed under a confocal microscope. We anticipate that there will be many applications for this technique, but to our knowledge, the methodology does not work reliably in embryos, even with enhanced GFPs.

### Immunocytochemical Staining of NMJs

Protocol 11.1 describes the immunocytochemical staining of NMJs from different stages:

- *Staining of NMJs in 0–15-hour-old Embryos.* This procedure was developed in the Bellen laboratory from standard protocols such as those described by Patel (1994). For original references for antibodies, see Table 11.1. Because some antigens require that the embryos be treated in specific ways, we refer the reader to the original publications.
- *Whole-mount Preparation for Staining NMJs of 15–22-hour-old Embryos.* This procedure is similar to the method for staining 0–15-hour-old embryos, but it adds a sonication step to allow penetration of the reagents through the cuticle. This protocol is based on a similar protocol described earlier by Patel (1994).
- *Staining of Dissected Embryos, and First-, Second-, and Third-instar Larval NMJs.* Immunocytochemical staining of NMJs from late embryos and from first-, second-, and third-instar larvae requires a dissection procedure. An alternative to pinning dissected body wall muscles to Sylgard (see K. Broadie, this volume) is the use of magnetic chambers. In this method, fillets are not pierced with sharp pins, but are stretched and held in place with appropriately sized and shaped metal rods attached to steel supports that adhere to a magnetic base (see Figure 11.3A). There are sever-

**Table 11.1.** Antibodies That Label NMJs

| Protein | Antigen[a] | Type of antibody | Source or references | Immunostaining embryos/larvae (dilutions)[a,b] | Westerns (dilutions) ECL[a,b] | Remarks |
|---|---|---|---|---|---|---|
| Synaptotagmin (SYT) | DSYT2 aa 134–474 | rabbit polyclonal | Littleton et al. (1993) | E/L 1:1000 +++ | 1:2000 +++ | presynaptic and synaptic vesicles and membrane |
| | DSYT1 aa 1–15 | rabbit polyclonal | Littleton et al. (1995) | E/L 1:1000 + | 1:1000 +++ | |
| Cysteine string protein (CSP) | DCSP-2 GST-FL | mouse monoclonal | Zinsmaier et al. (1994) | E/L 1:50 +++ | 1:1000 +++ | presynaptic and synaptic vesicles and membrane |
| | MAb 49 | mouse monoclonal | Zinsmaier et al. (1990) | | 1:1000 +++ | |
| Synaptobrevin (SYB) | NSYB1 aa 1–14 | rabbit polyclonal | van de Goor et al. (1995) | N | 1:1000 ++ | presynaptic and axonal |
| | NSYB2 aa 131–144 | rabbit polyclonal | van de Goor et al. (1995) | N | 1:1000 ++ | |
| | HV62 aa 33–94 | rabbit polyclonal | Sweeney et al. (1995) | L 1:500 | n.t. | |
| | anti-n-syb aa 120–141 | rabbit polyclonal | Deitcher et al. (1998) | E/L 1:250 + | 1:500 ++ | |
| Drab3 | N11440 MBP-FL | rabbit polyclonal | Schulze et al. (1995) | n.t. | 1:1000 ++ | n.t. |
| α-SNAP | ?? | rabbit polyclonal | L. Pallanck and B. Ganetzky (pers. comm.) | E 1:100 +++ | 1:1000 +++ | presynaptic |
| Synapsin (SYN) | SYNORF1 GST-SYN | mouse monoclonal | Klagges et al. (1996) | L 1:100 +++ | 1:100 +++ | presynaptic only |
| α-SAP47 | hybridoma library | mouse monoclonal | Reichmuth et al. (1995) | L 1:10 +++ | 1:100 +++ | presynaptic only |
| Ras opposite (ROP) | TrpE-ROP aa 51–492 GST-ROP | rabbit polyclonal | Salzberg et al. (1993) | E/L 1:500 + | 1:10,000 +/− | presynaptic and axonal |
| | 4F8 aa 179–379 | mouse monoclonal | Harrison et al. (1994) | E/L 1:2 +++ | 1:1000 +++ | |
| SNAP-25 | HIS-FL | guinea pig polyclonal | T. Lloyd, M. Wu, and H.J. Bellen (pers. comm.) | E 1:500 ++ | 1:5000 ++ | presynaptic and axonal |
| | 71.1 rat protein | mouse monoclonal | Chapman et al. (1994) | n.a. | 1:2000 ++ | |
| Syntaxin (SYX) | SYX 8C3 | mouse monoclonal | K. Zinsmaier (pers. comm.) Schulze et al. (1995) | E/L 1:10 ++ | 1:1000 ++ | presynaptic and axonal |
| | SYX44D5 | mouse monoclonal | Cerezo et al. (1995) | | | |
| | SYXI378 Rat Syx | rabbit polyclonal | T. Südhof (pers. comm.) | E/L 1:1000 +++ | 1:5000 +++ | |
| α-adaptin (Dα-Ada) | HIS-Cterm aa 580–940 | rabbit polyclonal | Gonzalez-Gaitan and Jäckle (1997) | E 1:100 + L 1:100 +++ | n.t. | presynaptic |
| | HIS-Cterm aa 606–715 | rabbit polyclonal | Dornan et al. (1997) | L 1:25 +++ | 1:2000 +++ | |

*(Continued on following pages.)*

**Table 11.1.** (*Continued.*)

| Protein | Antigen[a] | Type of antibody | Source or references | Immunostaining embryos/larvae (dilutions)[a,b] | Westerns (dilutions) ECL[a,b] | Remarks |
|---|---|---|---|---|---|---|
| Like AP180 (LAP) | GST-Aterm aa 1–301 | rat polyclonal | Zhang et al. (1999) | E 1:500 +/– L 1:500 ++ | n.t. | presynaptic |
| Dynamin (DYN) | SHI-3 aa 822–838 | rabbit polyclonal | Estes et al. (1996) | E n.t. L 1:200 | 1:2500 +++ | presynaptic type I boutons |
|  | MBP-Dyn Ab2074 aa 241–836 | rabbit polyclonal | Roos and Kelly (1998) | E n.t. L 1:200 +++ | 1:2500 +++ |  |
|  | GST-dyn Ab2073 aa 331–651 | rabbit polyclonal | Roos and Kelly (1998) | E n.t. L 1:200 +++ | 1:2500 +++ |  |
| Synaptojanin (SJ) | GST-SJPRD aa 976–1218 | rabbit polyclonal | R. Nolo, J. Roos, and H.J. Bellen (pers. comm.) | E 1:200 + L 1:200 ++ | 1:1000 ++ | presynaptic |
| Clathrin (CHC) | C8034 clathrin heavy chain | goat polyclonal | Zhang et al. (1999) Sigma | E n.t. L 1:400 ++ | n.t. | presynaptic membrane and cytoplasm |
|  | bovine anti-CHC |  | M. Ramaswami (pers. comm.) | L 1:25 + | 1:500 |  |
| Dap160 | GSTDap160 aa 264–520 Ab1703 Ab1704 | rabbit polyclonal rabbit polyclonal | Roos and Kelly (1998) | E n.t. L 1:200 +++ E n.t. L 1:200 +++ | 1:1000 +++ 1:1000 +++ | presynaptic membrane |
| StonedA | MBP-stonedA aa 27–350 | rabbit polyclonal | A. Marie Phillips (pers. comm.) Stimson et al. (1998) | L 1:500 ++ | 1:20,000 ++ | presynaptic terminals; subset of synaptic vesicles? |
| Discs Large (DLG) | GST-DLG aa 455–919 | rabbit polyclonal | Woods and Bryant (1991) Lahey et al. (1994) | E/L 1:250 +++ |  | type I boutons, pre- and post-synaptic SSR |
| Glutamate receptor (DGLUR-II) | DM2 aa 891–906 DS58 aa 537–547 | rabbit polyclonal rabbit polyclonal | Saitoe et al. (1997) | E/L 1:200 ++ | 1:1000 ++ | postsynaptic membrane |
| Small syn. bouton antigen (SSB) | anti-dunce aa 234–322 or aa 408–472 | rabbit polyclonal | Budnik and Gorczyca (1992) Nighorn et al. (1991) | E/L 1:200 | 1:1000 | type II boutons (antigen is not the Dunce protein) |

| | Immunogen | Type | Reference/Source | Immunostaining | Western | Localization |
|---|---|---|---|---|---|---|
| Antiglutamate | glutamate | rabbit polyclonal | Johansen et al. (1989b) Chemicon | E n.t. L 1:1000 ++ | n.r. | presynaptic, all motor endings |
| Antioctopamine | DL-octopamine | rabbit polyclonal | Monastirioti et al. (1995) J. Rapus and M. Eckert (pers. comm.) | L 1:1000 ++ | n.r. | presynaptic, type II boutons |
| Antiproctolin | Proctolin | rabbit polyclonal | Anderson et al. (1988) | | n.r. | presynaptic in many but not all endings |
| PACAP-38 | RIN 8920 | rabbit polyclonal | Zhong and Pena (1995) Peninsula Lab. | E n.t. L 1:200 ++ | 1:20,000 ++ | presynaptic |
| Fasciclin II | hybridoma library MAb 1D4 | mouse monoclonal | Schuster et al. (1996) G. Helt and C. Goodman (pers. comm.) | E/L 1:5 +++ | | pre- and post-synaptic SSR type I and II |
| Anti-HRP | horseradish peroxidase | mouse monoclonal mouse or goat polyclonal | Jan and Jan (1982) Cappel Lab. or Sigma | E 1:50 +++ L 1:50 +++ E/L:1:400 | 1:200 | plasma membrane |
| Mab 22C10 | hybridoma library | mouse monoclonal | Fujita et al. (1982) | E 1:50 +++ L 1:50 +++ | 1:500 + | axonal cytoskeleton (only a portion of the synapse) |
| Hiraku genki | His-HIG aa 35–295 | rat polyclonal | Hoshino et al. (1996) | A 1:10 ++ | ? | presynaptic and cleft in pupae and adult; NMJ M8 in larvae |
| Still life | His-SIF Abl3 aa 327–604 | rat polyclonal | Sone et al. (1997) | E/L/A 1:10 ? | ? | presynaptic |
| Frequenin | GST-FL | rabbit polyclonal | Pongs et al. (1993) | E/L/A 1:10 +++ | 1:500 +++ | presynaptic |

[a]Abbreviations: (FL) Full-length protein; (ECL) enhanced chemiluminescence; (E) embryos; (L) larvae; (A) adults; (n.r.) not relevant; (n.t.) not tested; (aa) amino acids; (Cterm) carboxyl terminus; (Aterm) amino terminus; (N) negative.

[b]Key for immunostaining embryos/larvae and Westerns: (+++) Excellent or very good; (++) good; (+) useful but not great; (+/–) difficult to use/poor.

al advantages to this method. First, evenly stretched preparations can be obtained with ease in 3–5 minutes for third-instar larvae. Second, using this method, it is possible to dissect preparations from late embryonic stages (after cuticle deposition) and early larval instars. Because the NMJs are covered only by the outside solution, images of higher resolution can be obtained (see, e.g., Gorczyca et al. 1994; Guan et al. 1996). Protocol 11.2 describes the fillet preparation of stage-17 embryos to third-instar larval stage and includes the construction of magnetic chambers. Protocol 11.3 describes the fillet preparation of embryos at stages prior to cuticle deposition. A fillet dissection of the embryo provides the clearest view of the internal structures and is especially important when viewing nerves to the body wall muscles.

## Labeling with FM1-43

The FM1-43 dye was first developed by Betz and Bewick (1992) and the methodology was adapted for the third-instar larval NMJs by Ramaswami et al. (1994). Several other similar dyes are now available, each with slightly different binding and kinetic properties (see, e.g., Klingauf et al. 1998). Protocol 11.4 describes labeling third-instar larvae with FM1-43 and can probably be used in second-instar larvae as well.

## Visualizing the Ultrastructure of Synaptic Terminals by Transmission Electron Microscopy

In addition to labeling NMJs, we have included procedures for transmission electron microscopy (TEM) preparation of NMJs in embryos and first-, second-, and third-instar larvae (see Protocol 11.5). These protocols are not described in detail elsewhere, and the ultrastructure of the NMJ can provide very valuable information about, for example, the number and morphology of active zones, size and number of synaptic vesicles, size of boutons, and alignment of pre- and postsynaptic densities. This information cannot be obtained in any other fashion than by TEM.

PROTOCOL 11.1*

## Immunocytochemical Staining of NMJs

### Materials

#### Media

Standard grape or apple juice agar plates (see Appendix 3)

#### Supplies and Equipment

Artist's brush
Squirt bottle of $H_2O$ or PBT
Nitex screen (1 inch sq. pieces)
Falcon tubes (50 ml), each with the bottom cut off and a hole cut in the cap
Scintillation vials (10 ml), for fixation
Pasteur pipettes
(*Optional*) Multiwell tissue culture plate (Fisher)
Microcentrifuge tubes (1.5 ml; Eppendorf)
Dark brown microcentrifuge tubes for fluorescently labeled antibodies

*For all protocols, $H_2O$ indicates glass distilled and deionized.

Dissecting microscope
Glass well
Hot plate set at 60–70ºC
Kimwipes
Coverslips (22 × 22 mm or 22 × 40 mm)
Ring-mounted slides
Cap tubes (15 ml)
Sonicator

## Solutions and Reagents

**Bleach** (full-strength Clorox)

*n*-**Heptane**

**Methanol** (100%)

**Ethanol** (95%)

**Glycerol**

Vectashield (Vector Laboratories)

Vectastain kit (Vector Laboratories); type of kit will depend on primary antibody

Tris-Cl (0.12 M, pH 7.6)

**Xylene**

Permount (Fisher)

**DAB** (3,3′-diaminobenzidine tetrahydrochloride) solution, 1 mg/ml in 100 mM Tris-Cl (pH 7.6); prepared from a 10 mg/ml stock solution of DAB in $H_2O$, passed through a 0.2-μm, 25-mm Nalgene disposable syringe filter; DAB is light-sensitive

**$H_2O_2$** (30%)

### 10× PBS

| | |
|---|---|
| NaCl | 87.6 g |
| **$NaH_2PO_4$** | 1.56 g |
| **$Na_2HPO_4$** | 12.35 g |
| $H_2O$ | to make 1000 ml |

Adjust pH to 7.6.

### PBT

1× PBS (pH 7.6)
0.1% Tween-20

### Fixative

Mix 2.0 ml of 10% **formaldehyde** (methanol-free; Polyscience 04018), 2.5 ml of 1× phosphate-buffered saline (PBS), and 0.5 ml of $H_2O$.

### Blocking Solution 1

PBT
3% bovine serum albumin (BSA)
2% normal goat serum (NGS)
If the secondary antibodies are derived from species other than goat, block with serum from appropriate species diluted 1:100 or 1:200 in PBT with 2% BSA and 2% NGS.

### Blocking Solution 2

PBT
5% NGS

*BBS*

    PBT

    0.1% BSA

    2% serum

*PBT/BSA*

    PBT

    2% BSA

*Tris-Cl/Tween-20 Solution*

    0.12 M Tris-Cl (pH 7.6)

    0.1% Tween-20

## Antibodies

Primary antibodies (see Table 11.1)

Secondary antibodies. The following types of secondary antibodies can be used, corresponding to the appropriate species of primary antibody: (1) fluorescently labeled, (2) directly linked to enzyme, or (3) conjugated to biotin for amplification using the avidin/biotin-HRP system.

**CAUTION: bleach, DAB, ethanol, formaldehyde, *n*-heptane, $H_2O_2$, methanol, $Na_2HPO_4$, $NaH_2PO_4$, xylene** (see Appendix 4)

## Method

### Staining of NMJs in 0–15-hour-old Embryos

1. Collect embryos using an artist's brush, or a spray of $H_2O$ or PBT from a squirt bottle. Transfer the embryos onto a Nitex screen, which is attached to the top of a modified 50-ml Falcon tube by screwing on the cap.

2. (*Dechorionation*) Perform the following steps by dipping the cap of the modified Falcon tube in a small container, and squirting solution on top of the embryos.

    a. Cover the embryos with bleach (full-strength Clorox) for 2 minutes.

    b. Rinse three times in PBT.

    c. Rinse once in $H_2O$.

3. Unscrew the cap and place the Nitex screen with the embryos in a regular 10-ml scintillation vial containing 2.5 ml of fixative and 2.5 ml of *n*-heptane. Wash off the embryos into the fixative and retrieve the Nitex screen. Fix at room temperature for 20 minutes on a shaker platform (e.g., ~100 rpm on a New Brunswick Model G2 gyratory shaker).

4. Pipette off the lower phase (fixative). The embryos should be floating on top of the fixative, at the fixative/heptane interface.

5. Add 2.5 ml of 100% methanol and shake vigorously for approximately 25 seconds.

    *Note:* This ruptures the vitelline membrane. Devitellinized embryos will sink to the bottom of the vial.

6. Aspirate the upper heptane phase and most of the methanol. Rinse once in methanol.

7. Use a pasteur pipette to transfer the embryos to a 24-well tissue culture plate (Fisher) or to microcentrifuge tubes. Rinse twice in methanol and three times in 95% ethanol.

    *Note:* The embryos can now be stored at –20ºC for months if needed.

8. Rehydrate the embryos by adding half the volume of $H_2O$. Remove the $H_2O$ and repeat rehydration. Rinse three times with PBT.

9. Incubate embryos in Blocking Solution 1 at room temperature for 1 hour.

   *Note:* Alternatively, BBS can be used instead of Blocking Solution 1.

10. Incubate the embryos in primary antibody (diluted 1:X [see Table 11.1] in Blocking Solution 1) at 4ºC overnight.

    *Note:* For double labeling, treat embryos with the first primary antibody, then perform steps 11–12a (i–ii) or steps 11–12b (i–ii). Rinse five to six times in PBT. Incubate in second primary antibody (appropriately diluted) at 4ºC overnight. Proceed with steps 11–12.

11. Rinse four times in PBT over 1 hour.

12. Treat embryos with secondary antibody in either of the following ways:

    a. If the secondary antibody is labeled with a fluorescent compound (e.g., Cy3, Cy5, FITC), perform the following steps:

       i. Add 200 µl of secondary antibody (1:200 dilution in Blocking Solution 1). Incubate at room temperature for 1–2 hours in the dark (we use dark brown microcentrifuge tubes).

       ii. Wash four times in PBT for 15 minutes (each wash) in the dark.

       iii. Remove PBT, add 100 µl of glycerol, and clear. Repeat until glycerol does not fluoresce under the microscope. Mount sample in Vectashield (Vector Laboratories).

    b. For permanent staining, use either secondary antibody directly linked to enzyme (e.g., affinity-purified goat anti-rabbit IgG, HRP conjugate [Biorad] if using a primary antibody from rabbit) or secondary antibody conjugated to biotin, for subsequent amplification using the avidin/biotin-HRP system (e.g., biotinylated, affinity-purified goat anti-rabbit IgG [Vectastain kit; Vector Laboratories] if using a primary antibody from rabbit).

       i. Apply 200 µl of secondary antibody (1:100 dilution in Blocking Solution 1) and incubate at room temperature for 1 hour.

       ii. Wash four times in PBT for 15 minutes (each wash) in the dark.

       iii. Rinse four times in PBT over 1 hour. (During this step, prepare the avidin-biotin-HRP solution as described in step iv, below.)

       iv. (*Optional, if using a biotin-conjugated secondary antibody*) Approximately 30 minutes before use, prepare a solution of 1:200 avidin and 1:200 biotinylated-HRP (Reagents A and B in the Vectastain kit) in PBT/BSA. Prepare 200 µl for each well or tube. Incubate the embryos in this solution at room temperature for 1 hour.

       v. (*Optional*) Wash three times in PBT over 30 minutes.

       vi. (*Optional*) Wash three times in Tris-Cl/Tween-20 solution over 30 minutes.

       vii. Incubate in 0.5 ml of 1 mg/ml DAB for 5–10 minutes. Mix 10 µl of 30% $H_2O_2$ in 1 ml of $H_2O$. Add 10 µl of the diluted $H_2O_2$ and watch under a dissecting microscope. Wait until NMJs become visible or background is relatively high under the dissection microscope, and rinse two to three times in Tris-Cl (pH 7.6).

       viii. Remove half the volume of Tris-Cl and add half-volume 95% ethanol. Wash three times with 95% ethanol. Transfer embryos to a glass well (very clean and prewiped with ethanol). Wash twice with ethanol. Remove all ethanol and add xylene.

    ix.  Cut 5 mm off a yellow pipette tip, wet the pipette tip with xylene, and transfer embryos to a small drop on a slide. Remove excess xylene with a Kimwipe and allow to dry. Add 2–4 drops of Permount. Embryos should stay where they are (use a pasteur pipette to bring embryos to middle). Cover with a 22 X 22-mm or 22 X 40-mm coverslip and then press with a ring-mounted-slide (moving the coverslip will destroy the embryos). Remove excess Permount and blow air on slides until no air is sucked under the coverslip. Put weights on slides and leave them on a hot plate at 60–70°C overnight.

### Whole-mount Preparation of 16–22-hour-old Embryos

1. Follow steps 1–3 of above method.

2. Remove as much fixative (lower layer) as possible. Retain the embryos at the interface and those below the interface. Make sure heptane is still remaining with the embryos in the container.

3. Follow steps 5–8 of above method.

4. Place approximately 50–100 µl of packed embryos into 0.5 ml of PBT in a 15-ml cap tube.

5. Sonicate the sample and place embryos on ice for 15–30 seconds between rounds of sonication to prevent embryo disintegration. After sonication, keep the embryos on ice until the next step.

   > *Notes:* The optimal settings for a given sonicator should be determined empirically. We use a Sonifier Cell Disruptor, a probe tip sonicator, set at 100% duty cycle and setting #7. At these settings, we sonicate for 2 seconds, four times.
   >
   >     The antisynaptotagmin anticysteine string antibodies are useful to optimize the conditions. Remember to include unsonicated embryos as a control.

6. Wash two times in PBT for 5 minutes (each wash).

7. Proceed with steps 9–12 of above method.

### Staining of Dissected Embryos, and First-, Second-, and Third-instar Larval NMJs

1. Dissect and fillet first-, second-, and third-instar larvae as described in Protocol 11.2. For fillet preparation of embryos, see Protocol 11.3.

2. Incubate preparations in fixative for 20 minutes.

3. Wash three times in PBT for 5 minutes (each wash); and then wash in PBT for 30 minutes. Transfer the embryos or larvae to a 1.5-ml microcentrifuge tube.

4. Incubate in 400 µl of Blocking Solution 2 for 30 minutes.

5. Add primary antibody (diluted 1:X [see Table 1] in Blocking Solution 2) and incubate at 4°C overnight.

6. Wash three times in PBT for 10 minutes (each wash).

7. Incubate in Blocking Solution 2 for 30 minutes.

8. Treat embryos or larvae with secondary antibody in either of the following ways:

a. If the secondary antibody is labeled with a fluorescent compound (e.g., Cy3, Cy5, FITC), perform the following steps:

   i. Add secondary antibody, diluted 1:200 in Blocking Solution 1. Incubate for 1–2 hours.

   ii. Wash four times in PBT for 15 minutes (each wash) in the dark.

   iii. Remove PBT, add 500 µl of glycerol, and clear. Repeat until glycerol does not fluoresce under the scope. Mount sample in Vectashield (Vector Laboratories).

b. For permanent staining, use secondary antibody conjugated to biotin for subsequent amplification using the avidin/biotin-HRP system (e.g., biotinylated, affinity-purified goat anti-rabbit IgG [Vectastain kit; Vector Laboratories], if using a primary antibody from rabbit).

   i. Apply secondary antibody (1:200 dilution in Blocking Solution 1) and incubate at room temperature for 1 hour.

   ii. Wash three times in PBT for 10 minutes (each wash) in the dark. (During this step, prepare the avidin-biotin-HRP solution as described in step iii, below.)

   iii. Approximately 30 minutes before use, prepare a solution of 1:200 avidin and 1:200 biotinylated-HRP (Reagents A and B in the Vectastain kit) in PBT/BSA. Prepare 400 µl for each well or tube. Incubate the embryos or larvae in this solution at room temperature for 1 hour.

   iv. Wash three times in PBT for 5 minutes (each wash).

   v. Wash three times in 0.12 M Tris-Cl (pH 7.6) for 5 minutes each wash.

   vi. Transfer to a 16-well plate or microcentrifuge tube.

   vii. Incubate in 0.5 ml of 1 mg/ml DAB for 5–10 minutes. Mix 10 µl of 30% $H_2O_2$ in 50 ml of $H_2O$. Add 500 µl of the diluted $H_2O_2$ and watch under a dissecting microscope. Wait until NMJs become visible under the dissection microscope, and rinse two to three times in 0.12 M Tris-Cl (pH 7.6).

   viii. Clear in glycerol for 24 hours and mount sample in Vectashield.

## PROTOCOL 11.2

# Fillet Preparation for Stage-17 Embryos to Third-instar Larval Stage Using Magnetic Chambers

## Materials

### Supplies and Equipment

Magnetic flexible strip with adhesive backing (Magnet Sales and Mfg. Co. 0648A, 11248 Playa Court, Culver City, California; Phone: 800-421-6692; Web site: http://www. magnetsales.com)

Glass slides (2" x 3", available from any major scientific supplier)

Paper-cutting tool and board

Silicone sealant

Flat toothpicks

Stainless-steel pins; size 000 for stage-17 embryos or first instar and size 00 for second- to third-instar preparations (Carolina Biological Supply Sciences)

Fine pliers or sturdy forceps

Epoxy glue (5 minute)

Metal signal tabs, available from any office supply store (Nu-vise projecting signals; Labelon/Noesting Co., Canandaigue, New York 14424; Phone: 800-428-5566)

Silicon-carbide paper (grit 220 and grit 400–600)

Dissecting tools

- Micro-dissecting scissors (Roboz RS5618). We carefully file down the tip of the scissors with a series of 600–1500 grit paper to sharpen the tips and reduce their profile. Extreme care is advised because improper filing may completely ruin these expensive scissors.
- (*Optional*) Sharp tungsten needle
- Forceps (#5 forceps biological grade, Inox). Nonmagnetic forceps are not necessary and are discouraged because of their softer tips (Roboz). File down the tips with grit paper series 600–1500.
- Dissecting microscope

## Solutions and Reagents

Saline, either $Ca^{++}$-free or containing 0.1 mM $Ca^{++}$ (below). $Ca^{++}$-free saline contains 1 mM EGTA.

### Saline

128 mM NaCl

2 mM **KCl**

4 mM **MgCl$_2$**

0.1 mM **CaCl$_2$**

35.5 mM sucrose

5 mM HEPES (pH 7.2)

**CAUTION: CaCl$_2$, KCl, MgCl$_2$** (see Appendix 4)

## Method

### Construction of Magnetic Chambers

Figure 11.3B illustrates the construction of a magnetic chamber.

1. Cut the magnetic strip to the same size as a 2 × 3-inch glass slide; a heavy-duty paper cutting board does nicely.

2. Punch out a hole in the center of the magnetic strip with an appropriate size hole-punch, about 3/4 inch. Alternatively, cut a hole with a scissors, scalpel, or Exacto knife using a medium size coin for size.

3. Peel off the white backing of the strip, and then place a bead of silicone sealant (on the sticky side of the magnetic strip) around the perimeter of the magnet and at the rim of the cut hole (see Figure 11.3B). This is to keep saline from leaking out.

4. Place the glass slide over the sticky side of the magnet and press. Clean up excess sealant and form sealant against the rim edge with flat toothpicks.

### Preparation of Pins

1. Bend the head of insect pins in the shape shown in Figure 11.3C-1.

   *Note:* For each preparation, six pins will be needed; two for the center and four for the corners.

2. Use quick-setting epoxy to glue pins to metal tabs as shown in Figure 11.3C-2. Raise the pin tip approximately 3 mm; set the tip of the pin on a small piece of cardboard for this purpose.

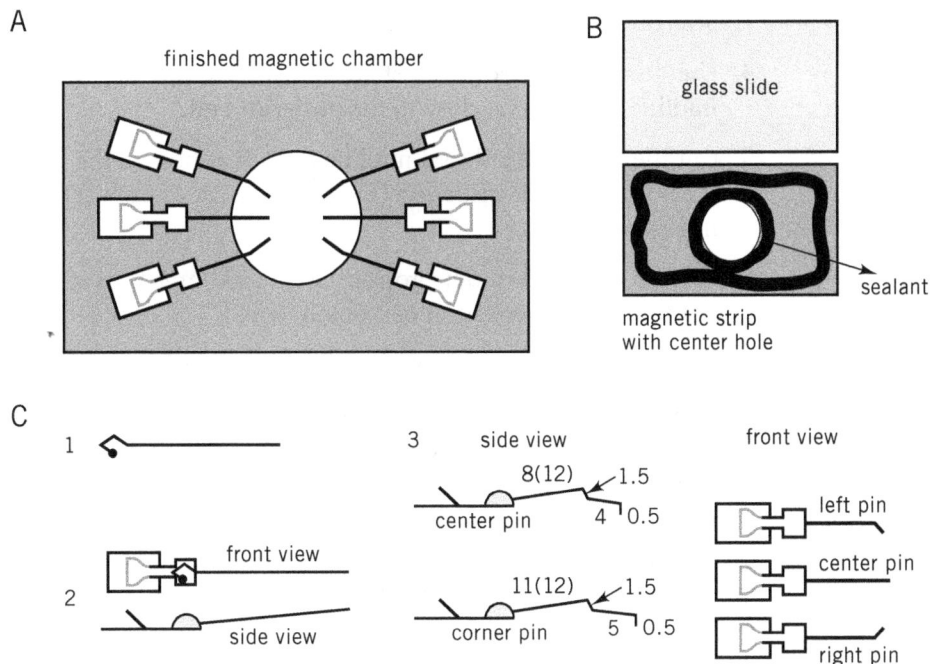

**Figure 11.3.** Design of magnetic chambers for late stage-17 embryos to third-instar larval stage. (*A*) Finished magnetic chamber and pins; (*B*) glass slide and bottom view of the magnetic strip showing the approximate placement of the sealant; (*C*) construction of the pins (see text for details).

> *Note:* Raising the level of the pin will prevent the liquids in the chamber from spreading all over the magnet via capillary action.

3. After the glue has hardened, cut the tips of the two center pins to approximately 1.4 cm and the four corner pins to approximately 1.8 cm. File the pin tips flat and thin with silicon carbide paper (grit 220). Finish with grit 400–600 paper.

4. Bend the pins with fine pliers or sturdy forceps as in Figure 11.3C-3 (dimensions are in millimeters; numbers in parenthesis correspond to the measurements for first instar dissection chambers).

> *Notes:* The pin tips should press down onto the glass and flex a bit, to ensure that they will hold onto the stretched specimen. The corner pins should be bent from the side as shown in Figure 11.3C-3.
>
> All dimensions are approximate and will depend on the chamber and purpose. The tips of the pin can be filed down further to adjust to the stage of the preparation (i.e., for younger larvae, the tips need to be thinner).

## *Dissection*

1. Place a larva (Figure 11.4A) in the dissecting chamber over a small drop of $Ca^{++}$-free saline (Figure 11.4B-1; in general, dissect body wall muscles in $Ca^{++}$-free or 0.1 mM $Ca^{++}$ to minimize contraction of the muscles during dissection). For stage-17 embryos, use a tungsten needle and fine-tipped forceps to free them from the chorion and vitelline membrane (see Protocol 11.3, steps 2–6). Then proceed as described for larvae.

2. Using the center pins and with the dorsal side of the larva facing up, pin the rear (between the posterior spiracles) and then the front (near mouth apparatus) of the larva (Figure 11.4B-2). Place the pins as close as possible to both ends of the larva for best results.

3. Stretch the larva taut using the center pins and fill the chamber with saline.

4. Use dissection scissors or a sharp tungsten needle to make a small hole at the dorsal midline of the larva, close to the posterior end.

5. Cut the larva at the dorsal midline from that hole to the anterior end and then from the hole to the posterior end (Figure 11.4B-3). At each end, cut two notches as shown in Figure 11.4B-3.

6. Use forceps to clean out protruding internal organs as much as possible, taking care not to damage the central nervous system.

7. Wash forcefully with drops of saline to get rid of organ debris as much as possible.

8. Gently stretch the posterior sides of the larva, and then the anterior sides using the corner pins until the preparation reaches the shape shown in Figure 11.4B-4 and Figure 11.4B-5. Wash with saline.

> *Note:* After dissection, fix the preparation in the dissection chamber. After 5–10 minutes in fixative solution, the extended shape of the body wall muscles will remain intact, and at this point the preparation can be transferred for further manipulation to another container.

## PROTOCOL 11.3

## Fillet Preparation of Embryos at Stages before Cuticle Deposition

### Materials

### *Supplies and Equipment*

Syringe (1 ml)
Silicone sealant
Microscope glass slide
Double-stick tape
Tissue paper (e.g., Kimwipes)
Forceps
Dissecting tools
- tungsten needle or sharp forceps
- dissecting microscope
Blunt needle

### *Solutions and Reagents*

**Ethanol** (95%)
Saline (see p. 188)

**CAUTION: ethanol** (see Appendix 4)

### Method

### *Preparation of Chambers*

1. Fill a 1-ml syringe with silicone sealant.

2. Use the syringe to dispense sealant onto the glass slide to make a rectangular well of approximately 1 × 3/4 inch (see Figure 11.5).

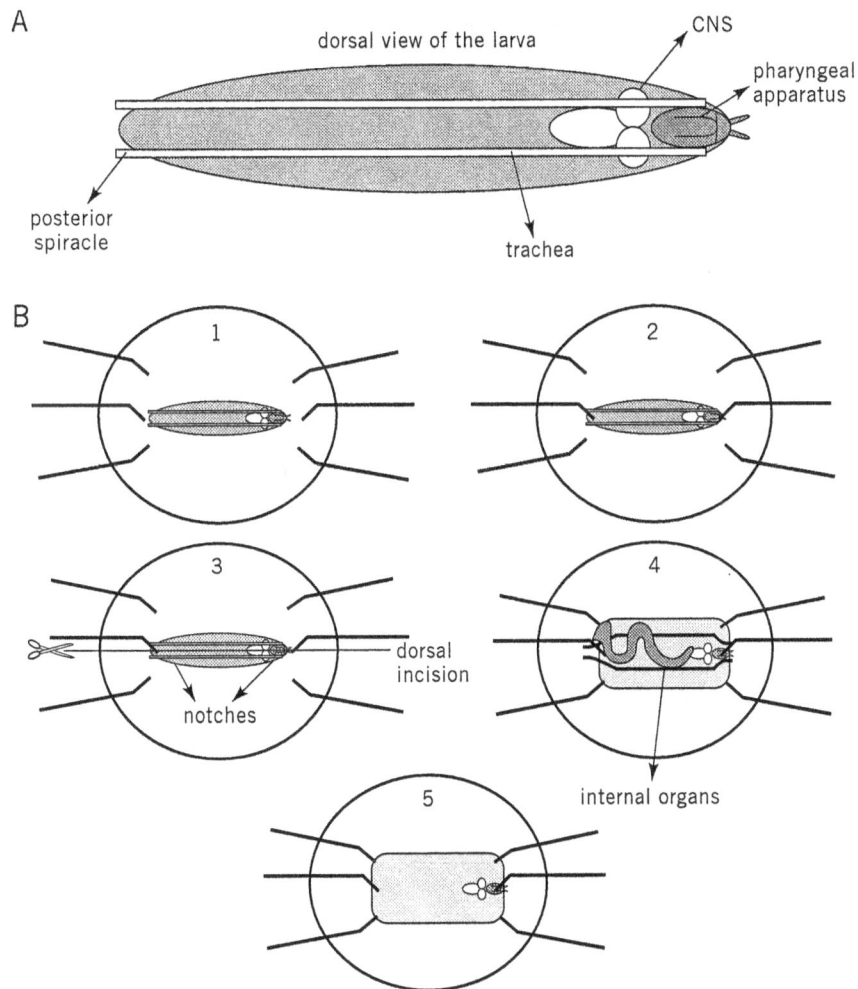

**Figure 11.4.** Dissection of a larval body wall muscle preparation using magnetic chambers. (*A*) Dorsal view of a larva; (*B*) steps of the dissection procedure (see text for details).

3. Before beginning the dissection procedure, place a small strip of double-stick tape outside of the well and a thin strip inside the well (see Figure 11.5).

## *Dissection*

1. Use 95% ethanol and tissue paper to clean the glass inside the well by vigorously rubbing.

   *Note:* This step is critical, as embryos will not stick to poorly cleaned glass.

2. Transfer about 20 well-staged eggs from egg-laying plates to the double-stick tape outside the well.

3. Use the back of the forceps tip to roll the embryo gently from side to side to manually dechorionate it.

4. Gently transfer the embryo to the double-stick tape inside the well (see Figure 11.5).

   *Note:* This can be done with the forceps by simply touching the embryo, which will usually stick to the forceps.

5. When 10–15 embryos have been transferred, gently fill the well with saline.

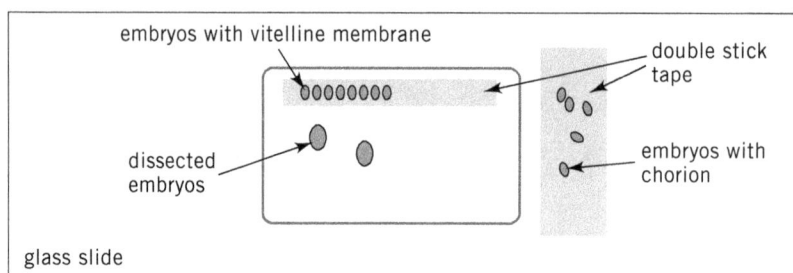

**Figure 11.5.** Arrangement of well and double-stick tape strips for the embryonic fillet preparation.

6. Use the tip of a sharp tungsten needle or very sharp forceps to puncture the vitelline membrane at the anterior end of the embryo. Note that there is a small empty space between the embryo and the vitelline membrane at the anterior end, allowing the membrane to be punctured without damaging the embryo.

7. Use the back of the forceps tip to push the embryo forward from the posterior end. At this point, the embryo will be extruded from the vitelline membrane, in a manner akin to toothpaste leaving its tube.

8. Use the needle or forceps to very gently float the embryo off the tape and onto the glass. Gently position the embryo so that the dorsal side is facing up (this can be performed by identifying the central nervous system, which at this stage is clearly seen at the ventral side).

9. Place the embryo onto the glass slide. At this point, the embryo sticks to the glass and cannot be moved for reorientation.

10. Gently rub the tip of the needle up and down along the dorsal midline. This will effectively cut along the dorsal midline of the embryo.

11. With a fairly blunt needle, gently press down the lateral edges onto the glass.

12. Use sharp forceps to gently pull out the gut and pieces of fat bodies, taking care not to damage the central nervous system or muscles.

13. Fix by gently exchanging the saline with fixative solution. *Important! Do not empty out the chamber, as the embryos will become unstuck.*

14. Gently wash three times (10 minutes each wash) in PBS buffer. Then proceed with immunocytochemical staining (Protocol 11.1) with the embryos stuck to the glass. To mount embryos, remove by gently teasing with a blunt needle before mounting.

> *Note:* For immunocytochemical staining of fillet preparations of embryos, Blocking Solution is seldom needed. Also, step 8a (iii) on p. 187 using glycerol can be omitted.

PROTOCOL 11.4

# FM1–43 Dye Uptake Labeling

## Materials

### *Supplies and Equipment*

Dissecting tools
- fine forceps
- dissecting microscope

Confocal microscope. These experiments require a confocal microscope. Alternatively, use a scope with epifluorescence and a CCD camera with appropriate software (Kuromi and Kidokoro 1998). These experiments also require a 40X water-immersion lens (e.g., Zeiss or Olympus, 40(/0.75NA or 40x/0.80NA). A 63x water-immersion lens is also useful.

### Solutions and Reagents

FM1-43 (4 μM, Molecular Probes)
HL3 containing $Ca^{++}$ (for preparation, see Table 15.1 in Broadie, this volume)
$K^+$ HL3 (60 mM)
$Ca^{++}$-free HL3

## Method

### Loading the Terminals with FM1-43 Dye

1. Fillet third-instar larvae to expose the muscles as described in Protocol 11.2.

2. Gently remove the viscera while keeping the central nervous system attached.

    *Note:* These preparations show spontaneous activity when incubated in HL3 containing $Ca^{++}$.

3. Load sample with FM1-43 dye. The simplest protocol uses high $K^+$, e.g., 60 mM $K^+$ HL3 containing 4 μM FM1-43 dye for 5 minutes.

    *Note:* Alternatively, stain a small number of NMJs by stimulating a single nerve in HL3 with dye (and no high $K^+$) using a suction electrode. As the nerve innervates only a few muscles in a single abdominal or thoracic domain, only few NMJs are labeled.

4. Wash the preparation twice rapidly, and then four to six times in $Ca^{++}$-free HL3 for 5 minutes, to remove surface-bound dye. The NMJs are now ready for analysis by epifluorescence or confocal microscopy.

### Unloading the FM1-43 Dye

1. Add HL3 with high $K^+$ for 5 minutes.

2. Wash in $Ca^{++}$-free HL3 and view with confocal microscope or a microscope equipped with epifluorescence.

## PROTOCOL 11.5

# Preparation of Samples for Transmission Electron Microscopy

## Materials

### Supplies and Equipment

Chamber (3-welled)
Spurr's **resin** (low-viscosity embedding media; Electron Microscopy Sciences 14300; prepare according to manufacturer's instructions using the "hard" modification)
Parafilm
Embedding molds
Insect pins, sharpened
Surgical carbon-steel razor blade
Ultramicrotome, for sectioning
Compound microscope
Microscope slides
Wooden stick with wire loop for transferring sections

Slide warmer set at 60–80ºC

Stainless-steel Teflon-coated blade

Copper grids type 460 Hex thin mesh; Electron Microscopy Sciences GT400H-Cu (plain copper grids will also work)

Forceps for handling grids

Paper tissue (e.g., Kimwipes)

Petri dishes (2)

Double-stick tape

Coverslips

Glass needles

Razor blade splinter

### Resin Blocks

To construct the blocks, place a thin layer of Spurr's resin at the bottom of the embedding molds and incubate at 70ºC for 3 hours.

## Solutions and Reagents

CBS (0.1 M, pH 7.2; see below)

**Osmium tetraoxide** ($OsO_4$) (2%)

Aqueous **uranyl acetate** (2%) (can be used for 1 month after preparation; prepare at least one night before use)

**Ethanol** series (30%, 50%, 70%, 85%, 95%, absolute)

Propylene oxide

**Acetone**

**NaOH** pellets

Dilute NaOH solution (~3 drops of 0.02 M NaOH in 30 ml of $H_2O$)

(*Optional*) Phosphate buffer (0.05 M, pH 7.2)

### Modified Trump's Universal Fix

Dissolve 0.28 g of **cacodylic acid** in 12.5 ml of $H_2O$. Adjust pH to 7.2. Add 5 ml of 16% **paraformaldehyde** (Electron Microscopy Sciences; 1-ml sealed ampules) and 2.5 ml of 8% **glutaraldehyde** (Electron Microscopy Sciences; 1-ml sealed ampules). Use as a freshly prepared solution.

### 0.2 M CBS

Dissolve 2.76 g of cacodylic acid and 9.04 g of sucrose in sufficient $H_2O$. Adjust volume to 100 ml with $H_2O$. Adjust pH to 7.2 with NaOH.

### 0.2 M CB

Prepared as for 0.2 M CBS, but without sucrose.

### Toluidine Blue Stain

1% toluidine blue

1% borax

### Lead Citrate Solution

Dissolve 5 mg of **lead citrate** in 1 ml of 4.5 mM NaOH. Centrifuge in a microcentrifuge at 12,000*g* for 5 minutes just before use.

**CAUTION: acetone, CaCl$_2$, cacodylic acid, ethanol, glutaraldehyde, KCl, lead citrate, MgCl$_2$, NaOH, osmium tetraoxide, paraformaldehyde, propylene oxide, resins, uranyl acetate** (see Appendix 4)

## Method

### Third-instar Larvae

1. Dissect body wall muscles in saline as described in Protocol 11.2. Stretch the larval fillet preparation for electron microscopy less than would be stretched for immunohistochemistry to prevent sample curling during dehydration and embedding.

2. Fix in Modified Trump's Universal Fix in the dissection chamber for approximately 10 minutes, and then transfer samples to a three-welled chamber and fix under mild agitation, at room temperature for 2 hours, and then at 4°C overnight.

   *Note:* The addition of 1 mM $MgCl_2$ to the fixative may improve the preservation of terminals.

3. Wash three times in 0.1 M CBS for 10 minutes (each wash).

4. Mix equal parts of 2% $OsO_4$ solution and 0.2 M CBS (~50 μl per sample) and postfix samples in this solution for 30 minutes.

5. Wash three times in 0.1 M CBS for 10 minutes (each wash).

6. Wash three times in $H_2O$ for 10 minutes (each wash).

7. Stain samples en bloc in aqueous 2% uranyl acetate for 20–30 minutes.

8. Dehydrate in ethanol series: 30%, 50%, 70%, 85%, 95%, and absolute for 3 minutes each. Repeat twice in absolute ethanol for 10 minutes.

   *Note:* We change the ethanol solution gradually between each ethanol step. For example, we exchange just a third of the previous ethanol step during the first minute, one half during the second, and the entire solution during the third.

9. Infiltrate in Spurr's resin as follows:

   Propylene oxide for 30 minutes.
   Propylene oxide:Spurr's resin (3:1) for 2 hours (or overnight).
   Propylene oxide:Spurr's resin (1:1) for 2 hours (or overnight).
   Propylene oxide:Spurr's resin (1:3) overnight. Cover the vial with Parafilm and make holes to allow evaporation of the propylene oxide.
   Spurr's resin, at least 3 hours twice.

10. Mount in resin blocks (prepared embedding molds). Carefully place the samples and Spurr's resin in the blocks. If the samples curl up, straightened them by pinning them down into the molds using sharpened insect pins.

    *Note:* Resin blocks should be constructed before placing the samples, otherwise the preparation sinks to the bottom of the mold, making block trimming somewhat difficult.

11. Bake at 70°C overnight.

12. Trim the block around the preparation using a surgical carbon-steel razor blade cleaned with acetone to approximately 2.5 × 1 mm and until the desired segment to be sectioned is reached.

13. Cut thick sections (0.5–1 μm) and place on a microscope slide using a wooden stick with a wire loop at the end (~2 mm in diameter). Add a drop of $H_2O$ and leave on a slide warmer at 60–80°C until the $H_2O$ completely evaporates to eliminate wrinkling. Add several drops of toluidine blue stain and incubate at 60–80°C for 1–2 minutes. Prevent the stain from drying out by placing a dish containing water close to the slide.

14. Wash the sections by squirting them with $H_2O$ and observe under a compound microscope.

*Note:* At this point, the muscles to be sectioned can be identified according to their profiles and position in cross-section. Also, it is sometimes possible to determine if there are terminals in the region being sectioned. Terminals are seen as small irregularities slightly lighter than their surroundings and close to the visceral surface of the muscles. Sometimes terminal-rich regions can also be identified if a portion of the nerve is seen close to the muscles.

15. Once a desired region to be sectioned is identified, trim the block further (to ~0.5 × 0.5 mm) using an acetone-cleaned, stainless-steel Teflon-coated blade.

16. Cut sections of approximately 50–80-nm thick (silver gray) and float them onto a clean grid.

17. To stain the grids, line the bottom of two petri dishes (number them 1 and 2) with Parafilm. Place two or three pellets of NaOH in the sides of petri dish 2 and keep covered. Place one drop/grid of 2% uranyl acetate in petri dish 1, and one drop/grid of lead citrate in petri dish 2. Then float the grids (section side down) in each drop of uranyl acetate for 10–30 minutes.

18. Pick up grids with forceps and blot excess solution with paper tissue. Wash the grids in $H_2O$ by picking them up with forceps and quickly dipping them down and up about 20 times in a 30-ml beaker filled with $H_2O$.

19. Use paper tissue to remove excess $H_2O$ and then float the grids in the drops of lead citrate for 5–8 minutes.

20. Wash the grids in a dilute NaOH solution. Blot excess solution and store until viewing. For long-term storage, keep the grids in a desiccator.

### Embryos

The overall procedure for TEM preparation of embryos is similar to that of third-instar larvae; the steps that differ from the method above are noted below. The method for embryos is based on the procedure of Prokop et al. (1996).

1. For embryonic preparations, dechorionate embryos of appropriate stages, and stick them in a row to a strip of double-sided tape fixed to a coverslip. Inject the embryos with 5% glutaraldehyde in 0.1 M CB (pH 7.2) using glass needles.

    Note: CB can be replaced by 0.05 M phosphate buffer (pH 7.2) with good results (A. Prokop, pers. comm.).

2. Cut off the anterior and posterior tips of injected specimens using a razor blade splinter.

3. Postfix in CB- or phosphate-buffered 2.5% glutaraldehyde for 1 hour.

4. Wash briefly in 0.1 M CBS.

5. Proceed as for third-instar preparations (steps 4–20).

### First- and Second-instar Larvae

The method for early larval instars has been published by Guan et al. (1996). We find that one of the most critical steps in obtaining good terminal morphology in dissected first- and second-instar fillets is the time of aldehyde fixation before osmication. In general, fixation in Modified Trump's Universal Fix is performed at room temperature for 30 minutes to 1 hour for early first-instar preparations and for 1–2 hours for second-instar preparations (see step 2 [Third-instar larvae]). The rest of the procedure remains the same as for the third instar.

## ACKNOWLEDGMENTS

We thank Mani Ramaswami, Jack Roos, Haig Keshishian, Andreas Prokop, and Michael Gorczyca for comments on the table and personal communications; Richard Atkinson, Tom Lloyd, and Jay Bhave for comments on the manuscript; and Salpy Sarikhanian for secretarial assistance.

## REFERENCES

Anderson M.S., Halpern M.E., and Keshishian H. 1988. Identification of the neuropeptide transmitter proctolin in *Drosophila* larvae: Characterization of muscle fiber-specific neuromuscular endings. *J. Neurosci.* **8:** 242–255.

Atwood H.L., Govind C.K., and Wu C.F. 1993. Differential ultrastructure of synaptic terminals on ventral longitudinal abdominal muscles in *Drosophila* larvae. *J. Neurobiol.* **24:** 1008–1024.

Betz W.J. and Bewick G.S. 1992. Optical analysis of synaptic vesicle recycling at the frog neuromuscular junction. *Science* **255:** 200–203.

Brand A.H. and Perrimon N. 1993. Targeted gene expression as a means of altering cell fates and generating dominant phenotypes. *Development* **118:** 401–415.

Broadie K.S. and Bate M. 1993. Development of the embryonic neuromuscular synapse of *Drosophila melanogaster*. *J. Neurosci.* **13:** 144–166.

Broadie K., Prokop A., Bellen H.J., O'Kane C.J., Schulze K.L., and Sweeney S.T. 1995. Syntaxin and synaptobrevin function downstream of vesicle docking in *Drosophila*. *Neuron* **15:** 663–673.

Budnik V. and Gorczyca M. 1992. SSB, an antigen that selectively labels morphologically distinct synaptic boutons at the *Drosophila* larval neuromuscular junction. *J. Neurobiol.* **23:** 1054–1065.

Budnik V., Koh Y.-H., Guan B., Hartmann B., Hough C., Woods D., and Gorczyca M. 1996. Regulation of synapse structure and function by the *Drosophila* tumor suppressor gene *dlg*. *Neuron* **17:** 627–640.

Callahan C.A. and Thomas J.B. 1994. Tau-β-galactosidase, an axon-targeted fusion protein. *Proc. Natl. Acad. Sci.* **91:** 5972–5976.

Cerezo J.R., Jimenez F., and Moya F. 1995. Characterization and gene cloning of *Drosophila* syntaxin 1 (Dsynt1): The fruit fly homologue of rat syntaxin 1. *Brain Res. Mol. Brain Res.* **29:** 245–252.

Chapman E.R., An S., Barton N., and Jahn R. 1994. SNAP-25, a t-SNARE which binds to both syntaxin and synaptobrevin via domains that may form coiled coils. *J. Biol. Chem.* **269:** 27427–27432.

Davis G.W. and Goodman C.S. 1998. Synapse-specific control of synaptic efficacy at the terminals of a single neuron. *Nature* **392:** 82–86.

Deitcher D.L., Ueda A., Stewart B.A., Burgess R.W., Kidokoro Y., and Schwarz T.L. 1998. Distinct requirements for evoked and spontaneous release of neurotransmitter are revealed by mutations in the *Drosophila* gene *neuronal-synaptobrevin*. *J. Neurosci.* **18:** 2028–2039.

Dornan S., Jackson A.P., and Gay N.J. 1997. α-adaptin, a marker for endocytosis, is expressed in complex patterns during *Drosophila* development. *Mol. Biol. Cell.* **8:** 1391-1403.

Estes P.S., Roos J., van der Bliek A., Kelly R.B., Krishnan K.S., and Ramaswami M. 1996. Traffic of dynamin within individual *Drosophila* synaptic boutons relative to compartment-specific markers. *J. Neurosci.* **16:** 5443–5456.

Fujita S.C., Zipursky S.L., Benzer S., Ferrus A., and Shotwell S.L. 1982. Monoclonal antibodies against the *Drosophila* nervous system. *Proc. Natl. Acad. Sci.* **79:** 7929–7933.

Giniger E., Wells W., Jan L.Y., and Jan Y.N. 1993. Tracing neurons with a kinesin-β-galactosidase fusion protein. *Roux's Arch. Dev. Biol.* **202:** 112–122.

Gonzalez-Gaitan M. and Jäckle H. 1997. Role of *Drosophila* α-adaptin in presynaptic vesicle recycling. *Cell* **88:** 767–776.

Goodman C. and Shatz C.J. 1993. Developmental mechanisms that generate precise patterns of neuronal connectivity. *Cell* (suppl.) **72:** 77–98.

Gorczyca M., Augart C., and Budnik V. 1993. Insulin-like peptide and insulin-like receptor are localized at neuromuscular junctions in *Drosophila*. *J. Neurosci.* **13:** 3692–3704.

Gorczyca M., Phillis R., and Budnik V. 1994. The role of *tinman*, a mesodermal cell fate gene, in axon pathfinding during the development of the transverse nerve in *Drosophila*. *Development* **120:** 2143–2152.

Guan B., Hartmann B., Koh Y.H., Gorczyca M., and Budnik V. 1996. The *Drosophila* tumor suppressor gene, *dlg*, is involved in structural plasticity at a glutamatergic synapse. *Curr. Biol.* **6:** 695-706.

Halpern M.E., Chipa A., Johansen J., and Keshishian H. 1991. Growth cone behavior underlying the development of stereotypic synaptic connections in *Drosophila* embryos. *J. Neurosci.* **11:** 3227–3238.

Harrison S.D., Broadie K., van de Goor J., and Rubin G.M. 1994. Mutations in the *Drosophila rop* gene suggest a function in general secretion and synaptic transmission. *Neuron* **13:** 555–566.

Hoshino M., Suzuki E., Nabeshima Y., and Hama C. 1996. Hikaru genki protein is secreted into synaptic clefts from an early stage of synapse formation in *Drosophila*. *Development* **122:** 589–597.

Jan L.Y. and Y.N. Jan. 1982. Antibodies to horseradish peroxidase as specific neuronal markers in *Drosophila* and in grasshopper embryos. *Proc. Natl. Acad. Sci.* **79:** 2700–2704.

Jia X., Gorczyca M., and Budnik V. 1993. Ultrastructure of neuromuscular junctions in *Drosophila*: Comparison of wild type and mutants with increased excitability. *J. Neurobiol.* **24:** 1025–1044.

Johansen J., Halpern M.E., and Keshishian H. 1989b. Axonal guidance and the development of muscle fiber-specific innervation of *Drosophila* embryos. *J. Neurosci.* **9:** 4318–4332.

Johansen J., Halpern M.E., Johansen K.M., and Keshishian H. 1989a. Stereotypic morphology of glutamatergic synapses on identified muscle cells of *Drosophila* larvae. *J. Neurosci.* **9:** 710–725.

Keshishian H., Chiba A., Chang T.N., Halfon M.S., Harkins E.W., Jarecki J., Wang L., Anderson M., Cash S., Halpern M.E., and Johansen J. 1993. Cellular mechanisms governing synaptic development in *Drosophila melanogaster*. *J. Neurobiol.* **24:** 757–787.

Klagges B.R.E., Heimbeck G., Godenschwege T.A., Hofbauer A., Pflugfelder G.O., Reifegerste R., Reisch D., Schaupp M., Buchner S., and Buchner E. 1996. Invertebrate synapsins: A single gene codes for several isoforms in *Drosophila*. *J. Neurosci.* **16:** 3154–3165.

Klingauf J., Kavalali E.T., and Tsien R.W. 1998. Kinetics and regulation of fast endocytosis at hippocampal synapses. *Nature* **394:** 581–585.

Kuromi H. and Kidokoro Y. 1998. Two distinct pools of synaptic vesicles in single presynaptic boutons in a temperature-sensitive *Drosophila* mutant, shibire. *Neuron* **20:** 917–925.

Lahey T., Gorczyca, M., Jia X., and Budnik V. 1994. The *Drosophila* tumor suppressor gene, *dlg*, is required for normal synaptic bouton structure. *Neuron* **13:** 823–835.

Lee H.M. and Wyse G.A. 1991. Immunocytochemical localization of octopamine in the central nervous system of *Limulus polyphemus*: A light and electron microscopic study. *J. Comp. Neurol.* **307:** 683–694.

Littleton J.T. and Bellen H.J. 1995. Presynaptic proteins involved in exocytosis in *Drosophila melanogaster*: A genetic analysis. *Inv. Neurosci.* **1:** 3–13.

Littleton J.T., Bellen H.J., and Perin M.S. 1993. Expression of synaptotagmin in *Drosophila* reveals transport and localization of synaptic vesicles to the synapse. *Development* **118:** 1077–1088.

Littleton J.T., Upton L., and Kania A. 1995. Immunocytochemical analysis of axonal outgrowth in synaptotagmin. *J. Neurochem.* **65:** 32–40.

Manzanares M., Marco R., and Garesse R. 1993. Genomic organization and developmental pattern of expression of the engrailed gene from the brine shrimp Artemia. *Development* **118:** 1209–1219.

Monastirioti M., Gorczyca M., Rapus J., Eckert M., White K., and Budnik V. 1995. Octopamine immunoreactivity in the fruit fly *Drosophila melanogaster*. *J. Comp. Neurol.* **356:** 275–287.

Nighorn A., Healy M.J., and Davis R.L. 1991. The cyclic AMP phosphodiesterase encoded by the *Drosophila dunce* gene is concentrated in the mushroom body neuropil. *Neuron* **6:** 455–467.

Patel N.H. 1994. Imaging neuronal subsets and other cell types in whole-mount *Drosophila* embryos and larvae using antibody probes. *Mol. Cell Biol.* **44:** 445–487.

Pongs O., Lindemeire J., Zhu X.R., Theil T., Engelkamp D., Krah-Jentgens I., Lambrecht H.G., Koch K.W., Schwemer J., and Rivosecchi R. 1993. Frequenin—A novel calcium-binding protein that modulates synaptic efficacy in the *Drosophila* nervous system. *Neuron* **11:** 15–28.

Prokop A., Landgraf M., Rushton E., Broadie K., and Bate M. 1996. Presynaptic development at the *Drosophila* neuromuscular junction: Assembly and localization of presynaptic active zones. *Neuron* **17:** 617–626.

Ramaswami M., Krishnan K.S., and Kelly R.B. 1994. Intermediates in synaptic vesicle recycling revealed by optical imaging of *Drosophila* neuromuscular junctions. *Neuron* **13:** 363–375.

Reichmuth C., Becker S., Benz M., Debel K., Reisch D., Heimbeck G., Hofbauer A., Klagges B., Pflugfelder G.O., and Buchner E. 1995. The *sap47* gene of *Drosophila melanogaster* codes for a novel conserved neuronal protein associated with synaptic terminals. *Brain Res. Mol. Brain Res.* **32:** 45–54.

Roos J. and Kelly R.B. 1998. Dap160, a neural-specific Eps15 homology and multiple SH3 domain-containing protein that interacts with *Drosophila* dynamin. *J. Biol. Chem.* **273:** 19108–19119.

Saitoe M., Tanaka S., Takata K., and Kidokoro Y. 1997. Neural activity affects distribution of glutamate receptors during neuromuscular junction formation in *Drosophila* embryos. *Dev. Biol.* **184:** 48–60.

Salzberg A., Cohen N., Halachmi N., Kimchie Z., and Lev Z. 1993. The *Drosophila Ras 2* and *Rop* gene pair: A dual homology with a yeast Ras-like gene and a suppressor of its loss-of-function phenotype. *Development* **117:** 1309–1319.

Schulze K., Broadie K., Perin M., and Bellen H.J. 1995. Genetic and electrophysiological studies of *Drosophila* syntaxin-1A demonstrate its role in nonneuronal secretion and neurotransmission. *Cell* **80:** 311–320.

Schuster C.M., Davis G.W., Fetter R.D., and Goodman C.S. 1996. Genetic dissection of structural and functional components of synaptic plasticity. I. Fasciclin II controls synaptic stabilization and growth. *Neuron* **17:** 641–654.

Sone M., Hoshino M., Suzuki E., Kuroda S., Kaibuchi K., Nakagoshi H., Saigo K., Nabeshima Y.-I., and Hama C. 1997. Still life, a protein in synaptic terminals of *Drosophila* homologous to GDP-GTP exchanges. *Science* **275:** 543–547.

Stimson D.T., Estes P.S., Smith M., Kelly L.E., and Ramaswami M. 1998. A product of the *Drosophila stoned* genetic locus regulates neurotransmitter release. *J. Neurosci..* **18:** (in press).

Sweeney S.T., Broadie K., Keane J., Niemann H., and O'Kane C.J. 1995. Targeted expression of tetanus toxin light chain in *Drosophila* specifically eliminates synaptic transmission and causes behavioral defects. *Neuron* **14:** 341–351.

Thomas U., Kim E., Kuhlendahl S., Koh Y.H., Gundelfinger E.D., Sheng M., Garner C.C., and Budnik V. 1997. Synaptic clustering of the cell adhesion molecule fasciclin II by discs-large and its role in the regulation of presynaptic structure. *Neuron* **19:** 787–799.

van de Goor J., Ramaswami M., and Kelly R. 1995. Redistribution of synaptic vesicles and their proteins in temperature-sensitive *shibire* (ts1) mutant *Drosophila. Proc. Natl. Acad. Sci.* **92:** 5739–5743.

Woods D.F. and Bryant P.J. 1991. The *discs-large* tumor suppressor gene of *Drosophila* encodes a guanylate kinase homolog localized at septate junctions. *Cell* **66:** 451–464.

Wu M.N. and Bellen H.J. 1997. The dissection of synaptic transmission in *Drosophila. Curr. Opin. Neurobiol.* **7:** 624–630.

Yoshihara M., Rheuben M.B., and Kidokoro Y. 1997. Transition from growth cone to functional motor nerve terminal in *Drosophila* embryos. *J. Neurosci.* **17:** 8408–8426.

Zhang B., Koh Y.H., Beckstead R.B., Budnik B., Ganetzky B., and Bellen H.J. 1999. Synaptic vesicle size and number are regulated by a synaptic adaptor protein required for endocytosis. *Neuron* **21:** 1465–1475.

Zhong Y. and Pena L.A. 1995. A novel synaptic transmission mediated by a PACAP-like neuropeptide in *Drosophila. Neuron* **14:** 527–536.

Zinsmaier K.E., Eberle K.K., Buchner E., Walter N., and Benzer S. 1994. Paralysis and early death in cysteine string protein mutants of *Drosophila. Science* **263:** 977–980.

Zinsmaier K.E., Hofbauer A., Heimbeck G., Pflugfelder G.O., Buchner S., and Buchner E. 1990. A cysteine-string protein is expressed in retina and brain of *Drosophila. J. Neurogenet.* **7:** 15–29.

# CONTENTS

# Histological Techniques for the *Drosophila* Eye Part I: Larva and Pupa

**Tanya Wolff**
*Washington University School of Medicine*
*Department of Genetics*
*St. Louis, Missouri 63110*

T HE *DROSOPHILA* EYE IS ESSENTIALLY A LIVING 96-well plate which can be manipulated using molecular and genetic tools. The fly eye has been used to study a broad range of topics, such as signal transduction, cell-fate specification, morphogenesis, and cell death. A number of features have led to the popularity of the eye for studying these and other topics. For example, cellular analysis of the eye is straightforward due to its relatively simple cellular architecture: Not only are the ommatidia, or unit eyes of the compound eye, composed of a small number of cells (19), but these cells can also be readily identified on the basis of their shape and location within an ommatidium. In addition, because the eye is a monolayer epithelium, all cells can be viewed in whole-mount preparations in both the larval eye imaginal disc and the pupal eye. Mosaic analysis is a powerful genetic tool that is routinely used in the eye to determine a given cell's requirement for a gene of interest. This type of analysis is possible because ommatidia are not clonal units; rather, cell fates are determined on the basis of positional cues. Furthermore, pattern formation proceeds as a wave of morphogenesis sweeps across the eye disc, and thus all early patterning events are laid out in a single imaginal disc preparation; because there are 750 ommatidia per eye, each developmental event occurs 750 times within a single preparation. A final attractive feature of the eye is that it is dispensable for viability and development, and therefore a number of tools have been developed for its manipulation.

Chapters 12 and 13 provide a collection of widely used protocols for study of the *Drosophila* eye. Chapter 12 covers protocols for preparation of larval and pupal tissue, and Chapter 13 includes protocols for preparation of adult tissue. Important notes often follow each protocol; the reader is encouraged to check these notes before starting a given procedure.

The methods outlined in this chapter have been modified by many anonymous individuals, so it is impossible to acknowledge all those who deserve credit for their contributions. Dissection instructions for the retrieval of eye tissue from larvae and pupae are provided in Protocol 12.1 (below). A list of the protocols presented in this chapter as well as the most common use for each is given in Table 12.1. Tables 12.2 and 12.3 identify fre-

**Table 12.1.** Index of Protocols and Their Primary Use

| Protocol | Type of staining | Primary use |
|---|---|---|
| 12.2 | antibody staining | identification of specific cell type |
| 12.3 | β-galactosidase detection | identification of specific cell type |
| 12.4 | BrdU labeling | identification of proliferating cells |
| 12.5 | in situ hybridization | analysis of gene expression |
| 12.6 | cobalt sulfide | highlights cell outlines |
| 12.7 | acridine orange | cell-death analysis (unfixed tissue) |
| 12.8 | TUNEL staining | cell-death analysis (fixed tissue) |
| 12.9 | phalloidin | highlights cell outlines |
| 12.10 | DAPI | nuclear marker |
| | double-labeling techniques | identification of several cell types in one preparation |
| 12.2 | using fluorescent antibodies | |
| 12.2 | using HRP-conjugated antibodies | |
| Protocol 12.3 Notes | staining with antibodies and X-gal | |
| Protocol 12.5 Notes | X-gal staining and in situ hybridization | |

quently used antibodies and enhancer trap lines, respectively. Chapter 13 provides a procedure for the preparation of larval eye-antennal imaginal discs and pupal retinas for transmission electron microscopy. For an overview of eye development and pattern formation, see Wolff and Ready (1993), and for a molecular review of eye development, see Dickson and Hafen (1993).

**Table 12.2.** Commonly Used Antibodies

| Antibody | Cell type | Localization | References |
|---|---|---|---|
| Elav | all neurons | nucleus | Robinow and White (1991) |
| BP104 | all peripheral neurons | membrane | Hortsch et al. (1990) |
| Sev | R3, R4, R7, CC mystery cells | apical region of cell | Tomlinson et al. (1987) |
| Boss | R8 (third instar), all R (adult) | membrane | Krämer et al. (1991) |
| Scabrous | R8 (after initial expression in clusters in furrow) | secreted | Baker et al. (1990); Mlodzik et al. (1990); Lee et al. (1996) |
| Rough | R3, R4, R2, R5, broad in furrow | nucleus | Kimmel et al. (1990) |
| Bar | R1, R6, primary pigment cells | nucleus | Higashijima et al. (1992) |
| Prospero | R7, CC | nucleus | Kauffmann et al. (1996) |
| Cut | CC | nucleus | Blochlinger et al. (1993) |
| Sparkling | CC, primary pigment cells, bristles | nucleus | Fu and Noll (1997) |
| Rh3, Rh4 | nonoverlapping sets of R7 cells (adults) | membrane | Feiler et al. (1992) |
| Rh1 | R1–R6 | membrane | Suzuki and Hirosawa (1991) |
| Atonal | broad stripe in furrow, then R8 | nucleus | Jarman et al. (1994) |
| Hairy | stripe anterior to Atonal | nucleus | Brown et al. (1991) |
| Dachshund | stripe anterior to furrow; R1, R6, R7, CC | nucleus | Mardon et al. (1994) |

This table lists the commonly used antibodies, subsets of cells in which they are expressed, their subcellular localization, and references describing the antibodies.

**Table 12.3.** Commonly Used Enhancer-trap Lines

| Enhancer trap | Cell type | Cytology |
|---|---|---|
| BGT-A141* | R3, R4/R7, AC, PC | 22D1 |
| G90* | R3, R4/**CC** | 34A1–2 |
| X38* | R1, R6/AC, PC/EQC, PLC | 86E16–19 |
| Z45* | R1, R6, R7 | 34C4–5 |
| AB64* | **R3, R4, R7** | 44C1–2 |
| P82* | R3, R4, R7 | 44C |
| BGH123* | R3, **R4** | 88F |
| AE127*/H162*(sev-up; Mlodzik et al. 1990) | **R3, R4**/R1, R6 | 87B4–5 |
| rK519* | R8 | 66A1–2 |
| H214* | R7 | 94D1–4 |
| l(2)07056 (Star) | R8, R2, R5 | 21E2–3 |
| R122* | equatorial, posterior to furrow | 53C1–2 |
| Rh4-*lacZ* | R7 axons | |
| Rh1-*lacZ* | R1–R6 axons | |
| dpp | furrow | |
| wg | dorsal/ventral margins | |
| E4 (gl-*lacZ*–Kevin Moses) | photoreceptor axons | |

This table lists the commonly used enhancer-trap lines, their expression patterns, and cytology, if known. Check these lines to make sure staining is as advertised before embarking on a complicated genetic scheme requiring many crosses. Bold indicates that the enhancer trap is expressed strongly in this subset of cells.

*These lines can be obtained from the Rubin Lab, LSAB Room 539, Dept. of MCB, HHMI, University of California, Berkeley, California 94720.

## PROTOCOL 12.1*

# Dissection Techniques

### Materials

### *Supplies and Equipment*

Dissecting microscope

Dissecting Tools

- Forceps. Dumont #5 (Ted Pella 505-NM; or Fine Science Tools 11252-30) or Dumont #55 (Fine Science Tools 11255-20) forceps are recommended; the #55 forceps have lighter and finer shanks than the #5 forceps.
- Dissecting scissors. Pascheff-Wolff microscissors by Moria are expensive but work well (Fine Science Tools 15371-92). Less expensive models also work and are available from Fine Science Tools.
- Dissection needles. For preparation, see below.

Sylgard dissection dish. This is an indispensable tool for dissections, as the soft base prevents the dissecting tools from getting damaged. It is also useful for performing the wash steps in many of the protocols outlined in this chapter. To prepare the dissecting dish, fill a 14-mm glass petri dish approximately half way with Sylgard (recipe is included with the product; Sylgard 184 Silicone Elastomer Kit [Dow Corning]; K.R. Anderson and Co. [distributor], 2800 Bowers Avenue, Santa Clara, California 95051; Phone: 408-727-2800).

Slides and coverslips

Double-stick tape

*For all protocols, H$_2$O indicates glass distilled and deionized.

### *Preparation of Dissection Needles*

Requirements:

> Tungsten rod, 0.005 x 4 inches (A-M Systems 7161)
>
> Monoject 200 hypodermic needles with aluminum hub, 27 gauge, 1/2 in. length. Manufactured by Kendall Medical (15 Hampshire St., Mansfield, Massachusetts 02048; Phone: 1-800-325-7472).

Insert a piece of tungsten rod into an aluminum hub hypodermic needle, leaving ~1 cm extending out from the tip of the needle. Crimp the needle in two spots, one to stabilize the tungsten rod and the other to bring the needle back to a parallel position relative to the syringe. Use a 1-cc syringe as a "handle" for the needle. Alternatively, pin vises can be ordered from Ernest Fullam, Inc. (54270).

The needles can be sharpened either electrolytically or chemically. If using a Variab (110V; Fisher 09-521-110), dip the needle in a 1 M **KOH** solution. Start with the rheostat set to a low voltage and carefully dial the voltage up until bubbles start to form. The chemical alternative is to dip the needle in a solution of boiling **sodium nitrite** contained in a nickel crucible. (Be cautious! A vat of boiling sodium nitrite is very dangerous! This procedure should be carried out in a chemical fume hood.)

## Solutions and Reagents

Mounting medium; the type of mounting medium depends on the staining protocol

### DISSECTION BUFFERS

Several buffers are suitable for dissection. Three commonly used buffers (1x PBS, *Drosophila* Ringer's solution, and 0.1 M sodium phosphate buffer) are given below:

#### *10x PBS*

| Component and final concentration | Amount to add to make 1 liter |
|---|---|
| 1.3 M NaCl | 75.97 g |
| 70 mM **$Na_2HPO_4 \cdot 2H_2O$** | 12.46 g |
| 30 mM **$NaH_2PO_4 \cdot 2H_2O$** | 4.8 g |

Combine all components in less than 1 liter of $H_2O$ and stir to dissolve. Adjust pH to 7.2 and bring to final volume of 1 liter with $H_2O$. Sterilize by autoclaving. Use as a 1x solution (final concentration: 130 mM NaCl, 7 mM $Na_2HPO_4 \cdot 2H_2O$, 3 mM $NaH_2PO_4 \cdot 2H_2O$).

#### Drosophila *Ringer's Solution (Tübingen and Düsseldorf)*

| Component and final concentration | Amount to add to make 1 liter |
|---|---|
| 182 mM **KCl** | 13.6 g |
| 46 mM NaCl | 2.7 g |
| 3 mM **$CaCl_2 \cdot 2H_2O$** | 0.33 g |
| 10 mM Tris-Cl | 1.21 g Tris base |

Combine all components in less than 1 liter of $H_2O$ and stir to dissolve. Adjust pH to 7.2 with 1 N HCl and bring to a final volume of 1 liter with $H_2O$. Pass through a 0.47-μm filter and sterilize by autoclaving. This is a 1x solution.

#### *Sodium Phosphate Buffer*

Prepare the following 2 stock solutions:

> (A) 0.2 M $Na_2HPO_4 \cdot$ (35.61 g/liter $Na_2HPO_4 \cdot 2H_2O$)
> (B) 0.2 M $NaH_2PO_4 \cdot$ (31.2 g/liter $NaH_2PO_4 \cdot 2H_2O$)

To prepare 0.1 M sodium phosphate (pH 7.2), combine the following:

| | |
|---|---|
| 0.2 M $Na_2HPO_4$ | 36 ml |
| 0.2 M $NaH_2PO_4$ | 14 ml |
| $H_2O$ | 50 ml |

## FIXATIVES

The following fixatives are used in protocols discussed in this chapter.

### 2× PEM

| | |
|---|---|
| PIPES (dipotassium salt) | 7.56 g |
| EGTA | 152 mg |
| 0.1 M **MgSO$_4$** | 2 ml |
| H$_2$O | to make 100 ml |

Adjust to pH 7.0. Filter-sterilize and store at 4°C for up to 1 year.

### Formaldehyde

Do not use commercial **formaldehyde** for histological preparations as it is unstable. Prepare formaldehyde from **paraformaldehyde** and store as a 20% stock at 4°C for 1–2 weeks, depending on the preparation.

To make a 20% stock solution, add 200 mg of EM-grade paraformaldehyde per milliliter of H$_2$O and heat at 60°C on a stir plate (in a ventilated chemical fume hood) to dissolve. Add a trace of **NaOH** to help dissolve the paraformaldehyde (no more than 1 ml of 1 N NaOH to 100 ml of H$_2$O). Do not overheat; if the solution boils or turns brown, discard and start over. Adjust pH with 1 N **HCl** (pH should be ~7.0). Formaldehyde is typically diluted to 4% in the desired buffer.

### Glutaraldehyde

Use EM-grade **glutaraldehyde** (Ted Pella 18426). For most uses (EM is an exception), open ampules of glutaraldehyde can be used for several months if kept covered with Parafilm and stored at 4°C.

### PLP

Prepare fresh immediately before use; it is not necessary to pH the final solution. PLP is a gentle fix and should be used on ice. The composition of PLP is:

0.075 M lysine

0.037 M sodium phosphate (pH 7.2)

2% formaldehyde

0.01 M **sodium m-periodate** (NaIO$_4$)

To make PLP, add 0.36 g of lysine to 10 ml of cold H$_2$O, 7.5 ml of cold 0.1 M NaH$_2$PO$_4$ (pH 7.2), and 2.5 ml of cold 0.1 M Na$_2$HPO$_4$ (pH 7.2). Immediately before use, mix 15 ml of this buffered lysine solution with 5 ml of cold 8% formaldehyde (freshly prepared) and 50 mg of NaIO$_4$.

CAUTION: CaCl$_2$, formaldehyde, glutaraldehyde, HCl, KCl, KOH, MgSO$_4$, Na$_2$HPO$_4$, NaH$_2$PO$_4$, NaIO$_4$, NaOH, paraformaldehyde, sodium nitrite, sodium phosphate (see Appendix 4)

## Method

### Larval Eye Imaginal Discs

Perform the following dissection with the larva immersed in dissection buffer.

1. To remove the eye-antennal imaginal discs from the larva (Figures 12.1B and 12.3A), use one pair of forceps to gently hold the larva at approximately one third of its length from the interior end. With a second set of forceps, grab a firm hold at the base of the mouth hooks and then pull the mouth parts away from the rest of the body (Figure 12.1A).

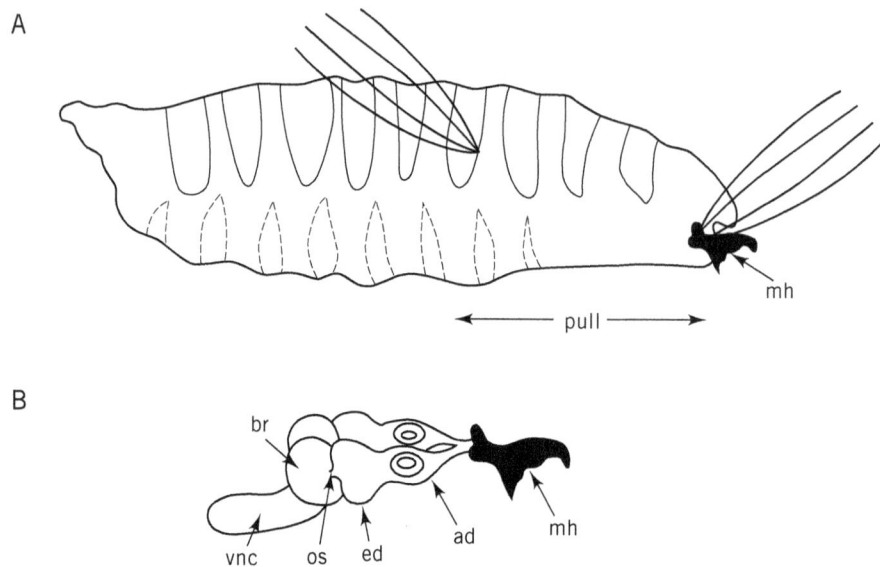

**Figure 12.1.** Larval dissection. For details, see Protocol 12.1. (*A*) Grasping third-instar larva with forceps. (*B*) Isolated CNS/eye-antennal disc complex. (ad) Antennal disc; (br) brain; (ed) eye disc; (mh) mouth hooks; (os) optic stalk; (vnc) ventral nerve cord.

*Note:* Typically, the brain with attached eye-antennal imaginal discs and salivary glands, as well as other tissues, will be removed as a single mass.

2. Remove all extraneous tissue, leaving the brain with attached discs connected to the mouth hooks (Figure 12.1B).

   *Note:* The brain can be removed before transferring the complex into fixative; however, this procedure is more easily performed while the tissue is in the fixative (step 4, below).

3. Transfer the mouth parts with attached discs into the fixative.

4. To remove the brain, use a tungsten needle to sever the optic stalks (Figure 12.1B) and then gently pull the brain away from the eye-antennal disc. Alternatively, if it is difficult to sever the optic stalks, brace the middle of the brain down on the dissecting dish with forceps and then pull on the mouth hooks.

   *Notes:* The disadvantage of the alternative approach is that the eye discs sometimes remain attached to the brain and are consequently "lost."

   The mouth parts serve as a convenient means of transferring the eye discs from one solution to the next throughout the staining procedure, so it is preferable to leave them attached until the final mounting stages. To transfer the eye-antennal imaginal discs, spear the mouth hooks with a tungsten needle.

5. (*Optional*) The peripodial membrane (ppm), which lies above the apical surface of the disc epithelium, can limit the accessibility of some antibodies to the tissue and interfere with viewing of some samples; in these cases, it will be necessary to remove the ppm. This membrane is most easily removed while the discs are in fixative. There is a gap between the ppm and the imaginal disc epithelium at the junction between the eye and antennal discs. Use a pair of tungsten needles to anchor the eye disc by holding the antennal disc against the base of the dissecting dish. Slip the second tungsten needle beneath the ppm at the junction between the eye and antennal discs and gently peel the membrane back towards the optic stalk.

*Note:* Although it is not necessary to completely remove the ppm at this stage, it is necessary to make a significant hole so that antibodies can readily access the underlying epithelium. Any remaining ppm can be removed following staining and before mounting the discs on the slide.

6. Stain samples according to the type of preparation and procedure of choice (Protocols 12.2 through 12.10).

7. To mount the discs, separate the eye-antennal discs from the mouth parts and transfer to mounting medium. Transfer the discs (in mounting medium) to a slide using a P20 Pipetman. To prevent the tissue from folding over on itself during mounting, use the minimal amount of mounting medium required to cover the area beneath the coverslip.

### *Pupal Retina*

1. Place a piece of double-stick tape in a Sylgard dissecting dish and tape the pupa, ventral-side down, on the tape. If the pupal case is wet, lay the animal on a Kimwipe to wick off excess moisture before placing it on double-stick tape.

2. Remove the operculum ("trap door"; Figure 12.2A) and then use forceps to grasp the dorsal surface of the cuticle and peel it away, a bit at a time (Figure 12.2B).

3. Once the animal is exposed, gently remove it from the cuticle and place it on the double-stick tape at a 45° angle, posterior-side down (Figure 12.2C).

4. Immerse the animal in a drop of dissecting buffer. Use the dissecting scissors to gently snip the "nose," and then cut the inner cuticle up to its "forehead" and then down to its "chin" (Figure 12.2C).

5. Use a P20 Pipetman to gently suck out the contents of the head. Within these contents are found radar dish-shaped structures, each attached to a round ball; these are the eyes attached to the brain (Figure 12.2D). Transfer tissue to fixative.

   *Note:* Once the eyes have been fixed, the brain complex provides a convenient means of transferring the eyes from one solution to the next, so leave the eyes attached to the brain during processing. Pupal eyes are most easily transferred by spearing the brain with a tungsten needle.

6. Stain samples according to the type of preparation and procedure of choice (Protocols 12.2 through 12.10).

7. Following staining, transfer eyes/brain to a drop of mounting medium in a Sylgard dish. Remove the eyes from the brain using tungsten needles to gently tease the eyes away from the brain.

8. Transfer eyes to a slide using a P20 Pipetman. Mount in a small volume (~10 μl) of mounting medium or enough to coat the coverslip.

PROTOCOL 12.2

## Antibody Staining

Staining tissue with antibodies is generally quite straightforward, although fixation and permeabilization conditions must be altered for some antibodies (see Protocol 12.2 Notes). Ideal concentrations will have to be determined for each antibody, but a general rule of thumb is to dilute monoclonal supernatants 1:1 and ascites or sera 1:250 to 1:5000.

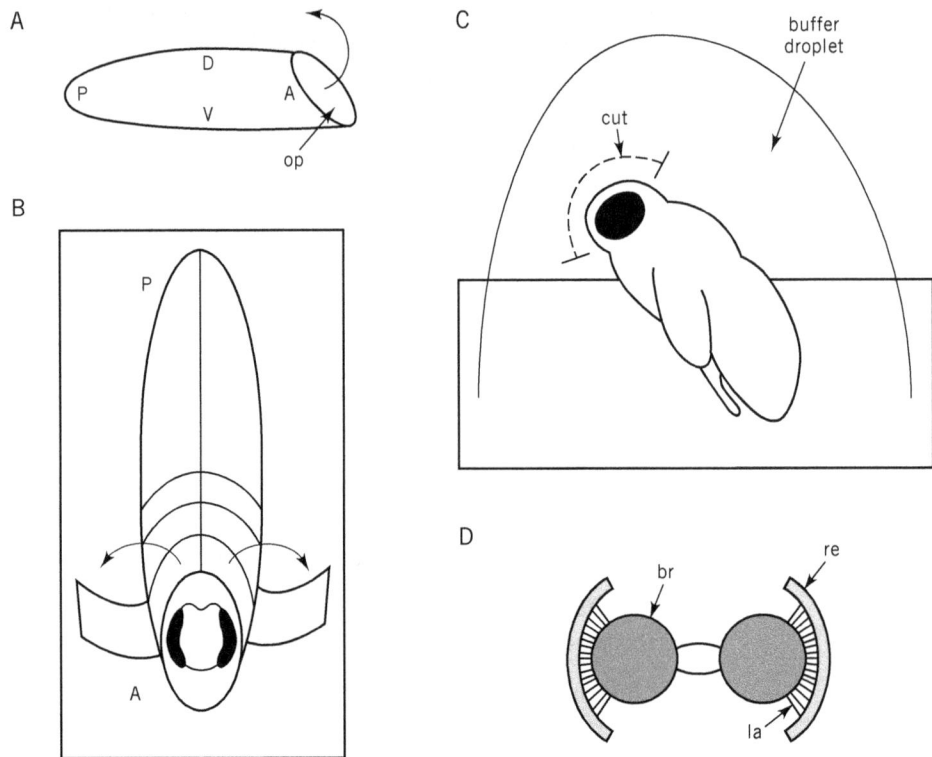

**Figure 12.2.** Pupal dissection. For details, see Protocol 12.1. (*A*) Removal of operculum from pupal case, lateral view. (*B*) Removal of pupal case, dorsal view. Box indicates surface of double-stick tape. (*C*) Dissection of pupal head, lateral view. (*D*) Isolated brain/retina complex. (A) Anterior; (br) brain; (D) dorsal; (la) lamina; (op) operculum; (P) posterior; (re) retina; (V) ventral.

All steps, with the exception of the antibody incubations and 3,3′-diaminobenzidene (DAB) reactions, can be performed in Sylgard dissecting dishes. Antibody incubations and DAB reactions can be performed in 60-well microwell plates (Nunclon).

Note that DAB is a potent carcinogen and should be handled with extreme care. Soak all materials that become contaminated with DAB overnight in bleach and dispose of them according to standard regulations.

## Materials

### *Supplies and Equipment*

Dissecting tools and equipment, Sylgard dissection dishes, and other materials (see p. 203)

Microwell plates (60-well, Nunclon). These plates are used for antibody incubations and DAB reactions. Up to five sets of discs or pupal eyes can be placed in each well. Antibody incubations can be performed in a minimal volume of antibody (12–15 μl). To prevent evaporation of the antibody, place a folded, damp Kimwipe at one end of the plate (preferably the end opposite the tissue).

Slides and coverslips

### *Solutions and Reagents*

Dissection buffer, e.g., 1x PBS, *Drosophila* Ringer's solution, or 0.1 M **sodium phosphate** (pH 7.2) (for preparation, see p. 204)

**Formaldehyde** (4%) in 0.1 M sodium phosphate (pH 7.2; for preparation, see p. 205)
Sodium phosphate (0.1 M, pH 7.2)
Mounting medium

- *If using a fluorescently tagged secondary antibody:* Antibleach mounting medium (Vectashield, Vector Laboratories).
- *If using an HRP-coupled secondary antibody:* 80% glycerol in dissection buffer. Low-purity glycerol contains contaminants that cause the horseradish peroxidase (HRP) reaction product to fade very rapidly. This problem can be avoided by using ultrapure glycerol (99.5% pure). Sigma and Boehringer Mannheim both carry this product. Low pH can also cause the HRP reaction product to fade, so make sure that the glycerol solution is at pH 7.2.

### Sodium Phosphate/TX

0.1 M sodium phosphate (pH 7.2)
0.3% Triton X-100

### Sodium Phosphate/TX/5% NGS

0.1 M sodium phosphate (pH 7.2)
0.3% Triton X-100
5% NGS (normal goat serum)
Inactivate the serum (GIBCO-BRL) by heating to 56ºC for 30 minutes. Sterilize by passing through a 0.2-μm filter and aliquot into sterile tubes. Store at –20ºC. Once thawed, store at 4ºC. If contaminants start to grow in aliquots stored at 4ºC, discard and use a fresh aliquot.

### DAB Staining Solution

0.5 mg/ml **DAB**
0.1 M sodium phosphate (pH 7.2)
0.003% $H_2O_2$
For HRP-signal intensification, add 1.5 mM **$NiCl_2$**.

## Antibodies

Primary antibody
Secondary antibody (e.g., fluorescein-conjugated or HRP-coupled secondary antibodies)

**CAUTION: DAB, formaldehyde, $H_2O_2$, $NiCl_2$, sodium phosphate** (see Appendix 4)

## Method

### Labeling with a Single Antibody

This is a standard antibody-staining procedure. For eye-antennal imaginal disc stained with anti-Elav antibody, a neuron-specific antibody, see Figure 12.3A.

1. Dissect tissue in a suitable buffer (e.g., 1x PBS, *Drosophila* Ringer's solution, or 0.1 M sodium phosphate [pH 7.2]) as described in Protocol 12.1.

2. Transfer to 4% formaldehyde in 0.1 M Sodium Phosphate and fix at room temperature for 15 minutes.

3. Rinse tissue three times in Sodium Phosphate/TX (10 minutes each wash).

4. Use a tungsten needle to transfer the tissue to a microwell plate containing the primary antibody diluted to the appropriate concentration in Sodium Phosphate/TX/5% NGS. Incubate at 4ºC overnight.

*Note:* Satisfactory staining can be obtained for many antibodies following incubation at room temperature for 2 hours.

5. Remove tissue from microwells using a P20 Pipetman and place in a drop of Sodium Phosphate/TX in a Sylgard dissecting dish. Rinse three times in Sodium Phosphate/TX (10 minutes each wash).

6. Use a tungsten needle to transfer tissue to secondary antibody in a microwell plate and incubate at room temperature for 2 hours.

7. Depending on the secondary antibody, perform either of the following steps:

   a. If a fluorescein-conjugated secondary antibody is used, treat tissue as follows:

      i. Rinse discs three times in Sodium Phosphate/TX (10 minutes each wash).

      ii. Mount in antibleach medium (Vectashield, Vector Laboratories).

   b. If an HRP-coupled secondary antibody is used, treat tissue as follows:

      i. Rinse tissue in 0.1 M sodium phosphate (pH 7.2; no detergent).

      ii. Develop in DAB Staining Solution, monitoring the reaction under a dissecting microscope. When the tissue is suitably dark, transfer tissue to a drop of 0.1 M sodium phosphate to stop the reaction.

      iii. Mount tissue in 80% glycerol in 0.1 M sodium phosphate (pH 7.2).

      > *Note:* The signal can be intensified with the addition of 1.5 mm $NiCl_2$ to the DAB staining solution; $NiCl_2$ gives a blue-black reaction product. If the antibody gives a high background or is weak, presoak the tissue in 0.5 mg/ml DAB in 0.1 M sodium phosphate (pH 7.2) for 10–15 minutes prior to addition of $H_2O_2$.

## Double Labeling with Fluorescent Antibodies

Before proceding, confirm that the two secondary antibodies to be used will not recognize each other. For example, do not use a combination of rat anti-mouse and goat anti-rat secondary antibodies. The problem of secondary antibody cross-reactivity can be avoided if secondary antibodies raised in a single species such as goat are used.

One limitation of fluorescent double labeling is that it generally does not work if both primaries were raised in the same species, unless the primaries are directly conjugated to fluorochromes. If labeling tissue using two primaries raised in the same species, follow the histochemical procedure outlined below (Double Labeling Using HRP-conjugated Antibodies).

For double labeling tissue with two primary antibodies raised in different species, perform the following steps:

1. Perform steps 1–3 as described in the method above (Labeling with a Single Antibody; cross-references to step numbers below refer to this method).

2. Incubate the tissue simultaneously in both primary antibodies (step 4).

3. Wash the tissue as described in step 5.

4. Incubate the tissue simultaneously in both secondary antibodies conjugated to fluorochromes (step 6).

5. Rinse tissue three times in Sodium Phosphate/TX (10 minutes each wash) and mount in antibleach medium (step 7a).

### Double Labeling Using HRP-conjugated Antibodies

For choice of secondary antibodies, see discussion on p. 209. Double labeling using DAB requires that the reaction product for one of the antibodies be intensified. The following factors should be considered when deciding which primary antibody to intensify.

- The intensification process is sensitive and should therefore be used to detect the less abundant antigen. Similarly, if one of the primary antibodies gives a high background or is widely expressed throughout the tissue, this signal should not be enhanced with $NiCl_2$.
- If one antigen is nuclear and the second is either cytoplasmic or on the cell surface, it is better to intensify the nuclear signal because the nickel-enhanced product tends to be granular and may obscure delicate outlines and details recognized by the cytoplasmic/cell surface antibody.
- If one antigen is less stable, for example, if it is extracted by exposure to detergent, then this antigen should be detected first.

Perform double labeling using HRP-conjugated secondary antibodies in either of the following ways, depending on the species in which the primary antibodies were raised:

1. If the primary antibodies were raised in different species, perform the following steps:

    a. Perform steps 1–3 as described in the first method above (Labeling with a Single Antibody; cross-references to step numbers below refer to this method).

    b. Coincubate the tissue in both primary antibodies, as in step 4.

    c. Following the Sodium Phosphate/TX washes, incubate the tissue first with one secondary antibody, and then develop using DAB with $NiCl_2$ (dark blue-black reaction product; steps 5, 6, and 7b, respectively).

    d. Repeat step 5 (the three 10-minute washes in Sodium Phosphate/TX). Incubate in the second secondary antibody and develop in DAB with no intensification (brown reaction product; steps 6 and 7b).

       *Note:* Be sure to perform the intensification reaction first, as indicated above, otherwise the two reaction products may turn an identical dark-blue–black color. (For a sample preparation, see Mardon et al. 1994.)

2. If the primary antibodies were raised in the same species, perform the following steps:

    a. Perform steps 1–3 as described in the first method above (Labeling with a Single Antibody; cross-references to step numbers below refer to this method).

    b. Follow steps 4–6 and 7b; i.e., perform the primary antibody and secondary antibody incubations sequentially. In other words, begin with the first primary antibody followed by the appropriate secondary antibody and perform the DAB reaction with $NiCl_2$ intensification.

    c. Stop the reaction by washing several times with Sodium Phosphate/TX and then wash the tissue three times in Sodium Phosphate/TX (10 minutes each wash).

    d. Incubate in the second primary antibody, wash, and incubate in the appropriate secondary antibody (i.e., repeat steps 4–6).

    e. Wash and develop in DAB with no intensification. Wash and mount (step 7b, iii).

Gentler fixation or permeabilization conditions are necessary to preserve the antigenicity of some antigens. For example, fixing and processing the tissue on ice can enhance the quality of staining. Alternatively, PLP is a good membrane fix (see p. 205). However, keep in mind that it may therefore not be suitable for staining nuclear antigens, as it may reduce the accessibility of antibodies to the nucleus. If using PLP, fix on ice for 45 minutes. Finally, some antigens are extracted from tissue by strong detergents, such as Triton X-100. Staining intensity of these antigens can be increased by substituting a gentler detergent, such as 0.1% saponin (or less), for Triton X-100.

To prevent shrinkage of tissue in a permanent mounting medium such as DPX (Fluka), it must first be dehydrated by taking it through a series of ethanol dehydration steps. Incubate the tissue in 30%, 50%, 70%, 90%, and absolute ethanol (10 minutes each), equilibrate in xylene for 10 minutes, and then mount in desired medium.

PROTOCOL 12.3

# β-galactosidase Detection

A number of screens have identified *lacZ* enhancer-trap lines that express β-galactosidase in specific patterns or subsets of cells. Table 12.3 gives a short list of commonly used lines. The presence of β-galactosidase activity can be detected using either X-gal (5-bromo-4-chloro-3-indolyl β-D-galactopyranoside) or an anti-β-galactosidase monoclonal antibody. For β-galactosidase activity detected using X-gal, see Figure 12.3A.

To double-label enhancer-trap lines using X-gal and antibodies, see Protocol 12.3 Notes, for modifications to this procedure.

## Materials

### Supplies and Equipment

Dissecting tools and equipment, Sylgard dissection dish, and other supplies (see pp. 203–204)

Microwell plates (60-well)

Slides and coverslips

### Solutions and Reagents

Dissection buffer, e.g., 1× PBS, *Drosophila* Ringer's solution, or 0.1 M **sodium phosphate** (pH 7.2) (for preparation, see p. 204)

Glutaraldehyde (1%) in 1× PBS (see p. 205)

1× PBS

Sodium phosphate/0.2TX (0.1 M sodium phosphate containing 0.2% Triton X-100)

Sodium phosphate/TX (0.1 M sodium phosphate containing 0.3% Triton X-100)

Sodium phosphate/TX/5% NGS (for preparation, see p. 209)

Glycerol (80%) in 1× PBS (for important note on glycerol, see p. 209)

DAB staining solution (for preparation, see p. 209)

### Staining Solution

| | |
|---|---|
| 0.2 M Na$_2$HPO$_4$ | 1.8 ml |
| 0.2 M NaH$_2$PO$_4$ | 0.7 ml |
| 5 M NaCl | 1.5 ml |
| 1 M MgCl$_2$ | 50 µl |
| 50 mM K$_3$[Fe(CN)$_6$] | 3 ml |
| 50 mM K$_4$[Fe(CN)$_6$] | 3 ml |
| H$_2$O | 40 ml |

This solution is stable for months when stored at room temperature in a foil-wrapped container.

### X-gal Stock Solution

Prepare 8% **X-gal** in *N,N*-dimethylformamide (DMF). Store in 50-µl aliquots at –20ºC.

### Preparation of X-gal Staining Solution

Prewarm the staining solution at 37ºC for 5 minutes. Add 8 µl of 8% X-gal stock solution per 300 µl of staining solution. Prewarming decreases precipitation of X-gal crystals.

## Antibodies

Primary antibody. The recommended source for the primary antibody, rabbit anti-β-galactosidase polyclonal antibody, is Organon Teknica. Preadsorb the primary antibody (1:50 dilution) using 0–2-hour fixed embryos at 4ºC overnight with rocking. Use the anti-β-galactosidase antibody at a final dilution of 1:5000 (i.e., a 1:100 dilution of the preadsorbed aliquot).

Secondary antibody (e.g., HRP-conjugated goat anti-rabbit secondary antibody)

**CAUTION: DAB, DMF, glutaraldehyde, K$_3$[Fe(CN)$_6$], K$_4$[Fe(CN)$_6$], MgCl$_2$, Na$_2$HPO$_4$, NaH$_2$PO$_4$, sodium phosphate, X-gal** (see Appendix 4)

## Method

### X-gal Staining of β-galactosidase Activity

For original reference, see Simon et al. (1985).

1. Dissect eye-antennal discs in a suitable dissection buffer (e.g., 1x PBS, *Drosophila* Ringer's solution, or 0.1 M sodium phosphate [pH 7.2]) as described in Protocol 12.1.

2. Fix in 1% glutaraldehyde in PBS at room temperature for 15 minutes.

3. Rinse tissue twice in Sodium Phosphate/0.2TX (10 minutes each rinse).

4. Incubate tissue in X-gal Staining Solution at 37ºC for 15 minutes to overnight (the length of time depends on the strength of expression of the *lacZ* line being used).

5. Once staining is complete, rinse tissue in PBS. Equilibrate and mount in 80% glycerol/PBS.

### Immunohistochemical Detection of β-galactosidase

Follow the single-antibody-labeling method given in Protocol 12.2.

## PROTOCOL 12.3 NOTES

*Double-labeling tissue with antibodies and X-gal.* To double label enhancer-trap lines using X-gal and antibodies, the following modifications to the above procedure must be made (modifications provided by U. Heberlein; see Figure 12.3A):

- Perform the antibody staining first, before the activity stain.
- Glutaraldehyde is not suitable for antibody staining; use either a formaldehyde-based fixative, e.g., 4% formaldehyde in 0.1 M sodium phosphate buffer or PEM. Fix tissue for 30–45 minutes on ice.
- Rinse the stained tissue in PBS or 0.1 M sodium phosphate buffer for 30 minutes following the DAB reaction (this is important for strong activity staining).

To improve activity staining of weak enhancer-trap lines, the following modifications must be made:

- Reduce the fixation time (as little as 20 minutes will work).
- Stain for activity with X-gal first, and then stain with antibody.
- Incubate in primary antibody for 1–2 hours only.
- Reduce the DAB reaction time to give a weaker antibody signal.
- Homozygose the enhancer-trap line.

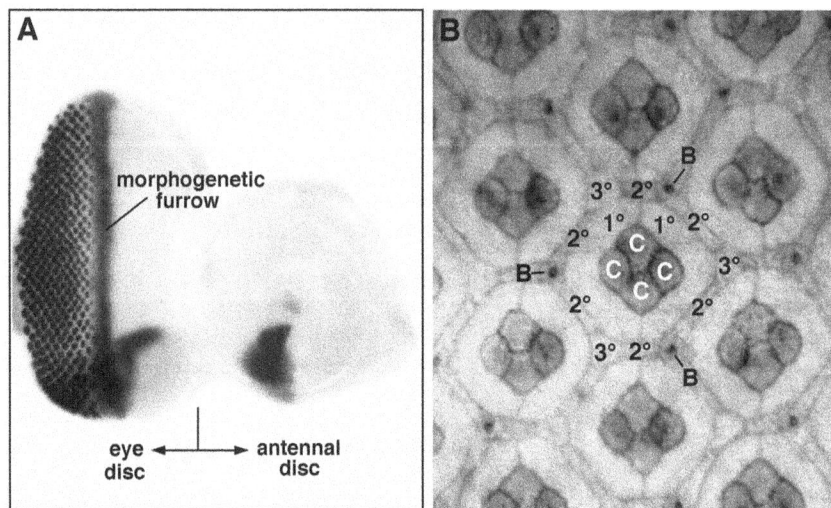

**Figure 12.3.** Histological preparations of third-instar eye-antennal imaginal disc (*A*) and pupal eye (*B*). (*A*) Third-instar eye-antennal imaginal disc stained with a nuclear neuron-specific antibody (anti-Elav) which stains the developing photoreceptors, visible as groups of cells posterior to the morphogenetic furrow. The morphogenetic furrow is highlighted using an enhancer trap (dpp) and is evident as a solid line. (This enhancer trap also identifies a group of cells in the antennal disc.) Anterior is to the right. (*B*) Apical surface of a midpupal eye stained with cobalt sulfide to outline cells. This eye has completed pattern formation, so all cells seen here will be present in the adult eye. (C) Cone cell; (B) mechanosensory bristle; (1°) primary pigment cell; (2°) secondary pigment cell; (3°) tertiary pigment cell. Anterior is to the right.

# BrdU Labeling

Bromodeoxyuridine (BrdU) is a nonradioactive thymidine analog, which is incorporated into the DNA of proliferating cells during S phase. Note that DAB is a potent carcinogen and BrdU is a mutagen, so both should be handled with extreme care. Soak all materials that become contaminated with DAB overnight in bleach prior to disposal.

## Materials

### Supplies and Equipment

Dissecting tools and equipment, Sylgard dissection dish, and other supplies (see pp. 203–204)
Slides and coverslips

### Solutions and Reagents

Dissection buffer, e.g., 1x PBS, *Drosophila* Ringer's solution, or 0.1 M **sodium phosphate** (pH 7.2) (for preparation, see p. 204)
**Formaldehyde** (5%) (for preparation, see p. 205)
**HCl** (3 M, freshly prepared)
Sodium phosphate (0.1 M, pH 7.2)
Glycerol (80%)/1x PBS (for note on glycerol, see p. 209)
**DAB** staining solution (for preparation, see p. 209)

#### BrdU

Prepare a stock solution of 7.5 mg/ml **BrdU** in 50% **ethanol**. Store at 4ºC. Dilute this stock BrdU solution to 75 µg/ml in 1x PBS.

#### PBS/TX

1x PBS
0.3% Triton X-100

#### PBS/TX/5% NGS

1x PBS
0.3% Triton X-100
5% NGS

### Antibodies

Primary antibody (anti-BrdU mouse monoclonal; Becton Dickinson)
Secondary antibody (HRP-conjugated anti-mouse)

**CAUTION: BrdU, DAB, ethanol, formaldehyde, HCl, sodium phosphate** (see Appendix 4)

## Method

1. Dissect tissue in a suitable buffer (e.g., 1x PBS, *Drosophila* Ringer's solution, or 0.1 M sodium phosphate [pH 7.2]) as described in Protocol 12.1.

2. Incubate in 75 μg/ml BrdU in PBS for 20–60 minutes.

3. Fix tissue in 5% formaldehyde at room temperature for 45 minutes.

4. Denature DNA by treating in 3 M HCl (freshly made) for 30 minutes.

5. Neutralize the acid by rinsing in several washes of PBS/TX for 30 minutes.

6. Incubate in anti-BrdU antibody (Becton Dickinson; 1:20 dilution in PBS/TX/5% NGS) at 4°C overnight.

7. Rinse three times in PBS/TX (10 minutes each wash).

8. Incubate in anti-mouse secondary antibody at room temperature for 2 hours.

9. Rinse three times in PBS/TX (10 minutes each wash) and once in PBS for 10 minutes.

10. Develop in DAB staining solution. When tissue is suitably dark, transfer it to a drop of 0.1 M sodium phosphate (pH 7.2) to stop the reaction.

    *Note:* 1.5 mM $NiCl_2$ can be added to the DAB solution for signal intensification; this treatment will yield a blue-black reaction product. The anti-BrdU antibody is very good so intensification is generally not necessary.

11. Mount tissue in 80% glycerol/PBS.

    *Note:* Alternatively, tissue can be dehydrated and mounted in DPX.

PROTOCOL 12.5

# In Situ Hybridization

This protocol has been through numerous generations of modifications by many laboratories, but it is based primarily on Tautz and Pfeifle (1989). All incubations and washes can be carried out in 1.5-ml microcentrifuge tubes. For X-gal staining followed by in situ hybridization, see Protocol 12.5 Notes.

## Materials

### Supplies and Equipment

Microcentrifuge tubes (1.5 ml, Eppendorf); for the preparation of RNA probes, use RNase-free tubes

G-50 Sephadex spin columns (Pharmacia) or NucTrap push columns (Stratagene)

Dissecting tools and equipment, Sylgard dissection dish, and other supplies (see pp. 203–204)

### Solutions and Reagents

Gel-purified DNA

Digoxigenin (DIG)-labeling kits (Boehringer Mannheim)

The Genius 1 kit for DNA-probe labeling includes the following:

- 10x hexanucleotide mix
- 10x dNTP mix
- Klenow
- Anti-DIG-alkaline phosphatase (AP) antibody (Boehringer Mannheim). Preadsorb the anti-DIG-AP antibody against a large volume of wild-type fixed embryos before use to reduce background staining.

- **NBT** (75 mg/ml) (nitroblue tetrazolium)
- X-phosphate (50 mg/ml) (**5-bromo-4-chloro-3-indolyl phosphate**)

The Genius 4 kit for RNA-probe labeling includes the following:

- 10x NTP labeling mixture
- 10x transcription buffer
- RNase inhibitor
- SP6 and T7 RNA polymerases
- RNase-free DNase I

EDTA (0.5 M)

Dissection buffer, e.g., 1x PBS, *Drosophila* Ringer's, or 0.1 M **sodium phosphate** (pH 7.2) (for preparation, see p. 204)

**Formaldehyde** (4%) in 1x PBS (for preparation, see p. 205)

Formaldehyde (4%) in 1x PBS containing 0.6% Triton

PBS/TX (1x PBS containing 0.3% Triton X-100)

(*Optional*) **Ethanol** series (25%, 50%, 75%, absolute)

Proteinase K (10 μg/ml) in PBS/Tween (see below)

### *PBS/Tween*

PBS
0.1% Tween-20

### *Hybridization Mix*

| Component and final concentration | Amount to add to make 5 ml |
|---|---|
| 50% **formamide** | 2.5 ml |
| 5x SSC | 1.25 ml of 20x |
| 100 μg/ml sonicated herring sperm DNA | 100 μl of 5 mg/ml |
| 50 μg/ml heparin | 5 μl of 50 mg/ml |
| 0.1% Tween-20 | 5 μl |
| 100 μg/ml tRNA | 25 μl of 20 mg/ml |
| $H_2O$ | to make 5 ml |

Prepare fresh before use.

### *Levamisole Solution*

| Component and final concentration | Amount to add to make 10 ml |
|---|---|
| 100 mM NaCl | 200 μl of 5 M |
| 50 mM **MgCl$_2$** | 500 μl of 1 M |
| 100 mM Tris (pH 9.5) | 1 ml of 1 M |
| 1 mM levamisole | 100 μl of 100 mM |
| 0.1% Tween-20 | 100 μl of 10% |
| $H_2O$ | to make 10 ml |

Store at –20ºC.

### *10x Glycine*

Prepare 20 mg/ml glycine in PBS/Tween. Store in 1-ml aliquots at –20ºC. To prepare 1x glycine, dilute an aliquot of the 10x stock with 9 ml of PBS/Tween.

**CAUTION: 5-bromo-4-chloro-3-indolyl phosphate, ethanol, formaldehyde, formamide, MgCl$_2$, NBT, sodium phosphate** (see Appendix 4)

## Method

Either DNA or RNA probes can be used with this in situ hybridization protocol. RNA probes tend to be more sensitive and give a stronger signal than DNA probes. The probes are made using the Genius digoxigenin-labeling kits (Boehringer Mannheim; see above). Additional tips can be found in the product sheets that accompany the kits.

### DIG-labeling of DNA Probe

1. Bring 50 ng of gel-purified DNA to 7.5 μl in H₂O. Denature by heating to 95ºC for 10 minutes and place on ice.

2. Add the following to the DNA.

   | | |
   |---|---|
   | 10x hexanucleotide mix | 1 μl |
   | 10x dNTP mix | 1 μl |
   | Klenow (2 units/μl) | 0.5 μl |

3. Incubate at 16ºC overnight (minimum incubation time is 4 hours).

4. Add 0.5 μl of 0.5 M EDTA to stop the reaction. Incubate at 65ºC for 15 minutes. Remove unincorporated nucleotides by filtration through G-50 Sephadex spin columns (Pharmacia) or NucTrap push columns (Stratagene).

5. Lyophilize the probe and redissolve in 25 μl of Hybridization Mix. This is enough probe for three hybridizations.

   *Note:* The probe can be saved and reused a second time with little loss of signal intensity.

### DIG-labeling of RNA Probe

1. Place a sterile, RNase-free microcentrifuge tube on ice and add the following:

   | | |
   |---|---|
   | gel-purified DNA | 1 μg |
   | 10x NTP labeling mixture | 2 μl |
   | 10x transcription buffer | 2 μl |
   | RNase inhibitor | 1 μl |
   | RNase-free H₂O | to 18 μl |
   | SP6 or T7 RNA polymerase | 2 μl |

2. Mix gently and incubate at 37ºC for 2 hours.

3. (*Optional*) To remove the DNA template, add 2 μl of RNase-free DNase I, and incubate at 37ºC for 15 minutes.

4. Add 2 μl of 0.5 M EDTA to stop the reaction.

### Tissue Preparation and Prehybridization

1. Dissect tissue in a suitable buffer (1x PBS, *Drosophila* Ringer's, or 0.1 M sodium phosphate [pH 7.2]) as described in Protocol 12.1.

2. Immediately transfer the tissue to 4% formaldehyde/PBS on ice. Fix for 15–20 minutes and then, for larval eye discs, remove the peripodial membrane.

3. Fix in 4% formaldehyde/PBS containing 0.6% Triton X-100 at room temperature for an additional 15–20 minutes.

4. Wash three times in PBS/TX with rocking (5 minutes each wash).

   *Note:* To store discs for future use, rinse in PBS, dehydrate through an ethanol series (25%, 50%, 75%, twice absolute, 10 minutes each), and store in ethanol at –20ºC. Storage in ethanol reduces nonspecific background staining.

5. Digest with 10 µg/ml proteinase K in PBS/Tween for 3–5 minutes. Gently invert the tube occasionally during digestion.

   *Notes:* Experiment with the digestion time; if not sufficiently digested, the probe will not get into the tissue and if left too long, the tissue will dissolve.

   Be excessively gentle with the tissue from this point on, as it is very fragile and the eye discs tend to break away from the mouth parts. When adding solutions to the microcentrifuge tube, let them drip down the wall of the tube to minimize disruption of the tissue.

6. Remove the proteinase K solution and wash twice in 2 mg/ml glycine in PBS/Tween. Rock at room temperature for 2 minutes.

7. Wash twice in PBS/Tween at room temperature with rocking (5 minutes each wash).

8. Postfix in 4% formaldehyde/PBS for 20 minutes with rocking.

9. Wash five times in PBS/Tween at room temperature with rocking (5 minutes each wash).

10. Rinse with 50% PBS/50% Hybridization Mix and then with 100% Hybridization Mix. For DNA probes, prehybridize at 45–48ºC for 1 hour. For RNA probes, prehybridize at 55ºC for 1 hour.

## Hybridization and Washing

1. Denature probe immediately before use by boiling for 10 minutes. Place on ice until ready to add to the Hybridization Mix.

2. Remove most of the Hybridization Mix and add 1 µl of diluted, denatured probe for every 10 µl of tissue. Mix gently and incubate overnight (for DNA probes, incubate at 45–48ºC; for RNA probes, incubate at 55ºC).

3. Perform the following washes at the hybridization temperature:

   a. Wash in 1 ml of Hybridization Mix for 20 minutes.

   b. Wash in Hybridization Mix:PBS/Tween (1:1) for 20 minutes.

   c. Wash five times in PBS/Tween (20 minutes each wash).

## Detection

1. Incubate the tissue in anti-DIG-AP antibody (1:1000 dilution in PBS/Tween) at room temperature for 1 hour. Use a volume equal to 10 times the volume of discs.

   *Note:* The antibody solution can be saved and reused up to seven times or more. The reaction time will take longer with each use, but background-staining levels will also decrease. Store antibody at 4ºC.

2. Wash tissue four times in PBS/Tween at room temperature with rocking (20 minutes each wash).

3. Rinse three times in Levamisole Solution (5 minutes each wash).

4. Add 4.5 µl of NBT and 3.5 µl of X-phosphate (for a 1-ml volume of levamisole) to the last wash. Mix gently. Keep tissue in the dark during the reaction, checking periodically to monitor staining.

5. Rinse the tissue four to five times with PBS/Tween to stop the reaction.

6. Mount in 80% glycerol.

## PROTOCOL 12.5 NOTES

For X-gal staining followed by in situ hybridization, perform the following (protocol provided by U. Heberlein):

1. Dissect imaginal discs in PBS as described in Protocol 12.1.

2. Immediately transfer to 4% formaldehyde/PBS on ice. Fix for 15–20 minutes and then remove peripodial membrane.

3. Rinse tissue three times in PBS (10 minutes each wash).

4. Stain with X-gal as described in Protocol 12.3.

5. Fix in 4% formaldehyde/PBS containing 0.5% Triton X-100 at room temperature for 15 minutes.

6. For in situ hybridization, proceed to step 4 (Tissue Preparation and Prehybridization above).

## PROTOCOL 12.6

# Cobalt Sulfide

Cobalt sulfide is a quick and reliable method for outlining cells and is primarily used for visualizing the apical surface of retinal epithelia. Although this technique cannot be used in conjunction with other staining protocols (e.g., antibodies), it is useful for phenotypic characterization (Melamed and Trujillo-Cenoz 1975; Cagan and Ready 1989). For an example of cobalt-sulfide-stained pupal eye, see Figure 12.3B.

All steps can be carried out in puddles in a Sylgard dissecting dish. Over time, a film of precipitate will form on the drop of cobalt nitrate and the ammonium sulfide will turn cloudy after several complexes have been developed (see method below). Change both of these solutions under these conditions. Transfer imaginal discs by spearing the mouth hooks with a tungsten needle and pupal eyes by spearing the brain with a tungsten needle.

### Materials

#### Supplies and Equipment

Dissecting tools and equipment, Sylgard dissection dish, and other supplies (see pp. 203–204)
Slides and coverslips

### Solutions and Reagents

Dissection buffer, e.g., 1x PBS, *Drosophila* Ringer's solution, or 0.1 M **sodium phosphate** (pH 7.2) (for preparation, see p. 204)

**Glutaraldehyde** (2.5%) in 0.1 M sodium phosphate (pH 7.2) (for preparation, see p. 205)

**$Co(NO_3)_2 \cdot 6H_2O$** (2–4%) in $H_2O$

**Ammonium sulfide** (2%) in $H_2O$. Dilute a 20% stock of ammonium sulfide (Fisher) to 2% just before use. Ammonium sulfide evolves hydrogen sulfide, which is poisonous, and the odor is unbearable, so handle the 20% stock in a chemical fume hood.

Glycerol (80%) in dissection buffer (for important note on glycerol, see p. 209)

**CAUTION: ammonium sulfide, $Co(NO_3)_2 \cdot 6H_2O$, glutaraldehyde, sodium phosphate** (see Appendix 4)

### Method

1. Dissect eye discs or pupal eyes in dissection buffer (e.g., 1x PBS, *Drosophila* Ringer's, or 0.1 M sodium phosphate [pH 7.2]) as described in Protocol 12.1.

2. Fix in 2.5% glutaraldehyde/0.1 M sodium phosphate for 10 minutes.

    *Note:* If staining eye discs, the peripodial membrane must be completely removed.

3. Rinse in $H_2O$.

4. Incubate in 2–4% $Co(NO_3)_2 \cdot 6H_2O$ in $H_2O$ at room temperature for 15 minutes.

5. Rinse *very briefly* in $H_2O$ (a quick in and out is sufficient; any longer and the staining intensity will be significantly diminished).

6. Develop in freshly prepared 2% ammonium sulfide in $H_2O$. The tissue will turn black immediately. Allow the tissue to sit in the ammonium sulfide for approximately 30 seconds, and then rinse in $H_2O$.

7. Transfer tissue to 80% glycerol, remove any residual peripodial membrane, mouth hooks, and brain (if working with pupal eyes). Mount in 80% glycerol.

PROTOCOL 12.7

# Acridine Orange

Acridine orange is a fluorochrome which can be used as a vital dye (Spreij 1971). Living cells are selectively permeable and therefore exclude the dye; however, dying cells lose their permeability barrier, so the dye molecules enter the cell and intercalate between the base pairs of the double-stranded DNA, causing it to fluoresce green (for representative examples, see Wolff and Ready 1991). Although this staining technique is quick and easy, two disadvantages are that it works only on unfixed tissue and the samples cannot be preserved, so they must be viewed and photographed immediately. The TUNEL staining method (see Protocol 12.8) is recommended if a permanent preparation is preferred. Wear gloves when handling acridine orange.

## Materials

### Supplies and Equipment

Dissecting tools and equipment, Sylgard dissection dish, and other supplies (see pp. 203–204)
Slides and coverslips

### Solutions and Reagents

*Drosophila* Ringer's solution (for preparation, see p. 204)

---

#### 1 mM Stock Solution of Acridine Orange

Dissolve 1.51 mg of **acridine orange** in 5 ml of absolute **ethanol**. Store in a foil-wrapped container at room temperature. Stock is stable for many months.

---

CAUTION: **acridine orange, ethanol** (see Appendix 4)

### Method

1. Dissect eye discs or pupal eyes in *Drosophila* Ringer's solution as described in Protocol 12.1. Alternatively, perform the dissection directly in the diluted solution of acridine orange (see step 2, below).

2. Incubate tissue for 5 minutes in a $1.6 \times 10^{-6}$ M solution of acridine orange (Aldrich) in *Drosophila* Ringer's solution. Immediately before use, dilute 1.6 µl of a 1 mM stock in 1 ml of Ringer's.

3. Rinse briefly in *Drosophila* Ringer's solution.

4. Mount in *Drosophila* Ringer's solution and cover with a coverslip.

   *Note:* It may be necessary to elevate the coverslip to prevent the tissue from being crushed; broken fragments of coverslips work well for this purpose. View samples immediately using fluorescence microscopy.

---

PROTOCOL 12.8

# TUNEL Staining

Although more labor intensive than the acridine orange method, the TUNEL-labeling technique (terminal deoxynucleotidyl transferase [TdT]-mediated dUTP nick end labeling) provides a permanent record for labeling dead cells. (From Gavrieli et al. 1992; White et al. 1996; protocol and additional modifications were provided by P. Kurada and Kristen White.)

### Materials

### Supplies and Equipment

Dissecting tools and equipment, Sylgard dissection dish, and other supplies (see pp. 203–204)
Microwell plates (60-well) for antibody incubations
Slides and coverslips

### Solutions and Reagents

**Formaldehyde** (2%) in 0.1 M sodium phosphate buffer (pH 7.4)
Proteinase K (10 µg/ml) in 1× PBS
PBTw (1× PBS containing 0.1% Tween-20)
Apoptag Kit (Oncor)

Triton X-100
Antibleach medium (Vectashield; Vector Laboratories)
Rhodamine-conjugated anti-DIG antibody (Boehringer Mannheim)

### *0.1 M Sodium Phosphate Buffer (pH 7.4)*

| | |
|---|---|
| 1 M $Na_2HPO_4$ | 77.4 ml |
| 1 M $NaH_2PO_4$ | 22.6 ml |
| $H_2O$ | to 1000 ml |

### *5× BSS (Balanced Salts Solution)*

| | |
|---|---|
| NaCl | 8 g |
| KCl | 7.5 g |
| **$MgSO_4$** | 4.5 g |
| **$CaCl_2·2H_2O$** | 1.725 g |
| tricine | 4.25 g |
| glucose | 9 g |
| sucrose | 42.5 g |

Adjust volume to 500 ml with $H_2O$, filter-sterilize, and store at 4ºC.

### *BSS/TX/NGS*

1× BSS
0.3% Triton X-100
5% NGS

**CAUTION: $CaCl_2$, formaldehyde, $MgSO_4$, $Na_2HPO_4$, $NaH_2PO_4$ (see Appendix 4)**

## Method

1. Dissect discs in 0.1 M Sodium Phosphate Buffer as described in Protocol 12.1.

2. Fix in 2% formaldehyde/0.1 M Sodium Phosphate Buffer (pH 7.4) at room temperature for 10 minutes.

3. Rinse twice in BSS (10 minutes each wash).

4. Treat with 10 μg/ml proteinase K in PBS at room temperature for 5 minutes.

5. Wash twice in PBTw (5 minutes each wash).

6. Postfix in 2% formaldehyde/0.1 M Sodium Phosphate Buffer (pH 7.4) for 10 minutes.

7. Wash five times in PBTw (5 minutes each wash).

8. Perform the TUNEL reaction using the Oncor Apoptag kit as follows:

   a. Incubate in equilibration buffer at room temperature for 1 hour.

   b. Mix the reaction buffer and terminal deoxynucleotidyl transferase (TdT) in a 2:1 ratio and add Triton X-100 to 0.3%. Remove the equilibration buffer from the sample and add enough of the enzyme mixture to just immerse the sample. Incubate at 37ºC overnight in a humid chamber.

   c. Prepare the stop-reaction mix by diluting stop/wash buffer in $H_2O$ in a 1:34 ratio. Remove the enzyme mixture from the sample and add stop-reaction mix. Incubate at 37ºC for 3–4 hours.

d. Wash three times in PBTw (5 minutes each wash).

9. Block in BSS/TX/NGS at room temperature for 1 hour.

10. To detect TUNEL, incubate tissue in rhodamine-conjugated anti-DIG antibody (diluted 1:200 in BSS/TX/NGS) at 4°C overnight.

11. Wash four times in BSS (20 minutes each wash).

12. Remove discs from mouth hooks and mount in antibleach medium (Vectashield).

# Phalloidin

Phalloidin is a toxin that binds filamentous actin, thereby stabilizing it. Phalloidin binds the abundant F-actin network along cell membranes and is therefore useful in highlighting cell borders. Phalloidin is extremely hazardous, so use caution when handling this toxin. (For an example of stained tissue, see cover of *Development* 113.)

## Materials

### Supplies and Equipment

Dissecting tools and equipment, Sylgard dissection dish, and other supplies (see pp. 203–204)
Microwell plates (60 well)
Slides and coverslips

### Solutions and Reagents

Dissection buffer, e.g., 1x PBS, *Drosophila* Ringer's solution, or 0.1 M **sodium phosphate** (pH 7.2) (for preparation, see p. 204)
**Formaldehyde** (4%) in 1x PBS (for preparation, see p. 205)
1x PBS
PBS/TX (1x PBS containing 0.3% Triton X-100)
Normal goat serum (NGS)
Antibleach medium (Vectashield, Vector Laboratories)

#### 20 µM Phalloidin Stock Solution

Sigma carries **phalloidin** conjugated to FITC, TRITC, and CPITC. It is shipped in rubber-stopper-sealed vials containing 0.1 mg. Inject 4 ml of absolute **ethanol** into a 0.1-mg vial of phalloidin to prepare a 20 µM stock.

CAUTION: ethanol, formaldehyde, phalloidin (conjugated to FITC, TRITC, and CPITC), sodium phosphate (see Appendix 4)

## Method

1. Dissect eye discs or pupal eyes in dissection buffer (e.g., 1x PBS, *Drosophila* Ringer's, or 0.1 M sodium phosphate [pH 7.2]) as described in Protocol 12.1.

2. Fix tissue in 4% formaldehyde in PBS at room temperature for 15 minutes.

3. Rinse tissue three times in PBS/TX (10 minutes each wash).

4. Prepare 1 μM fluorescently conjugated phalloidin by mixing together the following:

   | | |
   |---|---|
   | 20 μM phalloidin | 5 μl |
   | PBS/TX | 90 μl |
   | NGS | 5 μl |

5. Incubate tissue in 1 μM phalloidin at 4°C overnight or at room temperature for 2 hours in 60-well microwell plates (see p. 224).

6. Rinse three times in PBS (10 minutes each wash).

7. Mount in antibleach medium (Vectashield).

## PROTOCOL 12.9 NOTES

To double label tissue with an antibody and phalloidin, follow the standard procedure for antibody labeling (see Protocol 12.2 [Labeling with a Single Antibody]) and use a fluorescent secondary antibody (step 6). Following the secondary antibody incubation, wash tissue three times in PBS/TX (10 minutes each wash). Incubate in phalloidin, rinse in PBS, and mount as outlined in steps 3–6 above.

## PROTOCOL 12.10

# DAPI

DAPI (4′,6-diamidino-2-phenylindole) is a fluorescent dye used to stain DNA.

## Materials

### Supplies and Equipment

Dissecting tools and equipment, Sylgard dissection dish, and other supplies (see pp. 203–204)
Slides and coverslips

### Solutions and Reagents

Dissection buffer, e.g., 1x PBS, *Drosophila* Ringer's solution, or 0.1 M **sodium phosphate** (pH 7.2) (for preparation, see p. 204)
Formaldehyde (4%) in 1x PBS (for preparation, see p. 205)
PBS/TX (1x PBS containing 0.3% Triton X-100)
Antibleach medium (Vectashield, Vector Laboratories, Inc.)

### DAPI Staining Solution

Prepare a 50 mg/ml **DAPI** (Boehringer Mannheim) solution in 180 mM Tris-Cl (pH 7.5). Store at 4°C protected from light. Dilute to 1 mg/ml in 1x PBS.

**CAUTION: DAPI, formaldehyde** (see Appendix 4)

## Method

1. Dissect eye discs or pupal eyes in dissection buffer (e.g., 1x PBS, *Drosophila* Ringer's, or 0.1 M sodium phosphate [pH 7.2]) as described in Protocol 12.1.

2. Fix tissue in 4% formaldehyde/PBS at room temperature for 20 minutes.

3. Rinse tissue three times in PBS/TX (10 minutes each wash).

4. Incubate in 1 mg/ml DAPI in PBS for approximately 5 minutes or less.

> *Note:* If simultaneously staining with an antibody, tissue can be coincubated in the secondary antibody and DAPI (step 6 of Protocol 12.2 [Labeling with a Single Antibody]).

5. Rinse three times in PBS/TX (10 minutes each wash).

6. Mount in antibleach medium (Vectashield). View using a UV filter.

## ACKNOWLEDGMENTS

I am grateful to Nadean Brown, Graeme Mardon, and Jessica Treisman for their thoughtful comments and useful suggestions on this chapter. I also thank the many individuals who contributed specific protocols and modifications.

## REFERENCES

Baker N.E., Mlodzik M., and Rubin G.M. 1990. Spacing differentiation in the developing *Drosophila* eye: A fibrinogen-related lateral inhibitor encoded by *scabrous*. *Science* **250:** 1370–1377.

Blochlinger K., Jan L.Y., and Jan Y.N. 1993. Postembryonic patterns of expression of *cut*, a locus regulating sensory organ identity in *Drosophila*. *Development* **117:** 441–450.

Brown N.L., Sattler C.A., Markey D.R,. and Carroll S.B. 1991. *hairy* gene function in the *Drosophila* eye: Normal expression is dispensable but ectopic expression alters cell fates. *Development* **113:** 1245–1256.

Cagan R.L. and Ready D.F. 1989. *Notch* is required for successive cell decisions in the developing *Drosophila* retina. *Genes Dev.* **3:** 1099–1112.

Chou W.H., Hall K.J., Wilson D.B., Wideman C.L., Townson S.M., Chadwell L.V., and Britt S.G. 1996. Identification of a novel *Drosophila* opsin reveals specific patterning of the R7 and R8 photoreceptor cells. *Neuron* **17:** 1101–1115.

Dickson B. and Hafen E. 1993. Genetic dissection of eye development in *Drosophila*. In *The development of* Drosophila melanogaster (ed. M. Bate and A. Martinez-Arias), pp. 1327–1362. Cold Spring Harbor Laboratory Press, Cold Spring Harbor, New York.

Feiler R., Bjornson R., Kirschfeld K., Mismer D., Rubin G.M., Smith D.P., Socolich M., and Zuker C.S. 1992. Ectopic expression of ultraviolet-rhodopsins in the blue photoreceptor cells of *Drosophila*: Visual physiology and photochemistry of transgenic animals. *J. Neurosci.* **12:** 3862–3868.

Fu W. and Noll M. 1997. The *Pax2* homolog *sparkling* is required for development of cone and pigment cells in the *Drosophila* eye. *Genes Dev.* **11:** 2066–2078.

Gavrieli Y., Sherman Y., and Ben-Sasson S.A. 1992. Identification of programmed cell death in situ via specific labeling of nuclear DNA fragmentation. *J. Cell Biol.* **119:** 493–501.

Hay B.A., Wolff T., and Rubin G.M. 1994. Expression of baculovirus P35 prevents cell death in *Drosophila*. *Development* **120:** 2121–2129.

Higashijima S., Kojima T., Michiue T., Ishimaru, S., Emori Y., and Saigo K. 1992. Dual *Bar* homeo box genes of *Drosophila* are required in two photoreceptor cells, R1 and R6, and primary pigment cells for normal eye development. *Genes Dev.* **6:** 50–60.

Hortsch M., Patel N.H., Bieber A.J., Traquina Z.R., and Goodman C.S. 1990. *Drosophila* neurotactin, a surface glycoprotein with homology to serine esterases, is dynamically expressed during embryogenesis. *Development* **10:** 1327–1340.

Jarman A.P., Grell, E.H., Ackerman, L., Jan L.Y., and Jan Y.N. 1994. Atonal is the proneural gene for *Drosophila* photoreceptors. *Nature* **369:** 398–400.

Kauffmann R., Li S., Gallagher P.A., Zhang J., and Carthew R. 1996. Ras 1 signaling and transcriptional competence in the R7 cell of *Drosophila*. *Genes Dev.* **10:** 2167–2178.

Kimmel B.E., Heberlein U., and Rubin G.M. 1990. The homeo domain protein *rough* is expressed in a subset of cells in the developing *Drosophila* eye where it can specify photoreceptor cell subtype. *Genes Dev.* **4:** 712–727.

Krämer H., Cagan R.L., and Zipursky S.L. 1991. Interaction of *bride of sevenless* membrane-bound ligand and *sevenless* tyrosine kinase receptor. *Nature* **352:** 207–212.

Lee E.-C., Hu X., Yu S.-Y., and Baker N.E. 1996. The *scabrous* gene encodes a secreted glycoprotein dimer and regulates proneural development in *Drosophila* eyes. *Mol. Cell Biol.* **16:** 1179–1188.

Mardon G., Solomon N.M., and Rubin G.M. 1994. *dachshund* encodes a nuclear protein required for normal eye and leg development in *Drosophila*. *Development* **120:** 3473–3486.

Melamed J. and Trujillo-Cenoz O. 1975. The fine structure of the eye imaginal disc in *muscoid* flies. *J. Ultrastruct. Res.* **51:** 79–93.

Mlodzik M., Hiromi Y., Weber U., Goodman C.S., and Rubin G.M. 1990. The *Drosophila* seven-up gene, a member of the steroid receptor gene superfamily, controls photoreceptor cell fates. *Cell* **60:** 211–224.

Robinow S. and White K. 1991. Characterization and spatial distribution of the ELAV protein during *Drosophila melanogaster* development. *J. Neurobiol.* **5:** 443–461.

Simon J.A., Sutton C.A., Lobell R.B., Glaser R.L., and Lis J.T. 1985. Determinants of heat shock-induced chromosome puffing. *Cell* **40:** 805–817.

Spreij Th. E. 1971. Cell death during the development of the imaginal discs of *Calliphora erythrocephala*. *Neth. J. Zool.* **21:** 221–264.

Suzuki E. and Hirosawa K. 1991. Immunoelectron microscopic study of the opsin distribution in the photoreceptor cells of *Drosophila melanogaster*. *J. Electron Microsc.* **40:** 187–192.

Tautz D. and Pfeifle C. 1989. A non-radioactive in situ hybridization method for the localization of specific RNAs in *Drosophila* embryos reveals translational control of the segmentation gene *hunchback*. *Chromosoma* **98:** 81–85.

Tomlinson A., Bowtell D.D.L., Hafen E., and Rubin G.M. 1987. Localization of the *sevenless* protein, a putative receptor for positional information, in the eye imaginal disc of *Drosophila*. *Cell* **51:** 143–150.

White K., Tahaoglu E., and Steller H. 1996. Cell killing by the *Drosophila* gene *reaper*. *Science* **271:** 805–807.

Wolff T. and Ready D.F. 1991. Cell death in normal and rough eye mutants of *Drosophila*. *Development* **113:** 825–839.

_____. 1993. Pattern formation in the *Drosophila* retina. In *The development of* Drosophila melanogaster (ed. M. Bate and A. Martinez-Arias), pp. 1277–1325. Cold Spring Harbor Laboratory Press, Cold Spring Harbor, New York.

# CONTENTS

# 13

# Histological Techniques for the *Drosophila* Eye Part II: Adult

**Tanya Wolff**
*Washington University School of Medicine*
*Department of Genetics*
*St. Louis, Missouri 63110*

T HIS CHAPTER IS A CONTINUATION OF THE PREVIOUS CHAPTER, covering protocols used to study the adult eye. Important notes often follow each protocol and the reader is encouraged to read these notes before starting a given procedure. As with Chapter 12, the protocols provided here have been improved by many individuals. Protocol 13.1 describes the dissection procedure for retrieval of adult retinal tissue. Table 13.1 includes an index of protocols presented in this chapter and comments on their primary use. For an overview of eye development and pattern formation, see Wolff and Ready (1993) and for a molecular review of eye development, see Dickson and Hafen (1993).

## PROTOCOL 13.1*

## Adult Eye Dissection

### Materials

### *Supplies and Equipment*

Dissecting tools and equipment, Sylgard dissection dish, and other supplies (see Protocol 12.1, Materials list)

**$CO_2$ pad**

Scalpel blade (#11; Becton Dickinson)

**Table 13.1.** Index of Protocols and Their Primary Use

| Protocol | Type of staining | Primary use |
|----------|------------------|-------------|
| 13.2 | cryosectioning and immunocytochemistry | antibody staining of adult sections |
| 13.3 | β-galactosidase detection | identification of specific cell type |
| 13.4 | preparing thick sections of adult retinas | phenotypic and mosaic analysis |
| 13.5 | scanning electron microscopy (SEM): specimen preparation | high-magnification analysis of external structures |
| 13.6 | transmission electron microscopy (TEM): specimen preparation | ultrastructural analysis |

*For all protocols, $H_2O$ indicates glass distilled and deionized.

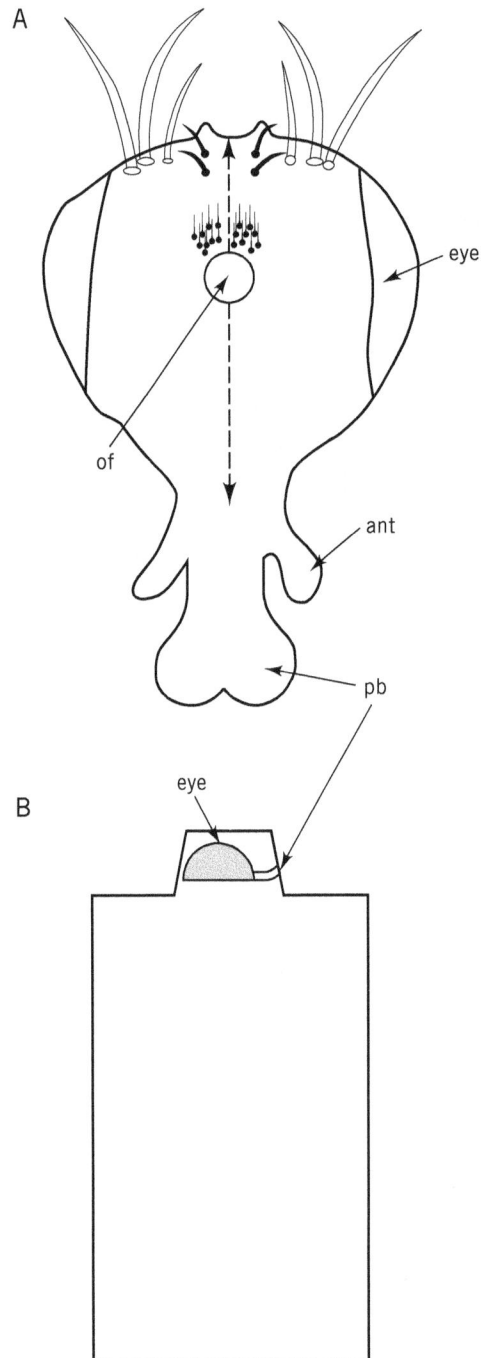

**Figure 13.1.** Adult dissection. (For details, see Protocol 13.1.) (*A*) Posterior view of adult head. (of) Occipital foramen; (ant) antenna; (pb) proboscis. (*B*) Laterial view of eye embedded in trimmed resin block.

### Solutions and Reagents

Dissection buffers and fixatives (see Protocol 12.1)

**CAUTION: CO$_2$** (see Appendix 4)

## Method

1. Anesthetize flies on a CO$_2$ pad.

2. Cut off the heads using a #11 scalpel blade (Becton Dickinson).

3. Use forceps to grab ahold of the proboscis and then insert the point of the scalpel into the occipital foramen, the hole through which the central nervous system tracks.

4. Gently slice the head in half, starting with either the dorsal or ventral half of the head (Figure 13.1A). Take care to prevent the eyes from bulging.

PROTOCOL 13.2

# Cryosectioning and Immunocytochemistry

This protocol is recommended for antibody staining of frozen adult eye sections.

## Materials

### Supplies and Equipment

Dissecting tools and equipment, Sylgard dissection dish, and other supplies (see Protocol 12.1)
Cryostat and accessory items
Razor blade
**Ethanol/Dry ice** bath
Drying plate set at 40ºC
Humidified chamber. To prepare a humidified chamber, place wet paper towels in an airtight Tupperware container. Balance the slides on risers, e.g., disposable 5-ml pipettes, arranged in parallel.
Coverslips
Coplin jars

#### Subbed Slides

Dissolve 5 g of gelatin (Sigma) and 0.5 g of chromium potassium sulfate in 500 ml of hot (80ºC) H$_2$O. Stir on a hot plate until dissolved (~2 hours).
Place slides in slide racks and wash in a warm solution of Haemo-sol. Rinse well in H$_2$O and then dip in the gelatin solution. Cover loosely and dry overnight.

### Solutions and Reagents

1x PBS (see Protocol 12.1)
**Formaldehyde** (3%) in 1x PBS
Formaldehyde (0.5%) in 1x PBS
Sucrose (12%) in 1x PBS

**O.C.T.** Tissue Tek Compound (Ted Pella 27050)

Graded ethanol series; 30%, 50%, 70%, 90%, and absolute

Mounting media: Either of the following can be used:

- 70% glycerol in 1x PBS. Low-purity glycerol contains contaminants that cause the HRP reaction product to fade very rapidly. This problem can be avoided by using ultrapure glycerol (99.5% pure). Sigma and Boehringer Mannheim both carry this product. Low pH can also cause the HRP reaction product to fade, so make sure that the glycerol solution is at pH 7.2.
- **DPX** (Fluka)

### Blocking Solution

1x PBS (phosphate-buffered saline)

1% BSA (bovine serum albumin)

0.05% Triton X-100

If a milder detergent is necessary, use 0.1% **saponin**.

### PBS/0.05% TX

1x PBS

0.05% Triton X-100

### PBS/0.1% Saponin

1x PBS

0.1% saponin

### PBS/1% BSA

1x PBS

1% BSA

### DAB Staining Solution

0.5 mg/ml **DAB** (3,3'-diaminobenzidene)

0.003% $H_2O_2$

1.5 mM $CoCl_2$

1.5 mM $NiCl_2$

The addition of $CoCl_2$ and $NiCl_2$ is optional. They intensify the reaction product and give a blue-black reaction product; in the absence of these compounds, the reaction product will be brown. See discussion in Protocol 12.2 (Double-labeling Using HRP-conjugated Antibodies) when deciding whether or not to intensify the signal.

Note that DAB is a potent carcinogen and should be handled with extreme care. Soak all materials that become contaminated with DAB overnight in **bleach** prior to disposal.

## Antibodies

Primary antibody

Secondary antibody (e.g., HRP-conjugated)

**CAUTION: bleach, $CoCl_2$, DAB, DPX, dry ice, ethanol, $H_2O_2$, formaldehyde, $NiCl_2$, O.C.T., saponin** (see Appendix 4)

## Method

1. Decapitate flies and remove the proboscis as described in steps 1 and 2 of Protocol 13.1.

2. Fix heads in 3% formaldehyde/PBS on ice for 60–90 minutes.

3. Rinse tissue in PBS.

4. Transfer heads to 12% sucrose in PBS and cut heads in half.

5. Infiltrate tissue with 12% sucrose in PBS at 4°C for 4–12 hours.

6. Transfer heads to O.C.T. Tissue Tek Compound and allow the compound to permeate the tissue at room temperature for 10–30 minutes.

7. Embed the tissue in O.C.T. compound as follows:

   a. Freeze a layer of O.C.T. onto the chuck by placing it in an ethanol/dry ice bath.

   b. When the first layer is almost frozen, remove it from the ethanol/dry ice bath and add another drop of O.C.T. to the frozen layer.

   c. Place up to 12 eyes in the unfrozen O.C.T. Use forceps to quickly orient the eyes (do this under the microscope) for either horizontal or transverse sections.

   d. Once oriented, put the chuck back in the ethanol/dry ice bath to freeze.

8. Place the chuck in the cryostat chamber and allow the block to equilibrate to cryostat temperature (–14°C to –18°C) for 20–30 minutes. Trim the edges of the block with a razor blade.

9. Cut sections 10–14-μm thick and transfer to subbed slides.

10. To dry sections onto the slides, heat slides on a 40°C drying plate for no more than 1 minute or air-dry slides at room temperature for 30 minutes. *Do not overheat or overdry.*

11. Fix the sections in 0.5% formaldehyde/PBS at room temperature for 20–60 minutes.

    *Note:* Slides can be stored in fix for several days at 4°C without adversely affecting the tissue.

12. Rinse slides three times in PBS (3 minutes each).

13. Block in Blocking Solution at room temperature for 30 minutes.

14. Wash slides several times in PBS/0.05% TX (or PBS/0.1% saponin).

15. Cover sections with primary antibody diluted in PBS/1% BSA and incubate in a humid chamber at room temperature for 30–60 minutes or at 4°C overnight. If the quantity of primary antibody is limited, place a coverslip on top of the antibody to distribute it evenly over the surface of the slide. Leave the coverslip in place during the incubation.

    *Note:* 100 μl of diluted antibody is adequate for a 22 x 40-mm coverslip.

16. Wash the slides five to seven times in PBS/0.1% Saponin at room temperature (5 minutes each wash) in Coplin jars.

17. Incubate in secondary antibody diluted in PBS/1% BSA at room temperature for 30–60 minutes.

18. Wash the slides five to seven times in PBS/0.1% Saponin at room temperature (5 minutes each wash) in Coplin jars. Rinse once in PBS.

19. (*Optional, if using HRP-conjugated secondary antibodies*) React in DAB as follows:

    a. Cover sections with DAB staining solution (0.5–1 ml per slide if coverslips are not used; 100–150 μl per slide if coverslips are used) for 5–30 minutes. Monitor the reaction under a microscope.

b. Once the reaction has reached completion, rinse slides in several changes of PBS in Coplin jars at room temperature.

20. Mount slides in either of the following ways:

a. Following washes, air-dry slides. Mount in 70% glycerol/PBS and observe immediately.

b. Dehydrate sections through an ethanol series (once in 30%, 50%, 70%, 90%, and twice in absolute; 10 minutes each) and mount in DPX (Fluka).

## PROTOCOL 13.3

## β-galactosidase Activity Staining of Frozen Adult Retinas

### Materials

#### Supplies and Equipment

Dissecting tools and equipment, Sylgard dissection dish, and other supplies (see Protocol 12.1)

Cryostat and accessory items

Razor blade

**Ethanol/dry ice** bath

Slides (subbed slides are not necessary)

Drying plate set at 40ºC

Coplin jars

Humidified chamber (for setup, see p. 231)

Coverslips

#### Solutions and Reagents

1x PBS (see Protocol 12.1)

**Glutaraldehyde** (2%; electron microscopy (EM)-grade) in 1x PBS

Graded ethanol series; 30%, 50%, 70%, 90%, and absolute

Mountants. Either of the following can be used:

- 80% glycerol in 1x PBS (see p. 232 for important note on glycerol)
- **DPX** (Fluka)

| *Staining Solution* | |
|---|---|
| 0.2 M **Na$_2$HPO$_4$** | 1.8 ml |
| 0.2 M **NaH$_2$PO$_4$** | 0.7 ml |
| 5 M NaCl | 1.5 ml |
| 1 M **MgCl$_2$** | 50 µl |
| 50 mM **K$_3$[Fe(CN)$_6$]** | 3 ml |
| 50 mM **K$_4$[Fe(CN)$_6$]** | 3 ml |
| H$_2$O | 40 ml |

This solution is stable for months when stored at room temperature in a foil-wrapped container.

### X-gal Stock Solution

Prepare 8% **X-gal** (5-bromo-4-chloro-3-indolyl β-D-galactopyranoside) in ***N,N*-dimethylformamide** (DMF). Store in 50-μl aliquots at –20°C.

### Preparation of X-gal Staining Solution

Prewarm the staining solution at 37°C for 5 minutes. Add 8 μl of 8% X-gal stock solution per 300 μl of staining solution. Prewarming decreases precipitation of X-gal crystals.

**CAUTION: DMF, DPX, dry ice, ethanol, glutaraldehyde, K₃[Fe(CN)₆], K₄[Fe(CN)₆], MgCl₂, Na₂HPO₄, NaH₂PO₄, X-gal** (see Appendix 4)

## Method

1. Dissect tissue as described in steps 1 and 2 of Protocol 13.1.

2. Embed up to approximately 12 heads in partially frozen O.C.T. Tissue Tek Compound on a cryostat chuck and orient accordingly using fine forceps as described in step 7 in Protocol 13.2.

3. Place chuck in the cryostat chamber and pretrim the block with a razor blade. Allow the block to equilibrate to cryostat temperature (–14°C to –18°C) for 20–30 minutes.

4. Cut sections (10–14-μm thick) and transfer to slides (subbed slides are not necessary).

5. To dry the sections onto the slides, heat slides on a drying plate set at 40°C for no more than 1 minute or air-dry at room temperature for 30 minutes. *Do not overheat or overdry.*

6. Fix the sections in 2% glutaraldehyde/PBS at room temperature for 15–20 minutes in a Coplin jar.

7. Wash the slides three times in PBS (5 minutes each wash).

8. Blot off excess PBS and place slides in humidified chamber.

9. Leave slides at room temperature until dry (5–10 minutes).

   *Note:* Slides can be left for several days at this point with no loss of enzymatic activity.

10. Cover sections with 50 μl of prewarmed X-gal Staining Solution and cover with coverslip. Incubate tissue in staining solution at 37°C from 15 minutes to overnight (the length of time depends on the strength of expression of the *lacZ* line being used).

11. Once staining is complete, rinse slides twice in PBS (5 minutes each wash). The coverslip will float off the slide in the first rinse.

12. Mount slides in 80% glycerol in PBS or dehydrate through a graded ethanol series (once in 30%, 50%, 70%, 90%, and twice in absolute; 10 minutes each) and mount in DPX (Fluka).

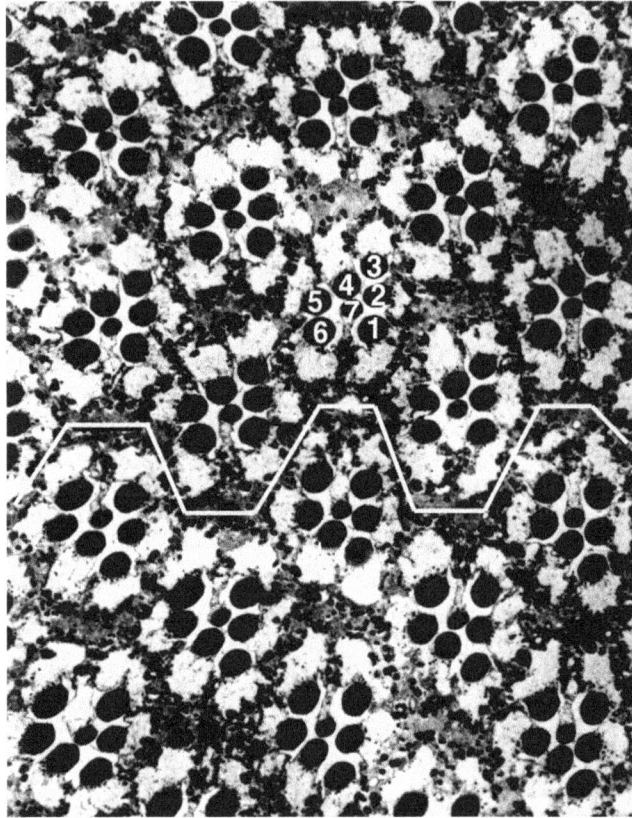

**Figure 13.2.** Thick section through an adult eye. The black circles labeled 1 through 7 are the photoreceptor rhabdomeres. The rhabdomere belonging to photoreceptor 8 lies below that of R7 and is therefore not visible in this section. The white line identifies the equator and divides the eye into mirror symmetrical dorsal and ventral halves.

PROTOCOL 13.4

## Preparing Thick Sections of Adult Retinas

This protocol is used for light microscopic analysis of the adult eye (Tomlinson and Ready 1987), and it provides nice preservation of pigment granules and photoreceptor rhabdomeres; therefore, it is primarily used for phenotypic and mosaic analysis of adult eyes (Figure 13.2). Because this fixation method does not preserve either cell membranes or subcellular structures, transmission electron microscopy (TEM) is recommended for visualizing these structures.

Osmium tetroxide ($OsO_4$), propylene oxide, and unpolymerized resin should be handled with extreme care. Gloves should be worn at all times as osmium (a volatile neurotoxin) and propylene oxide are particularly toxic, and Durcupan resin is carcinogenic when unpolymerized. Vinyl gloves should be worn for steps involving propylene oxide and resin because latex gloves are not impermeable to these reagents. Both osmium and propylene oxide are extremely volatile and all steps involving these reagents should be carried out in a well-ventilated chemical fume hood. Also note that $OsO_4$ vapor can damage the eyes. Although osmium solutions should not be stored in plastic, the incubation steps described below can safely be carried out in microcentrifuge tubes (Eppendorf). Any item that becomes contaminated with resin should be baked overnight before being discarded.

All items that come in contact with $OsO_4$ should be sealed in a Seal-a-Meal bag and disposed of according to institutional safety office guidelines. Be sure to follow proper disposal procedures for all reagents used in this protocol.

All steps can be performed in a single 1.5-ml microcentrifuge tube (Eppendorf) per genotype, by pipetting off the used solution and replacing it with the next solution. All incubations from steps 2–6 (Fixation and Embedding) should be performed on ice.

## Materials

### Supplies and Equipment

Dissecting tools and equipment, Sylgard dissection dish, and other supplies (see Protocol 12.1)

Microcentrifuge tubes (1.5 ml; Eppendorf)

BEEM flat embedding mold (package of 12; Ted Pella 111-22). The type of mold used may vary depending on the microtome model. This size and style works well for a Reichert-Jung microtome.

Sharpened (beveled) wooden stick or fine needle (e.g., microtools available from Ted Pella)

Teflon-coated razor blades (Ted Pella 121-3)

Subbed slides (see p. 231)

Heating block set at 80ºC

Coverslips

### Solutions and Reagents

Sodium phosphate (0.1 M, pH 7.2) (see Protocol 12.1)

**Glutaraldehyde** (2%) in 0.1 M **sodium phosphate** buffer (pH 7.2). Use freshly prepared fixative and high-quality glutaraldehyde or EM-grade. Glutaraldehyde from Ted Pella (18426) is available in sealed ampules. Cover opened glutaraldehyde vials with Parafilm and store for several months at 4ºC.

**$OsO_4$** (2%) in 0.1 M sodium phosphate (pH 7.2). Suspended $OsO_4$ is available from Ted Pella in sealed ampules (18459) and is available in powder form, but it is recommended that it be purchased already resuspended. Use opened vials of $OsO_4$ immediately and discard any unused portion (it oxidizes and is also too hazardous to keep open vials).

Graded **ethanol** series (30%, 50%, 70%, 90%, and absolute ethanol)

**Propylene oxide**

**DPX** mounting medium (Fluka 44581)

---

#### Durcupan Resin

Durcupan ACM is available as a set of four components (A, B, C, and D) or as separate components (Fluka 44610 for a set of all four components; 44611 [A]; 44612 [B]; 44613 [C], and 44614 [D]). Each component has a long shelf life.

Prepare as follows:

| | |
|---|---|
| Resin A | 54 g |
| Hardener B | 44.5 g |
| Accelerator C | 2.5 g |
| Plasticiser D | 10 g |

In a fume hood, combine all components in a disposable beaker. Stir until well mixed. Aliquot into 20-ml portions and use immediately or store at –20ºC (for 6 months) or at –80ºC (for several years) to slow polymerization. Aliquots can be freeze-thawed up to four times. Bake all waste at 70ºC for 24 hours before disposing!

*Toluidine Blue Staining Solution*

1% toluidine blue
1% borax

Dissolve components in $H_2O$. Filter immediately before use. The requirement for this solution is optional.

**CAUTION: DPX, resins (Durcupan resin), glutaraldehyde, $Na_2HPO_4$, $NaH_2PO_4$, $OsO_4$, propylene oxide, sodium phosphate** (see Appendix 4)

## Method

### Fixation and Embedding

1. Dissect adult eyes as described in Protocol 13.1.

2. Fix adult eyes in 0.25 ml of 2% glutaraldehyde in 0.1 M sodium phosphate buffer (pH 7.2) on ice. The eyes will float.

   *Note:* Eyes can be left in 2% glutaraldehyde until dissections are complete.

3. Add an equal volume of 2% $OsO_4$ in 0.1 M sodium phosphate (pH 7.2) and incubate on ice for 30 minutes.

4. Remove the glutaraldehyde/$OsO_4$ solution and rinse eyes with cold 0.1 M sodium phosphate (pH 7.2).

5. Add fresh 2% $OsO_4$ and incubate on ice for 1–2 hours.

   *Note:* Incubation for 1 hour is recommended if embedding eyes for clonal analysis; the tissue becomes darker with longer incubations, making it difficult to see the clones when orienting the eyes in the molds.

6. Remove $OsO_4$ and dehydrate the eyes through a graded ethanol series (once in 30%, 50%, 70%, 90%, and twice in absolute; 10 minutes each).

   *Notes:* Leaving enough ethanol in the tube to keep the eyes covered during these exchanges will prevent the cuticle from lifting away from the retina. If the cuticle does lift away, the eye will appear white; however, the underlying tissue has been preserved and can still be sectioned.
   Once the tissue is placed in absolute ethanol, subsequent steps can be performed at room temperature.

7. Equilibrate the eyes twice in propylene oxide (10 minutes each incubation).

8. Replace the propylene oxide with a mixture of 50% Durcupan resin/50% propylene oxide and incubate at room temperature overnight.

9. Remove the 50% resin/50% propylene oxide mixture and replace with 100% resin. Incubate for 4 hours to allow the resin to infiltrate the tissue.

   *Note:* The resin will polymerize at room temperature, so tissue should not be left "indefinitely" at this stage. Return resin to –20ºC during this 4-hour incubation step to minimize polymerization.

10. Embed eyes as follows:

    a. Fill molds with 100% resin and push the eye to the floor of the mold, close to the upper edge of the mold (Figure 13.1B; note this is a trimmed block), with the eye facing out toward the cutting surface.

    b. Orient the eyes using a sharpened wooden stick or fine needle.

c. Bake at 70°C for approximately 25 minutes or until the resin is slightly tacky. The exact time of baking will depend on the age of the resin and temperature of the oven, so watch it closely. As the resin heats up, it will become runny and the eyes will shift and need to be reoriented at the slightly tacky stage. The resin becomes very viscous as it cools, so work quickly to reorient the eyes once they have reached the slightly tacky stage.

d. Once the eyes have been reoriented, return them to the oven and bake at 70°C for 24 hours.

### Trimming and Sectioning Blocks

1. Use Teflon-coated razor blades (Ted Pella 121-3) to trim resin away to form a trapezoidal shape surrounding the eye.

   *Note:* The eyes should sit in a raised pedestal cut two to three times the depth of the half-head (Figure 13.1B). It should be tall enough so that the untrimmed edges do not bump into the knife during sectioning; if it is too tall, it will vibrate as the knife cuts through the resin, causing chatter.

2. Cut sections 1.5-μm thick (see Protocol 13.4 Notes, below).

3. Transfer sections to a subbed slide (see Protocol 13.4 Notes, below) by rolling sections first onto a beveled wooden stick and then off of the stick into a puddle of H$_2$O on the subbed slide.

4. Place the slide on a heating block at 80°C until the H$_2$O evaporates.

### Staining Slides

Staining slides with toluidine blue is optional. It is not necessary if viewing strongly pigmented eyes, but it will enhance the contrast of eyes with little or no pigment. Sections of eyes that will be used for mosaic analysis should not be stained, as it will be very difficult to distinguish between cells containing pigment granules versus those lacking pigment granules.

1. Place the slides containing the sections on a heating block at 70–80°C and allow the slides to warm up.

2. Incubate in enough toluidine blue to cover the sections for approximately 30 seconds.

   *Note:* The exact incubation time depends on the temperature of the block, the amount of toluidine blue, the thickness of sections, and the strength of w+ expression. As it is possible to overstain the tissue, it is best to understain and then restain to achieve an optimal level of staining.

3. Rinse in several changes of H$_2$O.

4. Examine under the microscope to assess the degree of staining. If necessary, repeat the staining procedure until the staining intensity is satisfactory.

### Mounting Slides

1. Dry the slides completely.

2. To mount both stained and unstained sections, place 1–2 drops of DPX on the sections and cover with a cover slip.

3. Store flat overnight to allow mounting medium to dry.

4. View sections using phase-contrast optics.

   *Note:* The image quality of unstained slides is very poor before mounting in DPX.

- If sectioning weak P[w+] lines, thicker sections (i.e., 2 μm) are preferable as it will be easier to identify the pigment granules.
- Subbed slides are not absolutely necessary, but because the subbed surface increases the surface tension of $H_2O$, the drop of $H_2O$ forms a bead rather than spreading out in a thin film across the slide. This is convenient because the sections are consequently confined to a smaller surface area on the slide.

PROTOCOL 13.5

# Preparation of Specimens for Scanning Electron Microscopy

Three protocols are provided, one for eyes that are to be critical-point-dried (CPD; Kimmel et al. 1990) and two alternatives if a critical point drier is not available. The major difference between the first two protocols is a preliminary fixation step. *Drosophila* eyes have a tendency to collapse when critical-point-dried, so fixing the animals before critical point drying improves the yield of good specimens.

## Materials

### Supplies and Equipment

Critical point drier or vacuum drier depending on method; both items are available in an electron microscopy laboratory.

Double-stick carbon conductive tabs (Ted Pella 16084-1; this size works for the specimen mounts given above. If ordering larger specimen mounts, be sure to order the appropriate size tabs).

Aluminum specimen mounts or stubs (15 × 10-mm cylinder; Ted Pella 16291). Stubs are coated with a thin film of grease that will contaminate the scanning electron microscope (SEM) chamber. To remove this grease, thoroughly rinse the stubs in 95% **ethanol** before use.

Microcentrifuge tubes (1.5 ml; Eppendorf)

### Solutions and Reagents

Graded ethanol series (25%, 50%, 75%, and absolute)

**Trichlorofluoroethane** (Aldrich)

Hexamethyldisilazane (HMDS; Sigma). HMDS is moisture-sensitive; open the ampule immediately before use.

### Fixative

1% **glutaraldehyde**
1% **formaldehyde**
1 M **sodium cacodylate** (pH 7.2)

**CAUTION: ethanol, formaldehyde, glutaraldehyde, sodium cacodylate, trichlorofluoroethane** (see Appendix 4)

## Method

### *Fixation, Dehydration, and Drying*

#### *Method 1: Critical Point Dry Method*

1. Immerse whole flies in fixative for 2 hours.

   *Note:* The addition of one small drop of 0.2% Triton X-100 or Tween-20 (diluted in $H_2O$) reduces the surface tension, so that the animals will become submersed in the fix.

2. Rinse in $H_2O$.

3. Dehydrate through an ethanol series (once in 25%, 50%, and 75% ethanol, and twice in absolute ethanol, each at room temperature for 12 hours).

   *Note:* Although 12-hour incubations are recommended, workers who find it necessary to cut corners have had success with 4-hour incubations. The exterior morphology of the eye is stable for at least 1 month in absolute ethanol.

4. Critical point dry (CPD) and sputter-coat.

#### *Method 2: Trichlorofluoroethane Method*

1. Pass samples through a graded ethanol series (once in 25%, 50%, and 75% ethanol, and twice in absolute ethanol, each at room temperature for 12 hours).

   *Note:* As stated in Method 1 (above), 12-hour incubations are recommended, but 4-hour incubations will work. These incubations can be performed in 1.5-ml microcentrifuge tubes.

2. Similarly, pass samples through a series of trichlorofluoroethane diluted in absolute ethanol (once in 25%, 50%, and 75% trichlorofluoroethane, and twice in 100% trichlorofluoroethane; each at room temperature for 12 hours). Then remove the trichlorofluoroethane and dry the samples under a vacuum.

   *Notes:* These steps also can be shortened to 4-hour incubations.
   Trichlorofluoroethane destroys the ozone layer and is therefore a less desirable option than CPD. Follow a proper disposal procedure for trichlorofluoroethane. An alternative low-surface tension solvent is HMDS (Braet et al. 1997; see below).

#### *Method 3: HMDS Method (Braet et al. 1997; modified by Naoto Ito)*

1. In 1.5-ml microcentrifuge tubes, dehydrate flies in 25% ethanol at room temperature overnight.

2. Dehydrate flies once in 50% and 75% ethanol, and three times in absolute ethanol at room temperature from 2 hours to overnight (each step).

3. Incubate flies in 0.25 ml of HMDS (Sigma; use a freshly opened ampule) at room temperature for 3 minutes. Repeat incubation in HMDS.

4. Dry flies overnight under vacuum.

### *Mounting Flies on Stubs*

Grease can damage the interior chamber of the SEM, so wear gloves when handling the stubs. Stick carbon conductive tabs onto aluminum specimen mounts and mount flies on the conductive tab. Up to eight flies can be mounted on each stub. Mount the animals so that the surface to be viewed is parallel to the stub surface (this minimizes the time-consuming and tedious task of having to orient the stub once it is in the microscope).

PROTOCOL 13.6

# Preparation of Specimens for Transmission Electron Microscopy

Standard transmission electron microscopy (TEM) procedures are described in great detail in McDonald et al. (this volume); therefore, this discussion is limited to a fixation protocol that works well for *Drosophila* eye tissue including larval imaginal discs and pupal and adult retinas. This procedure is adapted from Baumann and Walz (1989). All reagents used in this protocol are hazardous. They should be handled with care and disposed of according to standard regulations. Bake all resin overnight at 80°C before disposing. For additional precautions regarding special handling of $OsO_4$, propylene oxide, and resin, see Protocol 13.4. For details, see McDonald et al. (this volume).

## Materials

### Supplies and Equipment

Pulled glass micropipette attached to a syringe
Embedding molds (see p. 237)

### Solutions and Reagents

**Tannic acid** (1%) in fixative (see below)
**Sodium cacodylate** (0.1 M, pH 7.4)
**$OsO_4$** (2%) in 0.1 M sodium cacodylate
**Uranyl acetate** (2%) in $H_2O$
Graded **ethanol** dehydration series (30%, 50%, 70%, 90%, and absolute)
**Propylene oxide**
Durcupan resin (see p. 237 but use the following proportions: Component A, 50 g; component B, 50 g; component C, 1.75 g; component D, 0.75 g)

### Fixative

| Component and final concentration | Amount to add to make 10 ml |
|---|---|
| 0.1 M sodium cacodylate (pH 7.4) | 1 ml of 1 M stock |
| 4% **formaldehyde** | 2 ml of 20% stock |
| 3.5% **glutaraldehyde** | 0.7 ml of 50% stock |
| $H_2O$ | 6.3 ml |

Both formaldehyde and glutaraldehyde should be fresh. Prepare fixative fresh before each use.

A 20% stock of formaldehyde lasts up to 1 week if kept at 4°C. Important! Do not use a commercially available 37% stock formaldehyde; rather, make a 20% stock solution from paraformaldehyde (see recipe provided in Protocol 12.1).

Use EM-grade glutaraldehyde (available as a 50% stock in ampules; Ted Pella). Once opened, glutaraldehyde is good for 1 month if stored at 4°C.

**CAUTION: ethanol, formaldehyde, glutaraldehyde, $OsO_4$, propylene oxide, sodium cacodylate, tannic acid, uranyl acetate** (see Appendix 4)

## Method

1. Inject Fixative into larval, pupal, or adult head using a pulled glass micropipette attached to a syringe. Let the animal sit for 10 minutes.

   - *For larvae and pupae:* Remove retinal tissue from head, transfer to a fresh drop of Fixative, and continue to fix at room temperature for 4 hours.
   - *For adult eyes:* Place a drop of Fixative on top of the head after injecting the Fixative into the animal. Gently cut the head in half, taking care not to deform the heads, transfer to a fresh drop of Fixative and continue to fix for 4 hours at room temperature.

2. Fix in 1% tannic acid in Fixative overnight.

3. Wash in 0.1 M sodium cacodylate (pH 7.4) for 30 minutes.

4. Infiltrate with 2% $OsO_4$/0.1 M sodium cacodylate (pH 7.4) for 2 hours.

5. Rinse in $H_2O$.

6. Incubate in 2% uranyl acetate at room temperature overnight.

7. Dehydrate tissue through a graded ethanol series (30%, 50%, 70%, 90%, absolute; 5 minutes each).

8. Incubate in the following:

   - Propylene oxide for 5 minutes.
   - 33% resin in propylene oxide for 2 hours.
   - 66% resin in propylene oxide for 2 hours.
   - 100% resin overnight.

9. Embed in molds as described in Protocol 13.4, step 10.

## ACKNOWLEDGMENTS

I am grateful to Nadean Brown, Graeme Mardon, and Jessica Treisman for their thoughtful comments and useful suggestions on this chapter. I also thank the many individuals who contributed specific protocols and modifications.

## REFERENCES

Baumann O. and Walz B. 1989. Topography of Ca++-mediated sequestering endoplasmic reticulum in photoreceptors and pigmented glial cells in the compound eye of the honeybee drone. *Cell Tissue Res.* **255:** 511–522.

Braet F., DeZanger R., and Wisse E. 1997. Drying cells for SEM, AFM and TEM by hexamethyldisilazane: A study on hepatic endothelial cells. *J. Microsc.* **186:** 84–87.

Dickson B. and Hafen E. 1993. Genetic dissection of eye development in *Drosophila.* In *The development of* Drosophila melanogaster (ed. M. Bate and A. Martinez-Arias), pp. 1327–1362. Cold Spring Harbor Laboratory Press, Cold Spring Harbor, New York.

Kimmel B.E., Heberlein U., and Rubin G.M. 1990. The homeo domain protein *rough* is expressed in a subset of cells in the developing *Drosophila* eye where it can specify photoreceptor cell subtype. *Genes Dev.* **4:** 712–727.

Tomlinson A. and Ready D.F. 1987. Cell fate in the *Drosophila* ommatidium. *Dev. Biol.* **123:** 264–275.

Wolff T. and Ready D.F. 1993. Pattern formation in the *Drosophila* retina. In *The development of* Drosophila melanogaster (ed. M. Bate and A. Martinez-Arias), pp. 1277–1325. Cold Spring Harbor Laboratory Press, Cold Spring Harbor, New York.

# CONTENTS

# Preparation of Thin Sections of *Drosophila* for Examination by Transmission Electron Microscopy

**Kent L. McDonald**
*University of California at Berkeley*
*Electron Microscope Laboratory*
*Berkeley, California 94720-3330*

**David J. Sharp**
*University of California at Davis*
*Department of Molecular and Cellular Biology*
*Davis, California 95616*

**Wayne Rickoll**
*University of Puget Sound*
*Department of Biology*
*Tacoma, Washington 98410*

THIS CHAPTER IS FOR INDIVIDUALS who have not done electron microscopy (EM) before, but it also contains some new information about specimen preparation that can benefit experienced electron microscopists. It covers the materials and methods necessary to do *routine thin sections of embedded specimens*; however, it does not cover specialized EM techniques such as freeze-fracture, rotary shadowing, negative staining, scanning EM, and serial sectioning. Likewise, this chapter focuses mainly on embryos because a description of the fixation methods for all the various types of fly tissues would require an entire book. This chapter is divided into three main sections:

- *High-pressure Freezing and Freeze Substitution:* For those situations where the best possible preservation of ultrastructure and antigenicity is required.

- *Conventional Chemical Fixation:* For situations where optimal preservation is not required or when high-pressure freezing is not an option.

- *EM Immunolabeling:* Techniques for localizing specific antigens on thin sections.

Also included as a kind of checkpoint is an initial section that describes for would-be electron microscopists the problems encountered should they decide to perform EM.

EM has been around for so long that it would seem not to be necessary to have a special chapter on the methods. Why not just get a basic EM text or an EM paper and follow the steps laid out there? Why not just buy a kit and follow the directions? Well, it turns out that there is no single, simple procedure for fixing and embedding all tissues: The chemistry of different cell types is to some extent unique, and this affects the way each cell type reacts to the wide array of fixatives, buffers, organic solvents, and resins used in EM specimen preparation. This topic is thoroughly covered by Hayat (1981, 1989a) on fixation for electron microscopy, and the more recent book by Griffiths (1993) also contains an excellent review of the problems encountered with EM fixation.

A recurring theme in those organisms or cell types that are difficult to fix is the presence of a diffusion barrier that prevents the free diffusion of fixative and other chemicals in and out of the cell or tissue. This in turn means that fixation takes a relatively long time (measured in minutes or tens of minutes in some cases), during which the cells begin autolysis or are otherwise degraded from their original state. The reason *Drosophila* requires a special chapter on preparation methods for EM is that most fly tissues are surrounded by significant diffusion barriers. In the embryo, it is the vitelline envelope, and in larvae and adults, it is the cuticle. In this chapter, we discuss the methods that have evolved to cope with these barriers in order to achieve reasonable preservation of ultrastructure. We have tried to describe them in sufficient detail so that they can be reproduced as needed. However, before we get to that, let us consider some other issues that often are not thought of by the beginning microscopist until it is too late and the frustration level is high.

## CHECKPOINT

The following are some key questions that need to be considered before undertaking EM work. They are especially important if EM has never been done before, but even if it has, these questions are worth reconsidering as a reminder of what is involved in obtaining EM data.

### What Kinds of EM Resources Are Available?

EM resources are really of two different types: The equipment resources and the human resources. The types and availability of equipment are easy to ascertain, and in most institutions, there is still an electron microscope around and some support equipment such as microtomes and curing ovens for epoxies. In the best situation, a central Electron Microscope Facility will have all the necessary equipment and a mechanism in place for cost sharing or recharge. More commonly, a department (not necessarily one's own) or an individual researcher will have a microscope, with one person in charge of maintaining the equipment. Of the major pieces of equipment, the quality of the microtome is most important because sectioning is the most difficult skill to learn to perform EM. Try to find a Reichert-Jung (now Leica) Ultracut E or later model, or an RMC6000 or later. Obviously, earlier models and brands can work well, but the later models are so much better that they will shorten the learning curve significantly as well as cut better sections.

The human resource component is another critical issue when EM work must be done or the technique learned. If a central EM lab is available, someone there can teach these techniques and/or perform the work for a fee. If learning from someone who is a caretaker for a departmental or lab microscope, it will probably take longer and be more frustrating. These people are usually technicians, graduate students, or postdocs who have responsibilities other than maintaining EM equipment and thus working within their schedule may be a problem. In addition, the equipment will have to used when the primary users are not signed up.

## How Much Time and Money Should be Invested in the Project?

Two of us (K.M. and W.R.) supervise EM teaching and service labs and have many years of experience watching EM projects from their beginnings to (sometimes) completion. The one factor common to all of the projects is that they nearly always take longer and cost more than anticipated. At some point, investigators must decide whether it is more cost effective to have the work done or learn to do it themselves. It is a little bit like the old adage that if you give a person a fish, they will be fed for one day, but if you teach them how to fish, they can feed themselves indefinitely. But having said that, it is usually the best for all concerned to contract the work out. This is especially true when trying to get the answer to one specific question and have no intention of continuing to do EM. Learning to do microscopy that involves sectioning will take longer than many investigators will want. Sectioning is a physical skill involving hand/eye coordination and other motor skills, similar to learning to play a musical instrument. One may be able to pick out a tune, or cut a section after a week or so, but it takes lots of practice to make it routinely good. Learn how to section only if EM will be done routinely in the future. Perhaps the most workable solution in the short term is to learn everything but the sectioning and hire someone else to do that part. However, definitely learn how to operate the electron microscope because a tremendous amount can be learned about a research question by the time spent looking at sections. A person's mind will integrate much more information than can be recorded adequately on film, and after looking at many sections, the representative data can be determined and the appropriate images recorded.

Cost is an issue that varies considerably, depending on the situation at a given institution. Some EM facilities are heavily subsidized, and others not at all; however, a modest amount will probably have to be spent to get EM results. People who are willing to spend thousands of dollars each month for molecular biology or biochemistry projects often are reluctant to spend a fraction of the amount on EM. EMs are costly instruments to purchase and maintain, and in order to learn to section, a diamond knife may need to be purchased at a cost of $2000 to $3000.

In summary, if the EM project is a one-time effort, then the cost-effective solution is to hire someone to do the work, or find a collaborator with EM skills and the interest and time. If the investigator must do the work or wants to, then several months and several thousand dollars should be budgeted for the project. This time could be even longer if the equipment is old and sectioning must be learned.

## SAFETY CONSIDERATIONS _____

The electron microscopist routinely works around many dangerous chemicals and is, or should be, very safety conscious. If using a chemical that is not familiar, refer to the Material Safety Data Sheets available from chemical suppliers. If using cryotechniques, be aware of the problems with liquid nitrogen and cryogens such as propane. Hazards of reagents and chemicals used in EM fixation is given in Table 14.1; also see EM textbooks such as Bozzola and Russell (1999). Some safety precautions are given below:

- WEAR GLOVES when handling all chemicals. Latex gloves are generally useful, but vinyl gloves will disintegrate with acetone. Use double gloves when handling osmi-

**Table 14.1.** Hazards of Chemicals and Reagents Used in EM Fixation

| Type of chemical or reagent | Cautionary notes |
| --- | --- |
| Chemical fixatives and buffers | |
|    aldehydes such as glutaraldehyde and paraformaldehyde | tend to be reversible and not as dangerous as most other EM chemicals; however, use every precaution |
|    acrolein | basic ingredient in tear gas |
|    cacodylate buffer | an arsenic compound |
|    potassium ferri- or ferrocyanide | cyanide poisoning |
|    osmium tetroxide | very common and very dangerous; a heavy metal that will accumulate in the internal organs; when mixing with organic solvents for freeze-substitution, use extra caution |
|    uranyl acetate | heavy-metal poison; radioactivity usually depleted |
|    lead citrate | heavy-metal poison |
| Organic solvents | |
|    most common are ethanol and acetone, and sometimes methanol | can damage the liver if fumes are inhaled constantly |
| Epoxy and methacrylate resins | potentially carcinogenic with repeated skin exposure or inhalation of the accelerator component; contact dermatitis reaction common |
| Liquid nitrogen | can cause severe burns with prolonged skin contact; less evident is the danger from asphyxiation; liquid nitrogen expands almost 700-fold when converting from liquid to gas |

For additional information on each chemical or reagent, see Appendix 3.

um-acetone mixtures. If unsure about the suitability of the gloves, the catalogs usually have safety information for each type.

- WORK IN A CHEMICAL FUME HOOD! The hood should have a good draw. Even with a good hood, the volatility of some compounds such as Lowicryl resin is so great that it is hard not to smell it in the room. Remember to remove the gloves *in the hood* to avoid carrying toxic fumes around the lab.

- WEAR EYE PROTECTION WHEN WORKING WITH LIQUID NITROGEN! Use a shield that covers the whole face.

- WORK IN WELL-VENTILATED AREAS WHEN USING LIQUID NITROGEN! It is also important to work in a large room. Dumping large volumes of LN2 into the air in a small room is asking for trouble. For a more extended discussion of how to work safely with liquid nitrogen, see Sitte et al. (1987).

## HIGH-PRESSURE FREEZING AND FREEZE-SUBSTITUTION

The state of the art in fine-structure preservation for thin sectioning can be achieved by using fast-freezing technology followed by freeze-substitution and embedding in resin. The reason that fast freezing is so much better has to do with the speed of fixation, which in the context of freezing means stopping all molecular movements. It is estimated (Moor 1987) that samples prepared by high-pressure freezing (HPF) are "fixed" in 20–50 msec, as opposed to the seconds and even minutes that it takes for chemical fixatives to diffuse into the cell. Fast freezing also freezes every cell component regardless of its chemistry. In contrast, glutaraldehyde, paraformaldehyde, osmium, and other chemical fixatives are selective in their cross-linking (Hayat 1981).

Different ways to freeze tissues include impact against a cooled metal block, spraying onto a cooled surface, plunging into cooled cryogen, and cooling from two sides with cooled cryogen. The problem with most of these methods is that only a small volume of the sample can be well frozen, usually 5–10 µm from the cell surface, and up to approximately 40 µm with two-sided jet-cooling (Robards and Sleytr 1985; Echlin 1992). This makes these methods impractical for freezing large objects such as whole fly embryos or most adult tissues (except parts, e.g., wings, antennae, and surface bristles). But use of HPF (sometimes called hyperbaric freezing) can achieve good freezing up to depths of 200 µm and more, depending on the chemistry of the cells. It should be noted that all discussion of freezing depths and times refers to tissues that are not infiltrated by penetrating cryoprotectants such as sucrose or glycerol. For a more detailed discussion of how and why HPF works to freeze relatively large samples, see the articles by Moor (1987), Studer et al. (1989), and Kiss and Staehelin (1995) and books by Echlin (1992) and Robards and Sleytr (1985).

The procedure for high-pressure freezing is given in Protocol 14.1. The equipment and materials for HPF and the actual loading and operation of the machine are covered in detail in McDonald (1999). Rather than take up space here repeating the same details, we encourage reading that paper for specifics. Here, we review only the most essential of materials and also cover the specifics that deal just with *Drosophila* tissues. Information about freezing *Drosophila* can also be found in McDonald (1994).

Once frozen, tissues can be processed in a variety of ways before viewing in the EM, but in this chapter, we only discuss freeze-substitution and embedding in resin (Howard and O'Donnell 1987; Steinbrecht and Müller 1987). Resin-embedded samples lend themselves to thin sectioning, which is the most familiar way of viewing cell fine structure for morphology and immunolabeling.

Freeze-substitution (FS) is a method which dehydrates the cells at very low temperatures and replaces cell water with organic solvent at –80ºC to –90ºC. At this temperature, large molecules such as proteins are immobilized, yet smaller molecules such as water (ice) can be dissolved and replaced with organic solvents such as acetone or methanol. Chemical fixatives such as osmium tetroxide and glutaraldehyde will also diffuse into the cells at this temperature, but they do not actually produce much cross-linking. This occurs at warmer temperatures, thought to be approximately –50ºC for glutaraldehyde and approximately –30ºC for $OsO_4$ (Humbel and Müller 1986). However, because the chemical fixatives are effectively in place throughout the tissues when the cross-linking temperature is reached, there is effectively no diffusion-related fixation problem such as occurs by conventional room-temperature methods.

Protocol 14.2 describes the preparation of fixatives for freeze-substitution. Fixatives for freeze-substitution are made by dissolving either glutaraldehyde or osmium in an organic solvent. We also like to include a little uranyl acetate in these mixtures because it helps with membrane contrast. The freeze-substitution procedure is given in Protocol 14.3.

# High-pressure Freezing

## Materials

### Media

Agar (2%) in 60-mm plastic petri dishes (or fly food plates). This is a good surface for holding specimens because it keeps them moist, and for loading the specimen holders because they will not slip. Put a little yeast paste on one side to keep it moist and readily available.

Yeast paste. For preparation, see Sisson (this volume). Although the yeast paste used on food trays is suitable, it is even better to use a paste made up with 10% **methanol** rather than $H_2O$. This gives the sample some added cryoprotection and does not seem to affect the flies.

### Supplies and Equipment

High-pressure freezer (HPF). See Table 14.2 for a list of HPFs in the United States and information about how to find an HPF anywhere in the world.

Specimen holders. The commercially available specimen holders come in two designs. Type A (Ted Pella 39200) has a well on either side, one is 100-μm deep and the other is 200-μm deep. Type B (Ted Pella 39201) is flat on one side and has a 300-μm deep well on the other. By putting the wells and flat side in different combinations, six different chamber depths can be achieved. Specimen holders can also be obtained from Bal-Tec (Technotrade in the U.S.) and Leica (for addresses, see Table 14.2).

Dissecting microscope. Loading samples into the specimen holder works best if performed under a dissecting microscope. It is best to use one mounted on a pole stand with a large working area underneath.

Camel's hair paintbrush (size 00). This is the perfect tool for handling embryos and other fly parts. A little yeast paste on the tip of the brush picks up whatever is to be loaded into the specimen holder.

Fine forceps for picking up the specimen holder from the agar plate and transferring it to the specimen-holder carrier, which is the part that is put into the HPF for freezing. Forceps with angled tips (Ted Pella 5624) are more useful than the straight-tipped kind.

### Solutions and Reagents

**Liquid nitrogen** (LN2)
Freeze-substitution fixative frozen in LN2 (for preparation, see Protocol 14.2).

**CAUTION: liquid nitrogen, methanol** (see Appendix 4)

## Method

### Preparing Fly Tissues for Freezing

It takes approximately 5 minutes between freezing runs, so if doing a developmentally timed experiment such as looking at a particular stage of embryogenesis, timing will need to be taken into account when planning the experiment. Depending on the stage of interest, treat the samples as follows:

- *Embryos.* If using embryos, a 200-μm deep (type A) specimen holder holds up to 30 embryos. Multiple runs will be necessary, so plan the number of embryos accordingly. It is not necessary to remove the chorion layer, but we usually do if there are

*For all protocols, $H_2O$ indicates glass distilled and deionized.

**Table 14.2.** Locations of High-pressure Freezers

*U.S. Locations[a]*
Arizona State University, Tempe, Arizona
Cornell University, Ithaca, New York
Memphis State University, Memphis, Tennessee
Miami University, Oxford, Ohio
Shriners Hospital Research Lab, Portland, Oregon
University of California, Berkeley, California
University of Colorado, Boulder, Colorado
University of Georgia, Athens, Georgia
University of Massachusetts, Amherst, Massachusetts
University of Minnesota, Minneapolis, Minnesota
University of Wisconsin, Madison, Wisconsin
Wadsworth Center for Labs and Research, Albany, New York

*Other than U.S. (number of machines)[b]*
Australia (1)
Austria (1)
Brazil (1)
France (1)
Germany (7)
Japan (6)
Netherlands (1)
Sweden (1)
Switzerland (3)
United Kingdom (2)

[a]Contact information for individuals in charge of machines at U.S. locations can be obtained from: Technotrade International, 7 Perimeter Road, Manchester, New Hampshire 03103-3343; Phone: (603) 622-5011; Fax: (603) 622-5211.

[b]Information regarding specific contacts at locations outside the U.S. can be obtained from: BAL-TEC AG, Fohrenweg 16, FL-9496 Balzers, Liechtenstein; Phone: ++41-75-388-1212; Fax: ++41-75-388-1260 OR Leica AG, Hernalser Hauptstrasse 219, A-1171, Wien, Austria; Phone: 43 1 488 99-0; Fax: 43 1 485 21 42.

plenty of embryos. For dechorionating embryos, see step 1 in Protocol 14.4. Do not remove the chorion layer if using only a few embryos. In the latter case, just use the paintbrush with a little yeast paste to pick the embryos off the plate.

- *Larvae.* For larvae, use the 300-μm deep well of the type-B specimen holder, unless the larvae are small. If the larva is larger, then cut it or curl it up to fit.

- *Adult tissues.* These tissues may have to be dissected before freezing, or unwanted parts of the adult may have to be cut off.

### Important Considerations When Loading the Sample

Loading the sample into the specimen holder is without doubt *the most important step in the HPF process*. It is unlikely that the HPF will malfunction, so the quality of the final freeze can only be affected between the living cell and firing the HPF. Be careful about the following points:

- Do not allow the sample to dry out, but it should not be too wet either. Drying will simply ruin the tissue before it is even frozen, and if the sample is too wet, it may not freeze well. The reason for the latter is that when water freezes, it becomes an insulator, not a conductor of heat. The whole point of rapid freezing is to remove the heat from the sample as quickly as possible. Yeast paste protects the sample from drying out and is a good conductor of heat at the same time. Judging the correct consistency of the yeast paste takes a little practice; work quickly to keep things optimally moist.

- Shallower is better when it comes to specimen-carrier depth. Currently, the commercially available specimen holders come in depths of 100, 200, and 300 μm (all are 2 mm in diameter). Use the shallowest well depth for optimal heat transfer.

- The specimen carrier must be completely filled and without air spaces. Under pressure of 2100 bar, air spaces will cause the carrier wall to collapse and also the sample. If its wall is caved in toward the well after freezing, there was air space in the container. Before freezing, practice loading the specimen carriers until the volume is just right. This can be done anytime, but preferably some days before actually getting ready to freeze. Do not practice loading the specimen carriers after starting up the freezer because it is not as easy as it looks, but it is critically important.

### A Typical HPF Operation

1. Prepare the flies, taking timing into account, so that they are ready when the HPF is available.

2. Prepare yeast paste, specimen holders, and agar plates and have all necessary tools available to handle the particular specimen to be frozen.

3. Turn on the HPF (it takes ~30 minutes to be ready).

4. Collect flies and load into appropriately sized specimen holder.

5. Check that the specimen holder is full *with no air spaces.*

6. Load sample into specimen holder carrier and freeze (for step-by-step procedure, see McDonald 1999).

7. Put frozen sample into appropriate fixative vial under LN2.

8. Begin freeze-substitution (see Protocol 14.3) or store frozen samples for use at a later date.

## PROTOCOL 14.2

## Preparing Fixatives for Freeze-substitution

The method below is for the preparation of 50 ml of fixative.

### Materials

### Supplies and Equipment

Pencil (#2)

Nalgene cryovials (2 ml; Nalge 5000-0020). The catalog states that the cryovials should not be immersed in LN2, but we have been using them for years without problems. Alternatively, buy vials specifically to use with LN2.

Nalgene cryovial rack (Nalge 5030-0510)

Aluminum foil

Vial rocker
Bath sonicator
Conical tube (50 ml) with screw cap
Repeator pipettor to dispense up to 2-ml aliquots
Styrofoam box large enough to accept Nalgene cryovial rack

### Solutions and Reagents

**Uranyl acetate** crystals
Anhydrous **methanol**
Crystalline **osmium tetroxide** ($OsO_4$) (1 g; various EM vendors)
Pure **acetone**
**Glutaraldehyde** (10%) in acetone (Electron Microscopy Sciences 16530)
LN2 in 4-liter dewar

**CAUTION: acetone, glutaraldehyde, methanol, LN2, OsO4, uranyl acetate (solvent-based mixtures of glutaraldehyde, osmium, and uranyl acetate are highly hazardous)** (see Appendix 4)

## Method

REMEMBER! WEAR DOUBLE GLOVES FOR THIS PROCEDURE AND WORK WELL INSIDE A CHEMICAL FUME HOOD THAT DRAWS WELL. OSMIUM IN ACETONE IS VERY VOLATILE!

1. Use a #2 pencil to label approximately 35 Nalgene 2.0-ml cryovials as <u>2.1 - Os-UA</u> (which stands for 2% osmium tetroxide plus 0.1% uranyl acetate). If making a glutaraldehyde fixative, label the vial <u>.2.1 G-UA</u> (which stands for 0.2% glutaraldehyde plus 0.1% uranyl acetate). If using a freeze-substitution solvent other than acetone, indicate that also.

2. Put vials in vial rack with tops off.

3. Make up a stock solution of 5% uranyl acetate in methanol by adding 0.1 g of uranyl acetate to 2 ml of pure methanol, cover with aluminum foil, and put on rocker to dissolve (~10 minutes). If it has not dissolved after this time, just let it rock longer. Different vintages of uranyl acetate crystals seem to dissolve at different rates.

4. Sonicate an $OsO_4$ ampule (1 g) for 10 seconds to loosen crystals from glass.

5. Fill a 50-ml conical tube with 45 ml of pure acetone.

6. Add 1 g of $OsO_4$ crystals (or 1 ml of a 10% solution of glutaraldehyde in acetone) to the 45 ml of acetone, cover, and mix until completely dissolved.

7. Add 1 ml of 5% uranyl acetate stock to $OsO_4$-acetone, add pure acetone to make 50 ml, and mix well.

8. Use a repeater pipettor to dispense 1.5 ml into each cryovial. Work quickly, but not carelessly. Cap cryovials as soon as they are filled.

9. Add LN2 into a Styrofoam box to a level that will cover the cryovials in the rack.

10. Slowly lower the rack into LN2. Vials must remain upright so fixative will remain in the bottom of the vial.

11. When completely frozen, transfer vials to a LN2 storage device until ready to use.

PROTOCOL 14.3

## Freeze-substitution

The best way to do freeze-substitution is with a dedicated freeze-substitution device such as the Leica AFS system. These devices allow programming of the times and temperatures needed. When the time is up, simply unload the sample. Alternatively, if this equipment is not available, freeze-substitution can still be performed using items commonly found around the laboratory (see below).

### Materials

#### *Supplies and Equipment*

Freeze-substitution apparatus, such as the Leica AFS system
*OR* the following items (and reagents below):
Styrofoam box
Aluminum block with 13-mm holes (Fisher 11-718-14)
Cryotube with hole in the cap to admit the thermocouple probe
Type T thermocouple probe (Omega Engineering, 5SC-TT-T-30-36)
Digital thermometer (Cole-Parmer Instrument Model P-91100-40)
Rotary shaker
Elastic cord

#### *Solutions and Reagents*

**Acetone**
**Dry ice**

**CAUTION: acetone, dry ice** (see Appendix 4)

### Method

1. For a commercial freeze-substitution device, follow the instructions of the manufacturer.

2. If a freeze-substitution device is not available, load a Styrofoam box two-thirds full with dry ice.

3. Put the aluminum block on the ice to cool. Cover the top of the block with aluminum foil to prevent frost buildup on the surface.

4. Into one of the holes in the block, place a cryotube filled with acetone and into which has been inserted a thermocouple probe through the vial cap.

5. When the acetone has cooled to approximately –78ºC (it usually takes ~1 hour), transfer HPF-frozen samples in fixative from LN2 storage into the holes of the block.

6. Fill the remainder of the box with dry ice and cover the top of the aluminum block and the vials (e.g., with aluminum foil).

7. Put the lid on the box and place the whole assembly onto the rotary shaker (which must be able to take this much weight).

8. Use elastic cord to tie down the Styrofoam box on the shaker.

9. Start the shaker and rotate at least 100 rotations per minute.

10. Continuously shake until the ice has sublimed and the temperature of the vials has risen to the appropriate temperature for the kind of resin intended for use in infiltration and embedding. The entire process from initial transfer of samples onto dry ice to warm-up should take 3–4 days.

11. For epoxy embedding, e.g., using Epon or Epon-Araldite, warm to room temperature, rinse, infiltrate, and embed as usual (see steps 7–8, Protocol 14.4).

12. For low-temperature embedding, follow the manufacturer's instructions for resin:solvent exchange and polymerization (see also Newman and Hobot 1993).

## CONVENTIONAL CHEMICAL FIXATION

Many published works show that it is not necessary to use HPF-FS to preserve fly cells for ultrastructural analysis. Although it represents the state of the art in specimen preservation, not all projects require this level of precision. In addition, some tissues are just too large to fit into the HPF specimen carriers. Finally, some fly tissues such as eyes and ovaries do not freeze well in our experience, perhaps due to the fluids that surround the cells. Below, we present some conventional EM methods that will work for most fine-structure studies.

Because the vitelline membrane is impermeable to aqueous solvents, it is necessary to either mechanically disrupt it or render it permeable by treatment with organic solvents. Good ultrastructural preservation has been achieved by puncturing embryos immersed in fixative with extremely sharp tungsten needles, by microinjection of fixative (Prokop et al. 1998), and by permeabilization with heptane containing glutaraldehyde (Tepass and Hartenstein 1994). Note that even with chemical permeabilization, the vitelline membrane must still be subsequently removed manually to achieve infiltration with embedding resins.

Procedures for conventional fixation of embryos are given below. Protocol 14.4 describes a trialdehyde fixation procedure for embryos. The *n*-heptane permeabilization described in Protocol 14.5 uses simultaneous permeabilization with *n*-heptane and initial fixation with glutaradehyde. This procedure can produce excellent preservation. For an example of results using this fixation, see Tepass and Hartenstein (1994). These two methods for fixing embryos work well for us, but there are other procedures that can be used. One is described in McDonald (1994), and many more are found in the literature. If none of these give satisfactory results, consider fast-freezing and freeze-substitution.

When it comes to choosing a conventional fixation method for *Drosophila* tissues other than embryos, the best approach is probably to do an extensive literature search for EM papers on that tissue or something similar, and then try duplicating the published procedure.

PROTOCOL 14.4

# Trialdehyde Fixation of Embryos

## Materials

### Supplies and Equipment

If not noted otherwise, most items are routinely available from various EM vendors.

Dissecting microscope with zoom magnification and a fiber-optic light source if possible. We like to use a dissecting head mounted on a boom so that there is ample flat working space available.

Double-coated adhesive tape

Microscope slides

Chemical laboratory spot plate or depression-well slides. Ceramic plates with 12 depression wells (plates can be coated with a black glaze at a pottery studio for better visualization of embryos) make convenient wells for primary fixation. Depression-well slides can be substituted.

(*Optional*) Millipore vacuum filtration setup

(*Optional*) Fine paintbrush for transferring embryos

Pasteur pipettes (9 inches)

Specimen vials

Embedding molds

Wide-bore plastic pipette

Sharpened applicator stick

### Preparation of Tungsten Needles

The following is required:

12V power supply

**Sodium nitrite** or 10 N **KOH**

Alligator clip

Tungsten wire, 0.010-inch diameter (A.D. Mackay, Inc., P.O. Box 1612; Darien, Connecticut 06828; Phone: 203-655-7401, or Goodfellow, Cambridge, DB4 4DL, England)

Metal-handle needle and probe holder (Ted Pella 1350-4); these are convenient for handles for tungsten wire needles

Thermolyne Speci-Mix Test Tube Mixer (Fisher 12-814-2)

1. Take a 12V microscope power supply with continuous voltage regulation and remove insulation from one wire to make an electrode, which is submerged in a small beaker containing $H_2O$ saturated with **sodium nitrite** or 10 N **KOH.**

2. To the other lead, attach an alligator clip, which is used to connect the lead to the metal part of the tungsten needle holder. Then hold the needle and submerge the end of the wire into the solution.

3. Depending on the depth of submersion, make long sharp needles for puncturing (more submersion obviously) or short blunter needles for dissection. Needles can be rapidly sharpened this way (~1 minute usually) if the need arises to resharpen while working.

4. Another way to make the points is to attach the needle to a rocker assembly, e.g., a test-tube mixer like the one used to mix grids for immunolabeling (see p. 264), and arrange it so the needle dips in and out of the solution as the mixer rocks. This will make it easy to get tapered points.

### Solutions and Reagents

(*Optional*) **Bleach** (50%, Clorox)
Sodium cacodylate buffer (0.1 M, pH 7.4)
**Uranyl acetate** (2%) in $H_2O$
Graded series of **ethanol** (cold 30%, cold 50%, 70%, and 95%)
Absolute ethanol (use a freshly opened bottle)
**Propylene oxide** (100%)
**Epon-Araldite** embedding kit (Ted Pella)

### Fixative A

Prepare immediately before use as follows:

1. Prepare 0.2 M stock of **sodium cacodylate** buffer and pH to 7.2.

2. Prepare a fresh stock solution of 8% paraformaldehyde as follows: Add 8 g of paraformaldehyde to 100 ml of $H_2O$. Place on a hot plate at high setting *in a chemical fume hood*. Heat with stirring until steaming, but do not boil. Add 5–7 drops (may take a few more or less) of 1 N **NaOH**, one drop at a time until paraformadehyde goes into solution.

3. Open a new vial of 50–70% EM-grade glutaraldehyde (available in sealed ampules).

4. In *a chemical fume hood*, remove **acrolein** (Polysciences, Warrington, Pennsylvania) from bottle using a syringe. At a 2% concentration, it is obnoxious, but not harmful. It penetrates tissue rapidly and aides in the ability to rapidly dissect the embryo with minimal overall distortion. If possible, do the dissection in an exhaust hood. At the very least, work in a well-ventilated space where traffic is minimal. It is possible to eliminate acrolein from the fixative, and simply wait longer between puncturing and dissection, as paraformaldehyde also penetrates relatively rapidly. If acrolein is used, consult the Merck index or the Material Safety Data Sheet for acrolein.

5. Combine appropriate amounts of all components to obtain:

   2% paraformaldehyde
   2% glutaraldehyde
   0.13 M sodium cacodylate
   1 mM $CaCl_2$
   2% acrolein
   1.5% DMSO

6. Always check pH after all components mixed. The paraformaldehyde stock solution added will always increase pH variably. A pH-meter electrode dedicated to pH measurement of fixative is recommended.

### Fixative B

   1% $OsO_4$
   0.1 M sodium cacodylate buffer (pH 7.4)
   0.2 M sucrose

$OsO_4$ is available as crystals or 4% aqueous solution in sealed ampules. The 4% solution is more convenient, as the crystals normally take overnight to go into solution.

Mix 1 ml of 4% $OsO_4$, 1 ml of $H_2O$, and 2 ml of 0.2 M of cacodylate buffer. Add sucrose to 0.2 M.

**CAUTION: acrolein, bleach, $CaCl_2$, DMSO, glutaraldehyde, KOH, NaOH, propylene oxide, resins (Epon-Araldite), sodium cacodylate, sodium nitrite, uranyl acetate** (see Appendix 4)

## Method

1. Use either of the following methods for dechorionating embryos.

   a. *For manual dechorionation.* Assuming that the number of embryos needed is not too great, manual dechorionation can be used. Use a blunt dissection needle to transfer embryos individually from collection plate to double-stick tape on a microscope slide. Gently roll or nudge embryos with the end of the needle; the chorion will break open and the embryo, still enclosed in the vitelline membrane, is released. Try to keep the embryo resting on the broken chorionic fragments. Lift the embryo with the blunt end of the needle, and transfer to a second piece of double-stick tape in a depression-well slide or the well of a spot plate.

   b. *For dechorionation with bleach.* Embryos can be dechorionated in 50% Clorox. Using a small Millipore vacuum filtration setup makes the exchange of solutions efficient. Regulate the vacuum to provide a steady but not too rapid draining of solutions. Once embryos are dechorionated, remove the Millipore filter, group the embryos together with a fine paint brush, and transfer them by carefully applying the filter to a piece of double-sided adhesive tape, which can then be adhered to a depression well. Alternatively, transfer the embryos with a brush to a grape agar plate, where it is possible to visually select stages, group embryos according to desired stages, and transfer again with a piece of double-stick adhesive tape to a depresion well.

2. Prepare Fixative A immediately prior to fixation. Always use hundreds of times the volume of solvent to volume of tissue.

3. Perform primary fixation as follows:

   a. Cover embryos in a well with Fixative A.

   b. Puncture embryos carefully with sharp tungsten needle. Approximately 10 minutes after puncturing, gently tease embryos out of the vitelline membrane by stroking past the area where the original puncture was made. Blunter-ended needles work better for dissecting embryos out.

   c. Once a number of embryos are dissected out, transfer them with a 9-inch pipette, keeping embryos in the narrow part of the pipette, to a vial on ice with fresh fixative.

   d. Wash three times in 0.1 M sodium cacodylate (pH 7.4) for at least 20 minutes.

4. Postfix in Fixative B on ice for 2 hours (using the chemical fume hood!). Rinse three times in $H_2O$ and proceed to the step 5 *now*.

5. Stain en bloc in 2% aqueous uranyl acetate for 1 hour or leave at 4°C overnight. Cover the container because uranyl acetate is photosensitive. Rinse three times in $H_2O$.

6. Dehydrate the samples using the following graded series of ethanol, 5–10 minutes for each wash (Do not allow the tissue to dry out):

   30% (cold)
   50% (cold)
   70%
   95% (two changes)
   absolute (freshly opened bottle of glass-distilled ethanol; two changes)
   100% propylene oxide (two changes)

7. (*Infiltration*) There are different formulations of resins. Use the instructions that come with the Epon-Araldite kit. Add accelerator to all of the resin mixtures, but take care to ensure that the 100% resin does not become too viscous due to partial polymerization. If necessary, make fresh resin. Also, when mixing resin components, air bubbles often become trapped in the resin. Place the resin inside a vacuum desiccator and pull a vacuum over the resin for approximately 10 minutes to remove trapped air. Incubate the samples stepwise in the following:

> 100% propylene oxide:resin (1:1) for 1 hour
> 100% propylene oxide:resin (1:3) overnight
> 100% resin for 6–8 hours
> 100% resin overnight

*Note:* When working with resins, always use disposable plastic containers that will not be dissolved by propylene oxide (for Caution, see Appendix 4). Leftover resin can be polymerized and disposed of according to safety office guidelines. Material that comes into contact with resin is hazardous waste, as resins are carcinogenic.

8. (*Embedding*) Use a large-bore plastic pipette to carefully remove embryos from the vial. The embryos in resin can be placed in a plastic weigh boat. Transfer embryos individually with a sharpened applicator stick to freshly prepared resin in an embedding mold.

*Note:* There are many different types of embedding molds. Pyramid-tip molds (Ted Pella 10585) are convenient for orienting embryos for longitudinal sectioning. The pyramid tips can later be glued with Superglue to blank blocks polymerized in BEEM (size 00, Ted Pella 130) capsules.

## PROTOCOL 14.5

# *n*-Heptane Permeabilization

### Materials

### *Supplies and Equipment*

Dissecting microscope
Millipore filtration funnel attached to a flask with vacuum
Screw-cap vials (20 ml)
Siliconized glass slides
Double-coated adhesive tape
Petri plates (35 mm)
Tungsten needles (see p. 256)
Embedding molds
Embedding oven set at 65ºC
Specimen vials

### *Solutions and Reagents*

**Bleach** (50%, Clorox)
***n*-heptane**
**Osmium tetroxide** ($OsO_4$) (1%)
**Uranyl acetate** (2%)

Graded series of **ethanol** (cold 30%, cold 50%, 70%, and 95%)
Absolute ethanol
**Acetone** (100%)
**Epon-Araldite** embedding kit

### PBS-Triton

1x PBS (see Appendix 3)
0.1% Triton X-100

### Fixative A

25% **glutaraldehyde** (EM grade)
50 mM **sodium cacodylate** buffer (pH 7.0)

### Fixative B

2% glutaraldehyde (EM grade)
50 mM sodium cacodylate (pH 7.0)

### Fixative C

1% **osmium tetroxide**
2% glutaraldehyde
50 mM sodium cacodylate (pH 7.0)

**CAUTION: acetone, bleach, ethanol, glutaraldehyde, heptane, OsO$_4$, resins (Epon-Araldite), sodium cacodylate, uranyl acetate** (see Appendix 4)

## Method

The following method is used for fixing embryos:

1. Dechorionate embryos in 50% bleach (Clorox) using a Millipore filtration setup as described in step 1 of Protocol 14.4. Resuspend dechorionated embryos in a small volume of PBS-Triton and transfer to a 20-ml screw-cap vial.

2. (*Primary fixation*) Perform the following steps:

   a. Combine 2 ml of Fixative A with 8 ml of *n*-heptane. Shake vigorously, allow phases to separate, and use the upper (*n*-heptane) phase for fixation.

   b. Carefully remove the PBS-Triton and replace with *n*-heptane saturated with glutaraldehyde. Fix at room temperature for 15–20 minutes.

   c. Remove embryos from the fixative with a cut-off yellow micropipette tip and transfer to the siliconized glass slide.

   d. Just as the *n*-heptane completes evaporation, press a piece of double-coated adhesive tape to embryos and pick them up. Transfer the tape, with embryos up, to a 35-mm petri plate, and cover with Fixative B.

   e. Dissect embryos out of the vitelline membrane under a dissecting microscope using a tungsten needle. Once a sufficient number of embryos have been dissected out, transfer to Fixative C on ice for an additional 2 hours.

3. Wash the embryos two times in 50 mM sodium cacodylate. Postfix in 1% osmium tetroxide on ice for 2 hours. Then wash again in two changes of 50 mM sodium cacodylate. Transfer to 2% uranyl acetate on ice for 30 minutes.

4. Dehydrate in a graded ethanol series as described in step 6 of Protocol 14.4. After absolute ethanol washes, wash twice in 100% acetone. *Do not allow tissue to dry out completely.*

5. (*Infiltration*) Transfer embryos to a 1:1 mixture of Epon/Araldite and acetone and incubate overnight with mixing. Replace 1:1 mixture with 100% Epon/Araldite, incubate for 6 hours, replace with fresh Epon/Araldite, and incubate with mixing overnight.

6. (*Embedding*) Transfer embryos individually to fresh Epon/Araldite in embedding molds, orient embryos as required, and cure at 65°C for 48 hours to complete polymerization of the Epon/Araldite.

# EM IMMUNOLABELING

## Why ImmunoEM?

The main advantage of EM for immunolabeling is resolution, but there is also another aspect that is overlooked. For many investigators, the definition of an organelle includes an image and that image is the one generated by EM. It is especially true for membranous organelles, with the possible exception of the nucleus and plant vacuoles. When the words Golgi apparatus, smooth and rough endoplasmic reticulum, centriole, kinetochore, or mitochondrion are read or heard, what image comes to mind? It is usually the image in an electron micrograph. The components of the cytoskeleton also have characteristic structural features that are associated with their EM image. So when investigators see gold particles superimposed over the image of a microtubule (MT) or mitochondrion, it is somehow more convincing that they are labeling MTs or mitochondria than if they see bright dots or lines in the light microscope. This is especially true if the immunofluorescence image is of fixed cells. Light microscopy (LM)–fixation protocols are crude compared to those for EM when it comes to preserving cellular structural integrity close to the native state. This is why the development of probes for light microscopy, such as green fluorescent protein that can be used in living cells, is so exciting. The whole issue of LM fixation problems (cf. Melan and Sluder 1992) can be eliminated.

The ultimate solution to characterizing the distribution of particular gene products will be to use vital probes for LM in conjunction with selected EM labeling for confirmation and discrimination at higher resolution. Achieving adequate sample size with EM is a big problem, but it is not for light microscopy. The combination of the two is very powerful.

Space does not permit here an extensive presentation of the methods of EM immunolabeling. We provide an overview and some illustrations, but for further details, see McDonald (1994) and more general references such as the books by Griffiths (1993), Hayat (1989b,c, 1991), and Newman and Hobot (1993). For the most up-to-date articles on EM immunolabeling, although not usually on *Drosophila* tissues, consult recent issues of the *Journal of Histochemistry and Cytochemistry*.

## Which Type of ImmunoEM?

Two fundamentally different approaches to EM immunolabeling are known as *preembedding* and *postembedding* labeling. Within the category postembed are two main divisions, depending on whether *cryosections* or *resin sections* are used.

## *Preembedding Labeling*

In this method, cells are treated much the way they would be for LM labeling by fluorescent probes, except that instead of using a fluorescent secondary, one that is conjugated to an electron-dense probe visible in the EM is used. The big advantage of this technique is that once labeled, the cells can be processed for EM as they would normally. Furthermore, the electron-dense probes are throughout the final section, not just on the surface as with postembedding labeling techniques.

Two instances where preembed procedures are the methods of choice both involve living cells. The first is when the target to be labeled is on the cell surface. The second is when the investigator can take advantage of the cell's endocytotic pathway and present electron-dense particles such as ferritin or Nanogold to the cell, which then internalizes them. In the latter, endocytosis can be followed by fixing at selected time points and following the pathway of the electron-dense particles.

In general, we do not recommend the preembedding methods that require treating the cells with detergents or organic solvents to open up spaces large enough for the antibodies to penetrate into the cytoplasm. This is especially true if trying to label membranous components, or if the labeling target is liable to be part of a pool of soluble proteins. It is an acceptable alternative when labeling the cytoskeleton, which can be stabilized with either taxol for MTs or phalloidin for actin. However, even here the investigator must be certain that the stabilizing agent is not inducing polymerization of spurious filaments or only preserving a subset of polymers capable of reacting with the stabilizing agent. Careful comparisons of MT and actin distributions with and without taxol or phalloidin at the LM level must then be done.

If preembedding labeling can be used with the system, then it is necessary to consider what kind of probe to use to give the electron-dense localization signal. Although condensation compounds such as DAB have the advantage that they are small and can penetrate easily, they are not very discrete signals in the final image and can be difficult to interpret compared to gold. Our recommendation is Nanogold labeling followed by silver enhancement. Nanogold differs significantly from colloidal gold in being a defined chemical entity with specific side chains that can be modified for particular labeling challenges. In practice, the big difference between Nanogold and colloidal gold is that Nanogold will diffuse into tissues and colloidal gold will not. The exception to the latter is when tissues are so extracted that large holes allow the colloidal gold in, but such harsh treatment compromises the interpretation of the ultrastructure. A number of different products available from Nanoprobes are described in detail in their catalog, or on their Web site at http://www.nanoprobes.com.

## *Postembedding Immunolabeling*

*Cryosections.* Postembedding labeling essentially means labeling on sections, which is another way of gaining access to the interior of the cell, but without the harshness of detergent or ionic extraction as is done with preembed labeling. For purposes of EM immunolabeling, the term cryosections, or cryosectioning, has a very specific meaning. It refers to the technique first worked out by Tokuyasu (1973) where the cells are chemically fixed, then infiltrated with 3.2 M sucrose, frozen by plunging in liquid nitrogen, and then sectioned at low temperature, typically about –90°C. The cold sections are picked up, then thawed, rinsed, and labeled with primary and secondary antibody. They are then infiltrated with some compound to give them stability under the electron beam. An advantage of this method is that the cells never have to be dehydrated, a process that invariably leads to alteration in cell structure which is erroneously referred to as a "fixation artifact." Another advantage is that it is quick, and heavier labeling is obtained than on resin sections under the same conditions.

One of the disadvantages of cryosectioning is that it requires a lot of skill and experience to do it correctly. It is one of the techniques of EM that is more like an art than a science. When done correctly, the images are beautiful, but, more typically, the images are indistinct and difficult to interpret.

There have been some improvements in cryosectioning techniques since the original methods of Tokuyasu. Some of these involve better cryomicrotomes, such as the Ultracut T sold by Leica and the MTX sold by RMC. Other improvements have to do with the way the sections are made and subsequently handled. For the most recent information, see Griffiths (1993), Sitte (1996), and Webster (1999).

*Resin sections.* This is the method that we recommend the beginner to start with to carry out EM immunolabeling. First, investigators already familiar with routine EM-sectioning techniques find this method easiest to do. Simply use a different resin, and once the sections are obtained, label them following some simple procedures that are similar to those used when performing LM immunolabeling. Another reason is that the overall preservation of structure is best in resin compared to cryosections or preembed labeling. This is especially true for the cytoskeleton, although more problematical for endomembranes which sometimes show poor contrast. Finally, it is easy to do both LM and EM immunolabeling of resin sections.

A detailed discussion of EM immunolabeling procedures is beyond the scope of this chapter; however, most of the basic information can be found in McDonald (1994). Here, we include a brief summary of what is involved and add some information not included in the previous article. Two books are particularly useful for EM immunocytochemistry in general. The book by Griffiths (1993) is a good general reference covering a full range of topics from fixation for EM to stereological sampling of gold labels. The book by Newman and Hobot (1993) is more focused on methods for labeling of resin sections and has practical information on every aspect of that subject.

It is still true that the most critical component of immunoEM (iEM) is what primary antibody to use. As a general rule, we have found that polyclonal antibodies that are diluted 1:1000 or more have a reasonable chance of working at the EM level on sections of LR White, LR Gold, or Lowicryl HM20. Of these three resins, Lowicryl HM20 gives the best morphology, but we recommend starting with LR White because it is easier and less toxic. If good labeling is obtained with LR White, but not good morphology, switch to Lowicryl. In some cases, the antibodies are so reactive that it is possible to use Epon or Epon-Araldite as the resin, in which case, the morphology should be very good. In our hands, Spurr's resin gives poor morphology and little or no labeling.

For best results, use cryofixation (high-pressure freezing) followed by low-temperature embedding in Lowicryl HM20. If this is not possible, the next best approach is to use PLT (progressive lowering of temperature) dehydration (McDonald 1994), followed by infiltration and embedding with Lowicryl HM20. We do not recommend using Lowicryl K4M, even though it may give slightly better labeling density, because it is difficult to section and the morphological preservation is not as good as with Lowicryl HM20.

The antibody labeling procedure we now use is slightly different from the one described in 1994. Details are provided below (see Protocol 14.6). For other details regarding iEM procedures such as preparation of grids, sectioning, microscopy, and preparing images for publication, see McDonald (1994).

PROTOCOL 14.6

# Postembedding Immunolabeling

## Materials

### Supplies and Equipment

Petri dish (60 mm)

PFTE immunostaining pad (Ted Pella 10526-1). These pads have a 5 × 8 array of wells, so several will be needed when labeling many grids. If labeling five or fewer grids, two pads will suffice. Use 1 pad per 60-mm petri dish.

(*Optional*) Thermolyne Speci-Mix Test Tube Mixer (Fisher 12-814-2)

Small wire loop (2.5 mm in diameter)

### Solutions and Reagents

Primary and secondary antibodies

**Uranyl acetate** (1%) in 70% **methanol**

Graded methanol series (30%, 50%, and 70%)

**Lead citrate** (0.5%). The concentration for lead citrate seems to vary depending on the recipe. Most investigators use Reynold's lead citrate (Reynold 1963).

> ### Preparation of Solutions: Blocking Buffer (BB), Fixative, and PBS-Tween
>
> 1. Make up 50 ml of phosphate-buffered saline (PBS) (see Appendix 3).
>
> 2. To 25 ml of PBS, add 0.2 g of bovine serum albumin (BSA) (Sigma A 9647). Use a 50-ml Tripour beaker, add stirbar, and add powder gradually while stirring vigorously.
>
> 3. To 10 ml of PBS, add 100 µl of Tween-20. If using a micropipettor with disposable tips, cut off a few millimeters of the tip for a more accurate measurement. Put on rocker to mix. This is the Tween-20 stock solution.
>
> 4. Add 25 µl of cold-water fish gelatin (Sigma G 7765) to BSA mix.
>
> 5. Add 0.5 ml of Tween-20 stock solution (from step 3, above) to BSA mix (step 2 above), continue stirring. This is the Blocking Buffer (BB).
>
> 6. To prepare the fixative, add 0.2 ml of 25% **glutaraldehyde** solution to 10 ml of PBS.
>
> 7. Add 0.2 ml of Tween-20 stock solution (from step 3 above) to 10 ml of PBS. This is the PBS-Tween rinse.

**CAUTION: glutaraldehyde, lead citrate, methanol, uranyl acetate** (see Appendix 4)

## Method

1. Prepare a moist chamber by placing wet paper in the bottom of a 60-mm petri dish.

2. Place a PTFE immunostaining pad (Ted Pella 10526-1) on the wet paper.

   *Note:* The number of pads will depend on the number of grids to be labeled.

3. Fill the top row(s) of the pad with BB, using approximately 20 µl of solution. Float grids on the drops, section-side down, for 30 minutes. Place the top on the moist chamber. Agitate the solutions by putting the dish on a rocker such as the Thermolyne Speci-Mix Test Tube Mixer.

4. Fill the wells of the next row with primary antibody made up in BB. Transfer the grids from the BB to the primary with a small wire loop, 2.5 mm in diameter. With practice, this works much better than forceps. Cover the moist chamber, put on rocker, and incubate for at least 1 hour. For all subsequent steps, use the wire loop for grid transfer and the rocker for agitation.

5. Rinse in PBS-Tween for 2 minutes.

6. Rinse four times in PBS (2 minutes each).

7. Incubate in secondary antibody for 1 hour.

8. Rinse as in steps 5 and 6.

9. Fix in 0.5% glutaraldehyde in PBS for 5 minutes.

10. Rinse in PBS for 2 minutes.

11. Rinse four times in $H_2O$ (2 minutes each).

12. Poststain grids with uranyl acetate. We recommend using 1% uranyl acetate in 70% methanol for 5 minutes. Rinse with 70%, 50%, and 30% methanol; then rinse two times with $H_2O$.

13. Poststain with lead citrate for 3 minutes.

14. Allow the grids to air-dry, and then observe in microscope.

## EXAMPLES

In this section, we include illustrations of *Drosophila* tissues prepared for EM or iEM by some of the methods described above. Figures 14.1 and 14.2 compare the preservation of fine structure in cells prepared by HPF-FS (Figures 14.1a and 14.2a) with the same types of cells prepared by conventional chemical fixation (CF; Figures 1b and 2b), in this case, using Protocol 14.4. Figure 14.3 shows high-magnification details of cells prepared by HPF-FS, and Figure 14.4 illustrates examples of immunolabeled cells.

Figure 14.1 shows cells from a recently cellularized embryo at low magnification. The most noticeable difference is the contrast between nucleoplasm and cytoplasm in Figure 14.1b. This is probably due to the increased extraction of soluble proteins that occurs with room-temperature fixation and dehydration. A second point of comparison is the state of the membranes. Note that in CF cells, the outlines of the nuclear envelope are irregular (Figure 14.1b), whereas those in HPF-FS cells are smooth (Figure 14.1a). Closer inspection of the figures reveals that plasma membranes in HPF-FS cells are also smooth and continuous. For a higher magnification of cell membranes in this field, see Figure 1 in McDonald (1994). In general, when working at this magnification or lower, HPF-FS techniques are probably not needed. At this level of detail, different cell types can be distinguished, or organelles such as mitochondria can be counted, but the high-resolution information is not evident anyway, and CF techniques might as well be used. However, use HPF-FS technology if it is accessable, because it is actually quicker and easier than CF processing.

Figure 14.2 shows parts of epidermal cells that are in the process of extension during dorsal closure. Figure 14.2a is a medium magnification image of cells prepared by HPF-FS, and Figure 14.2b is of cells prepared by CF. Working in this magnification range allows

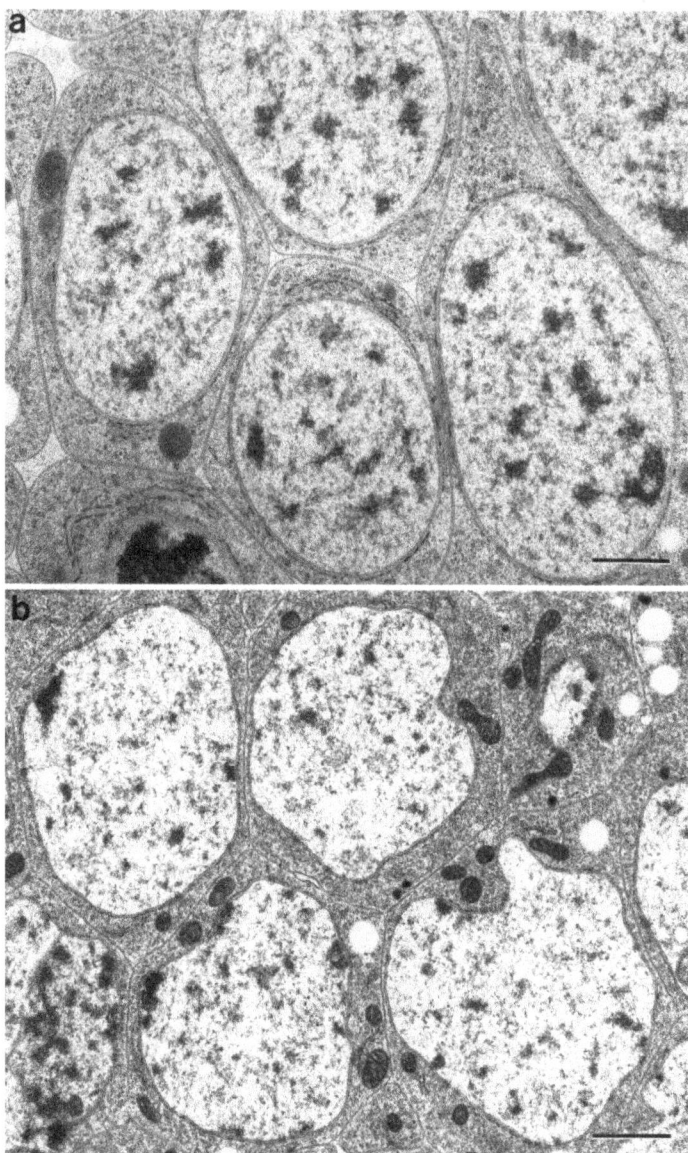

**Figure 14.1.** Low-magnification views of epidermal cells processed by HPF-FS (*a*) or conventional fixation (*b*). See text for details. Bars, 1 μm.

subcellular details such as MTs to be seen (larger straight arrows, Figure 14.2a and b), and in the case of the HPF-FS cells in Figure 14.2b, microfilaments as well (smaller straight arrows). Microfilaments are not usually well preserved in CF material, although hints of structure can sometimes be seen when they are in bundles. Within the box area in Figure 14.2a are individual microfilaments in cross section, although this is easier to see if working at higher magnification (Figure 14.3a). When studying membranes, use HPF-FS to prepare the cells. Comparison of membranes in Figure 14.2, a and b, shows that in CF material (Figure 14.2b) the membranes between adjacent cells become unnaturally ruffled (curved solid arrow) and that blebbing (curved open arrow) occurs on the surface membranes. In the HPF-FS cells (Figure 14.2a), the membranes become indistinct in areas of high curvature (curved arrows).

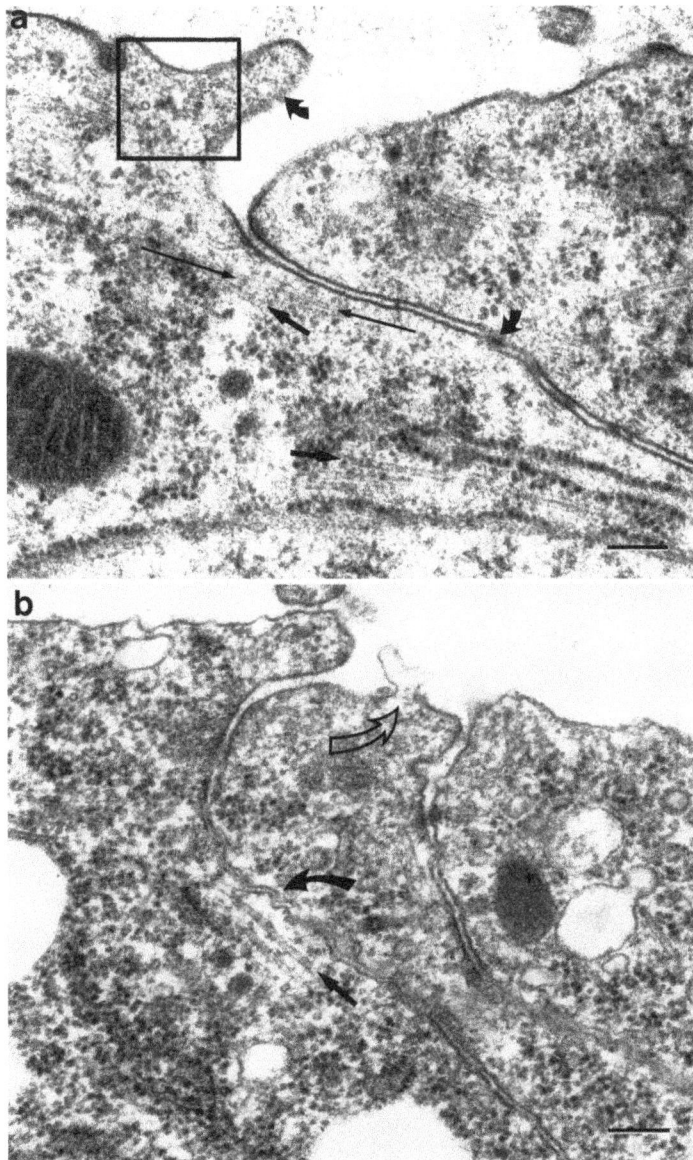

**Figure 14.2.** Medium-magnification views of epidermal cells prepared by HPF-FS (*a*) or conventional fixation (*b*). In panel *a*, the thicker straight arrows indicate microtubules, and the thinner straight arrows indicate microfilaments. Small curved arrows point out areas of membrane that are indistinct because of curvature. The box indicates the area of this cell shown at higher magnification in Figure 14.3a. In panel *b*, the straight arrow indicates a microtubule, the solid arrow shows how the membranes of adjoining cells are distorted, and the open curved arrow shows an area of the plasmalemma that is blebbing. Bars, 200 nm.

HPF-FS is best used when showing high-magnification images of fine structure. In Figure 14.3, we show how HPF-FS gives good preservation of both the cytoskeleton and cell membranes. Figure 14.3a is an enlargement of the area shown in the box in Figure 14.2a. At this magnification, it is easy to see the profiles of individual microfilaments in cross section (arrow). Figure 14.3b is also from an epidermal cell extension during dorsal closure, but showing microfilaments in longitudinal section. In addition, there are hints of cross-links between adjacent filaments (Figure 14.3b, arrows). Figure 14.3c shows two adjacent epidermal cell membranes and the dense material between them that is routine-

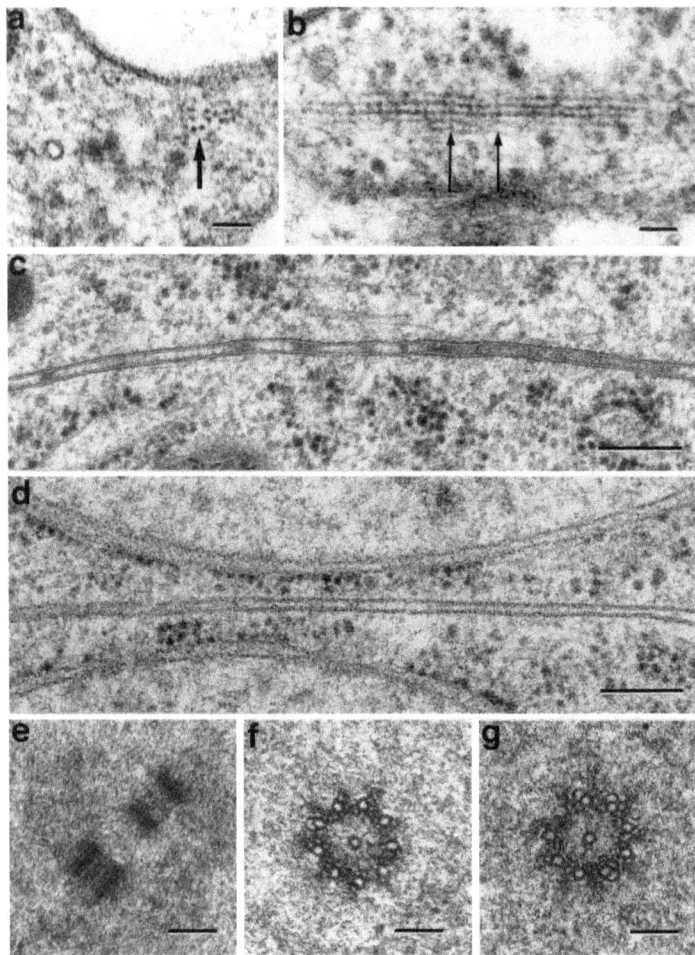

**Figure 14.3.** High-magnification views of organelles from cells prepared by HPF-FS. (*a*) Microfilaments (*arrow*) in cross section. Bar, 50 nm. (*b*) Microfilaments in longitudinal section. The arrows point out areas of apparent transverse elements in the bundle. Bar, 50 nm. (*c*) Membranes between adjacent cells in a postgastrulation embryo. Bar, 200 nm. (*d*) Membranes between cells in a pregastrulation embryo. Bar, 200 nm. (*e–g*) Centrioles from syncytial blastoderm mitotic spindle poles. Bars: (*e*) 200 nm; (*f,g*) 100 nm. See text for additional details.

ly found in postgastrulation embryos. In comparison, Figure 14.3d shows the membranes between adjacent epidermal cells in a pregastrulation embryo. This figure also shows parts of two nuclear envelopes that have smooth profiles expected from a round organelle (see also Figure 14.1). Finally, in the last three panels of Figure 14.3 (e–g), we show the high-resolution detail of centrioles from dividing nuclei in syncytial blastoderm nuclei (Sharp et al. 1999). Figure 14.3e shows a centriole pair at the pole of a metaphase spindle. In cross section (Figure 14.3f), the metaphase spindle centriole shows nine predominantly single MTs around a single central microtubule (see also Figure 11 in McDonald and Morphew 1993). There are weakly developed doublets at some positions. In late anaphase (Figure 14.3g), the doublets are more evident. Despite reports of triplet MTs in early embryos (Callaini and Dallai 1991), we have never seen these triplets.

In EM immunolabeling, the reactivity of the primary antibody is the most important ingredient for successful results. However, once it is established that the antibody works, there are several options to optimize the morphology while still retaining specific labeling.

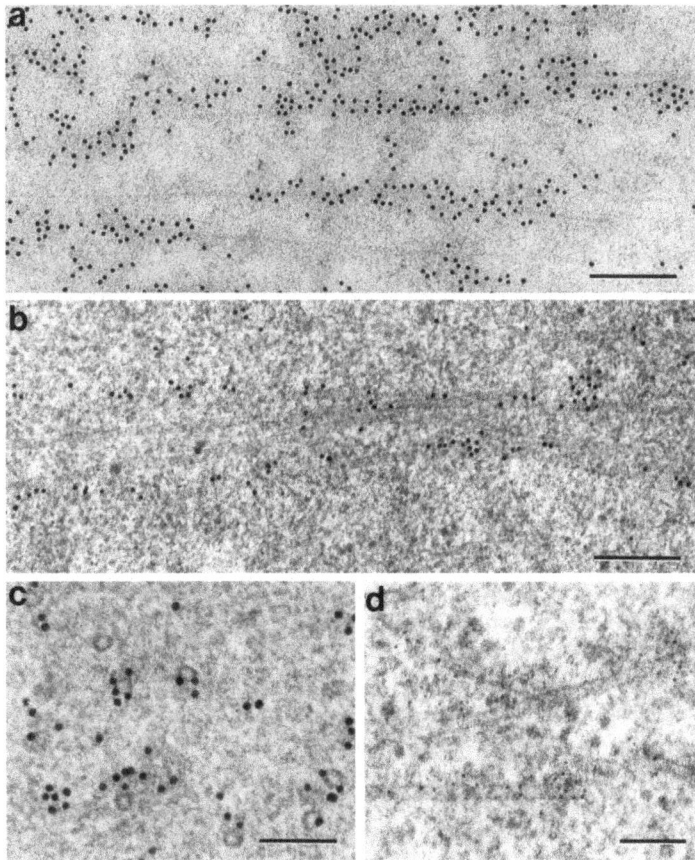

**Figure 14.4.** Immunogold labeling of mitotic-spindle–associated proteins. (*a*) KLP61F motor protein labeled with 10-nm gold on LR White sections. Bar, 200 nm. (*b*) Tubulin labeled with 10-nm gold on Lowicryl K4M sections. Bar, 200 nm. (*c*) Cross sections of spindle microtubules labeled with 10-nm gold on Lowicryl K4M sections. Bar, 100 nm. (*d*) KLP61F motor protein labeled with 5-nm gold on Epon-Araldite sections. Bar, 100 nm.

Figure 14.4a is a longitudinal section throughout the MT interzonal region in a metaphase spindle. It is difficult to pick out MTs, although the regions where they are located are shown by the 10-nm gold particles labeling the KLP61F motor protein in the interzone (Sharp et al. 1999). This spindle was embedded in LR White resin. A longitudinal section through a metaphase spindle embedded in Lowicryl K4M is shown in Figure 14.4b. In this resin, the MTs are more visible. A primary antibody to tubulin is labeled with a 10-nm gold secondary antibody. The same resin and antibody combination is shown in Figure 14.4c, where MTs in cross section are labeled with 10-nm gold secondary. This figure illustrates the fact that not all antigenic sites label with primary and that the gold does not always lie right over the molecule it is labeling. This is because the primary and secondary antibody combination may occupy a relatively large space (up to 20 nm long) and steric hindrance prevents saturation labeling of all possible antigen sites. Likewise, the distance between epitope and gold particle will vary according to the way the complex settles down on the resin surface during drying. In Figure 14.4d, 5-nm gold is used to label KLP61F motor protein in interzonal mitotic spindle MTs embedded in Epon-Araldite resin. This figure is included to illustrate the fact that if the antibody is good enough, resins such as Epon-Araldite can be used that normally do not support good gold labeling, although it gives the best morphology. Note also that 5-nm gold is difficult to see relative to 10-nm or larger gold.

## REFERENCES

Bozzola J.J. and Russell L.D. 1999. *Electron microscopy*, 2nd. edition. Jones and Bartlett, Boston, Massachusetts.

Callaini G. and Dallai R. 1991. Abnormal behavior of the yolk centrosomes during early embryogenesis of *Drosophila melanogaster*. *Exp. Cell Res.* **192:** 16–21.

Echlin P. 1992. *Low-temperature microscopy and analysis*. Plenum Press, New York.

Griffiths G. 1993. *Fine structure immunocytochemistry*. Springer-Verlag, Berlin.

Hayat M.A. 1981. *Fixation for electron microscopy*. Academic Press, New York.

———. 1989a. *Principles and techniques of electron microscopy*, 3rd edition. CRC Press, Boca Raton, Florida.

———, ed. 1989b. *Colloidal gold: Principles, methods and applications*, vol. 1. Academic Press, San Diego, California.

———, ed. 1989c. *Colloidal gold: Principles, methods and applications*, vol. 2. Academic Press, San Diego, California.

———, ed. 1991. *Colloidal gold: Principles, methods and applications*, vol. 3. Academic Press, San Diego, California.

Howard R.J. and O'Donnell K.L. 1987. Freeze substitution of fungi for cytological analysis. *Exp. Mycol.* **11:** 250–269.

Humbel B. and Müller M. 1986. Freeze substitution and low temperature embedding. In *The science of biological specimen preparation 1985* (ed. M. Müller et al.), pp. 175–183. SEM Inc., AMF O'Hare, Chicago, Illinois.

Kiss J.Z. and Staehelin L.A. 1995. High pressure freezing. In *Rapid freezing, freeze fracture, and deep etching* (ed. N.J. Severs and D.M. Shotton), pp. 89–104. Wiley, New York.

McDonald K. 1994. Electron microscopy and EM immunocytochemistry. *Methods Cell Biol.* **44:** 411–444.

———. 1999. High pressure freezing for preservation of high resolution fine structure and antigenicity for immunolabeling. *Methods Mol. Biol.* **117:** 77–97.

McDonald K. and Morphew M.K. 1993. Improved preservation of ultrastructure in difficult-to-fix organisms by high pressure freezing and freeze substitution. I. *Drosophila melanogaster* and *Strongylocentrotus purpuratus* embryos. *Microsc. Res. Tech.* **124:** 465–473.

Melan M.A. and Sluder G. 1992. Redistribution and differential extraction of soluble proteins in permeabilized cultured cells—Implications for immunofluorescence microscopy. *J. Cell Sci.* **101:** 731–743.

Moor H. 1987. Theory and practice of high pressure freezing. In *Cryotechniques in biological electron microscopy* (ed. R. A. Steinbrecht and K. Zierold), pp. 175–191. Springer-Verlag, Berlin.

Newman G.R. and Hobot J.A. 1993. *Resin microscopy and on-section immunocytochemistry*. Springer-Verlag, Berlin.

Prokop A., Martin-Bermudo M.D., Bate M., and Brown N.H. 1998. Absence of PS integrins or laminin A affects extracellular adhesion, but not intracellular assembly, of hemiadherens and neuromuscular junctions in *Drosophila* embryos. *Dev. Biol.* **196:** 58–96.

Reynolds E.S. 1963. The use of leach citrate at high pH as an electron-opaque stain in electron microscopy. *J. Cell Biol.* **17:** 208–212.

Robards A.W. and Sleytr U.B. 1985. Low temperature methods in biological electron microscopy. *Pract. Methods Electron Microsc.* **10:** 1–551.

Sharp D.J., McDonald K.L., Brown H.M., Matthies H.J., Walczak C., Vale R.D., Mitchison T.J., and Scholey J.M. 1999. The bipolar kinesin, KLP61F, associates with interpolar microtubule bundles within mitotic spindles of *Drosophila* early embryos. *J. Cell Biol.* **144:** 125–138.

Sitte H. 1996. Advanced instrumentation and methodology related to cryoultramicrotomy: A review. *Scanning Microsc.* (suppl.) **10:** 387–466.

Sitte H., Neumann K., and Edelmann L. 1987. Safety rules for cryopreparation. In *Cryotechniques in biological electron microscopy* (ed. R. Steinbrecht and K. Zierold), pp. 285–289. Springer-Verlag, Berlin.

Steinbrecht R.A. and Müller M. 1987. Freeze substitution and freeze drying. In *Cryotechniques in biological electron microscopy* (ed. R.A. Steinbrecht and K. Zierold), pp. 149–172. Springer-Verlag, Berlin.

Studer D., Michel M., and Müller M. 1989. High pressure freezing comes of age. *Scanning Microsc.* **S3:** 253–269.

Tepass U. and Hartenstein V. 1994. The development of cellular junctions in the *Drosophila* embryo. *Dev. Biol.* **161:** 563–596.

Tokuyasu K.T. 1973. A technique for ultracryotomy of cell suspensions and tissues. *J. Cell Biol.* **57:** 551–565.

Webster P. 1999. The production of cryosections through fixed and cryoprotected biological material and their use in immunocytochemistry. *Methods Mol. Biol.* **117:** 46–76.

# CONTENTS

# 15

# Electrophysiological Approaches to the Neuromusculature

**Kendal S. Broadie**
*Department of Biology*
*University of Utah*
*Salt Lake City, Utah 84112*

Drosophila is the most important genetic system for electrophysiological investigation. To date, it is the only system that combines systematic forward genetics with the ability to make detailed, high-resolution physiological recordings. The soil nematode *Caenorhabditis elegans* has the potential to become another powerful genetic model; however, its usefulness for many fields of physiology is limited because of its small size and the extreme simplicity of its body plan. The reverse genetics approach in mice is also very powerful, but remains limited in the type of genetic manipulation that may be attempted. It is therefore likely that *Drosophila* will remain the leading genetic model system for electrophysiological experimentation for the foreseeable future.

All stages of the *Drosophila* life cycle have been subject to electrophysiological recording: embryos, larval instars, pupae, and adults. In addition, cultures of both embryonic and larval cells are amenable to physiological experimentation. These systems have allowed detailed investigation of both developmental processes and mature neurological function. A variety of cell types have been the subject of study. In particular, a great deal of attention has been focused on the musculature and neuromuscular junction (NMJ) of the embryo, larva, and adult. This system allows study of the development and functional roles of ionic conductances in the somatic muscle and detailed examination of the glutamatergic NMJ: its synaptogenesis in the *Drosophila* embryo, the mechanisms of neurotransmission, and modeling of synaptic plasticity properties thought to underlie neuronal adaptation and modulation.

This chapter discusses all of the physiological techniques used to assay the *Drosophila* neuromusculature. First, a brief discussion on the requirements for *Drosophila* recording is provided, including a guide to the recording solutions, approaches to dissection, and an overview of the basic physiology equipment. Second, a more detailed discussion of the specific methods for each developmental stage is presented. This chapter describes recording configurations from the neuromusculature in culture, during embryogenesis, in the larva, and at the adult stage. Chapter 16 continues with outlines for recording techniques in the peripheral and sensory nervous system, including the adult retina and reflex circuits, and advances in central nervous system (CNS)-recording techniques, including cultured central neurons and in situ recording from the embryo and adult CNS.

## REQUIREMENTS

### Recording Solutions

#### *Haemolymph Bathing Solutions*

A variety of bathing solutions have been used for recording from *Drosophila*. These solutions come in two general classes (Table 15.1). First, the "standard" or "modified standard" bath salines are based on solutions commonly used for recording in other invertebrate systems. These solutions have been in use for several decades (Jan and Jan 1976a), and the bulk of published recordings has been performed in these or similar solutions. Second, the "hemolymph-like" (HL) bath solutions represent a compromise between the standard solutions and the ionic concentrations measured in the hemolymph of *Drosophila* and related Diptera (Table 15.2) (Stewart et al. 1994). Most experiments using these solutions have been performed in the most minimal HL saline, HL3, which has been reported to maintain both the morphology and physiological stability of the neuromusculature in a manner superior to that of the standard or modified standard solutions (Stewart et al. 1994).

None of the widely used bath salines accurately mimic the ionic concentrations measured in the hemolymph (Tables 15.1 and 15.2), which are found to be unfavorable for prolonged recording (Stewart et al. 1994). Most experiments are now being done in a variant of the modified standard salines or in HL3. The primary differences between these two solutions are $[Na^+]$ (115–140 mM for standard vs. 70 mM for HL3) and $[Mg^{++}]$ (4 mM for standard vs. 20 mM for HL3). The major advantage of the high $[Na^+]$ in the standard solution is to facilitate action-potential propagation and, in particular, the maintenance of normal, patterned synaptic transmission. The difference in $[Mg^{++}]$ alters the amplitude of synaptic transmission and shifts the $Ca^{++}$ dependency range upward in the HL3 solution (Stewart et al. 1994). Neither solution appears to be inherently better than the other. Careful control of osmolarity eliminates cellular vacuolation in both salines, and morphology appears to be equally well preserved. In addition, recording in both salines gives a similar starting resting potential, which is maintained equally well over 2 hours of recording (Stewart et al. 1994). The outstanding difference between these salines is the

**Table 15.1.** Representative External Bath Solutions

| (in mM) | A[a] standard | Modified standard B[b] | C[c] | D[d] | F[e] | G[f] HL3 | H[f] HL4 | I[f] HL5 |
|---|---|---|---|---|---|---|---|---|
| NaCl | 128 | 140 | 135 | 135 | 115 | 70 | 70 | 70 |
| KCl | 2 | 2 | 5 | 5 | 5 | 5 | 20 | 35 |
| CaCl$_2$•2H$_2$O | 1.8 | 5 | 1.8 | 0.5 | 6 | 1.5 | 1 | 1 |
| MgCl$_2$•6H$_2$O | 4 | 1 | 4 | 2 | 1 | 20 | 20 | 20 |
| NaHCO$_3$ | – | 4 | – | – | 4 | 10 | 10 | 10 |
| NaH2PO$_4$•2H$_2$O | – | – | – | – | 1 | – | – | 2 |
| L-glutamine | – | – | – | – | – | – | – | 5 |
| Trehalose | – | 5 | 5 | – | 5 | 5 | 5 | 5 |
| Sucrose | 35.5 | 100 | 36 | 36 | 75 | 115 | 115 | 50 |
| TES | – | – | 5 | 5 | 5 | – | – | – |
| BES/HEPES | 5 | 5 | – | – | – | 5 | 5 | 5 |
| pH | 7.1–7.2 | 7.2 | 7.15 | 7.4 | 7.3 | 7.2 | 7.2 | 7.2 |
| Osmotic pressure (mOsm) | 300±1.6 | 409±1.1 | n.d.[g] | n.d. | n.d. | 343±0.7 | 361±1.6 | 338± 0.7 |
| Membrane potential (mV)[h] | –53±9.8 | –63±6.2 | –56±7.3 | n.d. | n.d. | –59±8.3 | –54±7.1 | –49±5.0 |

References: [a]Jan and Jan (1976a); [b]Johansen et al. (1989); [c]Broadie and Bate (1993a); [d]Baines and Bate (1998); [e]Koenig and Ikeda (1983) and Trimarchi and Murphey (1997); [f]Stewart et al. (1994).

[g]n.d. indicates not determined.

[h]Recorded on third-instar larval muscle.

**Table 15.2.** Composition of Larval Hemolymph

| (in mM) | *Drosophila*[a,b,c] (total ion concentration; early measurements) | *Drosophila*[d] (ion selective electrodes; recent) | *Sarcophaga*[d] |
|---|---|---|---|
| Na$^+$ | 52–63 | 77±6.0 | 73±1.3 |
| K$^+$ | 36–55 | 40±2.0 | 22±2.2 |
| Ca$^{++}$ | 5–8 | 1.5±0.7 | 0.5±0.2 |
| Mg$^{++}$ | 20–33 | n.d.[e] | n.d. |
| Cl$^-$ | 30–42 | n.d. | n.d. |
| PO$_4^{-3}$ | 2.8 | n.d. | n.d. |
| Amino acids | 112.5–136.8 | n.d. | n.d. |
| Osmotic pressure | 360–396 | n.d. | 329 |

References: [a]Croghan and Lockwood (1960); [b]Begg and Cruikshank (1963); [c]Van der Meer and Jaffe (1983); [d]Stewart et al. (1994).
[e]n.d. indicates not determined.

time course of recorded synaptic currents and voltage changes; however, the physiological relevance of this difference is uncertain. The selection of one of these salines is thus primarily a personal choice that should reflect the question under study.

### Intracellular Recording Solutions

Standard intracellular electrodes are filled with a solution of 2–4 M KCl, potassium acetate, potassium citrate, or a combination of these salts. The various patch-pipette solutions are relatively similar, although some recipes are considerably more complicated (Table 15.3). For example, some recipes contain nucleotides (ATP, GTP) to maintain intracellular signaling and energy reserves, and some recipes attempt to balance osmolarity with the external bath. However, there is no clear evidence that any one of these recipes is superior, and a minimal patch pipette solution appears to be adequate for most purposes.

## Dissection

### Dissection Tools

Standard dissection tools can be adapted to all *Drosophila* physiological preparations. Fine forceps (#5) are required for all dissections, and fine iridectomy surgical scissors are use-

**Table 15.3.** Representative Patch Intracellular Solutions

| (in mM) | A[a] | B[b] | C[c] | D[d] | E[e] | F[f] |
|---|---|---|---|---|---|---|
| KCl | 120 | 140 | 144 | 70 | – | – |
| KF | – | – | – | 70 | – | – |
| CsCl | – | – | – | – | 130 | 158 |
| CaCl$_2$•2H$_2$O | 0.5 | – | 0.5 | 1 | – | – |
| MgCl$_2$•6H$_2$O | 4 | 2 | 1 | 2 | 4 | 1 |
| Trehalose | 5 | – | – | – | – | – |
| Sucrose | 36 | – | – | – | – | – |
| TES | 5 | – | – | – | 10 | – |
| BES/HEPES | – | 10 | 10 | 10 | – | 10 |
| EGTA | 5 | 11 | 5 | 11 | 0.1 | 5 |
| ATP | 4 | – | – | v | 2 | 1 |
| GTP | 0.4 | v | – | – | 0.4 | – |
| pH | 7.15 | 7.4 | 7.1–7.2 | 7.2 | 7.1–7.2 | 7.1 |

References: [a]Broadie and Bate (1993a); [b]Zagotta et al. (1988) and Baines and Bate (1998); [c]Zhao and Wu (1997); [d]O'Dowd and Aldrich (1988); [e]Hardie (1991); [f]Deitcher et al. (1998).

ful for dissection of late larval instars and adult stages. For detailed dissection, one of three types of tools is routinely used.

- *Hypodermic needles* are useful for making fine incisions, although frequent replacement is required as they quickly become blunt.
- *Sharpened tungsten dissection needles* have the benefit that they are relatively resilient and long-lasting. They can be used for both crude incisions and teasing apart delicate tissue (for preparation, see McDonald et al., this volume).
- *Glass electrodes* are useful because they are adaptable and easily replaceable. They can be pulled from either solid glass (these last longer) or standard thick-walled electrode glass and are useful for making incisions and the finest tissue dissection. Hollow glass electrodes can also be attached to plastic tubing and used for suction and expulsion of saline, useful in a variety of dissections.

### Mounting Preparations

The method of attaching the dissected preparation to a mounting platform is stage-specific and is described in detail for the embryo, larva, and adult (see below). In general, the following mounting approaches are used.

- *Embryos.* The early embryo (<16 hours after egg laying [AEL]) attaches directly to clean glass or glass coated with polylysine (poly-L-lysine hydrobromide; Sigma). Older embryos (>16 hours AEL), following cuticle formation, are attached to coverslips coated with Sylgard (Dow Corning). The embryos can be glued (cyanoacrylate glue; B. Braun Surgical D-34209 Melsungen, Germany), pinned (finest grade insect pins; 0.1 mm), or held in place with fibers (dental floss or spider silk, which is from *Nephia clavata*, a Japanese spider). The floss/silk is used with plastic lux 13-mm hermanox coverslips (Nunc). The strands are inserted into notches cut in the edge of the plastic coverslips (Yoshihara et al. 1997).
- *Larvae.* Young larvae (<L2) are best treated like mature embryos. Mature (wandering third instar) larvae are much easier to pin, but they also can be attached with cyanoacrylate glue.
- *Adults.* Adult flies are mounted with quick-setting glue/cements or with a soft, low-melting-point wax. The platform/chamber for adult recordings is adapted to the type of recording (see below).

## Equipment

### Microscopes

A good dissection microscope is required for most dissections (40x magnification suggested). For the embryo, it is suggested that 25x eyepieces be used to maximally increase magnification (1000x). For larger preparations (mature larvae, adult) the same (40x) dissection scope can be used for the recording session. For smaller preparations (embryo, cultured cells) or more detailed work, a compound microscope is required. For in vivo work, it is recommended that an upright microscope be used with a water-immersion lens (40–63x objective recommended; the lens can be of the kind that is used with a drop of water placed between lens and coverslip and is capable of being immersed into the open chamber containing culture medium, buffer, etc.) and fitted with differential interference contrast (Nomarski) optics. For culture or dissociated cell recording, an inverted microscope with Nomarski optics is recommended. Fluorescent capability may be useful if fluorescent vital dyes are to be used in labeling the preparation (Yoshikami and Okun 1984).

Detailed recording should be done with electrical and vibration isolation. Microscopes are routinely placed on an isolation table and protected with a grounded Faraday cage.

### Physiology Equipment

Standard electrophysiology equipment is used and a full description can be obtained from a variety of sources. Briefly, recording equipment appropriate to the experimental approach is required ranging from a simple amplifier for intracellular voltage recording, single- or two-electrode voltage-clamp amplifiers, patch-clamp amplifiers, or extracellular amplifiers. Stimulation equipment ranging from a dedicated stimulator to computer-generated stimulation will also be required for most procedures. Records can be monitored on an oscilloscope or directly on-line. Most experiments will be run through on-line computer software; several commercial packages are available.

Our lab uses Axopatch 1D or Axoclamp 2B amplifiers and pClamp 7 analysis hardware and software from Axon Instruments, and Grass stimulators (Astro-Med) and stimulus isolation equipment. However, there are a number of other commercially available options for both equipment and analysis software. The prospective physiologist is advised to shop around.

### Support Equipment

Equipment to make and modify glass microelectrodes requires first an electrode puller. A variety of pullers are available for intracellular and patch-clamp electrodes. Recommended is a versatile puller that is capable of pulling the full range of required electrodes. Several pullers are available that can be computer-programmed to pull a variety of shapes and sizes. The Brown-Flaming models (Sutter Instrument Co.) are widely favored. Second, a compound microscope (20–40x objective) should be available to inspect and visualize the modification of electrodes prior to use. Finally, for patch recording and stimulation electrodes, an electrode fire-polisher/microforge is required. This can be homemade from a simple heating element and variable transformer, or commercially purchased from a variety of sources.

## TECHNIQUES FOR STUDYING THE NEUROMUSCULATURE

The *Drosophila* neuromusculature has been subject to extensive physiological investigation for more than 20 years (Jan and Jan 1976a,b; Jan et al. 1997). It has been the primary tissue for the study of ionic conductances, synaptic transmission, and the output of central pattern generation. In all cases, recording has focused on the somatic muscles and the output onto these muscles by the glutamatergic NMJ. Discussed here are neuromuscular cultures derived from dissociated cells and whole-embryo culture, and recording techniques from the embryonic, larval, and adult neuromusculature in situ.

### Neuromusculature Cell Culture

Two protocols are given below for the generation and maintenance of neuromusculature cultures derived from embryos for electrophysiological recordings. Protocol 15.1 describes the culture of cells from dissociated embryos. Protocol 15.2 describes the culture of whole embryos and the conditions for maintaining these cultures for electrophysiology. For details on embryo collection and dechorionation, see Rothwell and Sullivan (this volume). For recordings for embryonic cultures, see p. 284 (Muscle Recording).

# Dissociated Embryonic Culture

## Materials

### Media

Apple juice agar plates (see Appendix 3)

Yeast paste (see Sisson, this volume)

Cell culture medium, e.g., original Schneider's *Drosophila* medium or modified M3 (MM3) medium (see discussion below [Culturing Conditions]).

### Supplies and Equipment

Dissecting tools, microscopes, and other equipment (see pp. 274–276)

Metal sieve

Double-stick Scotch tape

Plastic culture dishes

Humidified incubator at 25°C

Glass coverslips (untreated)

### Solutions and Reagents

Commercial **bleach** (undiluted) or a mixture of 75–95% ethanol and bleach (1:1)

**Ethanol**

Halocarbon oil (series 700)

**CAUTION: bleach, ethanol** (see Appendix 4)

## Method

### Embryo Collection and Dechorionation

1. Maintain breeding flies at 25°C in a small chamber containing apple juice agar plates with a dollop of yeast paste. Change the yeasted plates twice per day for at least 2 days before experimentation.

2. Collect eggs from a timed egg-lay (1–2 hours) into a metal sieve.

3. Dechorionate eggs manually either by rolling on double-stick Scotch Tape or by soaking in commercial bleach or a 1:1 mixture of 75–95% ethanol and bleach for 1–2 minutes.

   *Note:* Undiluted commercial bleach is used in this step; however, some investigators prefer to use diluted bleach.

4. Wash extensively in ethanol.

5. Place the dechorionated eggs in a plastic culture dish (under $H_2O$) for viewing.

### Selecting Embryos

Select embryos based on the appropriate developmental stage. Mid-stage gastrulae (2.5–3 hours AEL at 25°C) are selected based on morphological criteria (Campos-Ortega and Hartenstein 1985). A culture can be made by plating the cells from a single gastrula-stage

*For all protocols, $H_2O$ indicates glass distilled and deionized.

embryo (Seecof et al. 1971; Seecof 1979; Salvaterra et al. 1988; Zagotta et al. 1988; Tsunoda and Salkoff 1995; Chang and Kidokoro 1996).

### *Dissociated Neuromusculature Culture*

1. Place a dechorionated embryo under halocarbon oil (Series 700). Use suction to remove the entire contents of the embryo into the tip of a broken glass electrode (measuring ~50 μm at the tip).

2. Dissociate the contents of a single embryo in a small drop (5–20 μl) of culture medium on an untreated, sterile glass coverslip.

   *Notes:* The cultures may be incubated under halocarbon oil or in a humid chamber to prevent evaporation. Cultures can be grown in a variety of culture media including original Schneider's *Drosophila* medium or MM3 medium (Seecof et al. 1971; Salvaterra et al. 1988; Zagotta et al. 1988; Tsunoda and Salkoff 1995; Chang and Kidokoro 1996).
     Halocarbon oil can be used for the long-term incubation of cultures.

3. Incubate cultures at 25°C for 12 hours to several days (~48–72 hours) in a humidified incubator prior to electrophysiological investigation.

## PROTOCOL 15.1 NOTES

Cells from mid- to late-gastrula-stage embryos differentiate into several identifiable cell types, although these culture conditions favor the selective survival of muscle cells and neurons (Seecof et al. 1971; Zagotta et al. 1988; Tsunoda and Salkoff 1995; Chang and Kidokoro 1996). Myoblasts divide in vitro approximately 5 hours after the initiation of gastrulation or approximately 2 hours after plating. During the next 6–8 hours, myoblasts align and fuse to form myotubes, and neurons differentiate elongated processes. Functional, contractile myotubes differentiate within the first 12–24 hours, and functional NMJs are present (Seecof et al. 1972). Both muscle cells and neurons have been successfully patch-clamped (Seecof et al. 1971; Zagotta et al. 1988; Tsunoda and Salkoff 1995; Chang and Kidokoro 1996). Cultures remain viable for several days following plating.

## PROTOCOL 15.2

# Whole-embryo Culture

### Materials
### *Media*

Apple juice agar plates (see Appendix 3)

Yeast paste (see Sisson, this volume)

Cell culture media, e.g., original Schneider's *Drosophila* medium or modified M3 (MM3) medium. For preparation of the MM3 culturing medium, see Broadie et al. (1992). For media supplements, see p. 281 (Culturing Conditions).

### *Supplies and Equipment*

Dissecting tools, microscopes, and other equipment (see pp. 274–276)

Metal sieve

Plastic culture dishes

Humidified incubator at 25°C
Black tape
Poly-L-lysine-coated coverslips (poly-L-lysine hydrobromide; Sigma)

### Solutions and Reagents

Saline (see p. 283)
Commercial **bleach**
**Ethanol**
**CAUTION: bleach, ethanol** (see Appendix 4)

### Method

### Embryo Collection and Dechorionation

1. Maintain breeding flies at 25°C in a small chamber containing apple juice agar plates with a dollop of yeast paste. Change the yeasted plates twice per day for at least 2 days before experimentation.

2. Collect eggs from a timed egg-lay (1–2 hours) into a metal sieve.

3. Dechorionate eggs either manually or by soaking in commercial bleach for 1–2 minutes.

4. Wash extensively in ethanol.

5. Place the dechorionated eggs in a plastic culture dish (under $H_2O$) for viewing.

### Staging Embryos

Embryos viewed in reflected light with a dissection microscope are staged by morphological criteria (Campos-Ortega and Hartenstein 1985) to narrow (~15–30 minutes at 25°C), well-defined developmental time windows. Several developmental time windows can be used routinely:

- Narrow ventral furrow of gastrulation (2.75–3 hours AEL).
- Completion of germ-band retraction (8.5–9 hours AEL).
- Gut constriction (two-part gut, 12.5–13 hours AEL; three-part gut, 12.75–13 hours AEL).

Alternatively, embryos can be staged by relying entirely on time from fertilization. Under these conditions, embryogenesis lasts 21 +/– 1 hour at 25°C. Whole-embryo culture can be performed on dissected preparations from late, extended germ-band (7.5 hours AEL) onward; younger embryos (<7.5 hours AEL) can be removed intact from the vitelline membrane and cultured, but are difficult to dissect (Broadie at al. 1992). These very young stages are best examined with the dissociated culture technique (see Protocol 15.1).

### Dissection

1. Attach dechorionated, staged embryos (dorsal side up) to a small square (~0.25 cm²) of black tape on a poly-L-lysine-coated coverslip. Immerse the embryos in a drop of saline (Table 15.1).

2. Make a small incision in the vitelline membrane with a glass micropipette. Take care when choosing the membrane incision site and performing the incision: For dissec-

tion of older embryos (>8.5 hours AEL), use a dorsal incision to cut the membrane and embryo's dorsal midline simultaneously; for younger embryos, before germ-band retraction (<8.5 hours AEL), a small anterior incision is less likely to damage the embryo.

3. Remove embryos from the vitelline membrane by "blowing" the embryo free with a gentle stream of saline from a mouth pipette; the membrane remains attached to the tape.

   *Notes:* For dissection, the embryo is attached to the poly-L-lysine-coated coverslip in the desired orientation, usually dorsal surface up. For extended germ-band embryos (<8.5 hours AEL), the posterior part of the germ band can be gently reflexed, using a fine glass rod, to expose the internal, differentiating organs. The amnioserosa (after dorsal closure [~13 hours AEL] of the epidermis) is cut along the dorsal midline with a glass micropipette or sharpened tungsten needle, and the preparation is blown flat on the coverslip surface with a gentle stream of saline (Bate 1990).

   Care must be taken to maintain the embryo free-floating after removal from the membrane; contact with the air:solution interface or the poly-L-lysine-coated surface is disastrous at this stage (Broadie et al. 1992). Further dissection, e.g., removal of the yolk and internal organs to expose the neuromusculature (Figure 15.1), is performed by suction through a broken glass electrode as required.

   All manipulations to this point must be performed rapidly, as the poly-L-lysine-binding sites rapidly (<10 minutes) become saturated with yolk, cellular debris, etc.; however, with practice, several embryos (3–5) can be dissected on a single coverslip within this time.

4. Immediately after dissection, replace the saline with several washes of culture medium (see below).

## *Culturing Conditions*

Embryos are most successfully cultured in a MM3 medium (Shields and Sang 1977; Currie et al. 1988) at 25°C in a humid chamber. For preparation of the MM3 culturing medium, see Broadie et al. (1992). After filtering (0.22-μm Millipore filter), MM3 remains stable at 4°C for up to 3 months, although longer storage is not recommended. No striking difference has been detected in quality between MM3 stored at 4°C, –20°C, and –70°C. Fetal calf serum should be added several days (3–7) before use. Some considerations include:

- *Fetal calf serum (FCS).* FCS in the culturing medium saturates the poly-L-lysine-binding sites and can prevent adhesion between embryonic cells and the coverslip. To circumvent this difficulty, embryos are dissected in normal saline before addition of culture medium containing FCS. Batches of FCS serum can vary significantly in culturing quality, and thus, several batches should be tested before use. In general, noninactivated serum gives the best results; heat-inactivated serum (56°C, 30 minutes) is more stable, but may lower culturing quality (Broadie et al. 1992). It is recommended that 2–10% noninactivated serum be used and that serum be added to MM3 several days before use.
- *Glutamate.* Shields and Sang (1977) found that replacing KCl and NaCl with glutamate salts is beneficial to the differentiation of some cell types in dispersed cell cultures. During early embryogenesis (<13 hours AEL), this alteration can be used in whole-embryo culturing. However, during late embryogenesis (>13 hours AEL), high glutamate concentrations can perturb neuromuscular development, specifically the NMJ, possibly by saturation of the glutamate receptors at the developing synapse (Broadie and Bate 1993a). Replacing the L-glutamate salts with either equimolar D-glutamate salts or aspartate salts alleviates this difficulty without

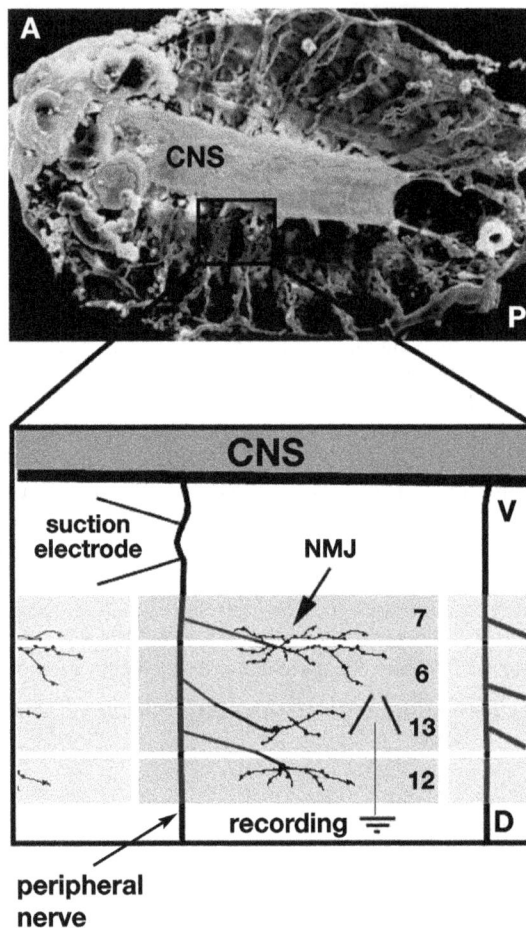

**Figure 15.1.** *Drosophila* embryonic neuromuscular preparation. (*Top panel*) Dissected *Drosophila* embryo viewed with a scanning electron microscope (SEM); anterior (A) to the left, posterior (P) to the right. The prominent CNS lies along the ventral midline. (*Bottom panel*) Schematic drawing of the four ventral longitudinal muscles in one segment. The CNS, peripheral nerve, and NMJ on these four muscles are drawn. The typical recording configuration involves whole-cell patch-clamp recording from muscle 6 and suction-electrode stimulation of the peripheral nerve.

apparent adverse effects. This modification can be routinely used when culturing late embryonic stages (>13 hours AEL).

- *Antibiotics.* Embryonic development of *Drosophila* is rapid, being complete by 21 hours postfertilization at 25°C. Since dissected preparations are cultured from extended germ band (>7.5 hours AEL) and most applications require culturing for only a limited period (<12 hours), antibiotics should not be necessary. If added, penicillin G (0.03 g/liter) and streptomycin sulfate (0.1 g/liter) can be used as in dispersed cell cultures (Shields and Sang 1977).

- *Morphological movements.* The *Drosophila* embryo undergoes gross morphological movements during development, including germ-band extension/retraction, the dorsal migration of epidermal cells during dorsal closure, and head involution (Campos-Ortega and Hartenstein 1985). When the dissected embryo is attached to a poly-L-lysine-coated substrate, these movements are inhibited or prevented, resulting in severe morphological abnormalities. However, if embryos are maintained free-floating in MM3, these movements can occur normally (Broadie et al. 1992). For this reason, younger embryos are gently freed from the substrate to continue development after the desired experimental manipulation has been performed. At a later stage, the embryo can be transferred to a fresh poly-L-lysine-coated coverslip and attached for observation or further experimentation. After 13 hours AEL, gross morphological movements of the epidermis are complete and the dissected preparation can be cultured continuously attached to the coverslip.

- *Culture configurations.* Dissected preparations deprived of oxygen develop poorly and much more slowly than the intact embryo. For this reason, embryos should be cultured in configurations to maximize the air-solution interface. Two configurations are routinely used: (1) For younger embryos (<9 hours AEL), the preparation in a small drop of MM3 is inverted over a well-slide constructed of two depression slides fastened together and the drop of MM3 is allowed to just touch both surfaces to form a column of solution. (2) For older embryos (>9 hours AEL), it is sufficient to culture the preparation flat in a small drop (20–100 µl) of MM3. In both cases, it is essential to culture in a humid chamber to avoid evaporation.

## Embryonic NMJ

An alternative to embryo culture is recording from acutely dissected *Drosophila* embryos (Broadie and Bate 1993a,b,c; Broadie et al. 1994, 1995, 1997). The main benefit of acute dissection is to avoid developmental perturbations that may be associated with culturing difficulties. The main drawbacks are that the preparation is not available for extended experimentation, and dissection of the late embryo (>16 hours AEL), following cuticle formation, is difficult and demands special techniques (Broadie et al. 1994). All experiments are performed on dissected whole-embryo preparations (Figure 15.1).

### Embryo Collection and Dechorionation

As described in Protocol 15.2, breeding flies are maintained on apple juice agar plates at 25°C, and eggs are collected from a timed egg-lay (1–2 hours). Eggs are dechorionated either manually or in commercial bleach and staged by morphological criteria to narrow (~15–30 minutes at 25°C), well-defined developmental time windows (Campos-Ortega and Hartenstein 1985). Under these conditions, embryogenesis lasts 21 +/− 1 hours.

### Dissection

*Younger Embryos (<16 Hours AEL).* As described in Protocol 15.2, embryos are removed from the vitelline membrane and transferred to poly-L-lysine-coated coverslips with the dorsal side up under saline. An incision is made along the dorsal midline with a glass electrode or sharpened tungsten needle. The embryo is blown flat to the coverslip surface with a gentle stream of saline from a glass pipette controlled by mouth (Bate 1990; Broadie and Bate 1993a).

*Older Embryos (>16 Hours AEL).* Embryos are transferred to coverslips coated with a thin layer of Sylgard (Dow Corning) under normal saline, and glued down at the head and tail with small drops of histoacryl tissue adhesive glue (Braun, Germany). The glue is delivered through a small glass pipette (10–20-µm inner diameter) with the flow controlled by mouth or with a syringe. An incision is made along the dorsal midline with a glass electrode or sharpened tungsten needle and the sides are glued down with more glue. The internal organs including the gut, fat bodies, and, optionally, trachea are then removed by suction applied to a glass pipette (Broadie and Bate 1993a,b,c).

Alternatively, the embryo is secured with small fibers, e.g., dental floss or spider silk, that are attached to the coverslip (Kidokoro and Nishikawa 1994). The dissected embryo is positioned so that a small bundle of fibers clamps the dorsal edges of the incision. The embryo should now be attached flat to the coverslip, epidermis down with the ventral CNS, peripheral nerves, and somatic musculature exposed for experimentation (Figure 15.1).

### Visualization of Embryos

For physiological experiments, the preparation is placed in a small recording chamber (<0.5 ml) and best viewed in transmitted light on an upright compound microscope fitted with differential interference contrast (Nomarski) optics and a 40–100x (typically 40x) water-immersion objective (Broadie and Bate 1993a).

Membrane-permeant fluorescent mitochondrial dyes, such as Rhodamine 123 (Rh 123; Sigma) or 4-Di-2-Asp (Molecular Probes), can be used to aid visualization of nerve terminals in the embryo (Yoshikami and Okun 1984). To label embryos, expose the embryos to the dye (5 μm) for 5 minutes and then remove the excess dye with several washes of saline. View the preparation with an epi-fluorescence attachment and appropriate filters.

### Muscle Recording

Recordings are typically made at or below room temperature (16–22°C) but become increasingly difficult at elevated temperatures (+23°C). Various recording techniques can be used as described below.

*Whole-cell Patch-clamp Techniques.* Patch-clamp recordings are made from the embryonic muscles using standard techniques (Broadie and Bate 1993a,b,c; Broadie et al. 1994, 1995, 1997; Nishikawa and Kidokoro 1995). Any of the standard patch-clamp variations—cell-attached, inside-out, outside-out—can be used to record single-channel activity in the muscle. Synaptic activity at the NMJ is monitored in whole-cell recording configuration (Figure 15.1). The muscles can be recorded in either voltage-clamp (typical holding potentials of –60 to –80 mV) or current-clamp configurations. Patch pipettes can be pulled from a variety of fiber-filled glasses (e.g., borosilicate glass, 1-mm outer diameter); the tips should be fire-polished to final resistances of 5–10 MΩ. Signals are amplified using a patch-clamp amplifier (e.g., Axopatch-1D, Axon Instruments), filtered with an 8-pole Bessel filter at 2–10 Hz, and either sampled on-line or stored digitally for later analysis.

In principle, any of the embryonic muscles may be recorded from in this preparation. In practice, most recordings have concentrated on the large, longitudinal muscles in the innermost muscle layer (Bate 1990). Records have been made from the dorsal, lateral, and ventral muscles of this muscle layer (Broadie and Bate 1993a,b,c; Kidokoro and Nishikawa 1994; Auld et al. 1995; Nishikawa and Kidokoro 1995). However, as in the larvae (see below), most experiments have focused on the group of four longitudinal, ventral muscles (6, 7, 12, 13; Bate 1990) in the anterior abdominal segments A2–A4 (Figure 15.1). In particular, electrophysiology has been performed on the identified NMJ on muscle 6 in the anterior abdomen (A2–A3) and the guideline measurements provided below come from this muscle (Broadie and Bate 1993a,b,c).

Very young embryos (<13 hours AEL) cannot be effectively voltage-clamped because of the extensive coupling between the embryonic myotubes (Broadie and Bate 1993a); this is usually not a problem during later (14+ hours AEL) developmental stages (Ueda and Kidokoro 1996). No enzymatic treatment is required prior to recording in young embryos (<16 hours AEL). However, a muscle sheath develops in the late embryo (>16 hours AEL) and must be removed with collagenase (collagenase IV, 1 mg/ml in divalent cation-free saline at room temperature for 0.5–1 minute) prior to patch recording. Seal resistances on the muscle are typically greater than 10 G$\Omega$. Whole-cell configuration is achieved easily with slight suction or an electrical "buzz" and input resistance of the myotube is in the range of 200 M$\Omega$ to 1+ G$\Omega$. Series resistances are typically 10–15 M$\Omega$. With maximum currents in the range of hundreds of picoamps to 1–2 nA (depending on age), series resistance errors (total current x series resistance) are usually less than 5 mV, but in mature embryos, they may be as high as 10 mV or greater and thus may require correction. Myotubes with these characteristics (average diameter of 10–20 $\mu$m and average length of 40–80 $\mu$m) should show reasonable space clamp. Cell capacitances are between 10 and 30 pF, increasing with developmental age, generating clamp time constants ($R_{series}$ x $C_{cell}$) averaging less than 0.25 msec.

*Perforated Patch-clamp Recording.* During whole-cell patch-clamp recording, dilution of the cellular cytoplasm, especially with long-time course recordings (up to 1 hour) or during sensitive developmental periods, might adversely effect the interpretation of results. Consequently, Nystatin-perforated patch-clamp recording techniques (Korn et al. 1991) can be used to verify records obtained with standard whole-cell techniques (Broadie and Bate 1993a). The perforation solution is prepared as follows: 10 mg of Pluronic F-127 (Molecular Probes) is dissolved in 20 ml of dimethylsulfoxide (DMSO) (Sigma); 10 mg Nystatin (Sigma) is dissolved in the Pluronic F-127 solution to make the stock; and 100 $\mu$l of the stock is added to 5 ml of filtered pipette solution to make the recording solution, which is used to back-fill the recording electrode.

Whole-cell currents are usually obtained less than 3 minutes after seal formation. Series resistance is increased in most instances to range from 20 to 40 M$\Omega$. Occasionally, cells developed an increased leakage current during prolonged recording (>15 minutes; Broadie and Bate 1993a), indicating that Nystatin had penetrated the cell, but this was usually not a problem.

*Isolation of Ionic Currents.* The whole-cell voltage-gated current in mature embryonic and larval muscle is composed of five prominent components (Wu and Haugland 1985; Singh and Wu 1990; Broadie and Bate 1993b): an inward calcium current ($I_{Ca}$) and four outward potassium currents. The potassium currents include two fast, inactivating currents, voltage-gated ($I_A$) and calcium-dependent ($I_{CF}$) and two delayed, noninactivating currents, voltage-gated ($I_K$) and calcium-dependent ($I_{CS}$). Ion-substitution experiments can be used to dissect these currents from the whole-cell response during voltage-clamp:

- *Calcium current ($I_{Ca}$):* CsCl is substituted for KCl in the patch pipette. Intracellular $K^+$ is replaced with $Cs^+$ via perfusion in whole-cell configuration for 5 minutes. In this configuration, no outward $K^+$ currents should be present. Thus, the inward $I_{Ca}$ can be recorded in the absence of all four outward potassium currents.
- *Rapid, voltage-gated potassium current ($I_A$):* External calcium is removed, eliminating both the inward calcium current and the calcium-dependent potassium currents, $I_{CF}$ and $I_{CS}$. $I_A$ and $I_K$ can be temporally separated. Alternatively, $I_A$ can be specifically eliminated in Shaker null mutants ($Sh^{KS133}$) and $I_A$ amplitude obtained by subtraction (wild-type–$Sh$). There is no significant difference in $I_A$ amplitude assayed with the two methods (Broadie and Bate 1993b).
- *Delayed, voltage-gated potassium current ($I_K$):* $Sh^{KS133}$ is used to eliminate $I_A$. External calcium is removed, eliminating both the inward calcium current and both calcium-dependent potassium currents ($I_{CF}$, $I_{CS}$). In this configuration, $I_K$ can be studied in isolation.

- *Rapid, calcium-dependent potassium current ($I_{CF}$):* $Sh^{KS133}$ can be used to eliminate $I_A$. $I_{CF}$ can be temporally separated from the delayed currents ($I_K$, $I_{CS}$); $I_{Ca}$ must be subtracted. Alternatively, $I_{CF}$ can be specifically eliminated using the slowpoke (*slo*) mutation and $I_{CF}$ amplitude obtained by subtraction (wild-type –*slo*). There was no significant difference in $I_{CF}$ amplitude assayed with the two methods (Broadie and Bate 1993b).
- *Delayed, calcium-dependent potassium current ($I_{CS}$):* A *Sh;slo* double mutant can be used to eliminate $I_A$ and $I_{CF}$, respectively. $I_{CS}$ is obtained by subtracting $I_K$ and $I_{Ca}$, both of which can be studied in isolation.

In these experiments, only $I_K$ and $I_{Ca}$ can be studied directly in the absence of all other ionic currents. $I_A$ can be studied in the presence of only the delayed $I_K$, which can be temporally resolved from the $I_A$ peak. Values for $I_{CF}$ and $I_{CS}$ are obtained only from multiple current traces; $I_{CF}$ can be specifically eliminated (*slo*), but $I_{CS}$ can be measured only in recordings in which the other current components are identified and subtracted as detailed above (Broadie and Bate 1993b).

### *NMJ Stimulation*

*Suction Electrode Motor Nerve Stimulation.* The most common way to stimulate synaptic transmission at the NMJ is with a glass suction electrode on the segmental nerve (Figure 15.1). Suction electrodes can be pulled from a variety of fiber-filled electrode glasses (1.0–1.5 mm outer diameter) and should be fire-polished to achieve the desired configuration (~5 µm inner diameter). A small segment (<50 µm) of the appropriate segmental motor nerve is drawn into the pipette with gentle suction to form a tight seal (Figure 15.1). It is crucial that the nerve have a tight fit in the suction electrode for adequate stimulation; too loose and the stimulation fails, too tight and it is not possible to stimulate an adequate section of the nerve. Stimulation can be applied with a variety of stimulators. Brief stimulation (0.2–1 msec) works best, and stimulation intensity (usually 1–5 V) will depend on the suction configuration and must be experimentally determined for each cell. Suprathreshold stimulation of the motor nerve is best achieved by setting the stimulus strength slightly above (1–2 V) threshold. Evoked NMJ responses are recorded from the patch-clamped muscle as described on p. 284. The shock artifact can be substantially decreased with the use of an isolated virtual ground.

It is common to stimulate the intact nerve by drawing in a loop of the nerve near the CNS exit point (Figure 15.1). This approach allows easy stimulation of the intact preparation, but has the disadvantage that spontaneous CNS-evoked activity can occur superimposed on the applied stimulation paradigm. An alternative is to stimulate the cut motor nerve. This is easy to achieve if the CNS is removed by suction, although this procedure tends to stretch the nerve and cause damage. The nerve may also be cut with a sharp glass microelectrode, but this procedure is tedious and difficult. It is recommended that records be made from the intact preparation.

*Direct CNS Stimulation.* An alternative to the standard suction electrode stimulation of the peripheral nerve is direct stimulation of the CNS (Nishikawa and Kidokoro 1995). A microelectrode filled with 3–4 M KCl or potassium acetate is inserted into the middle of the ventral ganglion and positive pulses of approximately 2 µA in intensity and approximately 2 msec in duration are delivered (Dietcher et al. 1998). Synaptic transmission is recorded in the patch-clamped muscle in the standard configuration.

*Iontophoresis.* Any small, charged molecule can be effectively iontophoresed onto the embryonic NMJ. Most experiments have been carried out with the NMJ neurotransmitter, L-glutamate (Broadie and Bate 1993c). A stock solution of 0.1 M L-glutamate (pH 8.0; monosodium salt) is used. Iontophoretic pipettes can be pulled for specific configurations to regulate the amount of glutamate delivered: high-resistance pipettes (100–200 MΩ) have been used for glutamate receptor (*gluR*) mapping in discrete membrane domains (Broadie and Bate 1993d), and lower-resistance pipettes (20–50 MΩ) have been used to estimate the total glutamate response at the embryonic synapse (Broadie et al. 1995, 1997). Glutamate can be iontophoresed for variable periods, although short pulses (0.1–1 msec) of negative current (~10 nA) work well. It is vital to prevent glutamate leakage between pulses with a small, positive backing current. The magnitude of this backing current will vary with pipette configuration and should be experimentally determined for each cell. Excessive movement in the dissected preparation upon approach of the iontophoresis pipette is an indication that glutamate is leaking from the tip and that the backing current must be increased.

For the purposes of *gluR* mapping, the muscle surface can be sampled by advancing the iontophoretic pipette short distances (e.g., 5–10 μm) with a micromanipulator and delivering a short pulse (e.g., 0.1 msec) of negative current (Broadie and Bate 1993d). Only the internal and lateral surface of the muscle is available to glutamate iontophoresis in this dissection; *gluR* maps of the external myotube surface can only be extrapolated from these exposed areas. Current responses in the voltage-clamped muscle and muscle grid position should be recorded simultaneously. In most cases, individual *gluR* openings can be resolved and counted; the number of *gluR*s in multiple *gluR* openings can be directly calculated by dividing the current amplitude by the amplitude of a single *gluR* opening (Broadie and Bate 1993d).

A commercially available arylamine spider toxin (Argiotoxin-636 pentahydrochloride; F.W. 819.13; Tocris-Cookson), which is a specific, noncompetitive, open-channel *gluR* blocker in both invertebrate and vertebrate preparations, has been used to block neuromuscular transmission (Broadie and Bate 1993a,c). Bath application of $10^{-4}$ to $10^{-6}$ M Argiotoxin-636 should completely block *gluR* openings within 5 minutes of application. Iontophoretic application of L-glutamate (as above) should elicit no response at junctional or extrajunctional sites. This blockage is apparently irreversible as no *gluR* openings are detected after repeated washing (>15 minutes; Broadie and Bate 1993a). The toxin can be dissolved in recording saline or MM3 culture medium and either (1) applied directly to a dissected preparation or (2) pressure-injected into an intact embryo with a bevelled micropipette. For pressure injection, the embryo should be slightly dehydrated (5–15 minutes in air) to reduce hemolymph volume, and toxin concentration and injection volume adjusted to approximately $10^{-5}$ M toxin in the hemolymph.

*Pressure Application of Noncharged Molecules.* A variety of noncharged molecules have been used to elicit synaptic transmission from the embryonic NMJ. Hyperosmotic saline (range 300–600 mOsm) triggers massive fusion of synaptic vesicles (Broadie et al. 1995). Black widow spider venom (BWSV), which contains latroinsectotoxin, also causes massive vesicle fusion with both $Ca^{++}$-dependent and -independent aspects (Broadie et al. 1995). Desiccated and frozen glands from *Latrodectus mactans* (black widow spider) may be purchased from several retailers (e.g., Sigma). The glands are homogenized with bath saline in a ground glass homogenizer placed in ice (4°C). The homogenate may be centrifuged at low speed for 5–10 minutes to remove tissue debris and gel-filtered over a Sephadex G-50 spin column at 4°C to isolate the high-molecular-weight latroinsectotoxins. This homogenate is a crude BWSV toxin and can be stored at –80ºC in small aliquots prior to use. For application, a pressure-ejection pipette (<5-µm inner diameter) containing the solution is positioned within 10 µm above the nerve terminal. The solution to be ejected is applied using a Picospritzer (General Valve Corp.) at 5–10 psi. During prolonged experiments, the recording chamber (volume <0.5 ml) should be perfused continuously (0.1–0.2 ml/sec) with normal bath saline to remove excess toxin.

## Early Larval (L1–L2) NMJ

Early larval instars, through L2, are dissected and recorded from in a manner analogous to the embryonic preparation. The dissection is identical and attachment with pins and other methods remains difficult through the second instar; it is recommended that the gluing technique be used.

Whole-cell muscle recordings are best achieved with the single-electrode patch-clamp technique (see p. 284) through the mid-L2 stage. However, the ability to successfully form seals on larval muscle decreases with age, and longer collagenase treatment (1–2 minutes) is required. Furthermore, the muscle grows rapidly, making space-clamp increasingly difficult with a single-patch electrode. Thus, the late L2 and L3 larval muscles are best recorded from using intracellular recordings (see p. 290 [Intracellular Muscle Recording]). Finally, it is best to stimulate and record from the intact neuromusculature through early L2 as the nerve-severing operation remains difficult until late larval stages.

## Late (Wandering Third Instar) Larval NMJ

### Dissection

Wandering third-instar larvae are selected based on size and active wandering behavior after leaving the food. Larvae can be dissected in culture medium (modified Schneider's medium), standard *Drosophila* saline, or HL saline. Some investigators prefer to dissect in culture medium followed by a slow series of solution changes (3:1, 2:1, 1:1, 1:2, 1:3) to the final recording saline (Stewart et al. 1994). However, most investigators dissect in low [Ca$^{++}$] or Ca$^{++}$-free saline to reduce muscular contractions, followed by addition of the appropriate Ca$^{++}$-containing saline, or simply dissect directly in the recording saline. There is no clear indication that any of these approaches is superior.

Wandering larvae are large (~5 mm) and are easily dissected (Figure 15.2). One common approach is to pin the head and tail using insect pins (0.1-mm pins), with the ventral surface down. A longitudinal incision is made with a pair of fine scissors along the dorsal midline (between the main dorsal trachea trunks), and then the animal is attached flat with pins along the cut dorsal edges.

Another approach is to secure the head and tail to Sylgard-coated glass coverslips with histoacryl tissue-adhesive glue (Braun, Germany) and glue flat to the coverslip. The internal organs (gut, fat body) are then removed by suction using a glass electrode attached to tubing (as above). A fine forceps is used to expose the CNS and muscles attached to the interior of the epidermis (Figure 15.2).

Under normal conditions, the segmental nerves of all preparations are severed with dissecting scissors near the VNC to eliminate CNS-mediated spontaneous transmission. For whole-cell recording, the preparation can be adequately viewed on a standard dissecting microscope (40X magnification). For detailed observation, the preparation is best visualized with a 40X water-immersion objective on an upright compound microscope using Nomarski optics.

## NMJ Stimulation

Nerve stimulation is similar to that described for the embryo (see p. 286). Usually, the segmental nerves of all preparations are severed with dissecting scissors near the CNS (see above), and the cut end of the appropriate segmental nerve is drawn into a fire-polished glass suction electrode (5–10-μm inner diameter) to form a tight seal (Figure 15.2). The nerve is stimulated with short pulses (0.2–1 msec) at a stimulus strength determined experimentally (as above). The suction electrode can also be used to record action potentials extracellularly (Wu et al. 1978; Ganetzky and Wu 1982). The signal-to-noise ratio of the neuronal action potentials recorded in this fashion is relatively poor, although a signal averager can be used to improve resolution.

*Iontophoresis.* Iontophoresis of charged molecules (e.g., L-glutamate) is performed essentially as described for the embryo (see p. 287). A stock solution of 0.1 M L-glutamate (pH 8.0) is iontophoresed from a glass pipette (10–20 MΩ) positioned directly over the NMJ synaptic boutons.

*Pressure Injection.* Pressure ejection of uncharged molecules (e.g., hyperosmotic saline) onto the larval NMJ can be done as described for the embryo (see p. 287). A pipette (5–10-μm inner diameter) filled with hyperosmotic bath saline (500 mM sucrose) is positioned 20–30 μm above the nerve terminal (Ranjan et al. 1998). The solution is ejected using a Picospritzer at 5 psi for 5–20 seconds. During experiments, the recording chamber (volume <0.5 ml) is perfused continuously (0.1–0.2 ml/sec) with normal bath saline.

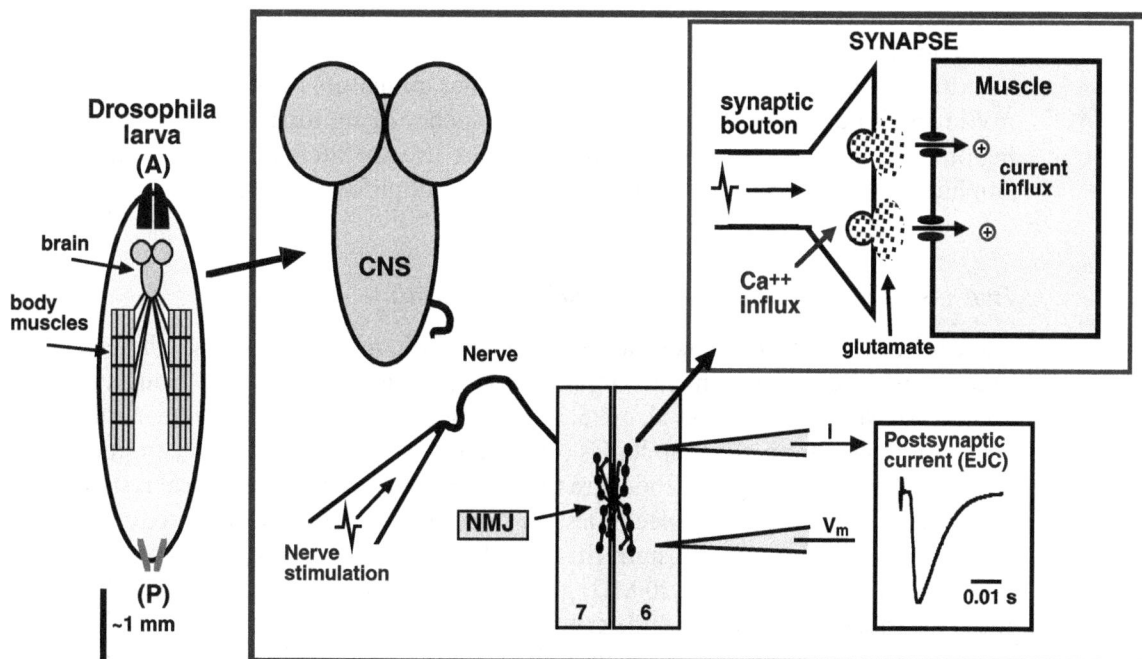

**Figure 15.2.** *Drosophila* larval neuromuscular junction system. A wandering third-instar larva is dissected open to reveal the ventral neuromusculature (see Figure 15.1). The peripheral nerve is severed and stimulated with a glass suction electrode. The muscle is recorded from in two-electrode voltage-clamp (TEVC) configuration. The postsynaptic excitatory junctional current (EJC) is recorded to assay synaptic transmission (*bottom inset*). (*Top inset*) Basis of synaptic transmission event being evoked by nerve stimulation and recorded via ion flux through muscle glutamate receptors.

BWSV can be pressure injected onto the larval NMJ in a manner similar to that for the embryo (Ranjan et al. 1998). Frozen glands from *L. mactans* (black widow spider) are commercially available (Sigma) and are used for preparation of BWSV as follows: The frozen glands are homogenized with 10 mM sodium phosphate (pH 7.2) in a ground glass homogenizer at 4°C (Ranjan et al. 1998). The homogenate is centrifuged at low speed for 5 minutes to remove tissue debris and gel-filtered over a Sephadex G-50 column at 4°C to isolate the high-molecular-weight latrotoxins. This homogenate is a crude BWSV toxin, which can be stored at –80°C in small aliquots prior to use (Ranjan et al. 1998).

### Intracellular Muscle Recording

In principle, any of the 30 muscles per hemisegment are available for recording, although the easiest targets are the most internal muscle layer, topmost in the dissected preparation (Bate 1990). Most electrophysiological records have been made from one of a group of four large, ventral longitudinal muscles (6, 7, 12, 13) in the anterior ventral abdomen (A2–A5; Figure 15.2). Similar records have been made from the pharyngeal muscles (Gorczyca et al. 1991). Records are typically made at room temperature (20–22°C), although many investigators prefer to record at a lower temperature (14–18°C) in a temperature-controlled room or using a Peltier battery to prolong the life of the dissected preparation and improve the stability of the recording.

Conventional intracellular voltage recording of synaptic potentials (EJPs [excitatory junctional potentials]; Jan and Jan 1976a) is the most traditional and simplest electrophysiological technique. The muscle is impaled with a single sharp electrode filled with 3 M KCl or potassium acetate (15–40 MΩ resistance). Nerve-evoked EJPs are recorded by stimulating the appropriate segmental nerve (0.2–1 msec) with a fire-polished suction electrode (see above).

In normal external [$Ca^{++}$] (1.8 mM), the evoked EJP has an amplitude of 25–40 mV in a normal resting potential of –60 mV. Spontaneous, miniature EJPs are approximately 0.5 mV in amplitude on average and occur at a frequency of approximately 1–5 Hz, depending on recording temperature. Voltage signals are usually filtered (1–2 kHz), and can be amplified and recorded with a simple intracellular amplifier using standard techniques.

### Two-electrode Voltage-clamp Muscle Recording

Standard two electrode voltage-clamp (TEVC) techniques allow the best whole-cell current recording conditions in the larval muscle (Figure 15.2) (Wu and Haugland 1985; Singh and Wu 1990; Davis et al. 1996; Rohrbough et al. 1998). Intracellular microelectrodes are pulled from glass capillaries (outer diameter 1 mm) containing an internal filament. Voltage recording electrodes are filled with 3 M KCl and have typical resistances of 20–40 MΩ. Current-passing electrodes are filled with a 3 M KCl or a mixture (3:1) of potassium acetate:KCl or potassium citrate:KCl and have typical resistances of 15–25 MΩ. Lower-resistance electrodes (<20 MΩ) give best results. Some investigators prefer to bevel the electrodes using silicon-carbide grit.

Current and voltage signals are amplified with a TEVC amplifier (Axon Instruments). Because the larval muscles are quite large (2000–5000 pF), proper clamp tuning is important, especially when recording large synaptic and voltage-gated currents (Wu and Haugland 1985). Excitatory junctional currents (EJCs) are typically recorded in the range of –60 to –80 mV by stimulating the appropriate segmental nerve (see above). The typical muscle cell resistance is 5–10 MΩ (Stewart et al. 1994).

In normal external [Ca$^{++}$] (1.8 mM), the nerve-evoked EJC amplitude is 100 nA or greater at the peak, with a duration of 30–40 msec (Figure 15.2) (Rohrbough et al. 1998). However, many studies of neuromuscular transmission are made in low [Ca$^{++}$] (0.2–0.4 mM), which greatly attenuate EJC amplitude. Low Ca$^{++}$ conditions greatly accenuate frequency-dependent forms of facilitation and potentiation (Zhong and Wu 1991; Zhong et al. 1992; Wang et al. 1994; Broadie et al. 1997; Rohrbough et al. 1998). In the absence of nerve stimulation, spontaneous miniature EJCs (mEJCs) are recorded with an average amplitude of about 0.5 nA and a frequency of 1–5 Hz, depending on temperature. Both evoked and spontaneous synaptic currents can be acquired and analyzed using several commercially available computer hardware and software packages.

## Focal Recording of Synaptic Boutons

Localized synaptic currents are recorded by placing an extracellular "macro-patch" micro-electrode over identified nerve terminals (Mallart et al. 1991; Kurdyak et al. 1994; Rivosecchi et al. 1994). This approach is used to examine release at single, identified presynaptic NMJ boutons. Dissected larvae are viewed with Nomarski optics through a 40–63× water-immersion objective on an upright compound microscope. Specific synaptic boutons are identified in living terminals either directly using Nomarski optics or with a vital fluorescent dye such as 4-Di-2-Asp (5 μM, 5-minute exposure). Loose-patch recordings can be made from any exposed NMJ terminal. However, synaptic arbors that allow direct, perpendicular approach are easier to obtain high-resistance seals. For this reason, muscle-13 terminals have been used in preference to muscles 6, 7, and 12, which have a more lateral innervation (Kurdyak et al. 1994).

Focal recording electrodes are pulled from glass capillary tubes (e.g., 75 μl, 1.5-mm outer diameter) and then fire polished and shaped on a microforge to allow perpendicular approach to the muscle for obtaining better seal resistance. The loose-patch electrodes are made with an inner diameter of 5–10 μm. When filled with the appropriate bath solution, they have resistances of 0.5–2 MΩ. Seals around the boutons are made by applying mild suction and usually increase the resistance two to six times (seal factor; Rivosecchi et al. 1994). Recordings are made with a loose patch-clamp amplifier and contain a calibration pulse to measure electrode series and seal resistances. These measurements are used to correct for attenuated current amplitudes at the electrode tip.

## Patch-clamp Recording of Synaptic Boutons

The perforated patch-clamp technique has been used to record both whole-terminal and single-channel currents from presynaptic NMJ boutons (Martinez-Padron and Ferrus 1997). The mature larval preparation is dissected as above and then extensively digested with collagenase (100 units/ml; type 1A; Sigma) for 8–10 minutes. High-resistance pipettes (10–20 MΩ) are pulled from thick-wall borosilicate glass pipettes, coated with Sylgard, and fire-polished. The tip of the electrode is filled with intracellular patch saline and then back-filled with the same solution containing a saturating concentration of Nystatin (200 μg/ml; Sigma) (Martinez-Padron and Ferrus 1997).

Seals are made by applying gentle suction to identified synaptic boutons visualized with Nomarski optics. Seal resistances of up to 30 GΩ form, and currents can be recorded in cell-attached, inside-out or whole-cell recording configurations (Martinez-Padron and Ferrus 1997). Records from enlarged, mutant type III boutons on muscle 12 (>2 μm in diameter) showed an average input resistance of 8–10 GΩ and mean bouton capacitance of approximately 1 pF. These boutons display a range of voltage-dependent ionic currents upon membrane depolarization (Martinez-Padron and Ferrus 1997).

### NMJ Synaptic Modulation Properties

Facilitation and potentiation are typically assayed in low external [$Ca^{++}$] (<0.4 mM, typically 0.2 mM $Ca^{++}$) because these synaptic properties are obscured by transmission fatigue at higher $Ca^{++}$ levels (Zhong and Wu 1991; Zhong et al. 1992; Wang et al. 1994; Broadie et al. 1997; Rohrbough et al. 1998). Four types of modulations are typically assayed: paired-pulse facilitation (PPF), short-term facilitation (STF), augmentation, and post-tetanic potentiation (PTP). In *Drosophila*, these synaptic modulation properties have been most extensively assayed in the mature NMJ.

Paired-pulse facilitation is assayed by delivering two paired stimuli in quick succession. PPF is typically first detectable at a 100-msec interval and becomes maximal at approximately 20 msec. Analysis of rapid PPF (<30-msec interval) is confounded by the duration of the recorded EJC event (30–40 msec). Short-term facilitation can be assayed by delivering short stimuli trains at increasing frequencies (e.g., 0.2–20 Hz). Frequency-dependent STF is typically triggered at approximately 3–5 Hz stimulation and increases with frequency. Long-term augmentation is assayed using a similar paradigm with the high-frequency stimulus train lasting 1 minute or longer. PTP is assayed by delivering a 5–10-Hz stimulus train ("tetanus") for a short period (e.g., 1 minute) and analyzing mean EJC amplitude before and following the tetanus (Zhong and Wu 1991; Wang et al. 1994).

## Adult NMJ

### Preparation and Dissection

Recordings from the adult muscles can be made in a fashion similar to that described in detail for the larvae (see p. 290–291). Most recordings make use of the large thoracic flight muscles (Wu et al. 1978; Salkoff and Wyman 1983; Elkins and Ganetzky 1988; Koenig et al. 1989). Young adult flies (1–4 days post-eclosion) are lightly anaesthetized with ether, cooling, or some other method. The flies are immobilized with glue (cyanoacrylate adhesive) or in a low-melting-point wax, such as myristic acid or Tackiwax, so that the lower half remains exposed to air and the thorax of the fly can be covered in saline. The orientation is adjusted to allow the passage and circulation of air in the tracheal system during recording. The thoracic and abdominal spiracles may be aerated as reported by Ikeda and Kaplan (1974).

The adult preparation is typically dissected and visualized with a standard dissection microscope. An incision is made in the dorsal or lateral thorax to dissect a passage to the dorsal longitudinal flight muscles (DLM). The DLM contains six identifiable fibers, each of which receives thousands of en-passant-type synapses from a single excitatory motor neuron (Ikeda et al. 1980; Ikeda and Koenig 1988). The motor neuron axons extend through the posterior dorsal mesothoracic nerve (PDMN). The dorsolateral surfaces of the DLM and the PDMN are exposed through dissection. The tergochanteral (TTM) jump muscle shares innervation with the DLM and can be recorded from in a similar fashion (Engel and Wu 1992, 1994). These preparations remain in good condition for several hours.

### Motor Nerve Stimulation

The PDMN may be cut with fine scissors approximately 20–40 μm from the thoracic ganglion. The entire nerve is sucked into a glass suction electrode (5–10-μm inner diameter) filled with bath saline. The nerve is stimulated with a 0.1–0.5-msec square pulse at just above threshold (1–5 V). Alternatively, the PDMN can be left intact, and an electrical stimulus applied to the cervical connective or directly to the brain. Stimulus is applied with a pair of insulated tungsten electrodes (10–30 mV; 0.1–0.5 msec) that are electrically isolated from ground (Wu et al. 1978; Elkins and Ganetzky 1988; Koenig et al. 1989). A tungsten electrode inserted in the abdomen measures reference potential.

### Muscle Recording

Excitatory junctional potentials (EJPs) are recorded via an intracellular electrode in a DLM fiber. A glass microelectrode with a resistance of 20–60 M$\Omega$ is inserted into the lateral surface of the DLM and voltage responses recorded during nerve stimulation (Wu et al. 1978; Koenig et al. 1989). The electrode may be beveled using silicon carbide grit to reduce resistance to 5–10 M$\Omega$ and for easier penetration. Healthy cells should have a resting potential greater than –70 mV. Additionally, two-electrode voltage and current-clamp can be used as in the larval preparation (Elkins and Ganetzky 1988). Voltage-clamp records have been made at 4°C (preparation cooled with a Peltier plate) or near room temperature (16–21°C).

## ACKNOWLEDGMENTS

I thank David Featherstone, Robert Renden, Emma Rushton, and especially Jeff Rohrbough for comments and input on this chapter. Jeff Rohrbough and Tim Fergestad contributed to the illustrations.

## REFERENCES

Auld V.J., Fetter R.D., Broadie K. and Goodman C.S. 1995. Gliotactin, a novel transmembrane protein on peripheral glia, is required to form the blood-nerve barrier in *Drosophila. Cell* **81:** 757–767.

Baines R.A. and Bate M. 1998. Electrophysiological development of central neurons in the *Drosophila* embryo. *J. Neurosci.* **18:** 4673–4683.

Bate M. 1990. The embryonic development of the larval muscles in *Drosophila. Development* **110:** 791–804.

Begg M. and Cruickshank W.J. 1963. A partial analysis of *Drosophila* larval haemolymph. *Proc. R. Soc. Edinb.* **68:** 215–236.

Broadie K. and Bate M. 1993a. Development of the embryonic neuromuscular synapse of *Drosophila melanogaster. J. Neurosci.* **13:** 144–166.

———. 1993b. Development of larval muscle properties in the embryonic myotubes of *Drosophila melanogaster. J. Neurosci.* **13:** 167–180.

———. 1993c. Activity-dependent development of the neuromuscular synapse during *Drosophila* embryogenesis. *Neuron* **11:** 607–619.

———. 1993d. Synaptogenesis in the *Drosophila* embryo: Innervation directs receptor synthesis and localization. *Nature* **361:** 350–353.

Broadie K., Skaer H., and Bate M. 1992. Whole-embryo culture of *Drosophila:* Development of embryonic tissues *in vitro. Roux's Arch. Dev. Biol.* **201:** 364–375.

Broadie K., Rushton E., Skoulakis E.C.M. and Davis R. 1997. Leonardo, a 14-3-3 protein involved in learning, regulates presynaptic function. *Neuron* **19:** 391–402.

Broadie K., Bellen H.J., DiAntonio A., Littleton J.T., and Schwarz T.L. 1994. The absence of Synaptotagmin disrupts excitation-secretion coupling during synaptic transmission. *Proc. Natl. Acad. Sci.* **91:** 10727-10731.

Broadie K., Prokop A., Bellen H.J., O'Kane C.J., Schulze K.L., and Sweeney S.T. 1995. Syntaxin and Synaptobrevin function downstream of vesicle docking in *Drosophila. Neuron* **15:** 663–673.

Campos-Ortega J. and Hartenstein V. 1985. *The embryonic development of* Drosophila melanogaster. Springer Verlag, Berlin.

Chang H. and Kidokoro Y. 1996. Kinetic properties of glutamate receptor channels in cultured embryonic *Drosophila* myotubes. *Jpn. J. Physiol.* **46:** 249–264.

Croghan P.C. and Lockwood A.P.M. 1960. The composition of the haemolymph of the larva of *Drosophila melanogaster. J. Exp. Biol.* **37:** 339–343.

Currie D., Milner M., and Evans C. 1988. The growth and differentiation in vitro of leg and wing imaginal disc cells from *Drosophila melanogaster. Development* **102:** 805–814.

Davis G.W., Schuster C.M., and Goodman C.S. 1996. Genetic dissection of structural and functional components of synaptic plasticity. III. CREB is necessary for presynaptic functional plasticity. *Neuron* **17:** 669–679.

Deitcher D.L., Ueda A., Stewart B.A., Burgess R.W., Kidokoro Y., and Schwartz T.L. 1998. Distinct requirements for evoked and spontaneous release of neurotransmitter are revealed by mutations in the *Drosophila* gene neuronal-synaptobrevin. *J. Neurosci.* **18:** 2028–2039.

Elkins T. and Ganetzky B. 1988. The roles of potassium curents in *Drosophila* flight muscles. *J. Neurosci.* **8:** 428–434.

Engel J.E. and Wu C.-F. 1992. Interactions of membrane excitability mutants affecting potassium and sodium currents in the flight and giant fiber escape systems of *Drosophila. J. Comp. Physiol. A* **171:** 93–104.

———. 1994. Altered mechanoreceptor response in *Drosophila* bang-sensitive mutants. *J. Comp. Physiol. A* **175:** 267–278.

Ganetzky B. and Wu C.-F. 1982. *Drosophila* mutants with opposing effects on nerve excitability: Genetic and spatial interactions in repetitive firing. *J. Neurophys.* **47:** 501–514.

Gorczyca M.G., Budnik V., White K., and Wu C.-F. 1991. Dual muscarinic and nicotinic action on a motor program in *Drosophila. J. Neurobiol.* **22:** 391–404.

Hardi R.C. 1991. Voltage sensitive potassium channels in *Drosophila* photoreceptors. *J. Neurosci.* **11:** 3079–3095.

Ikeda K. and Kaplan W.D. 1974. Neurophysiological genetics in *Drosophila melanogaster. Am. Zool.* **14:** 1055–1066.

Ikeda K. and Koenig J.H. 1988. Morphological identification of the motor neurons innervating the dorsal longitudinal flight muscles of *Drosophila melanogaster. J. Comp. Neurol.* **273:** 436–444.

Ikeda K., Koenig J.H., and Tsuruhara T. 1980. Organization of identified axons innervating the dorsal longitudinal flight muscle of *Drosophila melanogaster. J. Neurocytol.* **9:** 799–823.

Jan L.Y. and Jan Y.N. 1976a. Properties of the larval neuromuscular junction in *Drosophila melanogaster. J. Physiol.* **262:** 189–214.

———. 1976b. L-glutamate as an excitatory transmitter at the *Drosophila* larval neuromuscular junction. *J. Physiol.* **262:** 215–236.

Jan Y.N., Jan L.Y., and Dennis M.J. 1977. Two mutations of synaptic transmission in *Drosophila. Proc. R. Soc. Lond. B.* **198:** 87–108.

Johansen J., Halpern M.E., Johansen K., and Keshishian H. 1989. Stereotypic morphology of glutamatergic synapses on identified muscle cells of *Drosophila* larvae. *J. Neurosci.* **9:** 710–725.

Kidokoro Y. and Nishikawa K.-I. 1994. Miniature endplate currents at the newly formed neuromuscular junction in *Drosophila* embryos and larvae. *Neurosci. Res.* **19:** 143–154.

Koenig J.H. and Ikeda K. 1983. Characterization of the intracellularly recorded response of identified flight motor neurons in *Drosophila. J. Comp. Physiol. A* **150:** 295–303.

Koenig J.H., Kosaka T., and Ikeda K. 1989. The relationship between the number of synaptic vesicles and the amount of transmitter released. *J. Neurosci.* **9:** 1937–1942.

Korn S., Mary A., Connor J., and Horn R. 1991. Perforated patch recording. *Methods Neurosci.* **4:** 264–373.

Kurdyak P., Atwood H.L., Stewart B.A., and Wu C.-F. 1994. Differential physiology and morphology of motor axons to ventral longitudinal muscles in larval *Drosophila. J. Comp. Neurol.* **350:** 463–472.

Mallart A., Angaut-Petit D., Bourret-Poulain C., and Ferrus A. 1991. Nerve terminal excitability and neuromuscular transmission in T (X;Y)V7 and Shaker mutants of *Drosophila melanogaster. J. Neurogenet.* **7:** 75–84.

Martinez-Padron M. and Ferrus A. 1997. Presynaptic recordings from *Drosophila:* Correlation of macroscopic and single-channel K+ currents. *J. Neurosci.* **17:** 3412–3424.

Nishikawa K.-I. and Kidokoro Y. 1995. Junctional and extrajunctional glutamate receptor channels in *Drosophila* embryos and larvae. *J. Neurosci.* **15:** 7905–7915.

O'Dowd D.K. and Aldrich R.W. 1988. Voltage-clamp analysis of sodium channels in wild-type and mutant *Drosophila* neurons. *J. Neurosci.* **8:** 3633–3643.

Ranjan R., Bronk P., and Zinsmaier K.E. 1998. Cysteine string protein is required for calcium secretion coupling of evoked neurotransmission in *Drosophila* but not for vesicle recycling. *J. Neurosci.* **18:** 956–964.

Rivosecchi R., Pogs O., Theil T., and Mallart A. 1994. Implication of frequenin in the facilitation of transmitter release in *Drosophila*. *J. Physiol.* **474:** 223–232.

Rohrbough J., Pinto S., Mihalek R.M., Tully T., and Broadie K. 1999. *latheo*, a *Drosophila* gene involved in learning, regulates functional synaptic plasticity. *Neuron* **23:** 55–70.

Salkoff L. and Wyman R. 1983. Ion currents in *Drosophila* flight muscle. *J. Physiol.* **337:** 687–709.

Salvaterra P.M., Bournias-Vardiabasis N., Nair T., Hou G.,and Lieu C. 1988. In vitro neuronal differentiation of *Drosophila* embryo cells. *J. Neurosci.* **7:** 10–22.

Seecof R.L. 1979. Preparation of cell cultures from *Drosophila melanogaster* embryos. *Tissue Culture Assoc. Manual* **5:** 1019–1022.

Seecof R.L., Alleaume N., Teplitz R.L., and Gerson I. 1971. Differentiation of neurons and myocytes in cell cultures made from *Drosophila* gastrulae. *Exp. Cell Res.* **69:** 161–173.

Seecof R.L, Teplitz R.L., Gerson I., Ikeda K., and Donady J.J. 1972. Differentiation of neuromuscular junctions in cultures of embryonic *Drosophila* cells. *Proc. Natl. Acad. Sci.* **69:** 566–570.

Shields G. and Sang J. 1977. Improved medium for culture of *Drosophila* embryonic cells. *Drosophila Inform. Serv.* **52:** 161.

Singh S. and Wu C.-F. 1990. Properties of potassium currents and their role in membrane excitability in *Drosophila* larval muscle fibers. *J. Exp. Biol.* **152:** 59–76.

Stewart B.A., Atwood H.L., Renger J.J., Wang J., and Wu C.-F. 1994. Improved stability of *Drosophila* larval neuromuscular preparations in haemolymph-like physiological solutions. *J. Comp. Physiol. A* **175:** 179–191.

Trimarchi J.R. and Murphey R.K. 1997. The shaking-B mutation disrupts electrical synapses in a flight circuit in adult *Drosophila*. *J. Neurosci.* **17:** 4700–4710.

Tsunoda S. and Salkoff L. 1995. Genetic analysis of *Drosophila* neurons: Shal, Shaw, and Shab encode most embryonic potassium currents. *J. Neurosci.* **15:** 1741–1754.

Ueda A. and Kidokoro Y. 1996. Longitudinal body wall muscles are electrically coupled across the segmental boundary in the third instar larva of *Drosophila melanogaster*. *Invertebr. Neurosci.* **1:** 315–322.

Van der Meer J.M. and Jaffe L.F. 1983. Elemental composition of the perivitelline fluid in early *Drosophila* embryos. *Dev. Biol.* **95:** 249–252.

Wang J., Renger J.J., Griffith L.C., Greenspan R.J., and Wu C.-F. 1994. Concomitant alterations of physiological and developmental plasticity in *Drosophila* CaM kinase II-inhibited synapses. *Neuron* **13:** 1373–1384.

Wu C.-F. and Haugland F.N. 1985. Voltage clamp analysis of membrane currents in larval muscle fibers of *Drosophila*. *J. Neurosci.* **5:** 2626–2640.

Wu C.-F., Ganetzky B., Jan L.Y., Jan Y.N., and Benzer S. 1978. A *Drosophila* mutant with a temperature-sensitive block in nerve conduction. *Proc. Natl. Acad. Sci.* **75:** 4047–4051.

Yoshikami D. and Okun L. 1984. Staining of living presynaptic nerve terminals with selective fluorescent dyes. *Nature* **310:** 53–56.

Zagotta W.N., Brainard M.S., and Aldrich R.W. 1988. Single-channel analysis of four distinct classes of potassium channels in *Drosophila* muscle. *J. Neurosci.* **8:** 4765–4779.

Zaho M.-L. and Wu C.-F. 1997. Altercations in frequency coding and activity dependence of excitability in cultured neurons of *Drosophila* memory mutants. *J. Neurosci.* **17:** 2187–2199.

Zhong Y. and Wu C.-F. 1991. Altered synaptic plasticity in *Drosophila* memory mutants with defective cyclic AMP cascade. *Science* **251:** 198–201.

Zhong Y., Budnik V. and Wu C.-F. 1992. Synaptic plasticity in *Drosophila* memory and hyperexcitability mutants: Role of cAMP cascade. *J. Neurosci.* **12:** 644–665.

# CONTENTS

# Functional Assays of the Peripheral and Central Nervous Systems

### Kendal S. Broadie
*Department of Biology*
*University of Utah*
*Salt Lake City, Utah 84112*

$D$ROSOPHILA IS A HOLOMETABOLOUS INSECT that forms two radically different body plans, each controlled and directed by structurally distinctive nervous systems. The embryonic/larval nervous system has the advantage of simplicity; individually identifiable neurons can be reliably selected for repeated functional analyses. The adult nervous system, in contrast, is composed of hundreds of thousands of neurons and is correspondingly less tractable. However, its attractiveness is that it drives complex behaviors, including flight, courtship, learning, and memory. Thus, the adult nervous system holds the promise of allowing us to unravel the genetic and neuronal bases of complex behaviors.

Both the peripheral and central nervous systems (PNS and CNS, respectively) have been subject to functional analyses. The classic system for study of the PNS is the adult retina, through the recording of whole-eye electroretinograms (ERGs) and single-cell recordings from isolated photoreceptors. This preparation has allowed the detailed study of phototransduction and information processing in the visual system. Similarly, recordings have been made from neurons involved in sensory mechanoreception in the adult epidermis. Although more difficult, progress has also been made in recording from CNS neurons. Successful recordings have been made from central neurons in situ, in both the embryo and the adult, and also from cultured CNS neurons. Thus, *Drosophila* is a well-established preparation for sensory and PNS physiology and holds the promise for increasing investigation of central processes.

This chapter surveys the experimental approaches used to record from peripheral and central neurons in *Drosophila*. Most of this work has been accomplished in the adult fly, but recent progress has also been made on earlier developmental stages. For a discussion of recording solutions, dissection tools, and equipment, see Broadie (this volume).

## PERIPHERAL NERVOUS SYSTEM

### Adult Eye

Records from the adult eye are well-established models for photoreceptor physiology and phototransduction in *Drosophila*. The oldest and simplest technique is the recording of

ERGs, the measurement of whole-retina extracellular electrical response to light stimuli. The on- and off-ERG transients provide a crude massed measurement of synaptic activity downstream from photoreceptor activation. Records can also be made with intracellular electrodes of single photoreceptor responses in the intact eye. More recently, research has focused on single-cell recordings from isolated or semi-isolated photoreceptors in vitro.

## Preparation and Dissection

*Electroretinogram (ERG).* Young adult flies (1–4 days post-eclosion) are anesthetized with ether and immobilized on a stage or coverslip with low-melting-point wax or quick-setting cement/glue. The fly is embedded to prevent movement and to hold the head rigidly in place. The preparation is viewed with a low magnification (20–40X) dissection microscope.

*In Vivo Intracellular Photoreceptor Recordings.* Young adult flies (1–4 days post-eclosion) are immobilized on a stage with low-melting-point wax or quick-setting cement. Some investigators maintain the immobilized flies in a constant stream of moist oxygen (Hardie et al. 1991a,b). A small hole is cut in the cornea with a surgical scalpel or a vibrating razor blade and sealed with silicon grease. Alternatively, the entire head can be cut coronally to expose photoreceptors (Figure 16.1) (Niemeyer et al. 1996; Acharya et al. 1997).

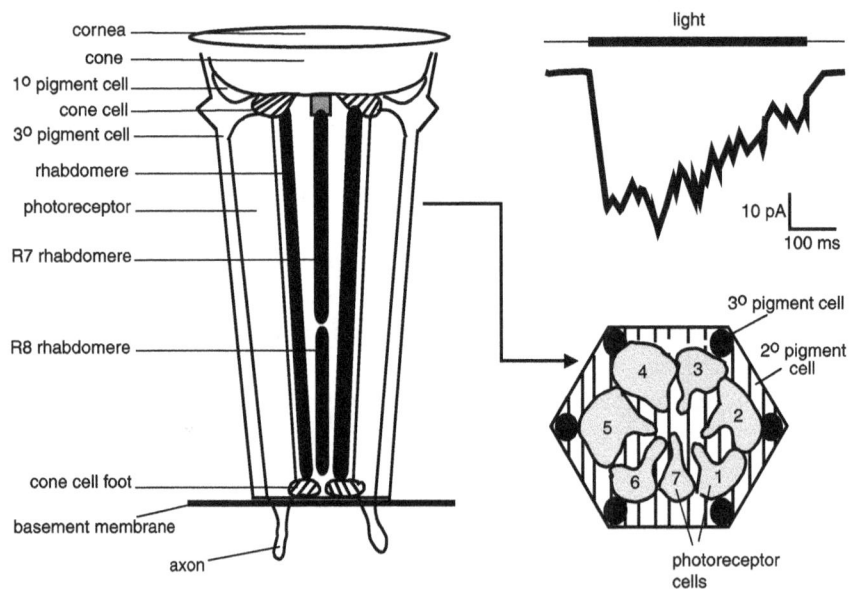

**Figure 16.1.** Schematic drawing of a *Drosophila* ommaditium and whole-cell light response. The pigment cells and photoreceptor cells are shown in longitudinal and transverse section. The pigment cell sheath is removed by the dissociation procedure allowing direct access to the photoreceptor membrane. The current inset shows a cartoon of a typical whole-cell response to a light stimulus. The photoreceptor is clamped at –60 mV.

*Isolated Photoreceptors.* Adult flies are immobilized by cooling or etherization and decapitated. The eyes are removed with a surgical scalpel or razor-blade chip in a chilled (4°C) saline solution. Retinas from either adults or pupae (after late stage p8; ~60 hours pupal development at 25°C; pupae are immobilized as described for adults) are dissected in normal bath saline solution and then quickly transferred to a bath solution containing 10% fetal calf serum (FCS saline; Hardie 1991a,b). Some investigators use enzymatic treatment (2 mg/ml collagenase) in divalent-cation-free saline at 22°C for 3–4 minutes to begin dissociation of the retina (Hevers and Hardie 1995). To mechanically dissociate ommatidia (Figure 16.1), the retina is gently triturated with an unsiliconized glass pipette, fire-polished to a diameter of approximately 100–150 μm. This treatment causes the ommatidia to break off at the basement membrane and pigment cells surrounding the photoreceptors to disintegrate (Figure 16.1) (Hardie 1991a,b). Thus, the photoreceptors are exposed, but lack their axonal terminals. A typical cell has a diameter of 5 μm and length of 80 μm. The dissociated cells can be recorded immediately or stored in FCS saline at 4°C for several hours before experimentation. A semi-intact preparation in which the retina is still attached to the lamina can be generated with milder treatment (Hevers and Hardie 1995). In both cases, the ommatidia are accessible for patch-clamp recording.

## Electrophysiology

*Electroretinogram (ERG).* Two fine tungsten-wire electrodes are used for recording. The recording electrode is inserted in the cornea and the reference electrode placed in the thorax. Light stimuli can be supplied by a variety of sources placed 2–20 cm from the fly eye. Electrical responses to light represent the summed extracellular activity from the photoreceptor field (Hotta and Benzer 1969; Pak et al. 1969; Kelly and Suzuki 1974).

*In Vivo Intracellular Photoreceptor Recordings.* Sharp glass electrodes with resistances in the range of 70–150 MΩ are back-filled with 2–3 M KCl. The electrode is lowered into the retina via the corneal hole or other photoreceptor exposure. Intracellular penetrations are signified by resting potentials of –40 to –60 mV (Hardie et al. 1991; Niemeyer et al. 1996; Acharya et al. 1997). The cells should display large, noisy responses to light (see Light Stimulation, below). Voltage-clamp experiments may be performed using a discontinuous, single-electrode voltage-clamp amplifier (Hardie et al. 1991).

*Isolated Photoreceptors.* Aliquots of ommatidia (~5–10 μl) are allowed to settle onto a clean, untreated coverslip and viewed with Nomarski optics on an inverted microscope. Recordings are made with patch pipettes pulled from borosilicate glass (fiber-filled) with resistances of 3–5 MΩ for whole-cell recordings, or 6–12 MΩ for isolated-patch recordings. Both whole-cell and isolated-patch recordings are made from pupal-stage photoreceptors using standard techniques (Hardie 1991a,b; Hardie et al. 1991). Seal resistances are typically 5–10 GΩ. In whole-cell records, series resistance was in the range of 6–25 MΩ and input resistance 100–2000 MΩ. Cell capacitance increases with age from approximately 3 pF at stage p8 to approximately 30 pF in stage p15 (mature pre-eclosion pupae; Hardie 1991a,b; Hardie et al. 1991).

Records can also be made from adult photoreceptors. However, the adult cells have relatively low input resistance (50–100 MΩ) and large currents (several nanoamperes), and so are difficult to voltage-clamp accurately (Hardie 1991a,b). Both late pupal and adult cells produce electrical response to light (Figure 16.1), but this light-activated current typically runs down within a few minutes. Relatively high resistance (10–25 M$\Omega$) pipettes are used to prolong recording time before rundown of the light response. Resting potential also tends to decay with time from approximately 50–60 mV to zero. Attempts to rescue resting potential or phototransduction machinery run down (with Mg-ATP or GTP) have been unsuccessful (Hardie 1991a,b; Hardie et al. 1991). However, voltage-gated currents appear normal under these conditions.

### Light Stimulation

Flies may be dark-adapted before preparation and dissection/recording done under dim red light (e.g., Schott OG630). Diffuse illumination can be provided by the collimated beam of a 12V 75–100W QI bulb passed through neutral-density filters and a yellow filter (e.g., Schott OG530) to ensure that only responses from one class of photoreceptor (R1–R6) are recorded. The light is focused on one end of a light-guide and the other end is placed approximately 2 cm over the bath (Hardie 1991a). In whole-cell configuration, the light response only remains stable for a limited period (<15–20 minutes) and often for only a few minutes.

## Adult Bristle Mechanosensory System

Mechanosensory bristle function in the adult fly can be used to assay the mechanotransduction machinery and characterize mutants defective in this mechanism or downstream function of the sensory neurons. Extracellular recordings of mechanically evoked bristle responses have been measured as a transepithelial potential (TEP), a voltage difference between the apical and basal sides of the sensory epithelium. The generation of the TEP serves as a general assay of mechanoreception.

### Preparation

Young adult flies are decapitated, the wings removed, and the body mounted on a pin through the thorax (Kernan et al. 1994). One electrode is placed over the cut end of a thoracic bristle (humeral or anterior notopleural macrochaete bristle). A second reference electrode is inserted into the thorax so that it contacts the hemolymph.

### Stimulation

To stimulate the bristle, the electrode is mounted on a piezoelectric microstage. Computer-generated voltage commands generate the mechanical stimulus, which deflects the bristle by 10–20 $\mu$m. Movement of the bristle by approximately 10° in the sagittal plane has been used to evoke response (Kernan et al. 1994).

### Recording

The bristle electrode is filled with high-potassium, low-calcium saline: 121 m$M$ K$^+$, 9 m$M$ Na$^+$, 0.5 m$M$ Ca$^{++}$, 4 m$M$ Mg$^{++}$, 35 m$M$ glucose, and 5 m$M$ HEPES. The reference electrode is filled with standard saline (see Chapter 15, Table 15.1). At rest, a positive TEP is maintained across the sensory epithelium. Mechanical stimulation causes inward ionic conductance in the bristle sensory neuron, reducing TEP (Kernan et al. 1994). This is measured by the recording electrode as a mechanoreceptor potential (MRP). The response is rapid, beginning within 3 msec of bristle deflection and is maintained for the duration of the stimulus.

## Adult Wing Campaniform Sensilla

The adult wing contains a set of eight campaniform sensilla, sensory structures that detect the mechanical deformation of the wing cuticle. The axons from these sensory neurons form two distinct tracts and branching patterns within the CNS (Palka et al. 1986). It is possible to record the extracellular activity from these axons as they run through the wing vein. Thus, the electrical properties of the mechanosensory campaniform neurons can be examined in response to mechanical stress applied to the wing (Dickinson and Palka 1987).

### Preparation and Dissection

A recording stage is prepared by applying a small amount of vaseline to two minutin pins situated on each side of a two-depression-well microscope slide (Figure 16.2) (Dickinson and Palka 1987). These pins will be used to fasten the wing to the stage. A young adult female fly (2–5 days post-eclosion) is anesthetized with $CO_2$. The distal tip of the wing is

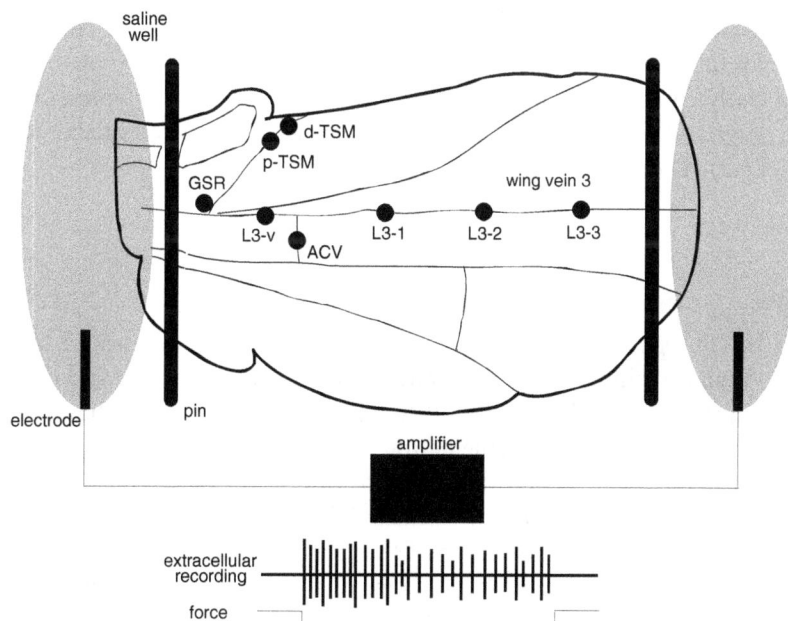

**Figure 16.2.** Preparation for extracellular recording from campaniform sensory neurons in the adult *Drosophila* wing. The wing is fastened in place using two greased insect pins. The electrical circuit is made between the hemolymph in wing vein 3 and the two saline wells. Extracellular spikes are recorded through electrodes placed in each well. The eight structurally similar campaniform sensilla are drawn. Force is supplied to selected sensilla and the extracellular response recorded as diagrammed.

removed with a fine dissection scissors; the entire wing is then removed from the thorax by cutting through the base of the wing so that a small portion of the thorax remains attached. A pair of forceps is used to hold the wing by the anterior margin (first wing vein), and the wing is placed dorsal side up across the prepared minutin pins on the recording stage (Figure 16.2) (Dickinson and Palka 1987). The proximal saline well is filled until the drop touches the cut thoracic tissue. With fine forceps, Vaseline is applied to plug all the cut distal veins, leaving only the third vein unplugged. The distal well is filled with saline until it touches the cut outer edge of the wing. The wing is then stretched between two drops of saline with an open passage through the third wing vein (Figure 16.2).

### Recording

A high-impedance, differential amplifier is used to record the extracellular activity of the campaniform neurons whose axons run through the third wing-vein (Dickinson and Palka 1987). One electrode is placed in each of the saline wells, and the amplifier output is monitored (Figure 16.2). One should be able to detect spontaneous activity from the campaniform neurons caused by the wing deformation produced by the mounting procedure. If not, it is likely that there is a fluid connection between the two saline wells. This shunts the electrodes and ruins the recording. In general, the signal-to-noise ratio improves with high electrical resistance between the two wells.

### Stimulation

A piezoelectric probe is positioned over the wing, and the tip is lowered until it just touches the wing surface (Dickinson and Palka 1987). The probe is depressed while monitoring for the activity of tonic neurons. The degree and duration of wing depression can be precisely controlled with a stimulator. A brief (500 msec) depression should yield a burst of large spikes at the onset, followed by tonic discharge of smaller amplitude events (Figure 16.2) (Dickinson and Palka 1987).

## Adult Reflex Circuit

In *Drosophila*, neural circuits underlying commonly assayed behaviors (e.g., olfactory/ visual learning and memory and courtship) are complex and difficult to assay. As a consequence, it has been difficult to analyze the effects of behavioral mutations on specific components of identifiable neural circuits. One system that has been used is a simple proprioceptive reflex controlling leg position in the adult fly (Jin et al. 1998). This reflex circuit involves a small set of identified femoral chordotonal sensory neurons with mapped CNS projections. The output of this circuit is measured in the tibia extensor muscle. This circuit displays several types of neuronal modulation including habituation.

### Preparation and Dissection

Mature adult flies (2–5 days post-eclosion) are anesthetized on ice for 5–10 minutes. The flies are decapitated with a sharp scalpel and left in a moist chamber for at least 1 hour to

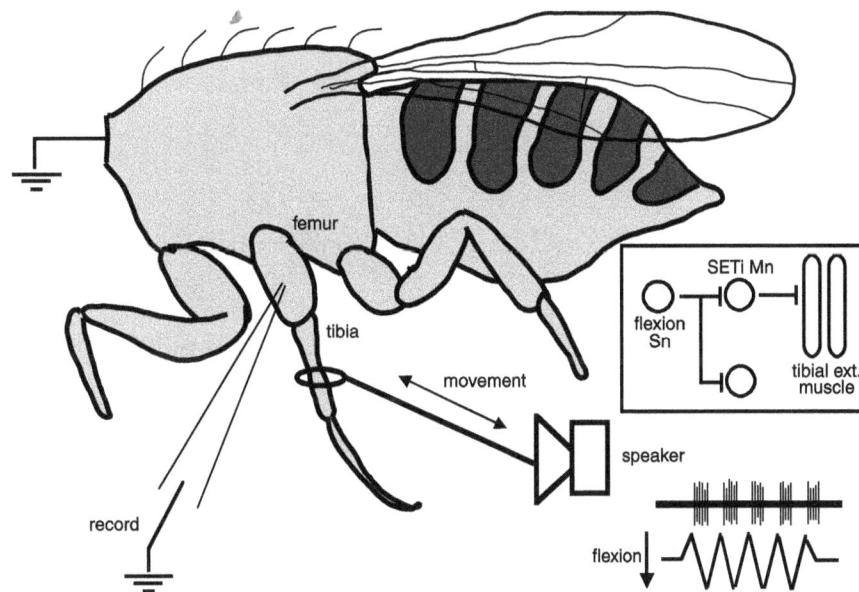

**Figure 16.3.** Schematic drawing of the preparation used for the adult leg reflex response. The fly is decapitated and the femur of the middle leg is immobilized. The tibia is rhythmically flexed by attachment to a movement generator (speaker) at a set frequency (e.g., 2 Hz). Myogram recordings are made from the tibial extensor muscle; the reference electrode is placed in the thorax. A cartoon of a typical recording session is shown. (*Inset*) Neural circuit underlying the resistance reflex. The sensory neurons of the chordotonal organ (flexion Sn) synapse on the motor neuron (Mn), which in turn synapse on the tibial extensor muscle.

stabilize following surgery. The decapitated fly is mounted on a dissecting microscope platform with soft wax, immobilizing the animal, but leaving the tibia and tarsi of the mesothoracic leg hanging free. To stabilize the leg femur, its distal end is waxed onto the platform (Jin et al. 1998).

### Muscle Recording

Excitatory junctional potentials (EJPs) are recorded from the tibia extensor muscle (Figure 16.3). A sharpened, tungsten recording electrode or glass electrode filled with intracellular recording solution is inserted into the tibia extensor muscle from the dorsal edge of the femur. A sharpened, tungsten ground electrode is inserted into the abdomen of the fly. Data can be recorded on-line and stored digitally for later analysis.

### Stimulation

To stimulate the femoral chordotonal organ, a fine metal loop is placed around the tibia and moved with a small speaker (Figure 16.3) (Jin et al. 1998). Flexing the femoro-tibial joint elicits a resistance response from the tibial extensor motor neurons. The frequency of EJPs in the tibia extensor muscle is monitored through a period of repetitive sinusoidal flexion-extension movements (Figure 16.3) (Jin et al. 1998). The tibia extensor muscle is innervated by two motor neurons; the fast extensor of the tibia (FETi) and the smaller slow extensor of the tibia (SETi; Trimarchi and Schneiderman 1994; Trimarchi and Murphey 1997). The FETi has a considerably higher threshold of excitation, and it is believed that only the SETi is actively involved in this reflex.

## CENTRAL NERVOUS SYSTEM

Recording from the *Drosophila* CNS is the most challenging of the electrophysiology techniques commonly used in this system. The reasons are the relatively small size of most CNS neurons (<5 µm), the tough sheath surrounding and protecting the CNS, and the relative complexity of the tissue. Consequently, most recordings from CNS neurons have been taken from dissociated cultures in the embryo and larvae. Recently, progress has also been made in recording from embryonic neurons in situ, suggesting that more extensive CNS investigation may be feasible in the future. Records from identified adult CNS neurons have been made successfully. However, such recording remains challenging because of the complexity of the adult thoracic ganglion and relatively small size of the CNS neurons. Most work has focused on the giant-fiber escape circuit and motor neurons with outputs onto the thoracic flight, haltere, and leg jump muscles.

### Culture of Embryonic Neurons

#### *Embryo Collection and Preparation for Cell Culture*

The culture method is similar to that described in Protocol 15.1 (see Broadie, this volume) for embryonic culture (Seecof et al. 1971; Seecof 1979; Salvaterra et al. 1988). Embryos are collected on agar plates in a timed lay (1–2 hours) and then incubated at 25ºC until gastrula formation (2–3 hours). Embryos are dechorionated in a 1:1 solution of bleach and 70–90% ethanol for 1–5 minutes, and gastrula-stage embryos are selected. After thorough washing, embryos are homogenized by grinding in a ground glass homogenizer in modified Schneider medium containing 20% FCS (some investigators use heat-inactivated FCS; O'Dowd and Aldrich 1988), 200 ng/ml insulin, 50 units/ml penicillin, and 50 µg/ml streptomycin (Saito and Wu 1991). A subpopulation of cells plated immediately will differentiate neuronal morphology with small soma (<5 µm). These small neurons have been successfully patch-clamped (see O'Dowd and Aldrich 1988; Leung and Byerly 1991).

Alternatively, dissociated cells can be collected by centrifugation (2500 rpm for 2 minutes), resuspended in medium containing 2 µg/ml cytochalasin B, and plated on glass coverslips. The cytochalasin acts to inhibit cell division and promote the growth of giant neurons; the drug can be removed after the second day of culture or left on with no ill effects (Wu et al. 1990; Saito and Wu 1991). The cultures are maintained in humidified chambers at room temperature (20–23ºC). Under these conditions, the majority of cells (>80%) differentiate neuronal morphology.

#### *Recording*

For electrophysiology, the culture coverslips are transferred to a bath of one of the standard recording solutions. Records are made in patch-clamp configurations (O'Dowd and Aldrich 1988; Wu et al. 1990; Leung and Byerly 1991; Saito and Wu 1991; Zhao and Wu

1997). Patch-pipettes are made from glass capillaries with a tip diameter of 1–2 μm and resistance of 2–5 MΩ for giant neurons and 5–15 MΩ for normal neurons or cell-attached patches. The pipettes can be coated with Sylgard (184 silicon elastomer, Dow Corning) to reduce stray capacitance and fire-polished on a microforge.

Whole-cell recordings have been made from both normal neurons (<5 μm in diameter) and the cytochalasin-treated giant neurons (9–15 μm in diameter) using standard techniques (Leung and Byerly 1991; Saito and Wu 1991; Zhao and Wu 1997). For the giant neurons, the average value of series resistance is approximately 20–25 MΩ and seal resistance is in the range of 5–50 GΩ prior to rupture of the patch membrane. Neurons have an input resistance of 1–4 GΩ at rest and a mean whole-cell capacitance of approximately 40 pF (Wu et al. 1990; Saito and Wu 1991). This capacitance value is nearly ten times greater than that predicted of a spherical cell of this size, suggesting that most of the membrane area is present in neurites. This complication makes it difficult to achieve reasonable space-clamp conditions.

## Culture of Larval CNS Neurons

### *Preparation and Dissection*

Neuronal cultures may be prepared from the CNS of mature larvae (Wu et al. 1983; Solc and Aldrich 1988; Kim and Wu 1991; Wright and Zhong 1995); 5–20 late third-instar larvae are collected and thoroughly washed in 70% ethanol for 1–2 minutes. Brain and the ventral nerve cord (VNC) are dissected from the larvae in phosphate-buffered saline (PBS; see Appendix 3) or any of the standard salines (see Table 15.1 in Broadie, this volume). The neurolemma surrounding the VNC is mechanically disrupted with fine forceps and then incubated in 0.5 mg/ml collagenase (Type I, Sigma) in divalent-free PBS for 1–2 hours (Wu et al. 1983; Solc and Aldrich 1988). The softened tissue is gently triturated in a glass pipette 50 times to completely dissociate the cells (Wright and Zhong 1995). The preparation is centrifuged at 3000–4000 rpm for 1–2 minutes and the supernatant discarded. The pellet is then resuspended in modified Schneider's medium (GIBCO-BRL) supplemented with 10–20% FCS and 1.5% penicillin/streptomyocin. The cells may be again triturated with a glass pipette until clumps are no longer visible and then plated on autoclaved untreated coverslips at a density of 0.5–1 VNC/coverslip (12-mm round). The cultures may be improved by placing a second coverslip on top of the cells (Solc and Aldrich 1988). The cultures are maintained at room temperature (19–22°C) for 1–12 days before electrophysiology experiments are conducted.

The distribution of cultured cells immediately after plating has been divided into three categories: large (>8–10 μm) round cells (type I neuroblasts), intermediate (5–10 μm) cells (type III neurons), and small (2–5 μm) cells (type II neurons) (Wu et al. 1983; Solc and Aldrich 1988). Type III neurons usually have multiple short, thick processes from one or both ends of a spindle-shaped cell body, whereas type II neurons have a single primary process extending from a round cell body (Wu et al. 1983; Solc and Aldrich 1988; Kim and Wu 1991). After a few days in culture, type III neurons differentiate round soma (8–12 μm) with a dense network of multipolar processes. These cells stain with anti-horseradish peroxidase (HRP) antibody, which recognizes a neuronal specific antigen in *Drosophila*. Type III neurons have been used for most electrophysiological studies (Wu et al. 1983; Solc and Aldrich 1988).

## Identification of Neurons

Expression of the reporter gene β-galactosidase has been used to identify specific neuronal cell types in culture prior to recording (Wright and Zhong 1995). The dissociated cells are resuspended in 100 μl of culture medium diluted 40–50% with $H_2O$ containing 0.2 mM of the fluorogenic β-galactosidase substrate analog 5-chloro-methylfluorescein di-β-D-galactopyranoside (CMFDG, Molecular Probes). After incubation for 1 minute, tonicity is restored by the addition of undiluted culture medium (250 μl), and the suspension is recentrifuged (Wright and Zhong 1995). The cells are washed several times in fresh medium and plated. Neurons expressing β-galactosidase will be marked with the fluorescent product and may be identified using a standard UV light source and FITC filters.

## Electrophysiology

Cultured neurons can be recorded from in whole-cell, attached patch, and cell-free patch recording configurations using standard techniques (Wu et al. 1983; Solc and Aldrich 1988). Patch electrodes are used with resistances of 3–20 MΩ. A typical cell has input resistance of 10–30 GΩ.

# Embryonic CNS

Recent progress has been made in recording from identifiable central neurons in situ in the *Drosophila* embryonic ventral nerve cord (VNC; Baines and Bate 1998). This advance is potentially important for a number of reasons. First, it allows the study of neuronal functional development in situ. Second, it allows study of central synaptic transmission to confirm and extend studies of normal and mutant transmission at the glutamatergic NMJ. Finally, it may allow access to central neural circuits for analysis of their developmental and functional properties.

## Dissection

The dissection is identical to that described in Chapter 15 (Broadie) for the study of the embryonic NMJ (Broadie and Bate 1993). In brief, the staged embryo is oriented with the ventral surface down and attached to a poly-L-lysine-coated coverslip (<16 hours AEL), or the head/tail is glued down with cyanoacrylate glue (Histoacryl; Braun, Germany) to a Sylgard-coated coverslip (>16 hours AEL) under dissection saline. A longitudinal incision is made along the length of the dorsal midline and the embryo is glued flat to the coverslip. Gut and fat body are removed to expose the VNC. The VNC is viewed using a 63X water-immersion lens with Nomarski optics.

## Enzymatic Treatment

A small section of the VNC neurolemma is ruptured using protease (1% Type XIV; Sigma) in external bath saline (Baines and Bate 1998). The protease saline is contained in a large-diameter (10–20 μm) patch-pipette which is brought into contact with a region of the neurolemma covering the dorsal surface of the VNC. A small portion of the neurolemma is drawn into the pipette using gentle suction and held for 2 minutes. Following this treatment, a small hole is made in the neurolemma by alternating suction and expulsion in the pipette. Any debris from this treatment is removed by suction with the pipette, which is then withdrawn. This treatment should expose neurons in at least two adjacent segments; however, the characteristic soma locations of the neurons are sometimes disrupted during the procedure, and therefore, neuronal identification based on position is not always possible (Baines and Bate 1998).

### Recording

Records are made using standard whole-cell patch-clamp. Patch electrodes are pulled from thick-walled borosilicate glass and fire-polished to final resistance of 15–20 M$\Omega$. Tight seals (usually >5 G$\Omega$) form readily, although obtaining stable whole-cell access is more problematic with 20–50% success rate in older embryos (>16 hours AEL) and only 10–20% success rate in younger embryos (<16 hours AEL). Recordings are best made in voltage-clamp mode with a holding potential of –60 to –80 mV. Recordings in current clamp have so far been unsuccessful due to large fluctuations in membrane voltage (Baines and Bate 1998).

### Stimulation

Evoked stimulation of synaptic transmission has not yet been attempted, although spontaneous endogenous transmission is reported (Baines and Bate 1998). Acetylcholine (ACh) iontophoresis is effective in generating whole-cell current responses in a development-dependent fashion. A 60-M$\Omega$ electrode containing 0.1 M ACh (HCl salt; Sigma) is positioned within 1–2 $\mu$m of the neuronal soma. A backing current of approximately –2 nA is used to prevent ACh leakage, and an ejection current of approximately +30 nA is used for the iontophoresis. ACh responses are first detected at 13 hours AEL with 25–30%

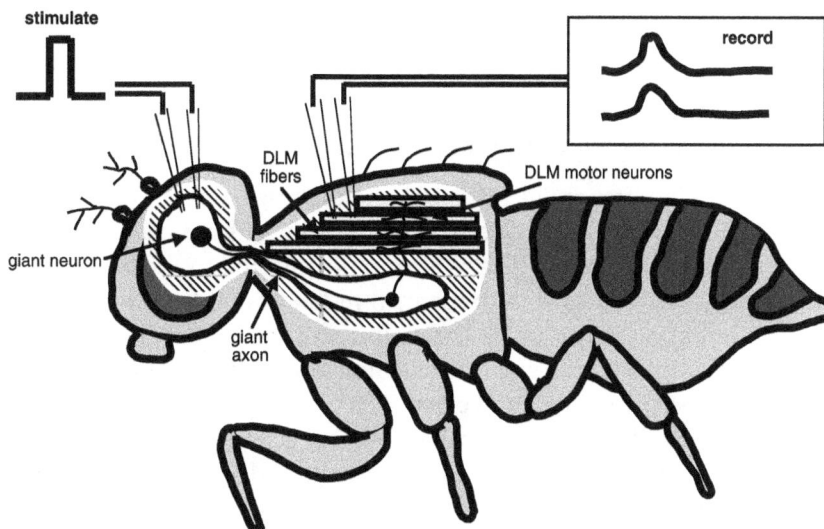

**Figure 16.4.** Giant fiber (GF) system of *Drosophila*. The schematic cut away of an adult fly shows the brain (in white) and the dorsal longitudinal muscle (DLM) fibers of the thorax. The soma of the giant neuron is found in the central brain and projects a giant axon into the thoracic ventral nerve cord (VNC) where it synapses on the DLM motor neurons. The giant neuron can be stimulated by electrodes placed in the brain, and the output of the circuit is monitored with intracellular electrodes in the DLM fiber.

of cells positive; after 16 hours AEL, 100% of cells show an ACh response (Baines and Bate 1998). Other agonists have yet to be assayed in this system.

## Adult CNS: Giant Fiber System

### Preparation and Dissection

Adult flies are mounted on a platform with a soft, low-melting-point wax usually with the ventral side up (Tanouye and Ferrus 1985). The ventral surface is bathed in saline, whereas the dorsal surface, specifically including the tracheal spiracles, is left dry and in contact with circulating air. The mounting wax is used to keep the fly immobile and maintain the air/water boundary. The legs, gut, and salivary gland may be removed to reduce movement, although particular care must be taken to maintain the tracheal system intact. Damage to the tracheal system results in compromised electrophysiological recordings, including reduced resting potential and smaller amplitude action potentials (Tanouye and Ferrus 1985).

The propreespisternum is removed to expose the cervical connective containing the axon of the giant fiber neuron (Figure 16.4). A small platform may be placed under the cervical connective to stabilize it and reduce movement. The dorsal longitudinal flight muscle (DLM) or the tergotrochanteral jump muscle (TTM) is exposed for impalement and recording (Figure 16.4). Flies prepared in this way remain stable for several hours at room temperature (Tanouye and Wyman 1980; Koto et al. 1981; Tanouye and King 1983; Tanouye and Ferrus 1985).

### Stimulation

The giant fiber may be stimulated in two ways (Tanouye and Ferrus 1985): (1) Most commonly, the brain is stimulated directly with electrical current through insulated extracellular tungsten electrodes (Figure 16.4). Stimulus voltage is kept just above threshold (typically in the range of 1–10 V) and duration of 0.1 msec. (2) The giant fiber can be stimulated intracellularly with depolarizing current passed through an impaling electrode (0.1–0.5-msec duration).

### Giant Fiber and DLM Recording

Recording sites are identified physiologically based on the fact that the giant fiber drives the DLM in a one-to-one fashion; the muscle is not driven one to one by any other cervical connective axon. Electrical stimuli delivered to the brain are used to drive the DLM; axonal recordings with the same threshold as DLM activation are considered giant-fiber recordings (Figure 16.4). The accuracy of this approach has been confirmed with dye-fill experiments (Ikeda et al. 1980; Tanouye and Wyman 1980; Koto et al. 1981; Tanouye and King 1983; Tanouye and Ferrus 1985).

Intracellular recordings of giant-fiber action potentials are made with sharp glass electrodes with resistances of 40–60 M$\Omega$. Recording from the DLM can be performed using insulated tungsten electrodes or glass microelectrodes with resistances of 5–40 M$\Omega$. The electrodes are filled with 3 M KCl. The giant-fiber resting potential is normally in the range of 60–80 mV, whereas adjacent cervical axons have substantially lower values (~30 mV). The action potentials average 60–70 mV and overshoot zero approximately 50% of the time (Tanouye and Ferrus 1985). Giant-fiber recordings may remain stable for 30 minutes or longer, but usually deteriorate within the first several minutes.

## Adult CNS: Motor Neuron Recording

### Preparation and Dissection

The adult preparation is similar to that of the giant fiber system (see above). Briefly, adult flies are anesthetized lightly using ether or ice (<5 minutes) and then waxed into an immobile position on a small platform, such as a small petri dish. The preparation is flooded with saline while maintaining an air interface (see above) or by trapping an air bubble around the fly in the saline (Trimarchi and Murphey 1997). The dorsal thoracic cuticle is dissected away to expose the fused thoracic ganglia to saline.

### Stimulation

A variety of circuits can be exposed to stimulation and recording (Tanouye and Wyman 1980; Koenig and Ikeda 1983; Trimarchi and Murphey 1997). Stimulation may be applied to the giant fiber (see above), or the afferents from the leg, wing, or haltere. Stimulation can be supplied with a tungsten or glass electrode (resistance of 2–5 M$\Omega$) directly to the afferent nerve. Afferents are stimulated with short pulses (20–100 $\mu$sec) at low to moderate frequencies (1–10 Hz; Koenig and Ikeda 1983; Trimarchi and Murphey 1997).

### Motor Neuron Recording

Using the contour of the ganglia and nerves as a guide, a number of motor neurons can be identified and impaled for intracellular recording. These cells include flight and leg motor neurons such as the dorsal longitudinal motor neurons, steering motor neurons (e.g., B1mn), the tergotrochanteral motor neuron (TTMn), and the fast extensor tibiae motor neuron (FETi; Ikeda and Koenig 1988; Trimarchi and Schneiderman 1994; Trimarchi and Murphey 1997).

Intracellular recording from the motor neurons is achieved with sharp glass electrodes of high resistance (100–250 M$\Omega$) filled with 3 M KCl. Records from the innervated muscle are often made in parallel using low-resistance (2–20 M$\Omega$) glass electrodes inserted into the appropriate muscle. The ability to successfully impale an identified motor neuron is relatively low (<20%), and so recording is usually coupled with an intracellular dye-injection technique that permits unequivocal identification of the cell following recording.

## ACKNOWLEDGMENTS

I thank David Featherstone, Robert Renden, Emma Rushton, and especially Jeff Rohrbough, for comments and input on this chapter.

## REFERENCES

Acharya J.K., Jalink K., Hardy R.W., Hartenstein V., and Zuker C.S. 1997. InsP3 receptor is essential for growth and differentiation but not for vision in *Drosophila*. *Neuron* **18:** 881–887.

Baines R.A. and Bate M. 1998. Electrophysiological development of central neurons in the *Drosophila* embryo. *J. Neurosci.* **18:** 4673–4683.

Broadie K. and Bate M. 1993a. Development of the embryonic neuromuscular synapse of *Drosophila melanogaster*. *J. Neurosci.* **13:** 144–166.

Campos-Ortega J. and Hartenstein V. 1985. *The embryonic development of* Drosophila melanogaster. Springer Verlag, Berlin.

Dickinson M.H. and Palka J. 1987. Physiological properties, time of development, and central projection are correlated in the wing mechanoreceptors of *Drosophila. J. Neurosi.* **7:** 4201–4208.

Hardie R.C. 1991a. Whole-cell recordings of the light-induced current in *Drosophila* photoreceptors: Evidence for feedback by calcium permeating the light sensitive channels. *Proc. R. Soc. Lond. B.* **245:** 203–210.

———. 1991b. Voltage sensitive potassium channels in *Drosophila* photoreceptors. *J. Neurosci.* **11:** 3079–3095.

Hardie R.C., Voss D., Pongs O., and Laughli S.B. 1991. Novel potassium channels encoded by the Shaker locus in *Drosophila* photoreceptors. *Neuron* **6:** 477–486.

Hevers W. and Hardie R.C. 1995. Serotonin modulates the voltage dependence of Shaker potassium channels in *Drosophila* photoreceptors. *Neuron* **14:** 845–856.

Hotta Y. and Benzer S. 1969. Abnormal electroretinograms in visual mutants of *Drosophila. Nature* **222:** 354–356.

Ikeda K. and Koenig J.H. 1988. Morphological identification of the motor neurons innervating the dorsal longitudinal flight muscles of *Drosophila melanogaster. J. Comp. Neurol.* **273:** 436–444.

Ikeda K., Koenig J.H., and Tsuruhara T. 1980. Organization of identified axons innervating the dorsal longitudinal flight muscle of *Drosophila melanogaster. J. Neurocytol.* **9:** 799–823.

Jin P., Griffith L.C., and Murphey R.K. 1998. Presynaptic calcium/calmodulin-dependent protein kinase II regulates habituation of a simple reflex in adult *Drosophila. J. Neurosci.* **18:** 8955–8964.

Kelly L.E. and Suzuki D.T. 1974. The effects of increased temperature on electroretinograms of temperature-sensitive paralysis mutants of *Drosophila melanogaster. Proc. Natl. Acad. Sci.* **71:** 4906–4909.

Kernan M., Cowan D., and Zuker C. 1994. Genetic dissection of mechanosensory transduction: Mechanoreception-defective mutations of *Drosophila. Neuron* **12:** 1195–1206.

Kim Y.-T. and Wu C.-F. 1991. Distinctions in growth cone morphology and motility between monopolar and multipolar neurons in *Drosophila* CNS culture. *J. Neurobiol.* **22:** 263–275.

Koenig J.H. and Ikeda K. 1983. Characterization of the intracellularly recorded response of identified flight motor neurons in *Drosophila. J. Comp. Physiol. A* **150:** 295–303.

Koto M., Tanouye M.A., Ferrus A., Thomas J.B., and Wyman R.J. 1981. The morphology of the cervical giant fiber neuron of *Drosophila. Brain Res.* **221:** 213–217.

Leung H.-T. and Byerly L. 1991. Characterization of single calcium channels in *Drosophila* embryonic nerve and muscle cells. *J. Neurosci.* **11:** 3047–3059.

Niemeyer B.A., Suzuki E., Scott K., Jalink K., and Zuker C.S. 1996. The *Drosophila* light-activated conductance is composed of the two channels TRP and TRPL. *Cell* **85:** 651–659.

O'Dowd D.K. and Aldrich R.W. 1988. Voltage-clamp analysis of sodium channels in wild-type and mutant *Drosophila* neurons. *J. Neurosci.* **8:** 3633–3643.

Pak W.L., Grossfield J., and White N.V. 1969. Nonphototactic mutants in the study of vision of *Drosophila. Nature* **222:** 351–354.

Palka J., Malone M.A., Ellison R.L., and Wigston D.J. 1986. Central projections of identified *Drosophila* sensory neurons in relation to their time of development. *J. Neurosci.* **6:** 1822–1830.

Saito M. and Wu C.-F. 1991. Expression of ion channels and mutational effects in giant *Drosophila* neurons differentiated from cell division-arrested embryonic neuroblasts. *J. Neurosci.* **11:** 2135–2150.

Salvaterra P.M., Bournias-Vardiabasis N., Nair T., Hou G., and Lieu C. 1988. In vitro neuronal differentiation of *Drosophila* embryo cells. *J. Neurosci.* **7:** 10–22.

Seecof R.L. 1979. Preparation of cell cultures from *Drosophila melanogaster* embryos. *Tissue Culture Assoc. Manual* **5:** 1019–1022.

Seecof R.L., Alleaume N., Teplitz R.L., and Gerson I. 1971. Differentiation of neurons and myocytes in cell cultures made from *Drosophila* gastrulae. *Exp. Cell Res.* **69:** 161–173.

Solc C.K. and Aldrich R.W. 1988. Voltage-gated potassium channels in larval CNS neurons of *Drosophila. J. Neurosci.* **8:** 2556–2570.

Tanouye M.A. and Ferrus A. 1985. Action potentials in normal and Shaker mutant *Drosophila. J. Neurogenet.* **2:** 253–271.

Tanouye M.A. and King D.G. 1983. Giant fiber activation of direct flight muscles in *Drosophila. J. Exp. Biol.* **105:** 241–251.

Tanouye M.A. and Wyman R.J. 1980. Motor outputs of giant nerve fiber in *Drosophila. J. Neurophysiol.* **44:** 405–421.

Trimarchi J.R. and Murphey R.K. 1997. The shaking-B mutation disrupts electrical synapses in a flight circuit in adult *Drosophila. J. Neurosci.* **17:** 4700–4710.

Trimarchi J.R. and Schneiderman A.M. 1994. The motor neurons innervating the direct flight muscles of *Drosophila melanogaster* are morphologically specialized. *J. Comp. Neurol.* **340:** 427–443.

Wright N.J.D. and Zhong Y. 1995. Characterization of K$^+$ currents and the cAMP-dependent modulation in cultured *Drosophila* mushroom body neurons identified by *lacZ* expression. *J. Neurosci.* **15:** 1025–1034.

Wu C.-F., Sakai M., and Hotta Y. 1990. Giant *Drosophila* neurons differentiated from cytokinesis-arrested embryonic neuroblasts. *J. Neurobiol.* **21:** 499–507.

Wu C.-F., Suzuki N., and Poo M.-M. 1983. Dissociated neurons from normal and mutant *Drosophila* larval central nervous system in cell culture. *J. Neurosci.* **3:** 1888–1899.

Zhao M.-L. and Wu C.-F. 1997. Alterations in frequency coding and activity dependence of excitability in cultured neurons of *Drosophila* memory mutants. *J. Neurosci.* **17:** 2187–2199.

# CONTENTS

# GFP and Other Reporters

**Tulle Hazelrigg**
*Department of Biological Sciences*
*602 Sherman Fairchild Center*
*Columbia University*
*New York, New York 10027*

Green fluorescent protein (GFP), encoded by the *Aequorea victoria gfp* gene, and β-galactosidase, encoded by the *Escherichia coli lacZ* gene, are popular reporters for studying the temporal and spatial patterns of gene expression in *Drosophila*. This chapter covers topics relevant to the use of each of these reporters, as well as CAT (bacterial chloramphenicol acetyltransferase) and firefly luciferase. Emphasis is placed on protocols designed for the detection of GFP and β-galactosidase by microscopy. In addition, quantitative protocols are presented for the detection of β-galactosidase, CAT, and luciferase.

## GREEN FLUORESCENT PROTEIN

GFP is a relatively small protein (~27 kD) and diffuses within the cell freely; when expressed in nerve cells, for instance, it readily fills and labels nerve processes (Brand 1995). When expressed as a fusion protein, GFP is an excellent tag for charting the subcellular localization of its fusion partner (Wang and Hazelrigg 1994). The most valuable attribute of GFP is that it can be imaged in live specimens, allowing observations by time-lapse video microscopy of movements and changes in the morphology of GFP-tagged cells (see Barthmaier and Fyrberg 1995; Yeh et al. 1995; Shiga et al. 1996; Edwards et al. 1997), as well as changes in the localization of GFP-tagged proteins (see Endow and Komma 1996, 1997; Theurkauf and Hazelrigg 1998).

## GFP: WHAT ARE ITS ADVANTAGES?

The ability to detect GFP is affected by several parameters, including the concentration of the protein, its site of localization within a cell, and the time required for formation of a fluorescent product (Hazelrigg 1998; Hazelrigg et al. 1998). In two studies, GFP and β-galactosidase were compared as reporters, by examining the expression patterns of GFP–β-galactosidase fusion proteins expressed under the control of a Gal4-inducible promoter (Shiga et al. 1996; Timmons et al. 1997). Timmons et al. (1997) characterized *gfp-lacZ* expression in more than 120 different Gal4 insertion backgrounds. Although both

GFP and β-galactosidase were detected in similar patterns in a broad range of tissues, several differences were noted. In some tissues, one or the other reporter was more readily detected. In general, β-galactosidase was a more sensitive reporter, because of the amplification of signal afforded by enzymatic detection, and β-galactosidase could often be detected earlier than GFP. However, these authors showed that the GFP$^{S65T}$ variant (see below) could alleviate these problems in GFP's sensitivity. A major advantage of GFP was, of course, the ability to detect its signal in living tissues.

## PROPERTIES OF GFP AND ITS VARIANTS

Wild-type GFP absorbs ultraviolet (UV) light (with a maximum peak at 395 nm) and blue light (with a lesser peak at 475 nm), and emits green light (maximally at 508 nm, with a shoulder emission at 540 nm). GFP variants possess shifted absorbance and emission spectra and may differ from wild-type GFP in other properties as well, such as the time required for chromophore maturation (for review, Tsien and Prasher 1998). Both wild-type and variant GFPs have been expressed successfully in *Drosophila* (Table 17.1).

**Table 17.1.** Examples of *Drosophila gfp* Transgenes

| Construct | Product | Reference |
| --- | --- | --- |
| p[Cas,NGE] | GFP-EXU | Wang and Hazelrigg (1994) |
| p[Cas,CGE] | EXU-GFP | Wang and Hazelrigg (1994) |
| pAct88F-GFP | GFP | Barthmaier and Fyrberg (1995) |
| UAS-GFP | GFP | Brand (1995) |
| UAS-*tau*-GFP | Tau-GFP | Brand (1995) |
| PUbnlsGFP | nlsGFP | Davis et al. (1995) |
| PUbGFP | GFP | Davis et al. (1995) |
| pMei-S332-GFP | GFP-MEI-S332 | Kerrebrock et al. (1995) |
| pUAS-GFP | GFP | Yeh et al. (1995) |
| α4 *tubulin-gfp-vasa* | GFP-Vasa | Breitweiser et al. (1996) |
| pCaSpeR/ncd-gfp | NCD-GFP | Endow and Komma (1996) |
| pCaSpeR/ncd-gfp* | NCD-GFP$^{S65T}$ | Endow and Komma (1996) |
| FI66GFP | GFP | Potter et al. (1996) |
| GMR-GFP | GFP | Plautz et al. (1996) |
| pUAS-*gfpn* | GFPN | Shiga et al. (1996) |
| pUAS-*gfpn-lacZ* | GFPN-β-galactosidase | Shiga et al. (1996) |
| pUAS-*gfpn-CAT* | GFPN-CAT | Shiga et al. (1996) |
| *hs-GFP-moe* | GFP-Moesin tail | Edwards et al. (1997) |
| α4 *tubulin-tau*-GFP | Tau-GFP$^{S65T/I167T}$ | Micklem et al. (1997) |
| UAS-GFP | GFP | Plautz et al. (1997a,b) |
| *dj*-EGFP | Don Juan-EGFP | Santel et al. (1997) |
| pUAST-β-gal/GFP | β-galactosidase-GFP | Timmons et al. (1997) |
| pUAST-GFP/βgal | GFP-β-galactosidase | Timmons et al. (1997) |
| UAS-*runt*-GFP | Runt-GFP$^{S65T/I167T}$ | Dormand and Brand (1998) |
| P[*gfp-bcd*] | GFP-Bicoid | Hazelrigg et al. (1998) |
| P[*hs-gfp-bcd*] | GFP-Bicoid | Hazelrigg et al. (1998) |
| *metchnikowin-gfp*$^{F64L}$ | GFP$^{F64L}$ | Levashina et al. (1998) |
| UAS-dj-EGFP | Don Juan-EGFP | Santel et al. (1998) |
| α4 *tubulin-mGFP6-stau* | mGFP6-Staufen | Schuldt et al. (1998) |
| pJM802, *drosomycin-gfp*$^{S65T}$ | GFP$^{S65T}$ | Ferrandon et al. (1998) |
| pJM804, *drosomycin-gfp*$^{S65T}$ | N-terminal Drosomycin-GFP$^{S65T}$ | Ferrandon et al. (1998) |
| UAS-*tau*-mGFP6 | Tau-GFP$^{m6}$ | Brand (1999) |
| UAS-*tau*-GFP S65T/I167T | Tau-GFP$^{S65T/I167T}$ | Brand (1999) |
| pB[Dmw,PUbnlsEGFP] | EGFP | Handler and Harrell (1999) |

One variant commonly used, GFP$^{S65T}$ (Heim et al. 1995), has properties that enhance its use for microscopy. Its excitation maximum is shifted to 488 nm, whereas its emission maximum remains close to wild type (511 nm), making it better suited to excitation with commonly used FITC (fluorescein isothiocyanate) filter sets and the 488 line of confocal microscopes equipped with a krypton/argon laser. This variant is brighter than wild-type GFP, and chromophore maturation occurs more rapidly. EGFP (enhanced GFP; Clontech) contains two mutations that increase its brightness and rate of chromophore formation, F64L and S65T (Cormack et al. 1996), as well as more than 190-bp changes to promote human codon usage. EGFP, which has an excitation maximum at 488 nm and an emission maximum at 507 nm, has been expressed in *Drosophila* (Santel et al. 1997, 1998; A. Handler, pers. comm.). Other variants that have been used in *Drosophila* include the GFP$^{F64L}$ variant (Levashina et al. 1998), GFP$^{S65T/I167T}$, and mGFP6 (Brand 1999). All of the variants just mentioned fluoresce green. The existence of variants that fluoresce other colors (see Tsien and Prasher 1998) makes possible double labeling in cells expressing both variants.

## VECTORS AND CONSTRUCTION OF *gfp* TRANSGENES

GFP has been expressed in *Drosophila* from a variety of promoters (Table 17.1; for review, see Brand 1995, 1999; Hazelrigg 1998; Endow 1999). Genes expressing GFP fusion proteins from the gene's natural promoter have been created by first mutating a target gene to contain appropriate restriction sites at the translation start or stop codons and inserting at these sites PCR-amplified *gfp* with these restriction sites appended. This is followed by cloning of the fusion gene into one of the available P-element transformation vectors (Wang and Hazelrigg 1994). In some cases, the P-element vector has supplied the promoter and polyadenylation signals, and the coding sequences for GFP or GFP fusion proteins have been cloned into available polylinker sites in these vectors. These vectors have provided promoter-specific patterns of GFP expression, including heat-shock-inducible expression from the *hsp70* promoter (Edwards et al. 1997; Hazelrigg et al. 1998) and maternal germ-line expression from the α4 *tubulin* promoter (Breitweiser et al. 1996; Micklem et al. 1997; Schuldt et al. 1998). Targeted expression to tissue types has been achieved with the Gal4 system (Brand and Perrimon 1993), using the PUAST transformation vector, which supplies an upstream-activated sequence (UAS)-regulated promoter and polyadenylation signal flanking a polylinker site for insertion of coding sequences (Brand 1995, 1999; Yeh et al. 1995; Shiga et al. 1996; Plautz et al. 1997a,b; Timmons et al. 1997; Dormand and Brand 1998; Santel et al. 1998).

## GFP BALANCERS

Balancer chromosomes expressing GFP are valuable research tools (Table 17.2). GFP acts as a genetic tag for the presence of the balancer, allowing the identification of homozygotes by virtue of their lack of GFP signal. For instance, in a heterozygous stock of genetic constitution m/GFP balancer (where m indicates a mutation of interest), m/m embryos can be identified by their lack of GFP fluorescence. The GFP balancers constructed in Thomas Kornberg's laboratory each bear a *Gal4* transgene expressed from the *Krüppel* (*Kr*) promoter, as well as GFP$^{S65T}$ (a bright variant) regulated by UAS elements (Casso et al. 1999). In animals bearing these balancers, GFP is expressed in a pattern reflecting the *Kr* promoter: GFP can first be detected at germ-band extension, as well as at later stages of embryogenesis, and in larvae, pupae, and adults. These balancers are available from the Bloomington Stock Center.

**Table 17.2.** GFP Balancer Chromosomes

*Kr-Gal4, UAS-GFP balancer stocks[a]*

    Df(1)JA27/FM7c, (P, w[+] KrGFP[10])

    Df(1)JA27/FM7c, (P, w[+] KrGFP[34])

    w ; L Pin/CyO, (P, w[+] KrGFP[19])

    w ; L Pin/CyO, (P, w[+] KrGFP[30])

    y w ; D gl[3]/TM3, Sb (P, w[+] KrGFP[4])

    y w ; D gl[3]/TM3, Sb (P, w[+] KrGFP[5])

*act5C-GFP balancer stocks[b]*

    FM7i-pAct-GFP: C(1)Dx, f/FM7, P[w+mC act::GFP = pActGFP]

    CyO-pAct-GFP: w; In(2LR)noc[4L],Sco[rv9R],b/In(2LR)CyO, P[w+mC act::GFP = pActGFP]

    TM3-pAct-GFP: w; Sb[1] / In(3LR)TM3 Ser,P[w+mC act::GFP = pActGFP]

These balancers are available from the Bloomington Stock center. A full description of these stocks, including all of the genetic markers present in each, is available from FlyBase. To access this information, go to the FlyBase main page at:

    http://flybase.bio.indiana.edu/.

Choose "Stocks," then select "Balancers search," and specify "GFP ('green balancers')" under "Carrying."

[a]The Kr-Gal4, UAS-GFP balancer stocks were constructed by D. Casso, F.-A. Ramirez-Weber, and T. Kornberg (Casso et al. 1999). Each balancer bears a *Gal4* transgene expressed from the *Krüppel* promoter, as well as a UAS-driven GFP$^{S65T}$ transgene.

[b]The act5C-GFP balancer stocks were constructed by Reichhart and Ferrandon (1998). Each bears an *actin5C*-driven GFP$^{S65T}$ transgene.

A second set of GFP balancers, constructed by J.M. Reichhart and D. Ferrandon (Reichhart and Ferrandon 1998), is also available from the Bloomington Stock Center. In these strains, GFP$^{S65T}$ expression is driven by the *actin5C* promoter, expressed both maternally and zygotically. The maternal expression complicates early scoring of nonbalancer embryos in heterozygous stocks. In the absence of maternal product, zygotic GFP expression is first detected at approximately 12 hours of development, and expression can be scored at later stages as well, including larvae and adults. A complete description of the GFP expression pattern is available at FlyBase (see legend to Table 17.2).

## GFP AS A MARKER FOR TRANSFORMATION

GFP has been used as an efficient marker for *Drosophila* transformation. In one set of experiments, *Drosophila* was transformed with a vector containing both a mini-*white* gene and a gene encoding an enhanced variant of GFP (EGFP) fused to a nuclear localization signal (nlsEGFP), expressed from the *Drosophila polyubiquitin* promoter (Handler and Harrell 1999). Transformants were screened either for the *white* marker in adults or by GFP fluorescence in larvae and pupae, using a fluorescent stereomicroscope (see below). These authors found that with these particular gene constructs, GFP was more efficiently detected than the *white* marker: In all cases where *white* expression was detected, GFP was also detected, but in many transformants GFP could be detected when *white* expression was not.

## MICROSCOPY

To augment the information presented here, see discussion by Endow and Piston (1998). Other valuable resources are Kiehart et al. (1994) and Davis (1999).

## Fluorescence and Confocal Microscopes

For confocal and fluorescence microscopy, microscope objectives should be chosen to maximize signal detection and image resolution. In general, objectives with a high numerical aperture (NA) should be used. Although Plan-Apochromat objectives are more highly corrected, Fluor or NeoFluar objectives may in some cases be preferable, because they contain fewer lens elements, allowing for a brighter image.

Many standard fluorescence microscope FITC filter sets are suitable for GFP use, and several filters have been designed specifically for GFP excitation and emission wavelengths. Some UV filter sets are also appropriate. The appropriate filter depends on the GFP variant being used; mutant GFPs differ from wild type in their excitation and emission spectra. Table 17.3 lists some commercially available epifluorescence FITC and DAPI (4′,6-diamidino-2-phenylindole) filter sets that are appropriate for wild-type GFP and the variants discussed in this chapter (see above and Table 17.1). Specific confocal filters are not listed here, due to the variable configurations and changing technology of confocal microscopes. Both the wild-type and variant GFPs discussed in this chapter can be viewed with confocal filters designed for FITC detection, utilizing excitation provided by the 488 line of a krypton/argon laser.

Filter sets must be chosen to maximize GFP signal while minimizing background signals from autofluorescence, the spectral characteristics of which will vary with different cell types. For instance, yolk in *Drosophila* oocytes fluoresces strongly when viewed with FITC filters. In our laboratory, we distinguish between yolk and GFP signals by using a long-pass emission filter instead of a band-pass filter; with the long-pass filter, the yolk appears yellow-green and can be distinguished from GFP's greener signal.

**Table 17.3.** Commonly Used Fluorescence Microscope Filter Sets for Imaging GFP

| Filter set | Excitation filter | Dichroic beam splitter | Emission filter |
|---|---|---|---|
| Zeiss 05 | BP 395–440 | 460 | LP 470 |
| Zeiss O9 FITC | BP 450–490 | 510 | LP 520 |
| Zeiss 10 FITC | BP 450–490 | 510 | BP 515–565 |
| Nikon V-2B | BP 380–425 | 430 | LP 460 |
| Nikon B-2A | BP 450–490 | 510 | LP 520 |
| Nikon B-2E | BP 450–490 | 510 | BP 520–560 |
| Leica K3 cube | BP 470–490 | 510 | LP 520 |
| Leica L4 cube | BP 470–490 | 510 | BP 515–560 |
| Chroma no. 31022 | BP 395+/–20 | 425 | BP 510 +/–20 |
| Chroma no. 41018 | BP 470+/–20 | 495 | LP 500 |
| Chroma no. 41017 | BP 470+/–20 | 495 | BP525+/–2 |
| Olympus U-MWIB | BP 460–490 | 505 | LP 515IF |
| Olympus U-MWIBA | BP 460–490 | 505 | BA 515–550 |

All of the filter sets can be used with wild-type GFP, which is excited by both UV (395-nm excitation maximum) and blue (475-nm excitation maximum) light. Both a band-pass (BP) emission filter and a long-pass (LP) emission filter are listed for the FITC filters (those with excitation wavelengths ≥450 nm). Several red-shifted green variants, including GFP$^{S65T}$, EGFP, and others, have their excitation maxima shifted to higher wavelengths, and the appropriate filter sets in the above list are the ones with excitation wavelengths ≥450 nm. Chroma (which also supplies the filters for several microscope companies) has designed filters tailored to the excitation and emission characteristics of several GFP variants, including GFPs that fluoresce blue, yellow, and cyan. A valuable resource about GFP filters is Chroma's GFP handbook, available at their website:

http://www.chroma.com/

Additional information can be obtained at their E-mail address (info@chroma.com).

Wild-type GFP is photoactivated upon exposure to UV or blue light (Shiga et al. 1996; see graphic demonstration in Endow 1999). It is believed that photoactivation of GFP results from isomerization of the molecule on exposure to excitation light, causing an increase in the 475-nm excitation peak (Cubitt et al. 1995). Thus, exposure to UV light can be used to increase brightness in response to blue excitation light (Chalfie et al. 1994). However, some commonly used variants of GFP, notably GFP[S65T], are not photoactivated.

## Dissecting Microscopes

GFP can be viewed with modular research stereomicroscopes equipped with recently developed epi-illuminators, available from Kramer Scientific (Elmsford, New York) and Micro Video Instruments (MVI; Avon, Massachusetts). These illuminators are designed to work with several existing brands of microscopes (the manufacturers offer guidelines for attaching the epi-illuminator); the filters are manufactured by Chroma Technology Corp. In our department, we have had success using both a Kramer GFP Fluorescence Illuminator attached to a Leica MZ microscope and an MVI STFL epi-illuminator attached to a Nikon SMZ microscope. Several microscope companies also sell stereomicroscopes already equipped with epi-illuminators, including Kramer, Leica, Nikon, Olympus, and Zeiss. The Kramer M$^2$ system has the advantage of allowing both macro- and microanalysis by providing stereo and high-power objectives on one instrument.

## Time-lapse Confocal Video Microscopy

GFP-tagged proteins can be analyzed in real time, in live cells, opening up many new, exciting applications. Analysis of live specimens can be done on either upright or inverted microscopes (Endow and Komma 1996, 1997; Theurkauf and Hazelrigg 1998). Confocal microscopes have software programs that allow time-series recordings, and imaging packages for CCD cameras also provide this capability. With the appropriate equipment, these images can be assembled and transferred to videotape. For a detailed protocol for time-lapse confocal imaging of *Drosophila* embryos, see Endow and Piston (1998).

# GFP PROTOCOLS

## Warnings and Cautionary Notes!

In GFP protocols, the following steps should be avoided, as they do *not* work:

- Do not seal coverslips with nail polishes, as many of these contain solvents that destroy GFP's fluorescence (Chalfie et al. 1994).
- Ethanol and methanol quench GFP's fluorescence (Ward 1998; Hazelrigg lab observations).
- Standard in situ hybridization protocols, involving hybridizations in formamide at 65ºC overnight, have been reported to quench GFP's fluorescence (Brand 1999; Hazelrigg lab observations).
- GFP is sensitive to both pH and temperature (Ward 1998).

## GFP Detection in Living Tissues: Preparation of Samples

Protocols 17.1–17.5 have been used successfully to image GFP in living tissues. Although each protocol is provided here in the context of specific tissues, in most cases, the protocol should also work for different tissue types. In some cases, alternative protocols are given for the same tissue (e.g., Protocols 17.2 and 17.3 are alternative protocols for imaging GFP in live embryos).

## GFP Detection in Fixed Tissues

Fixed embryos or tissues expressing GFP can also be labeled with antibodies to other proteins or with DNA stains or other compounds that bind specific cellular components, e.g., rhodamine-phalloidin to label F-actin (Wang and Hazelrigg 1994; Breitweiser et al. 1996; Edwards et al. 1997; Brand 1999). In our laboratory, we find that GFP's fluorescence is destroyed by extended exposure to methanol, so we omit steps that require methanol; but others report that short exposures to methanol are acceptable (Brand 1999). The methods given below (Protocols 17.6–17.8) all use formaldehyde fixation, but glutaraldehyde fixation also preserves GFP's fluorescence (Ward 1998).

PROTOCOL 17.1*

# Live Ovaries

This protocol was adapted from Theurkauf and Hazelrigg (1998).

## Materials

### Supplies and Equipment

$CO_2$ station
Microscope slides
Coverslips (22 × 40 mm and 18 mm$^2$)
Forceps (#5)
(*Optional*) Tungsten needle
Doublestick tape

### Solutions and Reagents

Halocarbon oil (#27, Sigma) or mineral oil

**CAUTION: $CO_2$** (see Appendix 4)

## Method

1. Collect and age females, and anesthetize with $CO_2$.

2. Place the females in a drop of halocarbon oil or mineral oil on a slide (for imaging on an upright microscope) or on a 22 × 40-mm coverslip (for imaging on an inverted microscope). Pull out the ovaries by grasping the tip of the abdomen with the forceps.

3. Tease apart the ovarioles with the forceps or a tungsten needle. Grasp the ovarioles at the germarium end and drag them along the surface of the slide or coverslip for well-separated egg chambers.

---

*For all protocols, $H_2O$ indicates glass distilled and deionized.

4. For analysis on an upright microscope, place small coverslips (18 mm²) on the slide at the sides of the oil to act as supports (secure with double-stick tape) and cover with a 20 × 40-mm coverslip. For analysis on an inverted microscope, place the coverslip on the microscope stand and proceed with imaging. Ovaries in this condition appear healthy for at least 1 hour, allowing time-lapse analysis.

PROTOCOL 17.2

## Live Embryos

This protocol was adapted from Davis et al. (1995) and Davis (1999).

### Materials

### *Media*

Egg collection plates

### *Supplies and Equipment*

Fine paintbrush for transferring embryos

Permeable membranes (Yellow Springs Instruments, Inc., Yellow Springs, Ohio; model 5793, standard membrane kit)

Custom-designed mounting slides. Use a razor blade to cut a square opening in a piece of colored tape on a coverslip (24 × 40-mm, #1.5). This opening acts as a well, and the sides of the tape provide support for an overlying permeable membrane. Coat the bottom of the well with heptane glue that is allowed to desiccate before embryos are applied to its surface.

### *Solutions and Reagents*

**Bleach** (50% Clorox) (dilute bleach 1:1 in $H_2O$)

$H_2O$ for washing embryos

Halocarbon oil series 95 (Halocarbon Products Corp., River Edge, New Jersey)

Heptane glue. Place a long strip of double-sided Scotch tape into a glass scintillation vial with 10–20 ml of **heptane** and shake overnight. Remove the heptane to a new vial. The glue strength will vary according to the amounts of tape and heptane used; thinner glue produces better imaging.

**CAUTION: bleach, heptane** (see Appendix 4)

### Method

1. Collect embryos on egg collection plates (for details, see Rothwell and Sullivan, this volume [Protocol 9.1]) and age appropriately.

2. Wash and dechorionate in 50% bleach (for details, see Rothwell and Sullivan, this volume [Protocol 9.2]).

3. Wash and blot dry.

4. Place the embryos in the center of the well of a custom-designed mounting slide. Cover with halocarbon oil. Place a permeable membrane over the embryos and the sides of the tape.

   *Note:* The membrane flattens and immobilizes the specimen but also allows oxygen transfer.

5. Image on an inverted microscope.

PROTOCOL 17.3

## Live Embryos

This protocol was adapted from Endow and Komma (1996) and Edwards et al. (1997).

### Materials

#### *Media*

Egg collection plates

#### *Supplies and Equipment*

Fine paintbrush for transferring embryos
Double-stick tape (Scotch 3MM)
Microscope slides and coverslips or Teflon-windowed chamber

#### *Solutions and Reagents*

(*Required*, if performing step 3a, below) halocarbon #27 oil (Sigma)
(*Required*, if performing step 3b, below) halocarbon oil mixture (a 1:1 mixture of halocarbon oils 27 and 700; Halocarbon Products Corp.)

> #### *$O_2$-saturated Halocarbon Oil (Optional, step 3a [below]; adapted from Endow and Piston 1998).*
>
> Use a size-E oxygen tank outfitted with a 0.25-7 liter min$^{-1}$ regulator. Attach Tygon tubing to the outlet nozzle. Insert a Pipetman tip at the end of the tubing and immerse in halocarbon oil in a 1.5-ml microcentrifuge tube. With the regulator opened at the lowest setting, bubble $O_2$ into the oil for about 30 seconds. Keep the tube tightly capped until the oil is used.

### Method

1. Collect eggs on egg collection plates (for details, see Rothwell and Sullivan, this volume [Protocol 9.1]) and age appropriately.

2. With a fine paintbrush, transfer the embryos to double-stick tape on a slide and gently roll to remove the chorions.

3. Use either a or b to prepare embryos for imaging.

    a. Transfer dechorionated embryos to a drop of halocarbon oil on a slide. Cover with a coverslip secured with layers of small strips of double-stick tape on both sides of the oil drop (Endow and Komma 1996).

       *Note:* The halocarbon oil can be saturated with oxygen (above) to help prevent anoxia of the embryos during long-term imaging.

    b. Alternatively, Edwards et al. (1997) used the following method for long-period observations. After dechorionation, cover embryos with halocarbon oil mixture in a Teflon-windowed chamber (for a description of how to construct this chamber, see Kiehart et al. 1994).

PROTOCOL 17.4

## Live Larval, Pupal, and Adult Tissues

This protocol is from Edwards et al. (1997). Intact larvae can be mounted live in halocarbon oil for observations of the whole animal (see Figure 9.5 in Hazelrigg 1998). For isolated tissues of postembryonic stages, the following protocol is appropriate.

### Materials

#### *Media*

Schneider's tissue culture medium (GIBCO)

> **EBR Medium**
> 130 mM NaCl
> 4.7 mM **KCl**
> 1.9 mM **CaCl$_2$**
> 10 mM HEPES (pH 6.9)

#### *Supplies and Equipment*

Forceps
Microscope slides
Vacuum grease
Coverslips
Filter paper

**CAUTION: KCl, CaCl$_2$** (see Appendix 4)

### Method

1. Dissect tissues in EBR Medium.

2. Mount on a slide in a drop of Schneider's medium enclosed by a circle of vacuum grease, to prevent dispersal.

3. Cover with a coverslip supported by a window frame of filter paper around the mounting medium. Leave a space between the paper and the medium to allow some oxygen to reach the sample.

PROTOCOL 17.5

## Using GFP Balancers to Genotype Embryos

This protocol was adapted from Casso et al. (1999) and is appropriate for genotyping embryos carrying the GFP balancers constructed in T. Kornberg's laboratory (see Table 17.2). GFP is also expressed in larvae, pupae, and adults bearing these balancers.

### Materials

#### *Media*

Egg collection plates

## Supplies and Equipment

Watch glasses (5-cm diameter wells). Fill each well with 1.2% (w/v) SeaKem agarose (FMC Corp.) dissolved in $H_2O$ and allow to solidify. Agarose has less autofluorescence than crude agar.

Dissecting stereomicroscope with epifluorescence illumination (see Microscopy above). The Kornberg laboratory uses the Leica MZ12 Fluorescent Stereo Microscope, equipped with a Chroma #41018 filter. For embryo observations, this long-pass filter enables differentiation of GFP from yolk fluorescence; GFP appears green against the yellow yolk.

## Solutions and Reagents

**Bleach** (50% Clorox) (dilute bleach 1:1 in $H_2O$)

**CAUTION: bleach** (see Appendix 4)

## Method

1. Collect embryos on egg collection plates (for details, see Rothwell and Sullivan, this volume [Protocol 9.1]) and age appropriately.

   *Note:* In the case of the KrGFP balancers, the embryos must be at the germ-band extension stage or older.

2. Dechorionate in 50% bleach (Clorox) and wash with $H_2O$ (for details, see Rothwell and Sullivan, this volume [Protocol 9.2]).

3. Place the embryos on agar beds in watch glasses.

4. Image with a fluorescent stereomicroscope.

---

PROTOCOL 17.6

# Fixed Ovaries

This protocol was adapted from Wang and Hazelrigg (1994). It can be used with dissected tissues of various types.

## Materials

### Media

Vials or bottles containing fly food supplemented with dry baker's yeast

## Supplies and Equipment

Forceps (#5)
Tungsten needle
Whatman no.1 filter paper, cut into small strips
Microscope slides and coverslips
$CO_2$ station

## Solutions and Reagents

PBS (phosphate-buffered saline; see Appendix 3)
Glycerol (60% v/v) in PBS

*Fixative A (From Theurkauf and Hawley 1992)*

8% **formaldehyde**
100 mM **potassium cacodylate** (pH 7.2)
100 mM sucrose
40 mM potassium acetate
10 mM sodium acetate
10 mM Na$_3$EGTA

*Fixative B*

Prepare 4% **paraformaldehyde** (w/v) in PBS. To dissolve paraformaldehyde, heat at 60°C in a chemical fume hood or sealed container (do not breathe fumes). Cool to room temperature before use. Prepare fresh daily.

CAUTION: CO$_2$, **formaldehyde, paraformaldehyde, potassium cacodylate**
(see Appendix 4)

## Method

1. Collect 0–1-day-old females and age for 2 days in vials or bottles containing fly food supplemented with dry baker's yeast, along with males to stimulate egg production.

   *Note:* The ovaries of 2–3-day-old females should contain egg chambers representing a range of developmental stages.

2. Anesthetize females with CO$_2$.

3. Dissect tissues in PBS.

4. Incubate in Fixative A or Fixative B for 10 minutes.

   *Note:* Fixative A is advised for cases requiring preservation of microtubule structures (Theurkauf et al. 1992).

5. Wash three times in PBS (10 minutes each).

6. Replace the PBS with 60% glycerol and allow the ovaries to equilibrate for at least 30 minutes.

7. Transfer in a drop of 60% glycerol to a slide and tease apart the ovarioles with forceps and a tungsten needle.

8. Apply a coverslip. To flatten the specimen, wick out excess mounting medium with strips of filter paper applied to the edges of the coverslip.

PROTOCOL 17.7

# Fixed Embryos

This protocol was adapted from Brand (1999). Antibody labeling of other proteins can be carried out after step 8.

## Materials

### Media

Egg collection plates

## Supplies and Equipment

Fine paintbrush for transferring embryos
Microcentrifuge tubes (1.5 ml)
Microscope slides
Coverslips (18 mm$^2$ and 22 x 40 mm)

## Solutions and Reagents

**Bleach** (50% Clorox) (diluted 1:1 with H$_2$O)
**Heptane**
**Methanol**
Vectashield mounting medium (Vector Laboratories)

### PBT

> PBS
> 0.1% Triton X-100

### Fixative

> 4% **paraformaldehyde**
> PBT
> Dissolve paraformaldehyde in PBT as described for Fixative B, Protocol 17.6.

**CAUTION: bleach, heptane, methanol, paraformaldehyde** (see Appendix 4)

## Method

1. Collect embryos on egg collection plates (for details, see Rothwell and Sullivan, this volume [Protocol 9.1]) and age appropriately.

2. Dechorionate in 50% Clorox bleach (for details, see Rothwell and Sullivan, this volume [Protocol 9.2]) for 3 minutes.

3. Wash with H$_2$O.

4. Transfer to a microcentrifuge tube containing 500 µl of heptane. Add 500 µl of fixative.

5. Fix for 30 minutes with constant gentle rocking.

6. Remove the aqueous (lower) layer and replace with methanol and shake vigorously for 30 seconds to remove the vitelline membranes.

   *Note:* Embryos that sink have their vitelline membranes removed; those that stay at the interface still have their vitelline membranes.

7. Remove as much of the heptane and methanol as possible, replace with fresh methanol, and wash rapidly (10 seconds). Keep methanol exposure to a minimum, as methanol will quench GFP's fluorescence.

8. Rehydrate in PBT.

   *Note:* Antibody staining can be done following this step.

9. Mount in Vectashield, using two small coverslips on the sides as supports, and a 22 x 40-mm coverslip to cover the embryos.

PROTOCOL 17.8

# Heat Shock Induction of GFP Expression

This protocol is from Hazelrigg et al. (1998). It was used to detect *hsp70*-driven GFP-Bcd in salivary glands, a polytene tissue. The method can be used for other tissues, but modifications to heat shock and recovery times may be required, especially for nonpolytene tissues, to compensate for lower levels of gene products. For a description of heat shock regimes for *hsp70*-driven expression of GFP-Moesin in several diploid tissues, see Edwards et al. (1997).

## Materials

### Media

Vials containing standard fly medium

### Supplies and Equipment

Water bath set at 37°C; ensure that the water level is just below the tops of the immersed
    vials (see step 1, below)
Forceps
Microscope slides and coverslips
Whatman no. 1 filter paper, cut into small strips

### Solutions and Reagents

PBS (see Appendix 3)
Fixative B (4% [w/v] **paraformaldehyde** dissolved in PBS; see Protocol 17.6)
Gycerol (60% v/v) in PBS

**CAUTION: paraformaldehyde** (see Appendix 4)

## Methods

1. Preheat vials by immersing them in a 37°C water bath until the temperature is equilibrated. Apply a weight to the tops of the vials to ensure full immersion of the food; take care not to allow water into the vials.

2. Place third-instar larvae in the heated vials, reimmerse in the water bath, and heat shock for 10 minutes.

3. Remove vials from the water bath and place them at 22–25°C to recover from heat shock.

   *Note:* In the case of an *hsp70-gfp-bcd* construct, 40 minutes recovery was sufficient for formation of a fluorescent GFP-Bcd product (Hazelrigg et al. 1998).

4. Dissect out the salivary glands in PBS (for details, see Kennison, this volume [Protocol 6.1]).

5. Place in Fixative B for 10 minutes.

6. Wash three times in PBS (10 minutes each).

7.  Replace PBS with 60% glycerol and allow the tissue to equilibrate for at least 30 minutes.

8.  Mount on a slide in 60% glycerol in PBS and cover with a coverslip. Flatten the specimen by wicking out excess mounting medium with filter paper strips applied to the edges of the coverslip.

## β-GALACTOSIDASE

β-galactosidase, the product of the *E. coli lacZ* gene, has seen extensive use as a reporter in *Drosophila* research. β-galactosidase catalyzes the hydrolosis of a variety of β-galactosides; this reaction can be assayed with appropriate substrates. Expressed from specific gene promoters, *lacZ* has been used to determine the spatial and temporal patterns of gene expression and to define *cis*-acting gene regulatory sequences (see, e.g., Hiromi et al. 1985; Glaser et al. 1986). Expressed from a minimal promoter in transgenics, *lacZ* has been a valuable reporter in "enhancer-trap" screens to identify genes expressed in specific cell types (O'Kane and Gehring 1987; Fasano and Kerridge 1988; Bier et al. 1989; Bellen et al. 1989; Grossniklaus et al. 1989; Wilson et al. 1989; Gönczy et al. 1992).

A large number of enhancer-trap transformant lines have been generated, as well as transformants expressing *lacZ* from defined promoters, providing an extensive collection of *Drosophila* stocks with specific patterns of *lacZ* expression. To search for such stocks in FlyBase, select "Genes" on the main page, and click on "Gene Expression Pattern Search" on the *Drosophila* Genes page. Specify "Reporters/Enhancers" after selecting a specific body part. In addition, many *UAS-lacZ* transgenes have been constructed for ectopic expression of *lacZ* using the Gal4 system, including UAS-*lacZ*, UAS-Nuclear-*lacZ*, UAS-*kinesin-lacZ*, UAS-*tau-lacZ*, and UAS-*gfp-lacZ* (for review, see Phelps and Brand 1998).

β-galactosidase fusion proteins are often localized like their endogenous counterparts. For example, fusions with the microtubule motor proteins Kinesin and Nod have been used to probe the polarity of microtubules in *Drosophila* cells (Clark et al. 1997). Although β-galactosidase by itself is cytoplasmic, fusion to a nuclear localization signal targets the protein to the nucleus (Noselli and Vincent 1991). In nerve cells, β-galactosidase remains in the cell body, but when fused to kinesin or the microtubule-binding protein Tau, the fusion proteins label nerve processes (Giniger et al. 1993; Callahan and Thomas 1994).

For microscopic analysis of *lacZ* gene expression, β-galactosidase can be detected histochemically by light microscopy, using a chromogenic substrate, or immunocytochemically, with labeled antibodies. β-galactosidase activity can also be measured quantitatively in tissue and cell culture homogenates (Lis et al. 1983; Lawson et al. 1984; Hiromi et al. 1985; Simon and Lis 1987; Ye et al. 1997). Fluorogenic substrates for β-galactosidase can be introduced into cells to fluorescently label *lacZ*-expressing cells, allowing the isolation of these cells from cell cultures and dissociated embryonic tissues (Krasnow et al. 1991; Minden 1996).

### Vectors and Construction of *lacZ* Transgenes

P-element vectors containing the *lacZ* reporter have been constructed by several laboratories (see Ashburner 1989a, chapter 33). These include vectors to accept promoters and their associated enhancer elements, with and without translation start codons provided by the vector (Kuhn et al. 1988; Molsberger et al. 1988; Thummel et al. 1988; Thummel and

Pirrotta 1992). In addition to driving expression from inserted promoter elements, some vectors are appropriate for creating β-galactosidase fusions with known *Drosophila* proteins. Maps of some of these vectors can be obtained at:

http://www-hhmi.genetics.utah.edu/thummel/pelement.html

## *lacZ* Balancers

Balancer chromosomes bearing *lacZ* genes facilitate the identification of homozygous mutant (nonbalancer) embryos. In FlyBase, sources and information about *lacZ* balancers can be found by choosing "Stocks" on the main page, clicking on "Balancers search" on the *Drosophila* Stocks page, and selecting "*lacZ* (blue balancers)" under "Carrying."

## β-Galactosidase Protocols

Histochemical methods for detecting β-galactosidase in fixed tissues are presented in Protocols 17.9–17.11. Although presented in the context of specific stages and tissues, each protocol can be adapted for different tissue types. These protocols all utilize the substrate X-gal (5-bromo-4-chloro-3-indolyl β-D-galactopyranoside), which produces a blue precipitate visible by light microscopy. Other substrates are available for assaying β-galactosidase activity, and β-galactosidase can also be detected by standard antibody-labeling methods, using commercially available antibodies.

## Warning: Endogenous β-galactosidase Activity

The patterns of endogenous *Drosophila* β-galactosidase activity in various tissues and developmental stages have been reported (Schnetzer and Tyler 1996). These results underscore the importance of including control tissues from wild-type *Drosophila*, to distinguish background endogenous β-galactosidase activity from that produced by expression of the transformed *E. coli lacZ* gene.

---

PROTOCOL 17.9

# Detection of β-galactosidase in Embryos

---

This protocol was adapted from Su et al. (1998) and can be used in conjunction with antibody staining or in situ hybridization. Embryos are first stained for β-galactosidase activity, and subsequently devitellinized before being labeled with antibodies or probes.

## Materials

### *Media*

Egg collection plates

### *Supplies and Equipment*

Scintillation vials (20 ml)
Fine paintbrush for transferring embryos
Microscope slides
Coverslips

## Solutions and Reagents

**Bleach** (30%) in $H_2O$
*n*-**Heptane**
**Methanol**
Absolute **ethanol**
$H_2O$
Permount (Fisher)

### NaCl-Triton Solution

0.7% NaCl
0.04% Triton X-100

### Fixative

0.1 M sodium phosphate (pH 7.5)
4% **formaldehyde**

### X-gal Staining Solution

10 mM **sodium phosphate** (pH 7.2)
150 mM NaCl
1 mM **MgCl$_2$**
3 mM **K$_4$[Fe(CN)$_6$]**
3 mM **K$_3$[Fe(CN)$_6$]**
0.3% Triton X-100

### X-gal Stock

10% **X-gal** in **DMF** (*N,N*-dimethylformamide)

**CAUTION: bleach, DMF, ethanol, heptane, formaldehyde, K$_4$[Fe(CN)$_6$], K$_3$[Fe(CN)$_6$], methanol, MgCl$_2$, sodium phosphate, X-gal** (see Appendix 4)

## Method

1. Collect embryos on egg collection plates (see Rothwell and Sullivan, this volume [Protocol 9.1]) and age appropriately.

2. Dechorionate in 30% bleach for 1–2 minutes.

3. Wash with NaCl-Triton Solution.

4. Wash briefly with $H_2O$ and transfer to a 20-ml scintillation vial containing 5 ml of Fixative and 5 ml of *n*-heptane. Fix, with shaking (350 rpm on a rotary shaker), at room temperature for 20 minutes.

5. Use a pasteur pipette to remove the fixative.

6. Wash with $H_2O$.

7. Remove the $H_2O$, and then all of the *n*-heptane with an aspirator. The embryos will stick to the sides of the vial.

   *Note:* It is critical to remove all of the *n*-heptane, because it inhibits β-galactosidase activity. Residual *n*-heptane can be removed by evaporation with forced air.

8. Add 10 ml of NaCl-Triton, rehydrate the embryos for 5 minutes, and wash once more with fresh NaCl-Triton.

9. Remove the NaCl-Triton with a pasteur pipette and replace with 1 ml of X-gal Staining Solution plus 20 µl of X-gal Stock.

10. Incubate at 37°C.

    *Note:* The incubation time must be determined empirically. It can range from several minutes to overnight, depending on the level of gene expression.

11. To devitellinize, remove the staining solution and add 5 ml of *n*-heptane, followed by 5 ml of methanol. Vortex for 20 seconds.

    *Note:* Devitellinized embryos sink to the bottom; embryos that retain their vitelline envelopes remain at the interface.

12. Use an aspirator to remove the *n*-heptane and methanol, including embryos that remain at the interface.

13. Wash with absolute ethanol.

    *Note:* At this point, the embryos can be stored in ethanol for extended periods.

14. Mount on a slide in Permount, cover with a coverslip, and seal the edges.

15. Alternatively, if antibody labeling or in situ hybridization is to be done, rehydrate the embryos and proceed with these methods.

## PROTOCOL 17.10

# Detection of β-galactosidase in Adult Testes

This protocol was modified from Gönczy et al. (1992) and was used to examine the expression patterns in the testes of enhancer-trap insertion lines. These investigators typically stained the testes for 12–16 hours; optimal staining times vary depending on levels of *lacZ* gene expression. This method should also be suitable for staining of dissected larval structures, including imaginal discs and salivary glands, and other adult structures.

### Materials

### *Supplies and Equipment*

Forceps
Microscope slides and coverslips
Nail polish

### *Solutions and Reagents*

X-gal stock (10% **X-gal** in **DMF**)
**Ethanol** series, diluted in PBS (50%, 70%, 90%)
Absolute ethanol
Canada Balsam:**methyl salicylate** (2:1), or other mounting medium of choice

*Drosophila Ringer's Solution (from Ashburner 1989b)*

182 mM **KCl**
46 mM NaCl
3 mM **CaCl$_2$**
10 mM Tris-HCl

*Glutaraldehyde Fixative*

1% **glutaraldehyde**
50 mM **sodium cacodylate**

*Prestaining Buffer*

7.2 mM **Na$_2$HPO$_4$**
2.8 mM **NaH$_2$PO$_4$**
1 mM **MgCl$_2$**
0.15 M NaCl

*Staining Buffer*

7.2 mM Na$_2$HPO$_4$
2.8 mM NaH$_2$PO$_4$
1 mM MgCl$_2$
0.15 M NaCl
5 mM **K$_4$[Fe(CN)$_6$]**
5 mM **K$_3$[Fe(CN)$_6$]**

*0.2% X-gal Staining Solution*

Dilute the X-gal stock in staining buffer.

*PBS/EDTA*

1 mM EDTA
PBS

**CAUTION:** CaCl$_2$, DMF, ethanol, glutaraldehyde, KCl, K$_3$[Fe(CN)$_6$], K$_4$[Fe(CN)$_6$], methyl salicylate, MgCl$_2$, Na$_2$HPO$_4$, NaH$_2$PO$_4$, sodium cacodylate, X-gal (see Appendix 4)

## Method

1. Dissect testes from 1-day-old adult males in *Drosophila* Ringer's Solution.

2. Fix in Glutaraldehyde Fixative for 15 minutes.

3. Wash three times with Prestaining Buffer.

4. Incubate in Prestaining Buffer at room temperature for 30 minutes.

5. Incubate in 0.2% X-gal Staining Solution at 37ºC.

   *Note:* Incubation times will vary with staining intensity and can range from minutes to many hours.

6. Wash three times in PBS/EDTA.

7. Dehydrate in increasing concentrations of ethanol in PBS, e.g., 50% ethanol, 70% ethanol, 90% ethanol, and absolute ethanol.

8. Mount under a coverslip in 2:1 Canada Balsam:methyl salicylate, or other appropriate mounting medium. Seal the edges with nail polish.

## PROTOCOL 17.11

# Detection of β-galactosidase in Adult Ovaries

This protocol was modified from Fasano and Kerridge (1988) and is also suitable for detecting β-galactosidase in other dissected adult, pupal, and larval tissues.

## Materials

### Supplies and Equipment

Forceps
(*Optional*) Tungsten needle
Microscope slides and coverslips
Whatman no. 1 filter paper, cut into strips
Nail polish

### Solutions and Reagents

*Drosophila* Ringer's Solution (see p. 331) or PBS
Glycerol (60% v/v) in PBS
X-gal stock (10% **X-gal** in **DMF**)

### Glutaraldehyde Fixative
2.5% **glutaraldehyde**
50 mM PIPES (pH 7.5).

### Staining Buffer
PBS
5 mM $K_4[Fe(CN)_6]$
5 mM $K_3[Fe(CN)_6]$

### 0.2% X-gal Staining Solution
Dilute the X-gal stock in staining buffer. Use as a freshly prepared solution, adding the X-gal fresh each time from a stock solution of 10% X-gal in DMF.

CAUTION: DMF, glutaraldehyde, $K_3[Fe(CN)_6]$, $K_4[Fe(CN)_6]$, X-gal (see Appendix 4)

## Method

1. Dissect ovaries out of adult females, in *Drosophila* Ringer's Solution or PBS. Separate ovarioles slightly (with a tungsten needle or forceps), but maintain connections to the ovaries.

   *Note:* To obtain egg chambers representative of all stages of oogenesis, it is best to use 2–3-day-old females that have been kept in well-yeasted vials with males to encourage continuous egg production.

2. Fix in Glutaraldehyde Fixative for 10 minutes.

3. Wash three times in PBS.

4. Incubate in 0.2% X-gal Staining Solution at 37ºC.

   *Note:* The timing will vary from several minutes to many hours, depending on staining intensity.

5. Wash three times in PBS.

6. Replace the PBS with 60% glycerol and allow the ovaries to equilibrate for at least 30 minutes.

7. Transfer to a drop of 60% glycerol on a slide and tease apart the ovarioles with forceps and a tungsten needle.

8. Apply a coverslip. To flatten the specimen, wick out excess mounting medium with strips of filter paper applied to the edges of the coverslip.

## PROTOCOL 17.12

# Quantitative β-galactosidase Activity Assay

This protocol was adapted from Lis et al. (1983).

## Materials

### Supplies and Equipment

Eppendorf microcentrifuge tubes (1.5 ml)
Microcentrifuge tube tissue homogenizers (plastic pestles to fit snugly inside microcentrifuge tubes)

### Solutions and Reagents

*Assay Buffer*
   50 mM **potassium phosphate**
   1 mM **MgCl₂**
   Adjust to pH 7.5.

*CPRG (Chlorophenol Red β-ᴅ-galactopyranoside) Solution*
   Prepare 1 mM CPRG in assay buffer.

CAUTION: MgCl₂, potassium phosphate (see Appendix 4)

## Method

1. Homogenize five adult flies in a microcentrifuge tube containing 100 μl of Assay Buffer, with approximately 20 strokes of the pestle.

2. Add 900 μl of Assay Buffer.

3. Mix by vortexing for 30 seconds.

4. Transfer an aliquot of the crude extract to a cuvette containing 1 or 2 ml of CPRG Solution.

5. Incubate at 37°C.

6. Read OD at 574 nm at intervals over a 2-hour time period, and calculate the β-galactosidase activity as a function of change in OD. To control for endogenous β-galactosidase activity, prepare parallel samples from control flies not expressing *E. coli lacZ*, and subtract the obtained control values from the experimental values.

## CAT AND LUCIFERASE

The bacterial CAT enzyme catalyzes the transfer of acetyl groups to chloramphenicol from acetyl coenzyme A (acetyl CoA). In a typical assay, this reaction is monitored with [14]C-labeled chloramphenicol: After separation by thin-layer chromatography (TLC), the acetylated and nonacetylated forms can be distinguished by autoradiography, and quantitation is achieved by isolating the forms and measuring their radioactivity in a scintillation counter. Quantitative CAT assays have been performed on *Drosophila* tissue culture and dissociated cell extracts (Di Nocera and Dawid 1983; Benyajati and Dray 1984; Thummel et al. 1988; Krasnow et al. 1989; Ye et al. 1997). CAT can also be detected with commercially available antibodies. In addition, a nonradioactive CAT assay exists that utilizes a fluorescent chloramphenicol derivative (Molecular Probes).

CAT is a very stable protein (its half-life has been estimated at 50 hours in mammalian cells; Thompson et al. 1991). This stability leads to perdurance of the protein, making it a problematic reporter for determining temporal changes in gene expression. However, stability is a desirable trait in situations where temporal changes are not being monitored.

Firefly luciferase catalyzes a bioluminescent reaction in which the substrate luciferin is oxidized, with light as a by-product. The light from this reaction can be measured quantitatively with a luminometer or a scintillation counter. The bioluminescence of luciferase has been measured in *Drosophila* tissue culture extracts (see Van Ohlen et al. 1997; Sawado et al. 1998) and in whole flies and isolated body parts (Brandes et al. 1996; Emery et al. 1997; Stanewsky et al. 1997, 1998; Plautz et al. 1997a,b). Luciferase has been a very valuable reporter for measuring circadian rhythms in living flies and isolated body parts.

Luciferase has a relatively short half-life, measured to be approximately 3 hours in mammalian cells (Thompson et al. 1991) and approximately 4 hours in flies (Plautz et al. 1997a). The relatively short half-life of luciferase is advantageous and necessary for certain situations where it has been used, such as the recording of 24-hour circadian rhythms.

### Vectors and Construction of Transgenes

Several vectors for expressing CAT in transformed flies and tissue culture cells have been constructed (see Ashburner 1989a, chapter 33). For primary references, see Mismer and Rubin (1987) and Thummel et al. (1988). A number of vectors for expressing firefly luciferase are available, including vectors that express *luc* from the *per, CRE, hsp70,* and *tim* promoters, and a UAS-responsive promoter (Lockett et al. 1992; Brandes et al. 1996; Emery et al. 1997; Plautz et al. 1997a,b; Stanewsky et al. 1997, 1998).

### CAT and Luciferase Protocols

Protocol 17.13 describes a quantitative CAT assay procedure using [14]C-labeled chloramphenicol. Protocol 17.14 has been used to monitor the circadian rhythms over periods of several days of *per-luc* gene expression in intact flies and isolated body parts (Plautz et al. 1997a,b; Stanewsky et al. 1997).

PROTOCOL 17.13

## CAT Assay on Cell Extracts

This protocol was adapted from Thummel et al. (1988), Ashburner (1989b; Protocol 134), and Ausubel et al. (1998; Protocol 1, unit 9.7).

### Materials

#### Samples

Harvested, washed tissue culture cells, transformed flies, or dissected tissues, expressing CAT

#### Supplies and Equipment

Microcentrifuge tubes (1.5 ml)
Microcentrifuge tube tissue homogenizers (plastic pestles to fit snugly in the microcentrifuge tubes)
**Dry ice/methanol** bath
Sonicator
Water bath (or heating block) at 65°C
Microcentrifuge maintained at 4°C
TLC sheets (plastic-backed, silica gel 1B; J.T. Baker)
TLC tank
Whatman 3MM filter paper
X-ray film
Film cassette

#### Solutions and Reagents

Tris-Cl (0.25 M, pH 7.8)
Bradford assay reagent
Acetyl CoA (4 mM)
Tris-Cl (1 M, pH 7.5)
[$^{14}$C]**chloramphenicol** (55 mCi/mmole)
Ethyl acetate
**Chloroform**:methanol (19:1)

> **CAUTION: chloramphenicol, chloroform, dry ice, methanol, radioactive substances** (see Appendix 4)

### Method

1. Treat samples as follows:

   a. *Whole flies or dissected tissues:* Add 100 μl of 0.25 M Tris (pH 7.8) to a 1.5-ml microcentrifuge tube. Transfer whole flies or dissected tissues to the microcentrifuge tube and homogenize with a plastic pestle.

   b. *Tissue culture cells:* Resuspend washed cells in 100 μl of 0.25 M Tris (pH 7.8).

2. Lyse cells (from step 1a *or* 1b) as follows:

   a. Freeze in a dry ice/methanol bath for 5 minutes, and then thaw at 37°C for 5 minutes.

   b. Sonicate in a cup sonicator for 20 seconds at full power.

      *Note:* Alternatively, use repeated freeze-thaw cycles to lyse cells: Freeze in dry/ice methanol for 5 minutes and then thaw at 37°C for 5 minutes; repeat twice for a total of three freeze-thaw cycles.

3. Heat the extracts at 65°C for 5 minutes.

4. Centrifuge in a microcentrifuge at 4°C for 10 minutes.

5. Transfer the supernatant to a new tube.

6. Determine the protein concentrations of the extracts using the Bradford assay.

7. In a microcentrifuge tube, set up the following reactions:

   | | |
   |---|---|
   | cell extract (adjust quantity; see note below) | 20 µl |
   | cocktail | 130 µl |

   Prepare the cocktail (per sample) as follows:

   | | |
   |---|---|
   | [$^{14}$C]chloramphenicol | 0.2 µCi |
   | 4 mM acetyl CoA | 20 µl |
   | 1 M Tris-Cl (pH 7.5) | 32.5 µl |
   | H$_2$O | to make 130 µl |

   *Note:* Different quantities of cell extract will have to be assayed: Adjust the volume of the cocktail so that the final concentration of Tris-Cl is 0.25 M, and the total volume is 150 µl. Aim for between 5% and 50% conversion to the acetylated form to be in the linear range of the assay.

8. Incubate at 37°C for 15 minutes to 2 hours.

   *Note:* The length of incubation time must be determined empirically.

9. Add 1 ml of ethyl acetate and mix by vortexing.

   *Note:* The nonacetylated chloramphenicol (CAM) and acetylated chloramphenicol (acetyl-CAM) are partitioned into the ethyl acetate.

10. Centrifuge in a microcentrifuge at room temperature for 1 minute.

11. Remove the top phase and dry in a microcentrifuge tube overnight in a hood, or for approximately 45 minutes in a rotary evaporator.

12. Resuspend in 15 µl of ethyl acetate.

13. Spot onto TLC sheets (5-µl spots), and chromatograph for 1.5–2 hours, in chloroform:methanol (the solvent should be ~2–3 cm from the top).

14. Air-dry and autoradiograph with X-ray film.

15. To quantify, perform the following:

    a. Cut spots from the TLC sheet and place each spot in a scintillation vial containing scintillation fluid.

       *Note:* Four spots should be present in each lane: a spot at the origin, followed by the lowest-mobility spot, CAM, followed by two forms of mono-acetyl-CAM. If five spots are present due to the presence of diacetylated chloramphenicol (the highest-mobility form), the reaction is not in the linear range and should be repeated with more dilute samples (step 7, above) or reduced reaction time (see step 8, above; Ausubel et al. 1998).

b. Count in a scintillation counter.

c. Calculate the percentage of acetylated choramphenicol.

% acetylated = counts in acetylated species/total counts in
acetylated + nonacetylated species.

# Detection of Luciferase in Intact Flies and Cultured Body Parts

This protocol was adapted from Plautz et al. (1997a,b), Stanewsky et al. (1997), and information provided by Jeff Hall.

## Materials

### Samples

Transgenic flies expressing firefly luciferase (*luc*)

### Media

#### Luciferin Fly Food Medium (for Recording from Intact Flies)

1% agar

5% sucrose

15 mM luciferin (D-luciferin firefly potassium salt, Biosynth)

#### Luciferin Tissue Culture Medium (for Recording from Isolated Tissues or Body Parts; Percentages Are v/v)

85.9% Shields and Sang M3 Insect Tissue Culture Medium (Sigma)

12% fetal bovine serum (Sigma), heat inactivated just before use at 60°C for 30 minutes

1% antibiotic mixture (available from Sigma; 5000 units of penicillin, 5 mg of streptomycin, 10 mg of neomycin/ml)

1% luciferin stock solution (100 mM stock solution; final concentration is 1 mM)

0.1% insulin stock solution (1 mg/ml stock solution, available from Sigma)

*Note:* Care must be taken to promote sterile conditions. Prepare the medium in a chemical fume hood and sterilize sample plates, dissecting forceps, etc., with ethanol. Microbial growth during days of recording, as is required for circadian analysis, can pose a serious problem.

#### Preparation of Microtiter Plates

White or black 96-well microtiter plates can be used (see below). Depending on the sample, plates are prepared as follows.

*For single fly recordings:* Place 100 μl of Luciferin Fly Food Medium in every other well of a 96-well microtiter plate. By using every other well, signal contamination from adjacent wells is avoided.

*For multiple fly recordings:* Place 250 μl of 1% agar, topped off with 100 μl of Luciferin Fly Food Medium, in each of 48 wells.

*For recording from isolated tissues or body parts:* Place 100 μl of Luciferin Tissue Culture Medium in each of 48 wells.

### Supplies and Equipment

$CO_2$ station

Microtiter plates (96 well). Use either white (Optiplate, Packard) *or* black (Micro-FLUOR, Dynex Technologies) 96-well microtiter plates. White plates produce more efficient signals, i.e., more light reaches the detector. Black plates reduce signal leakage (cross-talk) between wells.

(*Optional*) Forceps

Plastic domes (MicroAmp Caps; PCR covers, eight caps per strip; Perkin Elmer) trimmed to fit into microtiter wells. Use a small syringe needle to poke two holes in each dome top. Use an 18-gauge × 1.5-inch sterile, single-use, nontoxic, nonpyrogenic needle (the Hall lab uses a Terumo needle, Terumo Medical Corp.). This gauge and size are strong enough to penetrate the dome, leaving a hole large enough to promote air flow, yet small enough so that flies do not escape.

Microtiter press-on adhesive sealing film (TopSeal-A for 96-well plates, Packard). Two types of rectangular films can be prepared for experiments on intact flies, depending on whether domed or nondomed wells are used (see step 2, below). Prepare the TopSeal film as follows:

- *For domed plates:* Remove the backing from the film and place each film (sticky-side down) over an empty plate, which serves as a template. Poke two small holes over the positions of each of the relevant 48 wells using the needle described above.
- *For nondomed plates:* Stick plastic discs (cut out of overhead transparency film with a hole-puncher) on the adhesive side of the TopSeal film, in positions corresponding to the relevant 48 wells. By applying the TopSeal film first to an empty plate, the positions of these wells can be determined by the impressions left in the film. These plastic discs keep the flies from sticking to the TopSeal film. Place the film with plastic discs over a supportive surface (such as a piece of cardboard), and poke two holes over each relevant well space, such that the holes go through the film and the disc remains adhered to it.

Razor blade

(*Optional*) Parafilm

Clear spacer plates; microtiter styrene flat-bottom plates (Dynex 9205 or Polyfiltronics 7701-9300)

Bioluminescence/scintillation counter

### Solutions and Reagents

(*Optional*) Sterile $H_2O$
(*Optional*) **Ethanol** (95%)

**CAUTION: $CO_2$, ethanol** (see Appendix 4)

### Method

1. Anesthetize flies.

2. Treat samples as follows:

   a. *To record from intact single flies:* Place one fly in each alternating well of a microtiter plate (containing 100 μl/well of Luciferin Fly Food Medium), and cover with plastic domes. The domes restrict movement of the flies relative to the photodetector. Cover with a TopSeal film for domed plates, and use a razor to trim off excess TopSeal that extends beyond the plate.

b. *To record from multiple flies:* If recordings from multiple flies are desirable, e.g., for mutant screens, where less analytical precision is allowed, use nondomed plates. Place two to three flies in alternating wells (containing 250 μl/well of 1% agar, topped off with 100 μl of Luciferin Fly Food Medium). Cover with a TopSeal film for nondomed plates, doing so row-by-row (to prevent escapers), as the wells are filled. Trim off excess TopSeal that extends over the side of the plate.

c. *To record from isolated body parts:* Dip the anesthetized fly briefly in sterile $H_2O$ and then in 95% ethanol. Dissect out the tissue, or separate the relevant appendage, in sterile Luciferin Tissue Culture Medium. Place the specimen (or >1) in alternating wells (containing 100 μl of Luciferin Tissue Culture Medium). Cover with TopSeal film, with no holes, and trim off excess TopSeal. Take care to minimize microbial contamination. Cover with TopSeal as each well is filled, and put Parafilm over medium-containing wells until they are filled and covered. Alternatively, load the body parts into the wells in a laminar flow hood.

3. Place the sample plates separated by clear spacer plates (and a "stop" plate; see manufacturer's instructions) in the scintillation counter. Several laboratories (J. Hall, S. Kay, P. Hardin, R. Stanewsky, A. Sehgal) use a Packard TopCount bioluminescence/scintillation counter.

> *Note:* The spacer plates provide light access to the sample plates, allowing light/dark cycle monitoring. To avoid jamming during recording, use spacers that fit exactly with the test plates. Some laboratories use custom-made light banks flanking the TopCount stacker, to maximize light input during alternating light/dark conditions.

4. Enter instructions for recording (using software provided by Packard), e.g., which wells are to be read, time-point information (number of recordings per hour, duration of each time point). The overall duration of the run need not be entered at the beginning, since the run must be stopped at the appropriate time, depending on the experiment. For circadian analysis, set the recording instructions for the machine so that a given sample plate is read at the same time point (or points) within a given hour. Program the machine to read every other plate (given the spacer plates), and every other well of each plate (see above).

5. *Data handling and analysis:* Upon completion of a run on the TopCount, compress the data, copy to a disc, and transfer to a Windows-based PC for processing.

> *Note:* Data processing includes plotting time points of each well, averaging information between wells, etc. The Hall lab uses the Import & Analysis (I & A) macro set for Microsoft Excel to facilitate data handling. For details concerning the quantitative analysis of this type of periodic data, see Stanewsky et al. (1997) and Plautz et al. (1997b). Both an old and a newer version of I & A can be found at http://highwire.stanford.edu/~jplautz/.

## ACKNOWLEDGMENTS

I am thankful to several people who generously provided their time and help during the preparation of this chapter. Kathy Matthews provided helpful information about *gfp*- and *lacZ*-bearing *Drosophila* stocks and guidance on how to access this information through FlyBase. Jeff Hall provided extensive information and advice for the luciferase protocol. Several people critiqued protocols and offered advice for other sections, and in some cases provided unpublished information, including Andrea Brand, David Casso, Al Handler, Dan Kiehart, Tom Kornberg, Jennifer Mansfield, and Carl Thummel. Bill Sullivan was a helpful editor, and provided useful suggestions for improvements.

## REFERENCES

Ashburner M. 1989a. Drosophila: *A laboratory handbook.* Cold Spring Harbor Laboratory Press, Cold Spring Harbor, New York.

———. 1989b. Drosophila: *A laboratory manual.* Cold Spring Harbor Laboratory Press, Cold Spring Harbor, New York.

Ausubel F.M., Brent R., Kingston R.E., Moore D.D., Seidman J.G., Smith J.A., and Struhl K. 1998. *Current protocols in molecular biology.* John Wiley and Sons, New York.

Barthmaier P. and Fyrberg E. 1995. Monitoring development and pathology of *Drosophila* indirect flight muscles using green fluorescent protein. *Dev. Biol.* **169:** 770–774.

Bellen H.J., O'Kane C.J., Wilson C., Grossniklaus U., Pearson R.K., and Gehring W.J. 1989. P-element-mediated enhancer detection: A versatile method to study development in *Drosophila. Genes Dev.* **3:** 1288–1300.

Benyajati C. and Dray J.F. 1984. Cloned *Drosophila* alcohol dehydrogenase genes are correctly expressed after transfection into *Drosophila* cells in culture. *Proc. Natl. Acad. Sci.* **81:** 1701–1705.

Bier E., Vaessin H., Shepherd S., Lee K., McCall K., Barbel S., Ackerman L., Carretto R., Uemura T., Grell E., et al. 1989. Searching for pattern and mutation in the *Drosophila* genome with a P-lacZ vector. *Genes Dev.* **4:** 1273–1287.

Brand A. 1995. GFP in *Drosophila. Trends Genet.* **11:** 324–325.

———. 1999. GFP as a cell and developmental marker in the *Drosophila* nervous system. *Methods Cell Biol.* **58:** 165–181.

Brand A.H. and Perrimon N. 1993. Targeted gene expression as a means of altering cell fates and generating dominant phenotypes. *Development* **118:** 401–415.

Brandes C., Plautz J.D., Stanewsky R., Jamison C.F., Straume M., Wood K.V., Kay S.A., and Hall J.C. 1996. Novel features of *Drosophila period* transcription revealed by real-time luciferase reporting. *Neuron* **16:** 687-692.

Breitwieser W., Markussen F.H., Horstmann H., and Ephrussi A. 1996. Oskar protein interaction with Vasa represents an essential step in polar granule assembly. *Genes Dev.* **10:** 2179–2188.

Callahan C.A. and Thomas J.B. 1994. Tau-β-galactosidase, an axon-targeted fusion protein. *Proc. Natl. Acad. Sci.* **91:** 5972–5976.

Casso D., Ramírez-Weber F.-A., and Kornberg T.B. 1999. GFP-tagged balancer chromosomes for *Drosophila melanogaster. Mech. Dev.* **88:** 229–232.

Chalfie M., Tu Y., Euskirchen G., Ward W.W., and Prasher D.C. 1994. Green fluorescent protein as a marker for gene expression. *Science* **263:** 802–805.

Clark I.E., Jan L.Y., and Jan Y.N. 1997. Reciprocal localization of Nod and kinesin fusion proteins indicates microtubule polarity in the *Drosophila* oocyte, epithelium, neuron and muscle. *Development* **124:** 461–470.

Cormack B.P., Valdivia R.H., and Falkow S. 1996. FACS-optimized mutants of the green fluorescent protein (GFP). *Gene* **173:** 33–38.

Cubitt A.B., Heim R., Adams S.R., Boyd A.E., Gross L.A., and Tsien R.Y. 1995. Understanding, improving and using green fluorescent proteins. *Trends Biochem. Sci.* **20:** 448–455.

Davis I. 1999. Visualizing fluorescence in *Drosophila*—Optimal detection in thick specimens. In *Protein localization by fluorescence microscopy: A practical approach* (ed. V.J. Allan). Oxford University Press, United Kingdom.

Davis I., Girdham C.H., and O'Farrell P.H. 1995. A nuclear GFP that marks nuclei in living *Drosophila* embryos; maternal supply overcomes a delay in the appearance of zygotic fluorescence. *Dev. Biol.* **170:** 726–729.

Di Nocera P.P. and Dawid I.B. 1983. Transient expression of genes introduced into cultured cells of *Drosophila. Proc. Natl. Acad. Sci.* **80:** 7095–7098.

Dormand E.L. and Brand A.H. 1998. Runt determines cell fates in the *Drosophila* embryonic CNS. *Development* **125:** 1659–1667.

Edwards K.A., Demsky M., Montague R.A., Weymouth N., and Kiehart D.P. 1997. GFP-Moesin illuminates actin cytoskeleton dynamics in living tissue and demonstrates cell shape changes during morphogenesis in *Drosophila. Dev. Biol.* **191:** 103–117.

Emery I.F., Noveral J.M., Jamison C.F., and Siwicki K.K. 1997. Rhythms of *Drosophila period* gene expression in culture. *Proc. Natl. Acad. Sci.* **94:** 4092–4096.

Endow S.A. 1999. GFP fusions to a microtubule motor protein to visualize meiotic and mitotic spindle dynamics in *Drosophila. Methods Cell Biol.* **58:** 153–163.

Endow S. and Komma D. 1996. Centrosome and spindle function of the *Drosophila* Ncd microtubule motor visualized in live embryos using Ncd-GFP fusion proteins. *J. Cell Sci.* **109:** 2429–2442.

———. 1997. Spindle dynamics during meiosis in *Drosophila* oocytes. *J. Cell Biol.* **137:** 1321–1336.

Endow S.A. and Piston D.W. 1998. Methods and protocols. In *GFP, green fluorescent protein: Properties, applications, and protocols* (ed. M. Chalfie and S. Kain), pp. 271–369. Wiley-Liss, New York.

Fasano L. and Kerridge S. 1989. Monitoring positional information during oogenesis in adult *Drosophila. Development* **104:** 245–253.

Ferrandon D., Jung A.C., Criqui M., Lemaitre B., Uttenweiler-Joseph S., Michaut L., Reichhart J., and Hoffmann J.A. 1998. A drosomycin-GFP reporter transgene reveals a local immune response in *Drosophila* that is not dependent on the Toll pathway. *EMBO J.* **17:** 1217–1227.

Giniger E., Wells W., Jan L.Y., and Jan Y.N. 1993. Tracing neurons with a kinesin-β-galactosidase fusion protein. *Roux Arch. Dev. Biol.* **202:** 112–122.

Glaser R.L., Wolfner M.F., and Lis J.T. 1986. Spatial and temporal pattern of hsp26 expression during normal development. *EMBO J.* **5:** 747–754.

Gönczy P., Viswanathan S., and DiNardo S. 1992. Probing spermatogenesis in *Drosophila* with P-element enhancer detectors. *Development* **114:** 89–98.

Grossniklaus U., Bellen H.J., Wilson C., and Gehring W.J. 1989. P-element-mediated enhancer detection applied to the study of oogenesis in *Drosophila. Development* **107:** 189–200.

Handler A. and Harrell R.A. 1999. Germline transformation of *Drosophila melanogaster* with the *piggyBac* transposon vector. *Insect Mol. Biol.* **8:** 449–458.

Hazelrigg T. 1998. The uses of green fluorescent protein in *Drosophila*. In *GFP, green fluorescent protein: Properties, applications, and protocols* (ed. M. Chalfie and S. Kain), pp. 169–190. Wiley-Liss, New York.

Hazelrigg T., Liu N., Hong Y., and Wang S. 1998. GFP expression in *Drosophila* tissues: Time requirements for formation of a fluorescent product. *Dev. Biol.* **199:** 245–249.

Heim R., Cubitt A.B., and Tsien R.Y. 1995. Improved green fluorescence. *Nature* **373:** 663–664.

Hiromi Y., Kuroiwa A., and Gehring W.J. 1985. Control elements of the *Drosophila* segmentation gene fushi tarazu. *Cell* **43:** 603–613.

Kerrebrock A.W., Moore D.P., Wu J.S., and Orr-Weaver T.L. 1995. Mei-S332, a *Drosophila* protein required for sister-chromatid cohesion, can localize to meiotic centromere regions. *Cell* **83:** 247–256.

Kiehart D.P., Montague R.A., Rickoll W.L., Foard D., and Thomas G.H. 1994. High-resolution microscopic methods for the analysis of cellular movements in *Drosophila* embryos. *Methods Cell Biol.* **44:** 507–532.

Krasnow M.A., Saffman E.E., Kornfeld K., and Hogness D.S. 1989. Transcriptional activation and repression by Ultrabithorax proteins in cultured *Drosophila* cells. *Cell* **57:** 1031–1043.

Krasnow M.A., Cumberledge S., Manning G., Herzenberg L.A., and Nolan G.P. 1991. Whole animal cell sorting of *Drosophila* embryos. *Science* **251:** 81–85.

Kuhn R., Schafer U., and Schafer M. 1988. pW-ATG-lac, P-element vectors for *lacZ* transcriptional gene fusions in *Drosophila. Nucleic Acids Res.* **16:** 4163.

Lawson R., Mestril R., Schiller P., and Voellmy R. 1984. Expression of heat shock-β-galactosidase hybrid genes in cultured *Drosophila* cells. *Mol. Gen. Genet.* **198:** 116–124.

Levashina E.A., Ohresser S., Lemaitre B., and Imler J.L. 1998. Two distinct pathways can control expression of the gene encoding the *Drosophila* antimicrobial peptide metchnikowin. *J. Mol. Biol.* **278:** 515–527.

Lis J.T., Simon J.A., and Sutton C.A. 1983. New heat shock puffs and β-galactosidase activity resulting from transformation of *Drosophila* with an *hsp70-lacZ* hybrid gene. *Cell* **35:** 403–410.

Lockett T.J., Lewy D., Holmes P., Medveczky K., and Saint R. 1992. The *rough (ro+)* gene as a dominant P-element marker in germ line transformation of *Drosophila melanogaster. Gene* **114:** 187–193.

Micklem D.R., Dasgupta R., Elliott H., Gergely F., Davidson C., Brand A., Gonzalez-Reyes A., and St Johnston D. 1997. The *mago nashi* gene is required for the polarisation of the oocyte and the formation of perpendicular axes in *Drosophila. Curr. Biol.* **7:** 468–478.

Minden J.S. 1996. Synthesis of a new substrate for detection of *lacZ* gene expression in live *Drosophila* embryos. *BioTechniques* **20:** 122–129.

Mismer D. and Rubin G.M. 1987. Analysis of the promoter of the ninaE opsin gene in *Drosophila melanogaster. Genetics* **116:** 565–578.

Molsberger G., Schafer U., and Schafer M. 1988. A new set of lacZ fusion vectors, pUCPlac, for studying gene expression in *Drosophila* by P-mediated transformation. *Gene* **63:** 147–151.

Noselli S. and Vincent A. 1991. A *Drosophila* nuclear localisation signal included in an 18 amino acid fragment from the serendipity delta zinc finger protein. *FEBS Lett.* **280:** 167–170.

O'Kane C.J. and Gehring W.J. 1987. Detection in situ of genomic regulatory elements in *Drosophila. Proc. Natl. Acad. Sci.* **84:** 9123–9127.

Phelps C.B. and Brand A.H. 1998. Ectopic gene expression in *Drosophila* using the GAL4 system. *Methods* **14:** 367–379.

Plautz J.D., Day R.N., Dailey G.M., Welsh S.B., Hall J.C., Halpain S., and Kay S.A. 1996. Green fluorescent protein and its derivatives as versatile markers for gene expression in living *Drosophila melanogaster*, plant and mammalian cells. *Gene* **173:** 83–87.

Plautz J.D., Kaneko M., Hall J.C., and Kay S.A. 1997a. Independent photoreceptive circadian clocks throughout *Drosophila. Science* **278:** 1632–1635.

Plautz J.D., Straume M., Stanewsky R., Jamison C.F., Brandes C., Dowse H.B., Hall J.C., and Kay S.A. 1997b. Quantitative analysis of *Drosophila* period gene transcription in living animals. *J. Biol. Rhythms* **12:** 204–217

Potter S.M., Wang C.-W., Garrity P.A., and Fraser S.E. 1996. Intravital imaging of green fluorescent protein using two-photon laser-scanning microscopy. *Gene* **173:** 25–31.

Reichhart J.M. and Ferrandon D. 1998. Green balancers. *Drosophila Information Service* **81:** 201–202.

Santel A., Blümer N., Kampfer M., and Renkawitz-Pohl R. 1998. Flagellar mitochondrial association of the male-specific Don Juan protein in *Drosophila* spermatozoa. *J. Cell Sci.* **111:** 3299–3309.

Santel A., Winhauer T., Blumer N., and Renkawitz-Pohl R. 1997. The *Drosophila* don juan (dj) gene encodes a novel sperm specific protein component characterized by an unusual domain of a repetitive amino acid motif. *Mech. Dev.* **64:** 19–30.

Sawado T., Hirose F., Takahashi Y., Sasaki T., Shinomiya T., Sakaguchi K., Matsukage A., and Yamaguchi M. 1998. The DNA replication-related element (DRE)/DRE-binding factor system is a transcriptional regulator of the *Drosophila* E2F gene. *J. Biol. Chem.* **273:** 26042–26051.

Schnetzer J.W. and Tyler M.S. 1996. Endogenous β-galactosidase activity in the larval, pupal, and adult stages of the fruit fly, *Drosophila melanogaster*, indicates need for caution in *lacZ* fusion-gene studies. *Biol. Bull.* **190:** 173–187.

Schuldt A.J., Adams J.H., Davidson C.M., Micklem D.R., Haseloff J., Johnston D.S., and Brand A.H. 1998. Miranda mediates asymmetric protein and RNA localization in the developing nervous system. *Genes Dev.* **12:** 1847–1857.

Shiga Y., Tanaka-Matakatsu M., and Hayashi S. 1996. A nuclear GFP/β-galactosidase fusion protein as a marker for morphogenesis in living *Drosophila. Dev. Growth Differ.* **38:** 99–106.

Simon J.A. and Lis J.T. 1987. A germline transformation analysis reveals flexibility in the organization of heat shock consensus elements. *Nucleic Acids Res.* **15:** 2971–2988.

Stanewsky R., Jamison C.F., Plautz J.D., Kay S.A., and Hall J.C. 1997. Multiple circadian-regulated elements contribute to cycling period gene expression in *Drosophila. EMBO J.* **16:** 5006–5018.

Stanewsky R., Kaneko M., Emery P., Beretta B., Wager-Smith K., Kay S.A., Rosbash M., and Hall J.C. 1998. The cryb mutation identifies cryptochrome as a circadian photoreceptor in *Drosophila. Cell* **95:** 681–692.

Su M.-T., Golden K., and Bodmer R. 1998. X-gal staining of *Drosophila* embryos compatible with antibody staining or *in situ* hybridization. *BioTechniques* **24:** 918–922.

Theurkauf W.E. and Hawley R.S. 1992. Meiotic spindle assembly in *Drosophila* females: Behavior of nonexchange chromosomes and the effects of mutations in the nod kinesin-like protein. *J. Cell Biol.* **116:** 1167–1180.

Theurkauf W.E. and Hazelrigg T.I. 1998. In vivo analyses of cytoplasmic transport and cytoskeletal organization during *Drosophila* oogenesis: Characterization of a multi-step anterior localization pathway. *Development* **125:** 3655–3666.

Thompson J.F., Hayes L.S., and Lloyd D.B. 1991. Modulation of firefly luciferase stability and impact on studies of gene regulation. *Genes* **103:** 171–177.

Thummel C.S. and Pirrotta V. 1992. New pCaSpeR P element vectors. *Drosophila Information Service* **71:** 150.

Thummel C.S., Boulet A.M., and Lipshitz H.D., 1988. Vectors for *Drosophila* P-element-mediated transformation and tissue culture transfection. *Gene* **74:** 445–456.

Timmons L., Becker J., Barthmaier P., Fyrberg C., Shearn A., and Fyrberg E. 1997. Green fluorescent protein/β-galactosidase double reporters for visualizing *Drosophila* gene expression patterns. *Dev. Genet.* **20:** 338–347.

Tsien R. and Prasher D. 1998. Molecular biology and mutation of green fluorescent protein. In *GFP, green fluorescent protein: Properties, applications, and protocols* (ed. M. Chalfie and S. Kain), pp. 97–118. Wiley-Liss, New York.

Von Ohlen T., Lessing D., Nusse R., and Hooper J.E. 1997. Hedgehog signaling regulates transcription through cubitus interruptus, a sequence-specific DNA binding protein. *Proc. Natl. Acad. Sci.* **94:** 2404–2409.

Wang S. and Hazelrigg T. 1994. Implications for *bcd* mRNA localization from spatial distribution of *exu* protein in *Drosophila* oogenesis. *Nature* **369:** 400–403.

Ward W.W. 1998. Biochemical and physical properties of green fluorescent protein. In *GFP, green fluorescent protein: Properties, applications, and protocols* (ed. M. Chalfie and S. Kain), pp. 45–75. Wiley-Liss, New York.

Wilson C., Pearson R.K., Bellen H.J., O'Kane C.J., Grossniklaus U., and Gehring W.J. 1989. P-element-mediated enhancer detection: An efficient method for isolating and characterizing developmentally regulated genes in *Drosophila*. *Genes Dev.* **3:** 1301–1313.

Ye X., Fong P., Iizuka N., Choate D., and Cavener D.R. 1997. Ultrabithorax and Antennapedia 5′ untranslated regions promote developmentally regulated internal translation initiation. *Mol. Cell. Biol.* **17:** 1714–1721.

Yeh E., Gustafson K., and Boulianne G.L. 1995. Green fluorescent protein as a vital marker and reporter of gene expression in *Drosophila*. *Proc. Natl. Acad. Sci.* **92:** 7036–7040.

# CONTENTS

# 18

# Quantitative Microinjection of *Drosophila* Embryos

**Daniel P. Kiehart, Janice M. Crawford,**
**and Ruth A. Montague**
*Duke University Medical Center*
*Department of Cell Biology, University Program in Genetics*
*University Program in Cell and Molecular Biology*
*and the Duke Comprehensive Cancer Center*
*Durham, North Carolina 27710*

MICROINJECTION OF *DROSOPHILA* EMBRYOS is a common technique used by a wide range of investigators. The most frequent use of microinjection is to introduce foreign DNA into the developing germ cells of the embryo to generate transgenic flies. Because the injected DNA constructs are designed to carry a dominant, easy-to-score marker, even crude strategies for holding, imaging, and injecting the embryos are adequate, provided they yield transgenic animals with some acceptable frequency or efficiency. Several excellent protocols are available (see, e.g., Santamaria and Gans 1980; Rubin and Spradling 1983) that describe these microinjection techniques. In contrast, there are other applications for microinjection approaches in *Drosophila* that require a more refined strategy for handling the embryos.

The syncytial blastoderm is an excellent target for pharmacological investigation of maternally loaded gene products that are only marginally accessible to genetic analysis. Microinjection of drugs such as cytochalasins, phalloidins, colchicine (Edgar et al. 1987; Planques et al. 1991; Foe et al. 1993), and aphidicholine (Raff et al. 1990) has contributed to our understanding of cytoskeletal function during early development. Microinjection of antibodies to myosin II (Kiehart et al. 1990), myosin VI (Mermall and Miller 1995), $\alpha$-tubulin (Warn et al. 1987), or nuclear matrix components (Johansen et al. 1996) has also been used to evaluate the contributions of specific proteins during early development. Similarly, the microinjection of manipulated versions of proteins or RNAs of interest into the syncytial blastoderm has yielded information on the regulation of pattern formation. For example, truncated and/or deleted forms of the Nanos protein were compared using a *nanos* RNA injection assay (Curtis et al. 1997); this assay was based on previous experiments in which the aberrant phenotype of mutant embryos was "rescued" by the injection of wild-type cytoplasm (Lehmann and Nüsslein-Volhard 1986, 1987, 1991). Microinjection of both toxins and antibodies determined a role for topoisomerase II during early embryogenesis (Buchenau et al. 1993). Likewise, a recent study that utilized the microinjection of both toxins and manipulated proteins revealed that the small GTPase Rho has a key role in maintenance of the actomyosin cytoskeleton during cellularization (Crawford et al. 1998). Together, these examples demonstrate that this pharmacological approach, which uses the microinjection of drugs, antibodies, toxins, and manipulated RNAs and proteins, provides a unique opportunity to analyze cellular function in the developing embryo. Moreover, microinjection methods provide spatial and temporal resolution that is not readily available through genetic studies.

The pharmacological approach to studying gene product function in these microinjection studies requires that observed effects reflect a dose-response relationship so that the data can be accurately interpreted. Thus, unambiguous interpretation of the data requires that the *dose injected*, defined here as product of the tip concentration of the agent of interest times the *volume injected*, be evaluated with some degree of accuracy. Previously, we estimated the injected volumes by calibrating each micropipette, a process that is both tedious and time-consuming (Kiehart 1982). Quantitative microinjections can be achieved much more readily with the addition of a fluorescent tracer to the solution to be injected (Lee 1989). Analysis of the resulting integrated fluorescent intensity following injection can then be used to determine the volume and hence the concentration of the solution injected. Finally, the use of appropriately designed chambers to mount and observe the embryos with high-NA (numerical aperture) objectives and condensers greatly facilitates analysis of the biological response to a given injected material. In the particular case of the *Drosophila* syncytial blastoderm, microinjection studies can broaden our understanding of cellular functions through spatially and temporally specific manipulation of a stage of development that is not readily amenable to genetic approaches.

This chapter outlines the procedures for the microinjection and quantification of aqueous solutions during high-resolution observation of early development in the fly embryo.

## GENERAL STRATEGY

The general strategy for quantitative injection of embryos is summarized here. Embryos are collected, dechorionated, aligned on a small agar pad, and picked up on a coverslip coated with embryo glue. Next, the coverslip is affixed to a micromanipulation slide (see below) and the embryos are desiccated. They are covered with halocarbon oil, and if an upright microscope is to be used for microinjection, a second coverslip is added, creating a chamber. Protocol 18.1 describes the preparation of embryos for microinjection. The slide is then mounted on the compound microscope. The embryos are brought into focus, positioned in the center of the microscope field, and then moved back and out of the way using movement along a single axis of the mechanical stage of the microscope. A micropipette is drawn, backfilled with fluorescently "doped" injection solution, and then inserted into a micropipette holder that allows pressure to be applied to the solution in the pipette. The pipette holder is mounted on a micromanipulator and, using the controls on the manipulator, the tip of the micropipette is positioned in the center of the microscope field, and then lowered so that it is just out of focus. The mechanical stage is used to bring the embryos back into view, the tip of the pipette is brought back into focus, and microinjection ensues by inserting the pipette tip into the embryo and gently expelling the solution.

To quantitate the injected volume accurately, injection solutions are doped with an inert, fluorescent dextran, and the injected volume is measured as a function of injected fluorescence. A standard curve that relates the fluorescence to volume is generated as follows. A series of various sized droplets of aqueous injection solution is released into the oil surrounding the embryos. Digital, epifluorescent images of the fluorescent spheres are recorded with a low-magnification, low-NA objective such that their intensity is within the linear range of the digital camera. It is essential to use a low-NA objective so that fluorochromes throughout the depth of the specimen are both illuminated and subsequently imaged. Background is subtracted, and the acquisition software (MetaMorph, Universal Imaging Corp.) is used to generate a standard curve that relates the total fluorescence in each droplet (obtained by summing the intensities recorded by the digital camera for each pixel in the image of a given sphere) to the volume of each droplet calculated from its measured diameter. To measure the amount of fluorescent dextran (and therefore aqueous solution) injected into an embryo, a low-magnification, low-NA image of each embryo is taken both before and immediately following injection, using the same objective and illumination conditions that were used to generate the standard curve. In practice, this process is quite simple: A background image is taken, the droplets are released into the oil surrounding the embryos, and an image of the droplets is acquired. A journal (or macro) in MetaMorph automatically subtracts background, identifies the individual droplets, calculates their diameter, sums the intensity of all the pixels in the image of each sphere, and transfers the data into an Excel spreadsheet (Microsoft Corp.). We generate a new standard curve for every set of embryos we inject. The fluorescence in the background-substracted image of the embryo is calculated, and the volume injected is determined from the standard curve. Protocol 18.2 describes the quantitative microinjection of embryos.

Injected embryos are followed using high-resolution, differential interference time-lapse microscopy, confocal microscopy, or are incubated, recovered, and fixed for subsequent localization of various components.

## DESIGN OF MICROMANIPULATION SUPPORT SLIDES

### Upright Microscopes

The slide we use is specifically designed for use with an upright microscope (Figure 18.1; for a mechanical drawing, see Kiehart 1982). It is fashioned from stainless steel, aluminum, or brass. Plexiglas slides tend to warp and are not desirable. Overall, the slide is designed to have the same length and width as a standard microscope slide and is therefore compatible with most microscope stage clips or other mechanical slide movements. The metal slide is thicker than a standard glass microscope slide because it is designed to hold a coverslip that has a row of target embryos glued to its underside (see below), 3–5 mm above the microscope stage. This design allows an excellent optical window through which the embryos can be observed during development and microinjection. It also provides clearance for the micropipettes and their holders (located perpendicular to the optic axis), so that they do not interfere with the microscope stage. A second coverslip is affixed parallel to the first and prevents the halocarbon oil that surrounds the embryos from dripping down onto the condenser. In determining the distance between the coverslips (as defined by the thickness to which the appropriate part of the slide is machined), there are two conflicting features to be considered. By mechanical considerations, the distance between the coverslips should be maximized within reason so that introduction of the

A. Metal injection slide

B. Assembled injection slide on stage with pipette

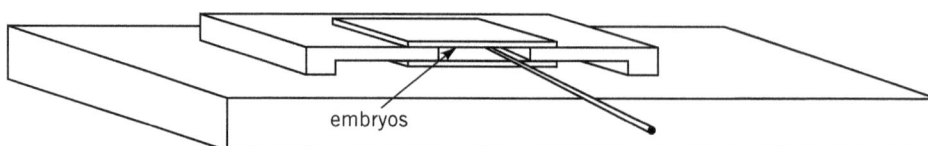

embryos

**Figure 18.1.** Micromanipulation slide designed for upright microscopes. (*A*) A perspective drawing of the metal microinjection support slide (for a mechanical drawing, see Kiehart 1982). (*B*) The assembled injection chamber showing top and bottom coverslips in place, the microscope stage, and part of a micropipette. The arrow points to the position of the embryos near the front edge of the bottom side of the top coverslip.

pipettes and their manipulation is facile, pipette tips are not readily broken, and pipette shafts are not dragged along the coverslip. In contrast, by optical considerations, the space between the two coverslips should be minimized so that it is little more than the 150-μm diameter of the embryo. With such a narrow chamber, optical qualities are maximized (one can achieve Köhler illumination with oil-immersion, high-NA, short-working distance objectives and condensers). Because it is usually messy to use immersion oil on the condenser, we typically use a high, dry condenser (0.9 NA, working distance ~2 mm). As a consequence, a fairly shallow chamber (0.8 mm) is a reasonable compromise. In practice, we have made a variety of slides of different thickness that we use for distinct purposes. In all cases, we adjust the pipette so that it is nearly perpendicular to the optic axis and is therefore in focus through a significant portion of its length. We also align embryos near the edge of the coverslip so that they receive a sufficient amount of oxygen to develop (see Kiehart et al. 1994).

For low-resolution work, we have used a low-NA, dry objective mounted on an upright microscope and have injected embryos mounted on a coverslip resting directly on a glass slide. This strategy is useful if injected embryos are subsequently observed on another microscope (e.g., an inverted, scanning confocal microscope).

## Inverted Microscopes

Inverted microscopes may be somewhat preferable to upright microscopes for microinjections when low-resolution imaging methods are employed because only a single coverslip is required, which allows greater access to the specimen. This coverslip provides both support for the embryos and the optical window through which they are viewed. Thus, a simple, rectangular metal slide provides adequate support for the coverslip in almost all applications. As described above, the metal support slide that holds the coverslip should have overall dimensions comparable to a glass microscope slide, with a hole for the optic axis and a slight indentation the size of a coverslip to keep the coverslip held up off the microscope stage. Embryos are mounted on the top of the coverslip, which is affixed to the bottom of the manipulation slide. The coverslip thus provides the window to the objective, and the halocarbon oil that surrounds the embryos is "held" in place by gravity. If high optical quality is not essential, the long-working-distance low-NA condenser that inverted microscopes are typically equipped with is often adequate for many imaging purposes. A clear advantage of this configuration is that pipettes can be used at fairly steep angles. Nevertheless, for highest resolution, a second coverslip is essential because the condenser should be oil-immersed. Under these conditions, the pipette must come in at a shallow angle and virtually all advantages of an inverted microscope are lost. In fact, we find that aligning the pipette almost perpendicular to the optic axis is desirable in almost all cases because this configuration allows the pipette shaft and tip to be simultaneously in focus along a significant part of their length.

PROTOCOL 18.1*

# Collection, Dechorionation, and Preparation of Embryos for Microinjection

### Materials

### *Media*

> #### *Standard Grape or Apple Juice Plates*
> Prepare grape or apple juice medium as described in Appendix 3. Pour into the bottoms of 60 × 15-mm polystyrene petri plates (Falcon 351007), which fit snugly into the top of the 100-ml polypropylene fly cages (see below). Dab each plate with yeast paste to facilitate egg laying (we use a mixture of dry yeast and $H_2O$).

### *Supplies and Equipment*

Filtration apparatus for embryo dechorionation (see Figure 18.2). The glass funnel assembly (Millipore XX10 025 00) with a Whatman filter paper (#1, #541, or 3MM) is mounted on a vacuum flask, which is attached to an aspirator, house vacuum, or a vacuum pump. This apparatus allows rapid exchange of solutions during dechorionation.

Artist's brush

Pasteur pipette

Dissecting microscope

Standard dissecting tools (steel probe with wood handle, applicator sticks, forceps, razor blades)

Clean glass coverslips (#1.5, 22 × 22-mm; this thickness is essential for optimal resolution)

Vacuum grease

Micromanipulation support slides. The type of support slide depends on whether the microinjection will be performed on an upright or an inverted microscope (see discussion on p. 347, Design of Micromanipulator Support Slides).

Desiccator; a screw-top specimen jar (~9 cm high × 9 cm diameter; e.g., Fisher 03-320-7E) with a 1–2-cm-thick layer of Drierite colored desiccant (Fisher)

*For all protocols, $H_2O$ indicates glass distilled and deionized.

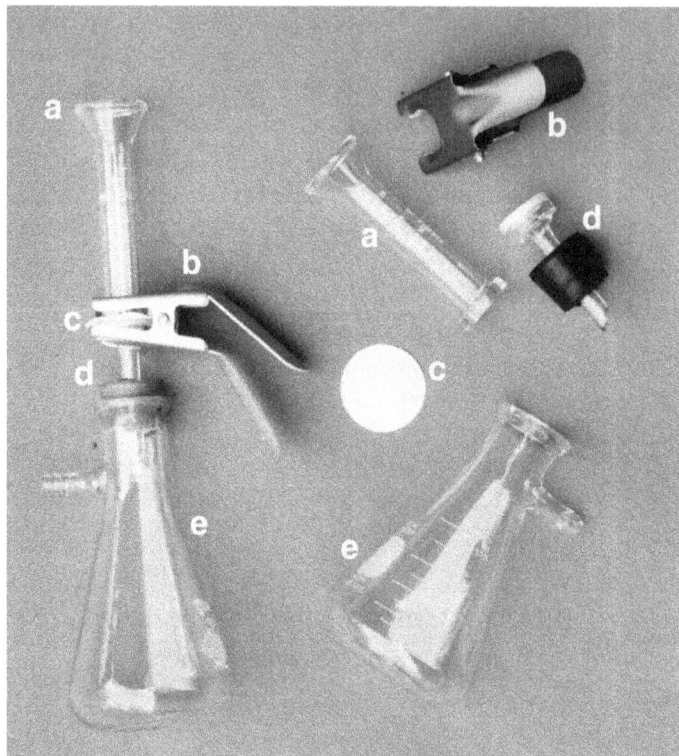

**Figure 18.2.** Filtration apparatus used for embryo dechorionation. Left side shows the assembled apparatus, right side shows the component parts. (*a*) Glass funnel; (*b*) clamp; (*c*) filter paper (preferably Whatman #1, #541, or 3 MM, 42.5-mm circles); (*d*) fritted glass base with a stopper that fits snugly into the opening of a 250-ml Erlenmeyer side-arm flask (*e*). Catalog number for *a*, *b*, and *d* is Millipore XX10 025 00. A piece of Whatman filter paper (*c*) is inserted between the upper (*a*) and lower (*d*) halves of the glass funnel apparatus and sealed into place with the clamp (*b*) that fits around the two parts of the apparatus. The embryos are washed from the agar collection plate into the filtration apparatus and are collected onto the piece of Whatman filter paper upon application of a gentle vacuum to the flask (*e*).

### Small Fly Cages

For "cytological" purposes (i.e., the injection of 10–100 embryos in a given experiment), 100-ml Tri-pour polypropylene beakers (Fisher 02-593-50B) fashioned into cages are excellent. Use a sharp razor blade to cut the bottom off of the beaker and heat the bottom until just melting using a Bunsen burner. Push the hot, softened plastic onto a sheet of fine-mesh, stainless-steel screening (Small Parts, Miami Lakes, Florida CX-100 [#100, opening size of 0.140 mm]) or nylon mesh (Small Parts CMN-300 [opening size of 0.3 mm]). This design is superior to putting holes in the bottom of a plastic beaker as the screening provides better exchange of air and more reliable control of humidity and temperature.

### Embryo Glue (Wieschaus and Nüsslein-Volhard 1986)

We dissolve the glue from approximately 7 cm of double-stick tape with 2 ml of **heptane** in a small glass vial with a screw cap.

## Solutions and Reagents

Squeeze bottle of $H_2O$
**Bleach** (50% Clorox; final concentration: 2.5% hypochlorite), prepare fresh each day
Halocarbon oil 700, to keep the embryos moist and oxygenated (Sigma H 8898)

**CAUTION: bleach, heptane** (see Appendix 4)

## Method

### Collecting Embryos

1. For best results, change plates frequently, even on days between experiments, by gently tapping down the flies and rapidly replacing each old plate with a fresh one. Collect embryos at desired stages. For injection into syncytial blastoderm stages, 0.5–1-hour collections at 25ºC are optimal.

   *Notes:* Changing the plates frequently will keep the flies well-fed and "conditioned" to being tapped to the bottom of the cage.

   The small cages and standard grape or apple juice plates described above are adequate for rapid collections of "cytological" numbers of embryos for quantitative microinjection studies (i.e., the injection of 10–100 embryos in a given experiment).

2. Wash embryos from a grape or apple juice plate directly into the assembled filtration apparatus using a stream of $H_2O$ from a squeeze bottle and the gentle application of an artist's brush. $H_2O$ is drawn through the filter paper into the flask below and the embryos remain on the filter paper.

   *Note:* We avoid washing excess yeast paste into the filtration apparatus, as it tends to clog the filter.

3. Wash the embryos three times by resuspending in fresh $H_2O$ (it should take ~15 seconds to drain 15 ml of $H_2O$ through the filter paper).

### Dechorionating Embryos

Several standard strategies for removing the chorion are available. We prefer chemical dechorionation by immersion in 50% commercial bleach (final concentration: 2.5% hypochlorite). Although in principle, chorions can be "cleared" by immersion in halocarbon oil, an intact chorion is tough on the tips of standard glass micropipettes and the use of fused-quartz capillary stock and a laser micropipette puller may be necessary if the chorion is left intact.

1. Resuspend embryos in 50% bleach solution and agitate by gentle stirring with a pasteur pipette. After 1 minute 15 seconds, draw off the bleach solution.

2. Wash the embryos three times with $H_2O$ as in step 3 (above).

   *Note:* The chorion-free embryos are fairly hydroscopic and stick to the side of the glass funnel. With some practice, the right combination of adding fresh $H_2O$ while removing the old wash will keep the embryos in suspension so that they can be drawn down onto the filter paper with minimal loss. We avoid the use of detergents (e.g., 0.1% Triton X-100) in the wash solutions because residual detergents may be detrimental to the embryos following penetration of the vitelline envelope by the micropipette.

3. Remove the funnel and place the filter paper with the dechorionated embryos on a fresh grape or apple juice plate (embryo-side down). Carefully remove the filter paper.

## *Aligning Embryos on Coverslips and Mounting Coverslips*

The next step is to arrange a row of dechorionated embryos on a glass coverslip. A carefully aligned row greatly expedites manipulations such as moving from embryo to embryo, subsequent injections, and the monitoring of injected embryos.

1. Transfer a small cluster of embryos onto a new, small grape or apple juice agar pad (~5 × 10 × 2 mm, cut from a grape or apple juice plate), and then use a metal probe or a sharpened wooden applicator stick to align them on the edge of the pad.

    *Notes:* Surface tension from the moist grape or apple juice agar holds the embryos gently in place, so that they can be oriented easily with respect to dorsal/ventral and anterior/posterior.

    We find it most convenient to inject through the posterior end of the embryo, but embryos can easily be arranged to facilitate injections into the lateral surfaces or anterior end. Embryos should be spaced at least one embryo diameter apart, especially if their long axes are parallel.

2. Place two to three drops (25–50 μl) of "embryo glue" near one edge of a clean 22 × 22-mm coverslip, tilt the coverslip on its side, and allow to air-dry. Prepare several coverslips in this manner for mounting embryos.

    *Note:* Glue is only applied to that region of the coverslip to which the embryos are affixed.

3. Pick up the line of embryos by gently lowering a coverslip coated with embryo glue onto the embryos until they are gently pressed into the agar pad. Upon lifting the coverslip, all of the embryos should be affixed in place on the coverslip.

4. Attach the coverslip to a stainless-steel micromanipulation slide designed for the upright or inverted microscope using a thin layer of vacuum grease.

## *Desiccating the Embryos*

Standard methods are used to relieve the turgor pressure that is normally (and explosively) released upon penetration of the vitelline membrane by a micropipette.

1. Place the mounted micromanipulation slide to which the embryos are affixed in a desiccator for 3–8 minutes.

2. Empirically determine the time for desiccation depending on temperature and humidity.

## *Assembling the Chamber*

1. Once the embryos are appropriately desiccated, cover them with halocarbon oil.

2. If subsequent manipulations are to be performed on an upright microscope, attach a second coverslip to the micromanipulation slide with a thin layer of vacuum grease. A second coverslip may not be necessary if an inverted microscope is used (see p. 348).

    *Notes:* The challenge is to get a sufficient amount of oil on the embryos so that it just fills the cavity between the two coverslips without overflowing. With a little practice, this is very straightforward.

    Several coverslips with embryos can be prepared and stored at 18°C to slow development until needed.

PROTOCOL 18.2

# Quantitative Microinjection of Embryos

## Materials

### Supplies and Equipment

Micromanipulation slides with mounted embryos, prepared as described in Protocol 18.1

Upright or inverted microscope

Capillary stock with a glass filament for micropipettes (e.g., 2-mm outer diameter and 0.68-mm inner diameter; A-M Systems)

Standard micropipette puller

Parafilm

(*Optional*) Hamilton syringe or a microloader pipette tip (Eppendorf) used for filling micropipette tips. Alternatively, a standard plastic p200 pipette tip can be hand-drawn into a thin microloader tip following heating over the pilot flame of a "Touch-O-Matic" Bunsen burner (Fisher).

Pressure-injection system

Micromanipulator; we use either a Leitz mechanical manipulator (Wetzlar, Germany) or a Narishige micromanipulator system (Model MMO-202N or MMO-202ND).

Software requirements

- MetaMorph imaging software (Universal Imaging) or similar acquisition and analysis software

- Graphing software, Microsoft Excel (Microsoft) or equivalent

### Solutions and Reagents

#### Injection Solutions

We have found that simple, low to medium ionic strength buffers are best for a variety of injection purposes. Two examples are given below.

> 0.5 mM **sodium phosphate** buffer (pH 7.5)
>
> 5 mM **KCl**

#### OR

> 150 mM NaCl
>
> 20 mM HEPES (pH 7.5)

If the pipette tips clog frequently, centrifuge the injection solution in a microcentrifuge at 14,000 rpm for 10–15 minutes to sediment offending particulates.

For microinjections, we dope the injection solution with rhodamine dextran, which we have found to be nontoxic and without effect on subsequent development when injected with control solutions. Prepare a 20 µg/µl stock solution of tetramethylrhodamine dextran, 70,000 molecular weight (Molecular Probes D-1819), sterilize by filtration, and store at –20ºC. Dilute the stock in injection solution to a final concentration of 1 µg/µl.

Antibody solutions can be prepared in 10 mM Tris-Cl or 10 mM phosphate buffer (pH 6.8–7.5) and concentrated to 5–50 mg/ml using a Centricon concentrator (Millipore). It may be worth rinsing the Centricon device with 1 mg/ml cytochrome C, ovalbumin, or bovine serum albumin before concentrating precious antibody, to avoid losing antibody through nonspecific absorption.

**CAUTION: KCl, sodium phosphate** (see Appendix 4)

## Method

### *Mounting the Chamber on the Microscope*

1. Place the assembled chamber on the microscope stage with the opening facing the micromanipulator.

2. Position the slide with the mechanical stage so that the first embryo in the row is in focus and centered in the field of view.

3. Back the slide out of the way and prepare the micropipette.

### *Preparation and Setup of Micropipettes*

1. Pull micropipettes from capillary stock with a glass filament (e.g., 2-mm outer diameter and 0.68-mm inner diameter; A-M Systems) using a standard micropipette puller.

   *Note:* Typically, pipettes with an appropriate tip diameter (~1 μm) can be pulled. However, if the tips are too small, they can be broken to an appropriate aperture later by gently pushing the loaded, mounted pipette against the edge of a coverslip while monitoring the process with the microscope.

2. Backfill the pipettes with injection solution.

   a. *If the pipettes are sufficiently clean*, introduce a small volume of the solution to the back end of the pipette by placing a drop of the injection solution on a clean piece of parafilm and touching the drop to the back of the pipette. Capillary action draws the solution into the pipette. Place the pipette tip-side down at a 45° angle to allow the tip to fill without air bubbles.

   b. *If the pipettes are at all dusty*, the solution will sweep the dust toward the tip and will very likely clog the pipette. To circumvent this dust problem, deposit a small drop of the injection solution directly into the neck of the pipette near the tip using either a Hamilton syringe or a microloader pipette tip.

3. Connect the pipette to a pressure injection system. For certain injection applications, use an air-filled system that consists of a 25- or 50-ml syringe connected via Tygon tubing (Norton) directly to the back of the pipette. For better control of injection, we use an oil-filled hydraulic system that consists of a screw-driven syringe attached through polyethylene tubing to a modified pipette holder that clamps the pipette in a silicon rubber chuck (described in detail in Kiehart 1982). Regardless of the nature of the pressure-injection system, clamp the pipette into a pipette holder that is in turn mounted on the micromanipulator.

4. To align the micropipette, adjust the attitude and position of the pipette near the center of the microscope field macroscopically. Use the micromanipulator controls (not the focus on the microscope) to center and focus the tip in the microscope field. Then lower the pipette tip so that it is just out of focus. Use the mechanical stage to move the micromanipulation slide into the field, thereby surrounding the pipette. Refocus the pipette tip using the micromanipulator controls and align the pipette tip near the embryo.

### Determining Volume as a Function of Fluorescent Intensity

Before microinjection of the embryos, collect the data necessary to generate a standard curve that relates fluorescence intensity (measured as the total pixel gray-scale value) to volume as follows:

1. Release a series of droplets of the injection solution (doped with rhodamine dextran) into the halocarbon oil adjacent to the row of embryos.

2. Acquire a digital image of the fluorescent spheres and a digital image of a region of oil with no droplets. Use a low-magnification, low-NA objective (see below, Figure 18.3a). Verify that the intensity in the digital image of the fluorescent spheres is not saturating the digital imaging system. Once these data are collected, the standard curve can be generated at any time.

3. Generate the standard curve as follows:

   a. From the image with droplets, subtract the background image with no droplets.

   b. Next, use the acquisition software to integrate the total pixel gray-scale value over the area of each sphere. In MetaMorph, this is accomplished easily as follows.

   - MetaMorph first identifies each sphere as a discrete object using a series of "object classifiers" (e.g., size, position, total pixel gray-scale value, etc.) that are defined by the user.

   - MetaMorph next sums the total pixel gray-scale value over each pixel in the image of each sphere, and determines the area (in number of pixels) defended by each sphere.

   - MetaMorph then uses the area defended to calculate a radius and then calculate the "equivalent sphere volume."

   c. Data for the calculated volume of the sphere and the total pixel gray-scale value are automatically dumped from MetaMorph to a spreadsheet in Microsoft Excel. Use the graphing software in Excel to plot total pixel gray-scale value as a function of volume to generate a standard curve that is linear over a fairly wide range (Figure 18.3e).

### Microinjecting the Embryos

1. Align the micropipette perpendicular to the region of the embryo to be microinjected.

2. Acquire a low-magnification, low-NA epifluorescent image of the embryo just before microinjection using the same objective and illumination conditions used to generate the standard curve (see above, Determining Volume as a Function of Fluorescent Intensity).

3. Without moving the microscope stage, use the micromanipulator controls to carefully insert the micropipette into the embryo at the desired location.

4. Apply gentle pressure to the air- or fluid-filled syringe attached to the micropipette to release the injection solution into the embryo. Use the micromanipulator controls to withdraw the micropipette from the embryo carefully but quickly to minimize the loss of cytoplasm. Appropriate desiccation allows injections without leakage.

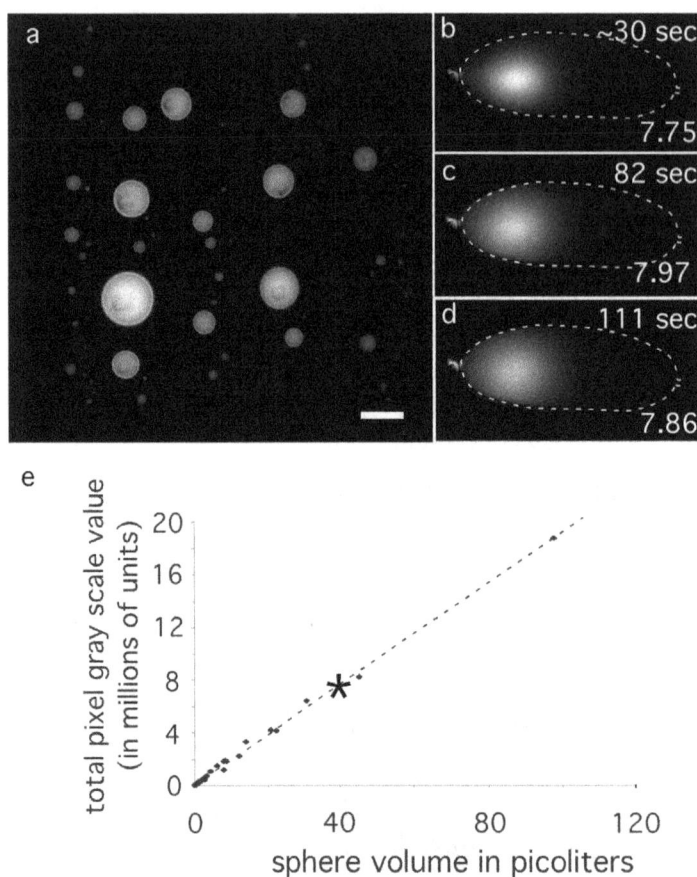

**Figure 18.3.** Determining the volume of solution microinjected as a function of fluorescent intensity. (*a*) An epifluorescent micrograph of droplets of injection solution doped with rhodamine dextran to a final concentration of 1 μg/μl and extruded from the microinjection pipette into the halocarbon oil that surrounds the target embryos. Bars: (*a–d*) 100 μm. (*b–d*) Epifluorescent micrographs of the same inject-ed embryo taken approximately 30, 82, and 111 sec after injection. (*Dashed white line*) Margin of the embryo (which is visible if the brightness and contrast of the digital image are altered). (*e*) Standard curve of total pixel gray-scale value vs. sphere volume in picoliters (pl; calculated from the data in panel *a*). Each diamond represents a fluorescent sphere in panel *a* (some spheres are too dim and/or too small to show up in *a* and the dotted line is fit by eye. Note that the total pixel gray-scale values for the embryo shown at the 3 time points in *b*, *c*, and *d* are 7.75, 7.97, and 7.86 (× 106 gray-scale units), respectively. This demonstrates that even though the injected material diffuses away from the site of injection, the esti-mated volume remains the same and within the total gray-value range of the calibration curve. The "star" symbol in panel *e* indicates the average of the three fluorescent intensities and indicates that the injected volume was approximately 40.5 pl or 2.7% of the estimated volume of the fly embryo (~15 nl; see Foe and Alberts 1983).

## Determining the Volume of Injected Material

1. Acquire another a low-magnification, low-NA epifluorescent image of the embryo immediately after microinjection. Since the microscope stage was not altered during the injection process, the "before injection" background image (not shown) aligns with and can be subtracted from the "after injection" image (Figure 18.3b,c,d). The resulting fluorescent intensity can then be integrated using MetaMorph or similar software.

*Note:* Although the microinjected material will diffuse, we found that four images taken 20 seconds apart showed an average change in fluorescent integrated intensity of only 1.5% (three time points are shown in Figure 18.3b,c,d). This indicates that the total pixel gray-scale value remains essentially constant at early time points after injection and is a reliable indicator of the microinjected volume.

2. To determine the microinjected volume, compare this total pixel gray-scale value to the standard curve of total pixel gray-scale vs. volume (see step 3 above, Determining Volume as a Function of Fluorescent Intensity).

## PROTOCOL 18.2 NOTES

The microinjected embryos can be imaged directly on the micromanipulation slide using high-resolution differential interference time-lapse microscopy. If the embryos express a green fluorescent protein (GFP) tagged marker or if the injected material is a fluorescent-

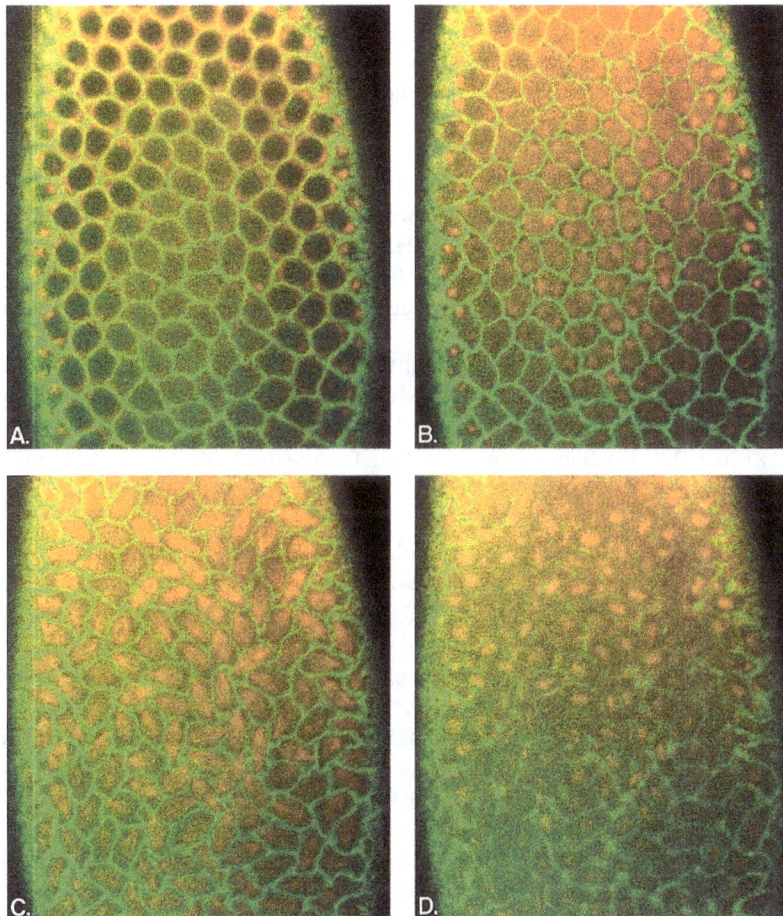

**Figure 18.4.** Confocal time-lapse microscopy of a microinjected *Drosophila* embryo. Dynamics of the actin (*green*) and microtubule (*red*) cytoskeleton are visualized during the syncytial mitotic divisions (*A–D*). An embryo that expresses the actin-binding protein GFP-moesin was microinjected with rhodamine-labeled α-tubulin (Molecular Probes), and the actin and microtubule dynamics were visualized as the mitotic cycle progressed from interphase (*A*) to anaphase (*D*).

ly tagged molecule (Figure 18.4), time-lapse images following microinjection can be obtained directly on the manipulation slide using confocal microscopy. Alternatively, following the desired length of incubation, the microinjection chamber can be disassembled and the embryos removed from the coverslip with a gentle stream of heptane released from a pasteur pipette. These embryos can then be fixed and stained for localization of components of interest.

# REFERENCES

Buchenau P., Saumweber H., and Arndt-Jovin D.J. 1993. Consequences of topoisomerase II inhibition in early embryogenesis of *Drosophila* revealed by in vivo confocal laser scanning microscopy. *J. Cell Sci.* **104:** 1175–1185.

Crawford J.M., Harden N., Leung T., Lim L., and Kiehart D.P. 1998. Cellularization in *Drosophila melanogaster* is disrupted by the inhibition of Rho activity and the activation of Cdc42 function. *Dev. Biol.* **204:** 151–164.

Curtis D., Treiber D.K., Tao F., Zamore P.D., Williamson J.R., and Lehmann R. 1997. A CCHC metal-binding domain in Nanos is essential for translational regulation. *EMBO J.* **16:** 834–843.

Edgar B.A., Odell G.M., and Schubiger G. 1987. Cytoarchitecture and the patterning of fushi tarazu expression in the *Drosophila* blastoderm. *Genes Dev.* **1:** 1226–1237.

Foe V.E. and Alberts B.M. 1983. Studies of nuclear and cytoplasmic behaviour during the five mitotic cycles that precede gastrulation in *Drosophila* embryogenesis. *J. Cell Sci.* **61:** 31–70.

Foe V.E., Odell G.M., and Edgar B.A. 1993. Mitosis and morphogenesis in the *Drosophila* embryo: Point and counterpoint. In *The development of* Drosophila melanogaster (ed. M. Bate and A. Martinez Arias). Cold Spring Harbor Laboratory Press, Cold Spring Harbor Laboratory, New York.

Johansen K.M., Johansen J., Baek K.H., and Jin Y. 1996. Remodeling of nuclear architecture during the cell cycle in *Drosophila* embryos. *J. Cell. Biochem.* **63:** 268–279.

Kiehart D.P. 1982. Microinjection of echinoderm eggs: Apparatus and procedures. *Methods Cell Biol.* **25:** 13–31.

Kiehart D.P., Montague R.A., Rickoll W.L., Foard D., and Thomas G.H. 1994. High-resolution microscopic methods for the analysis of cellular movements in *Drosophila* embryos. *Methods Cell Biol.* **44:** 507–532.

Kiehart D.P., Ketchum A., Young P., Lutz D., Alfenito M.R., Chang X.-J., Awobuluyi M., Pesacreta T.C., Inou S., Stewart C.T., and Chen T.-L. 1990. Contractile proteins in *Drosophila* development. *Ann. N.Y. Acad. Sci.* **582:** 233–251.

Lee G.M. 1989. Measurement of volume injected into individual cells by quantitative fluorescence microscopy. *J. Cell Sci.* **94:** 443–447.

Lehmann R. and Nüsslein-Volhard C. 1986. Abdominal segmentation, pole cell formation, and embryonic polarity require the localized activity of oskar, a maternal gene in *Drosophila*. *Cell* **47:** 141–152.

———. 1987. Involvement of the pumilio gene in transport of an abdominal signal in the *Drosophila* embryo. *Nature* **329:** 167–170.

———. 1991. The maternal gene nanos has a central role in posterior pattern formation of the *Drosophila* embryo. *Development* **112:** 679–691.

Mermall V. and Miller K.G. 1995. The 95F unconventional myosin is required for proper organization of the *Drosophila* syncytial blastoderm. *J. Cell Biol.* **129:** 1575–1588.

Planques V., Warn A., and Warn R.M. 1991. The effects of microinjection of rhodamine-phalloidin on mitosis and cytokinesis in early stage *Drosophila* embryos. *Exp. Cell. Res.* **192:** 557–566.

Raff J.W., Whitfield W.G., and Glover D.M. 1990. Two distinct mechanisms localize cyclin B transcripts in syncytial *Drosophila* embryos. *Development* **110:** 1249–1261.

Rubin G.M. and Spradling A.C. 1983. Vectors for P element-mediated gene transfer in *Drosophila*. *Nucleic Acids Res.* **11:** 6341–6351.

Santamaria P. and Gans M. 1980. Chimaeras of *Drosophila melanogaster* obtained by injection of haploid nuclei. *Nature* **287:** 143–144.

Warn R.M., Flegg L., and Warn A. 1987. An investigation of microtubule organization and functions in living *Drosophila* embryos by injection of a fluorescently labeled antibody against tyrosinated α-tubulin. *J. Cell Biol.* **105:** 1721–1730.

Wieschaus E. and Nüsslein-Volhard D. 1986. Looking at embryos. In Drosophila: *A practical approach* (ed. D.B. Roberts). IRL Press, Washington, D.C.

# CONTENTS

# 19

# Targeted Disruption of Gene Function in *Drosophila* by RNA Interference

**Leonie Misquitta and Bruce M. Paterson**
*Laboratory of Biochemistry*
*NCI, National Institutes of Health*
*Bethesda, Maryland 20892*

T O DETERMINE THE ROLE OF SPECIFIC GENES DURING DEVELOPMENT, the traditional strategy involves selecting or creating a mutation in the gene of interest. Because there is no targeted gene disruption involving homologous recombination that is applicable to *Drosophila*, mutations in a particular gene are usually obtained by a combination of genetic selection, P-element insertion, and deficiency analysis after γ-irradiation. These methods are very labor-intensive and time-consuming and, on occasion, do not yield a clear answer.

Recent studies in *Caenorhabditis elegans* have demonstrated elegantly that the introduction of double-stranded RNA (dsRNA) corresponding to all or a portion of the coding region for a particular gene into embryonic cells can interfere with the function of the endogenous gene to give a phenotype essentially equivalent to the known mutant phenotype (Fire et al. 1998). Moreover, the levels of dsRNA required to obtain a given phenotype suggested that a catalytic mechanism was involved. This phenomenon was termed dsRNA genetic interference or RNA-i. The precise mechanism of action for RNA-i is not understood, but studies in *C. elegans* strongly suggest that RNA-i targets the posttranscriptional process (Montgomery et al. 1998).

This chapter presents a procedure for the utilization of RNA-i in *Drosophila* embryogenesis. We have used this approach to successfully phenocopy a series of known mutations in *Drosophila*, including *twist, engrailed, daughterless, Dmef2*, and, to a lesser extent, *white* in the adult eye. On the basis of this success with known mutations, RNA-i was then used to demonstrate that the *Drosophila* MyoD homolog, *nautilus* (Michelson et al. 1990; Paterson et al. 1991), is required for somatic muscle formation in the embryo (Misquitta and Paterson 1999). The approach is very straightforward and involves improved methods for injection of the dsRNA product directly through the chorion of the embryo to minimize problems normally associated with desiccation of the dechorionated embryo and to facilitate postinjection analysis of gene expression. The same injection technique through the chorion is also used for the generation of trangenic *Drosophila* lines in our laboratory.

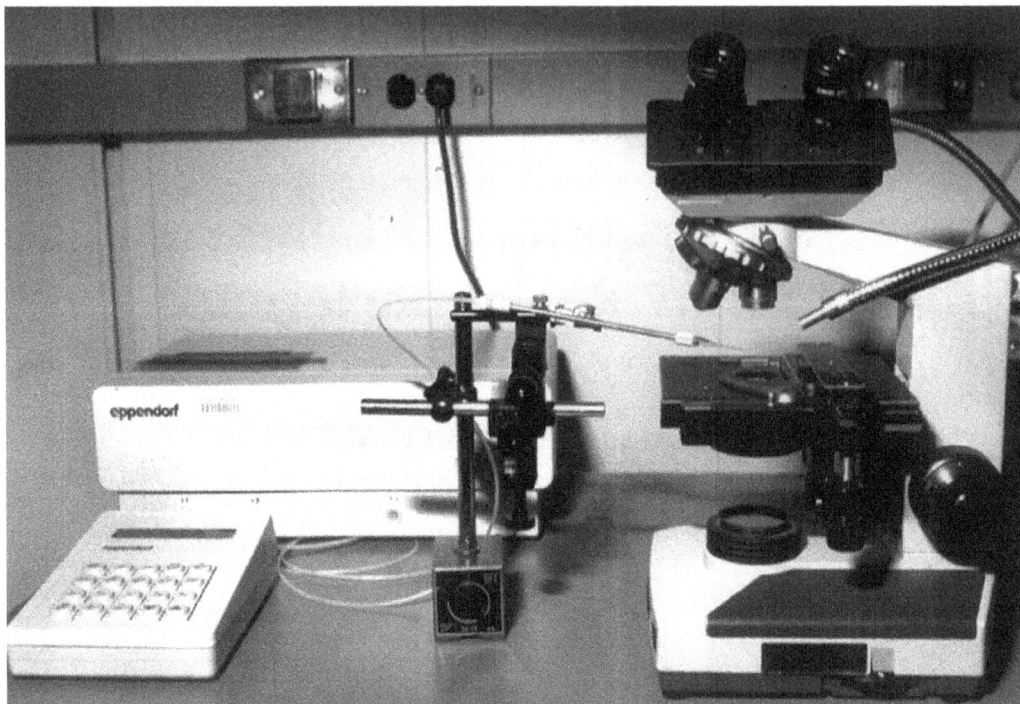

**Figure 19.1.** The injection setup with the Eppendorf Transjector Model 5246, showing the position of micromanipulator, the microscope, and the Transjector on the stone table.

Protocol 19.1 describes the preparation of dsRNA, by in vitro transcribing complementary strands of a cloned DNA fragment coding for all or a portion of the gene of interest, followed by annealing of the transcribed RNA. Embryos are collected for injection according to Protocol 19.2. Protocol 19.3 describes the setup and method for embryo injection (see also Figure 19.1). These procedures assume that the investigator has access to a working fly facility and is partially familiar with basic methods for injection and the analysis of gene expression in *Drosophila* embryos. If not, details for these basic procedures can be found elsewhere (Ashburner 1989a,b; for embryo collection, see also Rothwell and Sullivan, this volume; for microinjection procedures, see Kiehart et al., this volume).

## Protocol 19.1*

## Preparation of Double-stranded RNA

### Materials

#### Supplies and Equipment

Microcentrifuge tubes
Microcentrifuge
Water bath set at 90°C
Equipment for agarose gel electrophoresis

*For all protocols, $H_2O$ indicates glass distilled and deionized.

## Solutions and Reagents

Plasmid DNA. All or a portion of the coding region for the gene of interest is cloned into the BlueScript KS+ plasmid (Stratagene) containing T7 and T3 promoter sites flanking the multicloning site in the vector (see Protocol 19.1 Notes).

Appropriate restriction enzymes (see step 1 below)

In vitro transcription kits (Ambion, MEGAscript T7 Kit 1334 and MEGAscript T3 Kit 1338)

**Phenol:chloroform** for DNA extraction

**Ethanol**

Buffers and reagents for agarose gel electrophoresis (e.g., native 1–2% agarose gels in 0.5x TBE [see Appendix 3] containing 1 µg/ml **ethidium bromide**; appropriate DNA size markers)

DNase I

RNase-free H$_2$O

**Ammonium acetate** (3 M)

2-**Propanol**

Colored food dye. A standard food coloring kit can be purchased in the grocery store. Although any color will work, we use the green dye because it is dark and easy to spot in the embryo. Pass the dye through a 0.2-µm filter to remove any precipitate.

---

### Injection Buffer (Spradling 1986)

5 mM **KCl**

10 mM **NaH$_2$PO$_4$** (pH 7.8)

To prepare this buffer, use a 100 mM stock solution of NaH$_2$PO$_4$, adjusted to pH 7.8 with 1–5 N **NaOH**.

---

CAUTION: ammonium acetate, chloroform, ethanol, ethidium bromide, KCl, phenol, 2-propanol, NaH$_2$PO$_4$, NaOH (see Appendix 4)

## Method

1. Clone the DNA fragment to be transcribed first into BlueScript KS+ to take advantage of the T3 and T7 promoters at opposite ends of the multicloning site. Linearize a stock of 5 µg of plasmid with the appropriate restriction enzyme to generate a 5′ overhang at the run-off end of the transcribed DNA.

2. Extract the linearized DNA with phenol:chloroform and precipitate in ethanol.

3. Perform individual in vitro transcription reactions with T3 and T7 RNA polymerases as described by the manufacturer (Ambion MEGAscript Kits). For each reaction, use 1 µg of purified linearized DNA in a 20-µl reaction for 2–6 hours. (We generally use a 2-hour reaction.)

4. Treat each reaction with 2 units of DNase I at 37°C for 15 minutes.

5. To each reaction, add 115 µl of RNase-free H$_2$O and 15 µl of 3 M ammonium acetate. Extract with phenol:chloroform and precipitate the RNA with an equal volume of 2-propanol at –20°C for 1 hour. Aspirate the 2-propanol from the pellet and resuspend the RNA in 50 µl of injection buffer for annealing.

6. Analyze 1 µl of each single-strand reaction on a native 1–2% agarose gel (in 0.5x TBE with 1 µg/ml ethidium bromide in both the gel and the running buffer to check on the synthesis). The single-stranded RNA will run faster than the annealed RNA in the native gel.

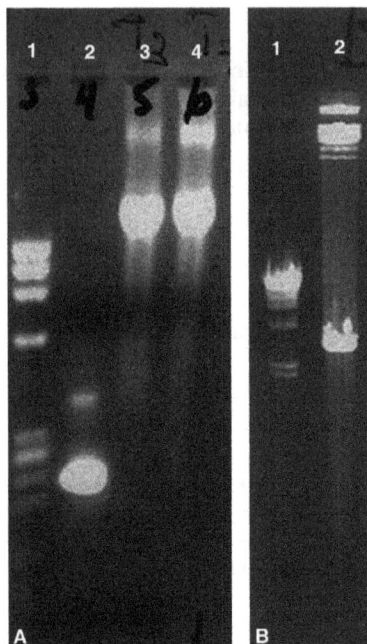

**Figure 19.2.** Native agarose gels showing the positions of single- and double-stranded RNA after synthesis and annealing, respectively. (*A*) Single-stranded β-galactosidase RNA synthesized with T7 (*lane 3*) and T3 (*lane 4*) RNA polymerase. Note that the single-stranded RNAs are slightly larger than the 1300-bp φX174 *Hae*III DNA marker (*lane 1*). (*Lane 2*) Single-stranded nautilus RNA is seen to migrate at around 310 bp. (*B*) Annealed β-galactosidase double-stranded RNA runs with the size of approximately 3 kb (*lane 2*) according to the *Hind*III λ DNA markers (*lane 1*). Note that there is aggregated RNA in the slot in lane 2, possibly due to incomplete solubilization of the single-stranded RNA or aggregation during annealing. This does not affect the activity of the dsRNA interference. We have noted that larger RNAs should be resuspended and annealed in a larger volume of injection buffer to minimize aggregation. Size markers: (Panel A) (*lane 1*) φX174 *Hae*III DNA marker (NEB): The fragment sizes from largest to smallest are (*1*) 1353 bp, (*2*) 1078 bp, (*3*) 872 bp, (*4*) 604 bp, (*5*) 310 bp, (*6b*) 281 bp, (*6a*) 271 bp, (*8*) 194 bp, (*9*) 118 bp, and (*10*) 72 bp. (Panel B) (*lane 1*) The marker is λ DNA-*Hind*III digest (NEB). The fragment sizes from largest to smallest are (*1*) 23,130 bp, (*2*) 9416, (*3*) 6557, (*4*) 4361, (*5*) 2322 bp, and (*6*) 2027 bp.

*Note:* Synthesis of β-galactosidase single-stranded RNA shown in Figure 19.2A runs just above the 1300-bp φX174 *Hae*III-cut DNA marker. Usually, the reactions are equally efficient for T3 and T7 RNA polymerase.

7. If the RNA is not concentrated enough, pool the reactions for each strand, add ammonium acetate to 0.3 M, and reprecipitate the RNA with an equal volume of 2-propanol at –20°C for 30 minutes. The RNA pellet should be clearly visible after centrifugation at 14,000 rpm (20,000g) for 10 minutes.

8. Drain the pellet well so that no obvious 2-propanol is visible. Resuspend the pellet directly in 80–100 μl of injection buffer to give a final concentration of 1–2 mg/ml dsRNA.

*Notes:* RNA concentration is determined based on absorbance at 260 nm (we assume 40–50 μg/ml dsRNA has an approximate $A_{260}$ = 1.0). We have used dsRNA samples of 0.1–2.0 mg/ml successfully.

The 80 μl of injection buffer may be a minimum volume for resuspension if the RNA transcripts are larger than 1 kb. Resuspension in less buffer causes the larger RNA transcripts to aggregate somewhat, but a substantial portion of the material is double-stranded after annealing and runs at the expected molecular weight.

**Table 19.1.** dsRNA Fragment Sizes Used to Inhibit Gene Function

| Gene | mRNA (bp) | CDS[a] (bp) | dsRNA region (bp) | % Mutant phenotype[b] |
|---|---|---|---|---|
| *daughterless* | 2813 | 150–2282 | 150–2282 (2132) | 85 |
| *Dmef2* | 2589 | 1039–2589 | 1039–2589 (1550) | 72 |
| *engrailed* | 2412 | 169–1827 | 169–1480 (1311) | 79 |
| β-*galactosidase* | 3350 | 100–3350 | 100–3350 (3250) | 80 |
| *nautilus* | 1410 | 263–1261 | 263–1261 (998) | 76 |
| nau-NH2 | | 263–703 | 263–703 (444) | 78 |
| nau-bHLH | | 704–902 | 704–902 (198) | 74 |
| nau-COOH | | 903–1261 | 903–1261 (358) | 76 |
| *twist* | 787 | 167–787 | 167–787 (620) | 86 |
| *white* | 2412 | 169–1827 | 169–1827 (1311) | 02 |
| *S59* | 2681 | 326–2305 | 326–2305 (1979) | 75 |

[a]Coding sequence.
[b]The percentage of injected embryos showing the mutant phenotype.

9. (*Annealing*) Heat the solubilized RNA in a 100–200-ml water bath to 90°C for 5 minutes. Allow to cool slowly to room temperature for 2 hours to overnight.

10. Analyze the annealed dsRNA on a native 1–2% agarose gel (see step 6, above) along with appropriate DNA markers to check for the expected dsRNA size.

    *Note:* The annealed dsRNA for β-galactosidase is now seen to run with a size of approximately 3 kb compared to the *Hin*dIII λ DNA markers (Figure 19.2B). Single-stranded RNA migrates much faster than its double-stranded form.

11. To prepare injection solution, mix 4.5 μl of dsRNA with 0.5 μl of filtered food dye.

## PROTOCOL 19.1 NOTES

We have used dsRNA representing the bHLH (basic helix-loop-helix) domain of *nautilus* (198 bp) with equal efficacy to the entire *nautilus*-coding region (998 bp) in disruption of *nautilus* function (Misquitta and Paterson 1999). Table 19.1 lists other examples we have tested. The lower size limit for effective disruption has not been determined.

## PROTOCOL 19.2

# Embryo Collection

### Materials

### *Media*

#### *Agar–Fruit Juice Plates*

Add 3 g of Bacto Agar (Difco) to 50 ml of $H_2O$ in a 200-ml beaker. Place the beaker in a microwave oven to dissolve the agar. Add 50 ml of prewarmed (60°C) fruit juice (grape or mango punch, any will probably work), 1 ml of a 10% solution (w/v) of methyl-*P*-hydroxy benzoate (Sigma H 6654) in **ethanol**, and 1 ml of **glacial acetic acid**. Pour into petri plates or collection dishes of choice to a depth of approximately 5 mm. The volumes can be scaled up as needed.

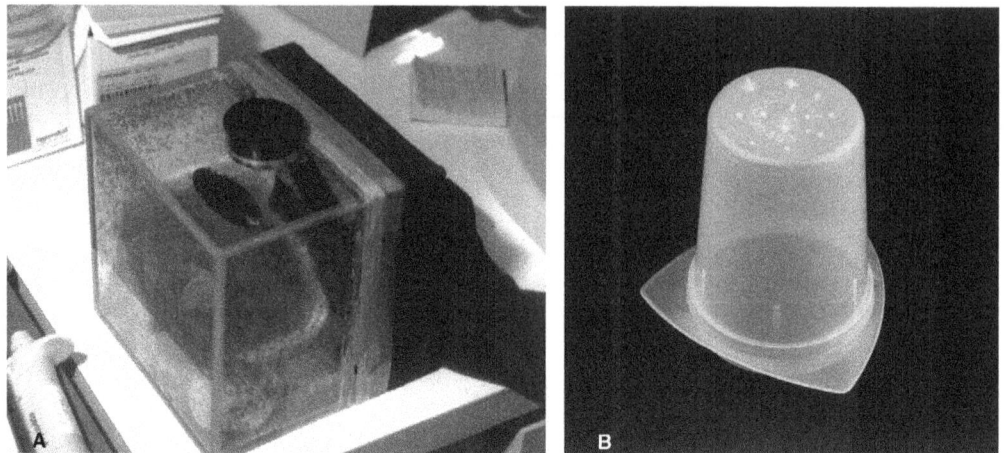

**Figure 19.3.** Types of collection cages for embryo injections. (*A*) Large collection cage (6.5 inches per side) that holds 15-cm collection plates. (*B*) 60-mm plastic beaker with holes in the bottom over a small petri plate filled with agar-fruit juice mixture for small collections.

## Supplies and Equipment

Glass slides (1 × 3 inches, 1-mm thick)

Stereomicroscope/dissecting microscope. We use the Reichert Stereo Star microscope, 0.7–4.2× magnification.

Fly cages for embryo collections. Place a small plastic beaker (60-mm diameter) with needle holes in the bottom over a 60-mm Agar–Fruit Juice Plate for small collections of embryos. Alternatively, use larger cages (6.5 inches on a side) depending on the convenience (see Figure 19.3). Collections are usually done at 24ºC (for timing, see step 1, below).

Nylon-mesh basket. The basket can be conveniently made from a 50-ml blue-top centrifuge tube cut in half. Cut a hole in the lid with a heated no. 15 cork-borer, and then fit the lid back onto the threaded end of the tube over a layer of Nitex filter cloth (Figure 19.4A). This basket is also required for collecting the embryos after microinjection.

Fine brush (#01)

Diamond pencil or fine black marker

**CAUTION: ethanol, glacial acetic acid** (see Appendix 4)

## Method

1. Set up cages at 24ºC using 2–4-day-old yw67c flies.

    *Note:* Agar–Fruit Juice Plates are changed every hour during the day to synchronize the egg collection over a 1–2 day period (*this is not done overnight!*).

2. Collect the eggs for 30–60-minute intervals just before injection.

3. Transfer the eggs into a nylon-mesh basket by washing with room temperature tap water.

4. Wash the eggs in room-temperature tap water. Use a fine brush to transfer approximately ten eggs at a time to a slide.

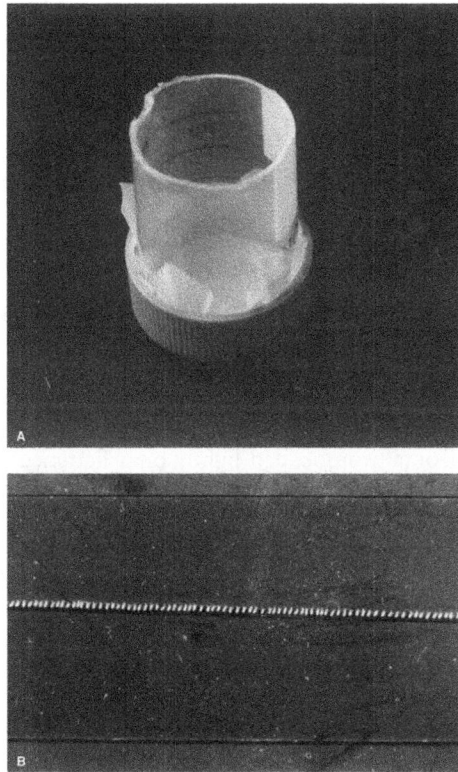

**Figure 19.4.** (*A*) Collection and wash basket made from a 50-ml blue-top disposable plastic centrifuge tube. (*B*) Fully chorionated embryos lined up for injection.

5. Line up the embryos side by side with their posterior ends perpendicular to a finely marked line running lengthwise down the center of the slide on the opposite side of the slide. Place approximately 50–60 embryos per slide (Figure 19.4B). The line can be made with a diamond pencil or a very fine black marker.

## PROTOCOL 19.2 NOTES

While lining up the embryos on the slide, keep the embryos relatively wet. This can be accomplished by dipping the brush lightly in water before the embryos are lined up. Approximately an embryo's width should be kept between each embryo on the slide to avoid clumping. Keeping embryos moist on the slide allows them to attach while drying to the slide (see Figure 19.4B). Normally, three to four slides can be lined up before injection if the embryos are kept in the moist chamber prior to injection. If the embryos dry, the chorion becomes too hard to inject (very important!). In our experience, pretreatment with an ethanol:water mixture did not improve injections through the chorion. Ethanol actually dried the embryos faster and made injection more difficult. Embryos should be injected within 30–60 minutes after collection.

# Embryo Injection

## Materials

### Supplies and Equipment

Glass capillaries for needles (Frederick Haer and Co., Bowdoinham, Maine 04008; 30-30-0-075). Note that these capillaries are fiber-filled, which helps in fluid movement within the needle.

Plastic back-loading pipette tips. Needles are back-loaded with Eppendorf Microloader tips (5242 956.003; 200 tips per box) using an Eppendorf P10 pipette.

Needle puller (Kopf Needle-Pipette puller, Model 730; David Kopf Instruments, Tujinga, California). For pulling needles, set heat at 12.5–13.0 and set the solenoid at 4. Store freshly pulled needles horizontally, imbedded lightly on a flattened narrow strip of modeling clay placed on the bottom of a 150-mm covered plastic petri dish.

Eppendorf Transjector Model 5246. Although primarily used for the injection of cultured cells, this Transjector is easily adapted for the injection of *Drosophila* embryos directly through the chorion. One advantage of this Transjector is that highly reproducible injection conditions can be defined with regard to the volume of material injected, while preventing backflow of material into the needle, which results in needle blockage. This is accomplished by varying the injection pressure, compensation pressure, and the time of injection. The Transjector was a key factor in being able to routinely inject approximately 500 embryos for each analysis. It also has a useful "clean button" that utilizes a burst of air pressure to clean the needle tip in case of blockage. This greatly reduces the number of times needles need to be changed. Figure 19.1 shows the Transjector and the injection setup. The Eppendorf Transjector model 5246 costs approximately $6000.

Injection setup (see Figure 19.1)
- Olympus CH-2 microscope (CHT-001A) for injection, or equivalent
- micromanipulator, Narishige model MN-153
- stone table for microscope and injection apparatus

Moist chamber such as a 150-ml agar collection plate with lid (required for storing slides after embryo injections). Five slides can be placed conveniently in the chamber.

Collection basket used in Protocol 19.2

Microcentrifuge tubes

## Method

### Filling the Needles

Back-load needles using the Eppendorf P10 pipette fitted with Eppendorf Microloader tips. Normally, we load 0.5–1.0 μl of dsRNA mixed with food coloring dye (from step 11, Protocol 19.1).

### Calibrating the Transjector

Based on the volume loaded into the needle (0.5–1.0 μl), each embryo receives approximately 100–200 pl of dsRNA at most. A typical loading, if the needle remains unbroken and unblocked, is good for the injection of approximately 1000 embryos or more.

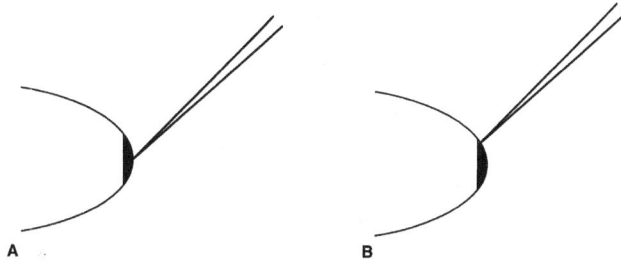

**Figure 19.5.** Correct injection position for the embryos. (*A*) Do not inject the embryos directly in the posterior center of the chorion, as the needles often break at this site due to the apparent increased hardness of the chorion. (*B*) Embryos are injected in the posterior end, slightly off-center, with much less needle breakage or blockage.

Adjust the settings for the 5246 Transjector as follows: set the Transjector to automatic injection; set the injection pressure (Po) to 450; set the compensation pressure (Pc) to 570; set the time of injections to 0.1 seconds. We have two Transjectors in the department and these settings work well on both instruments.

## *Injection*

Inject the embryos in the posterior end, slightly off-center (Figure 14.5), as the posterior tip of the chorion is very hard and needles often break in this position. Keep the injection room at 18°C to slow development.

1. Before injection, break the filled needle against the edge of the slide containing the embryos to create a sharp point. This is accomplished by moving the slide slowly toward the needle tip to tap it lightly while simultaneously depressing the "clean" button on the Transjector. The minute the tip is broken, a little of the dye will leak out back across the needle tip. Dye flow should stop when the "clean" button is no longer depressed. The sample is now ready for injection. A properly broken needle is shown in Figure 19.6.

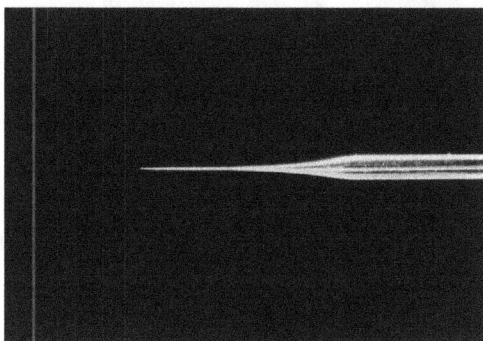

**Figure 19.6.** A properly broken needle ready for injection.

*Note:* If the tip is too large after breakage, the embryos will be destroyed during the injection. This will be obvious if the needle continues to drip dye when the "clean" button is not depressed. If dye flow stops after the "clean" button is released, then the needle is usually good.

2. Bring the needle tip and the posterior tip of the first embryo into the same focal plane and inject the embryo off-center (very important; see above and Figure 19.5). A small amount of food dye will appear as a small dot in the posterior end of the embryo. It takes approximately 1–5 seconds to inject per embryo.

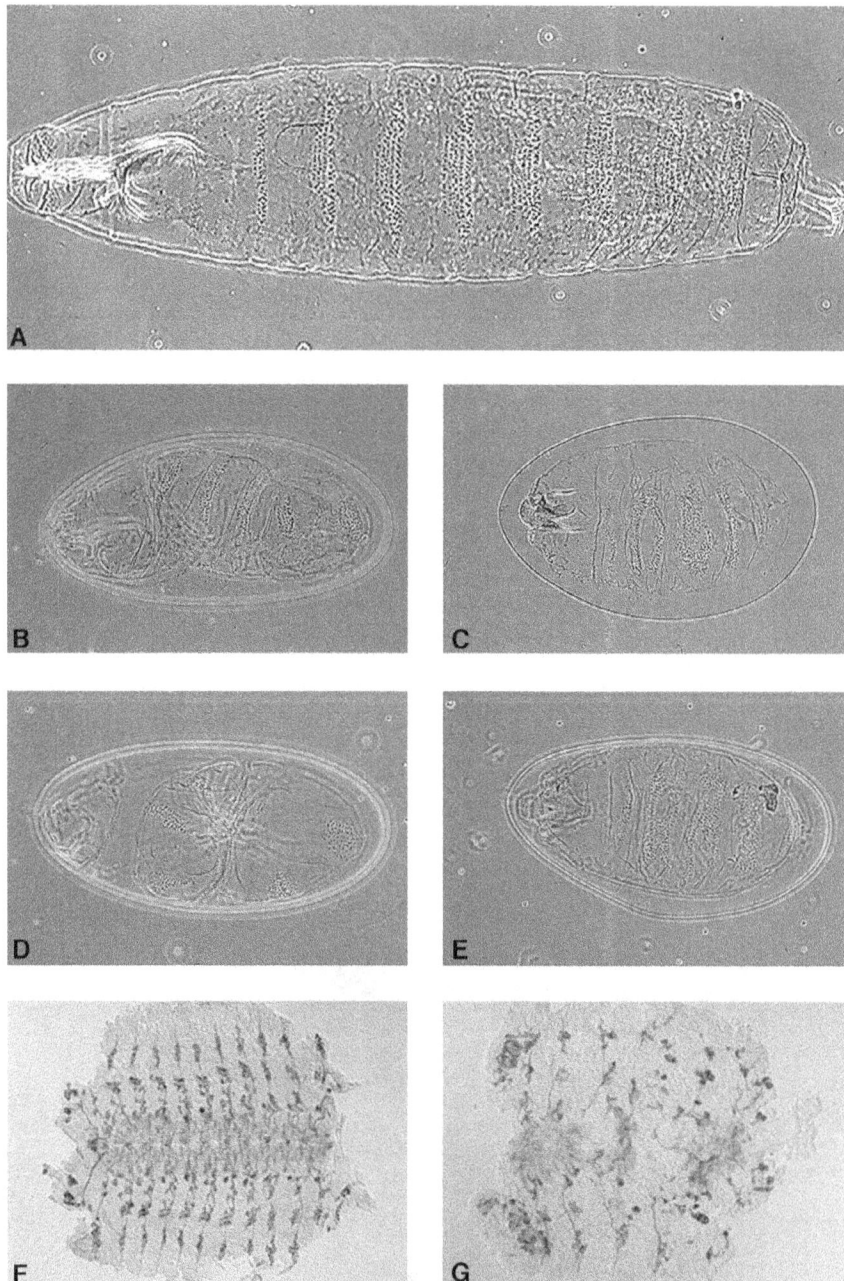

**Figure 19.7.** Phenotypes for RNA-i after the injection of dsRNA for *twist* (*D*), versus the *twist* 1096 mutant larva (*B*); *engrailed* (*E*) versus the *engrailed* IIB mutant larva (*C*); and *daughterless* showing the PNS and CNS in filleted embryos (*G*) versus the normal pattern (*F*). (*Panel A*) Wild-type larval pattern.

*Notes:* Embryos should be injected within 30–60 minutes after collection.

    The most common difficulty in injecting through the chorion is needle blockage. To minimize capillary backflow and blockage, do not push the needle too far into the embryo, but just far enough to penetrate the vitelline membrane. Make sure to inject off-center as shown in Figure 19.5. If blockage should occur on the Transjector, immerse the needle tip in drop of coverslip oil and hold down the "clean" button. When the needle clears, the food coloring will be visible in the oil. If blockage persists, change the needles.

3. Place the slide in a covered moist chamber at 18–22°C until completion of embryogenesis (~48 hours) or until the desired stage of development is reached.

    *Note:* It is critical to keep the embryos moist after injection if normal development is to occur. Punctured embryos tend to shrink and dry up without a humid environment. Likewise, excessive moisture, i.e., embryos floating in water, will also kill the embryos.

4. Wash the embryos off the slide into the original collection basket, transfer to a microcentrifuge tube, fix, and stain using standard protocols (Ashburner 1989a,b).

## PROTOCOL 19.3 NOTES

In our studies, typical efficiencies for generation of the mutant phenotype in the embryo ranged from 72% to 86%, but penetration of the *white* mutation to the adult eye was less than 3% (Misquitta and Paterson 1999). Figure 19.7 shows the patterns for both the authentic mutant and dsRNA-injected *twist* and *engrailed* IIB86 embryos along with the neuronal patterns for the central and peripheral nervous systems (CNS and PNS, respectively) after injection of *daughterless* dsRNA. Independent of our studies, RNA-i was also recently injected into dechorionated embryos to demonstrate that both *frizzled* and *frizzled 2* were in the *wingless* pathway (Kennerdell and Carthew 1998). For rescue of injected transgenic lines, apply a small amount of yeast paste to the slide and transfer the embryos from the paste to food vials.

## REFERENCES

Ashburner M. 1989a. Drosophila: *A laboratory handbook.* Cold Spring Harbor Laboratory Press, Cold Spring Harbor, New York.

———. 1989b. Drosophila: *A laboratory manual.* Cold Spring Harbor Laboratory Press, Cold Spring Harbor, New York.

Fire A., Xu S., Montgomery M.K., Kostas S.A., Driver S.E., and Mello C.C. 1998. Potent and specific genetic interference by double-stranded RNA in *Caenorhabditis elegans* (see comments). *Nature* **391:** 806–811.

Kennerdell J.R. and Carthew R.W. 1998. Use of dsRNA-mediated genetic interference to demonstrate that frizzled and frizzled 2 act in the wingless pathway. *Cell* **95:** 1017–1026.

Michelson A.M., Abmayr S.M., Bate M., Arias A.M., and Maniatis T. 1990. Expression of a MyoD family member prefigures muscle pattern in *Drosophila* embryos. *Genes Dev.* **4:** 2086–2097.

Misquitta L. and Paterson B.M. 1999. Targeted disruption of gene function in *Drosophila* by RNA interference (RNA-i): A role for nautilus in embryonic somatic muscle formation. *Proc. Natl. Acad. Sci.* **96:** 1451–1456.

Montgomery M.K., Xu S., and Fire A. 1998. RNA as a target of double-stranded RNA-mediated genetic interference in *Caenorhabditis elegans. Proc. Natl. Acad. Sci.* **95:** 15502–15507.

Paterson B.M., Walldorf U., Eldridge J., Dubendorfer A., Frasch M., and Gehring W.J. 1991. The *Drosophila* homologue of vertebrate myogenic-determination genes encodes a transiently expressed nuclear protein marking primary myogenic cells. *Proc. Natl. Acad. Sci.* **88:** 3782–3786.

Spradling A.C. 1986. P element-mediated transformation. In Drosophila: *A practical approach* (ed D.B. Roberts), pp. 178–197. IRL Press, Oxford.

# CONTENTS

# 20

# *Drosophila* Cell Culture and Transformation

**Lucy Cherbas and Peter Cherbas**
*Department of Biology*
*Indiana University*
*Bloomington, Indiana 47405*

PERMANENT *DROSOPHILA* CELL LINES DERIVED FROM MIXED EMBRYONIC TISSUES, including the most commonly used lines, S2 and Kc, have been available for approximately 30 years. More recently, lines derived from specific tissues, imaginal discs, and the larval central nervous system (CNS) have come into use. Although cultured cells were originally used by drosophilists mainly as convenient sources of DNA or carrier RNA, that situation has changed as an armamentarium of techniques for using the cells has slowly but steadily evolved. We have reviewed many of the techniques quite recently (Cherbas et al. 1994; Cherbas and Cherbas 1998). In this chapter, we provide an organized collection of pointers to published protocols, of comments based on more recent experience, and a few corrected or modified protocols.

Cherbas and Cherbas (1998) provides a brief account of the histories of the cell lines with speculations on the histological origin(s) of the embryonic lines and with a tabular account of much that is known about gene expression in various lines. For a broader biological review of the lines, see Schneider and Blumenthal (1978), Sang (1981), Cherbas and Cherbas (1981), and Echalier (1997). For cell maintenance and cloning protocols, see Cherbas and Cherbas (1998), and for transformation protocols, see Cherbas et al. (1994), supplemented, for the S2 expression system, by Kirkpatrick and Shatzman (1999).

Most investigators use *Drosophila* cell lines as hosts for transformation experiments. The goal may be to characterize a promoter, to investigate the role of a transcription factor, to overexpress a polypeptide, or to do something more novel. Figure 20.1 illustrates two novel and effective uses of stably transformed S2 cells.

Whatever the goal, it is important to appreciate that although the many embryonic cell lines are similar in properties, they are not identical. For example, S2 cells respond very poorly to the hormone ecdysone, but Kc cells exhibit a vigorous response. The first step is to choose a cell line that meets one's requirements.

There is, as yet, no central repository for *Drosophila* cell lines. S2 and Kc cells are widely used; obtaining starter cultures should present no problems. (Be aware, however, that

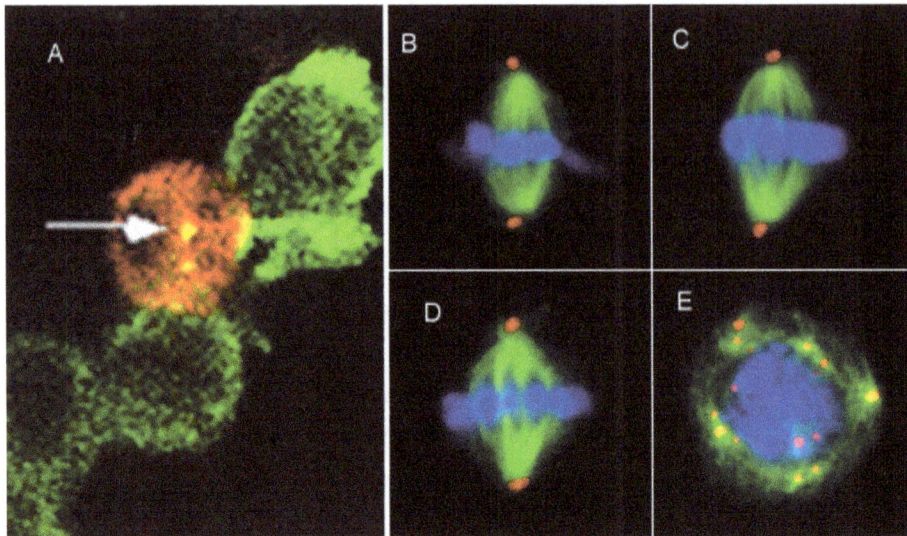

**Figure 20.1.** Examples of the uses of stably transformed S2 cell lines. (*A*) Mixed aggregate of cells from two stably transformed clonal lines, one transformed with a plasmid expressing Delta (Dl) from a metallothionein (Mt) promoter and the other transformed with a plasmid expressing Notch (N) from a Mt promoter. Cells from the two lines were mixed, treated with Cu⁺⁺, and immunostained for Dl (*green*) and N (*red*). Untransformed cells express no detectable N or Dl. The arrow indicates a vesicle within the N-expressing cell, which has taken up Dl protein. (*B–E*) Effect of high-level expression in mitotic S2 cells of the testis-specific *centrosomin* product CNN4 (Li et al. 1998). Stably transformed populations of S2 cells were selected following transfection with either the methotrexate resistance plasmid p8HCO alone (*B,C*) or p8HCO together with a plasmid in which CNN4 is expressed from a Mt promoter (*D,E*). (*C,E*) Cells were treated with Cu⁺⁺ to induce the Mt promoter. Cells were stained for centrosomes (*red*), tubulin (*green*), and DNA (*blue*). Untransformed cells, without or with Cu⁺⁺ treatment, are indistinguishable from the pattern seen in panels *B*, *C*, and *D*. Induction of the CNN4 protein leads to a proliferation of centrosomes and disorder of the tubulin array. (Panel *A* is courtesy of K. Klueg and M. Muskavitch and reprinted from Klueg et al. 1998 [© American Society for Cell Biology], and Panels *B–E* were kindly provided by T. Megraw.)

lab histories are sometimes vague and/or confused; e.g., the S2 and S3 lines have occasionally been interchanged—more reason for a preliminary check of a line's properties.) The newer lines can generally be obtained from their originators. If finding the necessary lines is a problem, an inquiry to bionet.Drosophila (see the Web site below) or to the authors of this chapter can sometimes help.

http://www.bio.net:80/hypermail/DROS/

A second step is to understand the virtues and limitations of the various transformation procedures available, which are detailed later in this chapter and outlined below:

- *Transient expression* provides a quantitative assessment of reporter gene expression. Cotransformation with expression vectors for other genes allows the assessment of their effects on the reporter gene's expression. The advantages of transient expression are that it is fast and statistically meaningful, as one is observing average behavior for a population of cells. The disadvantage is that only a minority of cells express—hence, one cannot look for losses of function or determine much about titers (e.g., of a coexpressed transcription factor) in the relevant population.

- *Stable transformation* generates cells containing high (>1000) numbers of copies of the vector in question or low-copy numbers. Either can be accomplished using either illegitimate recombination or transposition.

  *High-copy numbers* are of interest principally as a way to overexpress proteins. Usually, one uses S2 cells in which illegitimate recombination leads directly to long chromosomal direct repeats (arrays) containing the gene in question. When the gene is not toxic, this expression system is simple and relatively fast. For example, S2 cells can readily be made to express ADH (alcohol dehydrogenase) mRNA as 5% of the poly(A)$^+$ population. For a detailed account of applications, see Kirkpatrick and Shatzman (1999). Because of interactions within the array, these are not useful for gene-regulation studies. Transposition can also be used to generate high-copy numbers, but there is no published background.

  *Low-copy number* transformants can also be generated by illegitimate recombination, particularly in Kc cells. A better procedure is P-element-mediated transposition, which can readily yield single-copy transformants. In either case, one selects transformed cells using a selectable marker and works either with the population or with specific clones. The advantages of this transformation procedure are stable expression in the entire population and the ability to quantify expression levels when working with clones. The disadvantages include the time involved: Cloning requires 3–6 weeks depending on the health of the transformed cells. In addition, individual clones can vary in ways that are unrelated to the transformation, and thus a collection of clones must be studied.

- *Parahomologous targeting* occurs at high frequency when *Drosophila* cells are transformed with a plasmid containing 5–10 kb of chromosomal homology. It reveals itself as illegitimate insertions (containing vector sequences) within a few kilobases of the target site. This can be useful as a way to knock out genes.

## REQUIREMENTS FOR CELL CULTURE

### Culture Media

For a detailed discussion of culture media, see Cherbas and Cherbas (1998). The situation for casual cell users has been simplified by the fact that most common media for *Drosophila* cell culture are now commercially available. For example, most lines can be grown in M3 medium supplemented with 10% fetal calf serum, and M3 is available commercially both as liquid and as dry powder. For our larger-scale needs, we continue to manufacture M3 medium using the recipe in Schneider and Blumenthal (1978). Antibiotics are not in general necessary. For serum and its preparation by heat treatment and for more information about needles, see Cherbas and Cherbas (1998).

We have found that some of the commercial media designed for lepidopteran cells work satisfactorily for *Drosophila* cells. For example, the HyClone product CCM-3 is a convenient medium for Kc cells. It is particularly attractive in that its (proprietary) formulation obviates the need for serum. However, we have recently discovered that ecdysone-induced gene expression is unreliable in CCM-3. *The general point is not that one medium is universally best, but that the investigator cannot assume that the process of interest is insensitive to medium changes; it is important to test each medium for precisely those experiments it is intended to serve.*

## Saline

To wash cells, we use Robb's minimal saline (Robb 1969) formulated as follows:

### Solution A

| Reagent | Amount to add to make 1 liter |
|---|---|
| NaCl | 3.04 g |
| **KCl** | 2.98 g |
| Glucose | 1.80 g |
| Sucrose | 34.23 g |
| **$MgSO_4 \cdot 7H_2O$** | 0.28 g |
| **$MgCl_2 \cdot 6H_2O$** | 0.25 g |
| **$CaCl_2 \cdot 2H_2O$** | 0.15 g |
| $H_2O$ | to 1 liter |

### Solution B

| Reagent | Amount to add to make 1 liter |
|---|---|
| $Na_2HPO_4 \cdot 2H_2O$ | 0.356 g |
| **$KH_2PO_4$** | 0.050 g |
| $H_2O$ | to 1 liter |

Adjust to pH 6.75 with 1 N HCl. Autoclave Solutions A and B separately; when cool, mix in 1:1 ratio.

**CAUTION: $CaCl_2$, KCl, $KH_2PO_4$, $MgCl_2$, $MgSO_4$, $Na_2HPO_4$** (see Appendix 4)

## CELL GROWTH AND MAINTENANCE

### General Laboratory Practices

Although culture methods for *Drosophila* cells are, in many respects, quite forgiving, successful work does require disciplined laboratory hygiene, thoughtful attention to the ways in which routine maintenance techniques select for desirable or undesirable properties, and constant attention to the need for sterile conditions. For detailed recommendations, see Cherbas and Cherbas (1998) and any general manual on cell culture. Below, we highlight a few essential considerations, including:

- Maintaining sterility is more challenging for cell culture than for bacteriology. The medium is very rich and the cells of interest grow much more slowly than most likely contaminants. Although some investigators have managed to maintain cells for short periods at the laboratory bench, the best solution is to use a laminar flow hood and adopt high standards of sterile technique.
- Because *Drosophila* cells do not ordinarily harbor human pathogens, it is acceptable to work in a pharmacological laminar flow hood, rather than in the more awkward (and more expensive) biosafety cabinet.
- Media must be prepared using highest purity water. We have used a Milli-Q purification system with success; other investigators have used double-distilled water. In any event, avoid building distilled water that has been transported through metal pipes, at least for long-term work. Low-grade medium toxicity, whatever its cause, can take days or weeks to develop and even longer to diagnose; it is better to simply avoid forseeable problems.

- Spinner flasks are generally used for large-volume cultures. Since the presence of even a single bacterial or fungal spore leads to substantial waste of both medium and time, we take special precautions to sterilize these flasks: Fill the spinner with clean water. Autoclave, and leave at room temperature for 24 hours. Autoclave again. Pour off the water before use. The double autoclaving is sufficient to eliminate rare spores whose germination is induced by heat. The water leaches out chemical contaminants (e.g., soap residue) and detectably improves the health of the cultures.

## Equipment Needed

*Drosophila* cells grow in an air atmosphere; a $CO_2$ incubator is not required. We routinely grow our cultures in petri dishes (tissue-culture grade) stored in air-tight, plastic food storage containers to maintain high humidity. Because they grow optimally at 25ºC, it is possible to maintain *Drosophila* cells on the bench top. Be aware that some building temperature control systems are erratic enough to induce heat shock.

For large volumes, spinner flasks work well, but the cells grow somewhat more slowly, and appear more battered, than in stationary culture. It is important not to overfill spinner flasks; use no more than half the nominal capacity. The stirring bar should turn just quickly enough to maintain the cells in suspension, usually approximately 60 rpm. We maintain cell lines in petri dishes and use spinners only to expand populations for experiments.

Kc and S2 cells grow exponentially between approximately $5 \times 10^5$ and $10^7$ cells/ml and are healthiest if kept within this range. The cells double approximately every 24 hours, which generally means diluting them tenfold every 3–4 days. Cell density is easily determined using a standard hemocytometer. All of the *Drosophila* lines that we have worked with are either nonadherent or adhere to the substrate loosely enough so that they can be dislodged by blowing medium at the surface with a pasteur pipette. We do not use trypsin. For most purposes, it is adequate to simply dilute an aliquot of a dense culture in fresh medium. If desired, cells can be pelleted by gentle centrifugation ($\sim700g$ for 2 minutes), and then resuspended by pipetting up and down in fresh medium.

*Drosophila* cells appear to be much less subject to gross changes during culture than are mammalian cells; the karyotype is generally stable, and changes in cell properties with time in culture are usually subtle. Nevertheless, changes do occur. To minimize problems, we maintain many ampoules of frozen cells from which we restart our incubator cultures approximately every 6 months. Procedures for freezing and thawing viable *Drosophila* cells are given elsewhere (Cherbas et al. 1994; Cherbas and Cherbas 1998).

## MISCELLANEOUS TECHNIQUES

Procedures for the following techniques are given below:

- *Protein and nucleic acid labeling.* Newly synthesized proteins are readily labeled with a radioactive amino acid, using medium lacking the amino acid in question, fetal calf serum, and yeast extract (see Protocol 20.1). This procedure gives linear incorporation of $^3H$-labeled leucine for 1 hour, at a rate of 300 cpm/hour/$10^3$ cells (Savakis et al. 1980; Savakis 1981). Bieber (1986) labeled newly transcribed RNA to high specific activity in cells that had been minimally manipulated. He simply added [$^3H$]uridine to cells growing in complete culture medium. The rate of transcription was much higher in undisturbed cells than in cells that were concentrated before labeling. Furthermore, in undisturbed cells, early hormone-induced changes in transcription were detected that failed to occur when the cells were disturbed. Incorporation was linear for about 10 hours, at a rate of approximately $2 \times 10^4$ cpm/μg of RNA/hour (where the labeled uridine was present at 0.037 mCi/ml); the specific activity was proportional to the input label up to 0.1 mCi/ml.

- *Karotype analysis.* For karyotype analysis, we have used Protocol 20.2 successfully for a number of *Drosophila* cell lines. For an alternative protocol, see Echalier (1997).

- *Cell cultivation on polylysine-coated surfaces.* Most commonly used *Drosophila* cell lines adhere only loosely to the substrate. For some purposes (e.g., time-lapse photography and cytological examination of living cells), it is desirable to fix the growing cells more firmly to the substrate. This can be accomplished by treating the substrate with polylysine, with no obvious effect on the growth of the cells (see Protocol 20.3).

PROTOCOL 20.1*

## Protein Labeling

### Materials

### *Media*

Modified medium lacking leucine, yeast extract, and serum. M3 medium lacking leucine and methionine is available in powdered form from Sigma.

### *Supplies and Equipment*

Clinical centrifuge, tubes

### *Solutions and Reagents*

³H-labeled leucine (5 mCi/ml) in modified medium
Robb's saline (see p. 376) containing 1 mg/ml leucine

**CAUTION: radioactive substances** (see Appendix 4)

### Method

1. Harvest $10^8$ cells by centrifugation; dry the walls of the centrifuge tubes by blotting.

2. Resuspend the cells in 25 μl of modified medium (lacking leucine, yeast extract, and serum).

3. Incubate at 25°C for 5 minutes to deplete the intracellular leucine pool.

4. Add 75 μl of ³H-labeled leucine (5 mCi/ml) in modified medium.

5. Continue incubation at 25°C.

6. Stop the incorporation of ³H-labeled leucine by addition of 4 ml of Robb's saline containing 1 mg/ml leucine.

7. Harvest the labeled cells by centrifugation.

*For all protocols, $H_2O$ indicates glass distilled and deionized.

PROTOCOL 20.2

# Karyotype Analysis

## Materials

### Supplies and Equipment

Clinical centrifuge, centrifuge tubes
Microscope slides and coverslips

### Solutions and Reagents

**Vinblastine sulfate**, available from Sigma as a powder; prepare the stock solution (0.7 mg/ml) in ethanol.
Robb's saline (see p. 376)
**Acetic acid/ethanol** (3:1) fixative (freshly prepared)
**Acetic acid**/orcein/carmine (Ashburner 1989, p. 332) or 4% **Giemsa**

**CAUTION: acetic acid, ethanol, Giemsa, vinblastine sulfate** (see Appendix 4)

## Method

1. To a dish of exponentially growing cells, add vinblastine sulfate to a final concentration of 1.2 µg/ml, using the 0.7 mg/ml stock solution. Incubate at 25ºC for approximately 2 hours.

2. Centrifuge cells in a clinical centrifuge (~700g) for 2–3 minutes.

3. Resuspend in Robb's saline to a volume of 0.5 ml. Disperse cells with a pasteur pipette.

4. Add 0.5 ml of $H_2O$, mix (by swirling) very gently, and incubate at room temperature for 10 minutes.

5. Add 0.5 ml of $H_2O$, mix very gently, and incubate at room temperature for 5 minutes.

6. Add about 10 drops of acetic acid/ethanol fix. Mix very gently with a pasteur pipette.

7. Centrifuge in clinical centrifuge. Remove supernatant.

8. Add 2 ml of fix and gently dislodge the pellet by blowing over it with a pasteur pipette. Mix slightly more vigorously and leave for 10 minutes.

9. Mix thoroughly with a pasteur pipette. Add another 1 ml of fix.

10. Centrifuge the cells in a clinical centrifuge. Remove the supernatant.

11. Suspend cells in approximately 1 ml of fix to give a cloudy suspension.

12. Use a pasteur pipette to put one or two drops of cell suspension on a microscope slide held almost vertically; dry slide by vigorous shaking.

13. Stain chromosomes in acetic acid/orcein/carmine or in Giemsa.

PROTOCOL 20.3

# Treatment of Surfaces with Polylysine

## Materials

### Supplies and Equipment

Glassware, e.g., coverslips, slides, or petri dishes; alternatively, plasticware could be used
(see Protocol 20.3 notes, below).

### Solutions and Reagents

Poly-L-lysine hydrobromide (m.w. 150,000–300,000; Sigma P 1399), 1 mg/ml in sterile
H$_2$O. This solution can be stored at 4ºC and reused for at least 1 month.
**Nitric acid**, concentrated

**CAUTION: nitric acid** (see Appendix 4)

## Method

1. Clean glassware (e.g., coverslips, slides, or petri dishes) in concentrated nitric acid.

2. Rinse very well in tap water and then ten times in pure H$_2$O. Drain.

3. Soak for 15 minutes in poly-L-lysine stock solution.

4. Drain well (poly-L-lysine solution can be reused). Rinse ten times in pure H$_2$O.

5. Dry slides and autoclave or bake to sterilize.

## PROTOCOL 20.3 NOTES

To coat a plastic surface, the poly-L-lysine solution need only be 100 µg/ml. Since most
plasticware cannot be baked or autoclaved after coating, use filter-sterilized poly-L-lysine.

High concentrations of protein in the medium can inhibit cells from adhering tightly
to polylysine-coated surfaces. If this is a problem, allow cells to adhere to the treated surface in serum-free medium for 1–2 hours, and then replace medium with the normal
growth medium.

## PREPARATION OF DNA AND RNA

*Drosophila* tissue culture cells present no special problems for the isolation of DNA or
RNA. We use the method of Chomczynski and Sacchi (1987) for RNA preparation (yields
~200 µg total RNA/10$^7$ Kc cells) and the method of Maniatis et al. (1982) for DNA preparation (yields ~5 µg/10$^7$ Kc cells).

## CLONING CELLS

In our experience, any *Drosophila* cell line can be cloned by limiting dilution (in 96-well
plates) provided the growth of isolated cells is supported by the presence of a suspension
of X-rayed feeder cells. Some lines (notably S2) can be cloned in soft agar, at a considerable saving of labor and money. Detailed procedures for both techniques have been published recently (Cherbas et al. 1994; Cherbas and Cherbas 1998).

## TRANSFECTION METHODS

A number of techniques are available for the introduction of exogenous DNA into cultured *Drosophila* cells. The following are the most commonly used:

- *Calcium phosphate–DNA coprecipitation.* This technique is the oldest and probably the most commonly used. It requires no special equipment and is very inexpensive. The principal drawbacks of this technique are that (1) it works for only some cell lines and (2) it requires a fixed amount of DNA (20 μg/ml precipitate), thereby placing limits on experimental design.
- *Electroporation.* This technique is much less labor-intensive than calcium phosphate–DNA coprecipitation. It can be used for the introduction of a variety of molecules into cells, including dsRNA (double-stranded RNA), proteins, and small molecules. The technique confers no apparent restrictions on the amount of DNA being transfected and appears to work on all cell lines. The principal drawbacks are that (1) it requires a relatively expensive (several thousand dollars) piece of equipment and (2) the parameters for shocking the cells must be optimized for each cell line.
- *Lipofection.* This technique, like electroporation, apparently works for all *Drosophila* cell lines (V. Panin and K. Irvine, pers. comm.) and permits the introduction of a wide variety of types and concentrations of molecules. It requires no special equipment, but the reagents are relatively expensive. We have had no experience with the procedure, but others have found a variety of commercial lipofection systems to work well with *Drosophila* lines (e.g., Søndergaard 1996). Transfection of S2 cells with the lipofection reagent FuGene 6 (Roche Molecular Biochemicals) gives levels of reporter expression similar to those achieved with a similar amount of DNA delivered by the calcium-phosphate procedure (R. Flores-Saaib, pers. comm.). Lipofection is a relatively new technique, and the commercially available reagents for lipofection are rapidly decreasing in price and increasing in variety.

Each method causes at least some disruption to the normal physiology of the cells. In our experience, this is more of a problem with calcium phosphate than with electroporation. It is therefore a good idea to check the effect of transfection itself on the physiological process being studied.

Procedures for transfection by calcium phosphate–DNA coprecipitation and by electroporation are given in our earlier reviews (Cherbas et al. 1994; Cherbas and Cherbas 1998). The electroporation protocol for Kc cells (Cherbas et al. 1994) can be modified for S2 cells by increasing the voltage from 440 V/cm (Kc) to 715 V/cm (S2) (Cherbas and Cherbas 1998); the higher voltage also works well for S3 cells, for the *shibire* line EH34A3, and for the haploid line D (K. Klueg and M.A.T. Muskavitch, pers. comm.).

## TRANSIENT EXPRESSION

Transient expression is a simple, rapid procedure for examining the expression of exogenous DNA in cultured cells. DNA is introduced (by any of the transfection procedures described above), and its expression soon thereafter is monitored in the heterogeneous population of transfected cells. Typically, levels of expression in individual cells vary over a wide range, with a large fraction of the expression due to a small fraction of the cells. This is certainly true following calcium phosphate–DNA coprecipitation in Kc cells (Andres 1990), and preliminary data suggest that the pattern of expression is similar following electroporation in S2 cells (K. Klueg, pers. comm.). The following issues should be considered in designing a transient expression experiment:

- *Reporters.* The reporters used most commonly for mammalian cells all work for *Drosophila* cells: *Escherichia coli* β-galactosidase (*lacZ*), chloramphenicol acetyltransferase (CAT), and firefly luciferase. β-galactosidase is easily assayed without special equipment; however, a background activity of *Drosophila* β-galactosidase limits the sensitivity of the assay. The endogenous enzyme is particularly a problem in studies of ecdysone responses, because the *Drosophila* gene is induced by the hormone. CAT is a useful reporter because there is no background activity; like *lacZ*, it can be assayed without special equipment. However, the reagents for CAT assays are expensive, and the assay is labor-intensive. Luciferase is a very sensitive assay, with no background enzymatic activity, and is very easily assayed; it does, however, require access to a luminometer. Green fluorescent protein (GFP) is useful for some purposes; it is readily detected in transfected cells, but its expression is less readily quantifiable. For further information on these reporters and procedures for assaying their activity, see Hazelrigg (this volume).

- *Experimental design.* In designing transient expression experiments, it is important to bear in mind that most cells will express the exogenous gene product at low levels or not at all. For example, a transient experiment designed to test an antisense construct by inhibition of an endogenous gene is likely to disappoint. However, because cotransfection frequencies are very high, there are many successful strategies for using transients; these depend on ways of observing only the transfected cells. Although techniques exist for isolating transfected cells (cell sorting based on a cotransfected marker), the more common strategy is to use an identifiable reporter that can be detected by immunostaining (see, e.g., Fehon et al. 1990), by its intrinsic fluorescence (e.g., GFP), or by enzymatic assay. Because of the high efficiency of cotransfection, one can readily examine the effects of the product of one plasmid on the expression from a second plasmid (see, e.g., Swevers et al. 1996).

- *Promoters.* In general, investigators strive to achieve a high level of expression of the protein of interest in these experiments, which is best done by expressing the gene from a strong constitutive *Drosophila* promoter, such as actin5C or ubiquitin. The constitutive *copia* promoter is also useful, but it is much weaker than act5C or ubiquitin (L. Cherbas, unpubl.). The following vectors all make use of an act5C promoter and have been used extensively for expression in Kc or S2 cells:

  pPac (Krasnow et al. 1989) permits cloning of an open reading frame between an act5C promoter and an act5C polyadenylation region.

  Ract-Hadh (Swevers et al. 1996) was made by substituting an actin5C promoter fragment for the metallothionein promoter of pRmHa-1; pRmHa-1 contains a metallothionein promoter followed by a polylinker and a polyadenylation region from Adh (Bunch et al. 1988).

  pCMA (Hu 1998; Hu and Cherbas, in prep.) was made by adding an act5C promoter fragment to the expression vector pCMX (a gift from R.M. Evans); the resulting plasmid contains (in order) an actin5C promoter, a T7 promoter, a cytomegalovirus (CMV) promoter, a polylinker, and a polyadenylation region from SV40 and can be used for expression in *Drosophila* cells, in mammalian cells, and in vitro.

- *Quantification.* This is made easier if an internal control is used for transfection efficiency. The internal control is usually a second plasmid, carrying a second reporter expressed from a constitutive promoter, which is included at a constant concentration in each transfection. The same issues apply as for choosing the primary reporter. We have found a very useful combination to be firefly luciferase as the primary reporter and luciferase from the coelenterate *Renilla* as the secondary reporter. Promega sells reagents for assaying the two luciferases in succession from a single extract.

- *Effect of culturing conditions on levels of expression.* Treatment of the cells during the experiment can affect the level of a reporter in unexpected ways. For example, we have found that CAT activity, expressed from a constitutive promoter, is reduced approximately 30% by ecdysone treatment. Firefly luciferase activity is elevated about threefold by ecdysone, and *Renilla* luciferase is elevated about fivefold. We do not know the mechanisms of these effects, but we suspect that they are indirect effects of the hormone on stability of proteins or RNAs, or possibly on translational efficiency. It is advisable to check the properties of the reporter, expressed from a constitutive promoter, under the conditions of the experiment, and to correct the experimental results accordingly.

## STABLE TRANSFORMATION

## Strategies

### General Considerations

*Drosophila* cell lines may be stably transformed in several ways; the technique used should be chosen to fit the investigator's purposes. A traditional procedure, simple transfection with a supercoiled plasmid or combination of plasmids, leads to the generation of long head-to-tail arrays of the plasmid(s) inserted into the chromosome. In Kc cells, these arrays are fairly small (typically, 1–10 copies). In S2 cells, they are immense (>>1000 copies). This procedure is simple and can work well for making cell lines that produce large quantities of an exogenous protein. It is clear that reporter expression is not linear with copy number at high copy numbers, and we suspect that silencing and/or heterochromatization is occurring; this has not been studied. Two recent alternative procedures lead to the insertion of small numbers of copies of the plasmid sequence, either in random positions (by transposition) or in a targeted region (by parahomologous targeting).

Stably transformed populations must be selected following transfection; the procedure is relatively simple and rapid. However, it is often better to clone the transfected cells in the presence of the selective agent, rather than merely selecting a resistant population. There is a wide range in the properties of transformed cells resulting from a single transfection, including number of plasmid copies incorporated, expression per plasmid copy, and extent of both basal expression and inducibility of an inducible promoter. Bulk selection leads to a population consisting of those transformants that grow most rapidly, whereas those transformants whose properties are most desirable for the experiment may be lost. In our experience, the ease of finding transformed cells with suitable properties varies greatly with the protein being expressed; in some cases (e.g., CNN4, Figure 20.1B–E), a bulk-selected population is quite satisfactory, whereas in others (e.g., N, Figure 20.1A), cloning is essential.

### Promoters

The selectable marker must be expressed from a promoter whose strength is adequate to render the cells resistant to the selective agent. When transformation is expected to lead to the generation of long arrays, a relatively weak promoter may be adequate. Thus, for example, a methotrexate-resistant DHFR (dihydrofolate reductase) expressed from a *copia* promoter (pHGCO or its derivative pHCO, Bourouis and Jarry 1983) confers complete methotrexate resistance to S2 cells when present in at least 10 copies/haploid genome (Moss 1985; Cherbas et al. 1994). S2 cells transformed by array formation typically contain approximately 1000 plasmid copies per haploid genome. On the other hand, transposition and parahomologous targeting typically result in a small number of copies (1–5) of the plasmid sequence per cell; these procedures require that the methotrexate resistance marker be expressed from a stronger promoter. We have therefore substituted an actin5C promoter for the *copia* promoter in pHCO for use in the latter two procedures (Segal et al. 1996; Cherbas and Cherbas 1997).

When a plasmid that results in the expression of an exogenous protein is stably introduced, it is often necessary to regulate the expression of that protein with an inducible promoter to minimize its toxic effects while the line is maintained. This is usually done with a copper-inducible metallothionein (Mt) promoter, for which a number of vectors are available (Bunch et al. 1988). Figure 20.1 shows some examples of $Cu^{++}$-induced expression of exogenous proteins from the Mt promoter. It is important to realize that there is a low level of basal expression from the Mt promoter in cells that have not been treated with $Cu^{++}$; typically, the induction by $Cu^{++}$ is approximately 30-fold (Bunch et al. 1988). Both the basal level of expression and the extent of the induction may vary greatly from clone to clone for a given transfection (L. Cherbas and A. Mintzas, unpubl.).

When stably transformed cells are to be used for large-scale production of an exogenous protein, it is often helpful if the protein is secreted into the cell membrane or into the medium (see, e.g., Johanson et al. 1995). If the protein in question is not normally a secreted protein, its coding sequence can be cloned into one of a number of commercially available vectors which add a signal peptide to the expressed protein (Kirkpatrick and Shatzman 1999).

## Selectable Markers

A detailed description of selective systems can be found in an earlier review (Cherbas et al. 1994). The most commonly used selective systems for stable transformation of *Drosophila* cells are methotrexate, hygromycin B, and α-amanitin. Methotrexate resistance can be conferred by multiple copies of the DHFR resistance gene expressed from a *copia* promoter, or a single copy expressed from an actin promoter (see above). Resistance to α-amanitin can be conferred by a single copy of a genomic fragment containing a resistant form of the RNA polymerase subunit RPII215 (Jokerst et al. 1989; Segal et al. 1996); multiple copies cause no apparent toxicity (Thomas and Elgin 1988; L. Cherbas unpubl.). We know of no data on the efficacy of a single copy of the hygromycin resistance marker.

When cells are transformed by array formation, it is not necessary to incorporate the selectable marker into the plasmid of interest; cotransfection of two plasmids always leads to incorporation of both plasmids into a single array (for a discussion, see Cherbas et al. 1994). Indeed, it is preferable to use cotransfection. Not only does this approach minimize the labor in plasmid construction, it also permits more control over the number of plasmid copies incorporated in the transformed cells, since the arrays generally contain the cotransfected plasmids in the same proportions that were present in the mixture used for transfection.

## Selection Procedures

Following transfection, we allow 2 days for the cells to recover from the trauma of transfection and to express the selectable marker before adding the selective agent. Killing by the selective agent is always slower than it would be in an untransfected population, because a large number of cells that are not stably transformed nevertheless express the selective marker transiently. Procedures for methotrexate and amanitin selection, the two systems with which we are experienced, are given below. For a detailed protocol for hygromycin selection of S2 cells, see Kirkpatrick and Shatzman (1999).

### *Methotrexate Selection*

Two days after transfection, methotrexate is added to a final concentration of $2 \times 10^{-7}$ M. If the cells are to be selected as a population, simply spin the cells down every 4 days and resuspend in fresh medium containing methotrexate; after a few days, cell proliferation should be noticeably slowed, and after about 1 week, untransformed cells should begin to lyse. If the cells are to be cloned, allow 4 days of selection before cloning, during which growth of methotrexate-sensitive cells largely ceases; then clone in the presence of methotrexate. We clone S2 cells in soft agar, and Kc cells by dilution in 96-well plates. For cloning in 96-well plates, it is necessary to dilute the cells so that there will be approximately one viable cell per ten wells; this is obviously a problem when there is no estimate of the frequency of transformants in the partially selected population. We therefore generally clone a series of tenfold dilutions of the transfected cells ($10^2$ to $10^5$ intact cells/ml), and simply choose plates with an appropriate density of clones from which to pick transformants for further growth. Methotrexate-resistant lines that contain plasmid arrays should be maintained in the continuous presence of methotrexate, as a precaution against the (surprisingly rare) event in which the array is lost by homologous recombination.

### *α-Amanitin Selection*

α-Amanitin kills sensitive cells much more rapidly than does methotrexate. Since this reagent is very expensive and very toxic, we do not use continuous selection. Instead, α-amanitin is added 2 days after transfection (final concentration 5 μg/ml for S2 cells and 10 μg/ml for Kc cells), and the cells are left undisturbed for 1 week. By this time, all the sensitive cells have lysed, and it may be difficult to see the resistant cells in the debris of lysed cells. The cells are then cloned or grown up as a population, exactly as for methotrexate selection, except that the selective agent is not added to the medium used after the initial 1 week of selection. It is a good idea to add amanitin to the medium occasionally when maintaining cells containing arrays of plasmid, but it is not necessary (and for most labs, it is prohibitively expensive) to use continuous amanitin selection.

## Arrays

To generate arrays of plasmid, transfect cells by calcium phosphate–DNA coprecipitation, using 1 ml of precipitate for 3 ml of cells. The precipitate contains 20 μg of supercoiled plasmid DNA/ml, a mixture of the plasmid of interest, and a plasmid carrying a selectable marker. For S2 cells, if the composition of the plasmids does not lead to selection against their presence, incorporation of approximately 1000 plasmid copies per haploid genome will result, in the form of arrays containing the same ratio of plasmids that was present in the input mix. The products of this transformation procedure in other lines' cells are less completely characterized.

## Transposition

P-element transposition in Kc cells has been described (Segal et al. 1996). For this purpose, both the selectable marker and the fragment of interest must be included in a single transposon, because most cells will incorporate a single transposon or, at most, two to five transposons. Segal et al. (1996) used as a source of transposase the plasmid pUChsπΔ2-3, in which a partially spliced transposase transcript is expressed from a *hsp70* promoter. Using the basal expression of the heat shock promoter (i.e., no heat shock), electroporation of Kc cells with a mixture of 1–2 μg of the transposase plasmid and 11 μg of a 4.2-kb plasmid containing the transposon led to incorporation of an average of two to three copies of the transposon per cell; virtually all of the transformation occurred by P-element transposition, as shown by the absence of flanking vector sequences from the transformed cells.

We have continued to use this procedure in our laboratory and, although the data are still limited, we think that this will become the preferred procedure both for low-copy number experiments and for expression at high-copy numbers.

## Targeting by Parahomologous Recombination

Parahomologous targeting refers to clustered illegitimate recombination events that occur at high frequencies in *Drosophila* cells in the vicinity of a target sequence. We believe that it simply reflects high local concentrations of exogenous DNAs caused by the efficient pairing of plasmid DNAs with their chromosomal homologs. It is important to point out that such parahomologous events appear to occur during all *Drosophila* cell transformations; where the object is to obtain low-copy number transpositions, these events represent a serious problem that we are trying to eliminate.

Still parahomologous events are valuable for other purposes. Thus, they can be used to inactivate a target locus that is functionally haploid in the targeted line; the ploidy of a gene of interest is not always obvious from either karyotype or Southern analysis. Thus, we have successfully used parahomologous targeting to generate a Kc line that is deficient in *EcR* function; Kc cells are diploid for chromosome III (on which *EcR* is located), but they contain four copies of *EcR*, of which only one is functional (Cherbas and Cherbas 1997).

High frequencies of rearrangements in the targeted region (on the order of 50% of transformed clones) can be achieved. A linear fragment is prepared that contains both a selectable marker (methotrexate-resistant DHFR expressed from an actin5C promoter) and at least 4 kb of homology with the targeted chromosomal region. The DNA is introduced into Kc cells by electroporation; similar results were obtained using 2–60 μg of DNA per electroporation of fragments 8–13 kb in length. The resulting transformants are cloned in the presence of methotrexate, and the status of the targeted region is assessed by Southern analysis and/or phenotypic analysis.

## NOTE ADDED IN PROOF

A recent and exciting development is the demonstration by C.A. Worby and J.E. Dixon (Biological Chemistry, University of Michigan) that RNAi effectively abolishes expression of genes in several *Drosophila* cell lines. They have developed a simple protocol for its use and report success in blocking expression of each (of 10) signal transduction pathway gene tested.

## REFERENCES

Andres A.J. 1990. "An analysis of the temporal and spatial patterns of expression of the ecdysone-inducible genes *Eip28/29* and *Eip40* during development of *Drosophila melanogaster*." Ph.D. thesis, Indiana University, Bloomington.

Ashburner M. 1989. Drosophila: *A laboratory manual*. Cold Spring Harbor Laboratory Press, Cold Spring Harbor, New York.

Bieber A.J. 1986. "Ecdysteroid-inducible polypeptides in *Drosophila* Kc cells: Kinetics of mRNA induction and aspects of protein structure." Ph.D. thesis, Harvard University, Cambridge.

Bourouis M. and Jarry B. 1983. Vectors containing a prokaryotic dihydrofolate reductase gene transform *Drosophila* cells to methotrexate-resistance. *EMBO J.* **2:** 1099–1104.

Bunch T.A., Grinblat Y., and Goldstein L.S.B. 1988. Characterization and use of the *Drosophila* metallothionein promoter in cultured *Drosophila melanogaster* cells. *Nucleic Acids Res.* **16:** 1043–1061.

Cherbas L. and Cherbas P. 1981. The effects of ecdysteroid hormones on *Drosophila melanogaster* cell lines. *Adv. Cell Culture* **1:** 91–124.

———. 1997. "Parahomologous" gene targeting in *Drosophila* cells: An efficient, homology-dependent pathway of illegitimate recombination near a target site. *Genetics* **145:** 349–358.

———. 1998. Cell culture. In Drosophila: *A practical approach*, 2nd edition (ed. D.B. Roberts). IRL Press at Oxford University Press, United Kingdom.

Cherbas L., Moss R., and Cherbas P. 1994. Transformation techniques for *Drosophila* cell lines. *Methods Cell Biol.* **44:** 161–179.

Chomczynski P. and Sacchi N. 1987. Single-step method of RNA isolation by acid guanidinium thiocyanate-phenol-chloroform extraction. *Anal. Biochem.* **162:** 156–159.

Echalier G. 1997. Drosophila *Cells in Culture.* Academic Press, San Diego.

Fehon R.G., Kooh P.J., Rebay I., Regan C.L., Xu T., Muskavitch M.A.T., and Artavanis-Tsakonas S. 1990. Molecular interactions between the protein products of the neurogenic loci *Notch* and *Delta*, two EGF-homologous genes in *Drosophila. Cell* **61:** 523–534.

Hu X. 1998. "The mechanisms of activating the functional ecdysone receptor complex." Ph.D. thesis, Indiana University, Bloomington.

Johanson K., Appelbaum E., Doyle M., Hensley P., Zhao B., Abdel-Mequid S.S., Young P., Cook R., Carr S., Matico R., Cusimano D., Dul E., Angelichio M., Brooks I., Winborne E., McDonnell P., Morton T., Bennett D., Sokolski T., McNulty D., Rosenberg M., and Chaiken I. 1995. Binding interactions of human interleukin 5 with its receptor alpha subunit. Large scale production structural, and functional studies of *Drosophila*-expressed recombinant proteins. *J. Biol. Chem.* **270:** 9459–9471.

Jokerst R.S., Weeks J.R., Zehring W.A., and Greenleaf A.L. 1989. Analysis of the gene encoding the largest subunit of RNA polymerase II in *Drosophila. Mol. Gen. Genet.* **215:** 266–275.

Kirkpatrick R.B. and Shatzman A. 1999. *Drosophila* S2 system for heterologous gene expression. In *Gene expression systems* (ed. J.M. Fernandez and J.P. Hoeffler), pp. 289–330. Academic Press, San Diego.

Klueg K., Parody T., and Muskavitch M.A.T. 1998. Complex proteolytic processing acts on Delta, a transmembrane ligand for Notch, during *Drosophila* development. *Mol. Biol. Cell.* **9:** 1709–1723.

Krasnow M.A., Saffman E.E., Kornfeld K., and Hogness D.S.. 1989. Transcriptional activation and repression by Ultrabithorax proteins in cultured *Drosophila* cells. *Cell* **57:** 1031–1043.

Li K., Xu E.Y., Cecil J.K., Turner F.R., Megraw T.L., and Kaufman T.C. 1998. *Drosophila* centrosomin protein is required for male meiosis and assembly of the flagellar axoneme. *J. Cell Biol.* **141:** 455–467.

Maniatis T., Fritsch E.F., and Sambrook J. 1982. *Molecular cloning: A laboratory manual.* Cold Spring Harbor Laboratory, Cold Spring Harbor, New York.

Moss R.E. 1985. "Analysis of a transformation system for *Drosophila* tissue culture cells." Ph.D. thesis, Harvard University, Cambridge.

Robb J.A. 1969. Maintenance of imaginal discs of *Drosophila melanogaster* in chemically defined media. *J. Cell Biol.* **41:** 876–885.

Sang J.H. 1981. *Drosophila* cells and cell lines. *Adv. Cell Culture* **1:** 125–182.

Savakis C. 1981. "Ecdysteroid-inducible polypeptides and mRNAs in *Drosophila*." Ph.D. thesis, Harvard University, Cambridge.

Savakis C., Demetri G., and Cherbas P. 1980. Ecdysteroid-inducible polypeptides in a *Drosophila* cell line. *Cell* **22:** 665–674.

Schneider I. and Blumenthal A.B. 1978. *Drosophila* cell and tissue culture. In *The genetics and biology of* Drosophila (ed. M. Ashburner and T.R.F.. Wright), vol. 2a, pp. 265–316. Academic Press, London.

Segal D., Cherbas L., and Cherbas P. 1996. Genetic transformation of *Drosophila* cells in culture by P element-mediated transposition. *Somat. Cell Mol. Genet.* **22:** 159–165.

Søndergaard L. 1996. Efficiency of different lipofection agents in *Drosophila* S-2 cells. *In Vitro Cell. Dev. Biol. Anim.* **32:** 386.

Swevers L., Cherbas L., Cherbas P., and Iatrou K. 1996. *Bombyx* EcR (BmEcR) and *Bombyx* USP (BmCF1) combine to form a functional ecdysone receptor. *Insect Biochem. Mol. Biol.* **26:** 217–221.

Thomas G.H. and Elgin S.C.R. 1988. The use of the gene encoding $\alpha$-amanitin-resistant subunit of RNA polymerase II as a selectable marker in cell transformation. *Drosophila Inf. Serv.* **67:** 85.

# CONTENTS

# 21

# Generating Antibodies against *Drosophila* Proteins

**Ilaria Rebay**
*Department of Biology*
*Whitehead Institute/MIT*
*Cambridge, Massachusetts 02142*

**Richard G. Fehon**
*Developmental, Cell, and Molecular Biology Group*
*Duke University*
*Durham, North Carolina 27708-1000*

In RECENT YEARS, THE DIRECTION OF *DROSOPHILA* RESEARCH increasingly has turned toward understanding gene function at the level of individual cells. These studies almost invariably require the ability to detect specific proteins at multiple stages of development, either in situ or in cell extracts. Although recent advances in epitope tags, including green fluorescent protein (GFP), have provided other means for detecting specific proteins in situ and even in living cells, it is still true that specific antibody probes provide powerful tools for the analysis of protein localization, trafficking, stability, and function.

Numerous references, in particular the laboratory manual by Harlow and Lane (1988), describe in detail strategies for the production of antibodies. However, the *Drosophila* system provides some unique advantages and disadvantages for generating and characterizing specific antibody probes. In general, *Drosophila* cells are smaller than those of vertebrates, and they appear to express correspondingly smaller amounts of most proteins, making high-affinity antibodies crucial for immunolocalization studies. On the other hand, having access to null mutations in the gene against which antibodies are being produced provides a unique and very powerful control for determining the antibody specificity.

This chapter outlines strategies and methods that are useful for producing antibodies against *Drosophila* antigens. Excellent references on specific techniques for antibody production, such as that by Harlow and Lane (1988), as well as university and commercial facilities that provide animal handling services, are available to most investigators. Therefore, this chapter will not attempt to provide an exhaustive manual of all techniques necessary for the production of antibodies, but rather will provide strategies and techniques that should be of particular use to *Drosophila* researchers.

## CHOICE OF ANTIGEN

Assuming that antibodies are being made against a gene product whose coding sequence is known (surely the most common scenario), the two most likely sources of antigens for the production of antibodies are fusion proteins and synthetic peptides. Although there are some exceptions to this general rule (Baumgartner et al. 1996), synthetic peptide antigens can be used to produce antibodies suitable for Western blot analyses (Niemeyer et al. 1996) but only rarely yield antibodies suitable for in situ immunolocalization studies. The explanation for this empirical rule likely relates to the difficulty in producing a short primary peptide sequence that resembles the structure of a protein in an intact cell that has been fixed and permeabilized for staining. Thus, fusion proteins are almost always the preferred source of antigen for immunolocalization studies.

A wide variety of plasmid vectors are now commercially available for the production of fusion proteins in bacterial cells. Most are also designed to incorporate a tag that allows affinity purification of the expressed fusion protein from bacterial cell extracts. The most commonly used are vectors that incorporate a portion of the glutathione-S-transferase (GST) enzyme that is able to bind to immobilized glutathione and vectors that use a poly-histidine tag which binds immobilized nickel ions with high affinity. One useful strategy is to clone the same coding sequence into two different fusion-protein vectors. This increases the chances of obtaining a well expressed and soluble fusion protein. In addition, it provides a source of protein for testing serum or monoclonal antibodies, which will be recognized only by antibodies against the cloned protein and not by those that recognize the fusion tag itself.

The effect of ionic detergents, in particular SDS, on protein structure has important implications for the purification of fusion proteins. Although a number of protocols recommend SDS-polyacrylamide gel electrophoresis (SDS-PAGE) as a suitable method for the isolation of fusion proteins, empirical evidence indicates that some antigens isolated in this manner do not produce antibodies that recognize proteins in situ. For example, two groups working independently found that antibodies produced against SDS-PAGE-purified Notch fusion proteins would not work for immunolocalization in fixed tissues, although they did work for Western blot analyses (Kidd et al. 1989; R.G. Fehon, unpubl.). Anecdotal evidence suggests that this problem is less severe for nuclear antigens, but still the preferred method for fusion protein purification avoids altogether the use of ionic detergents, such as SDS or Sarkosyl. Interestingly, the converse argument—that antibodies raised against "native" fusion proteins will not recognize SDS-PAGE-treated antigens—is not usually valid. Antibodies raised against native fusion proteins that work well for immunolocalization in tissues generally work well for other techniques as well, including Western blots and immunoprecipitation (Fehon et al. 1991, 1994; O'Neill et al. 1994; Rebay and Rubin 1995; McCartney and Fehon 1996). Therefore, the best general strategy is to produce soluble fusion protein that can be affinity-purified from bacterial cell lysates.

Another important question to be addressed at the start of making antibodies is what portion of the protein should be expressed in the fusion protein? Although there are some theoretical guidelines and computer algorithms for the prediction of antigenic sites, empirical evidence indicates that this is a less than exact science. For example, our own experience with Notch showed that similar-sized fragments from different regions of the highly repetitive epidermal-growth-factor-related repeats in the extracellular domain of Notch produced very different results when used as immunogens (Fehon et al. 1991). Thus, it is difficult or impossible to predict which regions of a coding sequence will produce the best response. For this reason, the best general strategy is to prepare two or three (or more) fusion proteins expressing different regions of the coding sequence, in hopes that at least one of these will produce an acceptable response. With polymerase chain reaction (PCR) techniques, such subcloning is usually not difficult.

Why not just express the entire protein? If the coding sequence of interest is of the right size, this may be an option. Coding sequences greater than approximately 1.5 kb usually do not express as well as GST fusions. Conversely, small coding sequences usually express quite well but are not as effective as immunogens. Coding sequences of 0.8–1.2 kb in length appear to provide a reasonable compromise between expression and immunogenicity. If the available coding sequence is significantly smaller than this, it is possible to increase the size of the fusion protein by dimerizing or multimerizing the cloned portion within the fusion protein (McCartney and Fehon 1996).

Two protocols for expression and purification of GST fusion proteins are described below, one for soluble fusion proteins and one for insoluble fusion proteins (Protocol 21.1 and Protocol 21.2, respectively). The most commonly encountered problem in the preparation of fusion proteins is that the expressed protein becomes packaged into inclusion bodies and is therefore insoluble. One method for improving the yield of soluble protein is to induce expression in bacteria growing at lower than normal temperatures, from 25ºC to as low as 15ºC. Induction times at low temperature will range from 3 hours to overnight and should be determined empirically. Alternatively, it is possible to dissolve the inclusion bodies in urea and then dialyze out this denaturant before the affinity purification step. Using a combination of low growth temperature and the procedure for insoluble proteins (Protocol 21.2) will improve the yield of most difficult fusion proteins.

## Protocol 21.1*

## Preparation of Soluble GST Fusion Proteins

### Materials

#### Media

LB containing 100 µg/ml **ampicillin** or appropriate selection antibiotic for expression plasmid (~275 ml/sample)

#### Supplies and Equipment

Shaker incubator (37ºC, to grow broth cultures)
Sorvall centrifuge, GSA and SS34 rotors, or equivalent
Screw-cap centrifuge tubes (30 ml, which should be able to withstand immersion in liquid nitrogen)
Rocker placed at 4ºC
Water bath at 65ºC
Probe sonicator or French press
Polypropylene tubes (15 and 50 ml)
Clinical centrifuge
Microcentrifuge tubes (1.5 ml)
Microcentrifuge
Equipment for SDS-PAGE

*For all protocols, H$_2$O indicates glass distilled and deionized.

## Solutions and Reagents

**IPTG** (isopropyl-β-D-thiogalactopyranoside)

Lysozyme

Dithiothreitol (**DTT**, 1 M)

Triton X-100 (10% stock, v/v)

**Liquid nitrogen**

Reagents for determining protein concentration (e.g., **Bradford dye**)

Buffers and reagents for SDS-PAGE (**acrylamide/bisacrylamide, ammonium persulfate, TEMED, SDS**, etc.)

### STE (salty TE)

| | |
|---|---|
| 5 M NaCl | 3 ml |
| 1 M Tris-Cl (pH 8.0) | 1 ml |
| 0.5 M EDTA (pH 8.0) | 0.2 ml |
| $H_2O$ | to make 100 ml |

Use as an ice-cold solution.

### 10× PBS

| | |
|---|---|
| NaCl | 90 g |
| **$Na_2HPO_4$** | 20 g |
| **$NaH_2PO_4·H_2O$** | 8.3 g |
| $H_2O$ | to make 1 liter |

Filter-sterilize. Prepare 1× PBS daily, preferably with sterile $H_2O$.

### PBT

1× PBS

1% Triton X-100

### Preparation of Glutathione Agarose Resin

Hydrate the dry resin (Sigma G 4510) in freshly made PBT by rocking at 4ºC for at least 3 hours. Centrifuge in a clinical centrifuge at 1000 rpm for 15 seconds to pellet the resin. Rinse three times in ice-cold PBT. For the final resuspension, add an equal volume of ice-cold PBT to make a 50% slurry of the resin. Swelled resin can be kept at 4ºC almost indefinitely. For long-term storage, add **sodium azide** to 0.02%.

### Buffer A

0.1 M sodium acetate (pH 4.0)

0.5 M NaCl

### Buffer B

0.1 M Tris (pH 8.0)

0.5 M NaCl

### Elution Buffer

| | |
|---|---|
| 1 M HEPES (pH 7.0) | 2.25 ml |
| 5 M NaCl | 0.9 ml |
| 1 M DTT | 0.15 ml |
| Reduced glutathione (Sigma G 4251) | 0.14 g |
| $H_2O$ | to make 30 ml |

Adjust pH to 7.8. Prepare immediately before use.

**CAUTION:** acrylamide/bisacrylamide, ammonium persulfate, ampicillin, Bradford dye, DTT, IPTG, liquid nitrogen, $Na_2HPO_4$, $NaH_2PO_4$, SDS, sodium azide, TEMED (see Appendix 4)

## Method

1. Inoculate 25 ml of LB containing 100 μg/ml ampicillin with bacteria expressing the GST-fusion protein. Use of a protease-deficient strain such as BL21 is recommended. Grow at 37°C overnight.

2. Dilute the overnight culture 1:25 into larger culture (250 ml culture is adequate for most needs; if more protein is needed, scale up the volumes accordingly). Grow bacteria at 37°C with vigorous shaking until $OD_{600}$ reaches a minimum of 1.0.

3. Add 6 mg of IPTG per 250 ml of culture. Incubate at 37°C for an additional 1–3 hours.

   *Note:* For most proteins, induction is complete after 1 hour, although in some instances, a longer induction may improve the yield. Because long inductions may increase the risk of insolubility or degradation, the ideal time should be determined empirically for each fusion protein and the shortest time that gives a maximal induction should be selected.

4. Centrifuge cells in a Sorvall GSA or equivalent rotor at 5000 rpm for 7 minutes. Pour off the supernatant and rinse the pellet with a small volume of ice-cold STE.

5. Resuspend the pellet in 25 ml of STE. Add 2.5 g of lysozyme. Transfer to a 30-ml screw-cap centrifuge tube. Incubate on ice for 30 minutes.

6. Add DTT to a final concentration of 5 mM. Add Triton X-100 to 1% (dilute from a 10% stock). Mix well and rock at 4°C for 15 minutes.

7. Freeze in liquid nitrogen. Thaw rapidly in a 65°C water bath. Avoid letting the solution rise above 4°C during the thawing process by mixing frequently by inverting. If lysis does not occur (check viscosity of the solution), then repeat freeze/thaw three to five times until the solution becomes viscous from released bacterial DNA.

8. Use a sonicator or French Press to disrupt remaining cellular debris and released chromosomal DNA. Sonicate on ice in short bursts of 10–20 seconds at mid to high power until the viscosity of the solution is reduced to that of water (test with pasteur pipette).

   *Note:* Disruption using the French Press is reported to give greater protein yields than by sonication.

9. Centrifuge in a Sorvall SS34 or equivalent rotor at 10,000 rpm for 10 minutes to pellet cell debris. Transfer the supernatant to a 50-ml polypropylene tube. This is the soluble lysate. This pellet should be much smaller than the starting bacterial pellet. If the pellet looks like a typical pellet of whole cells, resuspend the pellet in STE and repeat lysis (steps 5–8).

10. Add 1 ml of a 50% slurry of prepared Glutathione-Agarose Resin to the soluble lysate.

11. Incubate with rocking at 4°C for 30 minutes to overnight.

12. Centrifuge in a clinical centrifuge at 1000 rpm for 15 seconds. Remove and save supernatant.

13. Resuspend beads in 10–15 ml of ice-cold PBT. (*Optional:* Add DTT to 5 mM.) Transfer to a 15-ml polypropylene tube. Centrifuge and remove supernatant. Repeat for a total of three washes.

14. Perform three additional washes with PBS without detergent. After the final wash, resuspend beads in 1 ml of ice-cold PBS and transfer to a 1.5-ml microcentrifuge tube. Centrifuge beads in a microcentrifuge at 3000 rpm for 15 seconds. Remove supernatant.

15. Resuspend beads in 1 ml of freshly prepared Elution Buffer and rock at room temperature or 4°C for 15 minutes to overnight. Centrifuge beads. Carefully remove and save the supernatant. This is the soluble protein solution.

16. Repeat elution (step 15) until no more protein is obtained. This can be quickly assayed using the "Bubble Test." Use a 1-ml pipette tip to *gently* blow air bubbles in the eluted fraction. If bubbles form easily and are stable, there is protein present. Do not blow more bubbles than necessary as this denatures the sample. Repeat the elution until any bubbles produced disappear almost instantly.

> *Note:* Most fusion proteins are eluted after one to two batches. If the yield is very high (5–10 mg or greater), up to five elutions may be necessary.

17. Determine protein concentration using the Bradford protein assay. If the Bradford assay is not used, make sure that the presence of glutathione in the sample does not affect the colorimetric assay being used. Run a sample of the eluted protein on an SDS-PAGE gel to confirm the protein is of the expected size and is not degraded.

18. Fusion proteins are usually quite stable in the glutathione Elution Buffer and can be injected into the selected animal host in this buffer. For short-term storage, keep the elution samples at 4°C. For long-term storage, freeze at –80°C.

19. Regenerate the glutathione-agarose beads before storing as follows:

    a. After eluting as much protein as possible, combine all available beads into a 15-ml polypropylene tube.

    b. Wash beads with at least three cycles of alternating pH. Each cycle consists of a wash in Buffer A, followed by a wash in Buffer B.

    c. Wash several times in PBT and store as a 50% slurry at 4°C.

PROTOCOL 21.2

# Preparation of Insoluble GST Fusion Proteins

## Materials

### Media

LB containing 100 µg/ml **ampicillin** or appropriate selection antibiotic for expression plasmid (~550 ml/sample)

### Supplies and Equipment

Shaker incubator (37°C, for growing cultures)
Sorvall centrifuge, GSA and SS34 rotors, or equivalent
Rocker placed at 4°C
Water bath at 65°C
Sonicator

Dialysis tubing
Polypropylene tubes (15 and 50 ml)
Clinical centrifuge
Microcentrifuge tubes (1.5 ml)
Microcentrifuge
Equipment for SDS-PAGE

## *Solutions and Reagents*

**IPTG**
Lysozyme
**Liquid nitrogen**
1x PBS
PBT (1x PBS/1% Triton X-100)
Solid **urea**
Urea (6 M)
Urea (3 M and 1 M) in 1x PBS
Glutathione-agarose beads (see p. 392)
Elution Buffer (see p. 392)
Reagents for determining protein concentration (e.g., Bradford dye)
Buffers and reagents for SDS-PAGE (**acrylamide/bisacrylamide, ammonium persulfate, TEMED, SDS**, etc.)

### *Phosphate-BME Buffer*

| | |
|---|---|
| 0.1 M **sodium phosphate** (pH 7.0) | 50 ml |
| β-mercaptoethanol | 100 µl |
| EDTA | 0.37 g |
| $H_2O$ | to make 100 ml |

**CAUTION: acrylamide/bisacrylamide, ammonium persulfate, ampicillin, β-mercaptoethanol, Bradford dye, IPTG, liquid nitrogen, SDS, sodium phosphate, TEMED, urea (see Appendix 4)**

## Method

1. Prepare a 50-ml overnight culture in LB containing 100 µg/ml ampicillin.

2. Dilute the overnight culture into 500 ml of LB containing 100 µg/ml ampicillin in a 1-liter flask. Incubate at 37°C for 1–2 hours with vigorous shaking until $OD_{600}$ reaches a minimum of 1.0.

3. Add 12 mg of IPTG to the culture.

4. Incubate at 37°C for 1 hour.

5. Cool on ice. Centrifuge in a Sorvall GSA or equivalent rotor at 5000 rpm for 7 minutes. Pour off supernatant.

6. Resuspend cells in 10 ml of Phosphate-BME Buffer with 0.5 mg/ml lysozyme and transfer to a screw-cap centrifuge tube. Incubate on ice for 30 minutes.

7. Freeze in liquid nitrogen for 5 minutes.

8. Thaw rapidly in the 65°C water bath as described in step 7 of Protocol 21.1.

9. Add solid urea to 6 M; invert to mix. *Be patient*—it will take 10–20 minutes for the urea to go into solution.

10. Sonicate two to four times on ice for 30 seconds, and then increase the volume to 20 ml with 6 M urea.

11. Centrifuge in a Sorvall SS34 or equivalent rotor at 12,000 rpm for 10 minutes. Decant the supernatant into dialysis tubing. There should be a clearish pellet at the bottom of the tube but little else, indicating that almost everything has dissolved.

12. Remove urea gradually by dialyzing at 4°C against the following solutions (minimum 3 hours each, but longer is fine):

    a. 3 M urea in PBS.

    b. 1 M urea in PBS.

    c. PBS (three times).

13. A large precipitate will form. Pour contents of dialysis tubing into a centrifuge tube. Centrifuge in an SS34 or equivalent rotor at 12,000 rpm for 10 minutes. Decant supernatant into a 50-ml polypropylene tube.

14. Add 1 ml of hydrated, glutathione-agarose beads. Proceed according to steps 11–17 of Protocol 21.1.

# IMMUNIZATION

## Choice of Host

Although rabbits are traditionally the organism of choice for producing antibodies, a number of investigators producing antibodies against *Drosophila* proteins have found that they often carry high levels of cross-reactivity to *Drosophila* proteins, even in preimmune serum. Useful antibodies have certainly been made in rabbit hosts (Young et al. 1993; Baumgartner et al. 1996), but many investigators have found that these sera require extensive affinity purification to remove cross-reactive species or specialized fixation and staining procedures to minimize nonspecific staining. On the other hand, other species including mouse, rat, and guinea pig have been found consistently to produce high-titer antisera that do not require affinity purification for use in immunolocalization, Western blot analysis, or immunoprecipitation. Thus, for the majority of antigens, the quantity advantage is more than outweighed by the disadvantages of high background staining and the necessity to affinity-purify antisera. In the event that a large volume of serum is required, guinea pigs are an excellent alternative to rabbits, as they provide both ample amounts of serum (30–40 ml from exsanguination) and very low, endogenous background reactivity against *Drosophila* proteins.

As a general rule, it is best to perform initial immunizations in mice for several reasons. First, mice can be maintained inexpensively and usually produce a readily detectable response within 6 weeks, thereby reducing the cost of initial antiserum production. Test bleeds of immunized mice typically produce 100–200 μl of serum, which at a titer of

1:1000 provides more than enough serum for initial immunolocalization and Western blot analyses of a particular gene product. In addition, although antigen from fusion proteins is not usually limiting, it can be in the case of poorly expressed or highly insoluble antigens. In such cases, mice are ideal because they require relatively little antigen to produce a strong response (75–150 μg of antigen per mouse total). Finally, mice have the advantage that if a strong response is elicited, it is then possible to obtain hybridoma cells that produce large quantities of highly specific monoclonal antibodies.

It is often desirable to perform double-labeled immunofluorescence experiments to compare the tissue or subcellular localizations of two proteins. It is worth considering this potential use in advance, because certain combinations of host species are better for this sort of experiment than others. In particular, combinations of mouse and rat antisera make a poor choice for double-labeling experiments. Because these two species are closely related, their immunoglobulin molecules are highly similar, and as a result, the producers of host-specific secondary antibodies must perform extensive purification to eliminate all cross-species reactivity. The result is secondary antibodies with relatively poor binding properties and correspondingly poorer fluorescent signals in staining experiments. In contrast, it is possible to obtain excellent specificity and staining activity using hosts that are more distantly related, such as mouse and guinea pig, together with secondary antibodies that have been specifically cross-reacted between these two species (e.g., those produced by Jackson ImmunoResearch Laboratories).

## Immunization Quantities, Schedule, and Route for Various Hosts

Recommended antigen amounts, adjuvants, routes, and schedules are listed below. Although repeated immunization can theoretically improve antibody titer, in practice, we have found that if a response is not seen after the fourth immunization, it is unlikely that a satisfactory antiserum will be produced by that animal. It is important to note that although Freund's has been the adjuvant of choice for some time, there is increasing pressure from governmental agencies to use less aggressive agents to stimulate the immune response. Commercially available alternatives, such as that from Ribi, appear to be less harmful to the host but have not been as extensively used, and their effectiveness is therefore less clear. Guidance in this issue should be available from local Institutional Animal Care and Use Committees, or commercial immunization services.

A number of commercial or university-based facilities are now available with trained personnel that perform all steps required for immunization and serum collection. These facilities are convenient, save time, and for the most part, free individual researchers from the increasingly burdensome bureaucracy associated with the use of vertebrate laboratory animals. However, they can be more expensive than the "do-it-yourself" approach, especially when overnight shipping expenses for antigen, serum, etc., are included. In practice, we have found that university immunization facilities, when available, offer a good compromise between convenience and expense. In either case, the facility will require information, including species to be immunized, antigen name, concentration, quantity of antigen to be used per animal, immunization route, and immunization schedule to be used. It is also generally a good idea to aliquot antigen in doses sufficient for each round of immunization, to explain explicitly how much antigen is to be used for each animal, and to check periodically during the process to ensure that procedures are being performed according to schedule.

### Schedule

*Mice:* Use 25–50 μg of protein per immunization in ≤250 μl of buffer.
*Number of animals:* Three

Day 0    Preimmune bleed
         Immunize in complete Freund's adjuvant, intraperitoneal (ip) injection
Day 21   Immunize in incomplete Freund's, ip
Day 35   Immunize in incomplete Freund's, ip
Day 42   First test bleed

*Note:* Immunization and test bleeds may be continued at 2-week intervals.

*Rats:* Use 50–100 μg of protein per immunization in less than 250 μl of buffer.
*Number of animals:* Three

Day 0    Preimmune bleed
         Immunize in complete Freund's adjuvant, ip
Day 21   Immunize in incomplete Freund's, ip
Day 42   Immunize in incomplete Freund's, ip
Day 49   First test bleed

*Note:* Immunization and test bleeds may be continued at 3-week intervals.

*Guinea Pigs:* Use 50–100 μg of protein per immunization in less than 250 μl of buffer.
*Number of animals:* Two

Day 0    Preimmune bleed
         Immunize in complete Freund's adjuvant, ip
Day 21   Immunize in incomplete Freund's, ip
Day 42   Immunize in incomplete Freund's, ip
Day 49   First test bleed

*Note:* Immunization and test bleeds may be continued at 3-week intervals.

### Preparation and Storage of Antiserum

Collected blood should be allowed to clot and centrifuged as described previously to separate the serum (Harlow and Lane 1988). Polyclonal antisera should be stable at 4ºC for several years if sodium azide (0.02%) is added to inhibit bacterial growth. However, it is usually best to aliquot precious reagents and store them at –80ºC. Working stocks should be kept at 4ºC, and repeated freeze-thaw cycles avoided. Some investigators prefer to partially purify immunoglobulins from serum using 50% ammonium sulfate (Harlow and Lane 1988).

## TESTING STRATEGIES TO DETERMINE ANTISERUM QUALITY _____

The sequence of tests described here is designed to establish conditions under which an antibody will work. It is a good idea to adopt a triage approach; use the fastest and easiest approach first to eliminate nonresponders quickly and then further test positives with more difficult and stringent tests. In other words, even if the final goal is immunoelectron microscopy, this should not be the first test of a new antibody. The list of tests presented below is not intended to be all-inclusive; other tests appropriate for the particular usage in mind may have to be developed.

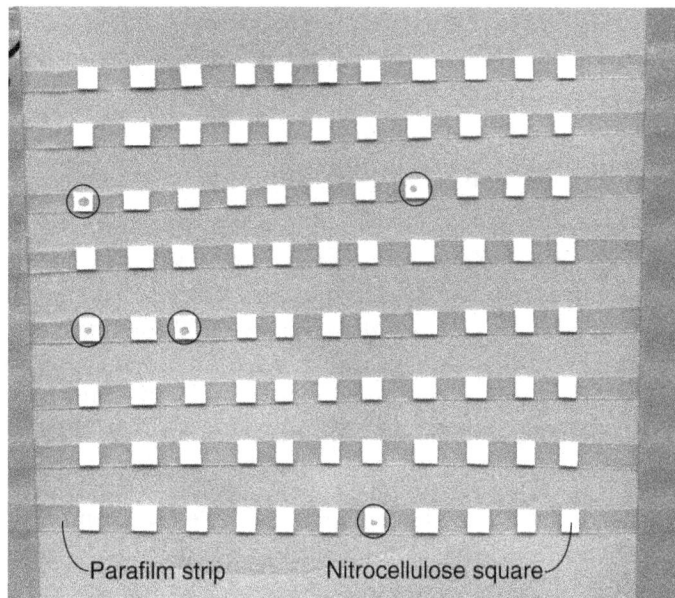

**Figure 21.1.** A dot blot used in a monoclonal antibody screen (Fehon et al. 1994). Five positive hybridoma supernatants are apparent (encircled), while the rest had no response to the fusion protein. The nitrocellulose squares for this dot blot were not arrayed to allow multichannel pipetting of primary supernatants.

## Test No. 1: Dot Blot

It is essential to determine first whether the antiserum recognizes the fusion protein against which it was raised. If a positive result is not obtained, it is futile to proceed to secondary tests. Continue boosting and retesting until a positive response is achieved, or try a different approach.

The dot blot will provide a qualitative estimate of antibody titer (Protocol 21.3). The same test can be done quantitatively by ELISA (enzyme-linked immunosorbent assay), but the technically simpler dot blot gives adequate information, uses less protein, and does not require an ELISA plate reader. If the same protein has been expressed in two different fusion-protein vectors, the test should be performed with the protein not used for immunization. If there is only one fusion protein, e.g., a GST-fusion, then a second dot blot can be done in parallel with just GST alone to compare the titers. If the antiserum recognizes the protein, there should be a lower titer of antibodies against GST alone as compared to against the entire fusion protein. A preimmune control is always advisable. Figure 21.1 shows a sample dot blot from a monoclonal antibody screen.

## Test No. 2: Staining of Transfected *Drosophila* S2 Cultured Cells

Once the antiserum has been determined to have a high titer against the fusion protein, the next step is to ask whether it can recognize the appropriate antigen in *Drosophila* tissues. Staining of transfected S2 cells for testing antisera is described in Protocol 21.4. There are several advantages to using transfected S2 cells as a first tissue test. First, the antibody only needs to penetrate a single cell. In addition, these cells are easy to fix and stain and require no dissection. Second, the overexpressed protein will be easier to detect than lower levels of endogenous protein. Third, background is generally less of a problem in S2 cells than in embryonic or imaginal tissues, allowing lower dilutions of the antiserum to be tested if the response is marginal. Finally, the untransfected cells form a useful built-in control. S2 cell stainings are also an easy way to determine the subcellular localization of a particular protein.

### Test No. 3: Staining of *Drosophila* Tissues—Helpful Hints

The most common reason for generating antibodies against a particular protein is to examine its pattern of endogenous expression during development. Once nice results are obtained with the transfected S2 cells, the next step is to try the antiserum on *Drosophila* tissues, usually embryos, imaginal discs, or ovaries. Providing protocols for each of these procedures is beyond the scope of this chapter. The intent is to provide a list of suggestions that will help optimize the use of the reagent. A streamlined embryo-staining protocol is provided that is useful when processing a large number of samples, e.g., when screening monoclonal antibodies (see below [Monoclonal Antibodies] and Protocol 21.5).

- *Experimental controls.* Remember to include the preimmune control in all initial experiments. A positive control using an antibody of a similar species is also recommended. This controls for the fixation, staining, and secondary antibody.
- *Incubation time.* Primary antibody incubations should be done at 4°C overnight with gentle rocking. Once nice results are obtained, it may be possible to reduce the time of primary antibody incubation. Cutting corners in the beginning is almost always counterproductive.
- *Antibody dilutions.* Recommended dilutions range from 1:1000 to 1:10,000. Occasionally, a very high-titer antiserum will require even greater dilution. Dilutions lower than 1:500 often have problems penetrating the tissue. Too high a concentration of antibody will cause nonspecific aggregation of the antibody over the surface of the tissue. In this case, the tissue will appear to glow on the outside, but there will be little or no signal internally. If this result is observed, dilute the antibody at least tenfold and repeat staining.
- *Detergents.* Try different detergents. If the antigen is nuclear, higher concentrations of detergents may be helpful (up to 0.3% Triton X-100 and 0.3% deoxycholate). If the antigen is membrane-associated, 0.1% saponin may work better.
- *Fixation.* Try different fixatives and different fixation times.
- *Detection of overexpressed protein.* Results can vary widely depending on the level of expression of the endogenous protein. If endogenous protein cannot be detected, staining of transgenic lines expressing the gene of interest will allow optimization of the protocols for detecting that particular protein. Once the optimal conditions are established for detecting overexpressed protein, it is likely that the endogenous protein will be easier to detect.
- *Methanol sensitivity.* If staining embryos gives no signal, it is possible that the antigen is sensitive to the methanol used for devitellinizing. To determine if this is the case, try staining imaginal discs or ovaries where standard protocols do not require methanol. Once a positive result is obtained, include a methanol fixation step and see if this prevents staining. If so, the antigen is methanol-sensitive, and for staining, embryos must be either hand-devitellinized or devitellinized using 85% ethanol in place of methanol (D. Kiehart, pers. comm).
- *Staining specificity.* Determine whether the staining is specific to the gene product of interest by staining populations of null mutant embryos. Use either a *lacZ*- or GFP-marked balancer chromosome so that the mutant class can be unambiguously distinguished. If the antibody is specific, then there should be no staining in a null

mutant, and staining should be visibly reduced in the heterozygous background. These experiments can be complicated if the allele is not truly null and still expresses the protein or if there is stable maternal contribution. If the problem is maternal expression, then the same test can be performed by generating clones and staining the imaginal discs. This is more laborious, but works nicely. If a null allele is not available, use a deficiency for the region instead. Western blots of single embryos from the mutant strain are also useful in determining specificity.

PROTOCOL 21.3

## Dot Blot

### Materials

### *Supplies and Equipment*

Clean glass plate (8 x 10 cm, or larger)
Double-stick tape
Nitrocellulose membrane (Schleicher & Schuell 0.2-µm membrane is recommended)
Clean razor blade
Parafilm
Clean petri dish
White labeling tape
Platform shaker

### *Solutions and Reagents*

Fusion protein (10–100 ng/µl)

*10x TBS*

| | |
|---|---|
| NaCl | 87.7 g |
| 1 M Tris-Cl (pH 7.5) | 100 ml |
| H$_2$O | to make 1 liter |
| Dilute to 1x. | |

*Blotto*

1x TBS
5% nonfat dry milk

### *Antibodies and Detection*

Appropriate HRP-conjugated secondary antibody
Diaminobenzidine (**DAB**; 10 mg/ml); DAB is a suspected carcinogen and should be handled with care.
**H$_2$O$_2$** (30%)

**CAUTION: DAB, H$_2$O$_2$** (see Appendix 4)

## Method

1. Cover one side of a clean glass plate (8 × 10 cm, or larger) neatly with double-stick tape.

2. Place a piece of nitrocellulose membrane on top of the double-stick tape.

3. With a clean razor blade and a straight edge, cut lengthwise (in the same direction as the strips of tape) 3-mm-wide strips. Use a fresh razor blade and change it frequently, because a dull blade will tear the nitrocellulose. Press firmly to cut through both the nitrocellulose and the double-stick tape, keeping the razor blade as close to parallel to the surface of the plate as possible. This takes a little practice, but once the technique is mastered, it takes only minutes to perform.

4. Gently and slowly peel the strips of nitrocellulose bound to the tapeoff the glass plate.

5. Attach 12 strips lengthwise in rows separated by approximately 3–5 mm on a 10 × 15-cm piece of Parafilm. Make sure that the strips lie flat and taut on the parafilm surface. If a 96-well microtiter plate is used as a template, the strips can be spaced precisely so that a multichannel pipettor can be used to add the primary antibody solutions. This saves time when screening monoclonal antibodies (see below), but for testing polyclonal sera, it is not important to be this precise.

6. Use a paper cutter to cut crosswise 3-mm-wide strips. The result will be strips of Parafilm onto which are attached 12 3-mm nitrocellulose squares.

7. Remove the backing from the Parafilm and array eight of these strips, Parafilm side down and each separated by approximately 5 mm, in a clean petri dish. Again make sure the strips lie flat and taut on the dish surface. Tape down the strips with a piece of white labeling tape (other colors may run during the staining process) at each edge of the Parafilm. The end result is a 96-well microtiter array of nitrocellulose squares.

8. Spot the fusion protein, at a concentration of 10–100 ng/μl, onto the nitrocellulose squares. When spotting the fusion protein, load a pipette tip with 5–10 μl of solution and then simply touch the end of the pipette tip (without depressing the plunger) onto a nitrocellulose square. A small (~1 mm) wet spot of fusion protein will be visible. This is sufficient. It is important that the spotted fusion protein not cover the entire square, as this would make it difficult to distinguish signal from background. Spot the protein five times for each antiserum so that several different antibody dilutions can be tested. Make note of the proper orientation of the plate by writing on the tape.

9. Air-dry at room temperature for 5–10 minutes or until the nitrocellulose is dry.

10. Block in Blotto with gentle agitation (50 rpm on a platform shaker) at room temperature for 5 minutes.

11. Quickly rinse three times with 1× TBS. Swirl each wash a few times and pour off immediately. Blot-dry with Kimwipes. Be thorough with the drying—if there is residual liquid, the antibodies will run together when they are added to the squares.

12. Prepare the following dilutions of antiserum in Blotto:

    1:100
    1:1000
    1:10,000
    1:100,000
    1:1,000,000

    Add approximately 10 ml (more or less depending on the size of the nitrocellulose squares) to a square.

13. Incubate at room temperature for 15 minutes.

14. Rapidly wash three times in 1× TBS. Then wash twice in 1× TBS at room temperature for 5 minutes (each wash) with gentle agitation on a platform shaker at 50 rpm.

15. Add the appropriate horseradish peroxidase (HRP)-conjugated secondary antibody, i.e., HRP-conjugated, goat anti-mouse antibody. For secondary antibodies obtained from Jackson ImmunoResearch Laboratories, dilutions of 1:5000 in Blotto are appropriate; 10 ml of solution will cover the entire plate. Swirl manually to make sure all the nitrocellulose squares are covered and then mix on a platform shaker (50 rpm) at room temperature for 15 minutes.

16. Wash as in step 14.

17. Develop in the following solution:

    | | |
    |---|---|
    | 1× TBS | 10 ml |
    | 10 mg/ml DAB | 100 µl |
    | 30% $H_2O_2$ | 1 µl |

    Store at room temperature for 5–10 minutes or until the signal is clear. A positive result will be seen as a dark brown or purple stain where the fusion protein was spotted. The remaining portion of the nitrocellulose square should be white. Preimmune spots should show no signal.

18. Dispose of DAB and wash the dot blot thoroughly in $H_2O$. Air-dry, preferably protected from direct light. Strips can be peeled off the plate and taped into a notebook.

## PROTOCOL 21.3 NOTES

The minimum recommended titer for proceeding to the next test is to have a 1:10,000 dilution of the antibody recognize the 10–100 ng/µl spot of fusion protein. A good response will have a much higher titer.

PROTOCOL 21.4

## *Drosophila* S2 Cell Staining

### Materials

#### *Supplies and Equipment*

Multiwell microscope slides (e.g., 10-well, Teflon-coated slides [Polysciences 18357] or 12-well multitest slides [ICN Biomedicals 6041205]). Treat the slides with poly-L-lysine to improve adherence of the cells (Harlow and Lane 1988). Cherbas and Cherbas (this volume; Protocol 20.3) describe the treatment of slides with poly-L-lysine.

Humid chamber (e.g., sealed plastic box lined with wet towels)

Cotton-tip applicators

Fluorescence microscope

#### *Solutions and Reagents*

1x PBS

---

#### *Paraformaldehyde Fix*

Prepare fresh daily. (Final volume ~10 ml.)

| | |
|---|---|
| 1x PBS | 10 ml |
| **paraformaldehyde** (Polysciences) | 0.2 (2%) to 0.4 (4%) g |

1x PBS can be preheated for a few seconds in microwave before the addition of paraformaldehyde. Place in a 60ºC oven or water bath for 0.5–1 hour, or heat on a stir plate for 10–15 minutes. *Do not leave longer than is necessary to dissolve the paraformaldehyde or it will evaporate.*

#### *PNT*

1x PBS

1% normal goat serum

0.1% Triton X-100

#### *Glycerol/n-Propyl-gallate Mounting Medium*

| | |
|---|---|
| glycerol | 9 ml |
| 1 M Tris-Cl (pH 8.0) | 1 ml |
| **n-propyl-gallate** | 0.05 g |

Heat to 60ºC for 10–15 minutes or microwave for 5–10 seconds to dissolve. Store in the dark at –20ºC.

---

#### *Antibodies and Detection*

Appropriate Cy3-conjugated secondary antibodies (Jackson ImmunoResearch)

**CAUTION: paraformaldehyde, *n*-propyl-gallate** (see Appendix 4)

### Method

1. Transfect S2 cells with an appropriate expression construct (refer to discussion in Cherbas and Cherbas, this volume).

*Notes:* With calcium phosphate transfection methods, typically 1–5% of cells will take up and express the transfected DNA. Thus, the remaining untransfected cells serve as negative controls.

Cotransfection with a GFP construct is recommended to control for transfection and to ascertain that the cells detected by the antiserum are actually of the transfected population. Thus, if the antiserum recognizes the correct overexpressed *Drosophila* antigen, bright staining should be detected in only a small fraction of the cells, and these same cells should also express the GFP construct. Even if the protein is endogenously expressed in S2 cells, there should still be an obvious difference in protein level between endogenous and overexpressed.

2. For antibody staining, place 50–100 µl of transfected S2 cells in each well of a multi-well microscope slide.

3. Let cells settle onto the slides at room temperature for 45–60 minutes in a humid chamber.

4. While the cells are settling, prepare fresh fix (2% paraformaldehyde in PBS).

5. When the cells are settled, *gently* flood the slide with fix (~1 ml). *Do not pipette the fix directly onto the cells.* Instead, pipette starting at the top of the slide away from the cells and let the slide flood. Pour off and replace with another 1 ml of fix.

6. Incubate at room temperature for 15 minutes in the humid chamber.

7. Pour off fix and wash three times with PBS by flooding the slide with approximately 2 ml of PBS and then pouring it off. Again, be gentle when adding each wash solution or the cells may be washed off the slide.

8. After the third wash, prop slides up so that they drain onto a Kimwipe or paper towel for 15 seconds. Then, carefully dry off the excess PBS with a cotton-tipped applicator stick. *Be careful not to disturb the cells.* The cotton often has fine threads that sweep behind, scraping off cells in their wake. Try to ball these strands up in the puddle of PBS at the bottom of the slide before trying to dry inbetween the wells.

   *Note:* There is a fine line between drying out the cells completely and leaving too much liquid on the slide. Drying out the cells will increase the background and in extreme cases may make the entire staining procedure fail. Too much liquid on the slide will cause the antibodies to mix between wells.

9. Add appropriate antibody dilutions to each well. The volume to be added to each well depends on the dryness of the slide. The usual range is 5–20 µl. Do not try to add too much solution to each well or there will be cross-contamination between wells.

   *Note:* Recommended dilutions for testing polyclonal antisera are 1:500, 1:1000, 1:5000, and 1:10,000. Dilutions should be made in PNT. If necessary, dilutions as low as 1:100 can be used on the S2 cells. The background is likely to be high, but the experiment will still be interpretable and will indicate whether there is any response at all.

10. Incubate cells in primary antibody at room temperature for 1 hour in the humid chamber.

11. Wash three times with PBS and dry. Add secondary antibody diluted in PNT. CY3-conjugated secondary antibodies from Jackson ImmunoResearch Laboratories work well at a 1:1000–1:5000 dilution.

12. Incubate at room temperature for 1 hour in the humid chamber.

13. Wash three times with PBS, dry, and mount in glycerol mounting medium with *n*-propyl gallate to prevent photobleaching. When adding mountant, cut off the tip of a P-200 pipette tip with a razor blade (or use a wide-bore tip) to facilitate pipetting the

glycerol solution. Depending on the number of slides being mounted, pipette 100–200 µl of mounting medium and place a small drop (~2–4 µl) on each slide well. For four 12-well slides, 200 µl should be sufficient.

14. Examine the results under a fluorescence microscope. A positive result, viewed with a 10X objective, will look like a dark field scattered with bright-red spots. Higher magnification (e.g., 63X oil-immersion objective) is recommended for visualizing subcellular localization and for digital or photographic recording.

> *Note:* A high-titer antiserum will show nice staining at a 1:1000–1:10,000 dilution. A minimal titer of 1:500 is generally required for success in staining of embryonic or imaginal tissues.

## MONOCLONAL ANTIBODIES

### Advantages of Monoclonal versus Polyclonal Antibodies

The advantage of making monoclonal antibodies is that they offer an unlimited supply of an antibody reagent. Especially when using mice, the volume of serum obtained from each bleed is small and will run out quickly. It is time-consuming and inconvenient to continually produce fusion protein, inject animals, and test the bleeds in order to maintain a supply of polyclonal antiserum. The production of monoclonal antibodies obviates the need for this.

Furthermore, monoclonal antibodies are frequently easier to work with for many experiments, including Western blotting and tissue staining, because background is rarely a problem and the reagent will work consistently in every experiment. On the other hand, polyclonal antibodies may work better than monoclonal antibodies for immunoprecipitation experiments or functional blocking experiments where having antibodies to different epitopes is likely to improve the efficiency of the technique. If the need for a large supply of polyclonal antibody is anticipated, guinea pigs are the animal of choice.

### When Should Monoclonal Antibodies be Made?

Monoclonal antibodies should be made only when the polyclonal antiserum gives satisfactory results at dilutions of 1:1000–1:10,000. If the titer is lower than this, the chances of a successful fusion are small. Mouse monoclonal antibodies are the easiest to make for technical reasons; rat monoclonal antibodies are possible but often less efficient.

The best monoclonal antibodies for most purposes are of the IgG class. Therefore, it is important to allow time for the animals' immune response to mature to produce primarily IgGs. If a high titer of antibody is found on the first test bleed, it is advisable to allow the mouse to age for 4–6 weeks without any further boosts prior to fusion. We have used mice that are over 1 year old and still obtained excellent results (R.G. Fehon and I. Rebay, unpubl.), although it is not recommended to push the time limit much beyond 1 year. When the decision is made to proceed, the mouse with highest titer and lowest background should be selected. The remaining mice should be kept as backups, without receiving further boosts, until the final cloned hybridoma lines are tested and frozen away.

### Immunization Prior to Fusion

Three to five days before the fusion date, the mouse of choice should be injected *intravenously* with 50–100 µg of the antigen without adjuvant. Intraperitoneal injection gives unsatisfactory results (I. Rebay, unpubl.).

## Hybridoma Fusion

The procedures for performing monoclonal fusions have been described elsewhere (Harlow and Lane 1988). It is important that the individual performing the fusion be adept at mammalian cell culture and have a record of successful monoclonal production. A good mouse is too precious to waste due to the incompetence of a monoclonal facility.

## Screening Strategies

Speed is critical when screening a monoclonal fusion. When the plates are ready to be screened, drop everything and focus entirely on the screening procedure. The faster the screening, the greater the chance that the selected monoclonal antibodies will be recovered and cloned. An ideal 3-day screening procedure is outlined below. If disaster strikes, an extra day of screening will probably be all right, but any longer than that the risk of losing the clones of interest increases drastically.

### Day 1

a. *Collect supernatants.* Usually 75–100 µl of supernatant will be available for screening. Most fusions will produce 10–20 microtiter plates (96-well) to be screened. Collecting supernatant from only those wells with cells growing may be more trouble than it is worth, particularly given the potential danger of mixing up samples.

b. *Perform a dot blot.* The first test is the dot blot (Protocol 21.3) shown in Figure 21.1. To avoid selecting monoclonal antibodies against the GST portion of the fusion protein, fusion protein produced in a different vector should be spotted. If this is not possible, spot two proteins on each nitrocellulose square with the GST-fusion protein in one corner and GST alone in the opposite corner. In this way, monoclonal antibodies recognizing the protein of interest can be distinguished from those recognizing GST. This also allows selection of an anti-GST monoclonal antibody that can be useful for many biochemical experiments. If the nitrocellulose spots are arrayed with 96-well microtiter spacing, supernatants can be added 8 or 12 at a time with a multichannel pipettor. This can save time and pipetting hand wear.

c. *Perform a single-plate trial run before screening all 20 plates!* Because it takes only one hour, a trial run with one plate is highly recommended. Include a positive control spot with the polyclonal antibody and a negative, medium-only control to ensure that all of the reagents are working. If the titer of the polyclonal serum was very high (1:10,000 or higher), it is not uncommon to find that all of the spots will light up, even from wells where there are no cells growing. This happens when the titer is so high that there are sufficient numbers of unfused spleen cells producing antibody, resulting in all wells containing fusion-protein-specific antibodies. If this is the case, then the collected supernatants must all be diluted. Dilutions of 1:10–1:100 in Blotto are recommended, depending on the strength of the signal. Then the trial dot blot should be repeated. A reasonable result is 1–10 strong positives per plate.

### Day 2

a. *S2 cell staining.* Screen only the positive clones from the dot blots on transfected S2 cells attached to multiwell slides. If there were too many positives to screen, then select the strongest ones. Using the multiwell slides, it is reasonable to screen 250 or more positives in half a day. Make sure to time everything properly to have sufficient transfected cells available. Supernatants should be used as 1:5–1:100 dilutions in PNT (see p. 404). Use low-end dilutions if straight supernatant was used for the dot blot and higher dilutions if diluted supernatant was used for the dot blot. Follow Protocol 21.4 using 45-minute to 1-hour incubations for both primary and secondary antibodies. A normal result is for approximately 10% of dot blot positives to be strong positives on transfected S2 cells.

b. *Embryo staining.* Screen S2 cell positives on embryos. Embryos are easiest for mass screening of multiple monoclonal antibodies. However, if necessary, disc or ovary stainings can be used instead. A streamlined version of the standard embryo-staining protocol works well and is provided below (Protocol 21.5). The primary antibody should be used at a dilution of 1:10–1:100 and incubated at 4°C overnight with gentle rocking. If the antibody is not intended for tissue staining, a Western blot screen against an appropriate tissue lysate can replace the embryo staining in the same amount of time.

### Day 3

a. Score embryo-staining results. Again, expect only 10% of S2-cell-positive clones to work well on embryos.

b. Select only the 5–10 strongest clones for expansion, freezing, and subcloning.

The net result after several months of subcloning and rescreening should be gorgeous monoclonal antibody reagents.

PROTOCOL 21.5

# Staining Embryos for Monoclonal Fusion Screening

## Materials

### Supplies and Equipment

Nitex embryo-collection baskets (for preparation, see Shermoen, this volume)
Squirt bottle of $H_2O$
Glass screw-cap vial or tube
Small paintbrush (for embryo manipulation)
Microcentrifuge tubes (1.5 ml)
Glass microscope slides
Glass coverslips (22 mm$^2$)
Whatman paper

### Solutions and Reagents

**Bleach** (50%)
**Heptane**
**Paraformaldehyde** (4%) in PBS (see p. 404)

**Methanol**
PNT (see p. 404)
Glycerol/*n*-**propyl gallate** mounting medium (see p. 404)


### *Antibodies and Detection*

Cy3-conjugated goat anti-mouse secondary antibody

**CAUTION: bleach, heptane, methanol, *n*-propyl gallate, paraformaldehyde** (see Appendix 4)


## Method

1. Collect embryos from an overnight collection in a Nitex basket. Rinse thoroughly with water.

   *Note*: For details on embryo collection and dechorionation, see Rothwell and Sullivan (this volume).

2. Dechorionate the embryos using 50% bleach for approximately 3 minutes. Rinse well with $H_2O$ until the smell of bleach is gone (be careful not to flush the embryos out of the Nitex basket). Use $H_2O$ (a squirt bottle is helpful here) to rinse the embryos off the sides of the basket and onto the mesh.

3. Use forceps to remove the Nitex mesh and dip into a glass screw-cap vial or tube containing a 50:50 mix of heptane:4% paraformaldehyde in PBS. Dunk the Nitex several times to allow the embryos to come off into the solution. Alternatively, use a paintbrush dipped in heptane to pick up the embryos from the Nitex and deposit them into the tube.

4. Fix at room temperature for 20–30 minutes with vigorous shaking.

5. Remove fix phase and replace with equal volume of methanol. Shake very vigorously for 15–30 seconds.

6. Let embryos settle to the bottom of the tube (these embryos are devitellinized). If interface is "sudsy," then shake or tap the tube to release embryos from the interface.

7. Remove heptane, any embryos that remain at the interface, and methanol but leave behind the embryos that have settled to the bottom of the tube.

8. Rinse three times with 1 ml of methanol.

9. Transfer the embryos in methanol to a microcentrifuge tube. Do not put more than 200 µl of embryos into each tube. Remove the methanol and fill the tube with 1.5 ml of PNT. Mix well by inverting the tube several times.

10. Rinse twice with PBS.

11. Block in PNT with gentle rocking until ready to use (at room temperature for 30 minutes or at 4°C for a longer period).

12. Aliquot embryos into as many 1.5-ml microcentrifuge tubes as needed to screen the monoclonal antibodies. If embryos are limiting, 50–100 embryos per tube are sufficient.

13. Incubate at 4°C overnight in PNT containing primary antibody.

14. Rinse twice in PBS.

15. Incubate at room temperature for 2 hours in CY3-conjugated, goat anti-mouse secondary antibody diluted 1:2000 in PNT.

16. Rinse once with PBS.

17. Transfer embryos to a microscope slide (two samples can be placed side by side on the same slide). Blot off the excess PBS with a strip of Whatman paper. Mount in a drop (~25 μl) of glycerol/*n*-propyl gallate mounting medium.

18. Cover with 22-mm² coverslip and examine.

## ACKNOWLEDGMENTS

We thank R. Ward for the protocol for preparation of soluble GST fusion proteins and B. Stevenson for teaching us the dot-blot procedure. We are grateful to the members of the Rebay and Fehon labs who have contributed through their experiments to the protocols found in this chapter. We also acknowledge S. Artavanis-Tsakonas in whose laboratory we initially developed most of the strategies and techniques described here.

## REFERENCES

Baumgartner S., Littleton J.T., Broadie K., Bhat M.A., Harbecke R., Lengyel J.A., Chiquet-Ehrismann R., Prokop A., and Bellen H.J. 1996. A *Drosophila* neurexin is required for septate junction and blood-nerve barrier formation and function. *Cell* **87:** 1059–1068.

Fehon R.G., Dawson I.A., and Artavanis-Tsakonas S. 1994. A *Drosophila* homologue of membrane-skeleton protein 4.1 is associated with septate junctions and is encoded by the *coracle* gene. *Development* **120:** 545–557.

Fehon R.G., Johansen K., Rebay I., and Artavanis-Tsakonas S. 1991. Complex cellular and subcellular regulation of Notch expression during embryonic and imaginal development of *Drosophila:* Implications for Notch function. *J. Cell Biol.* **113:** 657–669.

Harlow E. and Lane D. 1988. *Antibodies: A laboratory manual.* Cold Spring Harbor Laboratory, Cold Spring Harbor, New York.

Kidd S., Baylies M.K., Gasic G.P., and Young M.W. 1989. Structure and distribution of the Notch protein in developing *Drosophila. Genes Dev.* **3:** 1113–1129.

McCartney B.M. and Fehon R.G. 1996. Distinct cellular and subcellular patterns of expression imply distinct functions for the *Drosophila* homologues of moesin and the neurofibromatosis 2 tumor suppressor, merlin. *J. Cell Biol.* **133:** 843–852.

Niemeyer B.A., Suzuki E., Scott K., Jalink K., and Zuker C.S. 1996. The *Drosophila* light-activated conductance is composed of the two channels TRP and TRPL. *Cell* **85:** 651–659.

O'Neill E.M., Rebay I., Tjian R., and Rubin G.M. 1994. The activities of two Ets-related transcription factors required for *Drosophila* eye development are modulated by the Ras/MAPK pathway. *Cell* **78:** 137–147.

Rebay I. and Rubin G.M. 1995. Yan functions as a general inhibitor of differentiation and is negatively regulated by activation of the Ras1/MAPK pathway. *Cell* **81:** 857–866.

Young P.I., Richman A.M., Ketchum A.S., and Kiehart D.P. 1993. Morphogenesis in *Drosophila* requires nonmuscle myosin heavy chain function. *Genes Dev.* **7:** 29–41.

# CONTENTS

# 22

# Photoactivated Gene Expression for Cell–fate Mapping and Cell Manipulation

Jonathan Minden, Ruria Namba, and Sidney Cambridge
*Carnegie Mellon University*
*Department of Biological Sciences and*
*Center for Light Microscope Imaging and Biotechnology*
*Pittsburgh, Pennsylvania 15213*

$A$ LONG-STANDING GOAL OF DEVELOPMENTAL BIOLOGISTS is to track individual cells through development in order to create developmental fate maps. In *Drosophila*, a wide variety of fate-mapping methods have been developed, ranging from gynandomorph analysis (Garcia-Bellido and Merriam 1969; Janning 1978; Struhl 1981) to X-ray-induced mitotic recombination (Steiner 1970; Weischaus and Gehring 1976; Lawrence and Morata 1977) to cell ablation (Bownes 1975; Lohs-Schardin et al. 1979; King and Bryant 1982). These methods have generally dealt with marking relatively large areas of the embryo. To limit the number of marked cells, a number of methods for tagging individual cells have been developed, including single-cell dye injection (Technau and Campos-Ortega 1985; Bossing and Technau 1994) and single-cell heat shock by laser treatment (Halfon et al. 1997). We have developed a method that allows the expression of a wide variety of GAL4-UAS transgenes to be activated (Brand and Perrimon 1993) in embryos from nuclear cycle 12 to stage 12, in areas that range from single cells to large patches of cells (Cambridge et al. 1997). This method relies on caging the transcriptional activity of GAL4VP16, a potent transcriptional activator, with a photolabile compound, which can be removed with a brief exposure to long-wavelength UV light. Photoactivation is performed on a standard fluorescence microscope fitted with a standard mercury lamp and a pinhole aperture in the epifluorescence light path. The methods described in this chapter will be familiar to any *Drosophila* laboratory that does embryo injection. The preparation of caged GAL4VP16 requires a minimal amount of experience with protein purification. To photoactivate a single cell, the fluorescence microscope must be modified to produce a narrow beam of excitation light. Most microscope manufacturers sell these attachments at reasonable costs. We provide detailed descriptions of the various steps used for photoactivated gene expression, many of which are common to most *Drosophila* laboratories. The most critical components are caging of GAL4VP16, embryo injection, and the photoactivation microscope setup. Figure 22.1 shows a montage of UAS-*lacZ* embryos that were photoactivated at a variety of positions.

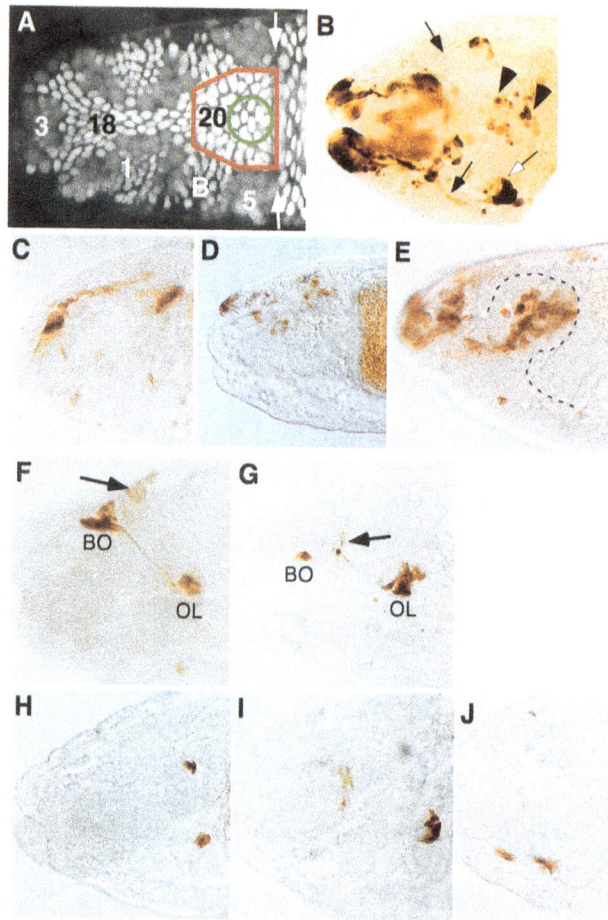

**Figure 22.1.** Examples of photoactivated embryos. (*A*) Individual cells of the dorsal head mitotic domains were visualized by the expression of nuclear GFP (*Ubi-GFPnls*) using confocal microscopy. δ20 is highlighted with a red border. The numbers indicate mitotic domains. Interphase nuclei appear as bright ovals with sharp edges, whereas mitotic cells are large and diffuse. The white arrows point to the cephalic furrow. The green circle indicates a typical size and location of the UV photoactivation beam. (*B–E*) Fates of mitotic domains were visualized by GAL4-dependent activation of *lacZ* using the photoactivated gene-expression system. All embryos are shown with anterior to the left. (*B*) Dorsal view of a stage-17 embryo with photoactivated δ20 cells. Cells in the posterior part of the brain (*white arrow*) and head peripheral nervous system (including axons projecting to the brain; *arrows*) are marked, as well as cellular debris (*arrowheads*). (*C*) Lateral view of a stage-17 embryo with photoactivated δ18 cells in the dorsal pouch. (*D*) Dorsal view of a stage-17 embryo with photoactivated δ5 cells at the anterior tip and in the brain. (*E*) Lateral view of a stage-15 embryo with photoactivated δB cells at the anterior tip and in the central nervous system, which is outlined with a dotted line. (*F–J*) Single-cell photoactivation of δ20. Single-cell photoactivation in the middle of δ20 marked cells in the developing larval visual system at stage 14 (*F*) and at stage 17 (*G*). They were confined to the Bolwig's organ (BO), the optic lobe (OL), and the dorsal pouch (arrows). (*H*) An embryo with marked cells in both optic lobes. (*I*) An embryo with marked cells in the optic lobe and the dorsal pouch. (*J*) An embryo with marked cells in the photoreceptor cells and both optic lobes.

## PREPARATION OF CAGED GAL4VP16 AND INJECTION INTO EMBRYOS

GAL4VP16 was purified according to the method of Chasman et al. (1989) from *Escherichia coli* that carried plasmid pJL2, which expresses GAL4VP16 under the control of the *lac* promoter. Protocol 22.1 generated enough GAL4VP16 protein to supply this laboratory for several years. Since GAL4VP16 was not commercially available at the writing of this chapter, the protein had to be purified or obtained from another laboratory.

The caging reaction involves covalent modification of approximately 60% of the lysine residues in GAL4VP16 with the photolabile, amine-reactive compound, 6-nitroveratryl chloroformate (NVOC-Cl). A typical preparative reaction (Protocol 22.2) caged 500 µl of 1.35 mg/ml GAL4VP16 in storage buffer. The extent of caging can be determined by fluorescamine labeling of free amines according to the method of Boehlen et al. (1973). This procedure is optional and is given in Protocol 22.2 Notes.

Protocol 22.3 describes embryo collection and preparation and injection with caged GAL4VP16.

## PROTOCOL 22.1*

## Purification of GAL4VP16

### Materials

### *Media*

#### *Standard Media*
1% Bacto tryptone
0.5% yeast extract
0.5% sodium chloride

### *Bacterial Strain*

*E. coli* carrying pJL2 (Chasman et al. 1989)

### *Supplies and Equipment*

Sorvall centrifuge and rotor, or equivalent
Sonicator (Bronson)
Glass or Teflon stirring rod
Column chromatography resins:
  • DE-52 column (2.5 x 16 cm [80 ml]; Whatman)
  • Heparin-Sepharose CL-6B column (2.5 x 7 cm [35 ml]; Pharmacia)
Centriprep-10 column (30 ml; Millipore)
Conductivity meter
Microcentrifuge tubes (1.5 ml)
Equipment for SDS-PAGE

### *Solutions and Reagents*

IPTG (isopropyl-β-D-thiogalactopyranoside) (200 mg/ml, 0.84 M)
Polyethyleneimine (10%)
**Ammonium sulfate** (solid)
**Bradford dye**
Reagents and buffers for SDS-PAGE (e.g., **acrylamide/bisacrylamide, ammonium persulfate, TEMED,** and **SDS**)

*For all protocols, H$_2$O indicates glass distilled and deionized.

*HEPES/NaCl Buffer (pH 7.5)*

20 mM HEPES-**NaOH** (pH 7.5)

0.2 M NaCl

*Buffer A*

20 mM HEPES-NaOH (pH 7.5)

10 mM **DTT** (dithiothreitol)

10 μM zinc acetate

2 μg/ml each **leupeptin, aproptonin**, and **pepstatin A**

Buffer A with 0.1 M, 0.2 M, 0.4 M, 0.6 M, and 0.75 M NaCl will be required.

*Storage Buffer*

50 mM HEPES-NaOH (pH 7.4)

0.2 M NaCl

1 mM DTT

30% glycerol

0.1 mM EDTA

10 μM zinc acetate

CAUTION: **acrylamide, ammonium persulfate, ammonium sulfate, aprotinin, bisacry-lamide, Bradford dye, DTT, IPTG, leupeptin, NaOH, pepstatin A, SDS, TEMED** (see Appendix 4)

## Method

### Bacterial Growth and Induction of GAL4VP16

1. Inoculate 50 ml of standard media with *E. coli* carrying plasmid pJL2. Grow to saturation at 37°C overnight.

2. Inoculate 12 4-liter flasks containing 2 liters each of standard media with 4 ml of saturated overnight culture. Incubate at 37°C for approximately 3 hours until the $A_{600}$ reaches 0.7–0.8.

3. To induce expression of GAL4VP16, add 2.4 ml of 200 mg/ml IPTG, to bring the final concentration of IPTG to 1 mM.

4. Incubate at 37°C for 3 hours.

### Harvesting and Lysis of Bacteria

1. Harvest cells by centrifugation at 10,000*g* at 4°C for 20 minutes.

2. Wash once by resuspending the bacterial pellet in 400 ml of HEPES/NaCl Buffer (pH 7.5). Centrifuge at 10,000*g* at 4°C for 20 minutes.

3. Resuspend the bacterial pellet in 400 ml of Buffer A containing 0.2 M NaCl. Lyse cells on ice by sonication until the $A_{600}$ reaches 10% of the starting value.

### Extraction and Column Chromatography

The following purification steps are performed at 0–4°C.

1. Centrifuge the crude lysate at 10,000$g$ at 4°C for 20 minutes.

2. Precipitate the GAL4VP16 from the crude lysate supernatant by the dropwise addition of 10% polyethyleneimine to a final concentration of 0.25% (w/v) with gentle stirring. Continue stirring for 30 minutes.

3. Centrifuge as in step 1.

4. Wash the pellet by resuspending in a small volume of Buffer A containing 0.4 M NaCl. When resuspending the pellet, use a glass or Teflon stirring rod to gently rub the surface of the pellet. Avoid making bubbles. Centrifuge as in step 1.

5. Resuspend the washed pellet with Buffer A containing 0.75 M NaCl. Centrifuge as in step 1. The supernatant contains the GAL4VP16 protein.

6. Precipitate the GAL4VP16 by adding solid ammonium sulfate to 35% saturation (0.194 g/ml) slowly with gentle stirring over 30 minutes and continue stirring for a further 30 minutes.

7. Centrifuge as in step 1.

8. Resuspend the ammonium sulfate pellet in Buffer A to a conductivity equal to that of Buffer A containing 0.1 M NaCl.

9. Purify the protein by column chromatography as follows:

   a. Load the sample (~20 ml) onto a 2.5 x 16-cm (80 ml) DE-52 column equilibrated with Buffer A containing 0.1 M NaCl.

   b. Wash with 10 column volumes of Buffer A containing 0.1 M NaCl.

   c. Elute with 1.5 column volumes of a continuous gradient of Buffer A containing 0.1–0.4 M NaCl.

   d. Collect 80 fractions (1.5 ml each).

   e. Assay for the presence of GAL4VP16 by SDS-PAGE. It should elute near the middle of the gradient.

   f. Pool the peak fractions. Load onto a 2.5 x 7-cm (35 ml) heparin-Sepharose CL-6B column equilibrated with Buffer A containing 0.2 M NaCl.

   g. Wash the column with 10 column volumes of Buffer A containing 0.2 M NaCl.

   h. Elute with one column volume of Buffer A containing 0.6 M NaCl.

   i. Collect 1-ml fractions.

   j. Assay the fractions by SDS-PAGE.

   k. Pool the peak fractions.

10. Equilibrate the pooled eluate (~10 ml) in storage buffer by two equilibration cycles in Centriprep-10 spin columns (30-ml capacity). These centrifugation steps take several hours because of the glycerol in the storage buffer.

11. Aliquot 500 µl of protein solution into 1.5-ml microcentrifuge tubes, and store at –80°C.

*Note:* Twenty-four liters of culture produced approximately 9 ml of GAL4VP16 protein with a protein concentration of 1.35 mg/ml as determined by Bradford assay. The final preparation should be greater than 95% pure as judged by SDS-PAGE.

# Caging of GAL4VP16

## Materials

### Supplies and Equipment

Centricon C-30 (3 ml) spin columns (Millipore)
Sorvall centrifuge and rotor, or equivalent
(*Optional*) Fluorimeter

### Solutions and Reagents

Sodium carbonate buffer (100 mM, pH 9.7)
Nitroveratryl chloroformate (1 mM; **NVOC**-Cl, Fluka) in **1,4 dioxane**, freshly prepared
Tris-Cl (100 mM, pH 1.75)
HEPES/NaCl buffer (pH 7.3; see p. 416)
(*Optional*) **Sodium phosphate** buffer (50 mM, pH 8.0)
(*Optional*) **SDS** (10%)
(*Optional*) Fluorescamine (30 mg/100 ml; Molecular Probes) in 1,4 dioxane

**CAUTION: 1,4 dioxane, NVOC, SDS, sodium phosphate** (see Appendix 4)

## Method

1. To 500 µl of 1.35 mg/ml GAL4VP16 in Storage Buffer, first add 500 µl of 100 mM sodium carbonate buffer (pH 9.7) and then 1 ml of freshly prepared 1 mM NVOC-Cl in 1,4 dioxane. The carbonate buffer should maintain the pH of the reaction at 9.5. Incubate the reaction at room temperature (20–21°C) for 30 minutes under dimmed room light.

2. Terminate the reaction by the adding 500 µl of 100 mM Tris-Cl (pH 1.75), which lowers the pH to 7.4.

3. Concentrate the sample by centrifugation in Centricon C-30 columns at 5000*g* at 4°C for 20–30 minutes.

4. Equilibrate the caged GAL4VP16 with 2 ml of HEPES/NaCl Buffer (pH 7.3) in the spin columns at 4°C for 40 minutes. Repeat this equilibration step once.

    *Note:* The final caged GAL4VP16 solution was adjusted to approximately 1 ml with a protein concentration of 0.4–0.7 mg/ml.

5. Aliquot 1–3-µl volumes and store at –80°C.

    *Note:* Caged GAL4VP16 is stable for more than 1 year under these conditions. Working solutions of caged GAL4VP16 can be kept at –4°C for up to 1 week.

## PROTOCOL 22.2 NOTES

To determine the extent of caging, parallel samples of approximately 30 µg of GAL4VP16 are treated with or without NVOC-Cl according to the standard caging protocol given above. The treated samples are equilibrated with 50 mM sodium phosphate buffer (pH 8.0). The samples are then brought up to a final volume of 1.5 ml containing 1% SDS and boiled for 3 minutes; 0.5 ml of 30 mg/100 ml fluorescamine in 1,4 dioxane is added and the fluorescence of the sample is analyzed with a fluorimeter (excitation at 390 nm/emission at 485 nm). Results of the caged protein samples are correlated to the control sample that was incubated without NVOC-Cl. There are 14 lysines per GAL4VP16 molecule. The target ratio of caging is 8 of 14 lysines modified by NVOC-Cl.

PROTOCOL 22.3

# Injection of Caged GAL4VP16

## Materials

### Media

Apple juice–agar plate for embryo collection (see Appendix 3)
Yeast paste, prepared by mixing dry yeast with a small volume of H$_2$O

### Supplies and Equipment

Standard, plastic fly bottle
Beaker (150 ml)
Stainless-steel filter baskets (150-mesh and 40-mesh screen; Bellco Glass) or baskets constructed as described below.
Nitex screen (130-mesh, Sefar America Inc., 4221 NE 34th Street, Kansas City, Missouri 64117; Phone: 800-283-8182)
Glass petri dish (50 mm)
Plastic petri dish (150 mm)
Fine brush
Stainless-steel probe for moving embryos around; a Dumont forceps (#5; Ted Pella) dulled with emery paper can be used as a probe.
Pasteur pipette
Kimwipes
No. 1 coverglass (20 x 20 mm and 20 x 50 mm)
Standard microscope slides
Drierite
Tape
Dissecting stereomicroscope. We have used the Leica Wild M3C with both transmitted light- and epi-illumination.
Injection microscope: Inverted tissue-culture microscope (CK-2, Olympus) fitted with a micromanipulator (MO-203, Narishige)
Syringe (50 ml) or a mechanical injector
Glass needles are formed from borosilicate glass capillaries (outer diameter 1.2 mm, inner diameter 0.94 mm, length 10 cm; Sutter Instrument Co.) having a thin, internal filament to aid filling on a horizontal needle puller (P-87, Sutter Instrument Co.). The needles are pulled to form a 3-mm tapered, sealed tip, which will be broken under halocarbon oil.
Glass syringe (10 µl; Gastight 1801, Hamilton Co.)

### Preparation of Filter Baskets

A filter basket can be prepared from a 50-ml screw-cap tube. Cut off the end of the 50-ml screw-cap tube and cut a large hole in the cap. Place a piece of 130-mesh Nitex screen between the cap and the tube, and screw in place to cover the hole.

## Solutions and Reagents

**Bleach** (50%)

Halocarbon oil (700 series, Halocarbon Products Corp.) in a 10-ml syringe fitted with an 18-gauge needle with a dulled tip

1x PBS (see Appendix 3)

### Egg Wash Solution

0.12 M NaCl
0.04% Triton X-100

### 2.25% Agar Blocks

Prepare 2.25% Bacto agar (Difco) in $H_2O$. Pour the agar into 100-mm plastic petri dishes, approximately 7-mm thick, and cut 2.0–2.5-$cm^2$ blocks with a razor blade. Cut a 45º bevel on the side on which the embryos will be aligned. This helps to view the embryos more clearly.

### Adhesive Solution

To prepare the adhesive solution, incubate double-side tape in **heptane** for several hours with constant mixing. Remove the heptane and centrifuge to remove any particular material. (See step 7, Notes.)

**CAUTION: bleach, heptane** (see Appendix 4)

## Method

1. Collect embryos for 3 hours on an apple juice–agar plate formed in the lid of a 35-mm plastic petri dish. Place the collection plate on top of a standard, plastic fly bottle containing only the flies, no food. Place a small dollop of yeast paste in the center of the collection plate to stimulate egg laying.

2. Wash and dechorionate embryos as follows (or as described in step 3, below):

   a. Rinse the embryos from the plate into a 150-ml beaker with a stream of Egg Wash Solution in a wash bottle. Typically, the volume of the wash solution is brought to 40–60 ml.

   b. Dechorionate the embryos by adding an equal volume of full-strength bleach (final concentration of 50%) to the beaker for 90–120 seconds with occasional swirling.

   c. Remove the embryos from the bleach by straining through a stainless steel filter basket (150 mesh). To remove any flies or large debris, place a second filter basket with a 40-mesh screen on top of the 150-mesh filter basket.

   d. Wash the embryos with three 100-ml volumes of $H_2O$, followed by a final rinse of Egg Wash Solution from the wash bottle.

   e. Transfer the embryos to a 50-mm glass petri dish with a stream of egg wash. First, rinse the embryos to one side of the collection basket with egg wash and then invert the basket over the petri dish and rinse the embryos out of the basket from the back of the basket. Tilt the basket at the 45º angle to direct the eggs into the dish.

3. Alternatively, washing and dechorionation can be performed as follows:

   a. Use a fine brush to collect embryos from the plate into a filter basket made from a 50-ml screw-cap tube. Rinse the embryos off the brush into the filter basket with $H_2O$.

   b. Dechorionate the embryos by immersing the filter basket in 50% bleach for 90–120 seconds.

c. Wash the embryos as in step 2d. Run Egg Wash Solution on the side of the basket to collect all the embryos to the bottom.

d. Unscrew the cap and remove the Nitex screen. Wash the embryos off the Nitex screen into a small, glass petri dish with a stream of Egg Wash Solution.

4. The following steps are best performed at 18–20°C, preferably at the lower temperature to improve survival rate and to slow down embryo development during manipulation. Examine the eggs under a dissecting stereomicroscope with transmitted-light illumination. Sweep the embryos into the center of the dish by gentle swirling. Form a pile of embryos that are between nuclear cycle 12 and 14 by gently pushing the embryos around with a stainless steel probe.

> *Note:* Stage 12–14 embryos have a distinct band of cleared cytoplasm surrounding the embryo. The older the embryo, the greater the nuclear density. Late-stage-14 embryos that are cellularizing have an additional thin dark ring within the cleared region. These are too old for cytoplasmic injection.

5. Remove the desired embryos from the dish with a pasteur pipette and deposit them on a 1.5-cm square of 130-mesh Nitex screen placed on top of a Kimwipe.

6. Rinse the embryos with a few drops of $H_2O$ to remove the Egg Wash Solution and transfer the Nitex screen to a block of 2.25% agar (Figure 22.2A). Gently move the embryos off the mesh with the stainless steel probe and line them up along the beveled edge with their anterior-posterior axes oriented parallel to the edge (Figure 22.2B). The agar block allows embryos to be oriented and minimizes embryo damage and premature drying. The embryos are rolled to the desired orientation. Photoactivation targets cells close to the glass coverslip. Therefore, position the surface to be activated away from the agar block surface.

7. Prepare a 20 x 50-mm no. 1 coverglass for mounting the embryos by making an adhesive stripe across the middle of the coverglass along the width with 2 μl of adhesive solution (Figure 22.2C).

> *Notes:* The consistency of the adhesive should be the same as pure heptane, and when 2 μl is applied to the coverglass, it should dry to form a transparent film. If the adhesive is too dilute, the embryos will not stick. If too concentrated, the film will have a rippled surface and cause some distortion when viewing the embryos through the microscope.
>
> Some batches of tape have been found to be toxic to embryos. Under the best conditions, expect a hatching rate of 80–90% for uninjected embryos.

8. Transfer the embryos from the agar block to the coverglass by very gently touching the adhesive strip on the coverglass to the surface of the embryos (Figure 22.2C). This is done by holding the corners of the coverglass between the two forefingers at one end and the thumbs at the other. Using the forefingers as a pivot, carefully lower the thumbs to allow the coverglass to touch the embryos. This step takes a bit of practice to master. If the embryos are moved during this procedure, gently move them into the desired position on the coverglass with the stainless steel probe.

9. To slightly dry the embryos, place the coverglass in a 150-mm plastic petri dish lined with a layer of fresh Drierite. We place a sheet of 130-mesh Nitex screen on top of the Drierite to keep dust off the coverslip. Drying the embryos takes 10–20 minutes. To monitor the extent of drying, view the embryos under the dissecting microscope with epi-illumination from an oblique angle. Move the lamp up and down in a small arc to observe the light reflecting off the vitelline membrane.

> *Note:* A fresh, undried embryo has a crisp, clear reflective surface. As the embryo is desiccated, the surface becomes less reflective and slightly dimpled. Overdried embryos are quite wrinkled. The slightly dimpled surface is the desired dryness.

A

B

C

D

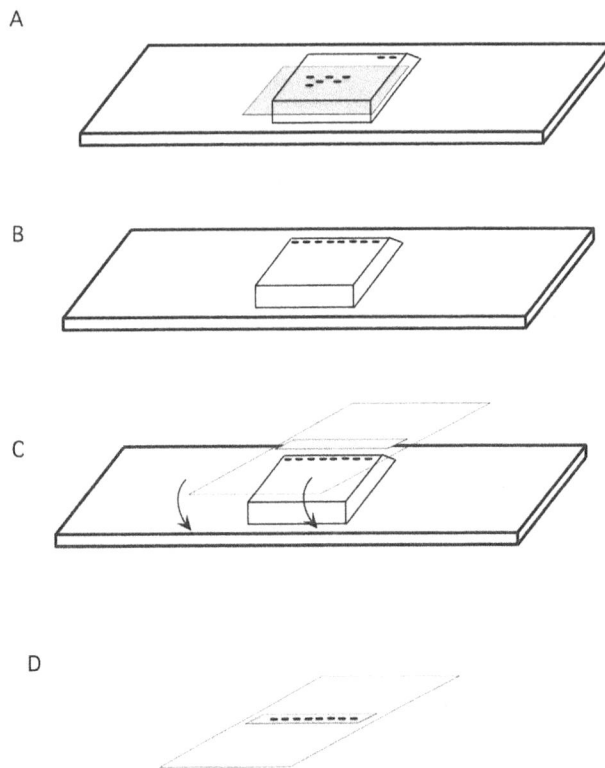

**Figure 22.2.** Transfer of embryos to a microscope coverglass. (*A*) An agar block placed on top of a microscope slide is shown. The embryos are transferred from the Nitex screen (shown in gray) to the beveled edge of the agar block. (*B*) Aligned embryos. (*C*) A coverglass with an adhesive stripe (shown in light gray) is placed over the embryos. The curved arrows show the rocking motion used to pick the embryos up off the agar block. (*D*) Final arrangement of the embryos on the coverglass.

10. Once the desired dryness is obtained, cover the embryos with halocarbon oil dispensed from a 10-ml syringe fitted with an 18-gauge needle with a dulled tip. Completely cover the embryos with halocarbon oil forming a 4–5-mm-wide stripe. For ease in handling and injection, affix the coverglass to a standard microscope slide with a small piece of tape.

11. Back-fill a glass capillary needle with caged GAL4VP16. Dilute the frozen aliquot of caged GAL4VP16 with 1–5 μl of PBS (as mentioned in Protocol 22.2, this working solution can be kept at –4°C for up to 1 week).

    *Note:* The optimal dilution of caged GAL4VP16 is determined empirically for each caging preparation. We dilute our preparation (0.42 mg/ml) fivefold with PBS.

12. Position a 20 x 20-mm coverglass next to the embryos. This provides a surface on which to break the needle tip. Place the slide, which carries the coverglass, in the slide holder on the stage of an inverted tissue culture microscope fitted with a 10x objective. Place the filled needle in the needle holder of a micromanipulator. The needle holder can be attached to either a 50-ml syringe or a mechanical injector (we use the 50-ml syringe).

13. Lower the needle into the halocarbon oil at one end of the line of embryos. Gently push the tip of the needle into the side of the 20 x 20-mm coverglass to break off a small portion of the tip. The goal is to create a 1–2-μm opening. Apply pressure to the needle to force a drop of the GAL4VP16 solution out of the needle and to remove any trapped air bubbles.

14. Position the needle adjacent to the site of photoactivation. The GAL4VP16 does not diffuse evenly throughout the embryo; it is limited to approximately one-quarter egg volume around the site of injection. Poke the needle against the side of the embryo to make sure that the tip is at the correct height at the midsection of the embryo. Plunge the needle into the embryo and apply enough air pressure to inject a small volume of GAL4VP16. A small clearing in the yolk plasm should be seen. If the embryo leaks, either too much was injected or the embryos were not dry enough. Do not use embryos with leakage for photoactivation.

15. Age the embryos to the desired stage at 18°C.

## SETUP FOR MICROSCOPY

Photoactivation is done on an inverted microscope equipped with epifluorescence illumination. To modulate the diameter and shape of the photoactivation beam, a conjugate image plane must be accessed in the epifluorescence light path. We use an Olympus IX70 inverted microscope fitted with a dual epifluorescence attachment from Olympus. This attachment has two mercury lamps connected to a "T-tube" with a beam splitter to select either source. One arm of the attachment is for full-field illumination and the other has a slider that can hold a variety of pinhole apertures (0.1–1 mm). Pinhole apertures can be purchased from either Melles Griot or Newport Corp. (1791 Deere Avenue, Irvine, California 92606; Phone: 800-222-6440). For a quick fix, a pinhole can be made from a piece of aluminum from a beverage can. There are a number of ways to access the epifluorescence light path. It is best to consult a microscopist or microscope sales representative. The following steps should be taken when setting up the microscope:

1. To direct the activation beam onto the embryos, select the DAPI or UV fluorescence filter set.

2. Align and focus the UV microbeam. Make a fluorescent-target slide by sealing a small volume of a fluorescent dye solution, such as fluorescein dextran, between a slide and a no. 1 coverglass. Use a 60X objective, which is what we generally use for photoactivation, to focus on a particle of dust in the dye. Adjust the length of the T-tube to bring the rim of the UV microbeam into focus. To maximize the light intensity of the microbeam, place the sensor of a power meter (Digital power reading 815 series, Newport) over the lens and adjust the position of the lamp with the condenser controls.

3. For UV irradiation of target cells, irradiation times necessary for photoactivation of caged GAL4VP16 in embryos are determined empirically for various size pinholes. For whole-embryo activation with full-field illumination using a 20X objective, set the center of the embryo in focus and illuminate the embryo for 3 seconds. For activation of selected cells, set the target cells in focus using a 60X objective. Irradiate small patches of cells for 7–10 seconds and single cells for 15 seconds. These times are determined empirically based on the success rate of photoactivation. We typically get greater than 50% activated embryos for patch photoactivation and 30% for single-cell photoactivation.

## VISUALIZATION OF PHOTOACTIVATED CELLS

### Immunostaining

Photoactivating caged GAL4VP16 in a selected group of cells in UAS-*lacZ* embryos allows the development of these cells to be traced using anti-β-galactosidase antibody staining. We use embryos that carry four copies of UAS-*lacZ*, which provides strong β-galactosidase expression. The β-galactosidase expression can be detected by this method from 1 hour after photoactivation until stage 16. Caged GAL4VP16 can be photoactivated from nuclear cycle 12 (~2 hours before gastrulation) to stage 12 (~6 hours after gastrulation). Protocol 22.4 describes the procedure we use for anti-β-galactosidase antibody staining, which has been adapted from Vincent and O'Farrell (1992).

## In Vivo Time-lapse Microscopy

The development of photoactivated cells can also be visualized in vivo using UAS-GFP (green fluorescent protein) embryos. This allows the movement of the selected cells to be followed by time-lapse microscopy. We use embryos carrying four copies of UAS-nlsGFP. After photoactivation, the embryos are transferred to a microscope capable of recording fluorescence, time-lapse movies. These recordings generally capture multiple focal planes spaced approximately 5 μm apart at 5–10-minute intervals. Our time-lapse microscope is fitted with a computerized stage that allows recording from up to 20 embryos per session.

PROTOCOL 22.4

# Visualization of Marked Cells by Antibody Staining

## Materials

### *Supplies and Equipment*

Razor blade
Glass screw-cap tube (8 ml)
Small funnel
Nitex mesh (130-mesh, see p. 419)
Falcon tube (15 ml)
Rubber tubing
Kimwipes
Plastic petri dish lined with double-stick tape
Forceps
Microcentrifuge tubes (1.5 ml)
Dissecting stereomicroscope

### *Solutions and Reagents*

**Heptane**
PBS
**Methanol**
**Ethanol** series: 30%, 70%, and absolute
Sodium azide (10%)

> *Fixative*
>
> 1x PBS:37% **formaldehyde** (7:3)
> *OR*
> 4% **paraformaldehyde** in PEM (0.1 M PIPES [pH 6.9], 1 mM EDTA, 1 mM MgCl$_2$)
>
> *PBT*
>
> 1x PBS
> 0.2% Triton X-100

### BSS

10 mM Tricine (pH 7)
40 mM NaCl
55 mM **KCl**
10 mM **MgSO$_4$**
1 mM **CaCl$_2$**
20 mM glucose
50 mM sucrose
0.2% bovine serum albumin (BSA)
0.02% **sodium azide**

### BST

BSS
0.1% Triton X-100

### BSN

BSS
5% normal serum of same species as secondary antibody
0.2% saponin

### PTW

10 mM **sodium phosphate** buffer (pH 7.5)
130 mM NaCl
0.1% Tween-20

## Antibodies and Detection Reagents

Primary antibody: mouse anti-β-galactosidase antibody (Sigma)

Secondary antibody: biotinylated horse anti-mouse antibody (Vector laboratories)

Vectastain Elite ABC Kit (Vector Laboratories). This is an immunoperoxidase signal detection system with increased sensitivity due to signal amplification by interaction of avidin and biotinylated horseradish peroxidase.

Diaminobenzidine tetrahydrochloride (**DAB**); aliquot and store stock solution (1–10 mg/ml) at –80ºC.

**Hydrogen peroxide** (30%)

Canada balsam:**methyl salicylate** (1:1)

(*Optional*) **Cobalt chloride** (0.3%)

**CAUTION: CaCl$_2$, cobalt chloride, DAB, ethanol, formaldehyde, heptane, hydrogen peroxide, KCl, methanol, methyl salicylate, MgSO$_4$, paraformaldehyde, sodium azide, sodium phosphate** (see Appendix 4)

## Method

1. Use the edge of a razor blade or another coverglass to remove as much as possible of the halocarbon oil from the coverglass. Wash off the residual oil from the coverglass by placing it in a small glass dish with heptane and gently rocking until the embryos begin to detach. Remove the coverglass from the dish and wash the embryos into an 8-ml glass screw-cap tube through a small funnel with a stream of heptane. Use several changes of heptane to remove residual oil and glue.

2. Fix the embryos in a two-phase mixture of 4 ml of heptane and 4 ml of fixative (PBS:37% formaldehyde [7:3]) by gently shaking for 30 minutes.

   *Note:* For some double-labeling experiments, vigorous shaking in heptane:4% paraformaldehyde in PEM (1:1) for 40 minutes yielded better staining results.

3. Devitellinize embryos as follows:

   a. Place a piece of Nitex mesh over the opening of a 15-ml Falcon tube. Cut off the end of the tube and hold the Nitex screen in place over the end with a ring made from a piece of rubber tubing. Insert a Kimwipe into the opening of the tube, pressed up against the Nitex screen; this will draw off the liquid once embryos are placed on the screen.

   b. Remove most of the heptane. Use a 200-μl pipetter with a tip clipped to make a wide opening to transfer the embyros on to the Nitex screen. Once the liquid is removed, immediately transfer the embryos onto double-stick tape lining the bottom of a plastic petri dish. Cover embryos with PBS.

   c. Devitellinize the embryos by gently pushing the ends with a forceps.

   d. Transfer the devitellinized embryos in PBS to a 1.5-ml microcentrifuge tube using a 200-μl pipetter with a tip that had been rinsed with PBT to prevent the embryos from sticking.

   e. Wash embryos several times with methanol.

      *Note:* Embryos that are not to be used immediately can be stored in ethanol at –20ºC for several days.

4. Antibody staining is performed with a slight modification of the method described by Bomze and Lopez (1994). Because the number of embryos to be processed is very small for a given experiment, the staining is typically done in a 200-μl volume in a 1.5-ml microcentrifuge tube:

   a. Rehydrate the embryos in four changes of BST over a 2-hour period.

   b. Block the embryos in BSN for 30 minutes.

   c. Incubate embryos with mouse anti-β-galactosidase (Sigma) diluted in BSN (1:1500) for 2–18 hours with gentle rocking.

      *Note:* Both primary and secondary antibodies are preincubated with an overnight collection of embryos to reduce background.

   d. Wash the embryos with BSN for 15 minutes. Repeat eight times.

   e. Incubate embryos with biotinylated, horse anti-mouse antibody (Vector Laboratories) diluted in BSN (1:2000) for 2–18 hours with gentle rocking.

   f. Wash embryos three times with BSN (15 minutes each).

   g. Wash embryos three times in PTW (15 minutes each).

   h. After the first PTW wash in step 4g, prepare the biotin-avidin signal amplification mixture by adding 5 μl each of reagent A and B (Vectastain Elite ABC kit, Vector Laboratories) in 1 ml of PTW; allow to mix for 30 minutes. After the third PTW wash, incubate the embryos in the signal amplification mixture for 30 minutes.

   i. To detect antibody binding, wash the embryos three times in PTW (15 minutes each). Dilute the stock solution of DAB to 0.5 mg/ml with PTW and incubate the embryos in 200 μl of this diluted DAB solution for 5 minutes. Add 2 μl of 3% hydrogen peroxide (freshly diluted from a 30% stock solution) to start the color reaction. This is typically done under a dissecting stereomicroscope in order to see the color develop on the embryos. Add 4 μl of 10% sodium azide to stop the reaction.

j. Wash the embryos three times in PTW (5 minutes each).

> *Note:* For detecting a second antigen after the first color reaction to detect β-galactosidase, perform the second round of immunostaining as described above (steps 4c–4j); however, the color reaction step is performed with DAB plus 0.3% cobalt chloride (step 4i). Typically, this reaction is much faster than the first one, so make sure that this is performed under the dissecting stereomicroscope with 10% sodium azide ready to be added to stop the reaction.

k. Dehydrate in an ethanol series (30%, 70%, and absolute; each for 5 minutes). Mount the dehydrated embryos in 1:1 Canada balsam:methyl salicylate.

# REFERENCES

Boehlen P., Stein S., Dairman W., and Udenfried S. 1973. Fluorometric assay of proteins in the nanogram range. *Arch. Biochem. Biophys.* **155:** 213–220.

Bomze H.M. and Lopez A.J. 1994. Evolutionary conservation of the structure and expression of alternatively spliced *Ultrabithorax* isoforms from *Drosophila. Genetics* **136:** 965–977.

Bossing T. and Technau G.M. 1994. The fate of the CNS midline progenitors in *Drosophila* as revealed by a new method for single cell labeling. *Development* **120:** 1895–1906.

Bownes M. 1975. Adult deficiencies and duplications of head and thoracic structures resulting from microcautery of blastoderm stage *Drosophila* embryos. *J. Embryol. Exp. Morphol.* **34:** 33–54.

Brand A.H. and Perrimon N. 1993. Targeted gene expression as a means of altering cell fates and generating dominant phenotypes. *Development* **118:** 401–415.

Cambridge S.B., Davis R.L., and Minden J.S. 1997. *Drosophila* mitotic domain boundaries as cell fate boundaries. *Science* **277:** 825–828.

Chasman D.I., Leatherwood J., Carey M., Ptashne M., and Kornberg R.D. 1989. Activation of yeast polymerase II transcription by herpesvirus VP16 and GAL4 derivatives *in vitro. Mol. Cell. Biol.* **9:** 4746–4749.

Garcia-Bellido A. and Merriam J.R. 1969. Cell lineage of imaginal discs in *Drosophila* gynandomorphs. *J. Exp. Zool.* **170:** 61–76.

Halfon M.S., Kose H., Chiba A., and Keshishian H. 1997. Targeted gene expression without a tissue-specific promoter creating mosaic embryos using laser-induced single-cell heat shock. *Proc. Natl. Acad. Sci.* **94:** 6255–6260.

Janning W. 1978. Gynandromorgh fate maps in *Drosophila*. In *Genetic mosaics and cell differentiation* (ed. W.J. Gehring), pp. 1–28. Springer Verlag, Berlin.

King B.H. and Bryant J.P. 1982. Developmental responses of the *Drosophila melanogaster* embryo to localized X irradiation. *Rad. Res.* **89:** 590–606.

Lawrence P.A. and Morata G. 1977. The early development of mesothoracic compartments in *Drosophila*. An analysis of cell lineage and fate mapping and an assessment of methods. *Dev. Biol.* **56:** 40–51.

Lohs-Schardin M., Cremer C., and Nüsslein-Volhard C. 1979. A fate map for the larval epidermis of *Drosophila melanogaster:* Localized cuticle defects following irradiation of the blastoderm with an ultraviolet microbeam. *Dev. Biol.* **73:** 239–255.

Steiner E. 1970. Establishment of compartments in the developing leg imaginal discs of *Drosophila melanogaster. Wilhelm Roux's Arch. Dev. Biol.* **180:** 1–30.

Struhl G. 1981. A blastoderm fate map of compartments and segments of the *Drosophila* head. *Dev. Biol.* **84:** 386–396.

Technau G.M. and Campos-Ortega J.A. 1985. Fate-mapping in wild-type *Drosophila melanogaster*. II. Injections of horseradish peroxidase in cells of the early gastrula stage. *Wilhelm Roux's Arch. Dev. Biol.* **194:** 196–212.

Vincent J.P. and O'Farrell P.H. 1992. The state of *engrailed* expression is not clonally transmitted during early *Drosophila* development. *Cell* **68:** 923–931.

Weischaus E. and Gehring W. 1976. Clonal analysis of primordial disc cells in the early embryo of *Drosophila melanogaster. Dev. Biol.* **50:** 249–265.

# Molecular Biology

# CONTENTS

# 23

# Recovery of DNA Sequences Flanking P-element Insertions: Inverse PCR and Plasmid Rescue

**Audrey M. Huang, E. Jay Rehm, and Gerald M. Rubin**
*Department of Molecular and Cell Biology, and*
*University of California at Berkeley*
*Howard Hughes Medical Institute*
*Berkeley, California 94720*

THE *DROSOPHILA MELANOGASTER* P-TRANSPOSABLE ELEMENT is a powerful and widely used research tool. The transposon can act as a mutagen to cause insertional alterations in proximally located genes (Rubin et al. 1982). P-elements are also a useful resource for local mutagenesis, as they have the tendency to transpose locally (within 200 kb; Tower et al. 1993; Zhang and Spradling 1993). A β-galactosidase reporter gene within a P-element can be expressed under the control of local endogenous promoters, thereby identifying patterns of expression associated with the genes at the site of insertion (Bellen et al. 1989; Bier et al. 1989; Wilson et al. 1989). Since P-elements have the tendency to insert into upstream promoter regions of genes (Spradling et al. 1995), they can be engineered to direct misexpression of endogenous genes at the site of insertion by containing a regulatable promoter (Rorth 1996). Expression of transposase can induce meiotic recombination in males at the site of a P-element insertion; such site-specific recombination can be used to map the location of a gene rapidly (Chen et al. 1998). Sequences flanking the P-element can be recovered and the site of insertion mapped to the nucleotide, to connect the genetic and physical maps and facilitate molecular analysis of the gene of interest.

The Berkeley *Drosophila* Genome Project (BDGP) has assembled a well-characterized collection of lethal mutations induced by single P-element insertions generated by a number of laboratories (see Spradling et al. 1995). The genomic DNA sequences adjacent to these insertions have been recovered by either plasmid rescue or inverse polymerase chain reaction (PCR); 95% of the flanking sequences are more than 25 bp long, and 86% of the flanking sequences are more than 200 bp long (Spradling et al. 1999; E.J. Rehm and G.M. Rubin, unpubl.). Figure 23.1 illustrates the general principle behind these two methods. Individual, experimentally modified P-elements can be mobilized in large genetic screens to generate thousands of stable mutant strains (see, e.g., Cooley et al. 1988). The combination of a complete genomic DNA sequence and relatively fast and easy molecular meth-

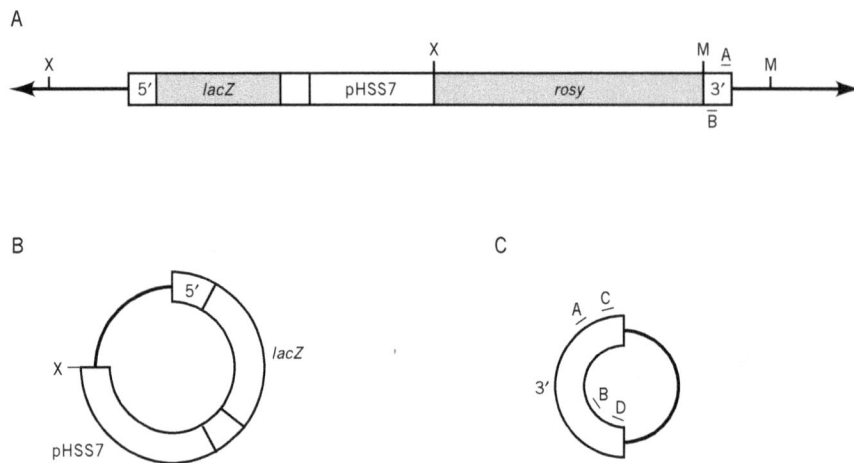

**Figure 23.1.** Rescue of sequences flanking P-element insertions. (A) A schematic diagram of P{PZ} (Mlodzik and Hiromi 1992). The P-element contains a *lacZ* reporter gene, plasmid rescue sequences (pHSS7) containing a bacterial origin of replication and a gene conferring kanamycin resistance, and the *rosy* gene for eye color selection. (X) *Xba*I; (M) *Msp*I; (A) Pry4 primer; (B) Pry1 primer. (B) Plasmid rescue off the "left" (5′) side of P{PZ}. Genomic DNA is digested with *Xba*I, which cuts within the P-element sequence downstream from the plasmid rescue sequences, and in the flanking genomic sequence, some distance away from the P-element insertion. This DNA is then self-ligated and transformed. Rescued flanking genomic DNA can then be sequenced with primers designed against the plasmid rescue sequences as well as the P-element end. (C) Inverse PCR off the "right" (3′) side of P{PZ}. Genomic DNA is digested with *Msp*I, which cuts within the P-element as well as the flanking sequence. This DNA is self-ligated and PCR-amplified with primers Pry4 (A) and Pry1 (B). This product can then be directly sequenced from both ends with primers Spep1 (C) and Sp3 (D).

ods for mapping P-element insertion sites to the nucleotide enhances the use of P-elements as tools in *Drosophila* research.

The inverse PCR protocol given below was developed to isolate genomic DNA sequences immediately adjacent to the insertion of large numbers of P-elements as part of the gene-disruption project of the BDGP. It has been successfully applied to various P-elements including the P{PZ} (Mlodzik and Hiromi 1992), P{lacW} (Bier et al 1989), and P{EP} (Rorth 1996) elements. Sequences and maps of these P-elements can be found on Flybase at:

http://flybase.bio.indiana.edu

This protocol can also be found at the BDGP Web site at:

http://www.fruitfly.org/methods/

Detailed procedures for isolating DNA flanking P-element insertions are given below.

## ISOLATION, DIGESTION, AND LIGATION OF GENOMIC DNA

Protocol 23.1 describes a standard fly miniprep that requires very few flies and produces high-quality DNA. This protocol can also be used to isolate RNA when RNase-free conditions are utilized; an extra step must be taken to rid the sample of genomic DNA (e.g., RNase-free DNase digestion). Genomic DNA prepared as in Protocol 23.1 is then digested with restriction enzymes (Protocol 23.2) and ligated (Protocol 23.3), and used for either plasmid rescue or inverse PCR as described below.

PROTOCOL 23.1*

# Quick Genomic DNA Prep

## Materials

### Supplies and Equipment

Microcentrifuge tubes (1.5-ml)

Disposable tissue grinders. Kontes Pellet Pestles can be purchased from Fisher (K749520-0000, pestles and tubes). The pestles fit into most (but not all!) microcentrifuge tubes. The better the fit, the better the homogenization. Kontes also makes a battery-operated, hand-held grinder, which makes processing of large batches much easier; this is also available from Fisher (K749540-0000).

Microcentrifuge

### Solutions and Reagents

**Isopropanol**
**Ethanol** (70%)
TE (see Appendix 3)

> ### Buffer A
> 100 mM Tris-Cl (pH 7.5)
> 100 mM EDTA
> 100 mM NaCl
> 0.5% **SDS**
> Store at room temperature.

> ### Buffer B
> 200 ml of 5 M potassium acetate
> 500 ml of 6 M **lithium chloride**
> Mix together and store at 4°C.

CAUTION: ethanol, isopropanol, LiCl, SDS (see Appendix 4)

## Method

1. Collect 30 anesthetized flies in a 1.5-ml microcentrifuge tube placed on ice.

   Note: Flies can be stored at –80°C indefinitely or DNA can be prepared immediately without freezing the flies.

2. Grind flies in 200 µl of Buffer A with a disposable tissue grinder. Add an additional 200 µl of Buffer A (total volume of 400 µl) and continue grinding until only cuticles remain (~1–2 minutes, grinding by hand).

3. Incubate samples at 65°C for 30 minutes.

4. Add 800 µl of Buffer B to each sample, mix well by inverting the tube multiple times, and incubate on ice for at least 10 minutes and up to a few hours.

---

*For all protocols, $H_2O$ indicates glass distilled and deionized.

5. Centrifuge in a microcentrifuge at 12,000 rpm at room temperature for 15 minutes.

6. Transfer 1 ml of the supernatant into a new microcentrifuge tube. Some of the precipitate will not pellet, but instead float on top of the supernatant; be extremely careful to avoid transferring any floating precipitate. Discard the pellet. If necessary, repeat the centrifugation step to get rid of any contaminating precipitate.

7. Add 600 μl of isopropanol to each sample, and mix well by inverting the tube several times.

8. Centrifuge at 12,000 rpm at room temperature for 15 minutes.

9. Discard the supernatant. Wash the pellet with 70% ethanol, air-dry, and resuspend in 150 μl of TE.

10. Store the DNA at –20ºC.

## PROTOCOL 23.2

# Restriction Enzyme Digestion

## Materials

### Supplies and Equipment

Microcentrifuge tubes (1.5 ml)
(*Optional*) Equipment for agarose gel electrophoresis

### Solutions and Reagents

Restriction enzymes; the following enzymes can be purchased from New England Biolabs:
    *Sau*3AI (Cat. no. 169), *Msp*I (Cat. no. 106), *Hin*P1I (Cat. no. 124).
Genomic DNA
10x Restriction enzyme buffer
RNase A (100 μg/ml stock solution; Boehringer Manneheim 109-169).
(*Optional*) Buffers and reagents for electrophoresis (e.g., 0.8% agarose gel in TAE)

### Method

1. Set up separate restriction digests for each enzyme. (See Table 23.1 for the appropriate enzyme to use for the particular method and P-element.) For each restriction digest, combine the following in a 1.5-ml microcentrifuge tube:

| | |
|---|---|
| genomic DNA (~2 fly equivalents) | 10.0 μl |
| 10 restriction enzyme buffer | 2.5 μl |
| 100 μg/ml RNase A | 2.0 μl |
| restriction enzyme | 10 units |
| sterile H$_2$O | to 25 μl |

2. Incubate restriction digests at 37ºC for 2.5 hours.

Table **23.1**. Restriction Enzymes for Inverse PCR and Plasmid Rescue

| | Plasmid rescue | | | | | |
|---|---|---|---|---|---|---|
| | P{PZ} | | P{lacW} | | P{EP} | |
| Inverse PCR | 5′ (left) | 3′ (right) | 5′ (left) | 3′ (right) | 5′ (left) | 3′ (right) |
| *Sau*3AI | *Xba*I | N/A[a] | *Eco*RI | *Bam*HI | N/A[a] | *Eco*RI |
| *Hin*P1I | *Xba*I + *Spe*I | | *Sac*II | *Bgl*II | | |
| *Msp*I | *Xba*I + *Nhe*I | | *Pst*I | | | |
| | | | *Xba*I | | | |

[a]Plasmid rescue will only work off the 5′ (left) side of P{PZ} and the 3′ (right) side of P{EP}.

3. Heat-inactivate restriction enzymes by incubating samples at 65ºC for 20 minutes for all enzymes except *Bam*HI, *Pst*I, and *Bgl*II; *Bam*HI and *Pst*I should be inactivated at 80ºC. Reactions containing *Bgl*II should be extracted with phenol/chloroform (1:1 v/v), precipitated with 1/10 volume of 3 M sodium acetate and 2.5 volumes of ethanol, and resuspended in 25 µl of TE.

4. (*Optional*) Check 10 µl of each digest on a 0.8% agarose gel in TAE. High-quality genomic DNA should run as a single high-molecular-weight band. After digestion, the DNA should run as an even smear.

PROTOCOL **23.3**

# Ligation

## Materials

### Supplies and Equipment

Microcentrifuge tubes (1.5 ml)
Microcentrifuge

### Solutions and Reagents

Digested genomic DNA
T4 DNA ligase (New England Biolabs, Cat. no. 202)
**Sodium acetate** (3 M, pH 5.2)
**Ethanol** (70%)

#### 10× Ligase Buffer

0.5 M Tris-Cl (pH 7.5)
100 mM **MgCl₂**
100 mM **dithiothreitol** (DTT)
10 mM ATP
0.25 mg/ml bovine serum albumin (BSA)
Store at –20ºC.

**CAUTION: DTT, ethanol, MgCl₂** (see Appendix 4)

## Method

1. Set up each ligation reaction by combining the following in a 1.5-ml microcentrifuge tube:

| | |
|---|---|
| digested genomic DNA (~1 fly equivalent) | 10 μl |
| 10x ligase buffer | 40 μl |
| sterile H₂O | 350 μl |
| T4 DNA ligase | 2 μl (2 Weiss units) |

*Note:* Performing ligations in a large volume minimizes intermolecular and favors intramolecular ligation events.

2. Incubate samples at 4°C overnight.

3. To each ligation reaction, add 40 μl of 3 M sodium acetate and 1 ml of ethanol. Mix well. Freeze sample at –20°C for at least 10 minutes.

4. Centrifuge in a microcentrifuge at 12,000 rpm at room temperature for 10 minutes.

5. Remove supernatant, wash the pellet with 70% ethanol, air-dry, and resuspend in 150 μl of sterile H₂O for inverse PCR and 10 μl of sterile H₂O for plasmid rescue. At this point, the ligated DNA can be stored at –20°C indefinitely.

## PLASMID RESCUE

The plasmid rescue sequences in each P-element have the respective dominant selectable markers: P{EP} and P{PZ} kanamycin resistance, and P{lacW} ampicillin resistance.

Transform the entire amount of the ligated DNA (one fly equivalent) by the method of choice. It is recommended that several transformants be screened by restriction digests of plasmid minipreps to check that only a single class of transformant is isolated. Isolation of multiple classes of transformants may indicate multiple P-element insertions in the starting line, sample contamination, or partial restriction enzyme digestion. See Table 23.2 for appropriate sequencing primers to use.

**Table 23.2.** Primer Combinations for PCR and Sequencing

| | | PCR primers | | Annealing | | |
|---|---|---|---|---|---|---|
| P-element | Rescued end | Forward | Reverse | temperature | Sequencing primers[a] | |
| P{Z} | 5′ | Plac4 | Plac1 | 60°C | Splac2 | Sp1 |
| P{Z} | 3′ | Pry4 | Pry1 | 55°C | Spep1 | Sp3 |
| P{lacW} | 5′ | Plac4 | Plac1 | 60°C | Splac2 | Sp1 |
| P{lacW}[b] | 3′ | Pry4 | Plw3-1 | 55°C | Spep1 | Sp6(*Sau*3AI) Sp5(*Hin*P1I) |
| P{EP} | 5′ | Pwht1 | Plac1 | 60°C | Sp1 | – |
| P{EP} | 3′ | Pry4 | Pry1 | 55°C | Spep1 | – |

[a]Where two primers are listed, the inverse PCR product can be sequenced from both ends by setting up sequencing reactions with the two different primers.

[b]For the rescue of sequences 3′ of P{lacW}, use the Sp6 primer on templates derived from *Sau*3AI digests, and Sp5 primer on templates derived from *Hin*P1I digests. The Spep1 primer works on both templates.

**Table 23.3.** Primer Sequences

| Primer | Length | Sequence (5′ to 3′) |
|---|---|---|
| (a) *PCR primers* | | |
| Plac4 | 27 mer | ACT GTG CGT TAG GTC CTG TTC ATT GTT |
| Plac1 | 24 mer | CAC CCA AGG CTC TGC TCC CAC AAT |
| Pry4 | 23 mer | CAA TCA TAT CGC TGT CTC ACT CA |
| Pry1 | 26 mer | CCT TAG CAT GTC CGT GGG GTT TGA AT |
| Pry2 | 28 mer | CTT GCC GAC GGG ACC ACC TTA TGT TAT T |
| Plw3-1 | 19 mer | TGT CGG CGT CAT CAA CTC C |
| Pwht1 | 29 mer | GTA ACG CTA ATC ACT CCG AAC AGG TCA CA |
| | | |
| (b) *Sequencing primers* | | |
| Splac2 | 25 mer | GAA TTC ACT GGC CGT CGT TTT ACA A |
| Sp1 | 22 mer | ACA CAA CCT TTC CTC TCA ACA A |
| Sp3 | 24 mer | GAG TAC GCA AAG CTT AA CTA TGT |
| Sp6 | 23 mer | TGA CCA CAT CCA AAC ATC CTC TT |
| Sp5 | 25 mer | GCA TCA CAA AAA TCG ACG CTC AAG T |
| Spep1 | 19 mer | GAC ACT CAG AAT ACT ATT C |

## INVERSE PCR AND SEQUENCING

Primer sets have been designed for each end of each P-element. The appropriate primer pairs and corresponding annealing temperatures are listed in Table 23.2, and the primer sequences are listed in Table 23.3. Generally, strong and unique bands, which can be directly sequenced without extensive purification, result from the PCR conditions given in Protocol 23.4. Use a cycle-sequencing protocol appropriate for the sequencing method. The BDGP is currently using Perkin Elmer ABI PRISM 377 automated sequencers with the PE Applied Biosystems Big Dye Terminator Cycle Sequencing kit. Under these conditions, 2.5 μl of the purified PCR product is sufficient for sequencing. The use of other thermal cyclers and sequencing strategies will require different conditions and protocols. Further information can be found at the BDGP Web site.

PROTOCOL 23.4

## Inverse PCR

### Materials

### *Supplies and Equipment*

PCR tubes and caps are available from several sources; e.g., the Strip-Ease strips of 12 0.2-ml tubes are easy to manipulate and can be purchased from Robbins Scientific (1044-50-0 tubes, 1044-72-0 caps).

Qiagen QIAquick PCR purification kit (kit of 50, Cat. no. 28104; kit of 250, Cat. no. 28106) for cleaning up PCR products

Thermal cycler

Equipment for agarose gel electrophoresis

### Solutions and Reagents

Ligated genomic DNA
dNTPs (Amersham Pharmacia Biotech 27-2035-01)
Forward and reverse primers
*Taq* DNA polymerase (Amersham Pharmacia Biotech T0303Y)
10x Pharmacia reaction buffer
Agarose gel (1%) in TAE

### Method

1. Set up reactions in thin-walled 0.2-ml PCR tubes placed on ice:

   | | |
   |---|---|
   | ligated genomic DNA | 10.0 µl |
   | 2 mM dNTP mixture | 2.0 µl |
   | 10 µM forward primer | 1.0 µl |
   | 10 µM reverse primer | 1.0 µl |
   | 10x reaction buffer | 5.0 µl |
   | sterile $H_2O$ | 31.0 µl |
   | 2 units *Taq* DNA polymerase | 0.4 µl |

2. Perform thermal cycling. The following parameters have been optimized for the MJ Research PTC-100 machine; other thermal cyclers have different cycling profiles, and adjustments may be necessary to achieve optimal results. Again, different primer combinations require different annealing temperatures; for the appropriate annealing temperature for the primers to use, see Table 23.2. These particular parameters require a run time of approximately 3 hours:

   | Cycle | Temperature | Time | Number of Cycles |
   |---|---|---|---|
   | | 94.0°C | 3 minutes | 1 |
   | denature | 94.0°C | 30 seconds | |
   | anneal | see Table 23.2 | 1 minute | 35 |
   | extend | 68.0°C | 2 minutes | |
   | | 72.0°C | 10 minutes | 1 |
   | | 4.0°C | hold | – |

3. Examine 5 µl (1/10) of each sample on a 1% agarose gel in TAE. If the resulting product is faint, reamplify the sample. If nonspecific bands appear, isolate the major product from the gel by standard gel purification techniques and directly sequence. Generally, this is not necessary.

4. Clean up the remainder (45 µl) of each sample with a Qiagen QIAquick PCR purification column to remove primers and nucleotides. Follow the protocol included with the kit and elute the PCR product from the column with 50 µl of sterile $H_2O$.

## REFERENCES

Bellen H.J., O'Kane C.J., Wilson C., Grossniklaus U., Pearson R.K., and Gehring W.J. 1989. P-element-mediated enhancer detection: A versatile method to study development in *Drosophila*. *Genes Dev.* **3:** 1288–300.

Bier E., Vaessin H., Shepherd S., Lee K., McCall K., Barbel S., Ackerman L., Carretto R., Uemura T., and Grell E. 1989. Searching for pattern and mutation in the *Drosophila* genome with a P-lacZ vector.

*Genes Dev.* **3:** 1273–1287.

Chen B., Chu T., Harms E., Gergen J.P., and Strickland S. 1998. Mapping of *Drosophila* mutations using site-specific male recombination. *Genetics* **149:** 157–163.

Cooley L., Kelley R., and Spradling A. 1988. Insertional mutagenesis of the *Drosophila* genome with single P elements. *Science* **239:** 1121–1128.

Mlodzik M. and Hiromi Y. 1992. Enhancer trap method in *Drosophila:* Its application to neurobiology. In *Methods in neurosciences* (ed. P.M. Conn), pp. 397–414. Academic Press. San Diego.

Rorth P. 1996. A modular misexpression screen in *Drosophila* detecting tissue-specific phenotypes. *Proc. Natl. Acad. Sci.* **93:** 12418–12422.

Rubin G.M., Kidwell M.G., and Bingham P.M. 1982. The molecular basis of P-M hybrid dysgenesis: The nature of induced mutations. *Cell* **3:** 987–994.

Spradling A.C., Stern D., Kiss I., Roote J., Laverty T., and Rubin G.M. 1995. Gene disruptions using P transposable elements: An integral component of the *Drosophila* genome project. *Proc. Natl. Acad. Sci.* **92:** 10824–10830.

Spradling A.C., Stern D., Beaton A.., Rehm E.J., Laverty T., Mozden N., Misra S., and Rubin G.M. 1999. The Berkeley Drosophila Genome Project Gene Disruption Project: Single P-element insertions mutating 25% of vital *Drosophila* genes. *Genetics* **153:** 135–177.

Tower J., Karpen G.H., Craig N., and Spradling A.C. 1993. Preferential transposition of *Drosophila* P elements to nearby chromosomal sites. *Genetics* **133:** 347–359.

Wilson C., Pearson R.K., Bellen H.J., O'Kane C.J., Grossniklaus U., and Gehring W.J. 1989. P-element-mediated enhancer detection: An efficient method for isolating and characterizing developmentally regulated genes in *Drosophila. Genes Dev.* **3:** 1301–1313.

Zhang P. and Spradling A.C. 1993. Efficient and dispersed local P element transposition from *Drosophila* females. *Genetics* **133:** 361–373.

# CONTENTS

# GAL4-mediated Ectopic Gene Expression in *Drosophila*

**Peter van Roessel and Andrea H. Brand**
*Wellcome/CRC Institute*
*University of Cambridge*
*Cambridge CB2 1QR, England*

T HE GENERATION OF GAIN-OF-FUNCTION PHENOTYPES by ectopic expression of known genes provides a powerful complement to the genetic approach, in which genes are identified through mutations that reduce or eliminate gene function. The GAL4 system (Brand and Perrimon 1993) is a method for ectopic gene expression in *Drosophila* that allows the selective activation of any cloned gene in a wide variety of tissue- and cell-specific patterns. Similar GAL4-based expression systems have been applied in mice (Ornitz et al. 1991), zebrafish (Scheer and Campos-Ortega 1999), and plants (Ma et al. 1988; J.P. Haseloff, pers. comm.). In addition to producing gain-of-function phenotypes, the GAL4 system can be used to interfere with gene expression by transcription of antisense RNA or expression of dominant-negative forms of a protein. Other applications of the GAL4 system include targeted cell ablation, performed by expression of cellular toxins or cell death genes (see Sweeney et al., this volume), and the expression of reporter proteins that enable the visualization of cell structure and function (Brand et al. 1995, 1999; Phelps and Brand 1998). Recent modifications of the GAL4 system have further enabled cell-fate mapping in embryos (Cambridge et al. 1997) and tissue-specific misexpression screens (Rørth 1996; Rørth et al. 1998).

In this chapter, we present an overview of the GAL4 sytem, including recent developments, and discuss methods for generating and visualizing transgenic flies that ectopically express GAL4.

## OVERVIEW

### How the GAL4 System Works

GAL4 is a transcription factor from yeast that can activate transcription in *Drosophila* (Fischer et al. 1988). GAL4 has no known target genes in *Drosophila* and can be expressed in various cells and tissues of the organism without observable effect. To express the yeast transcriptional activator GAL4 in different cell and tissue types, an enhancerless *GAL4*

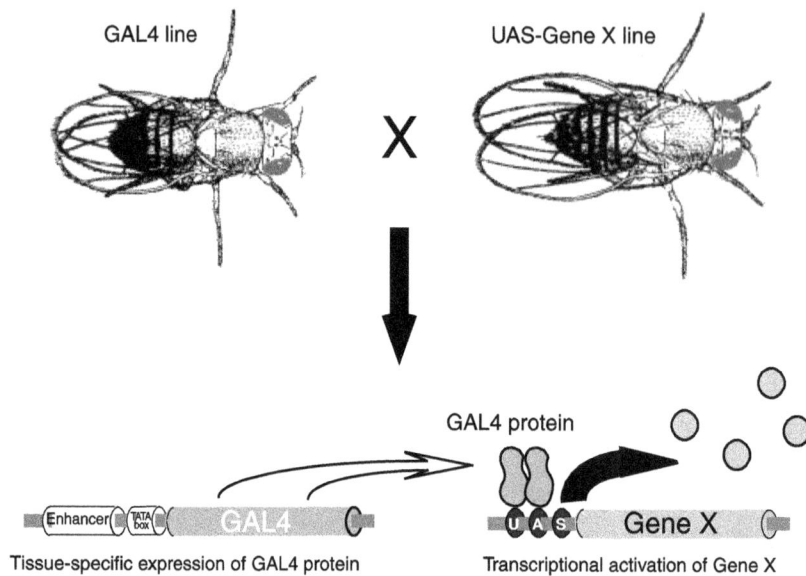

**Figure 24.1.** Targeted ectopic gene expression using GAL4. The *GAL4* gene, encoding a transcriptional activator from yeast, is introduced into the *Drosophila* genome as part of an enhancer-detection vector. *Drosophila* lines expressing the GAL4 protein in specific cells and tissues are crossed to lines carrying a target gene of interest (Gene X), subcloned downstream from five GAL4-binding sites (the UAS). The target gene is expressed in the progeny of such a cross, only in those cells and tissues where GAL4 is present.

gene is inserted randomly into the *Drosophila* genome by P-element transposition (Figure 24.1) (Brand and Perrimon 1993). Depending on its site of integration, GAL4 expression can be directed by any one of a diverse array of genomic enhancers. When a second gene containing GAL4-binding sites (the UAS, or upstream activation sequence) within its promoter is introduced into this background in a genetic cross, this gene is transcribed only in those cells where GAL4 is expressed (Figure 24.1). The effect of directed misexpression on development can then be followed.

Hundreds of lines expressing GAL4 in distinct and reproducible spatiotemporal patterns, each reflecting the native activity of the "trapped" enhancer, can be generated by this approach (enhancer detection or trapping; O'Kane and Gehring 1987). The lines are then screened for those that express GAL4 in a pattern appropriate to a given experimental goal. A second approach, in which a cloned promoter is used to drive expression of GAL4, can be taken if a characterized promoter known to drive gene expression in the desired pattern is available. GAL4 should then be expressed in a pattern similar to that of the endogenous gene from which the promoter was isolated (Brand and Perrimon 1993).

## Why Use GAL4?

The GAL4 system has at least two advantages over previous ectopic expression techniques. First, unlike expression from a heat shock promoter, the technique allows the rapid generation of individual strains in which ectopic expression can be directed to different tissues or cell types. To study the role of several different genes in a specific tissue, a single GAL4-expressing line can be crossed to an array of different UAS-target gene lines. Conversely, to study the function of a single gene in different cell and tissue types, one

UAS-target line can be crossed to a diverse array of lines expressing GAL4 in distinct cell and tissue-specific patterns. Second, the method separates the transcriptional activator from its target gene in two discrete transgenic lines. This ensures that the parental lines are viable: In one line, the activator protein is present, but it has no target gene to activate, and in the second line, the target gene is silent in the absence of the activator. The GAL4 system thus enables the propagation of transgenic lines carrying reporter genes encoding potentially toxic proteins. When the two lines are crossed, the target gene is turned on only in the progeny of the cross, allowing dominant phenotypes (including lethality) to be conveniently studied. Although transient expression of toxic proteins during transformation can prove lethal, such expression can be circumvented by transforming flies with a UAS construct containing an *FRT-transcriptional terminator-FRT* sequence before the target gene start codon, which is later excised with FLP-recombinase (Hidalgo et al. 1995; Lin et al. 1995; Smith et al. 1996).

GAL4-mediated ectopic expression is not readily inducible. The degree of expression is temperature-sensitive, however, with increased expression driven at 29ºC, and the insertional position of the UAS construct may affect expression of the target gene. These variables can be exploited to generate a series of increasingly severe phenotypes in overexpression experiments. Furthermore, specific temporal and spatial regulation of expression can be obtained using recently developed modifications of the GAL4 system. Laser-induced heat shock has been used to drive expression of GAL4 under the heat shock promoter in individual cells (Halfon et al. 1996), and caged GAL4 protein can be microinjected into syncytial embryos, after which it can be activated (uncaged) in individual cells by a fine beam of UV light (Cambridge et al. 1997). The FLP/FRT system can also be used to remove intervening transcriptional terminators from UAS reporter gene constructs. When FLP-recombinase is driven by the inducible heat shock promoter, the frequency of excision events—and hence, the number of cells that express a UAS target gene—can be controlled to some degree (Smith et al. 1996).

Until recently, it was not possible to drive ectopic expression with GAL4 early in embryogenesis. This obstacle has now been overcome by replacing the *hsp70*-derived 5′- and 3′-untranslated regions (UTRs) of the GAL4 driver with those of a maternally expressed gene and by fusing GAL4 to the viral transcriptional activator VP16. Using this driver line, GAL4-mediated expression is observed at the cellular blastoderm stage (Hacker et al. 1997; D. St. Johnston, unpubl.). Additionally, ectopic expression in the female germ line has been achieved by modifying the UAS vector to replace the *hsp70* promoter with the P-transposase TATA promoter and including the 3′UTR sequence from a maternally expressed gene (Rørth 1998).

A further variation on the GAL4 system uses GAL4 to misexpress unidentified genes in selected tissues (Rørth 1996). The UAS is first inserted at random positions within the genome, generating hundreds of independent "EP" (enhancer promoter) lines. EP lines are then crossed to a line that expresses GAL4 in a particular tissue. When ectopic expression gives an aberrant phenotype, the target gene can be isolated by plasmid rescue and characterized. This approach has enabled systematic "modular" gain-of-function genetic screens for genes that disrupt a particular developmental process (Rørth et al. 1998).

## TECHNIQUES

### Generation of Driver and Reporter Lines

For a description of plasmids pGaTB, pGawB, and pUAST, see Brand and Perrimon (1993), Brand et al. (1994), and Phelps and Brand (1998). Maps of these plasmids are shown in Figure 24.2 (pGaTB), Figure 24.3 (pGawB), and Figure 24.4 (pUAST).

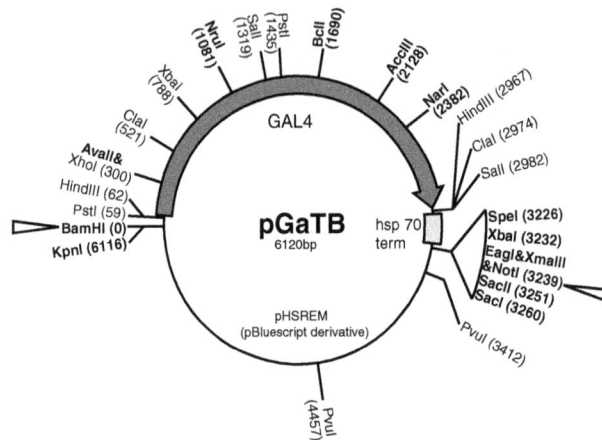

**Figure 24.2.** pGaTB. To construct promoter-*GAL4* fusions, the *GAL4*-encoding *Bam*HI/*Not*I fragment of pGaTB is subcloned downstream from the appropriate promoter sequence in a P-element vector. Unique restriction sites are presented in boldface type.

## Construction of Specific Promoter-GAL4 Fusions

To construct a promoter-GAL4 fusion, a three-way ligation can be carried out between the desired promoter sequence, the GAL4-coding sequence (excised as a *Bam*HI/*Not*I fragment from pGaTB; Figure 24.2), and an appropriate P-element vector (such as pCaSpeR 2, 3, or 4). Alternatively, the *Bam*HI/*Not*I fragment of pGaTB can be subcloned into a P-element vector, leaving *Bam*HI as a site into which promoter fragments can be cloned. The promoter fragment should ideally include sequence up to, but not including, the ATG of the native gene.

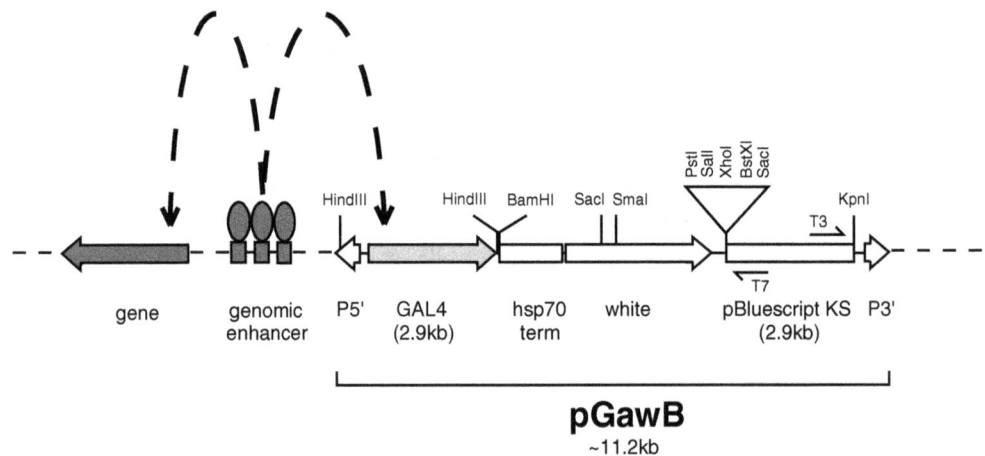

**Figure 24.3.** pGaWB. pGaWB is an "enhancer-trap" vector that makes use of neighboring genomic regulatory elements to drive expression of GAL4. The *white* gene gives eye colors ranging from pale yellow to red; an antibiotic resistance gene and bacterial *ori* within pBluescript allows plasmid rescue of neighboring genomic sequences. *Pst*I, *Sal*I, *Xho*I, *Bst*XI, and *Sac*I can be used to rescue downstream sequences, whereas the unique *Kpn*I site allows rescue of upstream sequences.

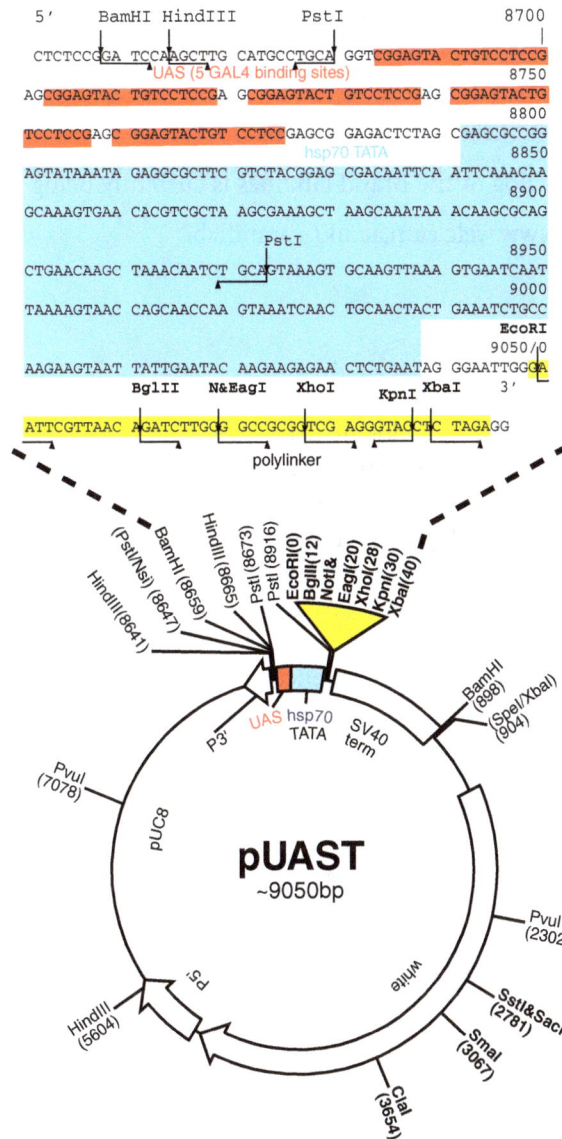

**Figure 24.4.** pUAST. pUAST contains the GAL4 UAS, a polylinker into which target genes can be sub-cloned, and a transcriptional terminator sequence, all within a transposable P-element vector. Restriction sites known to be unique are indicated in boldface type. Sequence data are provided for the UAS, *hsp70* TATA, and polylinker. GAL4-binding sites are highlighted in red; the *hsp70* sequence is highlighted in blue.

## Enhancer Trapping

To generate novel GAL4-expressing lines, an enhancerless GAL4 cassette (pGawB; Figure 24.3), can be mobilized by introducing P-transposase in a genetic cross. Different GAL4 expression patterns are obtained depending on the genomic enhancers or regulatory elements next to which GAL4 has inserted. Typically, approximately 65% of new insertions yield expression of GAL4 (Brand and Perrimon 1993). pGawB transposes less frequently, however, than other similarly sized enhancer trap constructs. This inefficiency may be due to alterations made in the 5′ P end of pGawB to enable the expression of GAL4 alone, rather than as a P-transposase-GAL4 fusion. A catalog of many GAL4 lines is available from FlyBase at:

http://www.ebi.ac.uk:7081/

and an on-line catalog of the Brand lab lines is currently being constructed; see

http://www.welc.cam.ac.uk/~brandlab/

### UAS-Reporter Constructs

To generate GAL4-responsive target genes, the coding sequence is cloned into the polylinker of vector pUAST, downstream from five GAL4-binding sites and the *hsp70* TATA and upstream of the SV40 terminator. The gene sequence should include the initiator methionine codon and minimal 5′UTR sequence, although it has been reported that in some cases, the 5′UTR can enhance expression. A map of the vector is presented in Figure 24.4, with sequence data for the UAS and polylinker. This sequence represents the most accurate map available for pUAST. Although a sequence for the vector is presented on FlyBase, the composite SV40 small t intron and polyadenylation sequence has not, to the best of our knowledge, been sequenced.

### Injection/Transformation Protocol

Transformant lines are generated using standard procedures (Spradling 1986). Embryos of strain *yw; +/+; Sb, P[ry+, Δ2-3]/TM6, Ubx* (Robertson et al. 1988) are injected with column-purified DNA (Qiagen Plasmid Midi Kit) at a concentration of approximately 0.6 μg/μl. Although 50% of the embryos die due to their genotype (*Sb/Sb* or *TM6/TM6*), the high efficiency of transformation obtained using P-transposase-expressing embryos outweighs this disadvantage. It is possible to obtain three or more independent insertion lines from 100 injected embryos. Furthermore, the lines with independent insertion sites can be generated by P-element transposition induced in the presence of P-transposase, for example, by crossing existing enhancer trap lines to the line *yw; +/+; Sb, P[ry+, Δ2-3]/TM6, Ubx*.

## GAL4 Expression Patterns

### Prediction and Visualization of the GAL4 Expression Pattern

The pattern of GAL4 expression should reflect that of native genes, both when using an isolated cloned promoter and when using unknown "trapped" gene regulatory sequences. Exact correspondence between GAL4 expression patterns and expression of native gene products cannot be assumed, however. Insertional position in the genome may affect spatial and temporal expression patterns. In addition, a period of transcription and translation is required before GAL4 accumulates to levels sufficient to activate the target gene, such that expression of the target gene lags behind that of expression of GAL4. The half-life of the GAL4 protein may differ from that of the protein regulated by the "trapped" enhancer. GAL4 may thus continue to drive target-gene expression in tissues when the endogenous gene products are no longer present.

Antibodies against GAL4 are the most direct means of characterizing GAL4 expression patterns. In our hands, however, neither commercially available nor in-house antisera have given satisfactory results. The use of reporter genes under the control of the GAL4-UAS provides a more accurate reflection of target gene expression. The expression pattern of such reporter genes should mirror that of other genes whose expression is driven by the same GAL4 line.

### Selection of a Reporter Construct

Many commonly used reporter constructs are based on the expression of β-galactosidase, encoded by the *lacZ* gene of *Escherichia coli*, or on green fluorescent protein (GFP), derived from the jellyfish *Aequoria victoria* (Chalfie et al. 1994). *lacZ*-based reporters can be detected using X-gal (Sigma) as a substrate for β-galactosidase activity, producing a dark blue precipitate in expressing tissues. The X-gal precipitate can diffuse away from the site of β-galactosidase activity, however, limiting the resolution of cellular and subcellular structures. A more precise approach is to characterize β-galactosidase expression by immunohistochemistry (antibody is available from Cappel, Organon Teknika).

GFP-based reporters can be viewed in live cells, noninvasively, without fixation or staining (see Hazelrigg, this volume). GFP-based target genes can thus be used to monitor GAL4 expression patterns in a single live embryo over time. GFP maintains its fluorescence after fixation in 4% formaldehyde, should fixation be required to characterize endogenous gene expression. Furthermore, GFP can be detected by immunostaining with a polyclonal anti-GFP antibody (available from Clontech). For immunochemical detection of both β-galactosidase and GFP, we perform antibody staining as described by Patel (1994).

Reporter constructs that are specifically targeted to, or excluded from, the nucleus, such as *UAS-nuclear GFP-lacZ* (Shiga et al. 1996), present a means by which cell nuclei can be resolved and counted. As cell morphology is instrumental in identifying cell types, reporter constructs that reveal subcellular structures, such as *UAS-tau GFP*, which labels microtubules (Brand 1995), are also advantageous. A further function of a reporter gene is to reveal the dynamic temporal expression pattern of GAL4. To achieve this, the reporter gene must require as little time as possible for transcription, translation, and posttranslational processing. Excessive "trancript-to-activity" lag time for the production of the reporter gene may cause GAL4 activity to appear delayed with respect to GAL4 expression. The lag time from transcription to fluorescence is several hours for wild-type GFP (Heim et al. 1994), but it is much reduced for modified forms of GFP that are now available (Brand 1999; Haseloff et al. 1999).

For available UAS reporters based on *lacZ*, *GFP*, and other genes, as well as a more extensive review of the GAL4 system, see Brand et al. (1994) and Phelps and Brand (1998). For descriptions of modified GFP variants and methods for the visualization of GFP in live and fixed embryos, see Brand (1999) and Haseloff et al. (1999).

## ACKNOWLEDGMENTS

The authors gratefully acknowledge the generous contributions of C. Phelps and J. Haseloff to the design of the figures.

## REFERENCES

Brand A.H. 1995. GFP in *Drosophila. Trends Genet.* **11:** 324–325.

———. 1999. GFP as a cell and developmental marker in the *Drosophila* nervous system. In *Green fluorescent proteins* (ed. K.F. Sullivan and S. Kay), pp. 165–181. Academic Press, San Diego.

Brand A.H. and Dormand E.L. 1995. The GAL4 system as a tool for unravelling the mysteries of the nervous system. *Curr. Opin. Neurobiol.* **5:** 572–578.

Brand A.H. and Perrimon N. 1993. Targeted gene expression as a means of altering cell fates and generating dominant phenotypes. *Development* **118:** 401–415.

Brand A.H., Manoukian A.S., and Perrimon N. 1994. Ectopic expression in *Drosophila*. In Drosophila melanogaster: *Practical uses in cell and molecular biology* (ed. L.S.B. Goldstein and E.A. Fyrberg), pp. 635–654. Academic Press, San Diego.

Cambridge S.B., Davis R.L., and Minden J.S. 1997. *Drosophila* mitotic domain boundaries as cell fate boundaries. *Science* **277:** 825–828.

Chalfie M., Tu Y., Euskirchen G., Ward W.W., and Prasher D.C. 1994. Green fluorescent protein as a marker for gene expression. *Science* **263:** 802–805.

Fischer J.A., Giniger E., Maniatis T., and Ptashne M. 1988. GAL4 activates transcription in *Drosophila. Nature* **332:** 853–856.

Hacker U., Lin X.H., and Perrimon N. 1997. The *Drosophila sugarless* gene modulates Wingless signaling and encodes an enzyme involved in polysaccharide biosynthesis. *Development* **124:** 3565–3573.

Halfon M.S., Kose H., Chiba A., and Keshishian H. 1996. Targeted gene expression without a tissue-specific promoter: Creating mosaic embryos using laser-induced single-cell heat shock. *Proc. Natl. Acad. Sci.* **94:** 6255–6260.

Haseloff J., Dormand E.L., and Brand A.H. 1999. Live imaging with green fluorescent protein. In *Confocal microscopy methods and protocols* (ed. S. Paddock.), pp. 241–259. Humana Press, Totowa, New Jersey.

Heim R., Prasher D.C., and Tsien R.Y. 1994. Wavelength mutations and post-translational autoxidation of green fluorescent protein. *Proc. Natl. Acad. Sci.* **9:** 12501–12504.

Hidalgo A., Urban J., and Brand A.H. 1995. Targeted ablation of glia disrupts axon tract formatiom in the *Drosophila* CNS. *Development* **121:** 3703–3712.

Lin D.M., Auld V.J., and Goodman C.S. 1995. Targeted neuronal cell ablation in the *Drosophila* embryo: Pathfinding by follower growth cones in the absence of pioneers. *Neuron* **14:** 707–715.

Ma J., Przibilla E., Hu J., Bogorad L., and Ptashne M. 1988. Yeast activators stimulate plant gene expression. *Nature* **334:** 631–633.

O'Kane C.J. and Gehring W.J. 1987. Detection in situ of genomic regulator elements in *Drosophila. Proc. Natl. Acad. Sci.* **84:** 9123–9127.

Ornitz D.M., Moreadith R.W., and Leder P. 1991. Binary system for regulating transgene expression in mice targeting *int-2* gene expression with yeast GAL4/UAS control elements. *Proc. Natl. Acad. Sci.* **88:** 698–702.

Patel N.H. 1994. Imaging neuronal subsets and other cell types in whole mount *Drosophila* embryos and larvae using antibody probes. In Drosophila melanogaster: *Practical uses in cell and molecular biology* (ed. L.S.B. Goldstein and E.A. Fyrberg), pp. 445–487. Academic Press, San Diego.

Phelps C.B. and Brand A.H. 1998. Ectopic gene expression in *Drosophila* using GAL4 system. *Methods* **14:** 367–379.

Robertson H.M., Preston C.R., Phillis R.W., Johnson-Schlitz D., Benze W.R., and Engels W.R. 1988. A stable source of P-element transposase in *Drosophila melanogaster. Genetics* **118:** 461–470.

Rørth P. 1996. A modular misexpression screen in *Drosophila* detecting tissue-specific phenotypes. *Proc. Natl. Acad. Sci.* **93:** 12418–12422.

———. 1998. GAL4 in the female germ line. *Mech. Dev.* **78:** 113–118.

Rørth P., Szabo K., Bailey A., Laverty T., Rehm J., Rubin G.M., Weigmann K., Milan M., Benes V., Ansorge W., and Cohen S.M. 1998. Systematic gain-of-function genetics in *Drosophila. Development* **125:** 1049–1057.

Scheer N. and Campos-Ortega J.A. 1999. Use of the GAL4-UAS technique for targeted gene expression in the zebrafish. *Mech. Dev.* **80:** 153–158.

Shiga Y., Tanakamatakatsu M., and Hayashi S. 1996. A nuclear GFP β-galactosidase fusion protein as a marker for morphogenesis in living *Drosophila. Dev. Growth Diff.* **38:** 99–106.

Smith H.K., Roberts I.J.H., Allen M.J., Connolly J.B., Moffat K.G., and O'Kane C.J. 1996. Inducible ternary control of transgene expression and cell ablation in *Drosophila. Dev. Genes Evol.* **206:** 14–24.

Spradling A.C. 1986. P-element mediated transformation. In Drosophila: *A practical approach* (ed. D. Roberts), pp. 175–198. IRL Press, Oxford.

# CONTENTS

# Functional Cell Ablation

**Sean T. Sweeney and Alicia Hidalgo**
*Department of Genetics*
*University of Cambridge*
*Cambridge, England CB2 3EH*

**J. Steven de Belle**
*Department of Biological Sciences*
*University of Nevada, Las Vegas*
*Las Vegas, Nevada 89154-4004*

**Haig Keshishian**
*Department of Molecular, Cellular and Developmental Biology*
*Yale University*
*New Haven, Connecticut 06520-8103*

CELL ABLATION IS A POWERFUL TOOL IN THE STUDY of eukaryotic developmental biology. The selective removal of cells by ablation may provide much information about their origin, fate, or function in the developing organism. In flies, three main methods have been used recently to ablate cells: chemical, genetic, and laser ablation. Each method has its own applicability with regard to developmental stage and the cells to be ablated, and its own limitations. In this chapter, we summarize the advantages and disadvantages of each approach and also describe instances where these procedures have been used successfully. We also outline protocols for the chemical ablation of adult brain structures and the assessment of such an ablation, a protocol for the assessment of apoptosis in the embryonic central nervous system (CNS), and protocols describing the setup and use of laser ablation. Many of the techniques that are applicable to this chapter, such as labeling cells with markers, ectopic expression, and antibody staining, are described in other chapters (see, e.g., Hazelrigg; van Roessel and Brand; both this volume).

## CHEMICAL ABLATION

### Hydroxyurea Ablation: Applications

Chemical ablation is an effective tool for studying nervous system development and function in *Drosophila*. The only substance currently used for this purpose in flies is hydroxyurea (HU), an inhibitor of ribonucleotide reductase that blocks DNA synthesis and kills

**449**

dividing cells (Timson 1975). The specificity of HU ablation is thus dependent on developmental events. In this respect, HU has been useful in determining temporal patterns of neuroblast proliferation and the origins of neuronal elements in flies (Broadie and Bate 1991; Ito and Hotta 1992; Prokop and Technau 1994; Ito et al. 1997; Stocker et al. 1997; Armstrong et al. 1998) and other insects (see, e.g., Truman and Booker 1986; Malun 1998).

In *Drosophila*, one especially fortuitous time window occurs at the end of embryonic development. For the first 8–12 hours after larval hatching (ALH), only five neuroblasts are proliferating in each brain hemisphere (Truman and Bate 1988). Four of these are found in the dorsal protocerebrum and give rise to the intrinsic elements (Kenyon cells [KCs] and glia) of the mushroom bodies (MBs) (Ito and Hotta 1992; Prokop and Technau 1994; Ito et al. 1997; Armstrong et al. 1998; Tettamanti et al. 1998). The remaining single neuroblast has an anterolateral position in the brain and is the progenitor of local interneurons (LocI) in the antennal lobe (AL) and a subset of lateral relay interneurons (RIl) in the inner antennocerebral tract (iACT) (Stocker et al. 1997).

HU treatment in newly hatched larvae (see below) results in adult flies with KCs and AL interneurons of embryonic origin only. At the level of the light microscope, MB ablation appears complete, the AL is reduced in volume by 32%, and the iACT is not obviously affected (deBelle and Heisenberg 1994; Prokop and Technau 1994). No LocI and not more than one or two of the approximately 40 RIl can be found in adult flies after ablation (Stocker et al. 1997).

Such HU-treated flies have explicit olfactory deficits (J.S. de Belle and M. Heisenberg; J.S. de Belle; both unpubl.) that can be attributed either to a loss of LocI contact among AL glomeruli or to missing RIl (Stocker et al. 1997). MB ablation also has a profound and specific effect on higher brain functions. HU-treated flies fail in associative learning tests, with odors that elicit normal avoidance responses (de Belle and Heisenberg 1994). Nonassociative odor learning, processing of visual signals, locomotion, flight, male courtship, circadian rhythmicity, and general vigor are not obviously affected by HU treatment (de Belle and Heisenberg 1994, 1996; Wittig et al. 1995; de Belle et al. 1998; O'Dell et al. 1998; C. Jones and T. Tully; R. Strauss and J. S. de Belle; both unpubl.).

## Advantages and Disadvantages of HU Ablation

The advantages of HU as an agent for chemical ablation are derived from its biological activity. Relatively low doses of HU can be administered orally (de Belle and Heisenberg 1994; see below) or by injection (Truman and Booker 1986; Prokop and Technau 1994). Within approximately 3 hours, HU accumulates to maximum concentrations in tissues and is metabolized to an inactive molecule approximately 1 hour thereafter (Timson 1975). The lethal effects of HU are only realized in dividing cells. All remaining cells are held synchronous in the $G_1$ phase of the cell cycle until metabolism of HU is complete.

The specificity of HU ablation is dependent on developmental events. This may be viewed as either an advantage or a disadvantage, depending on when cell proliferation takes place for the cells that go on to form the structures of interest being studied. There are not many time windows during which relatively few neuroblasts are dividing in the *Drosophila* CNS (see above; Campos-Ortega and Hartenstein 1985; Truman and Bate 1988; Prokop and Technau 1991, 1994; Ito and Hotta 1992; Younossi-Hartenstein et al. 1996). In newly hatched larvae, HU ablation is specific to a few identified cell types (de Belle and Heisenberg 1994; see above), whereas treatment during peak periods of neurogenesis and cell division, in general (e.g., molting and pupariation), has widespread or even lethal effects (Broadie and Bate 1991; Stocker et al. 1997; Armstrong et al. 1998; J.S.

de Belle, unpubl.). Despite a lack of specificity during much of development, chemical ablation can be a useful tool for studying isolated events in the nervous system. To date, large-scale descriptive studies of chemical ablation in flies have not been reported.

A further advantage to this technique is the ability to dissect structure and function by removing only subsets of neuron lineages derived from temporally related neuroblasts. Such partial ablation can be realized by administering reduced doses of HU (see below) (Ito et al. 1997; Armstrong et al. 1998). Progenitor cells and their lineages are either ablated or not. The effect is not consistent from one fly to the next and can range from complete ablation to none at all. One disadvantage of chemical ablation in studies of nervous system function is a lack of temporal control. Treatment has absolute effects that are realized from the moment HU is metabolized and cell division recommences. Thus, it is not possible to distinguish between immediate cellular processes and a neuronal network basis of function. In addition, neurons and structures that would normally associate with ablated tissues may undergo abnormal development as well. For example, it is not known whether ablation of MB KCs influences branching patterns of RIls in the iACT that survive HU treatment (Stocker et al. 1997).

Using enhancer-trap technology (O'Kane and Gehring 1987; Brand and Perrimon 1993), it is now possible to label cells efficiently in a highly reproducible fashion. An elegant study employing enhancer-trap labeling and HU ablation has led to the identification of the origins of two major classes of interneurons in the fly olfactory system (Stocker et al. 1997). This same combination also enables the assessment of the extent and efficiency of ablation of the mushroom bodies by HU (Armstrong et al. 1998).

## Procedures for Hydroxyurea Ablation of the Mushroom Bodies and Its Assessment

Newly hatched larvae are collected as described in Protocol 25.1 and treated with hydroxyurea as in Protocol 25.2. For dissection of adult brains following HU treatment, see Protocol 25.3. Procedures for assessing the extent of ablation by antibody staining (Protocol 25.4) or X-gal (5-bromo-4-chloro-3-indolyl β-D-galactopyranoside) staining (Protocol 25.5) are also given.

PROTOCOL 25.1*

## Collection of Newly Hatched Larvae

### Materials

*Media*

#### Preparation of Egg-laying Medium

Combine 16 g of Cream-of-Wheat (Nabisco) with 5 ml of molasses in 100 ml of tap water and heat until thickened. Add 1 ml of tegosept. Fill plastic tablespoons level with hot medium (note that the spoons must fit into 6-oz culture bottles), and ensure that there are no bubbles and that the surface is smooth (*important*). Refrigerate to store, and dry before use. Make a paste of yeast and add a small drop to the medium surface at the end of the spoon just before use. Cut holes in sponge stoppers so that spoon handles can protrude.

*For all protocols, H$_2$O indicates glass distilled and deionized.

### Supplies and Equipment

Glass or plastic fly culture bottles (~6 oz) and sponge stoppers
Dissecting microscope
Dissecting needle

### Method

1. For best results, use rhythmic flies that have been maintained under a specified light regime and allow flies to oviposit in the dark at 25ºC. Introduce spoons with egg-laying medium and yeast into bottles containing 200–250 2–5-day-old flies.

   *Note:* This initial oviposition period of at least 6 hours in duration (or conveniently overnight) is to allow females to lay eggs that might have already begun to develop.

2. Approximately 3 hours into the subsequent subjective day, exchange the first spoon for a new one and allow oviposition to continue for 22 hours (or less, if eggs on the first spoon are nearly confluent). The following day (22 hours after introducing the second spoon), remove the yeast drop.

3. Under a dissecting microscope, use a sharp needle (larvae will stick to it and to each other) to clear and dispose of any larvae that may have hatched.

4. Collect newly hatched larvae in subsequent 0–1-hour (or shorter) time intervals. Be sure to clear the spoon before the beginning of each interval.

   *Note:* Over the course of a day, it should be possible to obtain 200–500 timed larvae from one spoon, depending on the density of eggs. This is a two-person task when collecting larvae from ten or more spoons at a time. Larvae can be either treated immediately or maintained on standard culture medium for ablation later in development.

PROTOCOL 25.2

# Hydroxyurea Ablation

### Materials

### Media

#### Preparation of Ablation Medium

Heat (microwave) a paste of yeast and $H_2O$ until boiling. When cool, it should have the consistency of maple syrup. This yeast suspension is used as a control for ablation. For complete ablation, make a suspension of 10–50 mg of HU/ml of yeast paste at room temperature. For partial ablation, begin using 3 mg of HU/ml and vary the dose in subsequent experiments, depending on the result. The suspension must be mixed continually and made fresh on each day of treatment.

Fly culture medium

### Supplies and Equipment

Suction-filtration funnel and flask
Coarse filter paper (e.g., coffee filters)

Microcentrifuge tubes (2 ml)
Fine paint brush

## Solutions and Reagents

**Hydroxyurea** (Sigma); note that HU is toxic and care must be taken in its use; for appropriate disposal procedures, see institutional safety guidelines.

**CAUTION: hydroxyurea** (see Appendix 4)

## Method

1. Add 5 drops of either the HU suspension or yeast alone (control) to a 2-ml microcentrifuge tube. Introduce up to 500 newly hatched larvae to each tube with a very fine moist paintbrush. Allow larvae to feed at 25°C for 4 hours. Treat larvae in later stages of development in smaller numbers.

2. Following treatment, use tap water to rinse the microcentrifuge tube contents onto filter paper (cut to fit) in a suction-filtration funnel. Apply light suction and rinse continually until the yeast is no longer visible.

3. Transfer the larvae onto fresh culture medium using either of the following methods:

   a. Place the filter paper in the bottle and rinse with small amounts of tap water.

   or

   b. Collect individual larvae with a needle under a dissecting microscope.

      *Notes:* Although the second method is more labor-intensive, it ensures that each larva makes the transfer.

      The extent of ablation should always be inspected before further experimentation with HU-treated flies.

PROTOCOL 25.3

# Dissection of Whole-mount Adult Brains

## Materials

### Supplies and Equipment

Dissecting microscope
Sharpened size-5a watchmakers forceps; the forceps and sharpening stone can be purchased from Fine Science Tools.
Minuten pins (Fine Science Tools)

### Preparation of Sylgard Plates

Approximately 4–5 ml of Sylgard is required for a 50-mm petri dish. Combine Sylgard monomer and curing agent (supplied together, Dow Corning) in a 10:1 ratio and mix thoroughly. Using a Gilson P5000 pipette with the narrow opening of the plastic tip removed with a razor blade, add 4–5 ml of the Sylgard mixture to a 50-mm petri dish (or to a depth of 4–5 mm). Place the petri dish in a 60°C oven overnight to cure the mix. The plate is now ready for use.

### Solutions and Reagents

1x PBS (see Appendix 3)
**Ethanol**
**Paraformaldehyde** (4%) in 1x PBS, prepared from the 20% stock (see below)

---

#### 20% Paraformaldehyde

Combine 20 g of paraformaldehyde in 100 ml of $H_2O$. Heat overnight at 60°C with mixing in a chemical fume hood. Store as 1-ml aliquots at –20°C.

---

**CAUTION: ethanol, paraformaldehyde** (see Appendix 4)

## Method

1. Dip flies (usually 2 days old) quickly in ethanol.

   *Note:* This helps to stun the flies and also aids in immersing the flies in the fix/PBS solution by destroying the surface tension so that no air bubbles adhere to the fly.

2. Immerse the fly in 4% paraformaldehyde in PBS in a Sylgard dish. Use a minuten pin to pin the fly to the dish, ventral-side upward, through the abdomen.

   *Note:* If the adult CNS is to be immunocytochemically probed for neuropeptides, the dissection can be carried out in PBS/4% paraformaldehyde/1% EDAC (1-ethyl-3-(3-dimethylaminopropyl)-carbodiimide, a protein coupling reagent, Sigma).

3. Remove the wings and legs, taking care not to pull on the tissue at any point. *Do not hold the nervous system with the forceps.* Cutting is achieved by the aposition of one pair of forceps against the other.

4. Carefully pull the proboscis to the extended position and sever the oesophagus. This facilitates the removal of the proboscis by pulling in an anterior direction; the alimentary canal remains attached to the proboscis and is also removed in this movement by pulling it through the oesophageal foramen of the brain. Remove the large silvery tracheae within the head capsule (try to avoid damage to the brain and optic lobes).

5. One pair of forceps is then used to hold the flat section of cuticle that comprises the posterior of the head capsule ("collar") and the other pair of forceps is used to "peel" off the exterior (ommatidia) of the eye in one movement, not unlike the peeling of an orange. The posterior "collar" of the head capsule is removed by pulling equally from left and right sides, taking care not to place any lateral pressure on the cervical connective.

6. Use the forceps to cut along the sides of the sternal plates from the opening of the cervical connective to the abdomen, again taking care to only rupture the cuticle and cause as little damage to the underlying tissue. After this point, only the internal tissue is holding the dorsal thoracic cuticle to the ventral thoracic cuticle. Sever the abdomen from the ventral section of the thorax and remove the whole ventral thorax from the rest of the carcass. The carcass remains pinned and can be discarded at this point.

7. When working with the remaining ventral thorax with attached brain, hold the cuticle with one pair of forceps and work with the other forceps to remove as much muscular tissue as possible, leaving mainly the cuticle, gut, and heart, between which is sandwiched the ventral nerve chord (VNC). Remove the gut and heart; the VNC with the cephalic section still attached is levered off the cuticle. Remove remaining pieces of tracheae. The CNS with attached VNC is now ready for processing for antibody or X-gal staining.

PROTOCOL 25.4

## Antibody Staining of Adult CNS

### Materials

#### *Supplies and Equipment*

Microcentrifuge tubes (1.5 ml)

Pasteur pipettes (drawn out, short form), used for adding and removing solutions, minimize the risk of throwing out the CNS preparations with the wash.

Nutator

Watch glass

Binocular dissecting microscope

Microscope slides

#### *Solutions and Reagents*

1x PBS (see Appendix 3)

**Ethanol** series, 30%, 50%, 70% in H$_2$O

Absolute ethanol

Histoclear (National Diagnostics)

**Araldite**

Vectashield (Vector Laboratories)

#### *Permeabilizing Buffer*

1x PBS (phosphate-buffered saline)

4% **paraformaldehyde**

0.1% Triton X-100

0.1% deoxycholate

#### *PBT*

1x PBS

0.1% Triton X-100

#### *Blocking Solution*

1x PBS

0.1% Triton X-100

0.5% BSA (bovine serum albumin)

5% normal serum

It is advisable to match the donor species of the serum with that of the host of the secondary detection antibody. During the longer steps where the antibody is incubated with the CNS tissue, it is advisable to add 1 mM **sodium azide** to the blocking solution to prevent bacterial growth.

#### *Antibodies and Detection Reagents*

Primary antibody

Secondary antibody. We prefer to use a directly conjugated secondary antibody for detection, whether the detection is by diaminobenzidine (DAB) or by confocal fluorescence microscopy. To obtain a permanent preparation, we use HRP-conjugated goat anti-mouse or anti-rabbit antibody (at a dilution of 1:200; Bio-Rad). For visualization by confocal fluorescence microscopy, we use fluorophore-conjugated antibodies supplied by Jackson ImmunoResearch Laboratories, usually at a dilution of 1:200.

**DAB** solution; 1 tablet (5-ml size, Sigma) in 5 ml of PBT. When working with DAB, take strong precautions to avoid contact. DAB is a potent carcinogen.

**$H_2O_2$** (30%)

**CAUTION: DAB, ethanol, $H_2O_2$, paraformaldehyde, resins (araldite), sodium azide** (see Appendix 4)

## Method

1.  Immediately after dissection, place the CNS in Permeabilizing Buffer for 30 minutes.

2.  Wash the CNS three times in PBT at room temperature (5 minutes each wash), in a microcentrifuge tube on a nutator.

    *Note:* 1.5-ml microcentrifuge tubes are used for all incubation and washing steps. Usually, 0.5–1 ml of washing buffer/antibody solution is used. It is also useful to seal around the lid of the microcentrifuge tube with Parafilm to prevent leakage of solution during long incubations. At each wash/incubation step, remove the microcentrifuge tube from the nutator and allow the tube to sit upright for a minute or two so that the CNS preparations settle to the bottom of the tube.

3.  Incubate the CNS in Blocking Solution at room temperature for 1 hour.

4.  Add the primary antibody to the Blocking Solution at an appropriate dilution and incubate the microcentrifuge tube containing the CNS and antibody on a nutator at 4ºC for 48 hours up to 3 weeks.

    *Note:* Depending on the affinity/avidity of the primary antibody, the incubation period with this antibody can be increased up to 3 weeks (C. Helfrich-Forster, pers. comm.), for an antisera against a central brain neuropeptide. However, for a monoclonal or polyclonal primary antibody against antigens such as β-galactosidase or tetanus toxin light chain, 48 hours would usually suffice. Washing steps can also be increased accordingly.

5.  Wash the CNS five times with PBT at room temperature over a 10–12-hour period.

6.  Repeat step 3, above. Then, add the secondary antibody at an appropriate dilution to the block. Incubate at room temperature for 2 hours.

7.  Wash five times in PBT over a 3-hour period. Proceed to either step 8 or step 9.

8.  Treat the CNS as follows:

    a.  Visualize using DAB in a watch glass under observation:

        i.   Briefly, place the CNS in a watch glass in 3–5 ml of DAB solution for 10 minutes.

        ii.  While observing the CNS under a binocular dissecting microscope, add 1 μl of $H_2O_2$ to the solution in the watch glass and observe the resulting reaction.

        iii. Wash the CNS in PBT when it is deemed that the staining pattern in the CNS is sufficiently well developed. This wash step halts the reaction.

    b.  Dehydrate the CNS by passing the preparation through an ethanol wash series of increasing concentrations, 30%, 50%, and 70%, followed by three washes in absolute ethanol (20 minutes each).

c. Finally, immerse the CNS in Histoclear (National Diagnostics) for 30 minutes. Mount on a microscope slide in araldite.

9. Alternatively, after washes in step 7, mount the fly CNS directly in Vectashield and visualize under appropriate excitation, using a fluorescent confocal microscope. Excitation of 595–604 nm is used for visualization of Texas-Red-conjugated antibodies, whereas 494-nm excitation is used for FITC (fluorescein isothiocyanate)-conjugated antibodies.

PROTOCOL 25.5

# X-gal Staining of Adult CNS

## Materials

### Supplies and Equipment

Microtiter plate (96 well)

### Solutions and Reagents

**Ethanol** series, 30%, 50%, 70% (in $H_2O$)
Absolute ethanol
Histoclear
Araldite

#### Staining Solution

10 mM **sodium phosphate** buffer (pH 7.2)
150 mM NaCl
1 mM **MgCl$_2$**
10 mM **K$_4$[Fe$^{II}$(CN)$_6$]**
10 mM **K$_3$[Fe$^{III}$(CN)$_6$]**
0.1% Triton X-100
This solution should be frozen in 5–10-ml aliquots at –20°C in the dark. If background β-galactosidase activity must be kept to a minimum, a higher pH (e.g., pH 7.8) can be used. Shortly before use, 0.5–1 ml should be preincubated at 37°C.

#### 8% X-gal

Prepare 8% X-gal in dimethylsulfoxide (**DMSO**), store as 50–100-µl aliquots at –20°C in the dark. Shortly before use, preincubate a 12.5-µl aliquot at 37°C in the dark.

During the procedure, it is important to keep all solutions containing X-gal at 37°C. At lower temperatures, the X-gal will crystallize out of solution.

**CAUTION: DMSO, ethanol, K$_4$[Fe$^{II}$(CN)$_6$], K$_3$[Fe$^{III}$(CN)$_6$], MgCl$_2$, sodium phosphate** (see Appendix 4)

## Method

1. To 0.5 ml of staining solution, add 12.5 µl of 8% X-gal and maintain at 37°C. If precipitate forms, heat the solution to 65°C until the precipitate dissolves.

2. Place the dissected CNS in 150–200 μl of the staining solution prepared in step 1. We normally perform this step in a 96-well microtiter plate that is placed on a heating block and covered with parafilm. Cover the microtiter plate with a cardboard box or appropriate lid that excludes light. It may be convenient to maintain the heating block at 2°C or 3°C above 37°C, in order to maintain a temperature that will keep the X-gal soluble while adding and removing further CNS preparations.

3. Incubate at 37°C until staining is complete.

4. Dehydrate the CNS through an ethanol wash series of increasing concentrations, 30%, 50%, and 70%, followed by three washes in absolute ethanol (10–15 minutes each).

5. Incubate in Histoclear for 30 minutes. Mount on a microscope slide in Araldite.

## GENETIC ABLATION

### Advantages and Disadvantages

Genetic ablation consists of delivering a toxin or death-inducing gene that is under the control of a cell-specific enhancer (Kalb et al. 1993; White et al. 1996; Bergmann et al. 1998; Kurada and White 1998) or by means of the GAL4 system (Brand and Perrimon 1993; Hidalgo et al. 1995; Hidalgo and Brand 1997; Lin et al. 1995; Smith et al. 1996; Sweeney et al. 1995; Zhou et al. 1997; Booth et al. 2000; Hidalgo and Booth 2000). Because of the wide range of existing enhancers, both of known genes and of those identified by enhancer-trap techniques, toxins and death genes can be targeted to virtually any cell of choice. Hence, genetic ablation can be cell-type-specific. At least in the case of ricin-A chain, diphtheria-A chain, and the death-inducing genes, cell ablation is cell-autonomous; hence, the danger of damaging neighboring cells is eliminated (Kalb et al. 1993; Hidalgo et al. 1995; Lin et al. 1995; Smith et al. 1996; White et al. 1996; Hidalgo and Brand 1997; McNabb et al. 1997; Zhou et al. 1997; Bergmann et al. 1998; Kurada and White 1998). Furthermore, because expression is cell- or tissue-specific, every chosen cell type is ablated in, for instance, every segment of the embryo. This is important, because the phenotypic effects of ablation can vary depending on how many analogous cells are killed within a tissue. Finally, because ablation is technically extremely easy (it only involves setting up a genetic cross), large populations of individuals can be analyzed at once, which is important to validate the results. In summary, in contrast to laser ablation, genetic ablation is not labor-intensive; it is also very cheap, allowing the effects of eliminating every cell of a given type to be analyzed within an embryo; and it also allows the examination of populations rather than individuals.

The disadvantages of genetic ablation are that expression driven by enhancers may be irregular; i.e., the expression may not include every single cell of a given type in an embryo, the expression driven by the chosen enhancer may not be restricted to the chosen cell type, there may also be low levels of expression barely detectable but which will kill the expressing cells, and if the expression of the effector gene is GAL4 or enhancer-dependent, it may not allow choice of timing. Thus, if looking at adult or larval structures using a combination of GAL4 and germ-line UAS (upstream activation sequence) toxigene flies, care must be taken to ensure that the toxin-encoding gene is not expressed at nonrelevant developmental stages or in other cells in the embryo. Mosaic expression can circumvent this problem by allowing induction of ablation at a chosen developmental stage.

## Genetic Systems: Mosaic or Reiterative Ablation

Depending on the question pursued, ablation may be carried out, for instance, for every given member of a cell type (reiterative ablation) or for only a spatially localized group of cells (mosaic ablation). Methods are available for both mosaic and reiterative ablation, and the choice of method is important. For instance, mosaic ablation of neurons in the CNS can be bypassed by extending axons to result in a normal phenotype (Lin et al. 1995, using *Blue Death*), whereas ablation of the same set of neurons in a reiterative fashion causes severe loss of axon-glia bundles (Hidalgo and Brand 1997, using *UAS:ricin A*).

Considerable effort has been devoted to developing conditional means of delivering toxins to cells, e.g., by isolating temperature-sensitive alleles of such toxin-encoding genes. Conditional expression of the toxic products would be ideal, allowing induction of ablation at a precise developmental stage. So far, however, mosaic ablation is the only successful means of overcoming the noninducibility of the GAL4 system. When conditionality is not a requirement, use of straight *UAS:toxin* transgenic flies has been so far the most successful manner by which to achieve genetic ablation.

In the genetic ablation systems currently used, cell death is brought about by conditional or targeted expression of toxic gene products such as ricin-A chain or diphtheria-toxin-A chain (DT-A), both of which inhibit protein synthesis (Endo and Tsurugi 1988; Collier 1990). Alternatively, cell death can be induced by expression of gene products that activate apoptosis, such as *reaper* (Hay et al. 1995; White et al. 1996; McNabb et al. 1997; Kurada and White 1998; U.P. John et al., in prep.) and *hid* (Grether et al. 1995; McNabb et al. 1997; Zhou et al. 1997; Bergmann et al. 1998), which are often required in combination to elicit the required level of cell death (Zhou et al. 1997). Such expression can be controlled by direct use of the GAL4 expression system by cloning such genes directly downstream from a GAL4 responsive element, *UAS*. Actually producing a transgenic stock of a toxin-encoding gene, however, can be problematic. During transformation of *Drosophila* embryos, proteins encoded by the injected constructs may be transiently expressed (A.H. Brand, unpubl., for *UAS:DT-A*). This problem may be less severe for UAS constructs than for constructs employing the heat shock promoter for ectopic expression (G. Halder, pers. comm.). This problem has been avoided and transgenic toxigenes created by two slightly different strategies. Both of these strategies allow production of a straightforward *UAS:toxin* or they can be used to produce a conditional, mosaic expression of a toxin-encoding gene by a ternary control mechanism using the FLP/FRT (site-specific FLP recombinase–FRT recombination target) recombination system.

### *Blue Death: (Mosaic) Ablation with Diphtheria Toxin*

In the "Blue Death" system using the DT-A toxin (Lin et al. 1995), the coding region of the toxin is disrupted by the insertion of a heterologous gene, in this case, *lacZ* flanked by two FRT sites (Figure 25.1). This construct is placed downstream from the UAS sequence. Thus, in the presence of GAL4, *lacZ* is expressed. However, when a source of FLPase is added, an excision of *lacZ* is facilitated by the recombination of the FRT sites, allowing readthrough of the DT-A transcript, still under GAL4 control. FLP-mediated recombination, however, is not 100% efficient and a mosaic of DT-A expression is created. Monitoring of toxin function is achieved by assaying for *lacZ* activity. In the published study using the Blue Death system to ablate pioneer neurons (Lin et al. 1995), the addition of FLPase was achieved by the injection of FLP sense-strand mRNA in an effort to avoid the often detrimental effects seen when heat shock is used to drive FLPase under the control of a heat shock promoter. The requirement for heat shocks and the difficulty of working with RNA could be avoided by employing a *UAS:FLPase* source (Duffy et al. 1998).

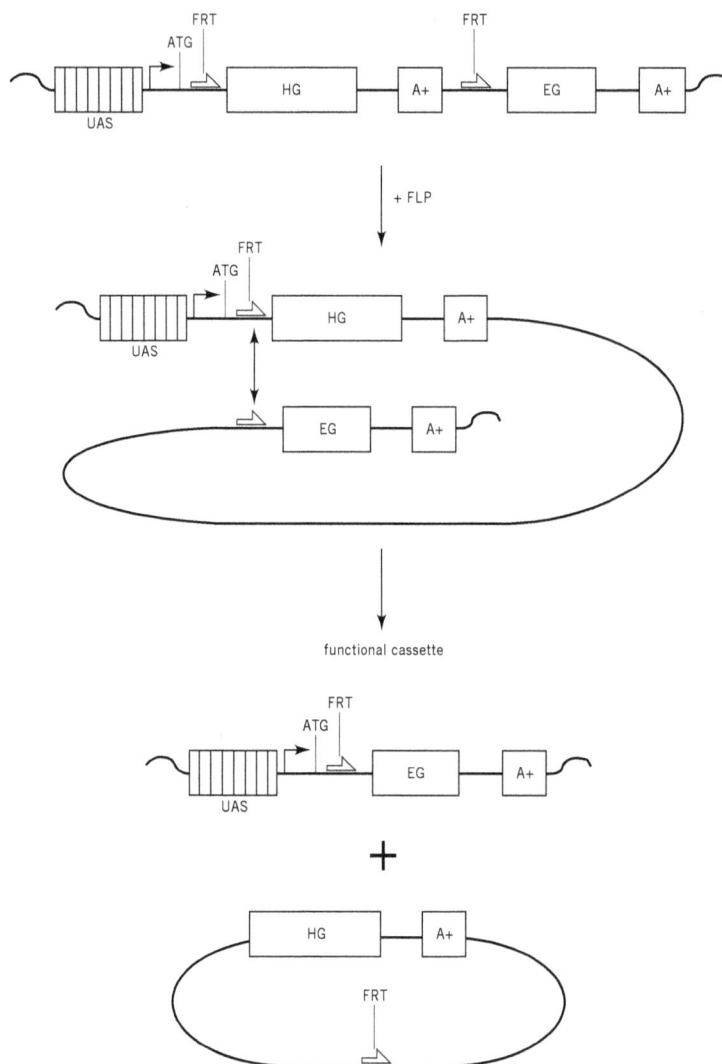

**Figure 25.1.** "FLP-out" method for regulating toxigene expression. (HG) Heterologous gene; (EG) effector gene. The effector gene is separated from the UAS sequence and the transcriptional start site (for Blue Death; Lin et al. 1995), or the UAS sequence and the TATA box (pUFWT; Smith et al. 1996), by the insertion of a heterologous gene (*lacZ*:Blue Death, *hsminiwhite*:pUFWT) flanked by FRT-recombination sites. Upon exposure of a source of FLPase, site-specific recombination removes the heterologous gene to generate the toxin-coding sequence.

## Reiterative and Mosaic Ablation with Ricin A

Flies transgenic for ricin A have been created following a method similar to that used for DT-A. Again, the toxic gene is brought under control of GAL4 after a FLPase-mediated recombination event (Figure 25.1). The toxic gene is separated from the UAS element by two FRT sites flanking a heat shock–*miniwhite*-poly(A) sequence (Smith et al. 1996). The poly(A) sequence of the *miniwhite* gene blocks transcription of any downstream sequences. If any readthrough transcription of the toxic gene should occur, the downsteam gene is the second gene in such a readthrough transcript and will be translated with low efficiency. This construct was used to produce a ternary control of ricin-A-chain expression in order to ablate neurons in the giant fiber escape circuit and also sections of the MBs (Smith et al. 1996). The advantage to this system, however, is that the vector, pUFWT, allows insertion of any gene downstream from the second FRT site to facilitate mosaic expression of almost any gene, toxic or not. The *miniwhite* gene from this construct was excised in the germ line to bring the *ricin-A* gene under direct control of the UAS sequence. This has been carried out successfully to create transgenic *UAS:ricin A* flies (K. Moffat and C. O'Kane, unpubl.) where direct injection of a *UAS:ricin A* construct would result in lethality for the injected embryos. In the presence of a heat shock FLP source and an appropriate GAL4 insert, the UAS-FRT–*miniwhite*–FRT-*ricin A* construct can be used to produce mosaics of ablation due to ricin-A expression at the required developmental stages of discrete heat shocks. Such heat shocks could feasibly be delivered using a laser setup such as that described below. However, care must be taken in using this system to choose an appropriate marker, because ricin A is an inhibitor of translation. For example, it might be preferable to use an endogenously expressed protein as a marker, rather than one that is ectopically expressed, as the translation of the latter would be inhibited by the action of the toxin.

## Choice of Cell Killing Agent

### Ricin A

Ricin A is the catalytic subunit of ricin toxin. It is capable of killing a cell, but not of crossing the cell membrane into neighboring cells, as it lacks subunit B, which is necessary for toxin internalization; and in the form employed for transgenic expression in flies, it lacks the secretory signal peptide (Moffat et al. 1992). The ricin-A chain kills cells by depurinating 28S rRNA and effectively halting protein synthesis. Used under GAL4 control in *UAS:ricin A* transgenic flies, it has been shown to be efficient, cell-specific, and cellautonomous (Hidalgo et al. 1995; Hidalgo and Brand 1997; Booth et al. 2000; Hidalgo and Booth 2000). Ablation appears to be fast, since absence of targeted cells is observed virtually at the same stage in which GAL4-driven *UAS:tau:lacZ* appears to be active as detected with anti-β-galactosidase antibodies (A. Hidalgo, unpubl.).

### Cold-sensitive Ricin A

Since GAL4-mediated expression does not allow temporal control of expression, an attempt was made to bring a conditional element into the control of ricin-A-chain action (Moffat et al. 1992). Cold-sensitive mutants of the ricin-A chain (RAcs) were selected for in yeast, which were then cloned into a construct allowing the expression of the toxin under the *sevenless* promoter. Flies bearing this construct reared at 29°C (control flies are raised at 18°C) show ablation of some photoreceptors, resulting in a visible eye defect. However, the expressivity of this ablation appears to be weak; many photoreceptors that should be killed by this treatment persist. Moffat et al. (1992) estimated the time for this mutant toxin to kill a cell from time of expression to be about 7 hours, 4 hours being sufficient to severely disrupt cell function. RAcs has also been placed under UAS control, the rationale being that because GAL4 expression is greatly elevated at higher temperatures, expression and action of RAcs would be complementary and possibly additive. However, use of *UAS:RAcs* has been shown to be impractical since damage to the expressing cells is observed only several days after expression of the toxin (Allen 1995).

## Diphtheria Toxin A Chain

The use of the transgenic DT-A subunit has been reported only in combination with the Blue Death method (see above; Lin et al. 1995) or with a specific promoter (Bellen et al. 1992; Kalb et al. 1993). DT-A is another toxin that interferes with protein synthesis, by ADP ribosylation of the translation factor EF2 (Collier 1990). Again, the B-chain of the toxin has been removed in order to render the effects of the toxin cell-autonomous. Published stocks of flies bearing a DT-A transgene are somewhat limited in their application, the expression of the toxin being either mosaic (Lin et al. 1995) or promoter-specific (Bellen et al. 1992; Kalb et al. 1993).

## Temperature-sensitive Diphtheria Toxin A Chain

In a strategy similar to that employed by Moffat and coworkers with the ricin-A chain, Bellen and coworkers (1992) isolated temperature-sensitive mutants of DT-A in yeast. These DT-A[ts] alleles were tested by driving their expression in photoreceptor cells R1–R6 using the *ninaE* promoter. A control strain expressing nonconditional DT-A has also been produced which gives rise to flies that exhibit rough eyes and abnormal ommatidia at all rearing temperatures (Bellen et al. 1992). For one particular DT-A[ts] allele expressed under the *ninaE* promoter at 25°C, examination of the eyes of such flies reveals abnormalities in as many as 5% of R1–R6 cells. Temperature shifting to 16°C for 6 hours for newly eclosed flies bearing this element can also be used to induce obvious defects in the eye. However, other DT-A[ts] alleles behave poorly, with problems of lack of expression of toxin and toxin effect, as well as pronounced position effects (Bellen et al. 1992). Clearly, from these two studies employing temperature-sensitive toxins, making temperature-sensitive alleles of toxins can lead to the attenuation of the enzymatic activity and therefore effectiveness of these toxins. Another potential problem is the possible "leaky" expression of toxin activity at nonpermissive temperatures, a problem that would need to be tightly controlled for.

## Diphtheria Toxin A Chain Containing an Amber Suppressor

An alternative mechanism for engineering conditionality into DT-A expression has been attempted by Kunes and Steller (1991). By changing a tyrosine codon to an *amber* (stop) codon in the coding sequence of DT-A, the expression of DT-A can be made conditional upon the presence of tRNA[Tyr] gene specifically modified to recognize an *amber* codon; this would allow the correct translation of the DT-A[am] message where the *amber* codon would normally be read by the translational machinery as a stop codon (Laski et al. 1989; Kunes and Steller 1991). DT-A[am] expression in the compound eye was mediated in this study using the photoreceptor-specific *chaoptic* promoter. Flies bearing this construct are crossed to flies bearing an amber suppressor tRNA[Tyr] element (Laski et al. 1989), and indeed, progeny can be recovered in which photoreceptor cell development had ceased by the end of pupal day 2, consistent with an ablation driven by the *chaoptic* promoter (Kunes and Steller 1991). Nevertheless, a number of transformants bearing only the *chp*-DT-A[am] element are not without defects in the photoreceptors, indicating that there is leaky suppression of the DT-A *amber* codon, even in the absence of the modified tRNA.

## Reaper *and* hid

Ablation has also been carried out successfully with *UAS:reaper* transgenic flies under GAL4 control, although this transgene must often be used in multicopy to be effective (U.P. John et al., in prep.) or in combination with *UAS:hid* (Zhou et al. 1997).

### Tetanus Toxin Light Chain

Although not an ablation in a literal sense, the blocking of neuronal communication can certainly be viewed as an ablation in a behavioral sense. Tetanus toxin light chain (TeTxLC) is a metalloproteinase, which cleaves a protein of the synaptic exocytosis machinery, synaptobrevin, leading to a block of evoked neurotransmitter release in the cell in which it is expressed (Schiavo et al. 1992). For the tetanus holotoxin, the heavy chain usually mediates its entry into a cell. Removal of the heavy chain renders expression of the light chain cell-autonomous in its action. Sweeney et al. (1995) expressed a synthetic, tetanus toxin–light chain gene under GAL4 control in the nervous system of *Drosophila* to find that evoked release of neurotransmitter is blocked and that no apparent developmental defects are observed as a consequence of this expression. The block of evoked release is due to the cleavage of the synaptically expressed, neuronal synaptobrevin (n-syb) protein; a ubiquitously expressed homolog of synaptobrevin (syb) is not cleaved by the toxin. The toxin has apparently no effect when expressed in nonneuronal tissues. Expression of TeTxLC in discrete patterns of the *Drosophila* nervous systems leads to behavioral defects (Sweeney et al. 1995; Reddy et al. 1997). A high-affinity monoclonal antibody has been raised against TeTxLC, and hence, the expression pattern of TeTxLC can be readily observed, even in thick tissue such as adult whole-mount brain preparations. Indeed, like the active form of the toxin, an inactive form of the TeTxLC (*UAS:IMPTNT*) works well as a cell marker so that the morphology of the cell in which the active or inactive toxin is expressed becomes evident upon detection (de Celis et al. 1996; S.T. Sweeney, unpubl.).

## Ablation Setup

For nonmosaic ablation, standard GAL4 methodology should be followed using *UAS:ricin A* (Hidalgo et al. 1995) or *UAS:reaper* (McNabb et al. 1997; Zhou et al. 1997; U.P. John et al., in prep.). Essentially, the GAL4 flies of choice are crossed to *UAS:ricin A, UAS:hid,* or *UAS:reaper* flies. Normal temperature should be 25°C. Lower temperatures can reduce GAL4-mediated expression and weaken the phenotype, presumably by reducing the number of cells being killed, which can, however, be exploited experimentally. For mosaic ablation, FRT-based methodology (see Smith et al. 1996 or Theodosiou and Xu 1998) or Blue Death methodology is followed (see Lin et al. 1995).

## Assessment of Ablation

Prior to ablation, in order to choose which cells to kill, the chosen GAL4 line can be crossed to *UAS:tau:lacZ* (Hidalgo et al. 1995; Hidalgo and Brand 1997) or *UAS:tau:GFP* (green fluorescent protein; Brand 1995; Dormand and Brand 1998), which allow visualization of the whole morphology of the cells. In the former case, *lacZ* is visualized with anti-β-galactosidase antibodies.

To assess whether the cells to be ablated have indeed been killed, reporter expression of *lacZ* or *GFP* under the control of the same GAL4 enhancer can be followed by building a recombinant chromosome by conventional genetics. Alternatively, cell-specific antibodies can been used that will reveal the absence of the cells being killed.

For *UAS:ricin A*, it is doubtful whether anti-ricin-A antibodies would be of much use because cells are extremely sensitive to ricin, and probably when the amounts of toxin are below the level of antibody detection. Ricin blocks protein translation; hence, absence of cell-specific markers reveals expression of ricin but not necessarily clearance of killed cells.

For *UAS:reaper*, methods normally used to monitor apoptosis can be exploited to monitor targeted ablation. Such methods include detection of *reaper* expression by in situ hybridization with *reaper* antisense DIG (digoxigenin)-labeled probes, apoptotic morphology (round, shrunken cells with nuclear condensation) of the cells, and TUNEL (terminal deoxynucleotidyl transferase [TdT]-mediated dUTP nick end-labeling). The advantage of TUNEL over other methods of detecting dying cells is that because it uses fixed material, it allows monitoring of antigen expression at the same time to confirm that the cells chosen for ablation are undergoing apoptosis.

Double-labeling *Drosophila* embryos using the TUNEL reaction is given in Protocol 25.6 and should be adapted to the selected antigen. Some antigens prefer the TUNEL reaction to be carried out first, whereas others prefer the TUNEL reaction to follow antigen detection (Booth et al. 2000). The reason for this may be because some antigens may not survive the 37°C incubation or the conditions of the reaction. Similarly, fixation times may be increased with better results for some antigens, but not for others.

## PROTOCOL 25.6

## TUNEL–Antibody Double-labeling Method for *Drosophila* Embryos

### Materials

#### *Media*

Apple juice–agar plates (see Appendix 3)
Yeast paste

#### *Supplies and Equipment*

Glass vials
Dissecting microscope

#### *Solutions and Reagents*

**Bleach**
**Formaldehyde** (4%) solution (in PBS):**heptane** (1:1)
Formaldehyde (4%) solution in PBT
**Methanol**
PBT (1x PBS, 0.1% Triton X-100)
Proteinase K solution (Boehringer Mannheim)
**Glycine** (10%)
Terminal transferase (TT) (25 units/µl; Boehringer Mannheim 220 582); the following solutions are also included with the enzyme:
- 5 TT buffer
- 25 mM **cobalt chloride**

Triton X-100 (10%)

DUTP (1 mM)

dUTP-FITC (25 nmoles/32 μl) or Biotin-16-dUTP (1 nmoles/μl); both work well.

### *Antibodies*

Primary antibody

Secondary antibody

- Texas-Red-conjugated secondary antibody (1:200), if labeling with dUTP-FITC
- FITC-conjugated secondary antibody (1:200) and avidin-Texas Red (1:200) if labeling with biotin-16-dUTP

**CAUTION: bleach, cobalt chloride, formaldehyde, glycine, heptane, methanol** (see Appendix 4)

### Method

1. Collect embryos on apple juice–agar plates, and wash off yeast with $H_2O$.

   *Note:* For details on embryo collection and dechorionation (step 2, below), see Rothwell and Sullivan (this volume).

2. Dechorionate embryos in bleach under a dissecting microscope.

3. Wash with abundant tap water, followed by $H_2O$.

4. Fix in 4% formaldehyde solution (in PBS):heptane (1:1) at room temperature for 20 minutes.

5. Remove the lower phase (fix), add methanol in equal volume to the remaining heptane (upper phase), and vortex briefly. Devitellinized embryos should sink down to the bottom of the vial in a few minutes.

6. Collect embryos and wash in methanol three times. Store at –20ºC if staining is to be performed at a later date.

7. Wash in PBT three times over a period of 15 minutes.

8. Treat with 2.5 μl of proteinase K solution for 1–3 minutes.

   *Note:* Proteinase K should first be titrated to find the optimal time of digestion that allows penetration of reagents but does not degrade the embryo. Normally, this is found to be between 1 and 3 minutes.

9. Rinse briefly in glycine (20 μl of 10% stock solution in 1 ml of PBT).

10. Rinse again in glycine for 2 minutes.

11. Wash in PBT five times over a period of 20 minutes.

12. Fix in 4% formaldehyde solution in PBT at room temperature for 20 minutes.

13. Wash in PBT at room temperature five times over a period of 15 minutes.

14. Wash in 50–100 μl of 5x TT buffer at room temperature for 15 minutes.

15. Set up the TT reaction (using dUTP-FITC) as follows:

| | |
|---|---|
| 5x TT buffer | 9 µl |
| 25 mM cobalt chloride | 4.5 µl |
| 10% Triton X-100 | 1.5 µl |
| dUTP (1:2 dilution with $H_2O$) | 0.5 µl |
| dUTP-FITC (1:4 dilution with $H_2O$) | 0.75 µl |
| Terminal transferase | 0.9 µl |
| $H_2O$ | 28.1 µl |

Final volume ~45 µl

Incubate embryos in TT reaction at 37°C for 3 hours.

16. Wash in PBT five times over a period of 20 minutes

17. Incubate in primary antibody overnight at 2–3x concentration normally used for HRP detection.

18. Wash in PBT eight times over a period of 1 hour.

19. Incubate with secondary antibody (in the dark to protect the fluorophore) as follows:

   a. If labeling is performed with dUTP-FITC, use Texas-Red-conjugated secondary antibody (1:200).

   b. If labeling is performed with bio-16-dUTP, use FITC-conjugated secondary antibody (1:200) and avidin–Texas Red (1:200).

20. Wash in PBT five times over a period of 1 hour.

21. Change PBT for Vectashield.

22. Mount in Vectashield.

## LASER ABLATION

### Advantages and Disadvantages of Laser Ablation

Lasers have been used for many years to ablate or manipulate *Drosophila* cells. Nevertheless, laser methods have remained the specialty of only a few laboratories. This is surprising given the widespread use of laser ablation in other genetic model systems, such as the nematode *Caenorhabditis elegans*. Part of the problem is that it can be difficult to image live cells in *Drosophila* embryos. However, with the introduction of vital fluorescent imaging methods, such as GFP, it is likely that laser-mediated cell manipulation methods will become more popular. The methods are in fact relatively straightforward, and in this chapter, we provide the necessary information for setting up and using a laser microscope.

Perhaps the most important advantage of using a laser rather than a genetic or gene expression method for manipulating cells is that the operations may be performed in essentially any cell pattern and, moreover, at any time in development. This makes it possible to do experiments that are currently impossible with methods such as GAL4/UAS-mediated toxin expression. For example, laser ablations can be performed on one side of an embryo, leaving the unoperated side as a control, or in even more elaborate patterns (Lin and Spradling 1993; Chang and Keshishian 1996; Farrell et al. 1996; Fernandes and Keshishian 1996, 1998). Even when suitable genetic methods exist, laser ablation can provide valuable supportive data (Nüsslein-Volhard et al. 1980; Cash et al. 1992; Chiba et al. 1993).

Laser methods are by no means limited to cell ablation. For example, subcellular structures have been targeted, such as the germinal vesicle of the developing oocyte (Montell et al. 1991), as well as individual chromosomes (Ponelies et al. 1989). On the other hand, lasers can be used for surgical manipulations, such as cutting nerves (Fernandes and Keshishian 1998). Lasers have also been used to disrupt molecular targets, as was done for the early morphogenetic gradients of the embryo (Lohs-Shardin et al. 1979a,b; Nüsslein-Volhard et al. 1980). A more specific effect can be achieved by targeting individual molecules using the CALI (chromophore-assisted laser inactivation) technique, where molecules are labeled with laser-light-absorbing tags (Jay and Keshishian 1990; Schmucker et al. 1994; Beermann and Jay 1994). Finally, a promising development is laser-mediated gene activation (Stringham and Candido 1993; Halfon et al. 1997). Appropriate heat shock promoters can be used to induce expression of heat shock transgenes or, with hsp-GAL4 drivers, to induce expression of UAS constructs in virtually any cell of the animal. The method can also be used to create marked cell clones and laser-activated genetic mosaics, by means of heat shock hsp-FLP/FRT constructs (Halfon et al. 1997).

Despite these attractive features, laser methodologies are not without their shortcomings. Cell ablation can be time consuming, especially when multiple cells are ablated in each animal. Genetic approaches will always generate more test animals than laser ablation. In addition, to ablate a cell, it must first be imaged in a live, undissected animal. This often requires video-based microscopy, sometimes with computer enhancement (Halpern et al. 1991; Inoue and Spring 1997).

Setting up and maintaining a laser microscope also involves time and expense. The simplest dye lasers suitable for cell ablation cost between $5000 and $10,000, which is above the cost of the microscope itself, assuming the investigator does all of the work to assemble the system. Fully configured laser microscopes are commercially available, but they can cost an order of magnitude more. Although laser confocal microscopes can also be used for cell ablation methods, they are cumbersome and usually require fluorescent cells (which is now possible with GFP expression, however). Thus, unless the investigator intends to use a laser extensively, the simple and economical methods based on toxin expression may be a wiser choice.

## Setup for Laser Ablation: Coupling of a Laser to the Microscope

### Microscope Requirements

Adapting a laser to a microscope is a relatively straightforward procedure. They can be adapted to either inverted or upright microscopes, and it is also possible to design their integration so that the microscope can still be used for other imaging purposes. Ideally, the investigator should use a microscope configured for simultaneous epi-illumination fluorescence and differential interference contrast (DIC or Nomarski) optics. All modern, inverted research microscopes have this capability, as do most upright research microscopes. It is also strongly recommended that the microscope have a video system capable of high-resolution real-time imaging. This is both for reasons of operator safety, and for the considerable image enhancement that results from video display of DIC/Nomarski images (Inouye and Spring 1997). The image enhancement is essential to identify cells located in the interior of live animals. Finally, it is essential that the microscope have plan-apochromats or oil-immersion fluorite objectives of the very highest available numerical apertures (NA).

## *Lasers*

Pulsed dye lasers are generally best suited for cell ablation. These devices pump a fluorescent dye to generate an intense output light pulse lasting a few nanoseconds. One advantage of these lasers is that they can be tuned to any wavelength, from the near ultraviolet (UV) to near infrared (IR). Because of their simple construction, they are compact and lightweight, relatively inexpensive, and very reliable.

The key parameters for consideration when selecting a dye laser are its pulse energy, pulse duration, and peak firing frequencies. A typical pulsed dye laser used for laser ablation studies (e.g., model VSL337 from Laser Sciences) generates 30 μJ of light energy per pulse, when pumping the laser dye Coumarin 440 (a dye often used for cell ablation). This laser has a firing rate of up to 20 Hz, and a life expectancy of 10 million pulses. Lasers are also often described in terms of their average and peak power. Average laser power refers to the light energy emitted each second. At its maximal settings, the VSL337 laser emits 0.6 mW average power (20 Hz × 30 μJ = 600 μJ/sec). By comparison, typical continuous lasers found on scanning confocal microscopes have outputs of 5–50 mW. A second parameter to consider is the laser's peak power. Although 30 μJ of energy per pulse may seem small, this is the energy emitted in 3 nsec. To emit light at that rate continuously (i.e., 10 μJ per nsec), the laser would have an output power of 10 kW. This value represents the laser's peak power, which is obtained by dividing the pulse energy by the pulse duration. The VSL337 dye laser is functionally equivalent to a continuous laser of 10 kW power, with an ultrafast 3-nsec shutter that opens 20 times a second. For Caution on lasers, see Appendix 4.

## *General Principles for Designing a System*

The easiest and safest way to direct laser light into a microscope is through the epi-illumination fluorescence port. In the simplest configuration, the laser replaces the mercury arc lamp. Unlike an arc lamp, the goal is not to fill the microscope field with light, but rather to focus the light to as small a spot as possible. To achieve this, a planoconvex lens having a focal length value usually between 50 mm and 125 mm (depending on the microscope) is positioned between the laser and approximately 5 cm from the entrance port of the epi-illuminator. The objective will then bring the laser light to a sharp, diffraction-limited spot. In practice, the appropriate laser-focusing lens and its exact location are determined by trial and error. Good-quality lens holders on a multi-axis micropositioner must be used, so that the lens can be critically adjusted. The focusing lens is also used to position the laser spot in the field of view.

The laser light is attenuated with neutral density (ND) filters mounted on manually adjusted filter wheels. Multiple wheels are used to control output light over a wide range. A good choice is two six-position wheels in series, one with five filters ranging up to log 0.5 ND in 0.1-unit increments, and the second with five filters up to log 2.5 ND in 0.5-unit increments. This allows laser attenuation up to log 3 absorbance (0.1% transmittance) in 0.1 ND steps. By adjusting laser-firing rates and the ND filters, the investigator can carefully control the amount of light energy focused on a cell.

In the epi-illuminator, the laser light is directed to the preparation using a standard, fluorescence filter set, appropriately matched to the laser wavelength (Figure 25.2). For example, Coumarin 440 emits light in the range of 430–470 nm, with a peak at 445 nm. A standard FITC filter set would therefore be suitable for this wavelength. To improve the laser throughput, the excitation filter is removed from the filter set. The dichroic mirror of the filter set reflects the laser light into the objective. Because wavelengths longer than 510 nm pass freely through the mirror, the investigator can view the ablation as it occurs.

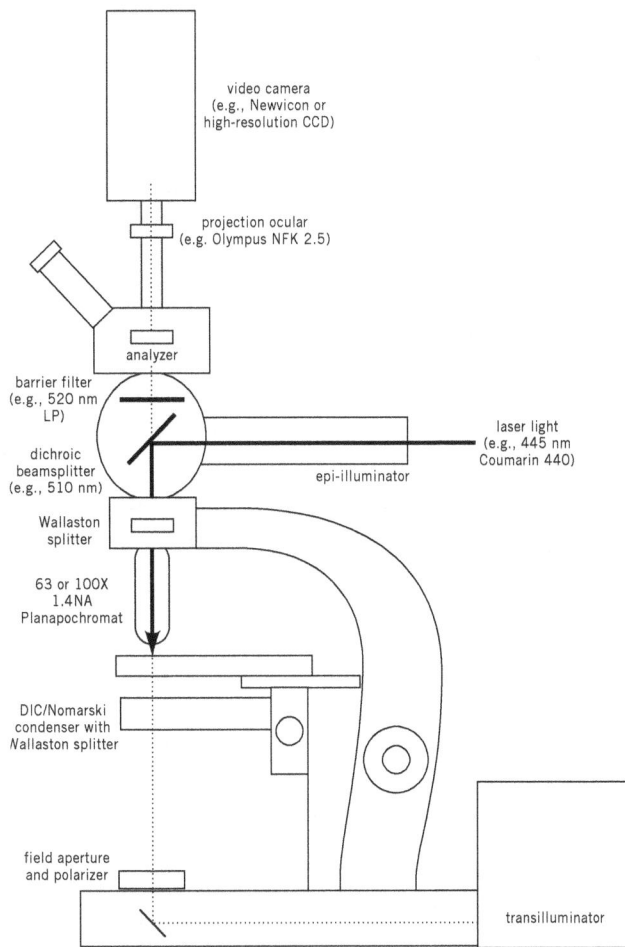

**Figure 25.2.** Side view of a laser microscope, illustrating the optical path taken by the laser light (*dark arrow*) once it has entered the epi-illuminator port. For a description of how the laser is coupled to the epi-illumination port, see Figure 25.3. In this example, the laser light is reflected into the preparation using a conventional GFP fluorescence filter set, from which the exciter filter has been removed. The rest of the optical train includes elements needed for DIC/Nomarski optics, and the coupling of a video camera. As in Figure 25.3, the microscope would be securely braced and mounted to a vibration-damping optical table.

The most important element is the microscope objective. Good success is obtained with objective with the highest possible NA. There are several reasons for this. First, the resolving power increases with the NA, allowing, with *Drosophila*, working close to the resolution limits of optical microscopy. Second, the light throughput increases with the square of the objective NA (for fluorescence, the brightness will vary by the fourth power of the NA). Finally, laser activation and ablation techniques depend on the light power transferred per unit volume. Laser light from the objective projects as a cone that converges at the focus point and diverges beneath it. Thus, it is desirable that the angle of the cone be as oblique as possible, something that occurs with objectives of high NA. This reduces out-of-focus damage to cells above and below the targeted cell. Oil-immersion objectives having an NA of 1.4 are available with magnifications of 63x and 100x, as are 1.3-NA/40x oil-immersion objectives. These objectives are recommended for laser manipulations.

Finally, one needs to see the target cell. Using DIC/Nomarski optics, the best way to visualize the target cell is to direct the image to a high-resolution monochrome video camera, capable of real-time video (there is little color information in living *Drosophila* embryos and larvae). This involves mounting the camera on the trinocular head of the microscope (or output port on an inverted scope) with the light projected by an ocular to the camera face plate. Although video camera mounts with internal lenses are available from microscope manufacturers, outstanding images can be obtained with the relatively inexpensive oculars, such as the NFK projection microscope oculars made by Olympus (available at 1.6x, 2.5x, 3.3x, and 5.0x). Unfortunately, inexpensive digital color cameras, which are popular for obtaining still images on microscopes, are usually inadequate for our purposes. Excellent CCD monochrome video cameras with 800 or better horizontal line resolution are widely available, and it is still hard to beat the richly detailed images obtained with the older technology Newvicon video tube cameras (e.g., Dage-MTI model 70 or the Hamamatsu model 2400 series). The field aperture should be kept in focus, the condenser aperture fully open, and sufficient transmitted light provided to drive the camera in its linear range. With the microscope adjusted in this way, *Drosophila* embryos and larvae can be observed with unparalleled detail and clarity. For further details and a physical explanation for the image enhancement obtained with video microscopy, see Inouye and Spring (1997).

Two approaches for coupling the laser to a microscope system are shown in Figure 25.3. Described in Figure 25.3A is a simple arrangement for laser ablation with DIC/Nomarski optics. Here, it is possible using a beam switcher to select light from either the laser or an arc lamp. This allows the microscope to be used for standard fluorescent epi-illumination as well as for laser methods, but not simultaneously. This is a good choice for a microscope that will be used for multiple purposes in a laboratory. In Figure 25.3B, the optics have been modified so that the laser ablation can be performed together with fluorescence excitation. This would permit, for example, the ablation of cells identified by their expression of a vital marker such as GFP. The rest of the microscope arrangement is shown in Figure 25.2. Here, the optical elements have been chosen assuming a 445-nm laser output wavelength, together with simultaneous imaging of GFP. In the latter case, a camera capable of low-light-level imaging would be used, such as a silicon intensifier target (SIT) or low-light-sensitive CCD camera.

### Mounting the Laser

It is essential to mount the microscope, laser, and the associated optical components to a vibration-damping optical table. Optical post and rail hardware are used (available from optics manufacturers, such as Newport Corp.) to position the laser and its associated optical elements. The microscope should be securely bolted to the table. The only element that needs to be finely adjusted is the focusing lens, which should be mounted in a 5-axis lens micropositioner.

Most upright microscopes have fixed turrets and moving stages. This means that during focusing, the epi-illumination port remains at a fixed height relative to the optical table. It is best to avoid using microscopes with fixed stages and moving turrets (upright microscopes that are modified for electrophysiology usually have this configuration). In this case, the epi-illuminator moves as the microscope is focused. A simple solution for this configuration is to launch the laser light into the epi-illuminator using an optical fiber (fiber-optic couplers for this purpose are standard accessories available from several laser manufacturers, such as Laser Sciences, Inc.). In this case, the laser rests on the optical

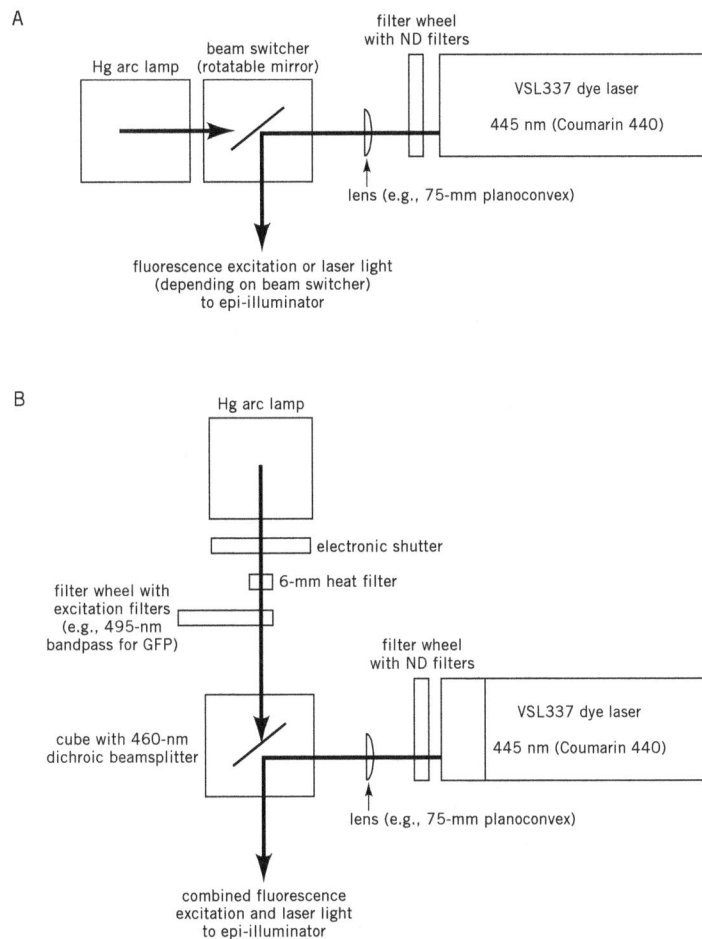

**Figure 25.3.** Recommended arrangements for coupling a laser to a microscope, as viewed from above. The light output (*dark lines and arrows*) in either example *A* or *B* (below) would be directed into the epi-illumination port of the microscope. The designs shown are suitable for either an upright or inverted microscope. (*A*) Simple arrangement for switching between a mercury arc lamp, for normal fluorescence illumination, and the laser. Light from a VSL 337 pulsed dye laser is filtered by the appropriate neutral-density filters mounted on a filter wheel. The light is focused by a planoconvex lens, whose focal-length value must be determined empirically and differs depending on the microscope used. The focused light is directed into a beam switcher, which is a standard item available from most microscope manufacturers. This allows the investigator to select either the laser or arc lamp as the epi-illuminator light source. The rest of the setup is shown for an upright microscope in Figure 25.2. (*B*) Arrangement permitting simultaneous fluorescence and laser excitation. The light from the arc lamp is controlled by a remote shutter and, after being heat-filtered, is passed through the appropriate excitation filter for the fluorophore of interest. In this example, a 495-nm bandpass filter is used for GFP, mounted in the filter wheel. Light from the laser is filtered and focused as in example *A* above. The laser light is reflected by a 460-nm dichroic beam splitter, mounted in a standard filter-cube holder. Because this beam splitter transmits wavelengths longer than 460 nm, the 495-nm GFP excitation light passes freely through it, while the 445-nm laser light is reflected. The output is directed into the epi-illuminator as illustrated in Figure 25.2. Not shown is the mounting hardware, which would include a vibration-damping optical table and micropositioners for the focusing lens. The arc lamp, filter wheels, beam switcher, filter cube, and laser should all be securely mounted and braced on optical platforms.

bench, and microscope movement relative to the table is no longer a problem. The cost of this solution is an inevitable amount of light loss, compared to the direct launch of the laser light into the epi-illuminator port, as shown in Figure 25.3.

**Figure 25.4.** Aligning the targeted cell with the laser. The method that is used when viewing a preparation with a video microscopy system is illustrated. The microscope is first focused onto the coverslip, which is cracked by a single pulse from the dye laser. The position of the crack is then marked on the video screen with an overlay. This can be as simple as a taping a small piece of transparent acetate film to the video monitor. More elaborate systems use computer-generated video overlays. In either case, the overlay marks the spot where the laser light is directed within the image field. To target a cell, the microscope is then focused onto the cell of interest, which is brought into position using the microscope stage controls. In practice, this has proven to be a far more accurate and effective technique than trying to steer the laser beam.

## Ablation Techniques

Figure 25.4 summarizes the method usually used in our laboratory for cell ablation. The first task is to find where the laser pulse is directed in the image field. An easy way to do this is to focus onto the coverslip and fire a single pulse from the laser. This will crack the slip, provided the light is transmitted unattenuated and a high-NA lens is employed. The crack is sufficient to identify the target spot. This is then marked on the video monitor with a target overlay. Computer-generated overlays are convenient, but a simple and effective alternative is to tape a small piece of acetate film onto which a target spot has been drawn. Once identified in this fashion, the cell of interest is brought into focus and aligned with the overlay by moving the stage.

When setting up the laser, it is important that the focused image on the video screen correspond exactly with the target spot. Errors can be corrected by careful adjustment of the tube length of the projection ocular (most microscopes provide adjustable barrels for this purpose). When properly aligned, the center of the target spot will be in focus on the video monitor.

PROTOCOL 25.7

## Embryonic Cell Ablation Using Lasers

### Materials

#### Supplies and Equipment

Slides and coverslips
Embryo glue (see Kiehart et al., this volume)
Plasticene
Microscope/Laser system described above

#### Solutions and Reagents

Halocarbon oil (#700)

### Method

1. Affix dechorionated *Drosophila* embryos to a slide (or coverslip, for inverted microscopes) using standard methods developed for microinjection (see Kiehart et al., this volume). An excellent glue for holding the embryos can be made by soaking double-stick tape in heptane. Spread a thin strip of glue on a slide and align embryos as desired.

2. Cover the embryos with #700 halocarbon oil. For upright microscopes, a coverslip must be used. Generally limit the time embryos are covered by the coverslip to approximately 20–30 minutes. Support the coverslip, if necessary, with a small amount of plasticene at each corner. The coverslip should be lowered as close as possible to the embryo, by pushing down gently on the plasticene at each corner. Do not compress the embryo itself.

3. Calibrate the laser power to a level suitable for cell killing. First, find the minimum power needed to crack the coverslip with a single pulse. If this can be done, there is ample power to ablate cells in the embryo. Start by attenuating the light power to 10–20% of this level (using the ND filters).

4. Fire single pulses at a test cell, using the alignment technique described in Figure 25.4. Moribund cells rapidly become granular or swollen.

5. Examine adjacent cells after 1–2 minutes to see if there is damage.

## PROTOCOL 25.7 NOTES

Too much laser power generates a small gas bubble in the cell, due to the cytoplasm boiling! Not all cells respond to laser light in the same way. Some seem to absorb light better, and the deeper targets generally need higher amounts of power. Perform a series of ablations at varying laser power levels, and examine the preparation acutely with a vital dye such as 1% Trypan blue in culture medium. Alternatively, an excellent fluorescent vital labeling system to detect dead and dying cells is available from Molecular Probes.

## ACKNOWLEDGMENTS

We are grateful for constructive comments on this manuscript from Cahir O'Kane, Paul Forscher, and Andrea Brand. S.T.S. thanks Sean McBride for his timely help. S.T.S. has been supported by a studentship from the BBSRC, by a BBSRC grant INS02992, a Wellcome Trust grant (048476/Z/96/Z, to Cahir O'Kane), and a Darwin College (Cambridge) Research Fellowship. A.H. was supported by a Wellcome Trust Career Development Fellowship.

## REFERENCES

Allen M.J. 1995. "Analysis of a cell ablation system in *Drosophila melanogaster*." Ph.D. thesis, University of Warwick, United Kingdom.

Armstrong J.D., de Belle J.S., Wang Z., and Kaiser K. 1998. Metamorphosis of the mushroom bodies; large scale rearrangements of the neural substrates for associative learning and memory in *Drosophila*. *Learn. Mem.* **5:** 102–114.

Beermann A.E. and Jay D.G. 1994. Chromophore-assisted laser inactivation of cellular proteins. *Methods Cell Biol.* **44:** 715–732.

Bellen H.J., D'Evelyn D., Harvey M., and Elledge S.J. 1992. Isolation of temperature-sensitive diptheria toxins in yeast and their effects on *Drosophila* cells. *Development* **114:** 787–796.

Bergmann A., Agapite, J., McCall K., and Steller H. 1998. The *Drosophila* gene *hid* is a direct molecular target of Ras-dependent survival signalling. *Cell* **95:** 331–341.

Booth G., Kinrade E., and Hidalgo A. 2000. Glia maintain follower neuron survival during *Drosophila* CNS development. *Development* **127:** (in press).

Brand A.H. 1995. GFP in *Drosophila. Trends Genet.* **11:** 324–325.

Brand A.H. and Perrimon N. 1993. Targeted expression as a means of altering cell fates and generating dominant phenotypes. *Development* **118:** 401–415.

Broadie K.S. and Bate M. 1991. The development of adult muscles in *Drosophila:* Ablation of identified muscle precursor cells. *Development* **113:** 103–118.

Campos-Ortega J.A. and Hartenstein V. 1985. *The embryonic development of* Drosophila melanogaster. Springer, Berlin.

Cash S., Chiba A., and Keshishian H. 1992. Alternate neuromuscular target selection following the loss of single muscle fibers in *Drosophila. J. Neurosci.* **12:** 2051–2064.

Chang T.N. and Keshishian H. 1996. Laser ablation of *Drosophila* embryonic motoneurons causes ectopic innervation of target muscle fibers. *J. Neurosci.* **16:** 5715–5726.

Chiba A., Hing H., Cash S., and Keshishian H. 1993. The growth cone choices of *Drosophila* motoneurons in response to muscle fiber mismatch. *J. Neurosci.* **13:** 714–732.

Collier R.J. 1990. Diptheria toxin: Structure and function of a cytocidal protein. In *ADP-ribosylating toxins and G-proteins* (ed. J. Moss and M. Vaughan), American Society for Microscopy, Washington, D.C.

de Belle J.S. and Heisenberg M. 1994. Associative odor learning in *Drosophila* abolished by chemical ablation of mushroom bodies. *Science* **263**: 692–695.

———. 1996. Expression of *Drosophila* mushroom body mutations in alternative genetic backgrounds: A case study of the mushroom body miniature gene (mbm). *Proc. Natl. Acad. Sci.* **93**: 9875–9880.

de Belle J.S., Wulf J., and Helfrich-Förster C. 1998. Mushroom bodies do not mediate circadian locomotor activity rhythms in *Drosophila*. *J. Neurogenet.* **11**: 150–151.

deCelis J.F., Barrio R., and Kafatos F.C. 1996. A gene complex acting downstream of dpp in *Drosphila* wing morphogenesis. *Nature* **381**: 421–424.

Dormand E.L. and Brand A.H. 1999. Runt determines cell fates in the *Drosophila* CNS. *Development* **125**: 1659–1667.

Duffy J.B., Harrison D.A. and Perrimon N. 1998. Identifying loci required for follicular patterning using directed mosaics. *Development* **125**: 2263–2271.

Endo Y. and Tsurugi K. 1998. The RNA N-glycosidase acivity of ricin-A chain. *J. Biol. Chem.* **263**: 8735–8739.

Farrell E., Fernandes J., and Keshishian H. 1996. Muscle organizers in *Drosophila*: The role of persistent larval fibers in adult flight muscle development. *Dev. Biol.* **176**: 220–229.

Fernandes J.J. and Keshishian H. 1996. Patterning the dorsal longitudinal flight muscles of *Drosophila*: Insights from ablation of larval scaffolds. *Development* **122**: 3755–3763.

———. 1998. Nerve-muscle interactions during flight muscle development in *Drosophila*. *Development* **125**: 1769–1779.

Girton J.R. and Berns M.W. 1982. Pattern abnormalities induced in *Drosophila* imaginal discs by an ultraviolet laser microbeam. *Dev. Biol.* **91**: 73–77.

Grether M.E., Abrams J.M., Agapite J., White K. and Steller H. 1995. The head involution defective gene of *Drosophila melanogaster* functions in programmed cell death. *Genes Dev.* **9**: 1694–1708.

Halfon M.S., Kose H., Chiba A., and Keshishian H. 1997. Targeted gene expression without a tissue-specific promoter: Creating mosaic embryos using laser-induced single-cell heat shock. *Proc. Natl. Acad. Sci.* **94**: 6255–6260.

Halpern M.E., Chiba A., Johansen J., and Keshishian H. 1991. Growth cone behavior underlying the development of stereotypic synaptic connections in *Drosophila* embryos. *J. Neurosci.* **11**: 3227–3238.

Hay B.A., Wassarman D.A., and Rubin G.M. 1995. *Drosphila* homologs of baculovirus inhibitor of apoptosis proteins function to block cell death. *Cell* **83**: 1253–1262.

Hidalgo A. and Booth G. 2000. Glia dictate pioneer axon trajectories in the *Drosophila* embryonic CNS. *Development* **127**: (in press).

Hidalgo A. and Brand A.H. 1997. Targeted neuronal ablation: The role of pioneer neurons in guidance and fasciculation in the CNS of *Drosophila*. *Development* **124**: 3253–3262.

Hidalgo A., Urban J., and Brand A.H. 1995. Targeted ablation of glia disrupts axon tract formation in the *Drosophila* CNS. *Development* **121**: 3703–3712.

Inoue S. and Spring K.R. 1997. *Video microscopy: The fundamentals*, 2nd edition. Plenum Publishing, New York.

Ito K. and Hotta Y. 1992. Proliferation pattern of postembryonic neuroblasts in the brain of *Drosophila melanogaster*. *Dev. Biol.* **149**: 134–148.

Ito K., Awano W., Suzuki K., Hiromi Y., and Yamamoto D. 1997. The *Drosophila* mushroom body is a quadruple structure of clonal units each of which contains a virtually identical set of neurones and glial cells. *Development* **124**: 761–771.

Jay D.G. and Keshishian H. 1990. Laser inactivation of fasciclin I disrupts axon adhesion of grasshopper pioneer neurons. *Nature* **348**: 548–550.

Kalb J.M., DiBenedetto A.J., and Wolfner M.F. 1993. Probing the function of *Drosophila melanogaster* accessory glands by directed cell ablation. *Proc. Natl. Acad. Sci.* **17**: 8093–8097.

Kunes S. and Steller H. 1991. Ablation of *Drosophila* photoreceptor cells by conditional expression of a toxin gene. *Genes Dev.* **5**: 970–983.

Kurada P. and White K. 1998. Ras promotes cell survival in *Drosophila* by downregulating *hid* expression. *Cell* **95**: 319–329.

Laski F.A., Ganguly S., Sharp P.A., RajBhandary U.L., and Rubin G.M. 1989. Construction, stable transformation and function of an amber suppressor tRNA gene in *Drosophila melanogaster*. *Proc. Natl. Acad. Sci.* **86**: 6696–6698.

Lin D.M., Auld V.J., and Goodman C.S. 1995. Targeted neuronal cell ablation in the *Drosophila* embryo: Pathfinding by follower growth cones in the absence of pioneers. *Neuron* **14:** 707–715.

Lin H. and Spradling A.C. 1993. Germline stem cell division and egg chamber development in transplanted *Drosophila* germaria. *Dev. Biol.* **159:** 140–152.

Lohs-Schardin M., Cremer C., and Nüsslein-Volhard C. 1979a. A fate map for the larval epidermis of *Drosophila melanogaster:* Localized cuticle defects following irradiation of the blastoderm with an ultraviolet laser microbeam. *Dev. Biol.* **73:** 239–255.

Lohs-Schardin M., Sander K., Cremer C., Cremer T., and Zorn C. 1979b. Localized ultraviolet laser microbeam irradiation of early *Drosophila* embryos: Fate maps based on location and frequency of adult defects. *Dev. Biol.* **68:** 533–545.

Malun D. 1998. Early development of mushroom bodies in the brain of the honeybee *Apis mellifera* as revealed by Brdu incorporation and ablation experiments. *Learn. Mem.* **5:** 90–101.

McNabb S.L., Baker J.D., Agapite J., Steller H., Riddiford L.M., and Truman J.H. 1997. Disruption of a behavioural sequence by targeted death of peptidergic neurons in *Drosophila*. *Neuron* **4:** 813–823.

Moffat K.G., Gould J.H., Smith H.K., and O'Kane C.J. 1992. Inducible cell ablation in *Drosophila* by cold-sensitive ricin A chain. *Development* **114:** 681–687.

Montell, D.J., Keshishian H., and Spradling A.J. 1991. Laser ablation studies of the roles of the *Drosophila* oocyte nucleus in pattern formation. *Science* **254:** 290–293.

Nüsslein-Volhard C., Lohs-Schardin M., Sander K., and Cremer C. 1980. A dorso-ventral shift of embryonic primordia in a new maternal-effect mutant of *Drosophila*. *Nature* **283:** 474–476.

O'Dell K.M.C., An X., Kaiser K., and de Belle J.S. 1998. Identification of centers for sex-specific behavior in the fly brain. *J. Neurogenet.* **11:** 181.

O'Kane C.J. and Gehring W.J. 1987. Detection of in situ genomic regulatory elements in *Drosophila*. *Proc. Natl. Acad. Sci.* **84:** 9123–9127.

Ponelies N., Bautz E.K., Monajembashi S., Wolfrum J., and Greulich K.O. 1989. Telomeric sequences derived from laser-microdissected polytene chromosomes. *Chromosoma* **98:** 351–357.

Prokop A. and Technau G.M. 1991. The origin of postembryonic neuroblasts in the ventral nerve cord of *Drosophila melanogaster*. *Development* **111:** 79–88.

———. 1994. Normal function of the mushroom body defect gene of *Drosophila* is required for the regulation of the number and proliferation of neuroblasts. *Dev. Biol.* **161:** 321–337.

Reddy S., Jin P., Trimarchi J., Caruccio P., Phillis R., and Murphey R.K. 1997. Mutant molecular motors disrupt neural circuits in *Drosophila*. *J. Neurobiol.* **33:** 711–723.

Schiavo G., Benfenati F., Poulain B., Rossetto O., DeLaureto P.P., DasGupta B.R., and Montecucco C. 1992. Tetanus and botulinum-B neurotoxins block neurotransmitter release by proteolytic cleavage of synaptobrevin. *Nature* **359:** 832–835.

Schmucker D., Su A.L., Beermann A., Jackle H., and Jay D.G. 1994. Chromophore-assisted laser inactivation of patched protein switches cell fate in the larval visual system of *Drosophila*. *Proc. Natl. Acad. Sci.* **91:** 2664–2668.

Shah E.M. and Jay D.G. 1996. Methods for ablating neurons. *Curr. Opin. Neurobiol.* **3:** 738–742.

Smith H.K., Roberts I.J.H., Allen M.J., Connolly J.B., Moffat K.G., and O'Kane C.J. 1996. Inducible ternary control of transgene expression and cell ablation in *Drosophila*. *Dev. Genes Evol.* **206:** 14–24.

Stocker R.F., Heimbeck G., Gendre N., and de Belle J.S. 1997. Neuroblast ablation in *Drosophila* P[GAL4] lines reveals origins of antennal target interneurons. *J. Neurobiol.* **32:** 443–456.

Stringham E.G. and Candido E.P. 1993. Targeted single-cell induction of gene products in *Caenorhabditis elegans:* A new tool for developmental studies. *J. Exp. Zool.* **266:** 227–233.

Sweeney S.T., Broadie K., Keane J., Niemann H., and O'Kane C.J. 1995. Targeted expression of tetanus toxin light chain in *Drosophila* specifically eliminates synaptic transmission and causes behavioral defects. *Neuron* **14:** 341–351.

Technau G.M. and Heisenberg M. 1982. Neural reorganisation during metamorphosis of the corpora pedunculata in *Drosophila melanogaster*. *Nature* **295:** 405–407.

Tettamanti M., Armstrong J.D., Endo K., Furokubo-Tokunaga K., Kaiser K., and Reichert H. 1998.. Analysis of mushroom body development by enhancer-trap expression patterns in the *Drosophila* brain. *Dev. Genes Evol.* **207:** 242–252.

Theodosiou N.A. and Xu T. 1998. Use of the FLP/FRT system to study *Drosophila* development. *Methods* **14:** 355–365.

Timson, J. 1975. Hydroxyurea. *Mutat. Res.* **32:** 115–132.

Truman J.W. and Bate M. 1988. Spatial and temporal patterns of neurogenesis in the central nervous system of *Drosophila melanogaster. Dev. Biol.* **125:** 145–157.

Truman J.W. and Booker R. 1986. Adult-specific neurons in the nervous system of the moth, *Manduca sexta:* Selective chemical ablation using hydroxyurea. *J. Neurobiol.* **17:** 613–625.

White K., Tahaoglu E., and Steller H. 1996. Cell killing by the *Drosophila* gene *reaper. Science* **271:** 805–807.

Wittig T., Dill M., and Heisenberg M. 1995.. Visual pattern discrimination learning in *Drosophila melanogaster* without mushroom bodies. In *Learning and Memory: Proceedings of the 23rd Göttingen Neurobiology Conference* (ed. N. Elsner and R. Menzel), p. 4. Stuttgart, Germany.

Younossi-Hartenstein A., Nassif C., Green P., and Hartenstein V. 1996. Early neurogenesis of the *Drosophila* brain. *J. Comp. Neurol.* **370:** 313-329.

Zhou L., Schnitzler A., Agapite J., Schwartz L.M., Steller H., and Nambu J.R. 1997. Cooperative functions of the reaper and head involution defective genes in the programmed cell death of *Drosophila* central nervous midline cells. *Proc. Natl. Acad. Sci.* **94:** 5131–5136.

# Genomics

# CONTENTS

# 26

# Preparation of DNA from *E. coli* Cells Containing Bacteriophage P1 Clones

Elena R. Lozovskaya

*Harvard University*
*Department of Organismic and Evolutionary Biology*
*Cambridge, Massachusetts 02138*

T HE ABILITY TO CLONE AND MANIPULATE LARGE FRAGMENTS of genomic DNA has created new research opportunities in virtually all areas of modern genetics. During the past 5 years, the bacteriophage P1 cloning system has proven to be a powerful tool in fields such as the construction of global physical maps of complex genomes (Smoller et al. 1991; Hartl et al. 1994; Vieira et al. 1997a) and in comparative studies of the evolution of genome organization (Lozovskaya et al. 1993; Segarra et al. 1995; Vieira et al. 1997b). For detailed descriptions of P1 vectors, construction of genomic libraries, and different applications of the system, see Hartl and Lozovskaya (1995).

## BACTERIOPHAGE P1 LIBRARIES

### *Drosophila melanogaster*

*Drosophila* P1 clones were produced with genomic DNA extracted from adult flies of an isogenic strain of genotype *y ; cn bw sp* according to the methods described in Smoller et al. (1991) and Lozovskaya et al. (1993). The *D. melanogaster* bacteriophage P1 library presently in use consists of 9216 clones. Approximately 40% of the clones are in the pNS582-tet14 Ad10 cloning vector (Sternberg 1990); the remaining (~60%) are in pAd10-sacBII (Pierce at al. 1992).

Figure 26.1 illustrates the size distribution of inserts of *D. melanogaster* DNA among 25 clones in pNS582-tet14 Ad10 and 20 clones in pAD10-sacBII. The average insert size in the former is 83.0 kb, and in the latter, it is 82.5 kb. These values are close to the theoretical expectations. A small proportion of clones in the pAD10-sacBII library, estimated at fewer than 1%, appear to lack inserts of *Drosophila* DNA; another small proportion of clones contain inserts smaller than 60 kb.

The size of *D. melanogaster* genome is 165 Mb. Approximately 21% of the entire genome is composed of simple satellite sequences which are not present in P1 libraries

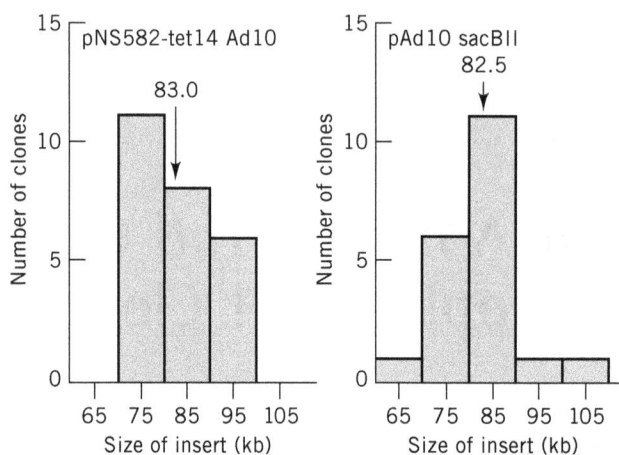

Figure 26.1. Size distribution of inserts among clones of DNA from *D. melanogaster* in pNS582-tet14 Ad10 and pAD10-sacBII.

(Hartl and Lozovskaya 1995). Assuming that the potentially clonable euchromatic sequences and more complex heterochromatic sequences in the *D. melanogaster* genome constitute the remaining 130 Mb, the clones in the library include about 5.9 haploid genome equivalents.

A total of 3104 clones were localized by in situ hybridization with the polytene salivary gland chromosomes. Among the localized clones, 388 hybridized with the chromocenter, the underreplicated mass consisting largely of the pericentromeric heterochromatin and the Y chromosome, and/or with multiple euchromatic sites (typically 10–100) without any apparent major euchromatic site of hybridization. A total of 64 clones yielded dual hybridizations (strong signals in two distinct euchromatic sites); these clones have not been investigated further, but some of them may represent chimeric clones containing ligated fragments from two different parts of the genome. An additional 191 clones were deliberate duplicates introduced into the workstream as blind controls in order to verify the accuracy of the procedures and the reproducibility of the cytological localizations.

The remaining 2461 localized clones yielded single major sites of euchromatic hybridization. Approximately 10% of these clones also exhibited multiple (typically 10–100) secondary sites of hybridization in the euchromatin; about half of this class also hybridized with the chromocenter. The multiple sites of hybridization are interpreted as resulting from transposable elements or other types of moderately repetitive, dispersed DNA contained in the cloned insert. With such clones, it is usually not difficult to identify a principal site of hybridization in the euchromatin in which the signal is intense and encompasses several bands, compared to which the multiple sites of hybridization are usually much weaker and present in single bands; the principal site of hybridization is the site to which the clone is assigned. (The distinction between the major site and secondary sites of hybridization can be difficult if the probe is excessive in amount or too heavily labeled.) A final class of clones, approximately 2% of those localized, hybridized with the chromocenter and also to one principal site of hybridization in the euchromatin, without detectable secondary sites of euchromatic hybridization.

Because the average insert size of the clones is approximately 82 kb, the mapped clones with unique major sites of hybridization include approximately 200 Mb of DNA, or the equivalent of about 1.8 copies of the haploid euchromatic genome. Assuming that all euchromatic sites are equally likely to be present in the clones, the proportion of euchromatic sites expected to be represented at least once among the clones is approximately 85%.

## *Drosophila virilis*

Our laboratory has been particularly interested in applications of genome analysis to evolutionary studies. To provide cloned material suitable for evolutionary applications in *Drosophila*, we also produced a bacteriophage P1 library from *D. virilis* (Lozovskaya et al. 1993). This species is in the subgenus *Drosophila* (*D. melanogaster* is in the subgenus *Sophophora*) and has been the object of substantial genetic and evolutionary study.

The source of genomic DNA was *D. virilis* strain 9, a wild-type strain collected in 1970 in Batumi, Georgia (former USSR), and maintained in laboratory culture since that time. High-molecular weight DNA was cloned into the bacteriophage P1 vector pAd10-sacBII, and clones were arrayed in the wells of 105 microtiter plates for a total of 10,080 clones.

The distribution of insert sizes among the *D. virilis* P1 clones is given in Figure 26.2. The average insert size is 65.8 kb, which is somewhat smaller than that in the *D. melanogaster* P1 clones. On the other hand, approximately 75% of the clones have inserts exceeding 50 kb, and approximately 25% have inserts exceeding 80 kb.

The DNA content of *D. virilis* is among the largest in the genus, approximately 313 Mb, 52% of which is composed of satellite sequences. Assuming a potentially clonable genome size of 150 Mb, the clones in the *D. virilis* library include approximately 4.4 haploid genome equivalents.

Working toward the ultimate objective of generating a framework bacteriophage P1 map of *D. virilis*, a sample of 732 randomly selected clones was mapped by in situ

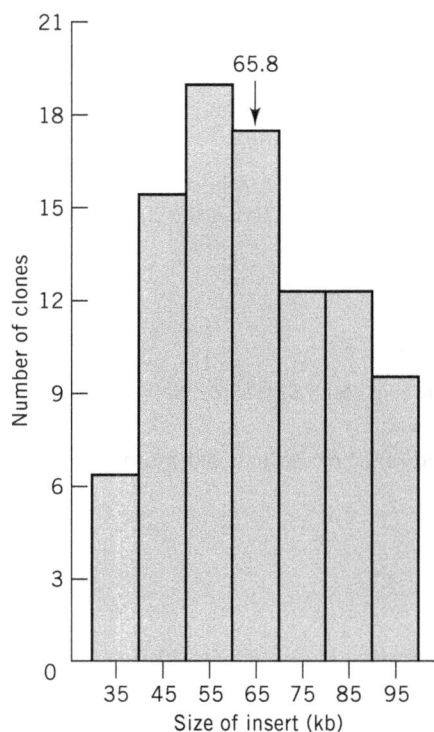

**Figure 26.2.** Distribution of insert sizes among P1 clones in pAD10-sacBII from *D. virilis*.

hybridization with the salivary gland chromosomes (Lozovskaya et al. 1993; Vieira et al. 1997a). Among these, 573 clones, or 78%, hybridized with a single euchromatic site each. As far as could be determined from this sample, no bias has been detected in the relative contribution of each large chromosome to the bacteriophage P1 clones. The only exception is chromosome 6 (the homolog of *D. melanogaster* chromosome 4) from which no clones have been recovered. The remaining 159 P1 clones from the sample contain repetitive DNA sequences that hybridized with multiple sites in the euchromatin and/or with the chromocenter. Interestingly, a significant fraction of these clones hybridized to multiple euchromatic sites but not to the chromocenter, which is a pattern of hybridization that is very rare among clones derived from *D. melanogaster*.

PROTOCOL 26.1*

# Preparation of DNA

This protocol is a modification of the procedure described by Birnboim and Doly (1979). For most bacteriophage P1 clones, the protocol yields a sufficient amount of DNA such that 15–20 µl can be used per restriction digest in a Southern blot. We typically use 1.5 µl per slide for in situ hybridizations with polytene chromosomes. DNA yield strongly depends on the specific bacteriophage P1 clone and may range from 3 to 15 µg.

## Materials

### *Media*

> **LB Medium (Luria-Bertani)**
>
> | | |
> |---|---|
> | Tryptone | 10 g |
> | Yeast extract | 5 g |
> | NaCl | 10 g |
> | $H_2O$ | 800 ml |
>
> Combine all components and stir to dissolve. Adjust the pH to 7.0 with 5 N **NaOH**. Add $H_2O$ to bring volume to 1 liter. Sterilize by autoclaving for 20 minutes at 15 psi on liquid cycle.

### *Supplies and Equipment*

Incubator shaker or water bath shaker

Refrigerated centrifuge and rotors (e.g., Sorvall GSA or GS3, or Beckman JA-10 rotors)

Corex tubes (15 ml)

Microcentrifuges, one refrigerated (4°C) or set in a cold room, and one at room temperature.

Microcentrifuge tubes (1.5 ml)

### *Solutions and Reagents*

Kanamycin stock solution (10 mg/ml) in $H_2O$; store at –20°C.

**Isopropanol**

---

*For all protocols, $H_2O$ indicates glass distilled and deionized.

**Ethanol**, 70% and absolute
TE buffer (pH 8.0; see Appendix 3)
**Chloroform**
(*Optional*) NaCl (4 M)
(*Optional*) **Polyethylene glycol** (13%) (PEG-8000)
(*Optional*) **Ammonium acetate** (5 M)

### 1 M IPTG

Dissolve 2.383 g of **IPTG** (isopropyl β-D-thiogalactopyranoside) in 8 ml of $H_2O$. Adjust the volume of the solution to 10 ml with $H_2O$ and sterilize by filtration through a 0.22-μm disposable filter. Store in 1-ml aliquots at –20ºC.

### Glucose-Tris-EDTA Solution

| Component and final concentration | Amount to add to make 100 ml |
|---|---|
| 50 mM glucose | 10 ml of 0.5 M |
| 25 mM Tris-Cl (pH 8.0) | 2.5 ml of 1 M |
| 10 mM EDTA | 2 ml of 0.5 M |
| $H_2O$ | 85.5 ml |

Autoclave for 15 minutes at 10 psi on liquid cycle. Store at 4ºC.

### NaOH-SDS (pH 4.8)

| Component and final concentration | Amount to add to make 100 ml |
|---|---|
| 0.2 N NaOH | 800 mg |
| 1% **SDS** | 10 ml of a 10% stock solution |
| $H_2O$ | to 100 ml |

Prepare weekly and store at room temperature.

### 3 M Potassium Acetate

| | |
|---|---|
| **Potassium acetate** | 60 ml of a 5 M stock solution |
| Glacial acetic acid | 11.5 ml |
| $H_2O$ | 28.5 ml |

The resulting solution is 3 M with respect to potassium and 5 M with respect to acetate. Store at room temperature.

### DNase-free Pancreatic RNase

For a stock solution, dissolve RNase A (GIBCO-BRL) in TE buffer at a concentration of 10 mg/ml. Heat at 100ºC for 15 minutes, dispense in aliquots, and store at –20ºC.

**CAUTION: ammonium acetate, chloroform, ethanol, glacial acetic acid, IPTG, isopropanol, NaOH, PEG, potassium acetate, SDS** (see Appendix 4)

## Method

1. Inoculate a bacterial culture containing the bacteriophage P1 clone of interest into 1 ml of LB medium containing 25 μg/ml of kanamycin. Incubate at 37ºC overnight with agitation.

2. Add 100 μl of the overnight culture to 10 ml of LB medium containing 25 μg/ml of kanamycin. Incubate with agitation at 37ºC to an optical density ($OD_{550}$) of 0.05–0.1. Growth to this density typically takes 1.0–1.5 hours.

*Note:* To allow thorough aeration in an incubator shaker or water bath shaker, the culture volume should not exceed 30% of the volume of the container.

3. Add IPTG to a final concentration of 1 mM and continue incubating the culture to an $OD_{550}$ of 0.5–1.5. This incubation typically lasts approximately 2 hours.

   *Note:* The bacteriophage P1 clone may be directly inoculated into 10 ml of LB medium containing 25 μg/ml kanamycin and 1 mM IPTG and grown at 37°C overnight with agitation. This approach requires less fussing than the procedure in steps 1–3. The disadvantage is that it does not work well with all P1 clones.

4. Centrifuge the culture at 6000*g* (6000 rpm in Sorvall GSA or GS3, or Beckman JA-10 rotors) at 4°C for 10 minutes. Glass tubes (e.g., 15-ml Corex) can be used. Discard the supernatant.

5. Add 200 μl of cold Glucose-Tris-EDTA Solution to the pellet. Resuspend the pellet making sure that it is completely dispersed in the liquid. Transfer the resulting suspension to a microcentrifuge tube.

6. Add 300 μl of NaOH-SDS Solution and mix well. Avoid vigorous shaking while mixing. Incubate on ice until the suspension clarifies (~5 minutes).

7. Add 300 μl of cold 3 M Potassium Acetate and mix gently until a whitish precipitate forms. Incubate on ice for a further 15 minutes.

8. Centrifuge in a microcentrifuge at 14,000 rpm at room temperature for 5 minutes. Transfer the supernatant to a clean microcentrifuge tube. Proceed to step 10.

9. (*Optional*) For a quick-and-dirty miniprep, perform the following:

   a. Add 0.6 volumes (~0.5 ml) of isopropanol and mix.

   b. Immediately centrifuge at 14,000 rpm at room temperature for 5 minutes. Remove the supernatant.

   c. Add 1 ml of 70% ethanol and shake.

   d. Centrifuge again as described in step b above. Discard the supernatant.

   e. Dry the pellet and resuspend in 50 μl of TE buffer.

10. Add 3 μl of DNase-free pancreatic RNase. Incubate at 37°C for 30 minutes.

11. Add an equal volume of chloroform and mix by vortexing for 1 minute. Centrifuge at 14,000 rpm at room temperature for 5 minutes. Transfer the aqueous layer to a clean microcentrifuge tube.

12. Add an equal volume of isopropanol and mix well. Centrifuge immediately at 14,000 rpm at room temperature for 10 minutes. Discard the supernatant.

13. Wash the pellet by adding 1 ml of cold 70% ethanol. Shake, centrifuge again as described in step 12 above, and discard the ethanol. Get rid of ethanol completely by centrifuging the tube briefly for a second time and removing the remaining drop with the Pipetman. Dry the pellet and resuspend in 40 μl of TE.

    *Note:* Do not overdry the pellet because it will be difficult to dissolve DNA.

14. (*Optional*) The DNA can be further purified as follows:

    a. To obtain higher-quality DNA preparations suitable for sequencing, perform the following steps:

  i. Add 10 µl of 4 M NaCl to the solution resulting from step 13. Mix and add 50 µl of 13% PEG-8000. Mix again and incubate on ice for 1 hour.

  ii. Centrifuge at 14,000 rpm at 4°C for 15 minutes (we perform this centrifugation in a cold room). Remove the liquid (at this stage the pellet is transparent, so be careful!).

  iii. Add 1 ml of 70% ethanol and vortex. The DNA should peel off the wall. If it does not, scrape it from the wall with a yellow tip.

  iv. Centrifuge at 14,000 rpm at room temperature for 5 minutes. Remove the liquid. Get rid of ethanol completely, dry the pellet, and dissolve it as described in step 13.

b. If DNA obtained in step 13 cannot be cut with restriction enzymes, the following purification step is used:

  i. Add 60 µl of TE or H$_2$O. Add 60 µl of 5 M ammonium acetate and mix to bring to a total volume of 160 µl.

  ii. Add 160 µl of ethanol, mix, and incubate at room temperature for 5 minutes.

  iii. Centrifuge at 14,000 rpm at room temperature for 10 minutes. Remove the supernatant.

  iv. Add 1 ml of 70% ethanol and shake.

  v. Centrifuge at 14,000 rpm at room temperature for 5 minutes. Remove the liquid. Get rid of the ethanol completely. Dry the pellet and resuspend in 50 µl of TE buffer.

## REFERENCES

Birnboim H.C. and Doly J. 1979. A rapid alkaline extraction procedure for screening recombinant plasmid DNA. *Nucleic Acids Res.* **7:** 1513–1523.

Hartl D.L. and Lozovskaya E.R. 1995. *The* Drosophila *genome map: A practical guide.* R.G. Landes, Austin.

Hartl D.L., Nurminsky D.I., Jones R.W., and Lozovskaya E.R. 1994. Genome structure and evolution in *Drosophila:* Applications of the framework P1 map. *Proc. Natl. Acad. Sci.* **91:** 6824–6829.

Lozovskaya E.R., Petrov D.A., and Hartl D.L. 1993. A combined molecular and cytogenetic approach to genome evolution in *Drosophila* using large-fragment DNA cloning. *Chromosoma* **102:** 253–256.

Pierce J.C., Sauer B., and Sternberg N. 1992. A positive selection vector for cloning high molecular weight DNA by the bacteriophage P1 system: Improved cloning efficacy. *Proc. Natl. Acad. Sci.* **89:** 2056–2060.

Segarra S., Lozovskaya E.R., Ribo G., Aquade M., and Hartl D.L. 1995. P1 clones from *Drosophila melanogaster* as markers to study chromosomal evolution of Mullers A element in two species of the *obscura* group of *Drosophila. Chromosoma* **104:** 129–136.

Smoller D.A., Petrov D.A., and Hartl D.L. 1991. Characterization of bacteriophage P1 library containing inserts of *Drosophila* DNA of 75–100 kilobase pairs. *Chromosoma* **100:** 487–494.

Sternberg N. 1990. Bacteriophage P1 cloning system for the isolation, amplification, and recovery of DNA fragments as large as 100 kilobase pairs. *Proc. Natl. Acad. Sci.* **87:** 103–107

Vieira J., Vieira C.P., Hartl D.L., and Lozovskaya E.R. 1997a. A framework physical map of *Drosophila virilis* based on P1 clones: Applications in genome evolution. *Chromosoma* **106:** 99–107.

———. 1997b. Discordant rates of chromosome evolution in the *Drosophila virilis* species group. *Genetics* **147:** 223–230.

# CONTENTS

# 27

# *Drosophila* Microarrays: From Arrayer Construction to Hybridization

**Kevin P. White**
*Department of Biochemistry*
*Stanford University Medical Center*
*Stanford, California 94305*

**Kenneth C. Burtis**
*Section of Molecular and Cellular Biology*
*University of California at Davis*
*Davis, California 95616*

Hʏʙʀɪᴅɪᴢᴀᴛɪᴏɴ ᴏꜰ ɴᴜᴄʟᴇɪᴄ ᴀᴄɪᴅ ᴘʀᴏʙᴇs ᴛᴏ ᴀʀʀᴀʏs of DNA comprising thousands of individual genes provides a powerful new technology for studying gene expression. The Berkeley *Drosophila* Genome Project (BDGP) is sequencing *Drosophila* cDNA clones, which will provide the raw material necessary for applying this technology to this model system. During the next few years, specific fragments of DNA (genomic and/or cDNA) representing the entire set of *Drosophila* genes will become available for experimental manipulation. Thus, it is an opportune time for the *Drosophila* community to begin to take advantage of the power of parallel approaches to the study of gene expression. To do so will require access both to the collection of genes and to the equipment necessary for the production and analysis of DNA microarrays.

In this chapter, we provide technical details regarding the construction of a robotic device capable of spotting DNA with the resolution required for microarray experiments, as well as protocols for preparation of the slides, spotting of the DNA, preparation of fluorescent probes, hybridization, and washing of slides. Available methods of scanning are also discussed.

It is important to note that microarray analysis is a newly emerging and rapidly evolving technology. It is thus probable that the technical details provided below will be superseded by later developments. It is essential that investigators be prepared to explore additional sources of information described in this chapter at the time they initiate their efforts. Currently, the cost of the microarray apparatus is approximately $30,000 and the cost of the scanner is about $50,000. These costs may come down in price over time.

The information provided in this chapter relies heavily on a remarkable Web site authored by Dr. Joseph DeRisi and others from Dr. Pat Brown's laboratory (Department of Biochemistry, Stanford Medical School), which can be found at:

http://cmgm.stanford.edu/pbrown/mguide/index.html

This group of investigators has provided public access to detailed instructions for construction of arrayer hardware, as well as to the software needed for operation of the arrayer. They have further established a public forum for discussions regarding advances and problems in microarray technology and continue to post updated technical protocols. This chapter makes frequent references to this Web site and avoids, insofar as possible, duplication of the information at this site. All information in the section of the chapter discussing arrayer construction reflects the content of this Web page as of May 1999; later updates may occur at any time in the future.

For those investigators who are planning to construct their own arrayer, the primary purpose of the first part of this chapter is to serve as an appendix to the "Mguide" Assembly Guide, which resides on the above Web page. Although the MGuide is remarkably comprehensive, there are certain areas in which additional information may prove useful. We will attempt to provide such information, in the context of the text in the MGuide. It would thus be advisable for the investigator to download and read the complete MGuide Assembly Guide, as well as this chapter before initiating construction of an arraying machine.

With respect to microarray experimental protocols, remember that this is a rapidly developing area, in which methods are being continuously invented and improved. In the spirit of the investigators who generously made information regarding the basic technology available on the Web, and in the hope of advancing the interests of the *Drosophila* community, we encourage investigators who develop further improvements to these techniques to share their ideas with the world by communicating them in a public forum.

## ARRAYER CONSTRUCTION

### Materials

The first step in construction is ordering the necessary parts. The major components are obtained from three principal suppliers:

- *Western Technology* (190 S. Whisman Road, Building G, Mountain View, California 94041; Phone: 650-968-6200). The Western Technology part numbers designated for robotic arms, amplifiers, and motion-control electronics are those indicated in the MGuide Parts List; however, the company representative for these items is now Vince Hodges (E-mail: vhodges@westerntechnology.com; Phone: 650-968-6200).
- *Newport Electronics Corp.* The breadboard and support stand from Newport Corp. are those listed.
- *Die-Tech, Inc.* (1721 Little Orchard St. #C, San Jose, California 95125; Phone: 408-279-3363; Fax: 408-279-3896). Die-Tech is a custom tool and die shop, which is experienced in production of items such as the slide platter and tip holders. Specifications and drawings for most of the necessary parts are provided in the MGuide if use of an alternative machine shop is preferred. The following are two significant considerations concerning the Die-Tech parts list:

  1. The Stanford arraying group has found that the original wash/dry clean station, with flowing water, is not necessary. The currently recommended alternative is to use the sonicator in place of the water rinse and a vacuum drying station. This requires a set of clean station brackets different from those originally specified, so that the drying station and sonicator can be placed in close proximity to the corner of the platter housing the microtiter plate. The alternative bracket set is

composed of four pieces: two angular brackets that attach to the wash station, and two straight brackets that attach both to the angle brackets and to the breadboard. We suggest contacting Die-Tech for details on these new parts.

2. The choice of tip holder is now dependent on the type of tips chosen. Although there are presently several alternative tip suppliers, we are most familiar with tips from two sources: Die-Tech and Majer Precision Engineering (236 West Lodge Drive, Tempe, Arizona 85283; Phone: 602-777-8222; Fax: 602-777-8244). Die-Tech 308 series stainless-steel tips are compatible with the 8- and 16-tip-holder assemblies listed in the MGuide Parts List. Majer presently sells two types of tips, both of which require use of a tip-holder assembly from Majer. The Majer tips are available in both 308L stainless steel and 17-4 series stainless steel, a new material with improved durability. A novel feature of the Majer tips and holder is the presence of a flat side on both the tip and the opening through which the tip passes. This prevents the tip from rotating about its long axis during a print run, which can lead to irregular spacing of spots in the array. Both 16- and 32-tip versions of the holder are available from Majer. Note that if printing with 32 tips, substitute a 32-hole drying station (from Die-Tech) for the 16-hole version.

Keep in mind that there may be a lag time of up to 1 month (or more) in delivery of the large parts from Western Technology and Newport, so be sure to plan ahead. The remaining large part to order is the PC required to run the arraying software. There are obviously multiple rapidly changing options available with respect to the computer hardware. Any computer meeting or exceeding the specifications in the MGuide is suitable. Considerations include the type of case (which must be compatible with the height of the controller card; some cases may be unsuitable) and convenience with respect to loading software (internal Zip drive may be convenient).

The remainder of the MGuide Parts List includes a variety of small items needed for construction or operation of the arrayer. Changes from the MGuide Parts List include the following:

- Because of the change in the method of cleaning and drying tips, the following items are no longer needed:
  PP Barbed Tube Fitting 1/2–3/4"
  PP Barbed Tube Fitting 1/8–1/4"
  Autoclavable Carboy PP, 50 liters
  Check Valve PP/5/16"
  Cal Pump, 180 gph, 20-ft cord
- The following items are needed:
  One SR Series AC Connector, Snap-in Mount
  Two 3-Conductor Power Supply cords (one for the vacuum pump and one to provide power to the relay box)
  Box of Metric 18-8 SS Socket Head Cap screws (M5 × 16 mm; McMaster-Carr 91292A126) to use in attaching the mounting bracket to the *Z* axis
  Mini-jumper (8 mm, open style, Newark Electronics, 16F1578) to modify the circuit board of the ICM-1900
- The following are additional useful items not presently listed in the MGuide Parts List:
  Helical Bundling Wrap (1/2"), black polyethylene (McMaster-Carr 7432K23), used to consolidate wiring and cables on completion of wiring
  Mobil Oil auto grease (1 lb.), Grainger No. 6Y788, used to lubricate the screws that drive the X and Y stages

Tri-flow with Teflon; a lubricant for the screw that drives the Z stage (available at most bicycle shops)

Stainless-steel shim stock sheets (McMaster-Carr 9011K82 and 9011K83), used to clean debris from the slots of the printing tips if they become clogged

## Construction

### *Assembly: Part 1 of the MGuide and Addendum*

Detailed instructions for construction of the microarrayer can currently be downloaded as three pdf files (Part I, Addendum, and Part II) from the MGuide Web site at:

http://cmgm.stanford.edu/pbrown/mguide/getguide.html

Before initiating construction, we suggest reading these documents in their entirety, as well as the information in this section. Although not everything will be meaningful at this point, it is a good idea to become familiar with the big picture at the outset in order to plan ahead during the actual construction process. As far as possible, the comments below are presented in the same sequence as relevant sections of the assembly guide.

While reading through the assembly guide and this section, keep in mind the types of tools required. Depending on how well stocked the investigator's departmental or personal toolbox might be, additional tools may have to be purchased. Standard sets of English and metric hex wrenches are required, as are normal tools for electrical wiring. A voltmeter will come in handy during the wiring phase. Keep a running list of tools needed while reading through guide for the first time. Check the list against the contents of the toolbox, and head for the hardware store.

Part 1 of the Assembly Guide involves initial setup and wiring. The arrayer will fit comfortably into a 10-foot square room, although it can be squeezed into a somewhat smaller space. The recent changes to the tip-washing procedures have eliminated the necessity of having water supply and drainage, so the only essential utility is electrical outlets. Less obvious but important considerations regarding the room are dust and noise. Arrays are sensitive to minute amounts of dust, and depending on the cleanliness of the ventilation system in the building, it may be necessary to add filtration systems to the room's air supply. A simple dust cover over the printing surface helps, but some groups have now developed more sophisticated systems involving forced delivery of HEPA-filtered air. Noise is a consideration for neighbors. The mechanical sound of the stages moving is not exceedingly loud, but the repetitive operation of the vacuum pump for drying the tips can be annoying. It would probably be best to identify a space that does not share a wall with someone's office.

Note that the current availability of a 4-meter cable for connecting the computer to the breakout box allows greater flexibility in relative placement of these two devices than was originally possible. The only critical consideration, particularly during the testing phase of construction, is that the computer keyboard and emergency stop button be in close enough proximity that the operator can immediately countermand a computer order if something goes wrong.

The isolation table from Newport is very heavy, and definitely requires the assistance of several individuals to assemble. Be sure to place it exactly where it is required, as moving the table after the arrayer is assembled will be problematic. Once the table is assembled, the next step is to start mounting parts to the breadboard. However, before attempting this, it is advisable to use a 1/4-20 tap to clean out the mounting holes that will be used. Although this should not be necessary, several groups have encountered debris in the threads of these holes, resulting in jammed screws.

Next mount the *X*-axis stage. Note that the guide was written from the perspective of someone sitting at a computer mounted close to the breakout box. From this position, the stage that moves the slide platter runs left and right, the normal orientation for a Cartesian *X*-axis. However, if observing from the "Front" of the breadboard (as indicated in the Assembly Guide drawing), the *Y*-axis runs left to right. This can be confusing, so it is best to simply remember that the X stage carries the platter, the Y stage carries the printing head, and the Z stage moves the printing head up and down.

Pay close attention to the screws that are used. The breadboard table is threaded using English units (1/4" holes), but several of the machined parts utilize metric sizes. If the screws do not seem to be going in easily, stop and check before using force. Screws have been purchased in a variety of diameters and lengths. Although diameters are specified in the Assembly Guide, lengths are not. For mounting the *X*-axis stage, use 1/4-20 ✕ 1" mounting screws.

Following the diagram in the MGuide, next position and mount the Y-stage uprights using 1/4-20 ✕ 1" screws. Before tightening these down, place the servo mounting plate in its final position, to make sure that everything will fit in the right place. If the Y stage came preassembled with the Z stage, it may be attached in the wrong orientation. Although there is some confusion on this issue, the situation is easily resolved by observing the polarity of the large stages in the MGuide pictures. The polarity of the long axis is defined by the placement of the motors, and the polarity of the other dimension is defined by the presence of a black rectangular box (the "outrigger") along one side. If everything is together properly, the outrigger should be up and the motor on the right when facing the mounted *Y*-axis from the front. In this case, the motor of the Z stage should be pointing up. If this is not the case, then detach and reattach the Z stage in the proper polarity. Once this issue has been resolved, and the uprights are securely fastened, mount the Y stage to the uprights using 1/4-20 ✕ 1/2" screws.

The servo plate can be mounted to the breadboard using 1/4-20 ✕ 1" screws. It is not necessary to mount this plate using all of the threaded holes provided; indeed, it is so heavy that there is probably no need to fasten it down to the table at all. If the plate is to be bolted down, one or two screws will suffice to keep it in place.

Now attach the amplifiers and circuit board to the servo plate, switching to metric (M5) screws (and metric Allen wrenches; 4 mm for these parts). Use the M5 ✕ 12 screws for mounting the circuit board, and the M5 ✕ 25 screws for mounting the amplifiers.

The installation of computer cards and software is beyond the scope of this chapter; however, if unfamiliar with this operation, we suggest seeking the assistance of a PC-knowledgeable person within the department. It is necessary to install the software that comes with the motion control system, as well as the latest versions of software from the MGuide Web site at:

http://cmgm.stanford.edu/pbrown/mguide/software.html

Updated versions of the arraying software may occasionally be posted to this site. It is advisable not to install extraneous software programs onto the PC that will be running the arrayer, as the activity of these programs at an inopportune moment may interfere with the operation of the arraying software in the middle of a print run, leading to unfortunate consequences.

## Wiring

The wiring of the arrayer, although appearing complex, is quite straightforward and well described in the MGuide. Work carefully and double-check at each step which color wire attaches to each terminal. It is much easier to be careful while wiring than to troubleshoot erroneous connections later. Try to work neatly; when working with the multistranded wire, leave as much as possible in the plastic casing so that the individual wires remain bundled.

Note that in the beginning, only the controller is wired to the limit switches, so that they can communicate with the computer. There is no power to the motors at this point. There is a good reason for this; the limit switches tell the computer when the carriage is getting too close to either end. The motors are deceptively powerful, and if the limit switches are not wired properly, and the signal is given to move the motors beyond the end of their travel limit, serious damage can occur. The initial tests are thus to ensure that when the carriages reach specific positions on each axis, the appropriate signal is sent to the computer, which will direct the motors to cease turning.

When there is no power to the motors, the X and Y stages can be moved manually by dragging the stages. It is easiest if one person moves the stages while another watches the computer screen for the numbers to change in the Jogger program window. The Z-axis can likewise be moved manually, not by dragging the stage, but rather by manually turning the coupler. If everything is wired correctly and working correctly, the numbers in the Jogger window should appear exactly as described in the MGuide, and the limit changes should occur at a safe distance before the carriage hits the limit of its mobility (i.e., bangs into the end). There are two conflicting influences here: The axes should have maximum extension to provide the largest possible working area for the arrayer, but the carriages should not run to their limit and suffer a "hard stop," which is potentially damaging to the motors and shaft. If the limit switches for the *X* and *Y* axes are not positioned by the factory far enough from the ends (as has been the case in some instances), use a small Allen wrench to remove the screws holding the rectangular covers of the outrigger boxes, to access the limit switches, which can then be repositioned.

It is likely (hopefully) that the only time the limit switches will ever be called into action is upon first testing the motors, using the Jogger program to move the carriages along the length of the axes. It is good to test the switches at least once, to make sure that they will work, but it is important to keep in mind that the arrayer must not be damaged in the process of testing it. One common error that several individuals have made is to perform the initial testing of the *X*-axis limit switch with the slide platter off, and then to test it again with the platter on. It is quickly discovered under these circumstances that the heavy platter, moving at a high velocity, has sufficient momentum to carry itself past the limit switch and continue moving, albeit without continued electrical power, until the carriage slams loudly into the end of the track. *Do not do this.* It is not an essential test, because under normal circumstances, the arrayer should never be moving for a long distance in this direction so close to the end of the *X*-axis. It is understandably fun to move the carriages around with the Jogger program when one has finished wiring the motors; just avoid doing this in close proximity to the limits of travel.

The next step is the wiring of the motors and amplifiers, but first, wire together the emergency stop switch. This switch should be kept in plain sight at all times and close to the computer. Practice hitting it a few times to get a feel for it. When starting to power the motors, be prepared to hit this switch instantly if anything seems wrong. It is much better to power down and then reboot everything than to have the motors destroy one of the axes.

Again, the wiring directions are explicit. The only addition to the current version of the MGuide is that when wiring the X-stage encoder, wire the shield sink (the bare twisted strands) to position 1 (GND) in the second bank of terminals as shown in the MGuide diagram.

In wiring the relay box, note that because there no longer is a water pump with the new system of tip washing, only two receptacles are needed rather than three. However, there is no harm in wiring the additional receptacle, and perhaps it will be useful to have an additional source of computer-controlled electrical power to handle some future innovation in arrayer technology. By trimming a few millimeters of plastic from the tabs on the sides of the inlet and outlet plugs, it is possible to snap them tightly into place in the PacTec box without the use of epoxy.

### Final Assembly: Part II of the MGuide

This section deals with final details of assembly. Mounting the slide platter is straightforward, but there are a few details not in the MGuide. The carriage assemblies moved by the X and Y axes have a back and front section, which are screwed together. The back half engages the drive shaft and is partially hidden behind the aluminum cover, whereas the front half is exposed. The front half of the assembly on the X-axis must be removed, and is not used in the arrayer. The slide platter thus mounts directly to the mounting holes on the sides of the back half. Problems with a metallic noise emanating from resonance vibrations in the slide platter can be alleviated by cutting small washers from a piece of thin rubber and placing them between the platter and the carriage. It may also be helpful to mount the platter using only four screws placed in the outside holes, rather than all six.

The MGuide Part II section on the cleaning station was written in the context of the original design rather than the current version. There is no longer a need for plumbing for water delivery or disposal, and no need for carboys. We have found that a better vacuum is drawn for the drying station by using black pressure tubing (Fisher 14-175-E) rather than Tygon. Note that the placement of the sonicator has also been altered as a consequence of its use as the wash station. Rather than sitting at the left side of the platter as in the MGuide illustrations, it is now placed immediately to the right of the dry station, which is located as close as possible to the platter.

The final step is mounting of the Z-axis bracket and tip holder. Note that attachment of the bracket requires the M5 × 16-mm screws mentioned in the addendum to the parts list. At this point, it is necessary to align the dry station with the printing tips, as described. The tips cannot be moved in the X-axis; therefore, this must be done by manually moving the dry station and retightening the brackets holding it in place. Once the bracket and tip holder are attached, and particularly if tips have been inserted, be extremely cautious about sending commands for movement of the axes. The motors will try very hard to move the tip assembly through an object left carelessly in its path, with resulting damage to the tips, bracket, and object. *Think before each move.*

### Problems

It is likely that the arrayer will be fully operational when construction is completed and small problems, if any, can generally be overcome with common sense and a little ingenuity. However, a few problems traced back to factory errors may require a bit more effort to resolve.

The most serious issue involves the alignment of the drive shaft and the motor, particularly in the X-axis. The shaft and motor are connected by an aluminum coupler,

**Figure 27.1.** Example of a misaligned drive shaft.

shown in the MGuide figure depicting motor attachment. This coupler is designed to break if, e.g., the carriage moves into an immovable object. The coupler is also designed to flex if the motor and drive shaft are *very slightly* out of alignment; however, a significant misalignment will cause failure due to wear, and the coupler will have to be replaced. This can be prevented by ensuring that the shaft and motor are aligned, which should be, but is not always, the case upon delivery from the factory. The degree of alignment can be assessed by observing the spacing of the fins in the coupler as the shaft turns: Misalignment causes the fins to flex as the shaft turns; perfect alignment allows the fins to remain parallel to one another with minimal flexing during rotation.

In minor cases, alignment can be corrected by removing the motor from its mount, loosening the screws that hold the motor mounting block, and then aligning the block with the shaft. An alignment tool helpful in this process can be obtained from Western Technology if necessary.

Misalignment of the drive shaft is a more serious problem that has been encountered several times, although hopefully the company is now dealing with this at the factory. Figure 27.1 shows this misalignment when the motor and mounting block have been removed. Note that the first concentric ring of metal inside the hole in the main block is distinctly off-center. In such cases, it is necessary to follow a procedure available from the manufacturer described in a document entitled 500000ET Series Tables: Realignment of the Ball Screw Ref: SB014. The protocol and a special tool required can be obtained through Western Technology. The staff at this company is quite willing to assist in dealing with such problems, so do not hesitate to contact them if necessary.

## OPERATION OF THE ARRAYER

Explicit instructions for operation of the arrayer are included in the documentation files that can be downloaded along with the ArrayMaker software. Read these carefully before initiating a print run. Remember that updates to the software and instructions may appear from time to time on the MGuide Web page, so be sure to visit often. On these Web pages can also be found the MicroArray Forum, which is a community bulletin board for array-related issues. Investigators can post problems here and communicate with other users about solutions to these problems. Be sure to post new modifications or techniques so others can benefit.

**Table 27.1.** cDNA Plasmid Libraries Used in Microarray Construction

| Library designation | Vector | Source tissue | Number of clones sequenced (as of 4/99) |
|---|---|---|---|
| LD | LD01001-LD21096 = pBS SK(+/–); LD21101 and greater = pOT2a | 0–22 hour embryos | 33,978 |
| HL | HL01001-HL07796 = pBS SK(+/–); HL07801 and greater = pOT2a | adult head | 3,004 |
| GM | GM01001-GM11496 = pBS SK(+/–); GM12101 and greater = pOT2a | ovary | 6,290 |
| GH | pOT2a | adult head | 21,773 |
| LP | pOT2a | larvae, pupae | 8,935 |
| SD | pOT2a | Schneider cells (S2) | 6,962 |
| Total | | | 80,942 |

# PREPARATION OF ARRAYS

## PCR Product Preparation

All *Drosophila* microarrays produced to date have utilized cDNA expressed sequence tag (EST) clones produced by the Berkeley/HHMI *Drosophila* EST Project. cDNA clones were sequenced from libraries constructed by Ling Hong and are listed in Table 27.1. More information on their construction is available at the BDGP Web site:

> http://www.fruitfly.org/

A unigene cDNA set is under production by the EST Project and can be used as a standard set for *Drosophila* array production (G. Rubin, pers. comm.). For purposes of polymerase chain reaction (PCR) amplification of the cDNAs, it is important to note that these sequences were cloned into either the pOT2a vector or the pBS SK(+/–) vector. The current unigene clone set is arranged such that clones containing pOT2a are grouped together in the same 96-well plates, as are the pBS-containing clones. These plates will be available commercially through Research Genetics; see the WWW site at:

> http://www.resgen.com/

For availability and updates on the contents of the sets, check the BDGP EST Web site. To amplify these cDNAs, vector-specific primers (Table 27.2) are used with either the pBS vector or the pOT2a vector. The PCR amplification procedure is described in Protocol 27.1 below.

**Table 27.2.** Primers Used with the pBS Vector or the pOT2a Vector

| Vector | Primer name | Sequence |
|---|---|---|
| pOT2a | OTf | AATGCAGGTTAACCTGGCTTATCG |
| pOT2a | OTr | AACGCGGCTACAATTAATACATAACC |
| pBS | ESTf | GAACAGCTATGACCATGATTACGCC |
| pBS | ESTr | CGGCCAGTGAATTGTAATACGACTC |

## Polylysine Slide Preparation

A number of methods are in use to link DNA to the surface of glass microscope slides for the purpose of microarray hybridization experiments. These include the production of polylysine-coated and aldehyde-coated slides for the direct attachment of DNA to the surface, and epoxysilynated surfaces that can be used to tether DNA containing amino linkages at its termini. Several commercial companies are currently experimenting with producing ready-to-print microscope slides, so preparation of slides for printing may be simplified in the near future. Currently, the most common method for preparing slides is polylysine treatment. Note that this process can result in significant variability in the quality of the slides that are produced. Therefore, it is important to make several independent batches in advance and to test a few from each batch before printing tens or hundreds of microarrays. Testing should include printing a few slides from each batch using DNA from a test plate (this plate generally consists of clones representative of the material to be printed, i.e., PCR-amplified cDNAs). The entire procedure is then carried out on this test batch, including post-processing, hybridization, and scanning as described below. The quality of slides is assessed by determining the ratio of signal to noise for each batch. Protocol 27.2 describes the preparation of polylysine-coated slides.

Protocols 27.2 through 27.5 are versions of the procedures that we have used in preparing *Drosophila* arrays. Alternative versions of these protocols can be found at the MGuide protocols Web page, http://cmgm.stanford.edu/pbrown/protocols/index.html, and in the methods paper by Eisen and Brown (1999).

PROTOCOL 27.1*

## PCR Amplification

### Materials

### *Supplies and Equipment*

Plates (96-well) for PCR amplification (Robbins Scientific 1038-00-0). Products can be precipitated directly in the PCR plates, although it is important to test the plates for DNA pellet-adherence characteristics before amplifying and precipitating *en masse*. We have found Robbins Scientific plates (1038-00-0) to work satisfactorily. Alternatively, PCR products can be transferred to Corning Costar V-bottom tissue culture plates (Costar 3894), which are known to work well for this technique.

Thermal cycler

Replicator (96-pin), re-usable (VP Scientific VP 408) or disposable (long pin type from Genetix X5051/L)

(*Optional*) Mushroom brush (VP Scientific VP 425 or a local kitchen-goods store)

(*Optional*) Lid of a pipette-tip box

Equipment for agarose gel electrophoresis

Centrifuge (e.g., Sorvall RC-3B or Beckman) equipped with a rotor capable of holding 96-well plates (e.g., Sorvall HL-2B).

PCR plate bases (Perkin Elmer N801-0531) for supporting multiwell plates during centrifugation

---

*For all protocols, $H_2O$ indicates glass distilled and deionized.

### Solutions and Reagents

Forward and reverse primers (OTf, OTr, or ESTf, ESTr)

**MgCl₂** (25 mM)

10x PCR buffer (various sources)

dNTP mixture (25 mM each of dATP, dGTP, dTTP, dCTP)

*Taq* DNA polymerase (various sources)

*Pfu* DNA polymerase

Template DNA either bacterial liquid cultures containing the cDNA EST clones or plasmid DNA

(*Optional*) **HCl** (0.5 N)

**Ethanol** (70% and absolute)

Buffers and reagents for agarose gel electrophoresis

Sodium acetate (pH 5.2)

**Isopropanol**

3x SSC (filter-sterilized)

**CAUTION: ethanol, HCl, isopropanol, MgCl₂ (see Appendix 4)**

### Method

1.  Set up PCRs in 96-well plates (reaction volumes of 100 μl per well) as follows:

    | | |
    |---|---:|
    | forward and reverse primers (OTf, OTr or ESTf, ESTr) | 0.2 μg each |
    | 2 mM MgCl₂ | 8 μl of 25 mM |
    | 10x PCR buffer | 10 μl |
    | 0.25 mM each dNTP | 1 μl of 25 mM dNTP mixture |
    | *Taq* DNA polymerase | 1.5 units |
    | *Pfu* DNA polymerase | 0.025 units |
    | H₂O | to 100 μl (adjust for volume of template DNA, see step 2 below) |

2.  Add template DNA to the reactions using either of the following methods:

    a.  Transfer 1–2 μl of saturated bacterial liquid culture containing the cDNA EST clone from a master plate to each well in the PCR plate using a re-usable 96-pin replicator (VP Scientific VP 408) or disposable replicators (Genetix X5051/L). To eliminate cross-contamination between transfers when using the re-usable replicator, perform the following steps:

        i.  After each transfer from bacterial culture to PCR, wash the replicator in H₂O while scrubbing the pins with a mushroom brush. Use a lot of water (this is done in a sink with a H₂O line).

        ii.  Pat dry H₂O on paper towels and transfer the replicator to a container (e.g., the lid of a pipette-tip box) with 0.5 N HCl. Let soak for 1–2 minutes.

        iii.  Wash again with plenty of H₂O.

        iv.  Pat dry and transfer to a container with absolute ethanol. Soak for a few seconds, and then flame the replicator.

        v.  Proceed to transfer from the next plate of cultures using the sterilized replicator.

b. Use plasmid DNA as template (<1 ng per reaction is sufficient).

*Note:* The quality of spots printed with DNA amplified from pure DNA template is less variable than the quality of spots printed with DNA amplified from bacterial culture. This may be due to carry over of some bacterial contaminant(s). However, it is a significant amount of work to prepare plasmid from each clone. During the initial phases of *Drosophila* microarray development, we have generally opted to amplify directly from bacterial culture. When the standardized unigene set of cDNA clones is completed, we expect that the preferred method will be to amplify directly from purified plasmid DNA because preparation of even a small amount will allow for hundreds of amplifications.

3. Perform thermal cycling using the following program:

   35 cycles:
   94°C for 30 seconds
   64°C for 30 seconds
   72°C for 5 minutes

4. After amplification, verify success of the reactions by electrophoresis of a small aliquot of each reaction on an agarose gel.

5. Precipitate and concentrate PCR products as follows (a number of kits are available for purifying PCR products in 96-well format; however, we have exclusively used isopropanol precipitation):

   a. Add 1/10 volume of sodium acetate (pH 5.2) and 1 volume of isopropanol to each reaction. Incubate at –20°C overnight.

   b. Centrifuge in a Sorvall (RC-3B) or Beckman centrifuge at 3000 rpm (~2000$g$) at 4°C for 1 hour.

      *Note:* The centrifuge must be equipped with a rotor capable of holding 96-well plates (e.g., Sorvall HL-2B). If PCR plates are used for precipitation, be sure to use PCR-plate bases to support the wells during centrifugation (Perkin Elmer N801-0531).

   c. Remove the liquid by decanting or by turning the plate upside down. In either case, be careful not to lose the pellets!

   d. Wash with 70% ethanol. Centrifuge again at 2000$g$ at 4°C for 5 minutes.

   e. Remove the ethanol and allow the pellets to air dry with the plates upright.

      *Note:* Do not put plates upside down to dry as the pellets can slide down the sides of the wells. Generally, pellets are allowed to air dry for several hours to overnight.

   f. Resuspend pellets in 30 µl of filter-sterilized 3x SSC. Typically, we resuspend at room temperature for 2–3 hours before transferring to the print plates.

## PROTOCOL 27.1 NOTES

Alternative protocols for cleanup of PCR products are described and compared in the MGuide Web page http://cmgm.stanford.edu/pbrown/protocols/ASH/pcrcleanupcomparison.htm. If DNA will be printed from 96-well plates, Falcon U-bottom plates (62406-220) work well, although we have printed directly from the Costar tissue culture plates. The DNA solution does tend to sneak up the walls of these plates a bit, so we generally transfer to the Falcon plates. In either case, evaporation during the print run can be significant with 96-well plates. This results in subsequent printings being suboptimal, although one can try to estimate the amount of evaporation and add $H_2O$ to replace the lost volume.

Alternatively, the sample can be completely dried down, stored, and rehydrated before the next print. Evaporation is less of an issue in 384-well plates (surface to volume is generally lower), and smaller volumes can be aliquoted into each plate (4–5 µl per well) allowing for up to five sets of print plates. If DNA will be printed from 384-well plates, Costar 6502 plates (Fisher 07-200-734) are recommended. At all stages, effort should be made to minimize evaporation from the plates. Plates should be kept covered when not in use. A convenient covering is adhesive-backed, pressure-sensitive film (Falcon 3073), which is easily removed and replaced. Depending on frequency of use, plates should be stored at 4ºC or –80ºC. It can be expected that at least two print runs will be obtained from each set of plates. Therefore, one round of PCR is enough to produce more than 1000 arrays.

## PROTOCOL 27.2

# Preparation of Poly-L-lysine-coated Slides

### Materials

### *Supplies and Equipment*

Glass microscope slides (usually 120 slides are prepared per batch)
Metal slide racks, each holding 30 slides
Glass staining dishes (350 ml; Shandon Lipshaw 121)
Orbital shaker
Centrifuge equipped with microtiter plate carriers
Vacuum oven for drying slides (vacuum is optional)
Closed slide box for storing slides

### *Solutions and Reagents*

#### *Slide-cleaning Solution*

Dissolve 35 g of **NaOH** in 140 ml of $H_2O$. Add 210 ml of 95% **ethanol** and stir. If the solution remains cloudy, add $H_2O$ until clear.

#### *Poly-L-lysine Coating Solution*

560 ml of $H_2O$
70 ml of poly-L-lysine (Sigma P 8920)
70 ml of 1x PBS (see Appendix 3)

**CAUTION: ethanol, NaOH** (see Appendix 4)

### Method

1. Place 30 slides in each metal slide rack and then place each rack into a 350-ml staining dish.

2. Fill each dish with Slide-cleaning Solution. Place dishes on an orbital shaker and agitate for 2 hours.

3. Wash slides thoroughly in $H_2O$. Rinse once quickly by plunging racks up and down, and then rinse at least four more times (10 minutes each rinse).

4. Transfer the slides, now free from the NaOH-ethanol, into Poly-L-lysine-coating Solution. Agitate slides for 15 minutes to 1 hour.

5. Transfer racks to clean staining dishes filled with $H_2O$. Plunge up and down five times to rinse.

6. Centrifuge slides on microtiter plate carriers (place paper towels below racks to absorb liquid) at 500 rpm for 5 minutes.

7. Dry slide racks in a 45°C vacuum oven for 10 minutes.

8. Store slides in a closed slide box.


## TREATMENT OF MICROARRAYS PRIOR TO HYBRIDIZATION

Following preparation of the microarrays, slides are rehydrated and UV-cross-linked (Protocol 27.3). The UV cross-linking step is optional; however, UV cross-linking has been observed to increase the amount of bound DNA to the slide and thus the signal intensity after hybridization (J. DeRisi, unpubl.).

Microarrays are then treated with succinic anhydride, which blocks the polylysine by acylation (Protocol 27.4). The result of this reaction is a negatively charged moiety that acts to reduce nonspecific binding of labeled DNA probes to the surface of the array during the hybridization step. This blocking step should be done in a chemical fume hood. 1,2-methyl-pyrrolidinone is volatile and toxic, and succinic anhydride is potentially carcinogenic, so be sure to take proper precautions and to dispose of the waste appropriately.


PROTOCOL 27.3

# Rehydration and Cross-linking

### Materials

### *Supplies and Equipment*

Styrene humidification chamber (Sigma H 6644)
Heating block set at 80°C
Saran wrap
(*Optional*) UV cross-linker (Stratagene UV Stratalinker 2400)
Metal racks


### *Solutions and Reagents*

4x SSC (see Appendix 3)

### Method

### *Rehydration*

1. Pour 100 ml of hot (50–60°C) 4x SSC into a styrene humidification chamber.

2. Place the microarray, DNA-side down in the chamber; there is no need to cover the chamber with the lid.

*Note:* When the spots become hydrated, they take on a glistening character. The time required during rehydration varies according to the size and density of the spots and the temperature of the 4x SSC; 5–10 minutes is typical for high-density arrays. Overhydration causes the spots to bleed together. Underhydration affects the shape of the spots and the quality of the subsequent hybridization.

3. As soon as all of the spots are glistening, remove the slide from the humidification chamber and snap-dry it face-up on a 80ºC heating block. This should only take a few seconds. A wave of evaporation flash across the surface should be seen. As soon as all the spots are dry, remove the microarray. Do not leave the array on the heating block for more than 6–8 seconds. Several arrays can be rehydrated simultaneously.

### Cross-linking

1. Place the snap-dried microarrays face-up on Saran wrap in a UV cross-linker.

2. Cross-link using 60 mJoules of energy.

3. Place arrays in a metal rack.

PROTOCOL 27.4

# Blocking with Succinic Anhydride

## Materials

### Supplies and Equipment

Glass beakers (500 ml and 4 liters)
Heating plate
Glass staining dishes (see p. 499)
Platform shaker
Stirrer and stir bar
Glass boat
Beckman table-top centrifuge
Plastic slide boxes (Applied Scientific AS-6051 and AS-6052) for storing slides

### Solutions and Reagents

**Succinic anhydride** (Aldrich 23,969-0)
**1-Methyl-2-pyrrolidinone** (Sigma M 6762)
Sodium borate (1 M, pH 8.0)
**Ethanol** (95%) (Note: do not use diluted absolute ethanol, as the benzene used to dehydrate this ethanol is fluorescent.)

**CAUTION: ethanol, 1-methyl-2-pyrrolidinone, succinic anhydride** (see Appendix 4)

## Method

1. Fill a 4-liter glass beaker with approximately 1.0 liter of $H_2O$ (enough to cover the slides), and place it on a heating plate to boil.

2. Measure 6.0 g of succinic anhydride into a 500-ml glass beaker with a stir bar.

3. Add 325 ml of 1-methyl-2-pyrrolidinone to the succinic anhydride and stir until completely dissolved.

   *Note:* The 1-methyl-2-pyrrolidinone should be clear when it is added, and the solution should assume a slight yellow hue once the succinic anhydride is dissolved.

4. Immediately add 25 ml of 1 M sodium borate (pH 8.0), and mix for 15–20 seconds. Act rapidly since the reactants of this solution are unstable.

5. Pour the resulting solution into a glass tray situated on a platform shaker.

6. Rapidly submerge the metal rack with the array slides into the solution.

   *Note:* It is very important to aggressively dunk the arrays up and down for a few seconds when they are first exposed to the succinic anhydride solution. This seems to eliminate diffusion and binding of the DNA away from the original spot (manifested as comet tails that emanate from the spots after the subsequent hybridization step; see Protocol 27.5).

7. Incubate the slides in the succinic anhydride solution for approximately 15 minutes. Shortly before the 15 minutes expires, turn off the heating plate with the boiling water.

8. When the bubbling of the heated water ceases, remove the rack of microarrays from the succinic anhydride solution and plunge it into the hot water. Incubate for 2 minutes. This step rinses the blocking solution from the arrays, and denatures the DNA.

9. Transfer the arrays into a glass boat containing 95% ethanol (see step 6 note above). Rinse for 30 seconds, and then centrifuge the arrays dry in a Beckman table-top centrifuge at 620 rpm for 5 minutes.

   *Note:* The slides are now ready for hybridization. These ready-to-hybridize microarrays should be stored in plastic slide boxes at room temperature. We have obtained satisfactory results with slides stored for up to 3 months.

## PROBE LABELING AND HYBRIDIZATION

Most experiments that we have performed involve isolating poly(A)$^+$ RNA from whole animals and directly incorporating fluorescently linked dUTP into cDNAs. There has also been some success isolating RNA from individual tissues and amplifying it using the protocol in Eberwine (1996). Protocol 27.5 describes the procedure that we use for directly labeling cDNA and hybridizing it to microarrays.

PROTOCOL 27.5

# Direct Labeling of cDNA Probes and Hybridization to Microarrays

### Materials

### *Supplies and Equipment*

Microcentrifuge tubes (1.5 ml)
Microcon-30 microconcentration tube
Microcentrifuge

Millipore UltraFree-MC columns
Heating block set at 100°C (or boiling water bath)
Hybridization chamber (custom made by Die-tech, specifications in MGuide)
Coverslips (22 x 22 mm)
Dissection forceps
Water bath set at 65°C
Metal racks
Beckman table-top centrifuge

## *Solutions and Reagents*

Poly(A)$^+$ RNA
Random hexamers
5x First-strand buffer for Superscript II
Dithiothreitol (**DTT**) (0.1 M)
Fluorolink Cy3 or Cy5 dUTP (Amersham)
dNTP mixture (25 mM each dATP, dCTP, dGTP; 10 mM dTTP)
Superscript II RNase H reverse transcriptase (GIBCO-BRL)
TE (pH 8.0; see Appendix 3)
Poly(A) DNA (10 mg/ml; Sigma P 9403)
20x SSC (see Appendix 3)
3x SSC
**SDS** (10%)

### *NaOH-EDTA Solution*
1 M **NaOH**
20 mM EDTA

### *Wash Solution 1*
330 ml of $H_2O$
20 ml of 20x SSC
1 ml of 10% SDS

### *Wash Solution 2*
350 ml of $H_2O$
1 ml of 20x SSC

**CAUTION: DTT, NaOH, SDS** (see Appendix 4)

## Method

### *Probe Labeling*

1. In each of two 1.5-ml microcentrifuge tubes placed on ice, set up the following annealing reaction:

   | | |
   |---|---|
   | poly(A)$^+$ RNA | 2–5 μg |
   | random hexamers | 3 μg |
   | $H_2O$ | to 15 μl |

2. Heat at 70°C for 5 minutes, and then return the tube to ice. Set up the following labeling reactions:

| | Reaction 1 | Reaction 2 |
|---|---|---|
| 5x First-strand buffer | 6 µl | 6 µl |
| 0.1 M DTT | 3 µl | 3 µl |
| Fluorolink Cy3-dUTP | 3 µl | – |
| Fluorolink Cy5-dUTP | – | 3 µl |
| dNTP mix (25 mM dATP, dCTP, dGTP; 10 mM dTTP) | 0.6 µl | 0.6 µl |
| Superscript II RNase H reverse transcriptase | 2 µl | 2 µl |

3. Incubate at 42°C for 60 minutes.

4. (*Optional*) Add 1 µl of Superscript II RNase H reverse transcriptase to each reaction and incubate at 42°C for an additional 60 minutes.

5. Base hydrolyze the RNA from the RNA:cDNA complex by adding 1.5 µl of NaOH-EDTA solution to each reaction.

6. Incubate at 65°C for 10 minutes.

7. Concentrate the labeled DNA as follows:

   a. Add 470 µl of TE and immediately load onto a Microcon-30 microconcentration tube. Add 2 µl of 10 mg/ml poly(A) DNA.

   b. Centrifuge in a microcentrifuge at 10,000 rpm until nearly all of the liquid has flowed through (~4–8 minutes).

   c. Add 470 µl of TE to the upper reservoir. Discard the flow-through from step b and centrifuge again at 10,000 rpm until nearly all of the liquid has flowed through.

   d. Repeat step c until the flow-through is free of color (usually once).

   > *Note:* After the final centrifugation step, there should be approximately 4 µl of colorful liquid in the upper reservoir. This is the labeled cDNA.

   e. Combine the labeled DNA from both the Cy3- and Cy5-labeled samples (final volume ~8 µl). To recover the labeled DNA, place the upper reservoir upside down in a new collection tube and centrifuge at 10,000 rpm for a few seconds.

   f. Add 2 µl of filtered 20x SSC to this sample to bring the total volume to 10 µl.

8. Prewet a Millipore UltraFree-MC with 5 µl of $H_2O$. Centrifuge at 10,000 rpm for 1 minute, and then remove any liquid from the bottom of the tube. Pipette the probe mix onto the side of the UltraFree chamber. Centrifuge the droplet through the column and collect the liquid (~10 µl).

## Hybridization

1. Immediately before setting up the hybridization to the array, add 0.3 µl of 10% SDS to the labeled probe. Do not refrigerate the probe after the SDS has been added.

2. Heat the probe to 100°C for 2 minutes, and then let stand at room temperature for 2–5 minutes.

3. Place a microarray slide in a hybridization chamber (Die-tech). Pipette 10 μl of 3x SSC on a corner of the slide on which the array is printed, away from the printed area of the microarray (this is to provide humidification). Pipette the probe directly onto the array approximately 5 mm below the top edge of the area where the microarrayed DNA is printed. Carefully place a dust-free coverslip (22 x 22 mm) onto the array by seizing the coverslip with a pair of dissection forceps and delicately easing it onto the array at an acute angle.

4. Seal the hybridization chamber and place it in a 65°C water bath for 6–18 hours. Longer hybridization times sometimes negatively affect the signal-to-noise ratio.

### Washing

1. Remove the arrays from the 65°C water bath and place them in a metal rack.

2. Submerge the arrays in Wash Solution 1 until the coverslips fall off. The arrays may have to be dunked up and down a few times, but do not be too aggressive as the coverslips could scratch the surface of the arrays.

3. Transfer the arrays to Wash Solution 2 for 30–60 seconds.

4. Centrifuge in a Beckman table-top centrifuge at 620 rpm for 5 minutes to dry. The slides are now ready for imaging.

## IMAGING

A number of commercial scanners are now available that produce high-quality images; therefore, we will not go into detail regarding scanning technology. Figure 27.2 depicts one quadrant from an array after hybridization with probes prepared from 0-hour white prepupal mRNA (displayed as green signal) and 12-hour prepupal mRNA (red signal). All scanners basically consist of dual lasers that excite the Cy3 and Cy5 dyes, and optics capable of producing confocal images of the surface of the slide. The commercial scanner with which we are most familiar is the GenePix 4000 from Axon Instruments, which produces high-quality images, includes an excellent software package, and is competitively priced; see the WWW site at:

www.axon.com/GN_Genomics.html

Alternative scanners are available that have been used successfully in other laboratories; before making a purchase, visit relevant Web sites and contact laboratories that have experience with specific products before making a purchase. An important consideration is the cost and quality of image analysis software available from each vendor, which varies widely. An excellent program for analysis of scanned images, written by Michael Eisen, is available free of charge to academics at:

http://rana.Stanford.EDU/software/

## FUTURE DIRECTIONS

We hope that the instructions and protocols provided here will encourage other investigators to begin analyzing transcription in *Drosophila* on a genomic scale. In the coming

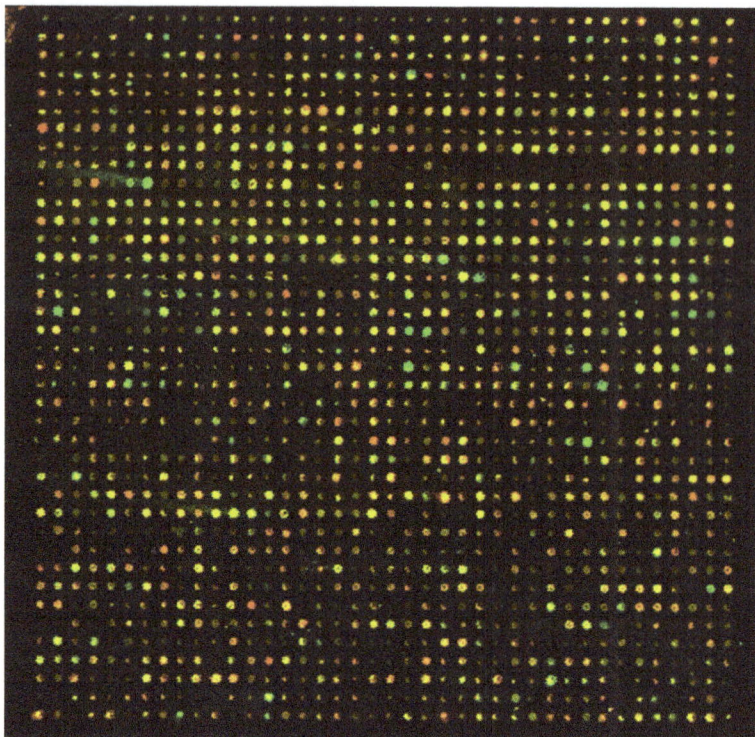

**Figure 27.2.** Hybridization of Cy3- and Cy5-dUTP-labeled probe mixtures to a microarray. An array with 8448 elements is shown following hybridization with probes prepared from mid-third larval instar mRNA (*green signal*) and 0-hour white prepupal mRNA (*red signal*). (Courtesy of Kevin White.)

years, we expect that a large amount of data will be produced describing the transcriptional behavior of the *Drosophila* genome in response to a wide variety of experimental situations. These types of data will allow investigators to group genes based on their regulatory profiles and eventually could provide a picture of the operations of most genetic regulatory networks in this organism.

It will be important to have a standardized way of producing and dealing with the data generated. Only a fraction of these data may be of immediate relevance to the investigator conducting the experiment. However, data accumulated in multiple array experiments may be of value to others in the fly community and should be stored in a manner that is accessible and useful. In the coming years, it will be a challenge for FlyBase to develop approaches to make this possible, and the community should participate by forwarding suggestions on this issue to the following E-mail address:

flybase-help@morgan.harvard.edu

Production of microarrays is an easily scaleable process, and we fully expect that sufficient *Drosophila* arrays will be produced to satisfy the demands of the community (either by academic investigators working together or by companies). It is also important to note that robotically printed DNA microarrays are not the only platform that will be available for analyzing gene expression on a genomic scale. Other technologies not yet available for *Drosophila* include photolithography-based oligonucleotide arrays and ink-jet type arrays. Because the data produced from any one of these platforms might have unique problems, for many experiments, it may be desirable to obtain gene expression information using different methods and to compare the results. The limiting factors at present are cost and availability, and robotically printed arrays are currently the best option for *Drosophila* research.

We expect that there may be a natural inclination for those initiating arrayer construction or array experiments to contact the authors of this article with further questions or problems. The section on arrayer construction was written by K.B. and that on preparation of arrays by K.W. Readers are encouraged to communicate with the relevant author (our addresses are listed in the FlyBase database) if they have further questions pertinent to these topics. Up-to-date protocols that we are following while working with *Drosophila* arrays are posted at K.W.'s Web Site:

http://cmgm.stanford.edu/~kpwhite

We assume that members of the *Drosophila* community contacting us will likewise be willing to help other members of the community who are in need of assistance in the future.

## REFERENCES

Eisen M.B. and Brown P.O. 1999. DNA arrays for analysis of gene expression. *Methods Enzymol.* **303:** 179–205.

Eberwine J. 1996. Amplification of mRNA populations using RNA generated from immobilized oligo(dT)-T7 primed cDNA. *Biotechniques* **20:** 584–591.

# CONTENTS

# 28

# Using *Drosophila* Genome Databases

**Sima Misra**
*University of California at Berkeley*
*Department of Molecular and Cell Biology*
*Life Sciences Addition*
*Berkeley, California 94720-3200*

**Madeline A. Crosby**
*Harvard University*
*Biological Laboratories*
*Cambridge, Massachusetts 02138*

**Rachel A. Drysdale**
*University of Cambridge*
*Department of Genetics*
*Cambridge CB2 3EH, United Kingdom*

WITH THE EXPLOSION OF GENETIC AND GENOMIC INFORMATION about the fly, comprehensive, integrated, and current databases are now crucial to *Drosophila* research. This chapter focuses on FlyBase, the core Internet resource of this kind. FlyBase includes information about the genome and genetics of *Drosophila* taken from genome projects and research publications, and practical information such as stock center catalogs, researcher contact information, and news. The data distribution efforts of the two major publically funded *Drosophila* genome projects, the Berkeley *Drosophila* Genome Project (BDGP) and the European *Drosophila* Genome Project (EDGP), have merged with FlyBase, increasing the amount of up-to-date information in FlyBase on genomic and cDNA clones and their sequences, annotation, and mapping information. Ultimately, FlyBase, BDGP, and EDGP data distribution will be integrated on the same public server (FlyBase). Until then, the genome centers will continue to maintain separate databases and Web sites; therefore, these databases (chiefly, the Berkeley Fly Database, or BFD) are described here as well. More specialized *Drosophila* databases and resources available via the World Wide Web (WWW) are described briefly at the end of the chapter.

By nature, databases and their query interfaces change constantly. Although the particulars of a display may change periodically, the underlying data structure in FlyBase is

**Table 28.1** Addresses on the WWW of the Major *Drosophila* Genome Databases

| Database | URL[a] |
|---|---|
| FlyBase Primary Server[b] | http://flybase.bio.indiana.edu/ |
| BDGP | http://www.fruitfly.org/ |
| EDGP | http://edgp.ebi.ac.uk |

[a]Uniform Resource Locator.

[b]For FlyBase, several international mirrors are available; their URLs can be found at the primary server (see "FlyBase mirrors" listed on the sidebar). FlyBase is also available via FTP (File Transfer Protocol) or Gopher (flybase.bio.indiana.edu).

rigorously organized and maintained, so that data classes themselves change very little. We focus on descriptions of basic query systems and the underlying data structure that are unlikely to become out of date. *We strongly suggest that this chapter be read at the computer, interacting with the various Web pages.*

## MATERIALS

The most efficient way to access FlyBase, BDGP, and EDGP is via the WWW, which requires a computer, a connection to the Internet, and browser software, such as Netscape or Internet Explorer. The addresses (or uniform resource locators, URLs) on the Web to access the databases are shown in Table 28.1.

A number of the interactive displays being developed by FlyBase-BDGP are Java programs, and thus more recent versions of browsers (versions 4.5 or better) are necessary to use them. Netscape Communicator 4.61 works well for these displays using computers with Windows (PCs), Macintosh OS7, or Unix operating systems, whereas Internet Explorer 4.5 with Apple's MRJ2.1 works well for Macintosh computers running MacOS8 or higher.

## WEB-SITE ORGANIZATION: HOW TO GET WHERE YOU WANT TO GO

### General Organization

FlyBase is the best starting point; *call up the FlyBase home page on the Web before reading further.* FlyBase contains the most comprehensive information on the following categories of data (*data classes*), which are described in further detail later in this chapter:

- genes and alleles
- molecular data and maps from the literature
- transposons, vectors, and transposon insertions
- aberrations and balancers
- fly stocks
- bibliographic references
- people (*Drosophila* researchers)

On the other hand, the BDGP (which can be accessed directly from the FlyBase home page) is the best source for up to date information on molecular data from the genome centers, including:

- genomic DNA clones
- genome maps (e.g., physical maps and molecular maps of genomic clones)

- cDNAs from the expressed sequence tag (EST) projects
- P-element insertions from the BDGP gene disruption and controlled misexpression projects

Finally, annotations of cosmid genomic clones on the *X* chromosome are best accessed from the EDGP (which can also be accessed directly from the FlyBase home page).

## Home Pages

The FlyBase home page lists the data classes, with links to pages containing information about each class and to different search forms for querying that particular class. In addition, a Quick Search tool on the home page allows users to immediately pose simple queries without going any further. Links to documentation and other general information are provided in the margin.

The BDGP home page has the major categories of information organized in a left side bar. One section has links to the three major BDGP efforts: the Genomic Sequencing Project, the BDGP/HHMI EST and Full-length cDNA Projects, and the P-element Disruption and Misexpression Projects. Another section provides links to the *D. melanogaster* BLAST server, other querying tools, and sequence analysis tools. The last section links to general information: obtaining clones, stocks, sequence databases, publications, and protocols.

Comprehensive listings of FlyBase and BDGP search tools are provided under the "All Searches" or "Searches" sections, accessed from the side bar or the top bar on FlyBase pages and from the side bar on the BDGP home page.

## Detailed Documentation

The FlyBase home page includes a link to the Reference Manual. This manual describes the searches, content, and format for each data section, with an explanation of the cross-links within FlyBase and between FlyBase and external databases. The Reference Manual is extensively hyperlinked within itself, to the files it describes, and to external resources. Each FlyBase data class provides a link, under "Documents," back to the relevant section of the reference manual.

The BDGP home page includes links to each BDGP project. Notes within each section provide experimental details and answers to Frequently Asked Questions (FAQs).

## Errors, Questions, and Comments

If all else fails, help is provided via E-mail (Table 28.2). For FlyBase, a separate address is provided for updates or corrections to data.

**Table 28.2** Contact Addresses

| Database | Concerning | Address |
|---|---|---|
| FlyBase | corrections and updates | flybase-updates@morgan.harvard.edu |
| FlyBase | questions and problems | flybase-help@morgan.harvard.edu |
| BDGP | general questions | bdgp@fruitfly.berkeley.edu |
| BDGP | EST/Full-length cDNA Project | EST@fruitfly.berkeley.edu |
| BDGP | Genomic Sequencing Project | drosophila@mhgc.lbl.gov |
| EDGP | questions and comments | edgp@ebi.ac.uk |

## QUERIES: HOW TO FIND AND USE THE APPROPRIATE SEARCH TOOL

### Text Searches by Data Class

#### General Principles

If the symbol, name, or specific characteristics of a data item (e.g., a gene, stock, EST) are already known, a *text* query is appropriate. Text queries can be simple or complex. In a *simple* query, words or numbers (a string of text) are entered, and many *fields* (repositories of information in the database) are searched. In a *complex* query, only certain fields or combinations of fields are searched. In addition, the words typed into a query form may be controlled or free text. For *free text*, any word (or keyword) can be typed and will be searched for. For fields using *controlled* terms, only certain words can be used, terms that are *identifiers* (such as symbol, name, line number, and identification number) or terms that are part of a controlled vocabulary.

A *controlled vocabulary term* is a previously determined and defined term that is used to describe the data type when it is originally curated in the database. The use of such precisely controlled terms allows more powerful and efficient queries for descriptive characteristics, such as phenotype and expression pattern. This can be helpful if the user has a general interest in a process, time in development, or tissue, but does not know the names of the genes involved, for example. Controlled terms can also be used to define different subsets or categories of a specific data class, e.g., dominant alleles of a gene, constructs useful as cloning vectors, or balancers carrying a reporter gene.

Another helpful trick is the use of a *wild card*, usually the "*" symbol, in place of text in a query. The asterisk stands for any letter or string of letters in a word, at the beginning, middle, or end, and is useful in text searches when users only know part of a name or want to obtain a list of related data types. (An example might be all of the genes on the second chromosome that have a gene symbol beginning with "ms" (male sterile); in which case, the wild-card query string "ms(2)*" is used in the symbol field in a simple search of genes in FlyBase.) Users must take care when searching descriptive material that the entry into a free text search does not correspond to a commonly used word. Plurals and other derivative forms of a term may be missed in a free text search, but this can be avoided by using a wild card at the end of a partial term.

Multiple terms can be entered in certain text queries. The default is "AND," which means that both words must be found associated with the report (e.g., learning and memory). However, users can also type "OR" between terms, which means that any of the words will be looked for independently (e.g., learning or memory) or "BUT NOT" (e.g., learning but not memory) to qualify the search.

Most FlyBase queries allow further refinements of a search, after an initial "hit list" is returned. Scrolling to the bottom of the page, the user will find a new query window, which can be set to search a specific field. The user can specify whether to use the AND, OR, or BUT NOT Boolean operators.

#### FlyBase

FlyBase text searches are accessible from many different entry points: the home page itself, the "All Searches" page, or the relevant data class page. In most cases, both simple and complex queries are available on the data class pages. FlyBase sections with searchable text include Genes, Aberrations, Clones, People, Stocks, Transposons, FlyBase documentation, News archives, and References (these data classes are described below, Scope of FlyBase Data Types).

On the FlyBase home page, the Quick Search facility uses a scrolling menu to define the data class(es) to be searched. "Symbol field" can be used for valid symbols or synonyms, and "All text" can be used to enter free text such as keywords and synonyms.

FlyBase makes extensive use of a controlled vocabulary of terms and qualifiers, to allow users to search the database more easily and retrieve data sets with particular properties. The terms are stored in a file, *controlled-vocabularies.txt*, in the Documents section of FlyBase. Because the term must match the controlled vocabulary entry exactly, the easiest way to perform these searches is by using the pull-down menus in FlyBase complex query forms, which provide all possible search terms for a field.

Most FlyBase search forms include hyperlinks to "Help" retrieve documents. For a full description of the searches available on FlyBase, see Reference Manual C, Using FlyBase on the Web under section C.3: FlyBase Search Tools.

### BDGP

Berkeley Fly Database (BFD) queries are accessible through the "Search Berkeley Fly Database" item on the BDGP home page and can be either text-based or graphical. The data types that can be queried in BFD include BDGP/EDGP data such as bacteriophage P1, BAC (bacterial artificial chromosome), cosmid, or YAC (yeast artificial chromosome) genomic clones, P-element insertion lines, STSs (sequence tagged sites), and BDGP and EDGP physical contigs (contiguous regions). All of these data classes are described below (Scope of FlyBase Data Types).

Data types can be queried by name in a field equivalent to the FlyBase symbol search; however, searches are case-sensitive and multiple terms cannot be entered. Queries can also be performed from this page by cytological location, for data that were explicitly localized by in situ hybridization. This means that although a number of bacteriophage P1 clones may map to a physical contig in a particular cytological region, via their overlaps with STSs, only those bacteriophage P1 clones that were themselves localized by in situ hybridization will be retrieved. Queries can also be qualified, e.g., to include only those clones or P-insertions for which sequence is available.

The result of a query will be a table of summarized results, including name, cytological location, accession number, or Bloomington stock number; each name hyperlinks to a full report (described below for each data type under Scope of FlyBase Data Types). The results can be saved as a text file for import into any common spreadsheet software that can interpret tab-delimited files (such as Excel or Filemaker), or as a FASTA formatted file of the sequences for sequence analysis.

EST queries can be accessed from the EST project pages (see ESTs and Full-length cDNA Clones below). "Search by identifier" allows users to query by identifier number or GenBank or dbEST accession number for any of the cDNA libraries. "Search by homology" allows users to search through the BLAST reports for sequence similarities of the consensus sequence of a group of ESTs to *Drosophila* and cross-species databases by keyword (e.g., by gene name).

## Searches by Cytological Location

### General Principles

If users are interested in a particular region of the genome, a *cytological location* query is appropriate. This can be done using a text query form, where the user specifies the cytological bands of interest , or with a *map viewer*, which allows users to browse genome maps interactively and query for data classes in particular regions. Java map viewers and queries are more interactive and highly visual, but they are still relatively slow (due to limitations of network speed and user machine memory). On the other hand, text-based displays are static, but faster, and include larger amounts of textual descriptive data. These considerations should influence the user's choice of query tools.

### FlyBase

The FlyBase CytoSearch query page can be used to create customized cytogenetic maps of any portion of the genome for any subset of genetic entities that have been localized on polytene chromosomes (e.g., Genes, Transposon insertions, and Aberrations). After checking off data types, the user enters the polytene interval of interest and can also add further refinements to the search with optional parameters explained on the CytoSearch help page. The resulting lists are hyperlinked to FlyBase data sets describing these genetic entities and to the stock centers.

The "Cytological maps" feature within the Maps section of FlyBase allows users to view different classes of genetic entities based on their determined or predicted cytogenetic map location. These classes of data include genes and various types of aberrations. After selecting cytological division and data class, the user can click on the symbol of an entity to open a hyperlinked text file describing that entity.

### BDGP

Queries for data types in BFD that have been localized cytologically can be performed using the text-based query tool described above. Such queries can also be run using the graphical query tools linked to the "Map Viewers" item on the BDGP home page, which allow interactive browsing of the genome at progressively higher resolution.

These Java map viewers can show data (including the physical map) for the entire *Drosophila* genome (ChromoView) or a chromosome arm (ArmView), allowing users to zoom in on a particular cytological band and to generate a table of results or link to CytoView, the next level of map viewer. CytoView diagrams how genomic clones are linked by STSs to form physical contigs, and links to text reports about the clones and STSs.

The best browser/computer platform combinations are described above in Materials. Images can be captured using graphics programs that allow users to take the image on the computer screen and save it as a file for printing.

## Searches by Sequence Similarity: BLAST and Pattern Search

If a user has a nucleotide or amino acid sequence in hand, the easiest way to identify *Drosophila* data classes (e.g., genomic, cDNA, and P-element insertion sites) with sequences identical or similar to it is with the *Drosophila* BLAST servers. FlyBase offers links to the BDGP and EDGP BLAST servers to search *D. melanogaster* sequence databases of different kinds using the BLAST2.0a suite of programs (W. Gish, unpubl.).

A BLAST help page describing the parameters of the program and the different BLAST programs links from the main BLAST query page. To use the BLAST query page, the user selects the BLAST program to be used from a pull-down menu. For nucleotide queries, BLASTN, BLASTX, or TBLASTX should be used; for amino acid queries, BLASTP or TBLASTN should be used. The user also selects from the various *D. melanogaster* sequence data sets (e.g., ESTs, genomic sequence, P-element insertion sites, and STSs). The default is "All *Drosophila*," which includes all of the other nucleotide databases except transposons and repeats. The user may then paste the query sequence in raw or FASTA format or upload a sequence file from their computer.

The results of a BLAST search will be a list of top matches to entries in the BFD, the EST database, or GenBank, and links to the alignment of the sequence at the score. If no results were found, the user has the option of putting the search into the "Request Service" where it will be repeated weekly with each update of the databases. When a match is found, the service automatically informs the user by E-mail.

To find exact matches of a short sequence pattern in the *Drosophila* databases, or in any of the GenBank databases from other species, the BDGP offers the Pattern Search tool under "Sequence analysis tools" on the home page. The user enters a nucleotide or amino acid query sequence (which can include N or X wild-card characters) and selects from a list of sequence data sets to search through. The result is a list with all database entries containing an exact match to the sequence pattern and the coordinates of the match.

## Searches by Body Part or Gene Expression Pattern

The FlyBase Gene Expression Pattern Search, an option on the Gene search page, provides information on the spatial and temporal distributions of transcripts and proteins from genomic and reporter genes, using the controlled set of hierarchical anatomical terms listed in controlled-vocabularies.txt (see Text Searches by Data Class, FlyBase above). The "Descend" option allows the user to move deeper into the hierarchy from the top level displayed on the Expression Pattern Search page. By clicking on "Search," the user receives tabular listings of proteins and transcripts from genes (and reporter constructs, e.g., *lacZ* and *GAL4*) expressed in any body part that is represented below that node in the hierarchy. This search tool, and a similar tool for queries based on phenotype (below), is being revised to utilize an image-based query interface.

The FlyBase Allele Search allows searching on the basis of anatomical structure affected by mutation. The Allele Query Form allows users to set the field to be searched to "Body Part." Query terms entered should conform to those listed in the controlled-vocabularies.txt file. The results of such a search is a list of alleles recorded as affecting the body part in question.

For the BDGP CK EST project (see ESTs and Full-length cDNA Clones below), "Search by expression pattern" allows users to select from a list of tissues and body parts to query for cDNAs expressed at particular times and places in the embryo, as assayed by RNA in situ hybridization.

## Tables of Data

Certain lists of information are available for browsing; these tables usually contain large amounts of data for commonly asked, large-scale queries. For example, the FlyBase genes directory includes the "Gene function and structure folder" containing the files function.html and structure.html. These files list the relevant terms in alphabetical order, linked to each gene whose product encodes that function or has that structure. The FlyBase Insertions page provides access to a listing of all transposon insertions by cytological location. The EST project page links to an "EST-Genes list" that allows users to browse all of the *Drosophila* ESTs that match known *Drosophila* genes, with appropriate hyperlinks. The BDGP genomic sequencing project page links to table of "Sequenced Genomic Clones," "Physical Contigs Sorted by Map Region," and "Physical Contigs Sorted by Size," to allow users to view all of the genomic sequence organized by clones or as larger sequence contigs.

## SCOPE OF FLYBASE DATA TYPES: WHAT YOU WILL FIND WHEN YOU GET THERE

The data in FlyBase concern the family Drosophilidae, including species other than *melanogaster*. This contrasts with the data from the BDGP and EDGP, which are exclusively from *D. melanogaster*, although limited genomic sequencing of *D. virilis* is planned.

The data in FlyBase are organized into several data classes. Within each data class, objects (such as genes) are uniquely identified by symbol and identifier number, and attributes are associated with each object. These attributes may be other named objects (such as an allele of a gene) or descriptions (such as phenotypic information for an allele). The identifiers for named objects are tightly controlled; they have the format FBxxnnnnnnn, where "xx" is a two-letter description of the data class and "nnnnnnn" is a unique integer identifier. The identifiers are linked to related objects, with the consequence that the different data sections of FlyBase are densely cross-referenced. These relationships are evident as the high degree of hyperlinking between items in FlyBase reports.

The symbol used for any given data object is unique and can be used with all of the text-based search options. Synonyms (invalid symbols used in the literature or FlyBase symbols that have become obsolete) can be used with the complex query options. FlyBase also maintains identifier numbers that have become obsolete as secondary identifiers; these can be used in free-text search fields.

Reports encompass a wide diversity of data and can be very large. For this reason, FlyBase offers the user alternative formats (listed after "Report Content") for displaying reports. These alternative subsets of the data all draw on the same "Full" report. For more complete descriptions of the data sections described below, see FlyBase on-line Reference Manual B: Detailed Descriptions of FlyBase Structure and Data.

## Gene and Allele Reports

### Genes

The Genes section of FlyBase includes information on *Drosophila* genes that have been defined genetically and/or molecularly and includes links to separate records for each allele of the gene. Each gene has a valid symbol, by which it is referred in all instances that it appears in FlyBase.

In addition to *Drosophila* chromosomal and mitochondrial genes, FlyBase includes other categories of genetic entities in the Genes section, such as natural transposons, transposon-borne genes including fusion genes and foreign genes, and chromosomal structural elements. For a full listing and explanation of the gene categories, see FlyBase Reference Manual B.1.3: Detailed Description of the Genes Fields.

FlyBase records and reports gene map data generated by chromosome in situ hybridization, complementation testing against defined chromosomal aberrations, and molecular mapping (with respect to other genes or to clones generated by the BDGP and EDGP).

### Alleles

Phenotypic information is captured in allele reports in several controlled vocabulary fields, as well as in a free-text format. The controlled fields include phenotypic class and anatomical structure(s) affected. Additional data captured for alleles include mutagen, molecular data, and, when appropriate, associated chromosomal rearrangement or associated transposon insertion.

In addition to alleles in the traditional sense, FlyBase names and curates further classes of alleles so that phenotypic/expression pattern data can be captured for in vitro construct alleles of both traditional and nontraditional genes, including alleles of reporter (e.g., *lacZ*), effector (e.g., *FLP1*), and toxin (e.g., *DT-A*) genes.

## Molecular Data in Gene Reports and Molecular Maps

### Proteins and Transcripts

This section provides data on specific transcripts and proteins, including which protein is encoded by which transcript, transcript exon structure, protein domains, and subcellular localization. Expression pattern data (including Northern and Western blot data as well as in situ hybridization and antibody staining data) are presented both in a tabulated form using a controlled vocabulary format and in a detailed free-text format. An image-based expression pattern search tool, which can access data captured in the controlled vocabulary format, is being developed.

### Sequence Data and Hyperlinks to Related Genes

DNA/RNA accession numbers are linked to the corresponding record in the DDBJ/EMBL/GenBank sequence databases. FlyBase receives updates of this information directly from DDBJ/EMBL/GenBank. Items in the "Protein accessions" field are linked to the corresponding record in the protein databases such as SwissProt and PIR. Homologs from other organisms are reported in the "Non-*Drosophila* homolog" field, which provides links to the homolog in the community database of the organism in question, wherever possible.

### Molecular Maps

A link to molecular map data curated from the literature by FlyBase is provided by the gene report; all maps that include the gene of interest are shown. Molecular maps are genomic maps of wild-type chromosomes. Each map represents the data presented in one publication and may contain information about gene structure, aberration breakpoints, mutations, regulatory elements, rescue fragments, and sites of transposon insertions. If known, the orientation of the map on the polytene chromosomes is indicated.

Map data are extracted from graphical maps in figures, from text, and from DNA sequence (either from the paper itself or from sequence database entries that cite the paper as a reference). Wherever possible, exact sizes or positions of entities are used as extracted from sequence data or from author statements. In the absence of sequence-based data, coordinates are estimated from graphical maps in figures.

A new feature found in an increasing number of genes reports is the "annotated reference sequence," represented by a small molecular map image at the top of the report. This is hyperlinked to a detailed sequence-level map that presents an integrated view of the molecular structure of the gene, plus molecularly characterized mutations and aberrations in the region of that gene. The entire map is available in the format standardized by the sequence databases and includes extensive annotation. In the future, computational annotation produced by the genome project will also be integrated into these maps.

## Transposons and Transposon Insertions

The Transposons section of FlyBase contains information on natural transposons and their component genes. It also includes information about engineered transposons and plasmid vectors, including cell culture vectors, annotated maps and compiled sequences of many of these transposons and constructs, and general information about them (e.g., selectable markers). Lists of specific types of constructs, such as P-element cloning vectors, enhancer traps, and reporter constructs, are provided, as are lists of insertions of specific constructs. Transposon insertion reports include cytological location, phenotypic data, and expression pattern data.

Data on transposons and constructs are derived from the literature, sequence database entries, and personal communications from investigators. Data concerning transposon insertions are derived primarily from the literature and from the BDGP.

Insertions generated by the BDGP can be accessed via the FlyBase Insertions query or using the BFD Search tool. These insertions were generated in two different projects: the Gene Disruption project (using the enhancer traps P{PZ} and P{lacW}) and the Controlled Misexpression project (using the UAS construct P{EP}). The BFD insertion reports include cytological location; links to the defined gene, allele, transposon, and insertion reports in FlyBase; genomic flanking sequence with exact insertion site and accession number; an image of the chromosome in situ hybridization data; and links to homologous sequence data determined by BLAST.

## Aberrations and Balancers

The Aberrations section contains reports on chromosomal rearrangements such as translocations, duplications and deficiencies, ring chromosomes, and inversions (including balancers). The system used for assigning symbols to aberrations is described in the Nomenclature guidelines (in the FlyBase Documents directory). Many aberrations are referred to in the literature by symbol(s) that do not correspond to the valid symbol. FlyBase stores these as synonyms.

Aberration reports include cytological localization of breakpoints; for all rearrangements with more than two breaks, the cytological sequence of the rearranged chromosome is presented. The "Genetic information" field presents map data conclusions drawn from genetic complementation tests. These data are used by FlyBase in constructing a consensus-order genome map. The origin of the rearranged chromosome is described in terms of mutagen, discoverer, and progenitor chromosome.

Balancer chromosomes are described in terms of associated aberrations, visible markers, and any other mutations or transposon insertions carried on the chromosome. Balancer reports can be accessed from within an aberration report or via the Balancer Search tool, which can be found within the Aberrations Query form. The Stocks section of FlyBase also provides access to Balancer Search.

## Stocks

The Stocks section of FlyBase includes subdirectories for stock lists from both public and private collections of *Drosophila*, order forms, quick and complex search facilities, and help files. Stock reports include a stock number, genotype, and comments. Alleles, aberrations, balancers, and transposon insertions appearing in stock reports are reciprocally hyperlinked to the full report for that item in the corresponding data section. If an item appearing in FlyBase is not available from one of the public stock centers, users can request a stock from the authors of a recent article.

The Stocks search allows querying by cytological location as well as by symbol. For comprehensive instructions for searching stock lists, see section C.3.3. of FlyBase Reference Manual C: Using FlyBase on the Web.

## Genomic Clones and STSs

The Clones section of FlyBase contains lists of genomic and cDNA clones, chiefly from the BDGP and EDGP, but complete reports about these clones can be obtained only from the genome project Web sites, most easily using the BFD search tools.

The Berkeley genomic sequencing project has a clone-based physical map of the *Drosophila* genome based on bacteriophage P1 and BAC genomic libraries. The European project is using cosmid and different BAC genomic libraries. YACs localized to the genome by in situ hybridization are also available.

STSs (sequence-tagged sites) are short sequences that were identified in order to construct the physical map. Most come from sequencing the ends of genomic clones, but others come from the sequence of known genes or regions flanking P-element insertions. Unlike a clone, an STS cannot be obtained, because an STS is simply a defined sequence.

Reports on all of the genomic P1, BAC, cosmid, or YAC clones and STSs can be found using the BFD text or graphical query tools. These reports also link to information and graphical displays of the physical contigs made up of these clones (CytoView, see below Genome Project Maps). When available, the sequence of a genomic clone is linked from the clone report page via the accession number of the submission in the DDBJ/EMBL/GenBank nucleic acid sequence data libraries. Sequences that can be joined to form large contiguous regions are also available in physical contig reports. Information on obtaining any of the clones is available at FlyBase and BDGP.

Reports for genomic clones that have been completely sequenced include links to a Java graphical display, CloneCurator, which shows the results of computational sequence analysis. Matches to *Drosophila* genes, ESTs, transposons, and so forth are indicated on the sequence, as well as BLAST-detected sequence similarities to other species. In addition, the results of gene prediction programs are displayed. For those clones inspected computationally and by a human curator, the most relevant results for each predicted and previously known gene identified in the sequence will be summarized and labeled with formal gene names or valid FlyBase symbols.

## ESTs and Full-length cDNA Clones

The EST (expressed sequence tag) section of FlyBase links to data from two different BDGP projects: the main BDGP/HHMI project sequencing the ends of full-length cDNAs and a small-scale project analyzing membrane-associated cDNAs (the CK library).

The full-length cDNA EST project generated the sequence of the 5′ ends of more than 80,000 cDNA clones made from high-quality *Drosophila* cDNA libraries. The long-term goal is to generate the full-length sequence of representative clones for each gene to generate a transcript map of the genome. These full-length sequences are available for querying and are submitted to DDBJ/EMBL/GenBank.

The CK library membrane-associated embryonic cDNA clones are in general not full length, but have been analyzed by whole-embryo in situ hybridization. Data from this library can be queried for clones expressed in particular tissues.

Users typically identify ESTs of interest by similarity or identity to their own sequence, e.g., by using the FlyBase BLAST server. The results of such a BLAST search will hyperlink to clot reports for the positive match, a *clot* being a group of clones with homologous sequences. A Java alignment viewer in the clot report allows users to inspect the alignment of EST sequences in detail. Because the clots are also analyzed for sequence similarities using BLAST, the results of these searches can also be browsed by querying using a key word (e.g., show me all the ESTs similar to the gene *Toll*).

## Genome Project Maps

The "Map Viewers" item on the BDGP home page links to Java displays, which allow interactive browsing of the genome at progressively higher resolution, described above (Searches by Cytological Location). ChromoView and ArmView show the physical map at any cytological position in the genome, with physical contigs shaded as to whether sequence information is available. These contigs link to the physical map viewer, CytoView, which diagrams the STS links between genomic clones that make up physical contigs. Users can move along the chromosome from contig to contig, and color-coding identifies clones that were localized by in situ hybridization and/or sequenced. Clones and STSs link to text reports, and the Clone reports link to the CloneCurator display of the annotated genomic sequence for that particular clone.

## References

The References section of FlyBase houses a searchable bibliography of publications about the genetics and biology of *Drosophila*, with a satellite file of journal and book abbreviations used in compiling the bibliography. FlyBase captures "personal communications" as publications, to accommodate otherwise unpublished information from the research community. Wherever possible, genetic and biological data presented in FlyBase are linked to the publication in which the data appeared.

A Reference report can be accessed using the Reference search tool or via a link from another object such as an allele or transposon. Reference reports include abstracts where available from PubMed and links to the article's full text where PubMed offers these. Reference reports include a tabular listing of genetic data recorded by FlyBase from that paper. References can be fetched in several formats including Refer/Endnote.

## People

The People directory of FlyBase stores contact information for *Drosophila* researchers. The information includes names, addresses, phone and FAX numbers, and E-mail and URL addresses. This information is primarily user-maintained, via the "Update" and "Add" facilities available directly from the FlyBase home page and in the People directory.

## HOW TO DOWNLOAD DATA

There are occasions when users may wish to download data files from FlyBase, BDGP, or EDGP. All data files are updated periodically. Users should be aware that newer versions will differ from previous versions not only by the addition of new data, but also as a consequence of error correction and editing of older data.

Smaller files (1 Mbyte or less) can be saved directly from the browser window under the "File" menu. Larger files are more reasonably obtained using FTP. For comprehensive instructions and FTP addresses for obtaining FlyBase files, see Reference Manual section D.2: FlyBase by FTP, and for field descriptions, see Reference Manual section B: Detailed Descriptions of FlyBase Structure and Data.

To download *D. melanogaster* sequence data sets that are searchable with the BDGP BLAST server in multiple fasta format, follow the link from the BDGP home page "Obtaining materials and sequences" and select "Sequence data download page." To download by FTP cosmid sequences or sequence datasets from the EDGP, follow these links from the EDGP home page.

## CITATION

Instructions for citing FlyBase are given in "Reference Manual I: The FlyBase Project." At the time of writing, the correct citation is:

The FlyBase Consortium 1999. The FlyBase Database of the *Drosophila* Genome Projects and Community Literature. *Nucleic Acids Res.* **27**: 85–88.

http://flybase.bio.indiana.edu/

Data generated by the BDGP or EDGP should be cited as such in any publication. Please contact the BDGP or EDGP for specific information on an appropriate citation for a particular case. Further information can be found on the WWW for the BDGP under "About the BDGP" on the home page.

## OTHER DATABASES

For the best comprehensive listing of *Drosophila* resources and databases, see *Drosophila Virtual Library*. A slightly expanded list is provided in Table 28.3.

**Table 28.3** List of *Drosophila* Resources and Databases

| Resource | URL | Notes |
| --- | --- | --- |
| The *Drosophila* Virtual Library | http://www.ceolas.org/fly/ | a helpful general resource for the beginner with links to most of the databases given below, as well as links to worldwide *Drosophila* labs and protocols |
| bionet.drosophila Newsgroup Archive | http://www.bio.net:80/hypermail/DROS/ | general or very technical questions answered in this forum by accessing old E-mail on the subject via the archives at bionet or in FlyBase/News, or by posting a message to the group by E-mail (dros@net.bio.net) |
| The Interactive Fly | http://sdb.bio.purdue.edu/fly/aimain/1aahome.htm | a guide to *Drosophila* genes and their roles in development, including narrative descriptions of the genes and non-linear developmental hierarchies and pathways |
| FlyView | http://flyview.uni-muenster.de/ | an image database of *Drosophila* expression patterns, including those from transgenic flies carrying enhancer-trap insertions; images and simple text descriptions of expression patterns can be searched for specific times during development |
| FlyBrain | http://flybrain.neurobio.arizona.edu/ | an atlas of neuroanatomical images and descriptions of the central and peripheral *Drosophila* nervous system |
| Virtual Fly Lab | http://vcourseware5.calstatela.edu/VirtualFlyLab/IntroVflyLab.html | a useful teaching tool to design "virtual" crosses of flies with various visible mutations, allowing users to analyze the results immediately |
| GIFTS (Gene Interactions in the Fly Transworld Server) | http://gifts.univ-mrs.fr/GIFTS_home_page.html | contains databases with information on various kinds of interactions: gene interactions during embryonic pattern formation; a knowledge base of networks of gene interactions; protein-binding sites identified on DNA sequences; and protein-DNA, protein-RNA, and protein-protein interactions. |
| DRES (*Drosophila*-related EST sequences) | http://www.tigem.it/LOCAL/drosophila/dros.html | a catalog of human cDNA clones that show homology with *Drosophila* genes |
| SegNet | http://www.csa.ru/Inst/gorb_dep/inbios/genet/s0sgnt.htm | diagrams and descriptions of gene networks involved in *Drosophila* embryonic segmentation |

## ACKNOWLEDGMENTS

The authors thank members of the FlyBase consortium, in particular Suzanna Lewis, as well as Ryan Baugh, David Tree, and Christian Siebel, for helpful comments and discussions about the manuscript.

# CONTENTS

# 29

# Preparation and Analysis of High-molecular-weight DNA in *Drosophila*

**Gary H. Karpen**
*Molecular Biology and Virology Laboratory*
*The Salk Institute*
*La Jolla, California 92037*

For many years, the upper limit of useful resolution for DNA electrophoresis was 23–50 kb. In the 1980s, new methods were developed for preparing, digesting, and analyzing high-molecular-weight (HMW) DNA (from 20 kb to 10 Mb). These methods are essential to any genomic analysis, because they allow characterization of the molecular structure and sequence composition of large chromosomal domains. Analyses of HMW DNA can be used effectively to identify polymorphisms, map rearrangement breakpoints and P-element insertion sites, and to compare restriction maps of large clones (e.g., bacterial artificial chromosomes [BACs]) to genomic DNA. The advantages of working with large chromosomal domains should be obvious; the more DNA covered in a mapping analysis, the greater the chance of discovering critical information (e.g., breakpoints and polymorphisms). These methods have been particularly effective in mapping regions that contain few widely separated restriction sites, most notably the centric heterochromatin. Surprisingly, HMW DNA technologies have been underutilized within the *Drosophila* community, most likely because of a general perception that the techniques are difficult to master. This chapter provides protocols which demonstrate that HMW DNA preparation and analysis can be as easy as standard electrophoresis, once misconceptions are overcome and some specialized equipment and supplies are obtained. These methods have applications to any project that involves DNA analysis, whether the goal is to understand genes involved in cell biology, development, neurobiology, and behavior or genomic regions involved in chromosome structure and function.

## PREPARATION OF HMW DNA

The key step in analyzing HMW DNA is to purify intact, chromosome-size molecules. Standard methods for extracting DNA from cells or organisms (e.g., phenol extraction and ethanol precipitation) produce fragments with an average size of 50–200 kb under optimal conditions. Frequent double-stranded breaks are produced by shearing forces applied to the DNA in solution during mechanical vortexing or mixing and pipetting. To better guard against such damaging forces, it is necessary to perform all extractions and manipulations on DNA that is embedded within a protective matrix. The overall strategy is to purify nuclei, gently embed them in molten agarose, and then extract proteins and perform other enzymatic reactions by transferring the solidified agarose block into the appropriate solutions. Salts, soaps, and enzymes act on the DNA by diffusing through the agarose matrix, while the matrix protects the DNA from shearing forces.

HMW DNA has been prepared from a variety of developmental stages, tissue culture cells, and tissues. The methods differ only in how nuclei or cells are isolated; all subsequent steps are identical, regardless of the DNA source. References for successful isolation from adults (Trapitz and Bunemann 1992), ovaries (Karpen and Spradling 1992), dissected larval tissues (Karpen and Spradling 1990; Glaser and Spradling 1994), and cultured cells (Locke and McDermid 1993) should be consulted as needed. Preparation of HMW DNA from embryos is described in detail below (Protocol 29.1), because in our hands, it is the simplest and most reliable protocol and can be used for large- or small-scale preparations (Le et al. 1995; Sun et al. 1997). For an outline of this protocol, see Flowchart 1.

## Protocol 29.1*

# Preparation of HMW DNA from Embryos

### Materials

### *Media*

#### *Preparation of Apple Juice–Agar Plates for Embryo Collection*

Collection plates are made from apple juice and agar media. Add 0.6 g of Nipigin hydroxy-benzoic acid (Sigma H 5501) and 10.0 g of sucrose to 250 ml of apple juice. Boil and then let cool. Separately, add 9.0 g of Bacto agar to 300 ml of $H_2O$ and boil. Mix the apple juice and agar solutions, cool to 65ºC, and pour into the lids of 35-mm petri plates (save the petri plate bottoms for collections; see Method below). This recipe makes 275 plates.

### *Supplies and Equipment*

Standard fly bottles

Weigh boats

Small, trimmed paint brush

Microcentrifuge tubes (1.5 ml) (matched with plastic pestles, Kontes 749520-0000)

Microcentrifuge (variable speed)

Plastic molds for forming agarose blocks/inserts are required and can be purchased from Bio-Rad, or made in a machine shop (Figure 29.1B).

Bent pasteur pipettes to manipulate the inserts. A Bunsen burner is used to seal the narrow end of the pipette. The end is then heated and twisted to form a U-shape.

Sequencing tape

Plastic tube (15 or 50 ml)

---

*For all protocols, $H_2O$ indicates glass distilled and deionized.

—— **FLOW CHART 1** ——

### HMW DNA PREPARATION FROM EMBRYOS

Fatten flies for at least 2 days on yeasted bottles.
↓
Precollect for 1–4 hours on a yeasted, apple juice–agar plate.
Collect overnight on a fresh yeasted apple juice–agar plate.
↓
Transfer embryos into weigh boat using NaCl/Triton and a paintbrush.
Collect embryos in collection sieve and rinse with NaCl/Triton.
↓
Dechorionate in 50% bleach for 90 seconds.
Rinse with NaCl/Triton, and then with 0.7% NaCl.
↓
Transfer to microcentrifuge tube with matching pestle (Kontes) with 1 ml of NaCl.
Centrifuge at 5000 rpm for 1 second.
↓
Estimate volume of eggs; 25 μl of eggs yields two inserts. Calculate amount of Homogenization
Buffer (40 μl per insert) and 1.5% InCert (FMC) agarose needed (store at 50ºC).
↓
Remove fluid. Add half of Homogenization Buffer to embryos (20 μl per insert).
↓
Homogenize, centrifuge at 5000 rpm for 3 seconds. Repeat.
Homogenize, centrifuge at 2000 rpm for 2 seconds.
↓
Gently transfer supernatant to 1.5-ml microcentrifuge tube (use a wide-bore, disposable tip)
↓
Add rest of Homogenization Buffer (20 μl per insert) to pellet.
Homogenize, centrifuge at 2000 rpm for 2 seconds.
↓
Combine supernatants, supplement with Homogenization Buffer as needed.
↓
Clean mold, seal one side with sequencing tape, label with genotypes.
↓
Place homogenate at 37ºC for 2 minutes.
Add an equal volume 1.5% agarose, and mix by pipetting gently.
↓
Pipette mixture into mold; cool at 4ºC for 10 minutes.
↓
Transfer inserts into tubes with NDS + Proteinase K (~2.5 ml per insert).
Incubate at 50ºC overnight with occasional mixing. Store at 4ºC.
↓
When ready to use, wash inserts with TE + PMSF at 25ºC for 1 hour. Repeat wash in TE +
PMSF for a second hour. Wash four times in TE, 20 minutes per buffer change. Store at 4ºC.

### Preparation of Embryo Collection Sieves

Embryos used to prepare HMW DNA are collected and rinsed in sieves. To construct a sieve, cut off the bottom of a plastic centrifuge tube and make a hole in the cap. Attach the stretched Nitex mesh (Tetko 3-130/43) onto the tube by screwing the modified cap over the Nitex (Figure 29.1A). Alternatively, superglue the mesh to the edge of the cut tube.

## Solutions and Reagents

**Bleach** (50%)

NaCl (0.7%)

FMC InCert agarose (1.5%) (FMC 50121), in 0.125 M EDTA (pH 7.5), previously boiled
   for 5 minutes in a screw-cap microcentrifuge tube and kept at 50ºC.

Proteinase K (EM Science 24568-2)

### NaCl/Triton Solution

0.7% NaCl
0.4% Triton X-100 (Boehringer Mannheim 789704)

### Homogenization Buffer

0.1 M NaCl
0.03 M Tris-Cl (pH 8.0)
0.05 M EDTA
0.5% Triton X-100
7.7 mM β-mercaptoethanol

### NDS

0.5 M EDTA
0.01 M Tris-Cl (pH 9.5)
1% Sarkosyl

**CAUTION: bleach, β-mercaptoethanol** (see Appendix 4)

## Method

### Embryo Collection

1. At least 2 days before embryo collection, fatten flies in bottles with yeast paste on one side of the bottle and a Kimwipe pushed into the food.

   *Note:* Depending on the health of the stock, combining adults from two to three stock bottles usually produces enough embryos for eight inserts, which is enough for 16 restriction digests or gel lanes. Move flies to new yeasted bottles daily to prevent humidity accumulation.

2. On collection day, spread one to two drops of dilute yeast solution over each apple juice–agar plate (two per bottle) and dry at 37ºC for approximately 5 minutes. For timed collections, remove older eggs or embryos retained by females by precollecting for 1–4 hours. Place flies in empty bottles with a Kimwipe taped to the bottom and a yeasted apple juice–agar plate over the bottle mouth. Cover with a 35-mm plate bot-

tom (to ensure that the collection plate creates a solid seal with the bottle) and tape to the bottle. Incubate upside down at 25°C in the dark.

3. Discard the precollection plate and collect embryos on the other apple juice–agar plate at 25°C overnight (~12 hours) in the dark. The collection plates can be aged 1–2 hours so that freshly laid eggs develop. Return flies to yeasted food bottles; subsequent collections can be performed depending on health of flies.

### Preparation of Embryos

1. Rinse embryos off apple juice–agar collection plates into weigh boats with NaCl/Triton Solution (dispensed from a squirt bottle). If necessary, use a small, trimmed paint brush to loosen embryos that are "stubbornly" embedded in the agar.

2. Pour embryos from the weigh boat into a collection sieve (Figure 29.1A). Rinse twice with NaCl/Triton Solution to remove yeast and bacteria.

3. Dechorionate embryos in 50% bleach for 90 seconds and transfer to a collection sieve. Rinse twice with NaCl/Triton Solution followed by three rinses with 0.7% NaCl. Transfer to a microcentrifuge tube (matched with a plastic pestle) with 1 ml of 0.7%

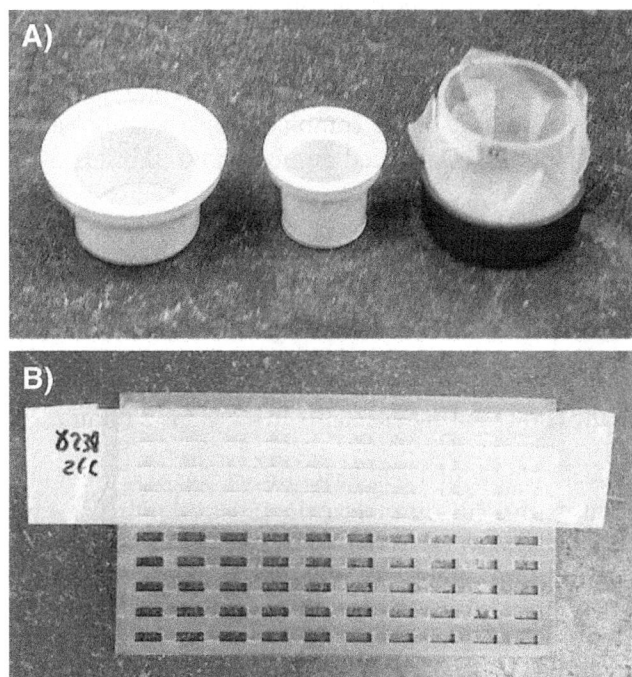

**Figure 29.1.** Equipment used in preparation of HMW DNA. (*A*) Sieves used in embryo preparations can be obtained commercially (*left*, 24-mm diameter [Costar 3479]; *middle*, 25-mm diameter [Costar 3477]) or made in the lab (*right sieve*) as described in the text. (*B*) Agarose block mold that can be produced by anyone familiar with small-unit plastic construction (e.g., at a machine shop). The overall dimensions of this mold are 10 cm (*length*) × 6 cm (*width*) × 6 mm (*depth*), but these can also be made to any specifications. In all cases, make sure that the individual mold wells are equal to the size of the wells in the PFGE gel (in this case, 2 mm × 6 mm × 6 mm). Plastic tape is used to seal one side of the mold and is folded onto itself at the edges to provide a surface for marking individual rows with genotype designations. Agarose blocks are made by carefully pipetting the molten agarose/nuclei mixture into the individual wells of the mold; be careful to avoid bubbles.

NaCl. Centrifuge in a microcentrifuge (e.g., Eppendorf 5415C) at 5000 rpm for 1 second. Estimate the volume of embryos; usually 25 μl of embryos produces two 80-μl agarose "inserts" (if this produces too much DNA per insert/lane, use 12.5 μl of embryos for two inserts).

### Homogenization of Embryos

1. Gently remove the NaCl "supernatant," and add Homogenization Buffer to the embryos; use 20 μl per insert. If the supernatant is more than 400 μl, split into more than one tube to avoid loss of fluid during subsequent steps (alternatively, for large preparations, use larger tubes and homogenizers, and adapt the protocol accordingly).

2. Homogenize embryos with the plastic pestle (~10 circular "strokes"), and then centrifuge at 5000 rpm for 3 seconds. Repeat homogenization twice in the same buffer. Remove cuticle and other contaminants by gentle centrifugation at 2000 rpm for 2 seconds.

3. Gently transfer the supernatant to a new tube with a cut-off pipette tip (increased diameter of tip decreases shearing forces).

4. Add another 20 μl of Homogenization Buffer (per insert) to the homogenate and rehomogenize. Centrifuge at 2000 rpm for 2 seconds and combine the supernatants. Adjust the total volume of the combined supernatants with Homogenization Buffer to yield the previously calculated number of 80-μl inserts.

### Preparation of Agarose Inserts

1. Prepare molds by taping one side with plastic tape, e.g., sequencing tape, and, if necessary, label rows to distinguish different genotypes (Figure 29.1B).

2. Prewarm the combined supernatants to 37ºC for 2 minutes.

3. Add an equal volume of 1.5% FMC InCert agarose (kept at 50ºC) to the combined supernatants with a cut-off pipette tip. Pipette gently three times to mix.

4. Pipette carefully into molds. Avoid bubbles by initiating mold filling from one side of the well, and try to fill from the bottom up.

5. Once inserts gel (at 4ºC for no more than 10 minutes), gently remove tape. Push inserts from the molds, using a bent pasteur pipette, into a 15- or 50-ml plastic tube that contains 2.5 ml per insert of NDS containing 0.1–1 mg/ml proteinase K.

   *Note:* Most published protocols stipulate 1 mg/ml proteinase K; however, we obtain identical results with 0.1 mg/ml, which is considerably less expensive.

6. Incubate overnight at 50ºC with occasional gentle mixing to remove membranes and protein, exposing the naked DNA.

   *Note:* Inserts can be stored indefinitely in NDS at 4ºC.

## RESTRICTION DIGESTION OF HMW DNA

Subsequent manipulations of the DNA are performed by transferring the agarose inserts into appropriate solutions. Protocol 29.2 describes methods for partial and complete restriction digests, but the same approach can be used for any type of DNA modification, such as ligation, polymerase chain reaction (PCR), and methylation; simply alter the buffers and incubation conditions in step 2 accordingly.

Many protocols include shaking tubes during the equilibration and incubation steps; shaking agarose inserts in solution can cause degradation of the DNA (see Chen et al. 1994) and therefore is not included in this protocol. Instead, gently invert tube once or twice to mix solutions.

PROTOCOL 29.2

# Restriction Digestion of HMW DNA

### Materials

### *Supplies and Equipment*

Bent pasteur pipette (see p. 526)
Large coverslips
Microcentrifuge tubes

### *Solutions and Reagents*

TE (pH 7.5; see Appendix 3)
**PMSF** (0.1 M, Sigma P 7626) in **isopropanol**
Appropriate 10x restriction buffer
Acetylated BSA (bovine serum albumin)
(*Optional*) Triton X-100 (Boehringer Mannheim 789704)
DNase-free (boiled) RNase A (Sigma R 5503)
Enzymes from a variety of different sources can be used successfully to digest HMW DNA. We find that the enzymes from New England Biolabs are the most reliable; this company also offers the largest variety of enzymes, especially rare cutters (recognizing >6-bp sites).
Filter-sterilized electrophoresis running buffer (0.5x TBE; see Appendix 3)

**CAUTION: isopropanol, PMSF** (see Appendix 4)

### Method

### *Removal of NDS and Proteinase K*

1. Transfer inserts to a tube containing TE (pH 7.5) and PMSF (add 1 µl of 0.1 M PMSF per milliliter of TE; PMSF inactivates proteinase K); use more than 3 ml per insert.

2. Incubate at room temperature for 1 hour with gentle shaking.

3. Replace TE/PMSF and repeat incubation (step 2).

4. Wash inserts with TE only to remove PMSF; use more than 3 ml per insert. Repeat for a total of four washes in TE (20 minutes each).

5. Store at 4ºC.

### Complete Restriction Digests

1. Remove inserts from TE with a bent pasteur pipette, and place on a clean glass surface (e.g., a large coverslip). It is best to view inserts against a black background.

2. Cut the inserts in half lengthwise (use a clean glass coverslip), and gently place in microcentrifuge tube. To prepare the restriction digestion buffer, mix together the following:

| Component | Final Concentration |
|---|---|
| 10x restriction buffer appropriate for the enzyme to be used in step 4 (below) | 1x |
| Acetylated BSA (if not already in the 10x buffer) | 100 µg/ml |
| Triton X-100 (use pure Boehringer Mannheim Triton X-100 diluted in $H_2O$) | 0.1% |
| DNase-free (boiled) RNase A (Sigma R 5503) | to 10 µg/ml |
| Sterile $H_2O$ | to make 500 µl |

*Note:* The addition of Triton X-100 is optional and is thought to improve access of some enzymes to the DNA.

3. Equilibrate inserts in approximately 250 µl of buffer without enzyme for about 15–30 minutes.

4. Pour off buffer and replace with 250 µl of buffer that includes 20 units of enzyme.

5. Preincubate on ice for 15–30 minutes to allow the enzyme to enter the matrix and contact the DNA before initiating digestion. Incubate at the appropriate temperature overnight, with periodic mixing by inversion.

   *Note:* Do not place agarose inserts at temperatures above 50ºC or the agarose will melt. Use 50ºC if enzyme specification calls for more than 50ºC.

6. If desired, add another aliquot of enzyme in the morning and incubate for 4 hours.

   *Note:* This step is optional; use if partial digests often occur or if trying a new enzyme.

7. After incubation is complete, pour off buffer while retaining insert. Add 500 µl of filter-sterilized electrophoresis running buffer (0.5x TBE) to the tube, and equilibrate for 15–30 minutes.

8. Place on ice, and then either run on gel or store at 4ºC for later use.

### Partial Restriction Digests

1. Perform steps 1–3 (complete restriction digests).

2. Pour off buffer and replace with 250 µl of buffer that includes different enzyme concentrations (e.g., 0, 1, 2, 5, and 20 units).

3. Preincubate at 4°C for 2 hours. Initiate digestion by transferring tubes to the appropriate temperature for 4 hours.

4. Halt digests by placing tubes on ice immediately.

5. Pour off buffer and add 500 µl of filter-sterilized 0.5× TBE.

6. Run on gel or store at 4°C.

## PULSED-FIELD GEL ELECTROPHORESIS

The most important advance in the analysis of HMW DNA was the development of electrophoresis methods for efficiently separating molecules ranging from 20 kb to 10 Mb or

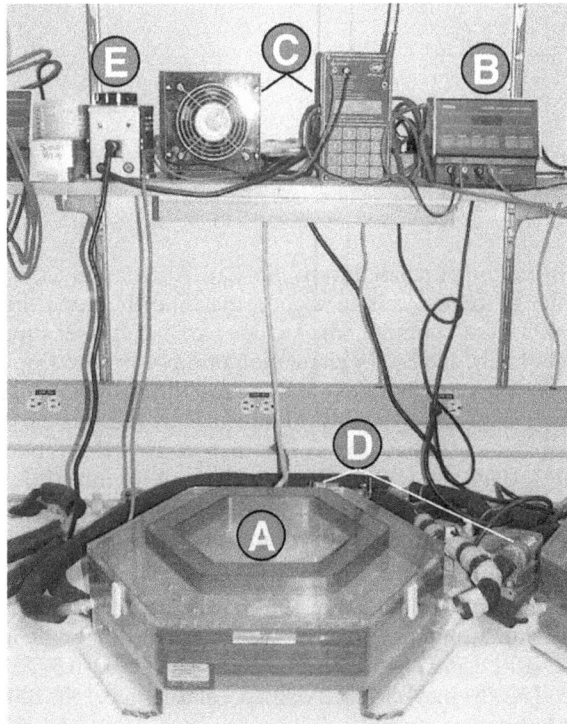

**Figure 29.2.** Components of this PFGE apparatus include the following: (*A*) CHEF gel box (CBS Scientific), (*B*) power supply (Bio-Rad 165-4761), (*C*) field-switch control unit; (*D*) circulation pump (Gorman-Rupp Industries [Belleville, Ohio 44813] 14925-005), and (*E*) circulation-pump controller. The Neslabs circulating chiller (RTE-211D) is not shown and is housed underneath the bench.

more. The seminal study of Schwartz and Cantor (1984) set the stage for subsequent modifications and improvements to the electrophoresis chambers, agarose gel matrix, and size markers. Only general information about pulsed-field gel electrophoresis (PFGE) and how to set up a PFGE gel are described here; for a comprehensive description of the theory and practice of PFGE, see the excellent book by Birren and Lai (1993).

Pulsed-field gel apparatuses come in a variety of shapes, including field-inversion gel electrophoresis (FIGE) and contour-clamped homogeneous electric field (CHEF; Chu et al. 1986). All PFGE approaches share one common theme: HMW DNA is separated by alternating the polarity of the electric field at precise intervals (for detailed information about how field inversion resolves HMW DNA and apparatus types, see Schwartz and Koval 1989; Birren and Lai 1993; Bustamante et al. 1993; Maule 1998). Methods for CHEF apparatuses, including general tips about PFGE, are described below. CHEF units are excellent for general use; they produce straight lanes, can resolve small and large fragments (up to 10 Mb), and are commercially available (see Protocol 29.3, Materials). This protocol is written for CHEF units manufactured in the late 1980s by CBS Scientific; currently, only Bio-Rad sells CHEF units. Therefore, defer to the manufacturer's instructions whenever differences arise.

## Protocol 29.3

## PFGE Technique

### Materials

### *Supplies and Equipment*

Casting chamber (15 × 15 cm)

Comb (18-well)

CHEF and FIGE units can be purchased from Bio-Rad. Alternatively, if handy with plastics and electronics or have access to a good machine shop, see the schematics in Birren and Lai (1993). Basic components include a gel box, power supply, field-switch control unit, cooling unit, and buffer circulation pump. For an example of the type of CHEF apparatus used in our laboratory and the specific components, see Figure 29.2. FIGE gels are somewhat easier to set up and run, but they are not described here because the upper limit of separation is less than with CHEF (about <250 kb; see Birren and Lai 1993, 1994, and contact the manufacturer for more information). FIGE can be run in a conventional gel box, with the addition of a field-switch unit, buffer circulation pump, and a cooling unit (or it can be run in a cold room).

HGT agarose (1%, FMC) in 0.5× TBE

Markers. Methods for making HMW markers in the lab are described in Birren and Lai (1993). Beginners should purchase commercially available markers, including the 1-kb ladder (GIBCO-BRL 15615-016), HindIII-digested λ DNA (0.5–23 kb), low-molecular-weight (0.5–150 kb) marker (NEB 350), λ concatamers (50–1000 kb) (Bio-Rad 170-3635), *Saccharomyces cerevisiae* chromosomes (Bio-Rad 170-3605), and *Schizosaccharomyces pombe* chromosomes (3–7 Mb).

Inserts (prepared as described above)

0.5× TBE (see Appendix 3)

Staining solution containing 500 μg/ml **ethidium bromide** in H$_2$O.

**CAUTION: ethidium bromide** (see Appendix 4)

**A) Uncut DNA**

**B) Restriction Digested DNA**

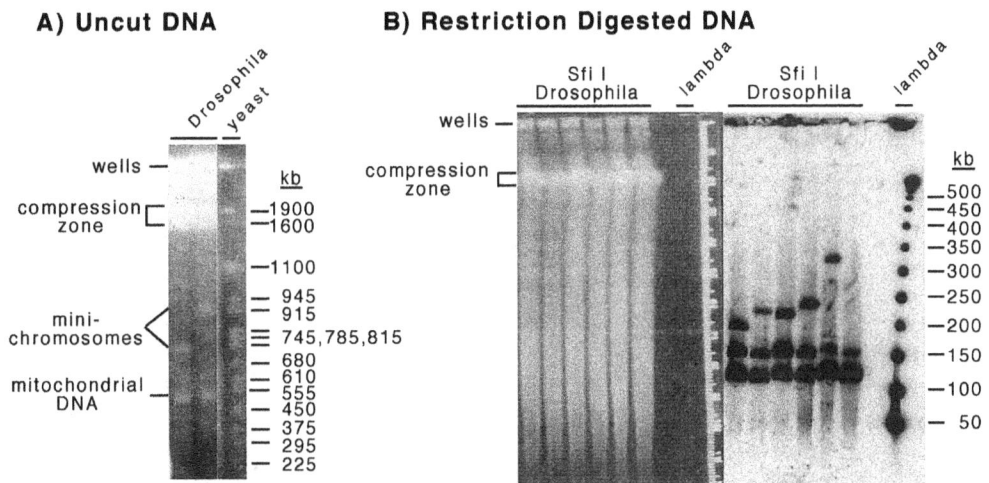

**Figure 29.3.** Examples of PFGE analysis of *Drosophila* DNA. (*A*) Uncut DNAs from two strains containing differently sized minichromosome derivatives (Le et al. 1995; Murphy and Karpen 1995; Sun et al. 1997), plus uncut *S. cerevisiae* DNA (size standard), were separated on a 1% HGT agarose CHEF gel. The size of each yeast chromosome is shown at the right. The pulse parameters were forward and reverse 60 seconds, increment 2 seconds, 30 repetitions, 28 hours at 180 V. Note the position of the wells, the compression zone, and the mitochondrial DNA. The migration and apparent size of the mitochondrial DNA can differ dramatically depending on the pulse conditions. In addition, uncut minichromosome derivatives of 720 kb and 915 kb (*left to right*) can be observed after ethidium bromide staining; band visualization does not always require Southern hybridization. Note that visible bands can be cut out of low-melting-point agarose gels, digested, and analyzed with PFGE (Sun et al. 1997). (*B*) *Sfi*I-digested *Drosophila* DNAs from strains with different minichromosome derivatives (Le et al. 1995; Murphy and Karpen 1995; Sun et al. 1997), plus concatemers of bacteriophage λ DNA, were separated on a 1% HGT agarose CHEF gel. Pulse parameters were forward and reverse 1 second, increment 1 second, 30 repetitions, 24 hours at 180 V. Sizes (*right*) were determined from the migration of the λ marker, which is barely visible in the ethidium-stained gel. The marker is visible on the blot because labeled λ DNA was included in the hybridization mix, along with an *X* chromosome probe that hybridizes to the *X* chromosome tip and the minichromosomes. All samples contain 125- and 150-kb fragments, and all except the minichromosome-less control (*right lane*) also contain an additional derivative-specific band ranging from 200 kb to 325 kb. Note that shorter pulse times result in greater resolution of lower-molecular-weight fragments, and failure to resolve higher-molecular-weight DNA, in this case, >500 kb (compare *A* and *B*). (Photos courtesy of Keith Maggert.)

## Method

### *Setup of Equipment*

Follow the manufacturer's instructions to set up the equipment. Become familiar with all plumbing and electrical components. It is particularly important to become knowledgeable about the programming capabilities of the unit. Start the circulating chiller, fill the gel box with buffer (for volume, see below), and circulate the buffer by starting the pump.

### *Experimental Plan*

Molecular weights are determined by comparison with mobility standards, so choose the marker(s) appropriately for the particular experiment. Put markers on both ends if possible, because gels may not always run straight. As for conventional gels, place samples in an order that will allow direct comparisons of controls and experimental samples.

Determine the appropriate pulse conditions that will be used for the gel according to the manufacturer's information and the size range that needs to be separated.

It is worthwhile to test new run conditions with markers only, to avoid sacrificing precious HMW DNA preparations. A Microsoft Word file that contains fields to enter the information that must be documented for each experiment can be downloaded from our Web Site at:

http://pingu.salk.edu/users/Karpen_web/Karpen_lab_home.html

### Gel Preparation, Loading, and Electrophoresis

1. To prepare the gel, pour 150 ml of 1% HGT agarose (FMC) in 0.5× TBE into a 15 × 15-cm casting chamber with an 18-well comb.

   *Note:* If the DNA will be isolated and purified from the gel after separation, use a low-melting-point agarose (we use GTG, FMC) and TAE buffer (Sun et al. 1997).

2. After the gel has solidified, place markers and inserts into the wells, and seal into position with a small amount of molten agarose.

3. Run gels in 1.8 liters of 0.5× TBE using the following parameters for the CHEF apparatus from CBS Scientific: buffer temperature 12°C, 180 V, approximately 170 mA, pulse conditions set to match the size range to be separated. Start the run.

4. Following separation, visualize the DNA by staining in 500 μg/ml ethidium bromide in $H_2O$ for 15 minutes, and then destaining in $H_2O$ for 15 minutes (Figure 29.3). Photograph the gels using standard methods, with a visible ruler next to the markers for molecular-weight determinations later (Figure 29.3B).

## SOUTHERN HYBRIDIZATION ANALYSIS

Once the HMW DNA is separated via PFGE, standard methods for DNA transfer to a membrane (Protocol 29.4) and Southern hybridization analysis are used to visualize specific HMW fragments.

PROTOCOL 29.4

## Transfer of DNA to a Membrane

### Materials

### Supplies and Equipment

Hybond N (Amersham RPN30313)
UV cross-linking device (e.g., UV Stratalinker 2400, Stratagene 400075)

### Solutions and Reagents

**HCl** (0.25 N)
5× SSPE (Sambrook et al. 1989)

> ### Alkaline Transfer Buffer
> 1.5 M NaCl
> 0.4 N **NaOH**
>
> ### 1× *Unblot*
> 0.2 M Tris-Cl (pH 7.5)
> 2× SSPE

**CAUTION: HCl, NaOH** (see Appendix 4)

## Method

1. Depurinate DNA in the pulsed-field gel (PFG) in 0.25 N HCl for 12 minutes. Rinse briefly in $H_2O$.

2. Incubate in Alkaline Transfer Buffer for 15 minutes with gentle shaking. Repeat incubation in Alkaline Transfer Buffer.

3. Transfer the DNA by capillary action to Hybond N (Amersham) overnight.

4. Neutralize in 1× Unblot and then in 5× SSPE (2–5 minutes each).

5. Cross-link the DNA to the membrane using the "autocross-link" setting of a UV Stratalinker 2400 (Stratagene 400075) or other UV cross-linking device.

## Probe Preparation and Hybridization

Probes are prepared by random-hexamer labeling of gel-purified fragments with [$^{32}$P]dCTP (ICN 33004X). We perform most hybridizations with heat-denatured probes (boiled for 10 minutes) for 2 hours in a Techne HB2 oven (VWR 35756-303) at 65ºC, using QuickHyb buffer (Stratagene 201220). Any hybridization device and buffers can be used, exactly as for conventional gels and blots. We perform hybridizations with probes that have a skewed AT:GC ratio (e.g., satellite DNAs, such as AATAT repeats) by nick-translating the probes, and then hybridizing overnight at a lower temperature (e.g., 48ºC for AATAT) in Church and Gilbert hybridization buffer (Church and Gilbert 1984). (Note: Quick-Hyb and other similar buffers become too viscous at temperatures lower than 60ºC.) Alternatively, probes can be produced using oligonucleotides corresponding to the satellite sequence, using terminal transferase to incorporate the label. Diluted, labeled marker DNA, such as λ concatemers (Figure 29.3B) and *S. cerevesiae* DNA (Figure 29.3A), can be included in the hybridizations to make the molecular-weight determinations more accurate. Signals can be observed after exposure to film (Kodak XAR-5, 2 Dupont Intensifying screens) or analyzed with a Molecular Dynamics or Fuji Phosphorimager.

## Data Analysis

The type of data analysis required is entirely dependent on the specific type of experiment being performed. Fragments identified by direct visualization of DNA with ethidium bromide or Southern hybridization are analyzed in the same manner as that for conventional electrophoresis. For example, fragment sizes are determined by comparing the length of fragment migration to the migration distances for the size standards (Figure 29.3). A

Microsoft Excel spreadsheet that automatically calculates fragment size after entry of the control and experimental migration distances can be downloaded from our Web Site at:

http://pingu.salk.edu/users/Karpen_web/Karpen_lab_home.html

As with any experiment, various problems can arise that lead to failure. Again, most of the problems are similar to what can happen with conventional electrophoresis and analysis, and the solutions are also similar. For example, unintentionally uncut or partially digested DNA appears as multiple bands, or the signal is only present in the wells or the compression zone (the area where all DNA that has a higher molecular weight than the limit of resolution accumulates) (Figure 29.3). Solutions include using more restriction enzyme from the start of the reaction, adding additional enzyme after a few hours or overnight, or making new DNA preparations. Degraded DNA appears as a smear of signal, which requires new DNA preparations, and perhaps making new buffers. Note that the low-molecular-weight markers can be run off the gel, depending on pulse conditions, which should be considered when analyzing the data; using a freshly mixed ethidium bromide solution can increase the lower limit of what can be seen on the gel. For solutions to problems observed with PFGE gels and blots, see the excellent book by Birren and Lai (1993).

## ACKNOWLEDGMENTS

I thank Hiep Le and Janice Wahlstrom for critical comments on this manuscript, and Keith Maggert for providing the photos used in Figure 29.3. Research in my laboratory is supported by National Institutes of Health grants R01-HG00747, R01-GM53435, and ACS-DB1200.

## REFERENCES

Birren B.W. and Lai E.H.C. 1993. *Pulsed field gel electrophoresis: A practical guide.* Academic Press, San Diego.

———. 1994. Rapid pulsed field separation of DNA molecules up to 250 kb. *Nucleic Acids Res.* **22:** 5366–5370.

Bustamante C., Gurrieri S., and Smith S.B. 1993. Towards a molecular description of pulsed-field gel electrophoresis. *Trends Biotechnol.* **11:** 23–30.

Chen L., Kosslak R., and Atherly A.G. 1994. Mechanical shear of high molecular weight DNA in agarose plugs. *Biotechniques* **16:** 228–229.

Chu G., Vollrath D., and Davis R.W. 1986. Separation of large DNA molecules by contour-clamped homogeneous electric fields. *Science* **234:** 1582–1585.

Church G. and Gilbert W. 1984. Genomic sequencing. *Proc. Natl. Acad. Sci.* **81:** 1991–1995.

Glaser R.L. and Spradling A.C. 1994. Unusual properties of genomic DNA molecules spanning the euchromatic-heterochromatic junction of a *Drosophila* minichromosome. *Nucleic Acids Res.* **22:** 5068–5075.

Karpen G.H. and Spradling A.C. 1990. Reduced DNA polytenization of a minichromosome region undergoing position-effect variegation in *Drosophila. Cell* **63:** 97–107.

———. 1992. Analysis of subtelomeric heterochromatin in the *Drosophila* minichromosome Dp1187 by single P element insertional mutagenesis. *Genetics* **132:** 737–753.

Le M.H., Duricka D., and Karpen G.H. 1995. Islands of complex DNA are widespread in *Drosophila* centric heterochromatin. *Genetics* **141:** 283–303.

Locke J. and McDermid H.E. 1993. Analysis of *Drosophila* chromosome 4 using pulsed field gel electrophoresis. *Chromosoma* **102:** 718–723.

Maule J. 1998. Pulsed-field gel electrophoresis. *Mol. Biotechnol.* **9:** 107–126.

Murphy T.D. and Karpen G.H. 1995. Localization of centromere function in a *Drosophila* minichromosome. *Cell* **82:** 599–609.

Sambrook J., Fritsch E.F., and Maniatis T. 1989. *Molecular cloning: A laboratory manual,* 2nd edition. Cold Spring Harbor Laboratory, Cold Spring Harbor, New York.

Schwartz D.C. and Cantor C.R. 1984. Separation of yeast chromosome-sized DNA by pulsed field gradient gel electrophoresis. *Cell* **37:** 67–75.

Schwartz D.C. and Koval M. 1989. Conformation dynamics of individual DNA molecules during gel electrophoresis. *Nature* **338:** 520–522.

Sun X., Wahlstrom J., and Karpen G. 1997. Molecular structure of a functional *Drosophila* centromere. *Cell* **91:** 1007–1019.

Trapitz P. and Bunemann H. 1992. Preparation of high molecular weight DNA from *Drosophila* adults for PFGE analysis. *Trends Genet.* **8:** 371–372.

# Biochemistry

# CONTENTS

# 30

# Culturing Large Populations of *Drosophila* for Protein Biochemistry*

**John Charles Sisson**
*Sinsheimer Laboratories*
*Department of Biology*
*University of California at Santa Cruz*
*Santa Cruz, California 95064*

Dᴜʀɪɴɢ ᴛʜᴇ ᴘᴀsᴛ 25 ʏᴇᴀʀs, ᴛʜᴇ *Dʀᴏsᴏᴘʜɪʟᴀ* ᴇᴍʙʀʏᴏ has proven to be a superb source for proteins involved in replication (Mitsis et al. 1995), transcription (Biggin and Tjian 1988; Soeller et al. 1988; Becker and Wu 1992), translation (Maroto and Sierra 1989), signal transduction (Liu et al. 1997), and extracellular matrix (Fessler et al. 1994) and cytoskeletal functions (Saxton et al. 1988; Miller et al. 1991; Cole et al. 1994; Field et al. 1996; Moritz and Alberts 1999). The success of these studies has depended on the production of gram quantities of embryos by large populations of adult flies.

Although the maintenance of a laboratory fly facility is essential for protein biochemistry, there are surprisingly few published procedures from which to choose. To a large extent, methods have been the subject of oral tradition and apprenticeship. However, the few published examples are very useful. Travaglini and Tartof (1972) describe an innovative method for culturing larva, Roberts (1986) provides helpful tips on constructing homemade fly containers, and Shaffer et al. (1994) present the most thorough protocol to date.

This chapter outlines a simple and efficient fly facility procedure perfect for small or large laboratories. The initial setup cost is approximately $4000. The facility can be run with 10–12 hours of work per week, at a minimum monthly maintenance cost of $200 at full capacity. Four fly houses will yield 40–60 g of Oregon-R embryos in 12 hours.

## SETUP AND MAINTENANCE OF FLY HOUSES

### Equipment and Supplies

Equipment and supplies required for maintaining four active fly houses each week include the following:
- *Biological Incubators.* A minimum of 40 square feet of total culturing shelf space is required in either two freestanding incubators or one walk-in incubator. In addition

*This chapter is dedicated to the memory of Ramon Mayo, fly facility manager and cherished member of the Department of Developmental Biology at Stanford University.

to a standard thermostat control, it is preferable to also have an automated humidification system. If the cooling system and/or condenser are mounted inside a free-standing-type incubator, they must have a phenolic coating, otherwise the propionic acid (which is used in many fly medium recipes as a mold inhibitor) will corrode them. A walk-in incubator that lacks a built-in humidification system equipped with a portable humidifier and dehumidifier (Kenmore brand) works quite well.

- *Instruments for Measuring Temperature and Humidity.* These include one handheld Digital Pyrometer (Fisher 13-904-50) and one Thermocouple Probe (Fisher 13-904-50A or 13-904-50C). The probe provides a digital display of temperature, relative humidity (RH), and dew point. It is very accurate and reads and recalls the above parameters during long monitoring periods.

- *Plastic Cylindrical Fly Houses (8).* The dimensions of each fly house are 45 cm (*length*), 30.6 cm (*outer diameter*), 29.5 cm (*inner diameter*) (see Figure 30.1). Drill a 6-cm hole into the middle of the surface for adding flies. Cut two 0.5-cm-wide and 0.2-cm-deep grooves along the circumference of each cylinder, 2 cm from each end; these grooves will help secure the cloth closures with large rubber bands or gaskets. Stabilizing plastic feet can be glued to the cylinders; however, they are not essential, since there is little trouble with the cylinders rolling and the feet tend to break. If necessary, ordinary clay bricks or large rubber stoppers are perfect low-tech alternatives. Plastic cylinders can be purchased, cut, and drilled by many commercial plastic companies.

- *Cloth Closures for Adult Fly Houses (8 sets).* Each set consists of one tubular and one square cloth closure (see Figure 30.1). The tubular closure should have a 8-cm wide band of durable cotton at one end sewn to a 65-cm long pliable nylon-mesh sleeve with a pore size of 200–300 μm. The tubular closure should fit snugly around the end of a cylindrical house. The square closure (37 × 37 cm) should be made of durable cotton. Purchase the materials at a fabric store and have the closures made by a competent sewer. Key considerations are durability and breathability.

- *Larval Condominiums (160).* Larval "condos" (see Figure 30.2) are made from 1-liter Nalgene polycarbonate, wide-mouth jars with lids (VWR 36318-646). Drill a 6-cm diameter hole in the middle of each Nalgene lid. Glue an 8-cm diameter piece of nylon screen on the *outside* of the lid over each hole using 100% silicone household adhesive (Dow Corning brand) (Figure 30.2). The nylon screen has a mesh size of 297 μm and is available in 30 × 30-cm sheets (Fisher 08-670-187).

- *Trays for Embryo Collection.* Styrofoam trays (~25.7 × 20.6 cm), available from most restaurant suppliers and at supermarket meat departments, are required.

- *Shelf Space.* One or two carts with a total of 14 square feet of shelf space are required.

- *Storage Space.* Approximately 30 cubic feet of cold room space is required to store collection trays, yeast paste, and spent condos and houses.

- *Miscellaneous Items.* These include (1) metal sieves with 850-μm, 425-μm, and 125-μm mesh sizes, respectively (VWR 57320-602, -680, and -828); (2) one soft, 1-inch wide paint brush or rubber kitchen spatula; (3) one large funnel with a 20-cm mouth diameter and wide neck; note that the small end of the funnel will have to feed into the 6-cm hole in the house; and (4) eight rubber or cork stoppers (size-12).

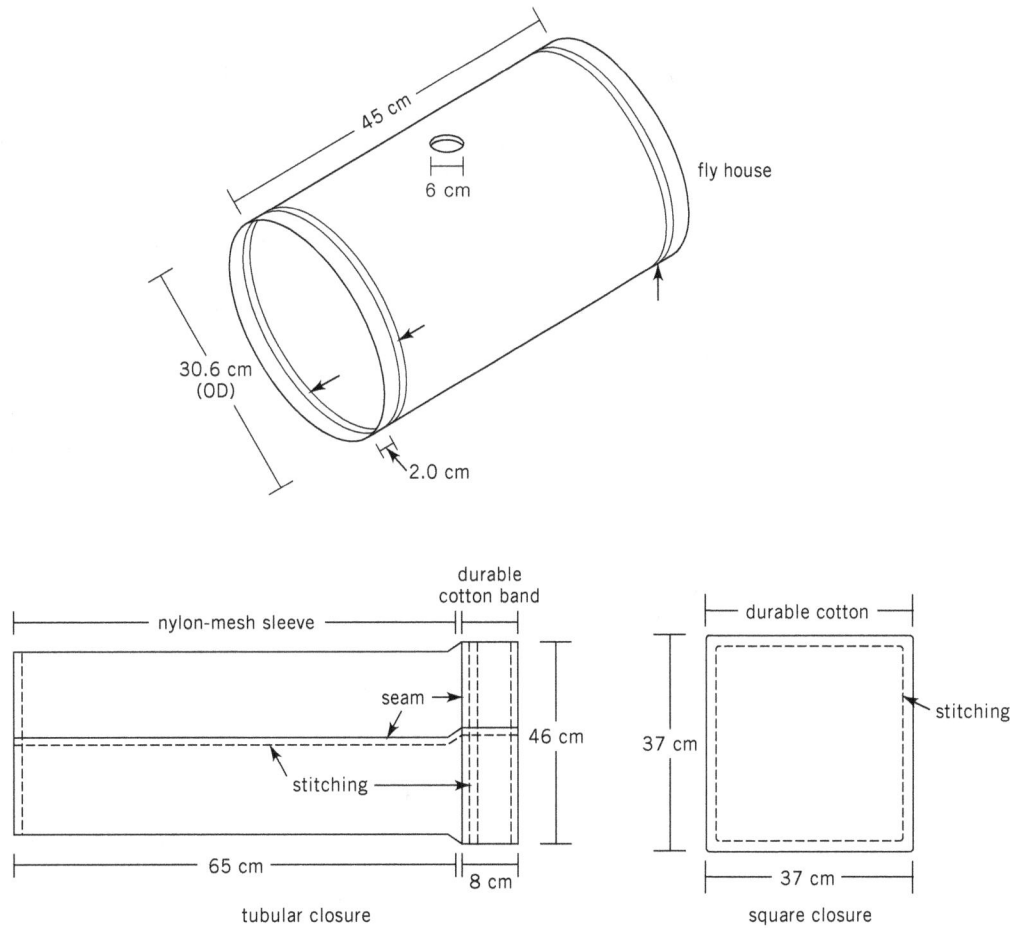

**Figure 30.1.** Diagram of a cylindrical fly house and a set of cloth closures. The dimensions are indicated in centimeters. The small arrows indicate the 0.5-cm wide, 0.2-cm deep grooves at either end of the house. OD is outer diameter. The tubular closure has a single seam that runs the length of the nylon-mesh sleeve and the full width of the cotton band. The dimensions for both cloth closures were taken while each was lying flat.

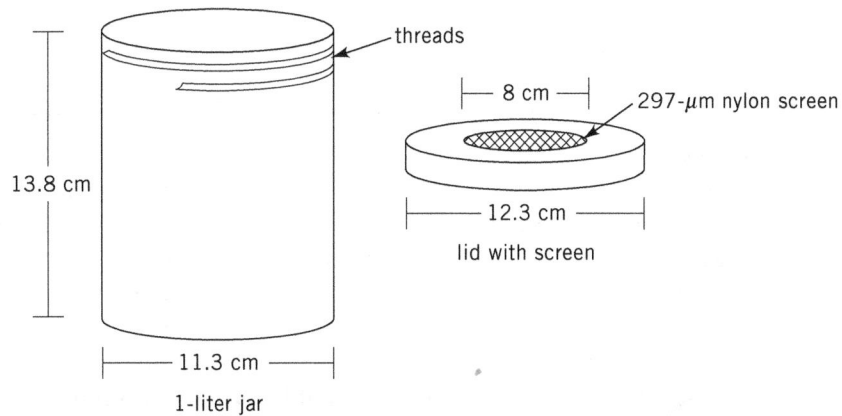

**Figure 30.2.** Diagram of one fly condo consisting of a 1-liter wide mouth jar and its screw-top lid. The lid is equipped with a 297-μm nylon screen for air flow.

## Media Preparation

For the preparation of larval condos, use a standard cornmeal-based fly media. Fill each condo with 100 ml of fly media. The food must be moist when the embryos are added. If the food is pulling away from the container walls, it is too dry; conversely, if the walls of the containers carry condensation, conditions are too wet. Find a way to store the food so that it is moist when needed. We make fly media the evening before embryos are added to the condos. After adding the media to each condo, loosely secure the screen-covered lids to the jars, cover the condos with plastic bags, and leave at room temperature overnight. When the embryos are added in the morning, the food is moist and the container walls are mostly dry.

For the preparation of embryo collection trays, we use grape juice–agar media. A less expensive alternative to grape juice–agar media is to use regular cornmeal-based media, provided the agar concentration is adjusted to 2.0%. When the agar content is lower, embryos become embedded in the media and fewer are washed from the collection trays, lowering yields. Grape juice–agar media is used largely for a slight enhancement in yield.

When determining how many embryo collection trays to prepare, consider the following: (1) An active fly house must always contain a collection tray for collections and to keep the flies well-fed between collections; (2) trays can be rinsed with warm tap water and reused once for noncollection, feeding periods; and (3) the shelf-life of the media at 4°C (several months for grape juice agar). It is usually a good idea to prepare more trays than may be needed. A 25.7 × 20.6-cm tray requires about 150 ml of egg-laying media.

> ### *Preparation of Grape Juice–Agar (1 liter)*
>
> 1. Add 20 g of agar to 750 ml of $H_2O$ and autoclave to melt the agar.
>
> 2. Combine 250 ml of concentrated (3:1) grape juice, 6.25 g of sucrose, and 1.5 g of p-hydroxybenzoic acid methyl ester (HAME, Sigma H 5501). Bring the mixture to a boil over a Bunsen burner or a hot plate; stir to dissolve the HAME and to prevent burning. Remove the mixture from the heat once the HAME has dissolved.
>
> 3. Mix the agar and grape juice solutions together once each has cooled to 70°C. Allow the Grape Juice–Agar solution to sit, allowing bubbles to rise.
>
> 4. Pour the media into the Styrofoam trays, forming a thin smooth layer (150 ml per 25.7 × 20.6-cm Styrofoam tray).
>
> 5. Once the agar has solidified, store media at 4°C. Store the Styrofoam trays in plastic bags, taking care to not gouge the media's surface.

## CULTURING CONDITIONS

*The most important features contributing to large embryo yields are the culturing conditions for the larvae and adults.* Keep the average temperature and average relative humidity at 25°C and 70%, respectively. Do not assume that the set and actual temperature and humidity coincide; they probably will not and will almost certainly display a range. Routinely monitor both variables with the handheld digital pyrometer and make the appropriate adjustments so that the desired average temperature and humidity are obtained.

Keep disturbances to the adults to an absolute minimum. Frequent disturbances will decrease embryo yields. For this reason, it is strongly recommended that the fly facility be separated from other lab stocks. This also reduces the probability of cross-contaminating stocks. It is a good practice to keep the incubator space clean—the embryo yields seem to improve in a clean environment.

Finally, although a regular light-dark cycle has been found to improve collection yields (Shaffer et al. 1994), we have not tried this and instead keep the flies in continuous low-level light.

## SCHEDULING

First, establish a regular weekly schedule. This will ultimately improve the yields of the collections. A sample schedule is provided in Figure 30.3. A full fly generation lasts 18 days, beginning when embryos are added to the condos and ending with the last collection day. Eleven days after the embryos are added to the condos, adults eclose (when cultured at 25ºC and 70% relative humidity). At this stage, transfer the flies from the condos to the embryo houses. Yields are best 3–4 days later and finally decline on the 18th day.

*Choosing a day of the week to add the embryos to the condos is the most important scheduling decision; it dictates the timing of everything else.* If restricting the fly maintenance work to weekdays is necessary, Friday is the best day to add the embryos. This maximizes the number of embryo collections that fall on weekdays. If the person maintaining the flies can work about 2.5 hours on Saturday or Sunday, then add the embryos to the condos on Wednesday or Thursday, respectively. This arrangement will place all the best embryo collections (days 14–16) on weekdays.

## A SOURCE FOR EMBRYOS

The quickest way to get started is to obtain a single overnight collection on two consecutive weeks from colleagues, maintaining an Oregon-R facility. By the third week, a large population of adults will exist from which to collect embryos (Figure 30.3). A minimum of 8 g is required from each collection to seed 80 condos.

Alternatively, expand a wild-type stock in bottles until approximately 10–20 g of adults are obtained and then combine them into a single house (see p. 547 [Day 11 procedure]). After 2–3 days of culturing these adults, an overnight collection should yield enough

| S | M | T | W | Th | F | S |
|---|---|---|---|---|---|---|
| | | | | | DAY 1 Add embryos to condos | DAY 2 |
| DAY 3 | DAY 4 Feed larva with dry yeast (optional) | DAY 5 | DAY 6 | DAY 7 | DAY 8 Absorb excess food moisture with paper towels | DAY 9 |
| DAY 10 | DAY 11 Transfer adults from condos to houses | DAY 12 | DAY 13 Wash dirty condos and houses | DAY 14 | DAY 15 Use an overnight collection to seed new condos | DAY 16 |
| DAY 17 | DAY 18 Put houses in the cold room | | | | | |

**Figure 30.3.** Recommended maintenance schedule for one full generation. Each day of the generation is indicated in the upper left corner and each task is noted. Cross-hatching indicates embryo collection days. On day 15, a new generation begins.

embryos to seed 40–80 condos. Pamper these adults, feeding them twice daily, so that a single long overnight collection can be taken on two consecutive weeks.

PROTOCOL 30.1*

# Maintaining a Population

For a recommended maintenance schedule, see Figure 30.3.

## Materials

### Supplies and Equipment

Metal sieves; paint brush or rubber kitchen spatula; one large funnel with a 20-cm mouth diameter and wide neck; and eight size-12 rubber or cork stoppers
Embryo collection trays and larval condominums (see p. 544 [Media preparation])
Metal spatula
Weigh boat
Squirt bottle of Embryo Wash Solution (see below)
Dry yeast
Fly houses and cloth closures (see pp. 541–542)
Large rubber bands or gaskets
Adhesive tape
Erlenmeyer flask (1 liter)
$CO_2$ station

### Solutions and Reagents

#### Dechorionating Solution

50% **Bleach**
0.7% NaCl
0.03% Triton X-100

#### Embryo Wash Solution

0.7% NaCl
0.03% Triton X-100

#### Inoculation Solution

14.4% (w/v) Sucrose
0.7% (w/v) NaCl
0.05% Triton X-100

#### Yeast Paste

In a 2-liter beaker, gradually mix active dry yeast (Red Star or Fleischmann's) and $H_2O$ with a sturdy kitchen spoon or spatula. Adjust the ratio of yeast and $H_2O$ until about 1 liter of paste is obtained, the consistency of cake frosting. (The amount of yeast required will depend on the amount of $H_2O$.) Leave the stirring implement in the beaker and cover the beaker with plastic wrap or foil. Place the beaker in a dish 1–2 inches deep and store it in the cold room. The yeast sometimes expands and pours out of the beaker and thus the dish will contain it.

**CAUTION: bleach, $CO_2$** (see Appendix 4)

*For all protocols, $H_2O$ indicates glass distilled and deionized.

## Method

*Day 1: Seed fly condos with 0.1 g of embryos (requires 2–2.5 hours) as follows:*

1. Stack the three metal sieves from largest mesh size to smallest (top to bottom) on the bottom of a sink. Rest a collection tray in the top sieve and rinse the embryos from the tray with $H_2O$. Use a paint brush or rubber spatula to clear the surface of residual embryos. Repeat this for each tray. Collect as many embryos as possible to the side of the bottom sieve with running $H_2O$ and then rinse the embryos with Embryo Wash Solution from a squirt bottle. From below the screen, blot the embryos dry with a paper towel.

2. Use a metal spatula to transfer the embryos to fresh Dechorionating Solution and incubate for 2 minutes to dechorionate. Pour the embryos into the small-mesh-size sieve and rinse away the bleach completely. After thoroughly rinsing the embryos, collect them to the side of the sieve and blot them dry as before.

   *Note:* Although dechorionating the embryos is not essential, it is recommended to ensure against mite infestations.

3. Transfer the embryos to a weigh boat and determine their total weight. Use 0.1 g to seed each condo.

4. Suspend the embryos in Inoculation Solution at a concentration of 0.05 g/ml and deliver 2 ml to each condo. Sprinkle some dry yeast into each condo, secure the screen-covered lids, and culture at 25°C, 70% relative humidity.

*Day 4: (Optional) Feed the larvae (requires 0.5–1 hour).*

Sprinkle approximately 1/4 teaspoon of dry yeast into each condo.

   *Notes:* In the interest of time, it is best to remove the lids of 10 condos at a time, add the yeast, and secure the lids.

   Feeding the larva can also be done on either day 3 or 5, instead of day 4.

*Day 8: Absorb excess liquid from the larval media if necessary (requires 1 hour).*

If the medium is wet, place a single absorbent paper towel into the food. Make sure that the paper towel is making good contact with the food, but not completely blocking the food's surface.

   *Notes:* This simple step can dramatically increase the yield of adults by reducing the number that become stuck in the wet food. This ultimately pays off with larger embryo yields.

   This job can also be done anytime between days 6 and 9.

*Day 11: Transfer adults from the condos to the houses (requires 2–2.5 hours) as follows:*

1. Affix the cloth closures to four fly houses. Stand one house on its end either on a clean benchtop or on the floor. Secure the square cloth closure to the upturned end with a large rubber band or gasket. Ensure that the cloth is sitting snugly against the house and that the rubber band lies in the groove. Turn the house over and attach the tubular closure to the opposite end. Slip the end with the durable cotton band around the circumference of the house. Secure the tubular closure with a rubber band as above. Twist the sleeve of the tubular closure into a rope and tie it into a loose

knot. Finally, secure the rubber bands (or gaskets) with masking tape by wrapping the tape around the circumference of each rubber band (or gasket). Repeat these steps with the remaining three houses.

2. Set up a $CO_2$ station that is relatively close to a benchtop where the fly houses are set up, and sufficiently out of the way so that other laboratory business is not disturbed.

3. Gather a 1-liter Erlenmeyer flask and the large funnel. Weigh the flask alone and record the weight (do not tare the scale with the flask, someone may need to retare it).

4. Move the condos from the incubator to the $CO_2$ station on one or two carts. Hold a condo in one hand with the lid angled toward the floor. Positioning the condo in this way prevents flies from getting stuck in the food. With the opposite hand, deliver $CO_2$ to the flies through the lid's screen using a $CO_2$ hose equipped with a clip at its end to regulate flow. When the flies fall from the condo wall, stop the flow of $CO_2$ and transfer the flies to the 1-liter flask. Place the large funnel in the mouth of the flask, position the condo above the funnel, and remove the condo's lid. As the flies are being transfered, keep the condo inverted. Tap both the jar and the lid on the funnel to knock additional flies loose. Pool the flies from 20 condos in this way, weigh them, and record the weight. As the flies are being pooled, periodically deliver $CO_2$ directly into the flask to keep the flies asleep. Add the flies from 20 condos to a single house through the 6-cm hole using the funnel, and then plug the hole with a rubber stopper. Repeat these steps for the remaining condos. When all the flies have been transferred, place the condos in the cold room.

   *Note:* Typically 20 condos yield 30–40 g of adults.

5. Place fresh trays with Yeast Paste into each house. Spread the Yeast Paste as a thin band (~14 X 4 cm) lengthwise down the middle of each collection tray. Place the yeasted trays together in pairs facing each other to keep the flies away. Untie the loose knot in the tubular closure and untwist the mesh sleeve. With one hand, hold the end of the sleeve and give it several shakes to keep the flies at bay. With the opposite hand, grab a yeasted tray and place it inside the house. To keep flies from escaping, draw the tubular closure around your wrist while delivering the tray. While removing your hand, shake the sleeve and then tie a loose knot in it as before. Repeat these steps for the other houses. Place them all at 25ºC and 70% relative humidity. If the houses are seeded in the morning, replace the morning trays with fresh trays with Yeast Paste before going home for the day.

*Days 13–17: During this period, collect embryos as follows:*

Females are the most productive between days 14 and 16, especially in the early morning hours. Therefore, collections that include the early morning hours of days 14–16 are recommended. Short collections (1–6 hours) are particularly good to acquire at this time.

1. Prewarm the collection trays and yeast paste. The night before a collection day, place the trays that are needed at room temperature in closed plastic bags. With an early morning collection, spread Yeast Paste onto the trays the night before the collection; for all other collections, spread the Yeast Paste onto the trays 1–2 hours before the collection.

*Notes:* A warm collection tray presents less of a disturbance to the flies than a cold one and helps increase embryo yields.

To keep flies off the collection trays and to keep the Yeast Paste from smearing, place the yeasted trays together in pairs facing each other, and then place them in a closed plastic bag.

2. When the collection trays are ready, untie and straighten the tubular closure and give it several shakes. Grab hold of the tray inside the house, turn it vertically, and tap it several times against the house to knock away the flies on its surface. Draw the tray into the mesh sleeve and blow onto its surface, through the mesh, to knock additional flies back into the house. Remove the tray, give the sleeve several more shakes, insert a new tray, and tie off the mesh sleeve. Throughout the collection period, keep the flies well fed. At a minimum, every house should get a collection tray with fresh yeast paste in the morning and again in the evening on days 12–17.

3. After making a collection, examine 50–100 dechorionated embryos under a microscope to verify the developmental staging (see Campos-Ortega and Hartenstein 1985).

4. Process embryos as described on p. 550 (Handling embryos for protein biochemistry).

**Day 13**: *Clean condos (requires 1 hour for two people or 2–2.5 hours for one person).*

Wash the condos that were placed in the cold room on day 11 with soap and water. This can also be done on day 12.

**Day 15**: *Start the next fly generation (requires 2–2.5 hours).*

Use an overnight collection from a single fly house to seed condos for the next fly generation. Follow the same procedure described for day 1.

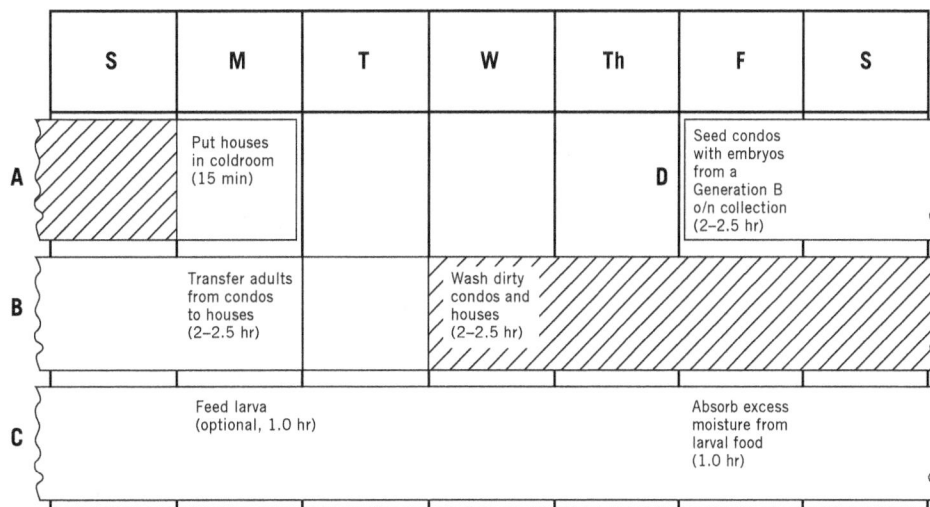

**Figure 30.4.** Sample weekly schedule. A portion of four fly generations (*A, B, C,* and *D*) overlaps with each week. The presence of a generation is indicated by a horizontal bar. Generation *A* ends on Monday, *D* begins on Firday, and both *B* and *C* run the length of the week. The tasks for each day are noted. The cross-hatching indicates embryo collection days. The approximate number of work hours required per task is indicated in parentheses. The time required to make condo fly media is not indicated.

*Day 18: Prepare for the next embryo collection.*

Remove the last collection trays from each house and place the four houses in the cold room. The houses and cloth closures can be washed anytime up to Sunday morning; the following Monday they will be needed. One full day is required for the cloth closures to air dry.

## COORDINATING OVERLAPPING FLY GENERATIONS

To sustain active fly houses each week, maintain three staggered fly generations each week. During any given week, adults, from which embryos are collected, larva, and embryos will be available. To see how the different tasks involved in maintaining these three generations interdigitate with one another each week, see Figure 30.4.

## HANDLING OF EMBRYOS FOR PROTEIN BIOCHEMISTRY

Embryos can be handled in one of several ways: (1) They can be dechorionated and processed immediately for protein biochemistry. (2) Individual collection trays can be placed in plastic bags or plastic wrap in the cold room until the desired number of collections is obtained and then processed all together. If the protein of interest is stable under these conditions, this approach is convenient and economical. Many transcription factors are stable under these conditions for up to 3 days (Biggin and Tjian 1988). (3) Embryos or their extracts can be frozen. To freeze embryos, place dechorionated embryos in a suitable tube, freeze in liquid $N_2$, and store at –80ºC. To freeze embryo extracts, homogenize the dechorionated embryos, and following centrifugation, freeze the supernatant and pellet in the same way. The proper method of storage for a given protein must be determined empirically.

## REFERENCES

Becker P.B. and Wu C. 1992. Cell-free system for assembly of transcriptionally repressed chromatin from *Drosophila* embryos. *Mol. Cell. Biol.* **12:** 2241–2249.

Biggin M.D. and Tjian R. 1988. Transcription factors that activate the *Ultrabithorax* promoter in developmentally staged extracts. *Cell* **53:** 699–711.

Cole D.G., Saxton W.M., Sheehan K.B., and Scholey J.M. 1994. A "slow" homotetrameric kinesin-related motor protein purified from *Drosophila* embryos. *J. Biol. Chem.* **269:** 22913–22916.

Fessler L.I., Nelson R.E., and Fessler J.H. 1994. *Drosophila* extracellular matrix. *Methods Enzymol.* **245:** 287–289.

Field C.M., Al-Awar O., Rosenblatt J., Wong M.L., Alberts B., and Mitchison T.J. 1996. A purified *Drosophila* septin complex forms filaments and exhibits GTPase activity. *J. Cell Biol.* **133:** 605–616.

Liu Z.P., Galindo R.L., and Wasserman S.A. 1997. A role for CKII phosphorylation of the *cactus* PEST domain in dorsoventral patterning of the *Drosophila* embryo. *Genes Dev.* **11:** 3413–3422.

Maroto F.G. and Sierra J.M. 1989. Purification and characterization of mRNA cap-binding protein from *Drosophila melanogaster* embryos. *Mol. Cell. Biol.* **9:** 2181–2190.

Miller K.G., Field C.M., Kellogg D.R., and Alberts B.M. 1991. Use of actin filament and microtubule affinity chromatography to identify proteins that bind to the cytoskeleton. *Methods Enzymol.* **196:** 303–319.

Mitsis P.G., Chiang C.-S., and Lehman I.R. 1995. Purification of DNA polymerase-primase (DNA polymerase α) and DNA polymerase δ from embryos of *Drosophila melanogaster*. *Methods Enzymol.* **262:** 62–77.

Moritz M. and Alberts B.M. 1999. Isolation of centrosomes from *Drosophila* embryos. *Methods Cell Biol.* **61:** 1–12.

Roberts D.B. 1986. Drosophila: *A practical approach.* IRL Press, England.

Saxton W.M., Porter M.E., Cohn S.A., Scholey J.M., Raff E.C., and McIntosh J.R. 1988. *Drosophila* kinesin: Characterization of microtubule motility and ATPase. *Proc. Natl. Acad. Sci.* **85:** 1109–1113.

Shaffer C.D., Wuller J.M., and Elgin S.C. 1994. Raising large quantities of *Drosophila* for biochemical experiments. *Methods Cell Biol.* **44:** 99–108.

Soeller W.C., Poole S.J., and Kornberg T. 1988. *In vitro* transcription of the *Drosophila engrailed* gene. *Genes Dev.* **2:** 68–81.

Travaglini E.C. and Tartof D. 1972. "Instant" *Drosophila*: A method for mass culturing large numbers of *Drosophila*. *DIS* **48:** 157.

# CONTENTS

# 31

# Preparation of Nuclear Extracts from *Drosophila* Embryos and In Vitro Transcription Analysis

**Michael J. Pazin**
*Cutaneous Biology Research Center*
*Massachusetts General Hospital*
*Charlestown, Massachusetts 02129*

THERE IS INCREASING AWARENESS OF THE ADVANTAGES of *Drosophila* as an organism for biochemical studies. The large base of genetic information currently available provides a conceptual framework for the investigation of mechanisms through biochemistry. Kilogram quantities of *Drosophila* embryos can be grown in incubators at relatively low cost. The similarity of vertebrate biochemical pathways to those in *Drosophila* facilitates generalizing results obtained with either experimental system.

Just as *Drosophila* cytoplasmic extracts are used by several laboratories in the investigation of chromatin (Becker and Wu 1992; Kamakaka et al. 1993; Pazin and Kadonaga 1998), *Drosophila* nuclear extracts are often used in the study of gene expression. Nuclear extracts can be used directly as a source of transcription factors, with the advantages that they are straightforward to prepare and highly active and appear to contain appropriate amounts of basal factors and coactivators. Nuclear extracts can also be fractionated, in some cases to purified proteins, with the advantage that the components of the reaction can be defined. Transcription experiments with nuclear extracts and fractionation studies of nuclear extracts have revealed many similarities between fly and human gene regulation (Kadonaga 1990; Wampler et al. 1990; Austin and Biggin 1996; Burke and Kadonaga 1997).

This chapter contains two protocols, one for making a *Drosophila* embryonic nuclear extract called soluble nuclear fraction (SNF), and one for RNA polymerase II transcription in vitro. SNF transcribes DNA and chromatin templates and functions with several *Drosophila* and mammalian promoters containing different sets of core promoter elements. SNF also functions with several activating proteins. This highly active extract is suitable for many in vitro transcription experiments.

## NUCLEAR EXTRACT PREPARATION

Protocol 31.1 presents the preparation of embryonic nuclear extracts, a method from the Kadonaga laboratory (Kamakaka et al. 1991; Kamakaka and Kadonaga 1994) with minor modifications. This extract is called SNF, for soluble nuclear fraction, to distinguish it from other methods. The now-standard SNF procedure involves a low-salt extraction of nuclei (Kamakaka and Kadonaga 1994), in contrast to other procedures that involve high-salt extraction (Heiermann and Pongs 1985; Soeller et al. 1988; Zhang et al. 1996). Low-salt extraction (0.1 M salt) probably explains the relatively low amount of nonspecific DNA-binding proteins, such as core histones and linker histones, present in SNF compared to some other extracts (0.4 M salt).

Embryos are grown in population cages as described in Sisson (this volume) and in Shaffer et al. (1994). Approximately 50–100 trays of 0–12-hour embryos are used in a preparation, with each tray containing 2.5–4 g of embryos, depending on culture conditions and timing within the fly cycle. It is standard to store the earliest trays in plastic wrap at 4°C for as long as 72 hours while further trays are accumulated. This extract is made with fresh, not frozen, embryos. A key to making highly active extracts is to work quickly to minimize protein degradation and leakage of proteins from the nuclei before extraction. It is also important to prepare all solutions from the highest-quality water available (glass-distilled and deionized, or Milli-Q-grade water). We do not treat the water with DEPC (diethyl pyrocarbonate), and we note that others have suggested that DEPC treatment may inhibit transcription (Zhang et al. 1996). Making SNF on this scale takes approximately 6 hours. As this procedure has been found to work with as little as 0.5 g of embryos (Kamakaka and Kadonaga 1994), it should be possible to make small-scale extracts from mutants if desired.

PROTOCOL 31.1*

## Preparation of SNF from *Drosophila* Embryos

### Materials

#### Stocks

*Drosophila* embryos, Canton S, 0–12 hours (~200 g)

#### Supplies and Equipment

Embryo-collection apparatus (CBS Scientific FC-100); similar to the device described in Shaffer et al. (1994).
Paint brush (1 inch) for collecting embryos
Plastic washtub for holding the lower level of the embryo-collection apparatus
Beckman ultracentrifuge and *SW28 swinging-bucket rotor (not SW28.1)
*SW28 polycarbonate tubes (1 × 3 inches, Beckman 355631)
Glass rod for stirring
Homogenizer (Yamato LH-21)
Nitex nylon screen; 500-μm mesh size (Sefar America 3-500/49) and 70-μm mesh size (Sefar America 3-70/43)
Funnel
Miracloth (Calbiochem 475855)
Sorvall centrifuge and *GSA rotor or equivalent
Centrifuge bottles for the Sorvall GSA rotor
Dounce tissue grinder (*40 ml) (Wheaton, Fisher 06-435C) and "B" pestle
Screw-cap polypropylene tube (50 ml)
Microcentrifuge tubes (1.5 ml)

*For all protocols, H₂O indicates glass distilled and deionized.

## Solutions and Reagents

**Bleach** (3 liters, 50% v/v). Combine 1.5 liters of $H_2O$ with 1.5 liters of bleach (5.25% sodium hypochlorite).

**Liquid nitrogen**

Solutions. The following are required for the preparation of the buffers given below:

0.5 M **DTT** (dithiothreitol; USB 15397). Prepare the 0.5 M stock and store at –20ºC.

0.5 M **sodium metabisulfite** (Sigma S 9000). Dissolve 0.95 g in 10 ml of $H_2O$; store at 4ºC. Use within 12 hours of preparation.

0.2 M **PMSF** (phenylmethylsulfonyl fluoride; Sigma P 7626). Dissolve in 100% **ethanol**; store at –20ºC.

0.5 M benzamidine (Sigma B 6506). Dissolve in $H_2O$; store at –20ºC.

### *Embryo Wash*

0.12 M NaCl
0.04% (v/v) Triton X-100
Prepare 1 liter.

### *Buffer 1*

Prepare 2 liters for 250 g of embryos:
15 mM HEPES, potassium salt (pH 7.6)
10 mM **KCl**
5 mM **MgCl₂**
0.1 mM EDTA
0.5 mM EGTA
350 mM sucrose
Store at 4ºC. Just before use, add the following (to 2 liters):
4 ml of 0.5 M DTT
4 ml of 0.5 M sodium metabisulfite
2 ml of 0.2 M PMSF
4 ml of 0.5 M benzamidine

### *Buffer AB*

Prepare 250 ml for 250 g of embryos:
15 mM HEPES, potassium salt (pH 7.6)
110 mM KCl
5 mM MgCl₂
0.1 mM EDTA
Store at 4ºC. Just before use, add the following (to 250 ml):
1.25 ml of 0.5 M DTT
0.625 ml of 0.5 M sodium metabisulfite
0.31 ml of 0.2 M PMSF
0.625 ml of 0.5 M benzamidine

### *HEMG20 containing 0.1 M KCl*

Prepare 25 ml for 50 g of nuclei:
25 mM HEPES, potassium salt (pH 7.6)
0.1 M KCl
12.5 mM MgCl₂
0.1 mM EDTA
20% (v/v) Glycerol

Store at 4ºC. Just before use add the following (to 25 ml):
    75 μl of 0.5 M DTT
    50 μl of 0.5 M sodium metabisulfite
    12.5 μl of 0.2 M PMSF
    50 μl of 0.5 M benzamidine
*Place these items in the cold room the night before the procedure is to be performed.

**CAUTION: bleach, DTT, ethanol, KCl, liquid nitrogen, MgCl$_2$, PMSF, sodium metabisulfite** (see Appendix 4)

## Method

Wear gloves for the entire procedure to prevent contamination by nucleases and proteases.

1. Collect embryos into an embryo collection apparatus by washing the embryos from the agar trays with tap water and a 1-inch paint brush.

    *Note:* The embryos pass through the upper layer of the apparatus with the 500-μm sieve and are collected on the 70-μm sieve (lower layer). Collection is done in a sink so that the running water can drain through the collection apparatus.

2. Remove the upper sieve containing flies and fly parts; discard the debris and wash the nylon (can be reused). Rinse the embryos on the lower sieve thoroughly with tap water to wash contaminating yeast through the sieve.

3. Place the lower level of the collection apparatus containing the embryos in a plastic wash tub. Soak for exactly 90 seconds in 50% (v/v) bleach, by pouring 3 liters of 50% bleach onto the embryos; swirl to mix.

4. Immediately rinse with 1 liter of Embryo Wash, by pouring Embryo Wash directly through the embryo collection apparatus; swirl to mix.

    *Note:* The rinsing in this step and step 5 are done in a sink to allow the collection apparatus to drain.

5. Rinse with H$_2$O to remove excess Embryo Wash and bleach (~1 liter of H$_2$O).

6. Dry embryos by wiping the bottom of the nylon mesh with paper towels to absorb excess water.

7. Weigh embryos in a large weigh boat (typically 100–250 g of embryos are used in an extract). *Perform all subsequent operations at 4ºC in a cold room.*

8. Combine 3 ml of Buffer I per gram of embryos, and stir with a glass rod to disperse the embryos evenly. Disrupt embryos with a single passage through the Yamato LH-21 homogenizer at 1000 rpm (setting = 100). Once the embryos are homogenized, work quickly to minimize amount of protein that leaks out of the nuclei into the buffer!

    *Note:* If a Yamato homogenizer is not available, other methods of homogenizing embryos may work (Heiermann and Pongs 1985; Soeller et al. 1988; Kamakaka and Kadonaga 1994).

9. Filter the homogenate through a funnel lined with Miracloth into a centrifuge bottle.

10. Wash the debris remaining in the Miracloth with additional Buffer I (2 ml/g of embryos).

11. To pellet nuclei, centrifuge at 10,000g in a Sorvall centrifuge using a GSA rotor at 4ºC for 15 minutes. Carefully decant supernatant (the pellets are very loose). Wipe off the white lipid coating from the walls of the centrifuge bottles with Kimwipes.

12. Resuspend the nuclei in Buffer I (3 ml/g of embryos); do not resuspend the small, solid yellow yolk pellet trapped under the larger, loose white nuclear pellet. Use a 40-ml Dounce tissue grinder with a B (loose) pestle to disperse the nuclei (three strokes), and transfer to clean centrifuge bottles.

13. Repellet the nuclei by centrifugation in the GSA rotor at 10,000g at 4°C for 15 minutes. Carefully decant the supernatant, and wipe the lipids from the walls of the centrifuge bottles with Kimwipes.

14. Resuspend the nuclei in Buffer AB (1 ml/g of embryos); do not resuspend the small yellow yolk pellet if present. Use three strokes with a 40-ml Dounce with a B pestle to disperse the nuclei. Transfer the entire suspension containing nuclei into one *preweighed* centrifuge bottle.

15. Centrifuge in the GSA rotor at 10,000g at 4°C for 10 minutes.

16. Decant the supernatant. Determine the total weight of the bottle and pellet. Determine the mass of the nuclei by subtracting the weight of the empty bottle from the total weight of the bottle and the pellet.

   *Note:* Typically 5 g of embryos yields approximately 1 g of nuclei. A lower yield may indicate incomplete homogenization at step 8.

17. For each gram of nuclei, add 0.5 ml of HEMG20 containing 0.1 M KCl. Resuspend the pellet by gently swirling the nuclei (do not use a Dounce homogenizer). Place the suspension on ice for 15–60 minutes.

18. Centrifuge the mixture in the SW28 rotor at 24,000 rpm (100,000g) at 4°C for 1 hour.

   *Note:* The SW28 tubes hold approximately 29 ml and do not have to be full during centrifugation.

19. After centrifugation, four distinct layers will appear (from top to bottom [a–d]) as follows:

   a. A thin opaque white lipid layer. Remove it with a spatula and discard.

   b. A clear, yellow-tinted liquid layer, which comprises about 50% of the total volume of the tube; this is the nuclear extract. Use a pipette to transfer the layer into a 50-ml screw cap polypropylene tube.

   c. A thin, cloudy gray liquid layer may be visible. Avoid this material—it will inhibit transcription.

   d. At the bottom, a solid off-white layer of nuclear material (DNA, scaffold, etc.).

20. Pool the nuclear extract, and quick-freeze in 300-μl aliquots in liquid nitrogen. Store at –80°C. A typical yield from 200 g of embryos is approximately 30 ml of SNF.

## IN VITRO TRANSCRIPTION

Protocol 31.2 for RNA polymerase II transcription with DNA templates is based on a method described by Kamakaka et al. (1991; Kamakaka and Kadonaga 1994) with modifications (Pazin and Kadonaga 1998). Although SNF was developed with DNA templates, it can also transcribe chromatin templates with high efficiency in the presence of transcriptional activators (Kamakaka et al. 1993; Pazin et al. 1994; Mizuguchi et al. 1997). The standard reactions are 50 μl containing 75 ng of DNA. The procedure consists of transcription and RNA purification (steps 1 through 7) followed by detection of the transcripts by primer extension with reverse transcriptase and a labeled oligonucleotide. Primer extension combines high sensitivity with confirmation that the transcripts are initiated from the promoter, rather than nonspecific initiation events occurring outside the promoter. Other methods for detecting transcripts have been reviewed recently (Mason et al. 1993; Zhang et al. 1996). Approximately 90 minutes are required for transcription and RNA purification, 2 hours for primer extension, and 2 hours for gel electrophoresis.

Many modifications of this basic protocol are possible, depending on the experimental question. The DNA can be preincubated with transcriptional regulators before adding SNF and rNTPs. Transcriptional start sites are mapped by running the primer extension products adjacent to a sequencing ladder made from the same template and primer used for transcription. The absolute amount of transcription is determined by quantitating the primer-extension products and standards with a phosphorimager. SNF may contain transcriptional activators and repressors in high enough concentration to alter the amount of transcription; this can be tested by comparing transcription with and without binding sites for the protein in question. Supercoiled plasmids are generally used as substrates, but relaxed and linear DNAs also appear to function. Increasing the DNA concentration generally results in lower efficiency (less transcription per template) and can actually decrease the total amount of transcription depending on the promoter (M.J. Pazin, unpubl.). Under the standard reaction condition of 20% (v/v) SNF, the RNA polymerase II machinery is limiting, as decreasing the amount of SNF per reaction reduces the amount of transcription (M.J. Pazin, unpubl.).

## PROTOCOL 31.2

## In Vitro Transcription Using SNF

### Materials

#### Supplies and Equipment

Microcentrifuge tubes (1.5 ml)

Microcentrifuge

Heating block set at 70–85°C

Heating block or water bath at 58°C (or appropriate annealing temperature)

Equipment for sequencing-gel electrophoresis

#### Solutions and Reagents

DNA diluted in DNA Dilution Buffer (see below) to a final concentration of 1 μg per 200 μl (i.e., 5 ng/μl).

HEPES (1 M), potassium salt (pH 7.5)

Potassium glutamate (3 M)

Transcription extract; SNF prepared as described in Protocol 31.1

**Polyvinyl alcohol** (10%) (m.w. 30,000–70,000; Sigma P 8136) dissolved in $H_2O$. Store a small aliquot at 4°C, the remainder at –20°C.

**PEG** (10%) (polyethylene glycol; m.w. 15,000–20,000; Sigma P 2263) dissolved in $H_2O$. Store a small aliquot at 4°C, the remainder at –20°C.

rNTP solution, containing 5 mM of each of the four rNTPs (ATP, GTP, CTP, UTP); made from 100 mM stocks (Promega, Fisher PR-E6000), store in 500-μl aliquots at –20°C.

Sodium acetate (0.3 M)

**Phenol:chloroform:isoamyl alcohol**

**Ethanol**

Labeled oligonucleotide primer. Typically, an oligonucleotide (28 nucleotides) is 5′ end-labeled with T4 polynucleotide kinase and [γ-$^{32}$P]ATP (Sambrook et al. 1989). The optimal annealing conditions will vary depending on the length and G+C content of the oligonucleotide. Primers are generally selected so that the extension product is about 100 nucleotides.

Polyacrylamide-urea sequencing gel (8% w/v), and buffer for electrophoresis

### DNA Dilution Buffer

10 mM HEPES, potassium salt (pH 7.5)

50 mM potassium glutamate

4.5 mM **MgCl$_2$**

10% (v/v) glycerol

### HEMG

25 mM HEPES, potassium salt (pH 7.6)

0.1 mM EDTA

12.5 mM MgCl$_2$

10% (v/v) glycerol

This solution can be stored at 4°C.

### Transcription Stop–Proteinase K Buffer

The composition of transcription-stop solution is:

20 mM EDTA, Na$^+$ (pH 8)

0.2 M NaCl

1% (w/v) SDS

0.25 mg/ml glycogen (Sigma G 0885)

Store at room temperature for up to 3 months, or at –20°C. If stored at –20°C, heating is necessary to redissolve the SDS. Prepare a stock solution of proteinase K (2.5 mg/ml; USB 20818) in TE. Store at –20°C. The solution can be freeze-thawed multiple times. Just before use, combine transcription stop and proteinase K in a 20:1 (v/v) ratio before adding to samples. Prepare fresh each time.

### Annealing Mix

10 mM Tris-Cl (pH 7.8)

1 mM EDTA

0.25 M **KCl**

0.83 nM labeled oligonucleotide primer

This mix can be prepared as a 5X stock without primer and stored at –20°C. Dilute with TE before use.

### Reverse Transcriptase–Primer Extension Mix

Prepare the extension mix as follows:

| | |
|---|---|
| 1 M Tris-Cl (pH 8.3) | 2.5 ml |
| 0.1 M **MnCl$_2$** | 500 µl |
| 2.5 mg/ml **actinomycin D** (USB 10415) in H$_2$O | 2 ml |
| 0.1 M solution of dNTPs (mixture of dATP, dGTP, dCTP, and dTTP; Promega U1240) | 132 µl |
| 0.5 M **DTT** | 1 ml |

Add the components in the order given. Add H$_2$O to bring the final volume to 40 ml. Store in 1-ml aliquots at –20°C in the dark.

Immediately before use, combine MMLV-reverse transcriptase (Fisher/Promega PR-M5301) with extension mix on ice. Add 10 units of reverse transcriptase (typically, ~0.05 µl) to 40 µl of extension mix for one reaction.

> ### *Formamide-NaOH Loading Buffer*
>
> Combine two volumes of formamide loading buffer with one volume of 0.1 M **NaOH.** Prepare immediately before use. The composition of formamide loading buffer is as follows:
>
> 80% (v/v) **formamide**
> 10 mM EDTA
> 1 mg/ml **xylene cyanol**
> 1 mg/ml **bromphenol blue**

**CAUTION: acrylamide, actinomycin D, bromophenol blue, CHCl$_3$, DTT, formamide, isoamyl alcohol, KCl, MgCl$_2$, MnCl$_2$, NaOH, PEG, phenol, polyvinyl alcohol, radioactive substances, urea, xylene cyanol** (see Appendix 4)

## Method

Wear gloves for the entire procedure to prevent contamination by nucleases and proteases.

1. Immediately before use, prepare the following cocktail. For each sample, the volumes of components are:

   | | |
   |---|---|
   | H$_2$O | 6.2 μl |
   | HEMG | 2.1 μl |
   | 1 M HEPES, potassium salt (pH 7.5) | 1.05 μl |
   | 3 M potassium glutamate | 0.69 μl |
   | 10% polyvinyl alcohol | 5 μl |
   | 10% PEG | 5 μl |
   | 5 mM rNTP stock | 5 μl |
   | SNF | 10 μl |

2. Add 15 μl of diluted DNA (~75 ng) to a 1.5-ml microcentrifuge tube, followed by 35 μl of the cocktail prepared in step 1. Mix gently by flicking the tube.

3. Incubate at room temperature for 20–60 minutes.

4. Add 100 μl of Transcription Stop–Proteinase K. Mix by vortexing. Incubate at 37°C for 15 minutes.

5. Add 0.3 M sodium acetate (250 μl), and then extract with phenol:chloroform:isoamyl alcohol (400 μl). Mix by vortexing for 2–5 seconds (excess vortexing may result in a large interface, and decreased RNA yield). Centrifuge in a microcentrifuge at 12,000–15,000g at room temperature for 5 minutes.

6. Transfer the aqueous phase to a 1.5-ml microcentrifuge tube containing ethanol (1 ml). Mix by vortexing. Centrifuge at 12,000–15,000g at room temperature for 15 minutes.

7. Remove supernatant. Centrifuge briefly. Remove the remaining supernatant and dry the pellet in a Speedvac. Samples can be stored at –20°C.

8. Resuspend the pellet in 10 μl of annealing mix. Incubate at 70–85°C for 2 minutes.

9. Incubate at 58°C for 40 minutes.

   *Note:* The optimal annealing conditions may vary depending on the oligonucleotide used.

10. Allow samples to cool to room temperature. Add 40 µl of Reverse Transcriptase–Primer Extension Mix, and mix by flicking the tube. Incubate at 37°C for 40 minutes.

11. Add 250 µl of ethanol, mix by vortexing, and centrifuge at 12,000–15,000g in a microcentrifuge at room temperature for 15 minutes.

12. Remove the supernatant, and dry the pellet in Speedvac.

13. Resuspend the DNA pellet in 5 µl of Formamide-NaOH Loading Buffer. Perform electrophoresis on a standard 8% (w/v) polyacrylamide-urea sequencing gel.

## ACKNOWLEDGMENTS

I thank Jim Kadonaga and Andrea Wurster for careful reading of this chapter. M.P. is supported by the Cutaneous Biology Research Center through the Massachusetts General Hospital/Shiseido Co. Ltd. Agreement.

## REFERENCES

Austin R.J. and Biggin M.D. 1996. Purification of the *Drosophila* RNA polymerase II general transcription factors. *Proc. Natl. Acad. Sci.* **93:** 5788–5792.

Becker P.B. and Wu C. 1992. Cell-free system for assembly of transcriptionally repressed chromatin from *Drosophila* embryos. *Mol. Cell. Biol.* **12:** 2241–2249.

Burke T.W. and Kadonaga J.T. 1997. The downstream core promoter element, DPE, is conserved from *Drosophila* to humans and is recognized by TAFII60 of *Drosophila. Genes Dev.* **11:** 3020–3031.

Heiermann R. and Pongs O. 1985. In vitro transcription with extracts of nuclei of *Drosophila* embryos. *Nucleic Acids Res.* **13:** 2709–2730.

Kadonaga J.T. 1990. Assembly and disassembly of the *Drosophila* RNA polymerase II complex during transcription. *J. Biol. Chem.* **265:** 2624–2631.

Kamakaka R.T. and Kadonaga J.T. 1994. The soluble nuclear fraction, a highly efficient transcription extract from *Drosophila* embryos. *Methods Cell Biol.* **44:** 225–235.

Kamakaka R.T., Bulger M., and Kadonaga J.T. 1993. Potentiation of RNA polymerase II transcription by Gal4-VP16 during but not after DNA replication and chromatin assembly. *Genes Dev.* **7:** 1779–1795.

Kamakaka R.T., Tyree C.M., and Kadonaga J.T. 1991. Accurate and efficient RNA polymerase II transcription with a soluble nuclear fraction derived from *Drosophila* embryos. *Proc. Natl. Acad. Sci.* **88:** 1024–1028.

Mason P.J., Enver T., Wilkinson D., and Williams J.G. 1993. Assay of gene transcription *in vitro.* In *Gene transcription: A practical approach* (es. B.D. Hames and S.J. Higgins), pp. 5–64. Oxford University Press, United Kingdom.

Mizuguchi G., Tsukiyama T., Wisniewski J., and Wu C. 1997. Role of nucleosome remodeling factor NURF in transcriptional activation of chromatin. *Mol. Cell* **1:** 141–150.

Pazin M.J. and Kadonaga J.T. 1998. Transcriptional and structural analysis of chromatin assembled *in vitro.* In *Chromatin: A practical approach* (ed. H. Gould), pp. 173–194. Oxford University Press, United Kingdom.

Pazin M.J., Kamakaka R.T., and Kadonaga J.T. 1994. ATP-dependent nucleosome reconfiguration and transcriptional activation from preassembled chromatin templates. *Science* **266:** 2007–2011.

Sambrook J., Fritsch E.F., and Maniatis T. 1989. *Molecular cloning: A laboratory manual,* 2nd edition. Cold Spring Harbor Laboratory Press, Cold Spring Harbor, New York.

Shaffer C.D., Wuller J.M., and Elgin S.C. 1994. Raising large quantities of *Drosophila* for biochemical experiments. *Methods Cell Biol.* **44:** 99–108.

Soeller W.C., Poole S.J., and Kornberg T. 1988. In vitro transcription of the *Drosophila* engrailed gene. *Genes Dev.* **2:** 68–81.

Wampler S.L., Tyree C.M., and Kadonaga J.T. 1990. Fractionation of the general RNA polymerase II transcription factors from *Drosophila* embryos. *J. Biol. Chem.* **265:** 21223–21231.

Zhang H., Iler N., and Abate-Shen C. 1996. Rigorous and quantitative assay of transcription in vitro. *Methods Enzymol.* **273:** 86–99.

# CONTENTS

# 32

# Preparation of Membrane Proteins from *Drosophila* Embryos

**Claire X. Zhang and Tao-shih Hsieh**
*Department of Biochemistry*
*Duke University Medical Center*
*Durham, North Carolina 27710*

Membranes and membrane proteins have key roles in many critical aspects of cellular functions, such as signal transduction, cell adhesion and movement, cytoskeleton organizations, and vesicle transport and targeting. *Drosophila* embryos provide a unique system for the study of membrane proteins because of an extensive database in both genetics and biochemistry. The recent efforts in identifying a large number of the *Drosophila* genes encoding membrane proteins promise to further accelerate progress in this area of research (Kopczynski et al. 1998). *Drosophila* early embryos are also unique in having an extraordinary peak in the activity of membrane biogenesis during development. The first 13 cycles of nuclear divisions occur in a syncytium without any cytokinesis; however, in the latter stage of the cycle-14 interphase, invaginating plasma membrane driven by cleavage furrows encircles approximately 6000 nuclei on the cortex of the embryo. Within a period of approximately 45 minutes, enough plasma membrane must be assembled and deposited in a highly ordered and coordinated manner, which accounts for a 23-fold increase in membrane area (Foe et al. 1993). There are two possible sources for the nascent membranes and their associated proteins: surface-membrane resorbtion resulting from flattening of the microvilli, and the transport of cytoplasmic pools of membranes, which exist as lamellar bodies (Fullilove and Jacobson 1971; Sanders 1975; Turner and Mahowald 1976). The latter possibility is also supported by the recent ultrastructural studies of cellular blastoderm (Loncar and Singer 1995) and the finding that a vesicle-docking protein, syntaxin, is essential for the cellularization processes (Burgess et al. 1997). The molecular mechanism by which these vesicles are transported, docked, and fused to the site of membrane growth remains to be elucidated.

To study the biochemical properties of membrane proteins, we use equilibrium sedimentation in a sucrose density step gradient (0.5 M, 2.0 M, 2.5 M) to purify the membrane fraction (Figure 32.1) following a procedure first developed by Strand et al. (1994). Protocol 32.1 describes this membrane-extraction procedure. Protocol 32.2 includes two useful methods to determine the membrane association of a protein of interest: (1) alkaline wash and (2) Triton X-114 phase separation. The integrity of the *Drosophila* mem-

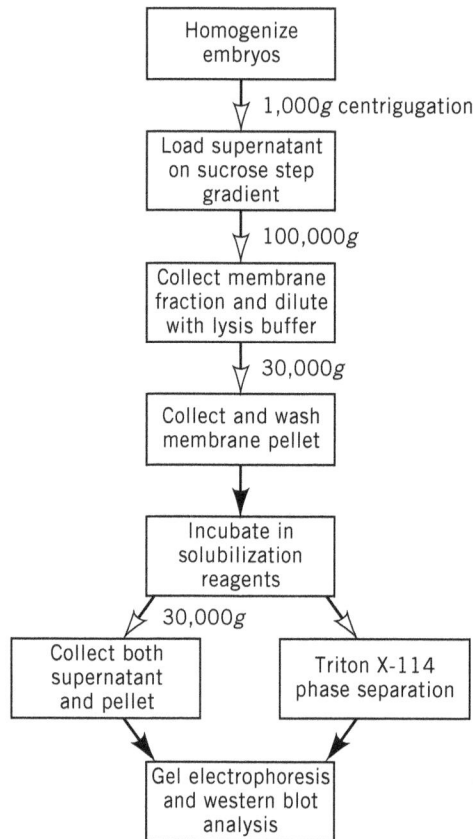

**Figure 32.1.** Flowchart of membrane fractionation. Centrifugation steps are indicated by open arrowheads.

brane can be maintained at pH 10 while many peripheral proteins are released from the membrane under such conditions (for review, see Hortsch 1994). Integral proteins can be solubilized by a nonionic detergent, e.g., Nonidet P-40 (NP-40) or Triton. The partitioning of integral membrane proteins into the detergent phase can also be used to study their membrane association. Triton X-114 solution is homogeneous at 0°C but separates into two phases, aqueous phase and detergent phase, at temperatures above 20°C. Integral proteins partition into the detergent phase, whereas hydrophilic proteins are in the aqueous phase (Bordier 1981). Here we use a transmembrane protein, neurexin, essential for septate junction formation (Baumgartner et al. 1996), as an example for the membrane fractionation and the various membrane treatments. We have also used this protocol to demonstrate that DAH protein encoded by a gene required for cleavage furrow formation (Zhang et al. 1996) is a membrane-associated protein.

## PROTOCOL 32.1*

# Membrane Extract Preparation

### Materials

### *Stocks*

*Drosophila* embryos of the desired stock

*For all protocols, H$_2$O indicates glass distilled and deionized.

## Supplies and Equipment

Dounce homogenization set (15 ml; Kimble/Kontes)

SS34 rotor, tubes, and adapters for Sorvall centrifuge

SW27 rotor, tubes, and adapters (or similar swinging-bucket rotor) for Beckman ultracentrifuge

Sterile cotton swabs

## Solutions and Reagents

All chemicals are reagent grade.

Sucrose solutions. 2.5 M sucrose solution is made in TKM (see below) and stored at
–20°C. 2.0 M and 0.5 M sucrose solutions are made on the day of the experiment, by
diluting 2.5 M sucrose in lysis buffer (see below). When performing the dilutions, keep
in mind that the lysis buffer contains 0.25 M sucrose.

Protease inhibitors. 20 mg/ml **leupeptin** and 5 mg/ml **pepstatin** (Boehringer Mannheim)
are dissolved in **methanol** and stored at –20°C. 100 mM PMSF (phenylmethylsulfonyl
fluoride) is dissolved in isopropanol and stored at –20°C. 1 M **DTT** (dithiothreitol) is
dissolved in sterile $H_2O$ and stored at –20°C.

Triton X-114 (10% v/v) (Sigma) is prepared in $H_2O$ and stored at 4°C.

### TKM

> 50 mM Tris-Cl (pH 7.5)
> 150 mM **KCl**
> 5 mM **MgCl$_2$**

### Lysis Buffer

> 50 mM Tris-Cl (pH 7.5)
> 150 mM KCl
> 5 mM MgCl$_2$
> 0.25 M sucrose
> 0.1 mM DTT
> 1 mM PMSF
> 2 μg/ml leupeptin
> 2 μg/ml pepstatin

**CAUTION: DTT, isopropanol, KCl, leupeptin, methanol, MgCl$_2$, pepstatin, PMSF** (see
Appendix 4)

## Method

1. Collect *Drosophila* embryos of desired stages, dechorionate, and store at –70°C (see
   Sisson, this volume).

2. Homogenize 1 g of frozen embryos in 2 ml of Lysis Buffer using pestle A (5 strokes;
   loose-fitting), followed by using pestle B (10 strokes; tight-fitting) in a 15-ml Dounce
   homogenizer placed on ice. The long tube of the 15-ml Dounce homogenizer prevents spilling during homogenization.

3. Centrifuge the extract at 1000*g* (3000 rpm for the Sorvall SS34 rotor) at 4°C for 10
   minutes. The debris and nuclei are pelleted down. Use a sterile cotton swab to pick up

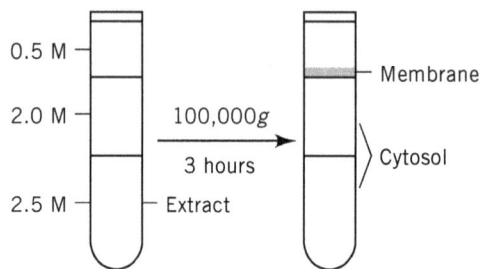

**Figure 32.2.** Sucrose step gradient sedimentation.

the white lipid layer on the top. Pipette out the supernatant carefully without disturbing the pellet. Save the supernatant for membrane preparation.

4. Add 15.2 ml of 2.5 M sucrose to a SW27 tube. Transfer the supernatant from step 3 (above) to the tube. Use a 5-ml glass pipette to mix the supernatant into the 2.5 M sucrose cushion by gentle stirring. Above this layer, slowly lay 12.5 ml of 2.0 M sucrose, and then overlay with 7 ml of 0.5 M sucrose (Figure 32.2). Six SW27 tubes will be required (i.e., 5 tubes are for balance). Make sure that all 6 tubes for the SW27 rotor are balanced prior to centrifugation at 100,000g (24,000 rpm) at 4°C for 2.5 hours.

5. The membrane fraction bands at the interface of the 0.5 M and 2.0 M sucrose layers. Take out the cloudy interface with a pasteur pipette. Add 2 volumes of lysis buffer, mix well, and centrifuge at 30,000g (16,000 rpm for SS34) at 4°C for 30 minutes. The membranes are pelleted down and ready for use or storage at –70°C.

## PROTOCOL 32.1 NOTES

For protein quantitation, the membrane pellet is resuspended in 0.2 ml of lysis buffer and the total protein concentration is determined by the Bradford dye-binding method (1976). Equal amounts of the total proteins in the membrane extract and the crude extract are loaded on SDS polyacrylamide gel and subject to Western blot analysis.

## PROTOCOL 32.2

# Extraction of Proteins from Membranes

### Materials
### *Supplies and Equipment*

Microcentrifuge tubes (1.5 ml)
SS34 rotor, tubes, and adapters for Sorvall centrifuge
Microcentrifuge

### *Solutions and Reagents*

Lysis buffer (see p. 565)
Phosphate-buffered saline (PBS; see Appendix 3)

**Na₂CO₃** $Na_2CO_3$ (0.1 M, pH 10.0)
NP-40 (1% v/v) in 50 mM Tris-Cl (pH 7.5)
Triton X-114 (10%)

### 1% (v/v) Triton X-114 Solution
10 mM Tris-Cl (pH 7.4)
150 mM NaCl
1% Triton X-114

### 6% (w/v) Sucrose
10 mM Tris-Cl, pH 7.4
150 mM NaCl
0.06% Triton X-114
6% sucrose

### 2% Triton X-114
10 mM Tris-Cl
150 mM NaCl
2% Triton X-114

### Tris-Cl/NaCl Solution
10 mM Tris-Cl (pH 7.4)
150 mM NaCl

### Laemmli Sample Buffer
2% SDS
100 mM DTT
50 mM Tris-Cl (pH 6.8)
10% glycerol
0.05% bromophenol blue

**CAUTION: Na₂CO₃** $Na_2CO_3$ (see Appendix 4)

## Method 1

### Alkaline Wash

1. Resuspend the membrane pellet in 3 ml of Lysis Buffer and aliquot into three micro-centrifuge tubes.

2. Centrifuge at 30,000*g* at 4°C for 30 minutes.

3. Resuspend each pellet in 0.5 ml of the following solutions:

   a. Phosphate-buffered saline (PBS)

   b. 0.1 M Na₂CO₃ $Na_2CO_3$ (pH 10)

   c. 1% (v/v) NP-40 in 50 mM Tris-Cl (pH 7.5)

   > *Note:* Because PBS will not dissociate any proteins from lipid bilayers, it serves as a negative control. Carbonate buffer will release most of the peripheral proteins from membranes, while leaving integral proteins on intact membranes. 1% NP-40 can solubilize most proteins including integral proteins from membranes and serves as a positive control.

4. Incubate the suspensions on ice for 30 minutes.

5. Centrifuge at 30,000$g$ at 4ºC for 30 minutes.

6. Save the supernatants and resuspend the membrane pellets in 0.1 ml of Laemmli Sample Buffer.

## Method 2

### *Triton X-114 Phase Separation (Cloudy-point Precipitation)*

1. Resuspend the membrane pellet in 0.25 ml of 1% (v/v) Triton X-114 solution.

2. Incubate on ice for 30 minutes.

3. In a 1.5-ml microcentrifuge tube, add 0.3 ml of 6% (w/v) sucrose cushion.

4. Carefully overlay with 0.2 ml of the Triton suspension (from step 2, above).

5. Incubate at 30ºC for 3 minutes. The Triton suspension becomes cloudy.

6. Centrifuge at 300$g$ (scale 2 for tabletop Eppendorf centrifuge, model 5415C) at room temperature for 3 minutes. The detergent-enriched phase appears like an oil drop at the bottom of the microcentrifuge tube after the centrifugation. The aqueous phase on top of the sucrose cushion is clear.

7. Repeat phase separation by removing the aqueous phase (0.2 ml) and adding a stock solution of 10% Triton X-114 to the aqueous phase to make a final concentration of 0.5%. Save the tube containing the sucrose cushion.

8. Incubate the mixture on ice until it clears. Lay it on top of the previous sucrose cushion (in the same tube), incubate, and centrifuge as in steps 4–6 (above).

9. After phase separation, the aqueous and detergent phases are processed separately. Remove the aqueous phase. Mix the aqueous phase with 0.3 ml of 2% Triton X-114, incubate, and centrifuge again (see steps 4–6) to remove any residual integral proteins. To recover the detergent phase (oil-drop in the bottom), use a capillary pasteur pipette to remove the sucrose cushion on its top. Resuspend the detergent phase in 0.35 ml of Tris-Cl/NaCl solution to make its volume similar to that of the aqueous phase.

## PROTOCOL 32.2 NOTES

All the samples are subjected to gel electrophoresis and Western blot analysis. The integral membrane protein neurexin was found to be highly enriched in the membrane extract and resistant to alkaline wash (Figure 32.3). Nonionic detergents, e.g., NP-40 and Triton, released the protein from the membrane. The protein partitioned into the detergent phase after Triton X-114 cloudy-point precipitation.

## ACKNOWLEDGMENTS

This work is supported by a grant from the National Institutes of Health (GM-29006). We appreciate a generous gift of neurexin antibody from Hugo Bellen (Baylor College of Medicine).

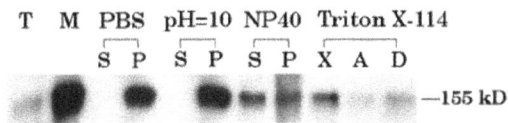

**Figure 32.3.** Membrane association of neurexin in 6–12-hour embryos. On a 7% SDS-polyacrylamide gel, 10 μg of total protein for lanes T (total extract) and M (membrane extract) was loaded. Approximately 3 μg of membrane protein is used in each membrane-treatment experiment. For the membrane-washing experiment (Protocol 32.2, Method 1), 21 μl of supernatant (S) and 4.2 μl of pellet (P) from PBS (pH 10), and NP-40 washes were loaded in parallel. For the phase-separation experiment (Protocol 32.2, Method 2), 10.5 μl of the Triton suspension (X), 21 μl of the aqueous phase (A), and 21 μl of detergent phase (D) were loaded. Western-blot analysis was performed using rabbit anti-neurexin serum (1:1000 dilution), according to the manufacturer's protocol for ECL (Amersham).

# REFERENCES

Baumgartner S., Littleton J.T., Broadie K., Bhat M.A., Harbecke R., Lengyel J.A., Chiquet-Ehrismann R., Prokop A., and Bellen H.J. 1996. A *Drosophila* neurexin is required for septate junction and blood-nerve barrier formation and function. *Cell* **87:** 1059–1068.

Bordier C. 1981. Phase separation of integral membrane proteins in Triton X-114 solution. *J. Biol. Chem.* **256:** 1604–1607.

Bradford M.M. 1976. A rapid and sensitive method for the quantitation of microgram quantities of protein utilizing the principle of protein-dye binding. *Anal. Biochem.* **72:** 248–254.

Burgess R.W., Deitcher D.L., and Schwarz T.L. 1997. The synaptic protein syntaxin1 is required for cellularization of *Drosophila* embryos. *J. Cell Biol.* **138:** 861–875.

Foe V.E., Odell G.M., and Edgar B.A. 1993. Mitosis and morphogenesis in the *Drosophila* embryo: Point and counterpoint. In *The development of* Drosophila melanogaster (ed. M. Bate and A.M. Arias), vol. 1, pp. 149–300. Cold Spring Harbor Laboratory Press, Cold Spring Harbor, New York.

Fullilove S.L. and Jacobson A.G. 1971. Nuclear elongation and cytokinesis in *Drosophila* montana. *Dev. Biol.* **26:** 560–577.

Hortsch M. 1994. Preparation and analysis of membranes and membrane proteins from *Drosophila*. *Methods Cell Biol.* **44:** 289–301.

Kopczynski C.C., Noordermeer J.N., Serano T.L., Chen W., Pendleton J.D., Lewis S., Goodman C.S., and Rubin G.M. 1998. A high throughput screen to identify secreted and transmembrane proteins involved in *Drosophila* embryogenesis. *Proc. Natl. Acad. Sci.* **95:** 9973–9978.

Loncar D. and Singer S.J. 1995. Cell membrane formation during the cellularization of the syncytial blastoderm of *Drosophila*. *Proc. Natl. Acad. Sci.* **92:** 2199–2203.

Sanders E. 1975. Aspects of furrow membrane formation in the cleaving *Drosophila* embryo. *Cell Tissue Res.* **156:** 463–474.

Strand D., Raska I., and Mechler B.M. 1994. The *Drosophila* lethal(2)giant larvae tumor suppressor protein is a component of the cytoskeleton. *J. Cell Biol.* **127:** 1345–1360.

Turner F.R. and Mahowald A.P. 1976. Scanning electron microscopy of *Drosophila* embryogenesis. 1. The structure of the egg envelopes and the formation of the cellular blastoderm. *Dev. Biol.* **50:** 95–108.

Zhang C.X., Lee M.P., Chen A.D., Brown S.D., and Hsieh T. 1996. Isolation and characterization of a *Drosophila* gene essential for early embryonic development and formation of cortical cleavage furrows. *J. Cell Biol.* **134:** 923–934.

# CONTENTS

# Preparing Cytoplasmic Extracts from *Drosophila* Embryos

### Michelle Moritz
*Department of Biochemistry and Biophysics*
*University of California at San Francisco*
*San Francisco, California 94143*

IN RECENT YEARS, *DROSOPHILA* HAS PROVEN TO BE AN EXCELLENT SOURCE for biochemical quantities of proteins involved in a variety of cellular processes. Extracts made from *Drosophila* embryos have been used successfully to isolate proteins involved in activities as diverse as cytoskeletal function (Kellogg et al. 1989; Miller et al. 1991; Saxton 1994) and RNA binding (Matunis et al. 1994).

The production of a high-quality *Drosophila* embryonic cytoplasmic extract for use in protein purification or biochemistry is relatively easy, providing population cages are available that produce at least 5 g of embryos during a 3-hour period (for methods to maintain population cages, see Sisson, this volume). Smaller quantities of embryos can be used to produce extracts that are useful for in vitro biochemical assays (see, e.g., Moritz et al. 1998), and thus it should be possible to make extracts from mutant stocks that could then be tested in vitro. Protocol 33.1 describes the collection and dechorionation of embryos for making cytoplasmic extracts. It is fairly easy to prepare even very concentrated cytoplasmic extracts from *Drosophila* embryos, with protein concentrations of 50–75 mg/ml (see Protocol 33.2).

PROTOCOL 33.1*

## Embryo Collection and Dechorionation

### Materials

#### Supplies and Equipment

Population cages for the collection of gram quantities of embryos. For methods in their maintenance, see Sisson (this volume).

Two sieves for the collection of embryos: no. 140, Tyler Equivalent 150 mesh, opening μm 106, 8 inches (diameter) × 2 inches (depth) (Fisher 04-881Z); no. 18, Tyler Equivalent 16 mesh, opening mm 1.0, 8 × 2 inches (Fisher 04-881L)

*For all protocols, H$_2$O indicates glass distilled and deionized.

Soft paintbrush or rubber kitchen spatula (~3-cm wide) for loosening embryos from the collection trays

Plastic container (e.g., Rubbermaid), for holding the lowest collection sieve

Dissecting microscope or compound microscope on low power (e.g., 10X) for staging embryos

### Solutions and Reagents

**Bleach** for dechorionating embryos. Prepare enough 50% bleach in $H_2O$ to submerge the embryos in their collection sieve. For dechorionating embryos, we place the sieve in a large Rubbermaid container, which requires 2 liters of 50% bleach. The bleach deteriorates fairly quickly, so prepare it fresh daily. It can be used for multiple collections on one day.

### Embryo Wash

0.04% NaCl

0.03% Triton X-100

Prepare a large stock (e.g., in a 5–20-liter carboy). Use a squirt bottle filled with Embryo Wash solution for performing the washing step.

**CAUTION: bleach** (see Appendix 4)

## Method

1. Collect embryos at the appropriate developmental stage.

   *Note:* To enrich for soluble proteins/protein complexes, and to minimize the amount of membranes and extraneous tissue in the extract, it is best to collect precellularized embryos. However, the best embryonic stage to use will depend on when the protein of interest is most abundant. The best collection period for the application should first be determined for a given set of population cages by checking a small amount of embryos under the microscope after dechorionation to ensure that most of the embryos are of the desired age. For example, to enrich for precellularized embryos, we collect for 2–3.5 hours (embryos are between 0 and 3.5 hours old).

2. Stack the collection sieves in a sink with the finer-meshed sieve on the bottom. The courser-meshed sieve is to separate any adult flies or large pieces of food from the embryos.

3. Use a squirt bottle to apply a pool of Embryo Wash to the food tray on which the embryos have been collected. Use a paint brush or rubber kitchen spatula to loosen the embryos from the surface of the food. Squirt the loosened embryos off the trays into the collection sieves.

4. When the embryos from all of the trays have been collected on the fine sieve, place the sieve in the plastic container containing 50% bleach solution for 2 minutes with occasional swirling.

5. Wash the dechorionated embryos in $H_2O$ until they no longer smell of bleach. Wash the embryos to one side of the sieve and blot dry from the bottom and top with paper towels (not many embryos will stick to the paper towels). Scrape the embryos into a weigh boat and weigh them.

## PROTOCOL 33.1 NOTES

Depending on the application, embryos may be quickly frozen in a tube plunged in liquid nitrogen and stored at –80ºC until needed. However, it is probably safest to prepare and freeze the extract as described in Protocol 33.2.

PROTOCOL 33.2

# Preparation of the Cytoplasmic Extract

## Materials

### Supplies and Equipment

Potter-Elvehjem-style plastic-coated tissue grinder with loose-fitting Teflon pestle "A," 30 ml (Wheaton 358011) or 55 ml (Wheaton 358013).

Overhead motorized stirrer with variable speed range of 1000–10,000 rpm (Wheaton 903475).

Miracloth (Calbiochem-Novabiochem 475855) for filtering the embryo homogenate

Plastic funnels (4–6, 65-mm diameter)

Sorvall centrifuge with SS34 or HB-4 rotor

Plastic centrifuge tubes to fit SS34 or HB-4 rotor

Plastic tubes for storing extracts at –80ºC

### Solutions and Reagents

**Liquid nitrogen** in a dewar flask

**PMSF** (100 mM) (phenylmethylsulfonyl fluoride). Prepare the stock in ethanol; store at –20ºC. The PMSF will come out of solution at –20ºC, so it must be redissolved just before use (place at 37ºC just until it dissolves).

### Homogenization Buffer

The buffer of choice will depend on the application of the extract. For work with cytoskeletal proteins, we typically use the following buffer:

50 mM HEPES, potassium salt (pH 7.6)

50–100 mM **KCl**

1 mM EGTA

1 mM **MgCl$_2$**

10% glycerol

### Protease Inhibitor Stock

10 mM benzamidine-HCl

0.1 mg/ml phenanthroline

1 mg/ml **aprotinin**

1 mg/ml **leupeptin**

1 mg/ml **pepstatin A**

Prepare the stock in ethanol and store at –20ºC (will not dissolve completely).

**CAUTION: aprotinin, KCl, leupeptin, liquid nitrogen, MgCl$_2$, pepstatin A, PMSF** (see Appendix 4)

## Method

1. Transfer the weighed embryos to a glass homogenizer placed on ice. Add the appropriate volume of buffer with final concentrations of 1:100 protease inhibitors and 1 mM PMSF.

   *Note:* The volume of buffer in which to homogenize the embryos depends on the proteins of interest and must be determined empirically. A good ratio of embryos to buffer to start with is 1:5 or 1:6 (weight to volume [g/ml]). We have used these ratios successfully for protein purification (Kellogg and Alberts 1992) and for the isolation of centrosomes (Moritz et al. 1995). However, some proteins may do better if prepared from a more concentrated extract; e.g., we have found that a 1:1 ratio of embryos to buffer works best for the isolation of the γ-tubulin ring complex, a 3-MD protein complex whose integrity is sensitive to dilution (Y. Zheng, K. Oegema, and M. Moritz, unpubl.). Alternatively, a 1:10 extract works well for the isolation of actin-binding proteins (Miller et al. 1991).

2. In the cold room, homogenize the embryos using five passes of the pestle mounted on the overhead stirrer. A complete cycle of one upward and one downward motion counts as one pass. Adjust the rotational speed of the stirrer during the course of each pass to achieve a balance between homogenizing effectively and keeping the extract from splattering out of the homogenizer.

3. Pour the extract through a funnel lined with Miracloth that is wetted with Homogenization Buffer. Collect the filtrate in a plastic tube suitable for centrifugation in an SS34 or HB-4 rotor.

4. Pass the filtrate from step 3 through a clean funnel lined with a second piece of wetted Miracloth. Collect the filtrate into a fresh tube.

5. Centrifuge the extract at 1500g (e.g., in a Sorvall centrifuge; 3500 rpm in an SS34 rotor or 3000 rpm in an HB-4 rotor) at 4°C for 10 minutes to remove nuclei, debris, and some of the yolk. Aspirate off the lipid layer at the top. Transfer the supernatant to a fresh tube for further clarification (step 6) or to plastic storage tubes for freezing (step 7).

6. (*Optional*) If desired, centrifuge the supernatant again as described in step 5 to further clarify the extract. Transfer the resulting supernatant to plastic storage tubes for freezing.

7. Plunge tubes into a dewar flask filled with liquid nitrogen. Store at –80°C.

   *Note:* We routinely freeze extracts in this way and store at –80°C for up to several months. For use, the extract is thawed quickly in a water bath at 37°C, and then a high-speed supernatant is prepared by centrifugation at 100,000g for 1.0–1.5 hours. The type of centrifuge and rotor will depend on the volume of extract to be spun, e.g., the extract may be centrifuged in the SW55 rotor in a Beckman L8-70M ultracentrifuge at 39,000 rpm. Whether the extract can survive freezing and thawing will, of course, depend on the protein of interest and must be determined empirically. However, we have not encountered many cases where fresh extract had to be used.

# REFERENCES

Kellogg D.R. and Alberts B.M. 1992. Purification of a multiprotein complex containing centrosomal proteins from the *Drosophila* embryo by chromatography with low-affinity polyclonal antibodies. *Mol. Biol. Cell.* **3:** 1–11.

Kellogg D.R., Field C.M., and Alberts B.M. 1989. Identification of microtubule-associated proteins in the centrosome, spindle, and kinetochore of the early *Drosophila* embryo. *J. Cell. Biol.* **109:** 2977–2991.

Matunis M.J., Matunis E.L., and Dreyfuss G. 1994. Isolation and characterization of RNA-binding proteins from *Drosophila melanogaster*. *Methods Cell Biol.* **44:** 191–205.

Miller K.G., Field C.M., Alberts B.M., and Kellogg D.R. 1991. Use of actin filament and microtubule affinity chromatography to identify proteins that bind to the cytoskeleton. *Methods Enzymol.* **196:** 303–319.

Moritz M., Zheng Y., Alberts B.M., and Oegema K. 1998. Recruitment of the γ-tubulin ring complex to *Drosophila* salt-stripped centrosome scaffolds. *J. Cell Biol.* **142:** 775–786.

Moritz M., Braunfeld M.B., Fung J.C., Sedat J.W., Alberts B.M., and Agard D.A. 1995. Three-dimensional structural characterization of centrosomes from early *Drosophila* embryos. *J. Cell Biol.* **130:** 1149–1159.

Saxton W.M. 1994. Isolation and analysis of microtubule motor proteins. *Methods Cell Biol.* **44:** 279–288.

# CONTENTS

# 34

# Immunoblotting of Proteins from Single *Drosophila* Embryos

**Tin Tin Su**
*Department of Molecular, Cellular, and Developmental Biology*
*University of Colorado, Boulder*
*Boulder, Colorado 80309-0347*

CELL CYCLE STUDIES IN *DROSOPHILA* EMBRYOS have led to increased understanding of how cell proliferation is regulated in a developmental context (for review, see O'Farrell et al. 1989; Edgar and Lehner 1996). Cell cycle progression involves and often relies on periodic changes in the properties of proteins in concert with passage through various phases of the cell cycle. Characterization of a protein of interest at specific points during a division cycle is often a prelude to understanding the function of the protein. As yet, however, we lack a way of synchronizing *Drosophila* embryos with respect to their cell cycle phase. Thus, it is not possible to prepare cell cycle stage-specific extracts and analyze the constituents. An alternate approach has been to analyze individual *Drosophila* embryos during the syncytial divisions. These divisions comprise embryonic division cycles 1–13 that occur within 2 hours and 10 minutes after egg laying (at 25°C; Foe et al. 1993). At these stages of development, nuclei within a given embryo divide metasynchronously in a common cytoplasm. Thus, each embryo at any given time represents a specific stage in the division cycle. The challenge then is to identify an embryo of desired division cycle stage.

This chapter describes a procedure for rapidly fixing precellular stage embryos and selecting those of the desired division cycle stage for further analysis. This procedure has been used to characterize changes in mitotic cyclin levels (Edgar et al. 1994), posttranslational modification on the mitotic kinase Cdk1 (Edgar et al. 1994), and posttranslational modification on a DNA replication factor (Mitsis 1995) during the cell cycle.

## Protocol 34.1*

# Preparation of Single *Drosophila* Embryos for Immunoblotting

### Materials

### *Media*

Grape juice–agar plates and yeast paste for embryo collection (see Sisson, this volume)

### *Supplies and Equipment*

Baskets for washing embryos (these are generally short cylinders of plastic that are sealed at one end with a nylon screen)
Soft brush for gathering embryos
Dissecting microscope
Microcentrifuge tubes (1.5 ml)
Compound fluorescence microscope
Glass microscope slides
Tungsten needle or a plastic pipette tip sealed at the end

### *Solutions and Reagents*

**Bleach** (50%) in $H_2O$
**Heptane**
**Methanol** (1 mM $Na_3VO_4$ may be added as a phosphatase inhibitor)
Embryo buffer (EB; see below) containing 4 μg/ml **bisbenzimide** (Hoechst 33258)
50%EB/50%glycerol

---

#### *2× SDS Sample Buffer (Sambrook et al. 1989, p. 18.53)*

100 mM Tris-Cl (pH 6.8)
200 mM dithiothreitol (DTT)
4% SDS (electrophoresis grade)
0.2% bromophenol blue
20% glycerol
This buffer without DTT may be stored at room temperature. DTT from a 1 M stock (Sambrook et al. 1989, Appendix B) should be added just before use and any unused portions discarded.

#### *Embryo Buffer (EB)*

10 mM Tris-Cl (pH 7.5)
80 mM sodium β-glycerophosphate (pH 7.5)
20 mM EGTA
15 mM **MgCl₂** (or 10 mM EDTA)
2 mM $Na_3VO_4$
1 mM benzamidine
1 mM **sodium metabisulfate**
2 mM **PMSF**
0.05% Tween-20

---

CAUTION: **bisbenzimide, bleach, heptane, MgCl₂, methanol, $Na_3VO_4$, PMSF, sodium metabisulfate, sodium orthovanadate** (see Appendix 4)

*For all protocols, $H_2O$ indicates glass distilled and deionized.

## Method

### *Harvesting and Dechorionating Embryos*

1. Collect embryos on grape juice–agar plates and wash them into baskets using $H_2O$ and a soft brush. Keep the embryos in the basket for subsequent steps until fixation. Rinse the embryos well with $H_2O$ to remove yeast.

    *Note:* Syncytial cycles 1–13 take 2 hours and 10 minutes at 25°C (Foe 1989; Foe et al. 1993). Therefore, a 3-hour collection of embryos should include all syncytial stages. Note that phases of a division cycle differ in their duration and will be represented accordingly. Thus, for example, Edgar et al. (1994) found 45% of cycle 2–7 embryos in interphase, 10% in prophase and prometaphase, 28% in metaphase, and 17% in anaphase and telophase.

2. Immerse embryos in 50% bleach for 2 minutes with occasional gentle swirling. This is usually achieved by placing the basket in a flat-bottomed dish that contains enough bleach solution to cover the embryos. Monitor dechorionation under a dissecting microscope; loss of the chorion is signified by the loss of dorsal appendages.

    *Note:* It is essential to remove the chorion completely for subsequent fixation procedures. The failure to dechorinate successfully usually occurs because the bleach solution is old.

3. Wash embryos with several milliliters of $H_2O$ to remove bleach completely, and pat dry with tissue.

### *Fixing Embryos*

1. Immerse embryos in heptane by placing the basket in a flat-bottomed dish that contains enough heptane to cover embryos.

2. Transfer the embryos into a 1.5-ml microcentrifuge tube containing 500 µl of methanol.

    *Note:* The transfer of embryos can be accomplished using a pipette and a plastic tip that has been cut off at the end to prevent drawing embryos through a small opening. Prior wetting of the pipette tip with liquid (heptane here and buffer [see below]) will prevent the embryos from sticking to the pipette tip. Heptane and methanol will form a double layer, with embryos resting at the interface. There should be an equal volume (~500 µl) of heptane on the methanol layer.

3. Fix for 5 minutes with gentle shaking (e.g., on a rocking platform set at low speed). Devitellinized embryos sink to the bottom of the tube while the others remain at the heptane-methanol interface. Discard most of the liquid and embryos that remain at the interface.

4. Rinse fixed embryos three times with ice-cold methanol.

5. All subsequent manipulations are carried out on ice using ice-cold solutions, except for sorting of embryos under the microscope, which can be done at room temperature.

### *Preparing Embryos for Visualization*

1. Rinse embryos twice with Embryo Buffer (EB).

2. Stain embryos with EB containing 4 µg/ml bisbenzamide (Hoechst 33258) for 5 minutes.

3. Rinse twice in EB.

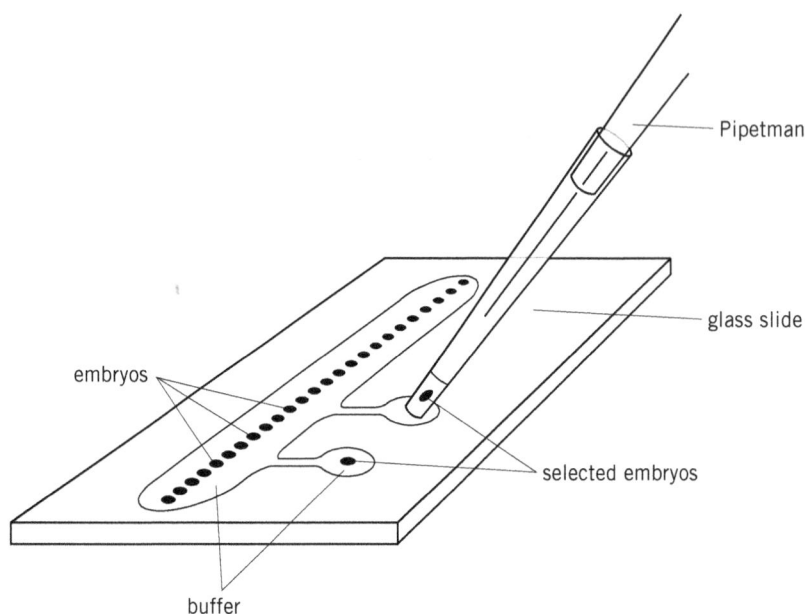

**Figure 34.1.** Selecting and manipulating single embryos from a line of embryos placed on a glass slide.

4. (*Optional*) Transfer to 50%EB/50%glycerol. The presence of glycerol in the buffer helps prevent embryos from drying out while being sorted.

> *Note:* Allowing embryos to sit on ice for up to 1 hour at this point helps "clear" the embryos and makes visualization of DNA easier.

## Visualization of Embryos

A compound fluorescence microscope using 10X or 20X objectives is used for the visualization of embryos. Transfer approximately 25 embryos onto a glass microscope slide such that embryos lie in a single file in a line of buffer (Figure 34.1). Visualize embryos using the DAPI channel to identify those of interest (excitation and emission maxima for Hoechst 33258 are 352 nm and 461 nm, respectively).

DNA morphology is used to determine cell cycle stage (Figure 34.2). To determine division cycle, use nuclei number (e.g., $n = 2$ for cycle 2, $n = 4$ for cycle 3, $n = 8$ for cycle 4, etc.) and nuclear position within the embryo (Foe et al. 1993). Divisions 1–8 occur in the interior of the embryo; nuclei migrate to the cortex during cycle 8 and nuclei reach the cortex in late cycle 9/early cycle 10; divisions 10–13 are cortical. Another useful marker for embryonic stage is the extent of pole cell formation. Pole buds form during cycle 9 and pole cells form during cycle 10 (Campos-Ortega and Hartenstein 1985). For cycles 9–13, it is impractical to count nuclei in the whole embryo. Instead, the number of nuclei per half embryo, or the number of cortical nuclei in the largest optical section, may be estimated (Figure 34.3). Note that an embryo in mitosis may look younger than an embryo in the interphase of the same cycle, especially under 10X objectives, because mitotic chromosome condensation leads to increased spacing among mitotic figures. Additionally, embryos that display pronounced asynchrony should be avoided.

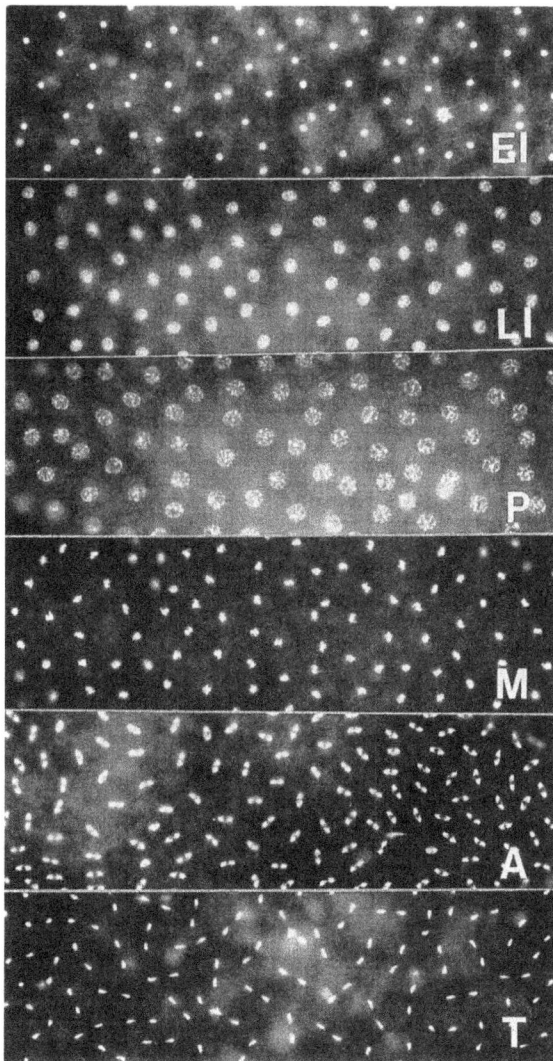

**Figure 34.2.** Division cycle stages in methanol-fixed, Hoechst 33258-stained embryos. (EI) Early interphase; (LI) late interphase; (P) prophase; (M) metaphase; (A) anaphase; (T) telophase. Cycle-10 embryos are shown. (Reprinted, with permission, from Edgar et al. 1994.)

## Embryo Selection

1. Gently push the embryo of interest out of the line of embryos using a tungsten needle or a plastic pipette tip with a heat-sealed end. Pipette and transfer the embryo to a microcentrifuge tube.

2. Add an equal volume of 2x SDS sample buffer to embryos (1–10 embryos for each division cycle stage) and boil for 10 minutes.

3. Proceed with SDS-PAGE and immunoblotting using standard procedures (Sambrook et al. 1989). Note also that Edgar et al. (1994) added 1 mM $Na_3VO_4$ to polyacrylamide gels to stabilize phospho-isoforms of Cdk1.

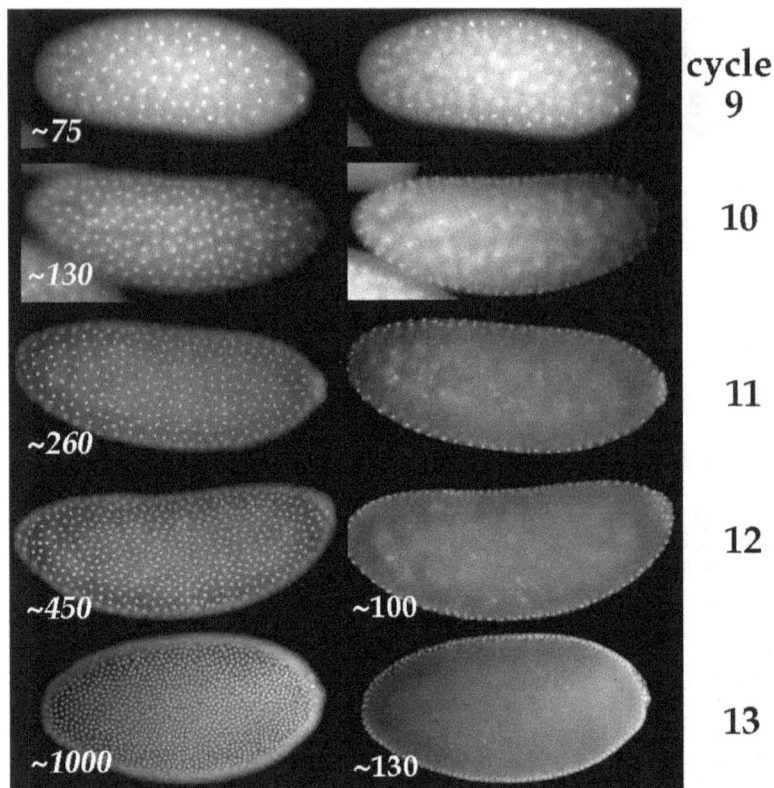

**Figure 34.3.** Approximate number of nuclei in cycles 9–13. (*Left panels*) Surface views and estimated number of nuclei contained within each view; (*right panels*) the corresponding optical cross sections and estimated number of cortical nuclei.

## OTHER SINGLE EMBRYO STUDIES

The protocol described above requires fixation in order to determine division cycle stage. Two methods have been used for studying native proteins from single *Drosophila* embryos and are summarized below. For detailed protocols, consult original papers.

The first method takes advantage of the cortical location of mitoses 11–13. Edgar et al. (1994) first identified live embryos that were undergoing mitosis by the disappearance of the nuclear envelope under DIC (differential interference contrast) optics. The embryos were then aged for appropriate time intervals to reach desired points in the following division cycle, prior to homogenization. Half of the extract from one embryo was used for H1 kinase assays and the other half was immunoblotted to correlate kinase activity with phosphorylation on Cdk1 and with cyclin levels.

In cycles 9 and earlier, nuclei are located in the embryo interior, making the mitotic disappearance of the nuclear envelope hard to visualize. To determine if Cdk1-associated kinase activity (Edgar et al. 1994) and the native size of MCM protein complexes (Su et al. 1996) change with progression through a syncytial cycle, extracts have been prepared and analyzed from randomly selected single embryos. Although the exact division cycle stage of each sample was not known, random selection and large sample size ensured that all stages were represented. Thus, despite its limitations, this approach can be used to ask a simple but fundamental question: Does the property of a native protein change within a syncytial cycle?

## SENSITIVITY OF DETECTION

The analyses described above require the detection of a protein of interest from a single embryo by Western blotting. The sensitivity of detection will depend on the abundance of the protein and the titer of antibodies. Note also that zygotic transcription does not begin until late syncytial divisions and that all proteins analyzed by single-embryo Western blots so far are maternally derived and present throughout precellular stages. Since syncytial cycles proceed entirely on maternally supplied products, most cell cycle functions should be amenable to analysis by methods described herein.

## ACKNOWLEDGMENTS

The procedures described in this chapter were developed by B.A. Edgar, P.H. O'Farrell, and colleagues (Edgar et al. 1994) and are based on a key technique that was suggested by D.R. Kellogg.

## REFERENCES

Campos-Ortega J.A. and Hartenstein V. 1985. *The embryonic development of* Drosophila melanogaster. Springer-Verlag, Berlin.

Edgar B.A. and Lehner C.F. 1996. Developmental control of cell cycle regulators: A fly's perspective. *Science* **274:** 1646–1652.

Edgar B.A., Sprenger F., Duronio R.J., Leopold P., and O'Farrell P.H. 1994. Distinct molecular mechanisms regulate cell cycle timing at successive stages of *Drosophila* embryogenesis. *Genes Dev.* **8:** 440–452.

Foe V.E., Odell G.M., and Edgar B.A. 1993. Mitosis and morphogenesis in the *Drosophila* embryo. In *The development of* Drosophila melanogaster (ed. M. Bate and A. Martinez Arias). Cold Spring Harbor Laboratory Press, Cold Spring Harbor, New York.

Mitsis P.G. 1995. Phosphorylation and localization of replication protein A during oogenesis and early embryogenesis of *Drosophila melanogaster. Dev. Biol.* **170:** 445–456.

O'Farrell P.H., Edgar B.A., Lakich D., and Lehner C.F. 1989. Directing cell division during development. *Science* **246:** 635–640.

Sambrook J., Fritsch E.F., and Maniatis T. 1989. *Molecular cloning: A laboratory manual.* Cold Spring Harbor Laboratory Press, Cold Spring Harbor, New York.

Su T.T., Feger G., and O'Farrell P.H. 1996, *Drosophila* MCM protein complexes. *Mol. Biol. Cell* **7:** 319–329.

# The Organism

# CONTENTS

# 35

# Laboratory Culture of *Drosophila*

**Michael Ashburner and John Roote**
*Department of Genetics*
*University of Cambridge*
*Cambridge, England*

Since its introduction to experimental biology more than 90 years ago, *Drosophila melanogaster* has proved to be an easily cultured and robust laboratory animal. Although culture techniques and the ways in which flies are handled have changed over the years, if he were to enter a fly room today, T.H. Morgan would clearly recognize what is being done, and why.

This chapter provides the basic methods for the laboratory culture of *D. melanogaster*. For a more detailed account, see Ashburner and Thompson (1978) and Chapter 8 of Ashburner (1989a). The recent book by Greenspan (1997) and that edited by Roberts (1998) are also very useful sources. The intelligent culture of *Drosophila* requires a basic understanding of the life cycle of this fly. In this chapter, we begin by describing the life cycle of *Drosophila* and then present information for setting up and maintaining a fly laboratory.

## LIFE CYCLE

*Drosophila* is a typical holometabolous insect, i.e., its life can be divided into four stages—embryo, larva, pupa, and adult—there being a complete metamorphosis of body form from larva to pupa. In the laboratory, *D. melanogaster* is usually cultured at 25°C or 18°C, and we provide timings of developmental stages appropriate to these temperatures only.

Eggs are deposited by females on, or inserted in, the surface of the culture medium. A female may lay approximately 100 eggs per day at her peak. Embryonic development takes approximately 24 hours at 25°C (~40 hours at 18°C); the first-instar larva begins to feed immediately. The larva passes through two molts: from first to second instar and from second to third instar. The first-instar larvae feed on the surface of the medium; second- and third-instar larvae burrow into the medium. The mature third-instar larva leaves the food medium at approximately 96 hours (25°C) and wanders, searching for a site to pupariate. Pupariation occurs at approximately 120 hours (25°C), and it is within the immobile confines of the pupal case (in fact, the tanned third-instar larval cuticle) that metamorphosis occurs. Eclosion of adult flies occurs from the pupal case about 9 days (25°C) (19 days at 18°C) after egg laying. Adult females will not mate until they are approximately 12 hours old; segregation of males and females during the first 8–10 hours of adult life is a convenient way of ensuring that the females are virgin (see p. 591).

**585**

## LABORATORY SETUP

### Supplies and Equipment

For the simple culture of *D. melanogaster* and routine genetic experiments, the following equipment and services are either necessary or to be greatly desired. Table 35.1 lists the specifications and sources of items discussed below. Much of the basic equipment is adapted, rather than being made, for fly culture. For this reason, only examples of sources can be given (see Table 35.1), and the best advice for those newly setting up a fly laboratory is to talk to groups in their locality. There are specialist companies that deal in equipment for fly laboratories; examples are LabScientific, Inc. and Applied Scientific.

**Table 35.1.** Sources of Items for Setting up a Fly Laboratory

| Item | Type or model and number | Supplier(s)/Manufacturer(s)[a] |
|---|---|---|
| Steam kettle | Joni High Tilting Cooking Kettle Model Euro 6 (60 liter), Euro 8 (80 liter), or Euro 10 (100 liter) | D.C. Norris and Co. Ltd.; WN Engineers Ltd. |
| Peristaltic pump | Accuramatic Mk 5 Peristaltic Pump Watson-Marlow 505Dz or Watson-Marlow 605Di Peristaltic Pump | Accuramatic Ltd. Watson-Marlow Ltd. |
| Glass fly bottles and vials | Powder bottles (8 oz) Glass or plastic vials (100 × 25 and 75 × 25 mm) | S.W. Richardson Ltd. T.P. Drewitt; LabScientific Inc.; Applied Scientific |
| Plastic fly bottles | ICRF design | Scientific Laboratory Supplies; LabScientific Inc. |
| Foam bungs (foam wads or closures) for fly bottles and vials | 3 × 3 cm and 5 × 5 cm | Astell Scientific Ltd; LabScientific Inc. Applied Scientific |
| Dissecting microscopes | MS5 Stereomicroscope; MZ6 Stereomicroscope; MZ FLII Stereo microscope (for GFP). Olympus SZ4045F Stereomicroscope | Leica Microsystems (UK) Ltd. Microscope Services and Sales |
| Carbon dioxide and related products | $CO_2$ cylinders, "non-dip"; 6 kg, 12 kg, or 34 kg Reducing/changeover valve No. 22 23 07 04 Multi-stage MSE-30 Regulator R600 Gas Regulator; M3000 Changeover Manifold | Messer UK Ltd. BIG Central Distribution Air Products |
| Porous polyethylene | Vyon D (3.2 mm thickness) | Power Utilities Ltd.; LabScientific Inc. |
| Digital thermometer | | Omega Engineering Inc. |
| Incubators | Series 1 Model 250, 303, or 305 ("for fly culture") Model JBH 800, JBH810, or JBH820 Forma Model 3919 large-capacity reach-in incubator, 32 cu ft; Forma Model 190281 Phenolic coated coils for above; Forma Model 224137 stainless-steel shelf kit for above | LMS Astell Scientific Ltd.; Percival Scientific Inc. Forma Scientific |
| Humidifiers | Microvap type PVP04/5 | Vapac Humidity Control; Novatron Ltd. |
| Liquid fly traps | Fly Dome Trap | Agrisense-BCS Ltd. |

[a]Addresses of suppliers and manufacturers are listed at the end of this chapter.

### Fly Food Kitchen

The facility for the production of fly food and for washing used vials and bottles can range from a very simple kitchen to one that is complex and expensive, depending on need. To maintain a handful of stocks, or set up a few crosses a week, all that is required is a domestic saucepan, a gas or electric ring, and a wooden spoon. We describe here a kitchen suitable for a reasonably large laboratory, producing 100 liters of food (enough for 10,000 vials) a week. Although this could be done entirely by hand, some investment in automation is well worthwhile. The most important equipment includes a steam kettle for preparing medium and a peristaltic pump for dispensing medium into bottles and vials.

*Bottles and vials:* D. melanogaster are usually cultured in 250-ml bottles, 40-ml vials (100 x 25-mm diameter), or 30-ml vials (75 x 25-mm diameter), which may be either plastic, and hence disposable, or glass, and hence reusable after washing. To keep the flies within the bottles or vials, a variety of stoppers can be used, which, again, may be disposable or washable and reusable (at least until they fall apart). The most commonly used plugs are made of polyurethane foam (3 x 3-cm or 5 x 5-cm diameter) or nonabsorbent cotton wool. Plugs can be washed and reused, but are not mite-proof, whereas cotton wool has the disadvantage that some people are irritated by cotton fibers in the air. Applied Scientific now supplies a disposable mite-deterring plug, call the Buzz Plug, made from cigarette filters (cellulose acetate). This has the advantage of not shedding but the disadvantage of cost. Stock or cross labels can be written on the bottle, or special tags (e.g., those available from LabScientific) can be used.

*Medium (fly food):* A number of different recipes for fly food are in use, and examples of the most popular, with general sources of their ingredients, are given below (see Fly Food Preparation).

### Microscopes

Although simple fly culture does not require microscopes, and males and females can readily be sexed with the naked eye after a little practice (especially by the young), any serious fly work requires many hours of careful observation of living flies under a dissecting microscope. Important considerations for the purchase of a dissecting microscope are (1) overall optical quality (which should be as good as one can afford); (2) magnification range (6–40x is adequate for most purposes); and (3) a light source (a "cold" light, such as from a fiber-optic source or a halogen lamp with a heat filter, is recommended). Many investigators prefer a microscope with a zoom, rather than fixed magnification settings. For the detection of GFP (green fluorescent protein) fluorescence in living flies, a microscope with a suitable UV light source is required.

### Autoclave

Old cultures should be autoclaved. For cultures in glass bottles or vials, this is essential before they are washed. Autoclaving is also recommended for cultures in plastic vials because it will prevent release of flies, in particular the release of transgenic flies. The size of autoclave used will depend on the volume of use.

### Anesthetization Equipment

Traditionally, fly pushers anesthetized their flies with diethyl ether, but because it is both safer and generally more pleasant, most (except the elderly) now use carbon dioxide.

Normal commercial-grade $CO_2$ as well as gas regulators for $CO_2$ tanks are readily available from many suppliers. To accommodate $CO_2$ tanks, a fly room should have racks that can be conveniently made or purchased, and automatic equipment for switching between tanks can save much grief. For any large fly laboratory, permanent piping of $CO_2$ at the bench is very worthwhile. $CO_2$ should be humidified before use, which can easily be done by passing the gas through a large flask of water. Our only recommendation here is to use a heavy-duty flask and bungs to avoid blowouts due to pressure. If an air-line is available in the laboratory, it can usually be easily adapted by a professional plumber to provide $CO_2$.

Each user should have a $CO_2$ pipeline from the humidified $CO_2$ source, which passes through a local valve that allows the supply pressure to be regulated and turned on and off. The pipeline then divides into two (using a Y-junction) arms, a meter or so from the end. One arm goes to the $CO_2$ plate and the other to a tube that can be inserted past the bung into a vial or bottle. The second arm should have a spring-clip regulator.

The $CO_2$ functions to deliver the gas to flies and to keep them anesthetized while being examined or handled. The plate is normally a shallow vessel that sits on the microscope stage. Gas is passed into this vessel and escapes through the surface of the plate, which is constructed of a white porous material, typically sintered glass or porous polyethylene. (If the surface of the $CO_2$ pad becomes clogged with the debris of dead flies it can be cleaned with sandpaper.) Plates can be purchased from fly equipment suppliers or constructed from Perspex by a competent workshop technician. Post a large notice—"HAVE YOU TURNED THE GAS OFF?"—that is visible to all as they leave the fly room.

### Controlled Temperature and Humidity Environment

Although *D. melanogaster* can be cultured happily at room temperature, any serious use requires a controlled temperature and humidity environment. These can either be stand-alone biological incubators, from any good manufacturer, or specially constructed rooms. Some incubators suffer from serious corrosion problems, especially if propionic acid is being used as a mold inhibitor. The cooling coils should be protected, for example, by enamel paint, or protected coils should be obtained at the time of purchase from the manufacturer. The temperature range should be between 18ºC and 29ºC (these being the extremes for most normal purposes) and the specification should demand a maximum deviation of +/–0.5ºC. Humidity is also important, but the requirement for humidity control will vary with geographical location and season. In general, 70% relative humidity (RH) is necessary.

Humidifiers should be of the type that releases a fine mist into the environment (rather than steam). It is preferable that constant temperature rooms also be air-conditioned, with the circulating air being prehumidified. It is difficult to maintain a high humidity in most laboratory incubators. Those marketed by LMS have, with *Drosophila* culture in mind, modified the cooling coils to reduce the loss of humidity normally caused by condensation on the coils.

Whatever controlled environment is used, some factors are critical. Most important of these is that the temperature control is fail-safe. The environment should be designed so that overheating can never occur. The cooling system must be such that it cannot, under any circumstances, freeze the flies. There must be continuous monitoring of temperature and humidity, noted in a permanent record. In any large facility, a hand-held, digital temperature probe should be regularly used to test for hot or cold spots in the controlled chamber. Finally, any serious fly laboratory must have its temperature-controlled environments wired to a commercial service that alerts users by phone at any time of day or night (usually, in our experience, at ~4.30 a.m.) if the temperature deviates from predetermined tolerances (typically more than 1°C above or 3°C below 25°C, with a decreased higher tolerance at 29°C, e.g., 0.5°C, and an increased lower tolerance at 18°C, e.g., 1°C).

Any enclosed space within which flies are cultured should be designed for ease of cleaning. This means an open floor space (e.g., shelving should end at least 12 inches above the floor), with rounded, rather than right-angular, junctions of walls and floor. There should be no nooks or crannies in which nasties can hide and the shelves should be of the type that are easily cleaned.

### Ancillary Equipment

Fly workers have their own favorite collection of equipment for handling flies, including a selection of fine (camel hair) brushes (e.g., grade 00), forceps (e.g., watchmaker's forceps, size 4 or 5), and perhaps carefully manicured feathers (flight feathers of geese are said to be the best) for pushing flies around a plate. In addition, small hand-held digital counters are useful for those with a poor short-term memory, and a morgue for dead and discarded flies is essential. Morgues are simply vessels filled either with a very-light oil or methylated spirits. A thick rubber mat is useful for deadening the sound of bottles being banged against the laboratory bench.

Flies escape into the fly room or culture room even in the best-run laboratories, and it is essential to trap as many of these as possible. Three methods can (and should) be used. Old-fashioned, sticky fly papers are remarkably effective. Liquid traps, of which there are a variety of beautiful Victorian designs, filled with a solution of live yeast and glucose are also effective. Finally, some laboratories use commercial fly electrocution devices, such as are typically sold to food shops, but these are not effective because the space between the wires is generally too large for *Drosophila*.

## Fly Food Preparation

A large number of different food media have been used for the culture of *D. melanogaster*. Several are described by Ashburner and Thompson (1978), Ashburner (1989b), Roberts and Standen (1998), and on the Bloomington Stock Center Web Site: http://fly.bio.indiana.edu/media.recipes.htm. Only two are described below: a general-purpose medium, and one suitable for growing larvae for polytene chromosome preparations. Table 35.2 lists the sources of ingredients used in the preparation of these media. In addition, instant media are available commercially (see Table 35.2).

**Table 35.2.** Sources of Ingredients for Fly Food Preparation

| Item | Supplier(s)[a] |
|---|---|
| Instant *Drosophila* medium | Carolina Biological Supply Co. Philip Harris Scientific Unilab Export |
| Yeast | available from baker's suppliers |
| Maize meal | available from whole-food shops |
| Dextrose monohydrate | Food Ingredient Technology Ltd. |
| Autolysed yeast powder | Food Ingredient Technology Ltd. |
| Food dye (dark blue is the preferred color) | available from food shops |
| Agar (No 3 [L13]) | Oxoid |
| Nipagin M (methyl hydroxy benzoate, methyl paraben) | Nipa Laboratories Ltd. |

[a]Addresses of suppliers are listed at the end of this chapter.

### General Purpose Medium

Mix A

| | |
|---|---|
| Water | 1300 ml |
| Agar | 18 g |
| Nipagin M (fungal inhibitor) | 50 ml of 10% (w/v) Nipagin M in 95% ethanol |

Mix B

| | |
|---|---|
| Dextrose | 150 g |
| Maize meal | 170 g |
| Dry yeast | 30 g |
| Water | 400 ml |

Combine components of Mix A and boil. Combine components of Mix B and blend to a smooth paste. Add Mix B to Mix A and then boil. Pour into vials and bottles, cover these with a cloth, and allow to cool. Dispense as for maize-meal medium (above). This medium, however, does not keep well and should be prepared daily. When cool and dry, seed surface of medium with a little dried yeast or live yeast suspension. This medium can be stored at 4ºC sealed in plastic bags for up to 10 days.

### Yeast-Glucose Medium

| | |
|---|---|
| Agar | 30 g |
| Glucose | 100 g |
| Dry yeast | 100 g |
| Nipagin M | 15 ml of 10% Nipagin M in 95% ethanol |

Mix the three dry ingredients in a blender and add tap water to 1 liter. Bring to a boil slowly and simmer to thicken. Then add the Nipagin M (fungal inhibitor).

## MAINTENANCE OF A FLY LABORATORY: GENERAL PROCEDURES

### Keeping Stocks

Most large fly laboratories maintain stocks that are not in every day use at 18ºC on a 4–5-week generation cycle. Stocks should be kept as 2–4 independent cultures, and it may be convenient to keep these on alternating generations, 2 weeks apart. Stocks are normally maintained in vials.

Most stocks can be kept by dump-transfer of flies to fresh vials. However, it is important to avoid too overcrowded cultures, and only 20 or so flies should be transferred. It is good practice to inspect the flies on transfer, to ensure that both sexes are present and that their phenotype is as expected. Fly laboratories may keep some stocks that require selection each generation, and it is important that the stock keeper knows of any special requirements to keep any stock (these should be entered on the stock database, see below). The "sick tray" is an inevitable part of any stock room; a place where sickly stocks, or stocks going though a crisis, are kept under special attention. It is very good practice to keep the old cultures for 2 weeks (at 18ºC) after transfer, so that they can be used as a backup should the new stocks fail for any reason.

## Collecting Virgins

Two general methods ensure that female *D. melanogaster* are virgin when used to set up a cross. These can be called the "biological" and "genetic" methods. Only the former is considered here; for genetic tricks useful for virgin collecting, see Chapter 12 of Ashburner (1989a).

Although some variation between stocks exists, the general rule is that females will not accept a male mate until they are 10–12 hours old (i.e., after eclosion from the pupa). Thus, flies can be collected during this window (or, better, between 8 and 10 hours after eclosion), anesthetized, separated into males and females, and stored until needed in yeasted vials. The females will then usually be virgin when used. As a preliminary check, the vials that were used for storing the virgin females should be kept and inspected 3 or 4 days later for any signs of larvae. If larvae are present, it is clear that at least one female in that vial was not virgin. Of course it does not matter too much if a single female is incorrectly stored with the males (as long as she is discarded); but a single male in the tube of females will play havoc. The rule for sexing for virgin collecting, especially when tired or rushed is: If in doubt it is a male. The following is a convenient schedule for virgin collection.

*Day 0:* Clear all flies from emerging cultures in the late afternoon or early evening (e.g., 5:00 p.m. to 6:00 p.m.). Discard these flies. Store emerging cultures at 18ºC in the dark.

*Day 1:* Put cultures at 25ºC in the light, first thing in the morning. Clear *all* flies from the cultures approximately 1 hour later, anesthetize, separate into males and females, and store these in separate vials at 18ºC until required. The young females, i.e., those that are relatively unpigmented and/or have unexpanded wings, will almost certainly be virgin. Check that the emerging cultures have *no* adult flies. Return the emerging cultures to 25ºC in the light and possibly collect virgins last thing in the evening. Keep the "female" vials after using the virgins and inspect 3–4 days later for larvae. If present, presume that any females from that vial were nonvirgin at the time of use. (Note that the presence of eggs in the female-holding vials is not evidence of nonvirginity, even virgin females will lay eggs, albeit at a low rate in comparison with mated females.)

In practice, fly workers develop their own protocols for virgin collection that suits not only the flies, but also social and other activities. But please bear in mind, nonvirginity is *by far* the most common reason for an "unexpected" result from a cross. It is therefore very good practice to design crosses so that nonvirgin progeny will be evident by their phenotype, especially if unexpected nonvirgin progeny could confuse the analysis of an experiment.

## Setting Up and Scoring Crosses

It is impossible to give any universal rules for setting up or scoring crosses, since the precise protocol will vary from experiment to experiment. Crosses can be set up readily with a single pair of parents, although the failure rate can be quite high; normally one would use four pairs for crosses in vials and between five and ten pairs for crosses in bottles. Crosses can be transferred to new vials or bottles after 2 days. When scoring crosses, it is usually important to score at least once a day and to continue scoring for 9 days after the first progeny have emerged, since many genotypes have a delayed development (and/or a short life span). (Scoring for longer than 9 days will cause confusion because $F_2$ flies may be emerging.)

## Mutagenesis

The three general techniques for mutagenesis in *Drosophila* are irradiation, chemical, and genetic (i.e., by transposon insertion). Only the former two are discussed here. The choice between irradiation and chemical mutagenesis is determined by the objective of the experiment. In *very* general terms, only about one third of irradiation-induced mutations will be associated with chromosome aberrations, whereas most mutations induced with either of the two chemicals discussed here will be "point" mutations (usually due to single-base-pair changes; see Chapter 9 of Ashburner [1989a] for a more systematic treatment, and Grigliatti [1998] for detailed protocols). In this section, we discuss only the general protocols for the mutagenesis itself, not the genetic schemes required to detect the desired mutations.

For all routine purposes, only 3–5-day-old males are mutagenized. After treatment, the males are mated immediately to harems of virgin females (usually as 20-pair bottles). These cultures should be transferred daily for 6 days. They can then be discarded or the males removed and the females further subcultured; this ensures that only postmeiotic stages are sampled and will avoid recovering clusters of identical mutations. Bottles should be labeled in such a way that identification of progeny from the same batch of parents is possible.

### Irradiation

Flies may be irradiated with either γ- or X-rays. Many small laboratories use whatever source is conveniently available (e.g., a machine otherwise used for therapy); for purchase, a small industrial X-ray machine is strongly recommended. X-ray equipment such as the Torrex TRX2800 and Torrex 120/150D, 24-inch cabinet, is available from Faxitron Corporation. γ-ray equipment is available from AEA Technology.

An X-ray machine has three advantages over a γ-ray source: (1) the relative biological effectiveness of X-rays is higher than that of γ-rays; (2) X-ray machines are safe when switched off, whereas γ-ray sources require expensive shielding that must be maintained; and (3) for γ-ray sources, the exposure time will need recalibration over time, since the source will be decaying (see Ashburner 1989a [p. 307]). Self-contained X-ray machines suitable for irradiation of flies, i.e., with an operating voltage of 100 kV or more and 5 mA, are available. These machines, designed for industrial use, need to be modified to remove damaging low-energy X-rays for irradiating flies. This is done by inserting a 1-mm filter of aluminum or Perspex between the source and the flies. For irradiation, male flies are placed either in small plastic vials or gelatin capsules (use capsules with a 3-mm-thick wall for $^{60}$Co γ-rays). For the routine induction of mutations, and chromosome aberrations, use a dose of 4 kR with X-rays and 5 kR with $^{60}$Co γ-rays. The dose rate of X-ray machines needs to be calibrated, which may be performed by the supplier, or a dosimeter can be used, for example, the Farmer Dosimeter 257D from NE Technology. A monitor should be used to test the machine for X-ray leakage, e.g., the Mini-monitor from Mini Instruments.

### Chemical Mutagenesis

*Ethylmethanesulfonate (EMS).* The most convenient chemical for routine mutagenesis is EMS, administered by feeding to adult males. The standard dose for EMS (available from Sigma BioSciences) is 25 mM (0.24 ml of EMS in 100 ml of 1% aqueous sucrose, dispersed by repeated aspiration with a 10-ml syringe or P1000 Gilson pipette [Anachem]). Males are placed in an empty bottle and starved for 12–24 hours before being allowed to feed for 12–24 hours on the EMS solution. Tissue paper (or paper towel) is fitted tightly to the floor and sides of the bottle, so that the flies do not get trapped in nooks and crannies. Freshly made up EMS solutions can then be conveniently dispensed on the tissue paper with a P5000 Gilson pipette. (Starvation of males can lead to a high death rate, and consequent loss of yield; this is especially true if the males are already genetically weak. If so, keep them overnight in a food vial without any additional yeast.)

> **CAUTION:** *EMS is a potent and dangerous mutagen and carcinogen. Always work in a well-ventilated fume hood, and wear gloves and eye protection. Decontaminate all material that has come into contact with EMS by treatment in a large volume of 10% (w/v) sodium thiosulfate. Have this solution prepared and available before opening the bottle of EMS, so that any spill can be immediately decontaminated. Do not mouth-pipette EMS under any circumstances. Keep the vials of flies being treated with EMS in the fume hood. Ensure that pipette tips are kept below the surface of fluid, so as to avoid creating an aerosol (see also Appendix 4 for Caution).*

*Ethylnitrosourea (ENU).* ENU is, like EMS, a very effective mutagen for *Drosophila*. Although EMS may induce chromosome aberrations, ENU is far less effective in this respect. Use as a 5 mM solution in 5% aqueous sucrose. ENU can be purchased from Sigma BioSciences as "Isopac" vials. Dissolve in 0.01 M sodium acetate buffer (pH 4.5), by injecting this solution into the vial; dilute in 5% sucrose to 5 mM (Grigliatti 1998). The ENU solution is used in the same way as if. ENU solutions should be freshly made up for use.

> **CAUTION:** *ENU is a potent mutagen and must always be used in a well-ventilated fume hood with proper personal protection (gloves and eye protection). Decontaminate all material that has been in contact with ENU using 1 N NaOH. (see also Appendix 4 for Caution).*

## Plagues and Diseases

One reason for the success of *D. melanogaster* as a laboratory organism is that it is relatively resistant to plagues and diseases. The two most common problems are molds overgrowing the medium and mites. There have also been disturbing reports recently of viral infections in some laboratories. The culprit is apparently a picornavirus (*Drosophila* C virus, DCV) and the symptoms of infection are black, dying pupae. The extent to which fly laboratories suffer from molds and mites varies greatly, but there is some general advice that can be given. The most important advice is that prevention is better (*far* better) than cure. There are two rules of prevention:

- *Keep a clean fly room, fly kitchen, and culture environment.* For cleaning surfaces in fly rooms, use alcohol or a spray disinfectant, e.g., "Astell D."
- *Quarantine all incoming stocks, no matter from what source.* All incoming stocks must be quarantined, even if the distributor swears that they are free of mold, mites, or viruses. A quarantine facility (e.g., a dedicated incubator) should be situated as distant from the normal fly and culture rooms as possible, and all materials, especially discarded vials, must be segregated from those in regular use. It is probably sufficient to quarantine for two generations, and only transfer the stocks to the regular facility when close inspection shows them to be free of infection or infestation. For flies brought to the laboratory straight from the wild, four generations of quarantine is recommended.

If an infection or infestation occurs, the first rule is to isolate all affected cultures to a quarantine facility. What is done next depends very much on the nature and extent of the problem.

### Bacteria

The most common bacterial problem is mucus-producing bacteria on the food, which often produce a reddish-brown pigment (e.g., *Acinebacter* sp.). The addition of antibiotics (streptomycin, tetracycline, or ampicillin; available from Sigma) to the food at a concentration of 250 mg/liter is usually sufficient to cure the problem within one generation. If the problem is recurrent, then investigate the possible sources of the contamination (e.g., the yeast). The use of dextrose, rather than sucrose, in the fly medium should prevent most bacterial growth. The routine use of antibiotics in the food medium is not recommended, as this will inevitably lead to resistance.

### Molds

Molds, usually species of *Penicillium* or *Aspergillus*, are a common problem, as fly medium is an ideal substrate for their growth. With healthy cultures, the flies normally outcompete these fungi, but they can prove to be serious for weak stocks or for cultures at low density. It is now routine practice to include mold inhibitors in medium. Those most commonly used are Nipagin M or propionic acid (see fly food recipes). For both bacteria and molds, persistent infections that are refractory to treatment can best be overcome by washing eggs in 70% alcohol.

### Mites

Several species of mites can infect cultures of *Drosophila* (see Chapter 38 of Ashburner 1989a). Broadly speaking, the mites may be interested either in the flies' food (food mites) or in the flies themselves. The fly mites are rarer than the food mites, but far more dangerous. Food mites often come in with the raw materials of fly medium (e.g., corn meal). For this reason, it is good practice to store bulk meal at –20ºC and to be scrupulous about cleaning up any spills. One of the commonest causes of a serious mite infestation is allowing old fly cultures to fester in culture rooms or the fly room. These rooms must be inspected regularly by someone with the authority to autoclave old cultures without question. This is not an issue where the liberal social attitude so characteristic of fly labs can be allowed to constrain effective management.

If mites are found, then the affected cultures must be immediately quarantined (even better, autoclaved, but this is not always acceptable). If foam or muslin cotton-wool bungs are used, then replace these immediately with bungs of nonabsorbent, tightly balled cotton wool; these should prevent the mites spreading further.

Rapid (i.e., daily) transfer of stocks or cultures can rid them of mites, but this can be dangerous for weak stocks. An alternative is to collect eggs and wash them free of any mites eggs before transfer to clean vials, or to collect pupae on paper inserts and wash them free of mites and mite eggs in 70% ethanol, again before transfer to clean vials.

Tedion has been found to be effective against some common species of food mite. Tedion is available from the Sigma Aldrich Library of Rare Chemicals and also from local suppliers of agrochemicals. Dilute the commercial product (usually 8% active compound) to 5000 ppm in acetone and soak 7-cm filter papers in this solution. Allow filters to dry completely and introduce one into each culture.

Serious endemic mite infestations should never be allowed to build up. If they do, then seek professional advice to combat them, since they will require complete fumigation of all fly-handling rooms and equipment (see Ashburner 1989a [p. 1218]).

### Viruses

In contradiction to the statements in Ashburner (1989a, p. 1192), infection with the double-stranded RNA virus DCV has been found to be a serious problem in a few fly laboratories in recent years. Its symptoms are the presence of dying black pupae, particularly in old cultures. In addition, DCV infection seems to block the induction of transgenes under the Hsp70 heat shock promoter (T. Tully, pers. comm.).

T. Tully (pers. comm) has developed protocols for eradicating viral infection.

## Obtaining and Sending Flies

Two general publicly funded stock centers for *D. melanogaster* are at Bloomington, Indiana and at Umeå, Sweden. In addition, there is a public stock center for P-element stocks at Szeged, Hungary and a center for species other than *D. melanogaster* at Bowling Green, Ohio. The stock lists of all of these Stock Centers (as are those of several individual labs) are available from FlyBase at:

http://flybase.bio.indiana.edu/stocks/

FlyBase provides both search tools for these lists and E-mail ordering forms. The Bloomington Stock Center requires a *Bloomington user number* (BUN) before ordering; these stocks are only available for a small handling charge.

A strong tradition of the fly community is to send stocks to those colleagues who request them. A condition of publication is, or should be, that any stocks described in a publication should be available to any colleague who requests them. *Period!* Conventions for stocks as yet unpublished are less well established. It would certainly be reasonable to enquire why the stocks are requested and a polite refusal should be accepted should this reason conflict with one's own work planned or in progress. Should an investigator supply such stocks to others, it would also not be unreasonable to ask that these not be passed on to a third party without the investigator's explicit permission, at least until they are described in a publication.

There is a regretable trend of requiring "Material Transfer Agreements" to be signed for some stocks, particularly those carrying certain transposons that are covered by patent; see the Web Site at:

http://fly.bio.indiana.edu/patent.htm

It is very good practice for any fly laboratory to maintain a database of all stocks (e.g., in Filemaker Pro). This database should include (1) the genotype of the full stock; (2) notes about particular stocks that will aid in their maintenance and use; and (3) a note of the origin of each stock, and to whom and when it was sent (this helps to recover a stock if one is unfortunate enough to lose it). If an investigator wishes to make the stock list available to a wider public, then it can be posted on FlyBase (E-mail to flybase-help@morgan.harvard.edu).

Stocks can usually be sent by first-class mail. If sent internationally, a customs form declaring them as "Research material" of "No commercial value" may be required. Under *no* circumstances describe them as "fruit flies," or else they will be thought to be Tephritid fruit flies and destroyed. Some countries (e.g., Australia) demand that an import licence number be quoted on the package, which should be provided by the intended recipient. The policy of courier services toward carrying flies seems to vary greatly; some are prepared to carry *Drosophila*, some not. If the investigator requests stocks to be sent by courier, it is a courtesy to provide the sender with an account number. The rules for the direct importation of flies into different countries vary. Importing flies into the United States (e.g., as hand baggage) requires a permit from the USDA and declaration to the Agriculture Inspector at U.S. Customs. Experience has shown that nondeclaration is not looked on kindly, and will inevitably lead to serious delays and seizure of the flies. The Bloomington Stock Center has some advice on these matters; see the Web site at:

> http://fly.bio.indiana.edu/import-export.htm

To send flies by mail or courier, set up a new culture and allow a day or so for egg laying. Wrap the vials well, so that they will not break (e.g., in "bubble wrap") and pack in a cardboard tube. Enclose a data sheet describing the stock fully (conveniently printed out from one's database). It is usually best to mail or otherwise send flies early in the week and not to send flies when the temperature is extreme.

## Address List

### *Stock Centers*

Bloomington *Drosophila* Stock Center
Department of Biology
Indiana University
Bloomington, Indiana 47405
Phone: (812) 855-5782
Fax: (812) 855-2577
E-mail: flystocks@bio.indiana.edu
Web site: http://fly.bio.indiana.edu/

European *Drosophila* Stock Center
Department of Genetics
University of Umeå
Umeå, S901 87, Sweden
Phone: +46 90 7865275
Fax: +46 90 7867665
E-mail: karin.ekstrom@genetik.umu.se
Web site: http://flybase.bio.indiana.edu/stocks/
    stock-centers/lk/umea/

Szeged *Drosophila* Stock Center
Department of Genetics
Jozsef Attila University
Kozepfasor 52
H-6726 Szeged, Hungary
Phone and Fax: +36 62 432 485
E-mail: sidonja@biocom.bio.u-szeged.hu
Web site: http://www.bio.u-szeged.hu/genetika/
    stock/

*Drosophila* Species Resource Center
Department of Biological Sciences
Bowling Green State University
Bowling Green, Ohio 43403
Phone: (419) 372-2742
Fax: (419) 372-2024
E-mail: kayoon@bgnet.bgsu.edu
Web site: http://flybase.bio.indiana.edu/stocks/
    stock-centers/lk/species-center/

FlyBase
Biological Laboratories
16 Divinity Avenue
Harvard University
Cambridge, Massachusetts 02138-2020
Phone: (617) 495-9925
Fax: (617) 496-1354
E-mail: flybase-help@morgan.harvard.edu
Web site: http://flybase.bio.indiana.edu/

## Suppliers and Manufacturers

Accuramatic Ltd.
Watlington
King's Lynn
Norfolk PE33 0JB, United Kingdom
Phone: +44 1553 777253

AEA Technology
Amersham
Buckinghamshire, United Kingdom
Phone: +44 1494 543810
Web site: http://www.aeat.com

Agrisense-BCS Ltd.
Pontypridd
Mid-Glamorgan, United Kingdom
Phone: +44 1443 841155
Fax: +44 1443 841152

Air Products Europe
Hersham Place
Molesley Road
Walton-on-Thames
Surrey, KT12 4RZ, United Kingdom
Phone: +44 1932 249200
Fax: +44 1932 249565
Web site: http://www.airproducts.com/

Air Products
7201 Hamilton Blvd.
Allentown, Pennsylvania 18195-1501
Phone: (610) 481-4911
Fax: (610) 481-5900

Anachem
Anachem House
20 Charles Street
Luton, Beds, LU2 0EB, United Kingdom
Phone: +44 1582 745040
Fax: +44 1582 483332
Web site: http://www.anachem.co.uk

Applied Scientific
West Harris Avenue
San Francisco, California 94080
Phone: (650) 244-9851
Fax: (650) 244-9866
E-mail: cutserv@appliedsci.com
Web site: http://www.appliedsci.com

Astell Scientific Ltd.
Powerscroft Rd., Sidcup

Kent, DA14 5DT, United Kingdom
Fax: +44 181 300-2247
Web site: http://www.astell.com

BIG Central Distribution
Newcut Industrial Estate
Woolston, Warrington
Cheshire, WA1 4AG, United Kingdom
Phone: +44 1925 811005
Fax: +44 1925 821735

Carolina Biological Supply Co.
2700 York Rd.
Burlington, North Carolina 27215
Phone: (800) 334-5551
Phone: (336) 584-0381
Fax: (800) 222-7112
Fax: (336) 584-7686
Web site: http://www.carolina.com/

D.C. Norris and Co. Ltd.
Great Gransden, Sandy
Beds, SG19 3AH, United Kingdom
Phone: +44 1767 677515
Fax: +44 1767 677956
E-mail: mail@dcnorris.co.uk
Web site: http://www.dcnorris.co.uk

Faxitron Corp.
225 Larkin Drive, Unit 1
Wheeling, Illinois 60090
Phone: (847) 465-9729
Fax: (847) 465-9740
Web site: http://www.faxitron.thomasregister.
  com/olc/faxitron/

Food Ingredient Technology Ltd.
HiTec House
Sand Road Industrial Estate
Great Gransden
Beds, SG19 3AH, United Kingdom
Phone: + 44 1767 677666
Fax: +44 1767 677966

Forma Scientific
Marietta, Ohio 45750
Phone: (800) 848-3080
Phone: (740) 373-4763
Fax: (740) 373-6770
Web site: http://www.forma.com/

LabScientific Inc.
114 W. Mt. Pleasant Ave.
Livingston, New Jersey 07039
Phone: (800) 886-4507
Phone: (973) 992-0850
Fax: (973) 992-0827
E-mail: labsupply@aol.com
Web site: http://www.labscientific.com/
    binder_fly.htm

Leica Microsystems (UK) Ltd.
Davy Avenue
Knowlhill, Milton Keynes
Beds, MK5 8LB, United Kingdom
Phone: +44 1908 666663
Fax: +44 1908 609992
Web site: http://www.leica.com

LMS
The Modern Forge
Riverhead, Sevenoaks
Kent, TN13 2EL, United Kingdom
Phone: +44 1732 451866
Fax: +44 1732 450127

Messer UK Ltd.
Station Road
Coleshill
Birmingham, United Kingdom
Phone: +44 1675 46884
Fax: +44 1675 467053
Web site: http://www.messergroup.com

Microscope Services and Sales
1 Whitehall Lane
Egham, Surrey, TW20 9NE, United Kingdom
Phone: +44 1784 432694
Fax: +44 1784 438594
E-mail: microscope@clara.net

Mini Instruments
8 Station Road
Burnham on Crouch
Essex CM0 8RN, United Kingdom
Phone: +44 1621 783282
Fax: +1 1621 783132
Web site: http://www.mini-instruments.co.uk

NE Technology Ltd.
Bath Road
Beenham
Berkshire RG7 5PR
United Kingdom

Phone: +44 118 971 2121
Fax: +44 118 971 2835
E-mail: enquiries@netech.co.uk
Web site: http://www.netechnology.co.uk

Nipa Laboratories Ltd.
Llantwit Fadre, Pontypridd
Mid Glamorgan, CF38 2SN, United Kingdom
Phone: +44 1443 205311
Fax: +44 1443 207746

Novatron Ltd.
Unit 34, Southwater Industrial Estate
Horsham
Sussex, RH13 7DU, United Kingdom
Phone: +44 1403 733012
Fax: +44 1403 733311
E-mail: novatron@BTinternet.com
Web site: http://www.novatron.co.uk/

Omega Engineering Inc.
One Omega Drive
Stamford, Connecticut 06907-0047
Phone: (203) 359-1660
Phone: (800) 826-6342
Fax: (203) 359-7700
Web site: http://www.omega.com/

Oxoid
Wade Rd., Basingstoke
Hants RG24 8PW, United Kingdom
Phone: +44 1256 841144
Fax: +44 1256 814626
E-mail: Oxoid@oxoid.com
Web site: http://www.oxoid.co.uk

Percival Scientific Inc.
1805 East 4th Street
Boone, Iowa 50036
Phone: (800) 695-2743
Phone: (515) 432-6501
Fax: (515) 432-6503
E-mail: percival@netins.net
Web site: http://www.percival-scientific.com/

Philip Harris Scientific
618 Western Avenue
Park Royal
London, W3 0TE, United Kingdom
Phone: +44 181 992-5555
Fax: +44 181 993-8020
Web site: http://www.philipharris.co.uk/scientific

Power Utilities Ltd.
Unit 5, Drome Road
Deeside
Flintshire CH5 2NY, United Kingdom
Phone: +44 01244 280408
Fax: +44 01244 280409
E-mail: alex@power.telme.com
Web site: http://www.power-utilities.ltd.uk

Scientific Laboratory Supplies
Unit 27, Wilford Industrial Estate
Ruddington Lane, Wilford
Nottingham, NG11 7EP, United Kingdom
Phone: +44 115 9821111
Fax: +44 115 9825275

Sigma Aldrich Library of Rare Chemicals
P.O. Box 335
Milwaukee, Wisconsin 53201
Phone: (414) 273-3850

Sigma BioSciences
Fancy Rd., Poole
Dorset, BH17 7NH, United Kingdom
Phone: +44 1202 733114
Fax: +44 1202 715460
E-mail: ukcustsv@eurnotes.sial.com
Web site: http://www.sigma-aldrich.com

S.W. Richardson Ltd.
Unit 14, Mowlem Trading Estate
Leeside Rd.
Tottenham, London N17, United Kingdom
Phone: +44 181 801-6077
Fax: +44 181 801-0541

T.P. Drewitt
15 Palace View Rd.
Chingford, London E4 9EN, United Kingdom
Phone: +44 181 529-2310

Unilab Export
1604 Walker Lake Rd.
Mansfield, Ohio 44906
Phone: (419) 747-1040
Fax: (419) 747-1041
E-mail: unilab@richnet.net

Vapac Humidity Control
Edenbridge, Kent, United Kingdom
Phone: +44 1732 863447
Fax: +44 1732 865658

Watson-Marlow Ltd.
Falmouth
Cornwall, TR11 4RU, United Kingdom
Phone: +44 1326 370370
Phone (USA): (800) 282-8823
Fax: +44 1326 376009
Fax (USA): (978) 658-0041
E-mail: support@watson-marlow.co.uk
E-mail: u/s.support@watson-marlow.com
Web site: http://www.watson-marlow.co.uk

WN Engineers Ltd.
Hoejeloftvaenge 165
DK-3500 Vaerloese
Denmark
Phone: +45 44 48 36 95
Fax: +45 44 48 96 95

## REFERENCES

Ashburner M. 1989a. Drosophila. *A laboratory handbook*. Cold Spring Harbor Laboratory Press, Cold Spring Harbor, New York.
———. 1989b. Drosophila. *A laboratory manual*. Cold Spring Harbor Laboratory Press, Cold Spring Harbor, New York.
Ashburner M. and Thompson J. 1978. The laboratory culture of *Drosophila* . In *The Genetics and Biology of* Drosophila (ed. M. Ashburner and T.R.F. Wright), vol. 2a, pp. 100–109. Academic Press, London.
Greenspan R. 1997. *Fly pushing*. Cold Spring Harbor Laboratory Press, Cold Spring Harbor, New York.
Grigliatti T. 1998. Mutagenesis. In Drosophila. *A practical approach*, 2nd edition. (ed. D. Roberts), pp. 55–83. IRL Press at Oxford University Press, United Kingdom.
Roberts D. 1998. Drosophila. *A practical approach*, 2nd edition. IRL Press at Oxford University Press, United Kingdom.
Roberts D. and Standen G.N. 1998. The elements of *Drosophila* biology and genetics. In Drosophila. *A practical approach*, 2nd edition (ed. D. Roberts). IRL Press at Oxford University Press, United Kingdom.

# CONTENTS

# 36

# Preparation of Larval and Adult Cuticles for Light Microscopy

**David L. Stern and Elio Sucena**
*Laboratory for Development and Evolution*
*University Museum of Zoology*
*Cambridge, England CB2 3EJ*

THE FINELY SCULPTED CUTICLE OF *DROSOPHILA* CARRIES A RICH ARRAY of morphological details. Thus, cuticle examination has had a central role in the history of genetics. Studies of the *Drosophila* cuticle have focused mainly on first-instar larvae (Figure 36.1) and adult cuticular morphology (illustrated in Hodgkin and Bryant 1978, for *D. melanogaster*; and Grimaldi 1990, for the family Drosophilidae). The cuticles of second- and third-instar larvae are, however, strikingly different from those of the first instar (Figure 36.2), but these differences have been poorly studied (Kuhn et. al. 1992).

Almost as many methods exist for preparing cuticles for light microscopy as there are *Drosophila* biologists. In this chapter, we offer some simple protocols that we use routinely and have found satisfactory for analysis of fine cuticular details. With experience, investigators will undoubtedly evolve their own variations.

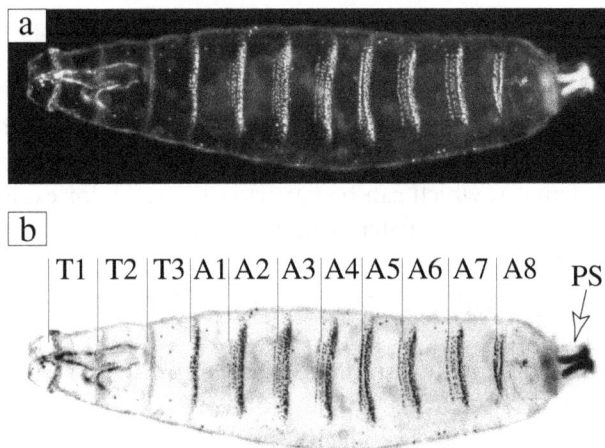

**Figure 36.1.** (*a*) Dark-field image of the ventral cuticle of a first instar larva of *D. melanogaster*. (*b*) Segments are labeled on a digitally inverted version of this image: (T1–T3) first, second, and third thoracic segments; (A1–A8) the eight abdominal segments; (PS) posterior spiracles.

## MOUNTING OF CUTICULAR PREPARATIONS

### Mountants

Hoyer's medium is a useful mountant for both larval and adult cuticles. The medium rapidly digests soft tissues, leaving the cuticle cleared for observation. In addition, samples can be transferred directly from water to Hoyer's medium. However, specimens mounted in Hoyer's medium will degrade over time. For example, the fine denticles on the larval dorsum are best observed soon after mounting; they begin to fade after 1 week, and they may disappear completely after several months. More robust features, such as the ventral denticle belts, will persist for a longer period of time. The preparation of Hoyer's medium is described in Protocol 36.1.

Since in some countries chloral hydrate is a controlled substance, we can recommend lactic acid:$H_2O$ (3:1), an alternative mountant, that also gives excellent results in the short term, although crystals form with time.

To prepare fine "museum-quality," permanent slides, it is best to mount specimens in Canada Balsam. Canada Balsam is usually supplied as a solution in xylene, although it is sometimes supplied solid and must be dissolved in xylene (~60–65% [w/v]) before use. With use, the concentration of xylene in a stock of Canada Balsam decreases, and it is necessary to add more xylene. It is difficult to give precise recipes for Canada Balsam, because every user seems to prefer a slightly different viscosity. We tend to use a rather dilute solution of Canada Balsam so that it spreads easily and does not dry too rapidly while mounting specimens. The disadvantage is that there is actually less Balsam in a "drop" of the solution, and when dried, it may contract from the sides of the coverslip, sometimes even disturbing the specimen. Unfortunately, there is no substitute for experience when using Canada Balsam.

### Slides and Coverslips

Slides and coverslips must be carefully cleaned before use, especially for observation with dark-field optics. Slides and coverslips can be cleaned by rubbing them with lens paper moistened with 70% ethanol.

## PREPARATION OF LARVAL CUTICLES

We distinguish between techniques for mounting larval cuticles of first-instar larvae that have not hatched from the egg (Protocol 36.1) from techniques for mounting larvae that have hatched and begun feeding (Protocol 36.2). Protocol 36.1 is for sampling all eggs laid by one or more females, which can be particularly useful, for example, when a mutation produces embryos that *cannot* hatch from the egg.

### Unhatched First-instar Larval Cuticles

Protocol 36.1 describes the preparation of cuticles from larvae that have not yet hatched from the egg. We have seen this simple protocol in many similar incarnations. Several methods have been proposed for collecting a large number of larvae that have not yet fed. One option is to allow the flies to lay eggs for 3 hours, and then let the embryos develop until they produce a cuticle (~18 hours later at 25ºC). If a longer egg-laying period is required, e.g., to collect eggs from a small number of flies, it is possible to allow the flies

a

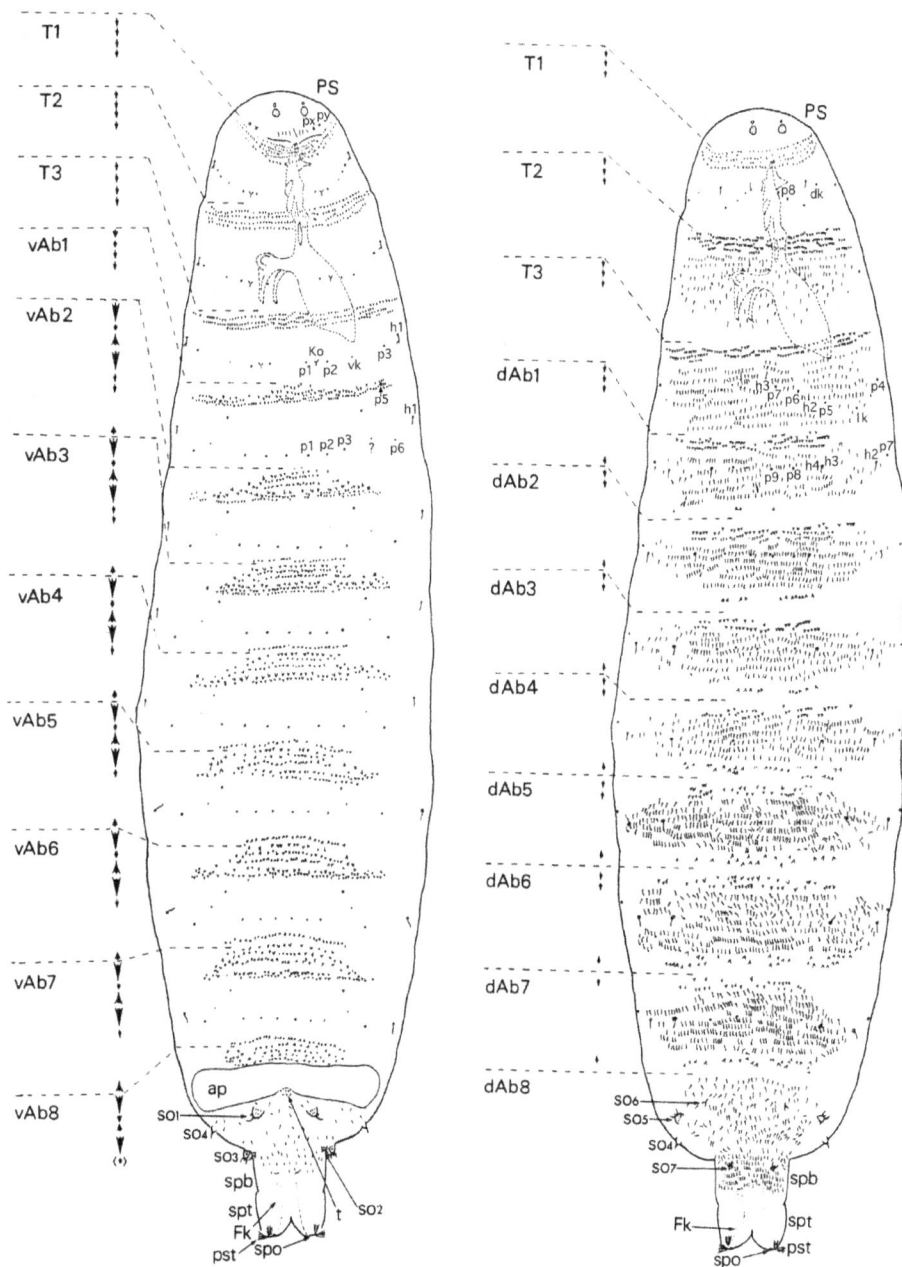

**Figure 36.2a.** (*See page 605 for legend.*)

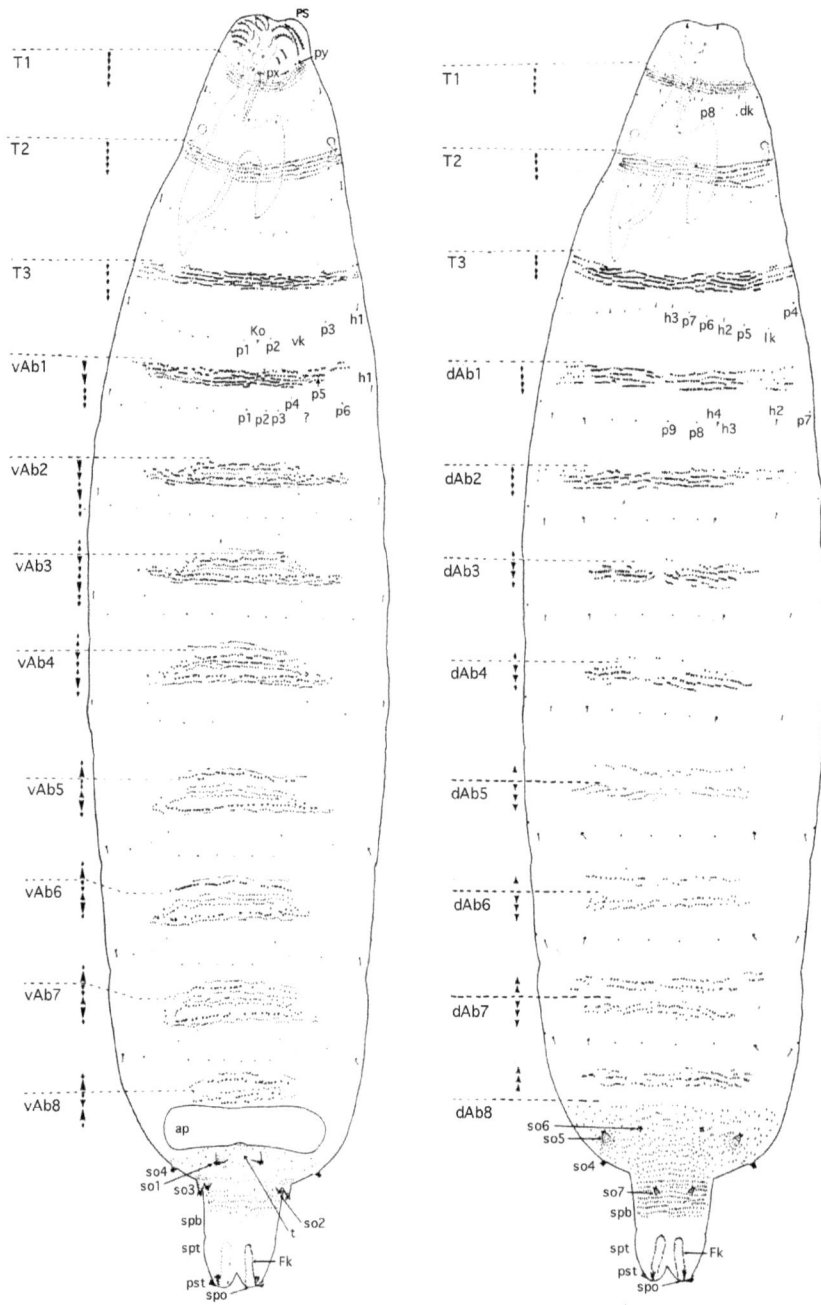

**Figure 36.2b.** *(See facing page for legend.)*

C

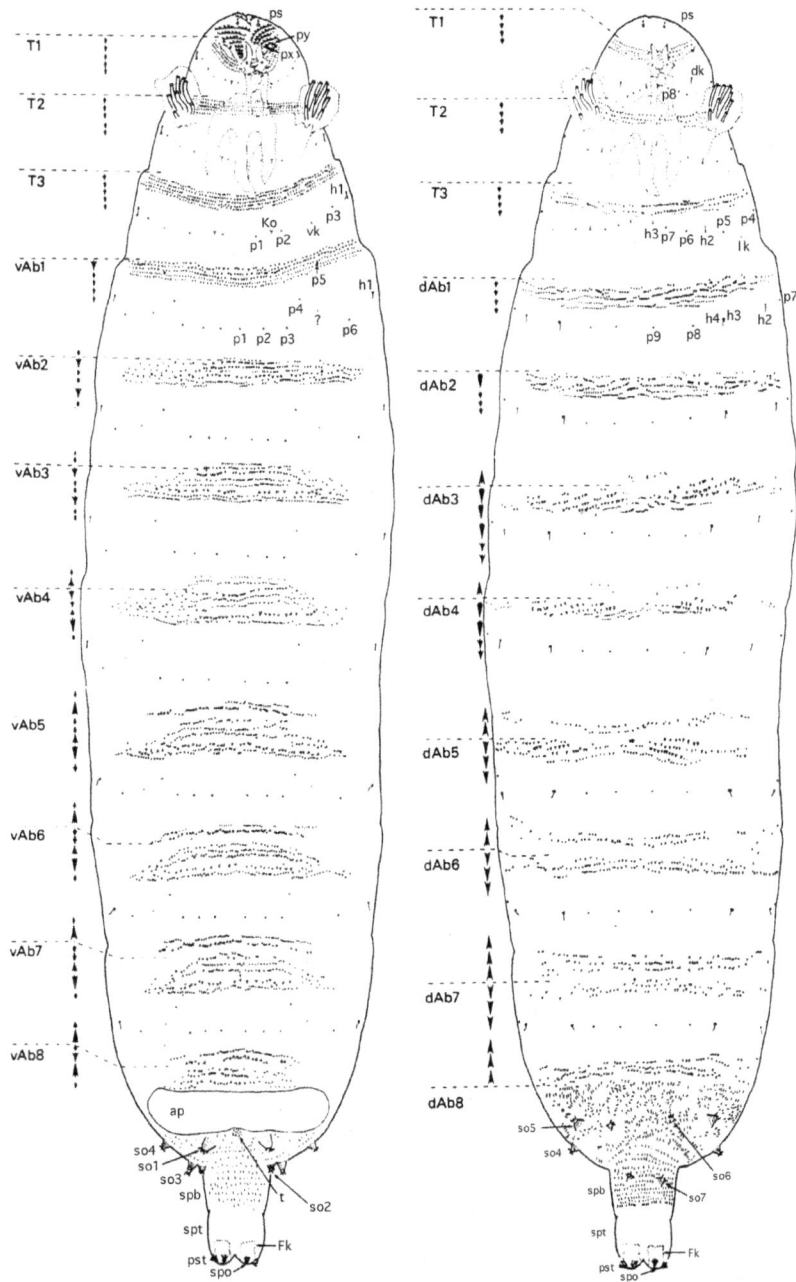

**Figure 36.2.** (*See pages 603–604 for parts a and b.*) Drawings of the cuticles from first-instar (*a*), second-instar (*b*), and third-instar (*c*) larvae. For each instar, the ventral view is on the left, dorsal on the right. Segment borders are identified by horizontal dashed lines. The orientations of spinules on the dorsal side and denticle bands on the ventral side are shown by arrowheads above and below the dashed lines. Other major features are labeled as follows: (ap) anal pad; (dAb1–dAb8) dorsal abdominal segments 1–8; (dk) dorsal *kolbchen*; (Fk) Filzkorper; (h1–h4) hair 1–4; (Ko) Keilin organ; (lk) lateral *kolbchen*; (p1–p9) papillae 1–9; (px–py) papillae x and y; (pst) posterior stigmatophore hairs; (so1–so7) sense organs 1–7; (spb) stigmatophore base; (spo) spiracular opening; (spt) stigmatophore top; (T1–T3) thoracic segments 1–3; (vAb1–vAb8) ventral abdominal segments 1–8; (vk) ventral *kolbchen*. (Reprinted, with permission, from Kuhn et al. 1992.)

to lay eggs for 12 hours (or overnight) and then let the embryos develop for a further 6 hours. The eggs are then dechorionated and washed (step 3 of Protocol 36.1) and allowed to develop for a further 18–24 hours in water (Wieschaus and Nüsslein-Volhard 1986). The protocol can then be continued from step 4.

## Cuticles of Feeding Larvae

To mount larvae that have already fed, it may be useful to remove the gut contents. Protocol 36.2 describes three methods for preparing cuticles from fed larvae. Van der Meer (1977) provides a commonly used procedure (Method 1), which involves manually pricking the larvae. A simpler method for preparing larval cuticles is to burst the larvae once they have been mounted (Method 2). This method is used for first- and second-instar larvae and does not require pricking; it removes the gut contents by "popping" the rear of the embryo using pressure from the coverslip. If just the right amount of medium is used, the coverslip will be pulled toward the slide, applying pressure on the samples. The larvae usually burst from their posterior ends. For both of these protocols, it is useful to first put larvae at 4°C for 10 minutes to allow the larvae to stretch out. For third-instar larvae, we recommend Method 1 (Van der Meer's protocol) or Method 3.

## PROTOCOL 36.1*

# Preparation of Cuticles from Unhatched First-instar Larvae

### Materials

### *Media*

#### *Apple Juice-Agar Plates*

1. In two separate conical flasks mix:

   Flask A: 18 g of agar in 600 ml of $H_2O$

   Flask B: 20 g of sucrose in 200 ml of apple juice

2. Place the conical flasks in a microwave on medium setting (7 on a scale of 0–10) for 30 minutes.

3. When dissolved, pour contents of Flask B into a 2-liter plastic measuring jug with a magnetic stirring rod.

4. Carefully (to avoid bubbles) add contents of Flask A to the jug (containing the apple juice solution) and place the mixture on a magnetic stirrer.

5. Allow the mixture to cool to at least 60°C, and then add 20 ml of 20% Nipagin M (dissolved in absolute ethanol; Nipa Laboratories). Stir for a few minutes and then pour plates. (This recipe will make 65 50-mm plates or 40 90-mm plates).

6. Wait for the medium to solidify. Store plates upside down at 4°C.

#### *Yeast Paste*

Commercial baker's or brewer's yeast can be used. Mix the yeast with tap water to make a soft paste and spread it in a thin layer on the plate.

*For all protocols, $H_2O$ indicates glass distilled and deionized.

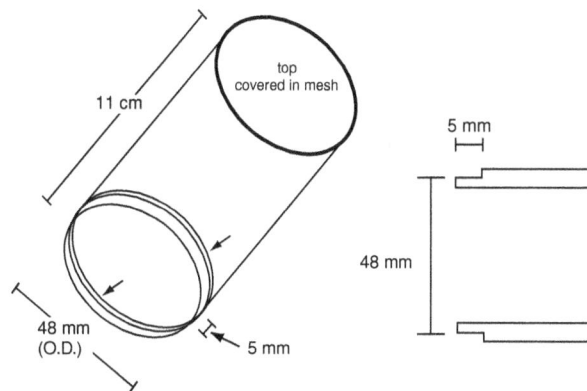

**Figure 36.3.** Diagram of a small egg-laying cage (*left*) and a cross-sectional view of the bottom of the cage (*right*), not to scale. The cage is constructed from 5-cm-wide, 3-mm-thick, plastic tubing, cut into 11-cm lengths. The bottom edge, indicated with arrows, is ground to a height of 5 mm so that the outer diameter equals 48 mm. This is just the right diameter to fit snugly within a 5-cm-wide petri dish. The top is covered with a mesh fine enough to exclude flies, which is attached to the tubing with tape. We normally use a fine plastic mesh with 0.5 x 0.5-mm holes, or cheesecloth. Larger egg-laying cages that fit 90-mm petri dishes can be constructed using 90-mm-wide, 5-mm-thick, plastic tubing with the following dimensions: 15 cm (*height*); 87.3 mm (*outer diameter of ground end*).

## *Supplies and Equipment*

Egg-collecting cages. The precise dimensions of the cages will depend on the size of petri dish used for apple juice–agar plates. We normally use two sizes of cages that fit 5-cm and 9-cm-wide petri dishes. A diagram with dimensions for a small cage is provided in Figure 36.3.

Egg sieves (Figure 36.4)

Small paintbrush

Dissecting microscope

Microcentrifuge tube (1.5 ml)

Slides and coverslips

Filter paper (or Kimwipes)

Fine forceps

Nail polish

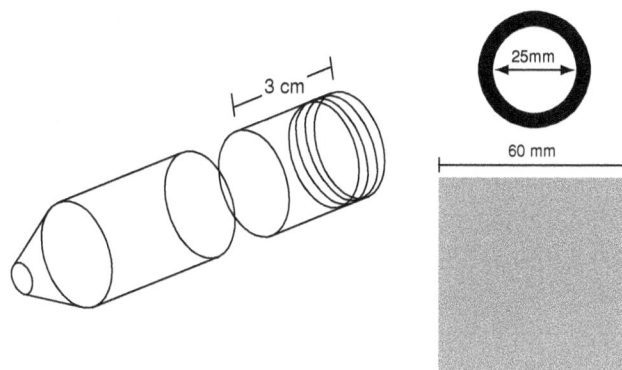

**Figure 36.4.** Diagram of an egg sieve constructed from a 50-ml plastic Falcon tube. First, cut the tube 3 cm from the top edge (*left*) and discard the bottom. Second, cut a 25-mm hole in the top of the Falcon tube (*upper right*). Third, cut a 60 x 60-mm piece of 100-μm polyester mesh (*lower right*). Place the mesh firmly over the top of the cut tube and screw on the cap. The eggs are placed within the tube.

### Solutions and Reagents

Sodium hypochlorite (12%) (commercial **bleach**:tap water [1:1] is satisfactory)
**Methanol**
**Heptane**
Triton X-100 (0.1%) in $H_2O$
Mountant, either Hoyer's medium or **lactic acid**:$H_2O$ (3:1)

---

#### Hoyer's Medium

1. In a chemical fume hood, add 30 g of **gum arabic** to 50 ml of $H_2O$ in a glass beaker with a magnetic stirring rod and dissolve overnight on a magnetic stirrer (heating to 60°C helps).

2. Add 200 g of **chloral hydrate** in small batches.

3. After the chloral hydrate has completely dissolved, add 20 g of glycerol.

4. Centrifuge for 30 minutes at 10,000 rpm (~10,000$g$) and then filter the supernatant through glass wool.

Store the solution at room temperature in a tightly sealed flask. We prefer to dilute the standard Hoyer's recipe with $H_2O$, to allow the solution to spread rapidly under the weight of a coverslip. Hoyer's medium itself may lose $H_2O$ and thicken over time, but it can be reconstituted with more $H_2O$. Hoyer's medium can be mixed 1:1 with **lactic acid** to increase contrast, although crystals will eventually form in these preparations.

---

**CAUTION: bleach (sodium hypochlorite), chloral hydrate, heptane, gum arabic, lactic acid, methanol** (see Appendix 4)

### Method

1. Allow the flies to lay eggs on apple juice–agar plates in the dark for 3 hours. *D. melanogaster* females 4–15 days old lay eggs at the highest rate and prefer to lay eggs in the dark, especially at "dusk" (Ashburner 1989). If it is important to maximize egg production, put females on a light/dark cycle and schedule egg collections for the 3-hour period starting 1 hour before the onset of darkness. Remove the flies and allow the eggs to develop at 25°C for a further 18 hours.

2. When the embryos are ready for collection, transfer them to an egg sieve as follows.

   a. Place enough tap water in the petri dish to just cover the eggs.

   b. Brush over the surface of the apple juice–agar plate vigorously with a small paintbrush, breaking up any clumps of yeast. The eggs will soon be seen swirling in the water. Pour this water into the egg sieve. If eggs remain on the plate, add more water (repeat the brushing if necessary), and pour the water through the sieve.

   c. When all the eggs have been collected, wash them thoroughly under a stream of tap water.

3. Dechorionate embryos as follows:

   a. Place the egg sieve in a small dish containing 50% bleach (12% sodium hypochlorite) for 2–3 minutes.

      *Note:* When the eggs are fully dechorionated, they appear shiny instead of dull white. The first few times it may be useful to observe dechorionation under a dissecting microscope.

b. Remove the egg sieve from the bleach; wash the eggs and container thoroughly with tap water.

> *Note:* As discussed on p. 606, at this stage, the egg sieve can be placed in a petri dish with water to allow the embryos to develop further.

4. Devitellinize embryos as follows:

a. Dry the mesh of the egg sieve by blotting the bottom of the sieve on a paper towel.

b. Dry the paintbrush on the paper towel and then swirl it around the bottom of the egg sieve. The embryos will stick to the paintbrush.

c. Shake the brush in heptane, i.e., in the top phase of a 1:1 mixture of heptane and methanol (500 μl each) contained in a 1.5-ml microcentrifuge tube. The eggs will fall off the brush to the interphase. If large numbers of embryos have been collected, simply use a larger tube and larger volumes of heptane and methanol (1:1).

d. Repeat steps 4b and 4c until all of the eggs are transferred.

e. Close the microcentrifuge tube and, with extreme vigor, shake the tube for 30 seconds. This will cause the vitelline membrane to pop and the embryos will fall to the bottom of the tube.

5. Remove the heptane and methanol with a glass pipette. Replace with 1 ml of methanol. Repeat methanol wash twice.

6. Replace the final wash of methanol with 0.1% Triton X-100. Repeat once.

7. Mount cuticles as follows:

a. Use a glass pipette to transfer the embryos with a small amount of liquid to a clean slide. Fold a piece of filter paper (or a Kimwipe) to a point and use this to wick up most of the liquid, carefully avoiding the embryos. Try not to let the embryos dry out, although irreparable damage will not occur if they dry for a brief period.

b. When most of the liquid is removed, place a drop of Hoyer's medium (or lactic acid:$H_2O$ [3:1]) onto the embryos. Spread out and arrange the embryos using a fine forceps. Place a coverslip onto the preparation.

c. Place the slide at 60ºC for at least 1 hour.

> *Note:* It may be preferable to let the slides cook for one to several days, as this hardens the Hoyer's mountant, but may overdigest the cuticles.

d. Seal the edges of the coverslip with nail polish.

PROTOCOL 36.2

# Preparation of Cuticles from Feeding Larvae

## Materials

### *Media*

Yeasted apple juice–agar plates (see p. 606)

### *Supplies and Equipment*

Egg cages (see p. 607)
Microcentrifuge tubes (1.5 ml)

Slides and coverslips
Nail polish
(*For Method 1*) Fine tungsten needle and fine forceps
(*For Method 2*) Filter paper (or Kimwipes)

### Solutions and Reagents

Mountant: Hoyer's medium (for preparation, see p. 608) or lactic acid:$H_2O$ (3:1)
(*For Method 1*) 70% glycerol (diluted in $H_2O$)
(*For Method 1*) glycerol:**acetic acid** (4:1) fixative
(*For Method 3*) 10% **KOH** (diluted in $H_2O$)

### PBS

130 mM NaCl
7 mM **$Na_2HPO_4 \cdot 2H_2O$**
3 mM **$NaH_2PO_4 \cdot 2H_2O$** (pH 7.0)
Sterilize by autoclaving.

### Perenyi Fix (For Method 2)

To 51 ml of $H_2O$, add 15 ml of 1% **chromic acid** (1 g/100 ml of $H_2O$), 4 ml of **nitric acid**, and 30 ml of 95% **ethanol**. Store at –20°C.

**CAUTION:** acetic acid, chromic acid, ethanol, KOH, $Na_2HPO_4$, $NaH_2PO_4$, nitric acid
(see Appendix 4)

## Method 1

### Modified Van der Meer Protocol

1. Allow females to deposit eggs for 24 hours on apple juice–agar plates daubed with yeast.

2. (*Stretching*) Remove the females and add more yeast as necessary to allow the larvae to grow to the desired stage. Use fine forceps to remove the larvae from the plate and place them in a 1.5-ml microcentrifuge tube containing 500 µl of PBS. Place the tube at 4°C for 15 minutes.

3. (*Pricking*) Place the larvae in a glass dish containing 70% glycerol. Prick the larvae with a fine tungsten needle.

4. Transfer the larvae to a 1.5-ml microcentrifuge tube containing fixative (glycerol:acetic acid [4:1]). Store at 60°C overnight, making sure that the tubes are closed tightly to prevent evaporation of the acetic acid.

5. Mount larvae as follows:

   a. Place a drop of Hoyer's medium (or lactic acid:$H_2O$ [3:1]) on a clean slide. Use fine forceps to place and position the larvae on the slide.

   b. Place the slide at 60°C overnight (up to 72 hours, depending on the size of the larvae).

   c. Seal the edges of the coverslip with nail polish.

## Method 2

### *Alternative Procedure for Larval Cuticle Preparation*

1. Collect larvae from a yeasted apple juice–agar plate into 500 µl of PBS.

2. (*Stretching*) Place at 4°C for 10 minutes.

3. Place the larvae in Perenyi Fix for 20 minutes.

   *Note:* Other fixatives will work, although the relatively fast action of Perenyi Fix is useful in this case.

4. Wash the larvae twice in H$_2$O.

5. Mount larvae as follows:

   a. Place a drop of H$_2$O containing the larvae on a slide. Dry the larvae as much as possible (as in Protocol 36.1, step 7a) with filter paper or Kimwipes.

   b. Drop 25–30 µl (for a 22 × 22-mm coverslip) of Hoyer's medium (or lactic acid:H$_2$O [3:1]) over the larvae. Arrange the larvae and place a coverslip over the preparation.

   c. Heat at 60°C until clear (16–48 hours, depending on the size/stage of the larvae).

   d. Seal the edges of the coverslip with nail polish.

## Method 3

### *Third-instar Larval Cuticle Preparation*

1. Collect larvae with forceps and transfer them to a microcentrifuge tube with 500 µl of PBS. Simultaneously, keep 500 µl of 10% KOH at 4°C.

2. (*Maceration*) Replace the PBS with the cooled 10% KOH. Place the tube at 70°C for 10–15 minutes.

3. Mount larvae as follows:

   a. Cut 5 mm off a 1-ml pipette tip, and pipette the larvae to a slide. Soak up most of the liquid, as described in step 7a (Protocol 36.1).

   b. Place a drop of Hoyer's mountant (or lactic acid:H$_2$O [3:1]) onto the larvae and spread out the samples. Drop a coverslip over the preparation.

   c. Place the slide at 60°C until the samples clear (24 hours or more).

   d. Seal the edges of the coverslip with nail polish.

## ADULT CUTICLES

Protocol 36.3 describes a procedure for mounting adult cuticles in Hoyer's medium; Protocol 36.4 uses Canada Balsam. Adult flies can be stored in 70% ethanol before preparation for mounting. However, tubes of ethanol often dry out. A useful solution, so to speak, is to add glycerol to absolute ethanol to a final concentration of 70% ethanol/30% glycerol. The glycerol never evaporates and protects samples for years. Flies can then be prepared for mounting by first washing them thoroughly in 70% ethanol.

# Rapid Mounting of Adult Structures in Hoyer's Medium

## Materials

### *Supplies and Equipment*

Dissecting tools
- forceps
- dissecting scissors
- dissecting microscope

Slides and coverslips

Nail polish

### *Solutions and Reagents*

Hoyer's medium (for preparation, see p. 608)

**Ethanol** (70%)

**CAUTION: ethanol** (see Appendix 4)

## Method

1. (*Dissection*) Since adults cannot be profitably mounted whole in Hoyer's medium, some, but surprisingly little, dissection is necessary. Samples stored in ethanol are easier to dissect than freshly killed samples. First transfer the specimen to 70% ethanol in a glass dish for dissection. Proceed with dissection.

   *Note:* The exact method of dissection will vary depending on the organs under study, but we provide some general guidelines. The head and abdomen can be pulled off the thorax with ease and usually require no further dissection. The thorax is more problematic as it is large and resilient and has many appendages. One option is to remove the appendages (wings and legs) by holding the thorax with one pair of forceps and pulling at the base of the appendage with another pair. The remains of the thorax can then be cut in two with dissecting scissors, either along the dorsal and ventral midlines or laterally, depending on the regions of greatest interest. Some investigators prefer to leave some or all of the appendages on the thorax and then cut laterally, between the wings and legs, and mount the top and bottom sections independently.

2. Transfer the dissected parts to a glass dish containing $H_2O$ and allow the specimens to rehydrate for several minutes (more time for larger pieces, less time for smaller pieces).

3. Mount as follows:

   a. Transfer the fly pieces directly to a drop of Hoyer's medium spread on a glass slide.

      *Notes:* Entire abdomens and heads can be positioned in the medium, and in this case, it is often best to include four samples per slide arranged at the vertices of a square. Then, when the coverslip is dropped on the medium, it will drop evenly onto the specimens.

      For smaller parts of the fly, such as legs, we routinely mount up to 16 or more legs per slide.

   b. Store the slides at 50°C until the soft tissues have cleared, usually several hours, or longer for hard mounts.

   c. Seal the edges of the coverslip with nail polish.

PROTOCOL 36.4

# Preparation and Mounting of Adult Structures in Canada Balsam

## Materials

### *Supplies and Equipment*

Round-bottomed watch glasses
Slides and coverslips
Dissecting microscope
Solutions and Reagents
**Ethanol** (70% and absolute)
**KOH** (10%)
Clove oil
(*Optional*) **Xylene**
Canada Balsam (see discussion on p. 602 [Mountants])
Nail polish

**CAUTION: ethanol, KOH, xylene** (see Appendix 4)

## Method

We usually perform all of the steps in round-bottomed watch glasses.

1. Wash entire adult flies in 70% ethanol. If they were stored with glycerol, repeat this wash twice.

2. Macerate as follows:

    a. Remove the ethanol and add enough 10% KOH to cover the specimens.

    b. Cover with a glass slide and store at 70ºC for approximately 45 minutes. Monitor this step with a dissecting microscope. When the thorax becomes transparent, i.e., when most of the thoracic muscles have been digested, proceed to step 3 (below). Overcooked flies become fragile. If some of the solution evaporates, add $H_2O$.

       *Note:* If the specimens were stored in ethanol for a long time (i.e., longer than 6 months), it may be necessary to cook them longer. Fresh specimens will require less time.

3. Let the specimen cool to room temperature. Remove the KOH and add enough $H_2O$ to cover the specimen. Return to 70ºC for 45 minutes. Do not add $H_2O$ to the warm watch glass or it may break.

4. Let the specimen cool to room temperature and remove as much of the water as possible, then add 70% ethanol. Dissect the specimens.

5. To dehydrate specimens, replace the 70% ethanol with absolute ethanol. Leave for 10 minutes, then remove the ethanol, and add fresh absolute ethanol.

6. After 10 minutes, remove the ethanol and add just enough clove oil to cover the specimens; alternatively, xylene can be used. Let the specimens sit for at least 15 minutes.

7. Mount as follows:

    a. Spread a drop of Canada Balsam onto a clean slide. Transfer the specimens from the clove oil and arrange as desired. Leave for a few minutes until the surface just becomes sticky when touched with forceps.

b. Place a clean coverslip in a well of xylene to wet the coverslip. Drop the coverslip onto the Canada Balsam at an angle.

c. Dry the slide at 50°C until hard (~3–5 days).

d. Ring the coverslip with nail polish.

## VIEWING CUTICLE PREPARATIONS

Dark-field illumination is the most common technique for viewing larval cuticles, particularly for viewing "large" features such as the mouthparts, ventral denticle belts, and posterior spiracles. A handy substitute for true dark-field illumination is to use low-power objectives with the condenser set in the phase-contrast position for a higher-powered objective (Wieschaus and Nüsslein-Volhard 1986). For example, use a 10X objective with the condenser in Ph3 position. Other larval features, e.g., the fine dorsal denticles of the first instar larva, are best observed using phase-contrast optics.

The adult cuticle presents a wider range of observable features. Large features, such as entire appendages, and the macrochaetes and microchaetes are often best observed with bright-field or DIC illumination. Microtrichiae, which are often erroneously referred to as trichomes, cover most of the body and can be difficult to observe with bright-field optics. Phase contrast often provides suitable viewing, but dark-field illumination is sometimes ideal. We often digitally invert the gray-scale values on images captured with dark-field illumination, to give images that have dark bristles and microtrichiae on a white background (see Figure 36.1).

# REFERENCES

Ashburner M. 1989. Drosophila: *A laboratory handbook*. Cold Spring Harbor Laboratory Press, Cold Spring Harbor, New York.

Grimaldi D.A. 1990. A phylogenetic, revised classification of genera in the Drosophilidae (Diptera). *Bull. Am. Mus. Nat. Hist.* **197:** 1–139.

Hodgkin N.M. and Bryant P.J. 1978. Scanning electron microscopy of the adult of *Drosophila melanogaster*. In *The genetics and biology of* Drosophila (ed. M. Ashburner), vol. 2c, pp. 337–358. Academic Press, London.

Kuhn D.T., Sawyer M., Ventimiglia J., and Sprey Th.E. 1992. Cuticle morphology changes with each larval molt in *D. melanogaster. Dros. Info. Services* **71:** 218–222.

Van der Meer J. 1977. Optical clean and permanent whole mount preparation for phase-contrast microscopy of cuticular structures of insect larvae. *Dros. Info. Services* **52:** 160.

Wieschaus E. and Nüsslein-Volhard C. 1986. Looking at embryos. In Drosophila: *A practical approach* (ed. D. B. Roberts), pp. 199–227. IRL Press, Oxford.

# CONTENTS

# 37

# Exposing *Drosophila* to Neuroactive Drugs

**Jay Hirsh**
*Department of Biology*
*University of Virginia*
*Charlottesville, Virginia 22903*

*D*ROSOPHILA IS BECOMING A USEFUL MODEL for studying responses to neuroactive drugs that affect vertebrates, including alcohol (Moore et al. 1998), volatile anesthetics (Campbell and Nash 1994; Lin and Nash 1996), nicotine (U. Heberlein, pers. comm.), and aerosolized crack cocaine (McClung and Hirsh 1998, 1999; Andretic et al. 1999). We describe methods of feeding drugs to living flies (Protocol 37.1) and a simple method for the direct application of drugs to the nerve cord of decapitated flies (Protocol 37.2). A rather complex device, the inebriometer, can be used for selecting and scoring flies with altered responses to volatile anesthetics and alcohol (Liebovitch et al. 1995; Moore et al.1998); however, a much simpler apparatus, which is described in Protocol 37.3, can be used for studing responses to cocaine, nicotine, and related aerosolized drugs.

Cocaine is an indirect aminergic agonist, blocking reuptake of amines and thus increasing their concentration in the synaptic cleft. In flies and vertebrates, it induces stereotypic motor behaviors that increase in intensity with repeated, intermittent exposures—a process termed sensitization. Simplified forms of these behaviors can be induced in decapitated flies by direct application of aminergic agonists or cocaine (Yellman et al. 1997; Torres and Horowitz 1998). Flies also respond to alcohol; although the direct alcohol target has not been determined, alterations in cAMP signaling affect the severity of alcohol responses (Moore et al. 1998).

A limited number of studies have been performed in which flies will respond to drugs added to food. The agents fed to flies include several biogenic amines such as octopamine (which can restore fertility in females sterile due to a genetic deficiency in octopamine; Monastirioti et al. 1996) and those such as tyrosine, tyramine, L-DOPA, and dopamine (C. McClung and J. Hirsh, in prep.) that can affect in vivo responses to aerosolized cocaine and activity levels. In addition, cocaine-HCl can be fed to flies. It does not have same behavioral consequences as exposing flies to aerosolized cocaine-OH, most likely due to the ability of animals to adapt to relatively constant concentrations of cocaine, but can be shown to be effective due to its effects on responsiveness of dopamine receptors (R. Andretic and J. Hirsh, unpubl.). Finally, an inhibitor of tyrosine hydroxylase, 3,5 diiodo-L-tyrosine, can also be fed to flies (Neckameyer 1996). This enzyme performs the rate-limiting step in dopamine biosynthesis. Similar effects can be obtained by feeding reserpine (P. Bartell and J. Hirsh, unpubl.), which depletes amines from presynaptic terminals. Both of these drugs result in hypoactivity in flies and can cause defects in ovarian development (Neckameyer 1996; R. Andretic et al., unpubl.). The major limitations on which drugs will be effective in flies are likely to be in vivo stability, stability of the drug in the food, and finally, expense—many costly drugs require large monetary outlays to reach effective levels.

## PROTOCOL 37.1*

## Feeding Drugs to Flies

### Materials

### *Media*

Instant fly food (Carolina Biologicals)

### *Solutions and Reagents*

Agents to be fed (all available from Sigma-Aldrich):
- Biogenic amines: **dopamine, serotonin, octopamine**, tyramine
- Drugs affecting biogenic amine metabolism: di-iodo-tyrosine, **reserpine**

### *Preparation of Agents*

Prepare the agents in $H_2O$ (unless otherwise indicated) at appropriate concentrations. Effective concentrations are

5–10 mM for amines

10 mg/ml for di-iodo-tyrosine (make up in 0.1 N **HCl**)

20 mM for reserpine

**CAUTION: dopamine, HCl, octopamine, reserpine, serotonin** (see Appendix 4)

### Method

Add the agents at the appropriate concentrations to instant fly food. Add 1–3-day-old adult flies to the food to observe behavioral effects.

The amines will make the flies hyperactive and will generally increase responsiveness to aerosolized cocaine, whereas the amine-depleting drugs will make flies hypoactive. Effects of amine-depleting agents on responsiveness to aerosolized cocaine can be complicated due to sensitization of amine receptors (P. Bartell and J. Hirsh, unpubl.). The pharmacological specificity for given drugs may be significantly different in flies than in vertebrates; this must be critically evaluated for each drug. All of these agents can be toxic to flies over one to several days. Care must be taken to perform behavioral measurements while the flies are still healthy; sick flies can show quite aberrant responses to drugs.

*For all protocols, $H_2O$ indicates glass distilled and deionized.

PROTOCOL 37.2

## Applying Neuroactive Drugs to the Nerve Cord of Decapitated Flies

### Materials

#### *Supplies and Equipment*

*High-tech option:* Microscissors (Fine Science Tools 15010-11), $CO_2$ dissecting stage, fine forceps

*Low-tech option:* Razor blade, aluminum block chilled to 4°C, fine forceps

Petri dish or other small container lined with moistened filter paper.

Dissecting microscope

Microscope accessories. If videotaping the responses, use a good high-resolution camera (Hitachi, KPM-1) attached to the dissecting microscope via a C-mount adaptor. A ring illuminator (various stock numbers to fit on the microscope objective, Edmund Scientific) fitted to a fiber-optic illuminator provides even and shadow-free illumination.

#### *Solutions and Reagents*

Drugs (available from Sigma-RBI or other suppliers):

Amines: **dopamine, octopamine, serotonin**

D1-like dopamine receptor agonists: SKF 82958, SKF 81297

Dopamine D1-like receptor antagonists: SCH 23390, SKF 85366

D2-like dopamine receptor agonist: quinpirole

D2-like dopamine receptor antagonists: eticlopride, raclopride

#### *Preparation of Drugs*

Prepare drugs in 5 mM **sodium phosphate** buffer (pH 7.0), containing 2% green food coloring, within 24 hours of use. Most drugs are used at approximately 1–5 mM. The food coloring used does not interfere with responses and is a useful marker to ascertain that the cut at the cervical connective is open and capable of taking up drugs.

**CAUTION: dopamine, octopamine, serotonin, sodium phosphate** (see Appendix 4)

### Method

1.  Decapitate flies using either of the following:

    *   A microscissors, on a porous polyethylene stage through which humidified $CO_2$ is bubbled as an anesthetic.

    *   A razor blade, anesthetizing the flies using an aluminum block chilled to 4°C (Torres and Horowitz 1998).

    In either case, minimize the amount of time the flies are anesthetized to ≤5 minutes. Use flies 1–3 days from eclosion.

2.  Place the decapitated flies on a moistened filter paper in a small petri plate or other container.

    *Note:* Within 5 minutes after removal from the anesthetic, the decapitated flies should regain their balance, stand upright, and show a basal level of grooming.

3. Drugs are applied to the cut nerve cord after at least 30 minutes recovery time. Apply drugs using a 20-µl pipettor, holding a small droplet of the drug solution against the cut at the cervical connective for approximately 5 seconds.

*Notes:* The flies remain alive for several days if kept in a humid environment, although healing and closing of the cut nerve cord occurs progressively during this time. To determine that a fly is responsive, check for a provoked grooming response by touching a sensory bristle on the thorax with the hair from a paintbrush. The fly should respond with a grooming response to the affected area. For amines and amine agonists, the decapitated flies should begin to show grooming and locomotor responses within approximately 15 seconds.

The most striking responses are seen with the D1-like antagonists, which cause akinesia with associated tremor, and quinpirole, which stimulates locomotion, hindleg grooming, and hyperactive behaviors (Yellman et al. 1997). Cocaine has also been applied to decapitated flies (Torres and Horowitz 1998), but we find the responses to cocaine are highly variable and difficult to quantitate (C. Yellman and J. Hirsh, unpubl.).

4. To quantitate responses, videotape the flies for 2 minutes after drug application, following locomotion on a grid of 1-mm graph paper.

*Note:* If quantitation of responses is important, keep the temperature and humidity of the testing room relatively constant, and only use flies showing an active provoked grooming response.

## PROTOCOL 37.3

# Exposing Living Flies to Aerosolized Crack Cocaine

### Materials

#### *Supplies and Equipment*

Dosing chamber (see Figure 37.1, and below for construction)

Viewing chamber. Viewing chambers can be made in a glass shop from rectangular glass tubing. We have used either 27 x 9-mm or 12 x 43-mm tubing (see Figure 37.2), with conical openings formed to facilitate fly transfer. The openings are plugged with cut foam from a foam stopper.

Camera setup. We routinely film behaviors using a high-resolution black and white CCD camera (Hitachi KP-M1U) equipped with a 12.6–70-mm video zoom macrolens (Navitar, Associated Microscope; Haw River, North Carolina).

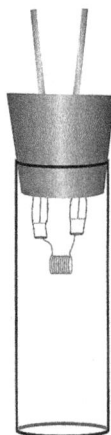

**Figure 37.1.** Fly dosing chamber.

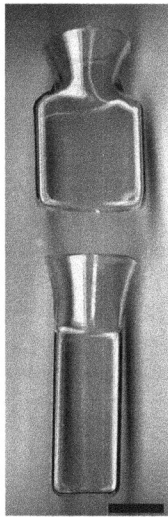

**Figure 37.2.** Two representative behavioral monitoring chambers. Bar, 2 cm.

## Construction of Dosing Chambers

A schematic diagram of a dosing chamber is shown in Figure 37.1. The following items are required for the construction of dosing chambers:

28-gauge Nichrome wire (McMaster-Carr 8880K25 [Phone: 908-329-3200])

telephone butt connectors (Radio Shack 64-3073A)

crimping tool (Radio Shack 64-410)

#4 rubber stopper

25 × 95-mm glass vial

insulated copper wire (~24 gauge, available at hardware stores for bell wiring)

8-gauge nail

Power supply capable of providing 2.2 V, 5 Amp. An inexpensive (~$70) yet adequate power supply is available from Parts Express (120-545; 725 Pleasant Valley Drive, Springboro, Ohio 45066; Phone: 800-338-0531). In the absence of a power supply, the voltage supplied by two C- or D-cell batteries will suffice.

A thermocouple gauge is useful to calibrate the filaments, but is not absolutely required (Cole Parmer 91000-00).

To construct a dosing chamber, tightly wrap ten turns of the Nichrome wire around the 8-gauge nail. Cut off approximately 1-cm leads and attach to the terminal block via the screw connectors. Bore two holes through the #4 stopper and insert the copper wires through the bores. Crimp the wires into one side of the butt connector, and crimp the filament onto the other side. Calibrate the filaments using the thermocouple gauge, determining the voltage required to attain 200°C in 15 seconds. In the absence of such a gauge, we find that 2.2 V applied to the filament will generally suffice. Cocaine base sublimes at 100°C and will pyrolize if heated over 200°C. This is not a large problem in this device since the cocaine rapidly leaves the filament once it sublimes into an aerosol.

## Solutions and Reagents

### Cocaine

Make a stock solution of **cocaine** base at 10 mg/ml in absolute ethanol. This solution is stable at 4°C for days. Apply the cocaine solution to the filament; typically 75 µg for males and

110 µg for females. Allow the ethanol to dry thoroughly. Even small amounts of residual ethanol remaining on the filaments will markedly decrease the motor responses to cocaine. Be aware that ethanol inside the butt connectors will take a rather long time to dry fully. Use the filament within 30 minutes of drying. The filaments lose potency over time, probably by loss of the cocaine crystals.

The active form of cocaine is (–) cocaine; the other isomer, (+) cocaine, is approximately 5 times less effective in flies (C. McClung and J. Hirsh, unpubl.), but can be a useful control. Both drugs are obtained as the free base and are available to basic research investigators through the National Institute on Drug Abuse (NIDA) Drug Supply Program (Phone: 301-443-6275). To receive and use either drug, the investigator or a colleague will need Federal and State Schedule II Drug Licenses.

If (–) cocaine-HCl is substituted for (–) cocaine free base, the cocaine-induced behaviors are markedly reduced, because the HCl salt requires a much higher temperature to sublime, such that most of the cocaine is pyrrolized.

**CAUTION: cocaine, ethanol** (see Appendix 4)

## Method

1. Sort flies in groups of 15–20 at least 12 hours before dosing; exposure to $CO_2$ anesthetic shortly before dosing will alter the responses.

   *Notes:* We routinely dose males and females separately because males are generally more sensitive than females.

   Flies that are old or not healthy can show altered responsiveness to cocaine.

2. Transfer the flies by tapping them into an empty vial without anesthetic. Place the filament assembly on the vial. Tap the flies to the bottom of the vial using the minimum force necessary.

3. Apply approximately 2.2 V to the filament; a smoke of aerosolized cocaine will be apparent from as little as 50 µg of cocaine. Remove the voltage after 10 seconds, and let the flies remain in the vial for another 50 seconds.

4. Transfer the flies to a clean viewing chamber. We routinely perform this transfer in a chemical fume hood even though no aerosol is apparent at 1 minute subsequent to the exposure. The cocaine aerosol rapidly condenses on cold surfaces, such that the duration of the exposure is probably limited by this factor. This factor also makes changing the geometry of the dosing apparatus a bit tricky.

## PROTOCOL 37.3 NOTES

The behaviors induced by cocaine in living flies are rather complex. Because different types of behaviors are induced by different cocaine exposures (McClung and Hirsh 1998), we have devised a behavioral scale to describe the behaviors, as shown in Table 37.1. In general, flies show behaviors that change incrementally along the numbered behavioral scale as they come under the influence of cocaine, and during recovery.

**Table 37.1** Behavioral Scoring of Flies following Exposure to Cocaine Free-base Vapors

| Scale | Behavior |
| --- | --- |
| 0 | Normal behavior: locomotion, flight with a basal level of grooming. |
| 1 | Intense, nearly continuous grooming and reduced locomotion. |
| 2 | Stereotyped locomotion, extended proboscis; some locomotion with simultaneous grooming. In this and higher behavioral scores, there is a loss of negative geotaxis and flight, with flies remaining at the bottom of the container. |
| 3 | Slow stereotypic locomotion in a circular pattern, extended proboscis. |
| 4 | Rapid twirling, sideways or backward locomotion sometimes accompanied by a front leg twitch. |
| 5 | Hyperkinetic behaviors including wing buzzing, erratic activity with flies often bouncing off the wall of the container. |
| 6 | Severe whole body tremor, no locomotion, usually overturned with legs contracted to body. |
| 7 | Total akinesia or dead. |

Adapted, with permission, from McClung and Hirsh (1998).

# REFERENCES

Andretic R., Chaney S., and Hirsh J. 1999. Circadian genes are required for cocaine sensitization in *Drosophila*. *Science* **285:** 1066–1068.

Campbell D.B. and Nash H.A. 1994. Use of *Drosophila* mutants to distinguish among volatile general anesthetics. *Proc. Natl. Acad. Sci.* **91:** 2135–2139.

Leibovitch B.A., Campbell D.B., Krishnan K.S., and Nash H.A. 1995. Mutations that affect ion channels change the sensitivity of *Drosophila melanogaster* to volatile anesthetics. *J. Neurogenet.* **10:** 1–13.

Lin M. and Nash H.A. 1996. Influence of general anesthetics on a specific neural pathway in *Drosophila melanogaster*. *Proc. Natl. Acad. Sci.* **93:** 10446–10451.

McClung C. and Hirsh J. 1998. Stereotypic behavioral responses to free-base cocaine and the development of behavioral sensitization in *Drosophila*. *Curr. Biol.* **8:** 109–112.

———. 1999. The trace amine tyramine is essential for sensitization to cocaine in *Drosophila*. *Curr. Biol.* **9:** 853–860.

Monastirioti M., Linn C.E. Jr., and White K. 1996. Characterization of *Drosophila* tyramine β-hydroxylase gene and isolation of mutant flies lacking octopamine. *J. Neurosci.* **16:** 3900–3911.

Moore M.S., DeZazzo J., Luk A.Y., Tully T., Singh C.M., and Heberlein U. 1998. Ethanol intoxication in *Drosophila*: Genetic and pharmacological evidence for regulation by the cAMP signaling pathway. *Cell* **93:** 997–1007.

Neckameyer W.S. 1996. Multiple roles for dopamine in *Drosophila* development. *Dev. Biol.* **176:** 209–219.

Torres G. and Horowitz J.M. 1998. Activating properties of cocaine and cocaethylene in a behavioral preparation of *Drosophila melanogaster*. *Synapse* **29:** 148–161.

Yellman C., Tao H., He B., and Hirsh J. 1997. Conserved and sexually dimorphic behavioral responses to biogenic amines in decapitated *Drosophila*. *Proc. Natl. Acad. Sci.* **94:** 4131–4136.

# Appendices

# Appendix 1

## 1989 Table of Contents from Drosophila: *A Laboratory Manual*

As INDICATED IN THE PREFACE, THIS EDITION MAKES NO ATTEMPT to be comprehensive. Instead, we took a rather democratic approach to its design and selected protocols that are likely to be widely used by the *Drosophila* community over the next 10 years. This unfortunately required sacrificing many excellent protocols that were included in the original edition. Although this edition has passed its tenth anniversary, most labs still have a copy, and the tattered and taped covers attest to its continued value as a reference. Consequently, this appendix reproduces the 1989 Table of Contents of the original edition by Michael Ashburner.

## CHROMOSOMES

*Protocol 1*
Preparation of mitotic chromosomes

*Protocol 2*
Preparation of mitotic chromosomes

*Protocol 3*
Preparation of mitotic chromosomes without colchicine treatment

*Protocol 4*
Staining mitotic chromosomes with Hoechst 33258

*Protocol 5*
Staining with quinacrine hydrochloride

*Protocol 6*
Staining with chromosomes of embryonic nuclei with DAPI

*Protocol 7*
C-banding mitotic chromosomes

*Protocol 8*
N-banding mitotic chromosomes

*Protocol 9*
Labeling chromosomes in S phase with BrdU in vivo

*Protocol 10*
Labeling nuclei with BrdU

*Protocol 11*
Labeling chromosomes with BrdU in vitro

*Protocol 12*
Preparation of meiotic chromosomes from testes

*Protocol 13*
Staining meiotic chromosomes with DAPI

*Protocol 14*
Feulgen/Giemsa staining of meiotic and mitotic chromosomes

*Protocol 15*
Whole mount technique for female meiosis

*Protocol 16*
Squash method for female meiosis

## MOLECULAR BIOLOGY

## TISSUE CULTURE

*Protocol 132*
Quantitative measurement of brown-eye pigments

## TRANFORMATION

*Protocol 133*
Quantitative assay for β-galactosidase

*Protocol 134*
Assay for chloramphenicol acetyltransferase (CAT)

*Protocol 135*
Plasmid rescue of integrated transposons

*Protocol 136*
Purification of DNA for transformation

*Protocol 137*
Isolation of RNA from transfected embryos

## APPENDICES

*A* Stains
*B* Fixatives and other histological reagents
*C* Mounting and embedding media
*D* Histochemical substrates
*E* Proteins and enzymes
*F* Hormones
*G* Oils
*H* Chemicals and reagents
*I* Materials
*J* Buffers
*K* Solutions
*L* Simple media
*M* Tissue and organ culture media
*N* Food media
*O* Suppliers list

# Appendix 2

## Anatomical Drawings of *Drosophila*

Appendix 2 consists of simple line drawings of many key aspects of *Drosophila* biology. We have included these because many of us have often spent a great deal of time searching the *Drosophila* literature for a basic drawing to make a transparency for a seminar or to use as a template for other purposes. Although they are available from a number of sources, one is often hard pressed to locate a given example 20 minutes before a group meeting. While not comprehensive, the following set of figures (roughly ordered in accord with *Drosophila* development) should be useful in this regard.

635

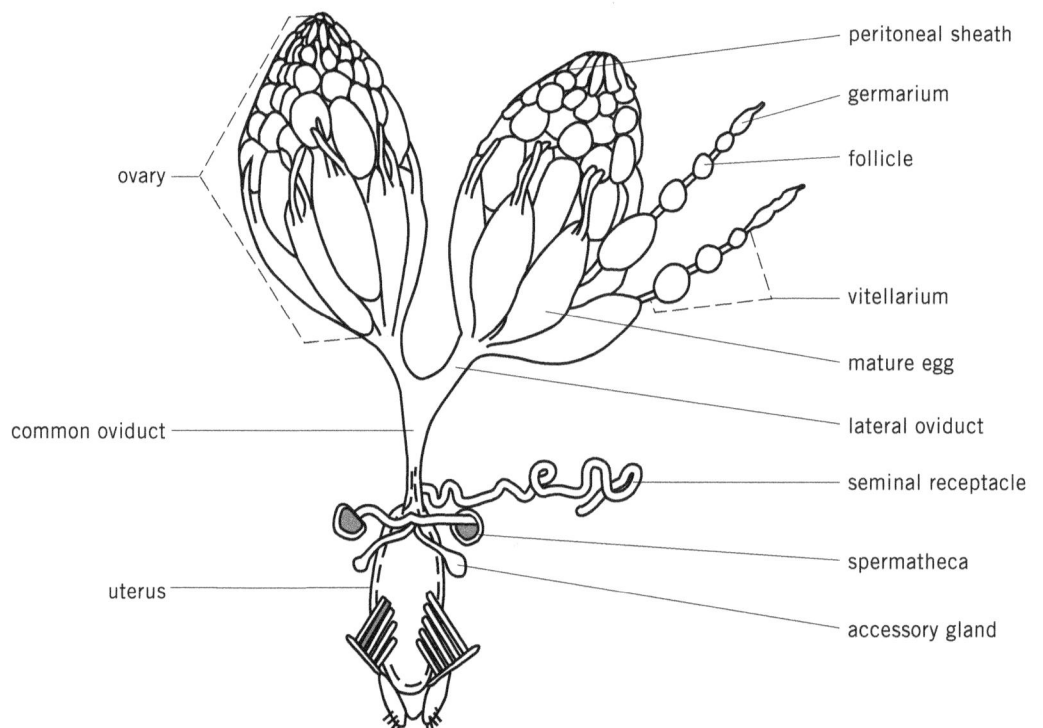

**Figure A2.1.** The female reproductive system contains two ovariole clusters that connect to a common oviduct. Eggs pass through the oviduct and are momentarily stored in the uterus until egg deposition. The broken line indicates the position of the mature egg in the uterus (the posterior end faces down). (Redrawn, with permission, from Miller 1950.)

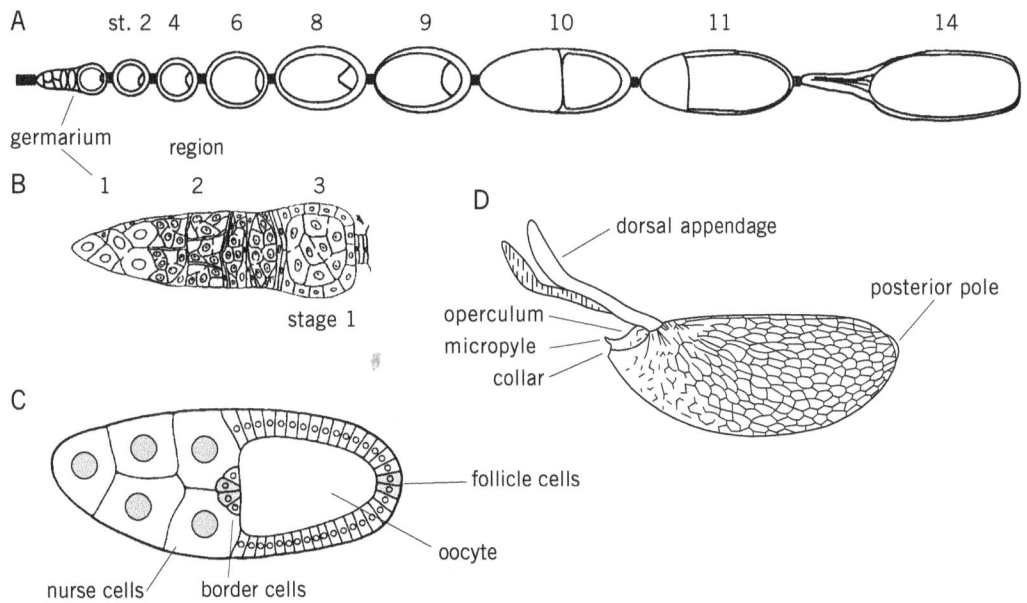

**Figure A2.2.** The *Drosophila* ovary consists of 15–20 functional units called ovarioles (see Figure A2.1). (*A*) Each ovariole consists of a germarium connected to series of developing egg chambers. (Redrawn, with permission, from Robinson et al. 1994 [©Company of Biologists Ltd.].) (*B*) The germarium contains stem cells (region 1) that produce 16 cell clusters connected through a series of intercellular bridges (region 2). As these clusters grow, they are encompassed by follicle cells (region 3). (Redrawn, with permission, from Robinson et al. 1994 [©Company of Biologists Ltd.].) (*C*) Egg chamber containing an oocyte, 15 nurse cells, and a layer of follicle cells (Redrawn, with permission, from Lawrence 1992 [©Blackwell Science Ltd.].) (*D*) Chorion-encased mature egg. (Redrawn, with permission, from Parks and Spradling 1987.)

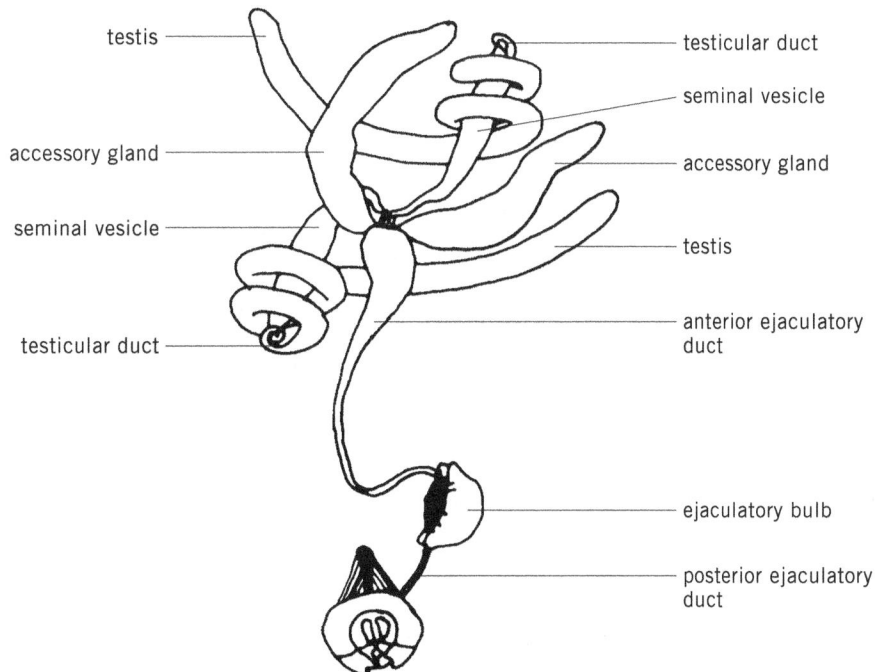

**Figure A2.3.** The male reproductive system. Germ cells, positioned at the anterior tip of the testis, produce cysts containing 64 spermatids. Growth and maturation of the spermatids occur within the testis. (Redrawn, with permission, from Miller 1950.)

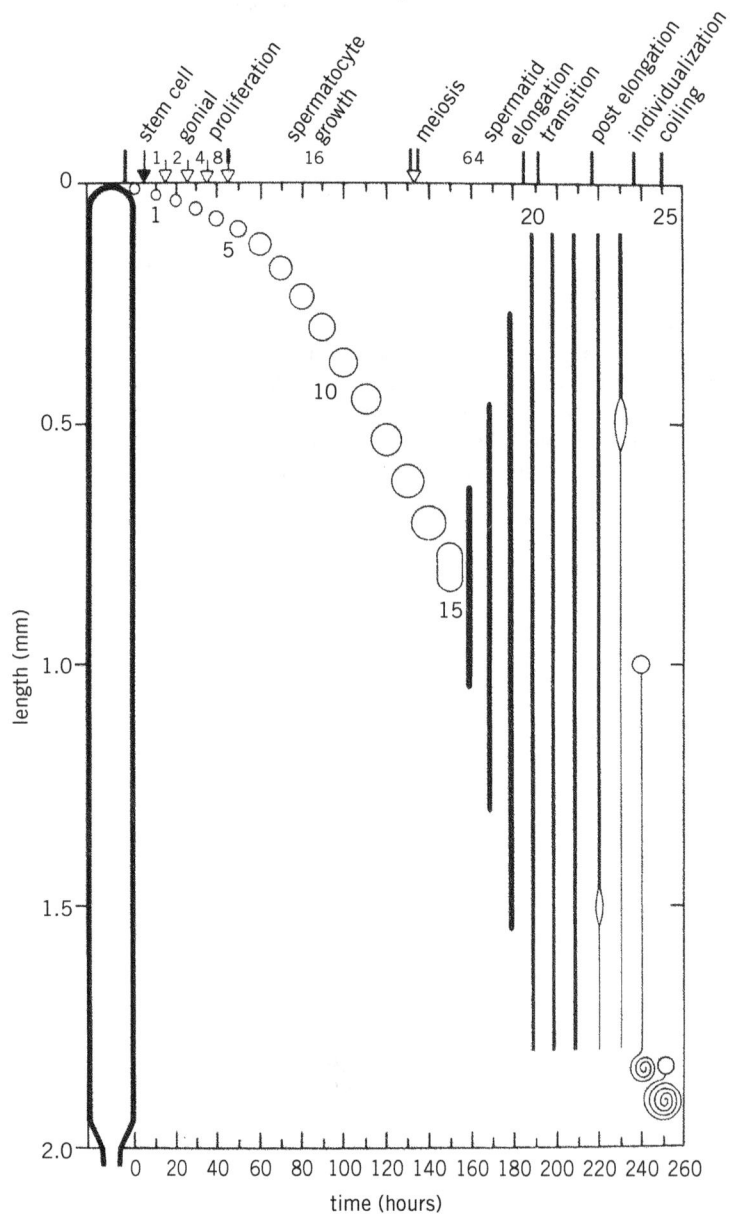

**Figure A2.4.** Cysts derived from a single stem cell are diagrammed according to their age and position in the testis. Each cyst contains 64 spermatids arising from the four synchronous mitotic divisions followed by two meiotic divisions. Because cytokinesis is incomplete in all of these divisions, the spermatids are connected by intracellular bridges. (Redrawn, with permission, from Lindsley and Tokuyasu 1980.)

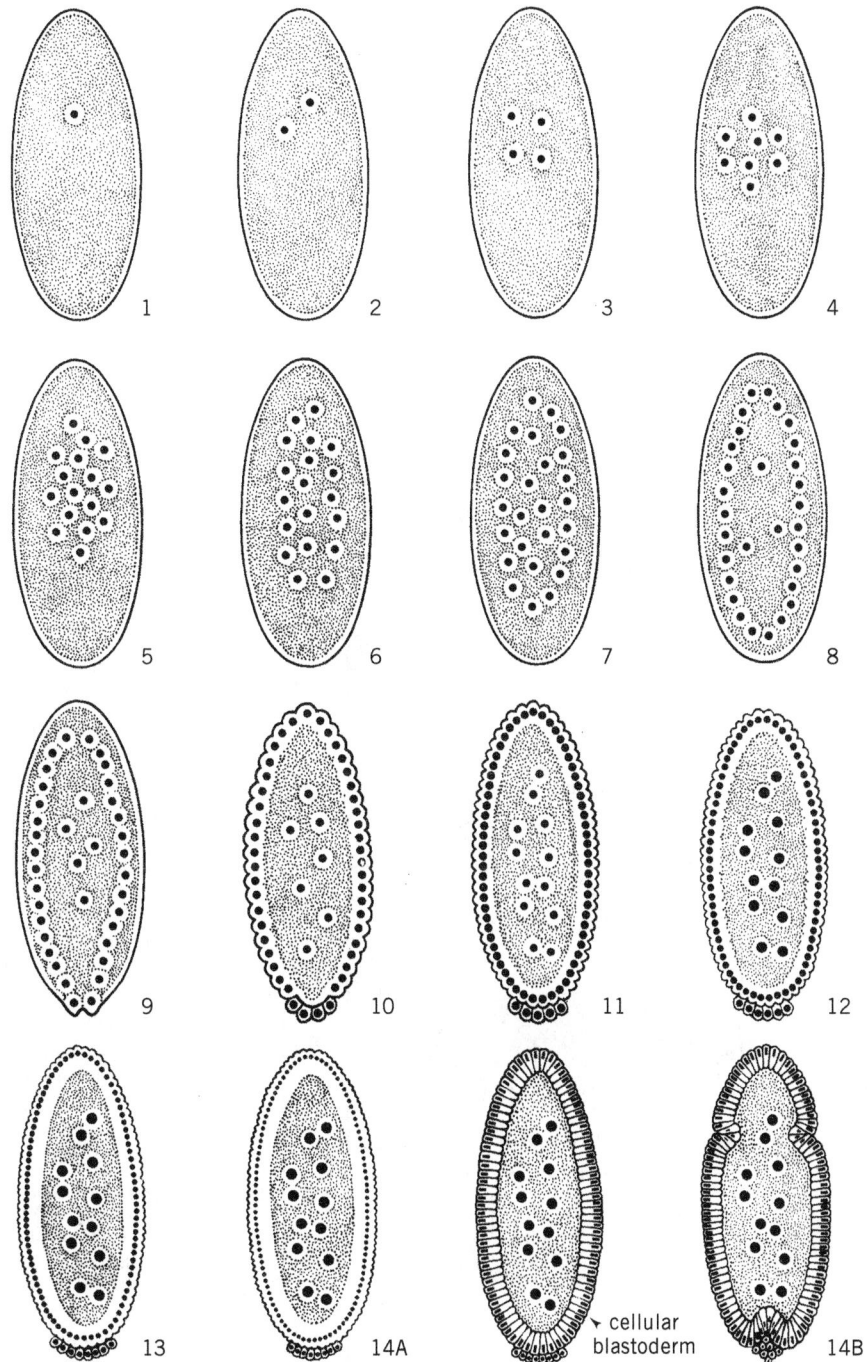

**Figure A2.5.** Early *Drosophila* embryogenesis consists of a series of rapid synchronous nuclear divisions without accompanying cytokinesis. Posterior localized pole cells, the germ-line precursors, form during nuclear cycles 9 and 10. Cellularization occurs during nuclear cycle 14 and is immediately followed by gastrulation. (Redrawn, with permission, from Foe and Alberts 1983 [©Company of Biologists Ltd.].)

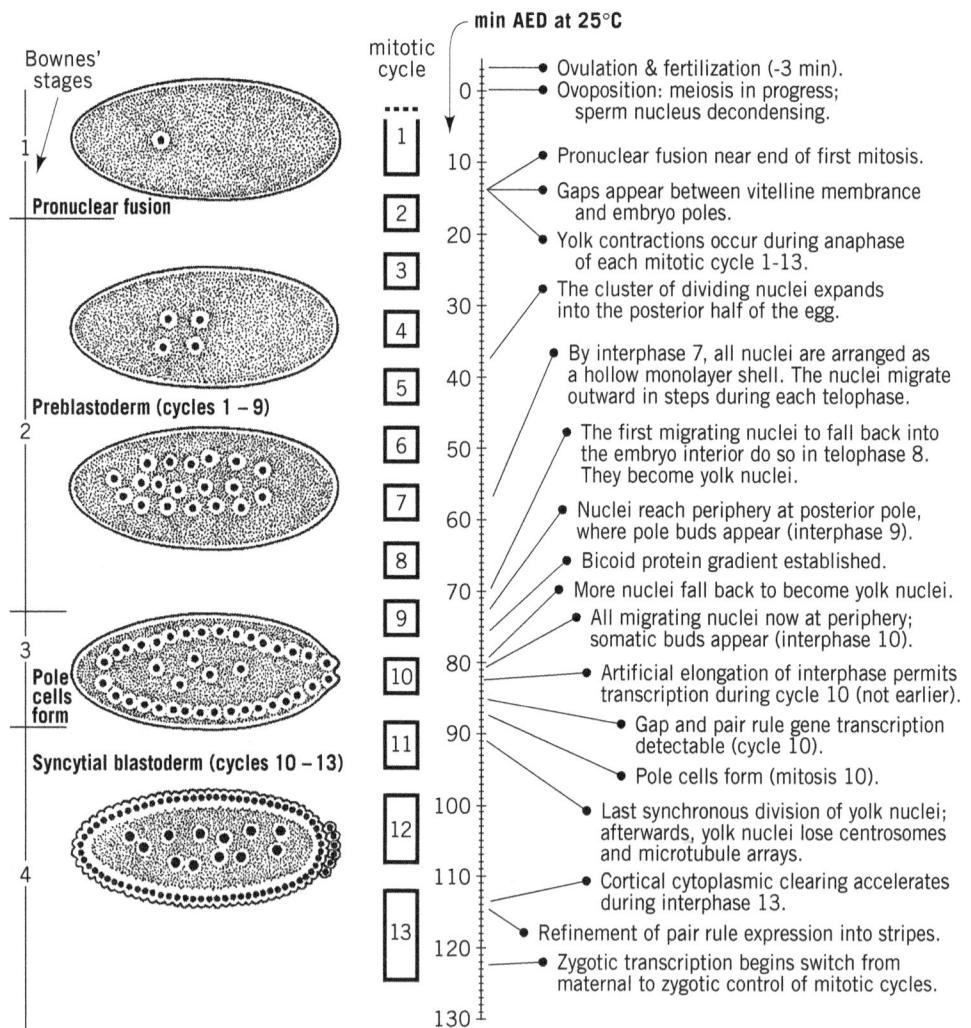

**Figure A2.6.** (*See facing page for legend.*)

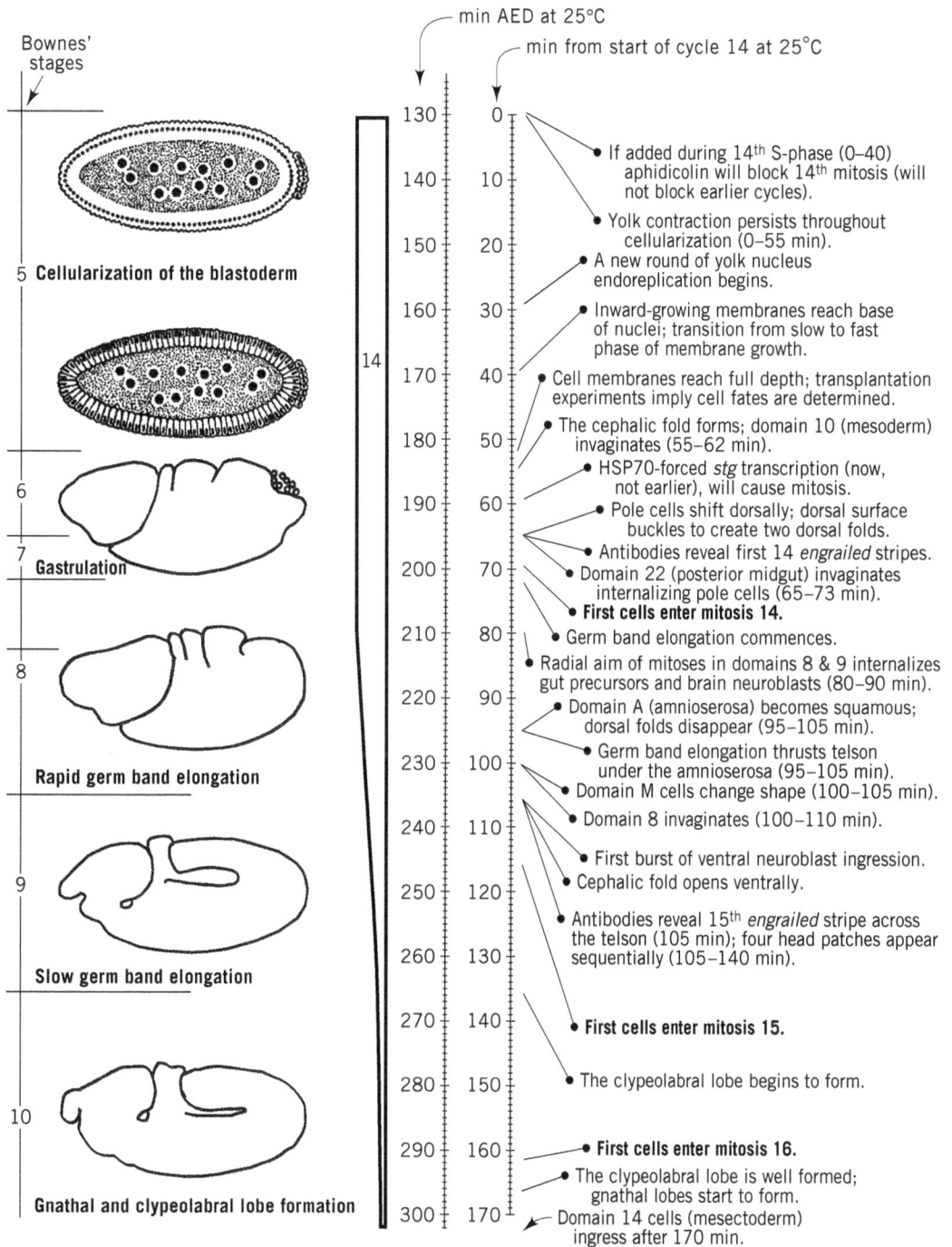

min AED at 25°C

min from start of cycle 14 at 25°C

Bownes' stages

5 **Cellularization of the blastoderm**

6

7 **Gastrulation**

8

**Rapid germ band elongation**

9

**Slow germ band elongation**

10

**Gnathal and clypeolabral lobe formation**

14

- If added during 14th S-phase (0–40) aphidicolin will block 14th mitosis (will not block earlier cycles).
- Yolk contraction persists throughout cellularization (0–55 min).
- A new round of yolk nucleus endoreplication begins.
- Inward-growing membranes reach base of nuclei; transition from slow to fast phase of membrane growth.
- Cell membranes reach full depth; transplantation experiments imply cell fates are determined.
- The cephalic fold forms; domain 10 (mesoderm) invaginates (55–62 min).
- HSP70-forced *stg* transcription (now, not earlier), will cause mitosis.
- Pole cells shift dorsally; dorsal surface buckles to create two dorsal folds.
- Antibodies reveal first 14 *engrailed* stripes.
- Domain 22 (posterior midgut) invaginates internalizing pole cells (65–73 min).
- **First cells enter mitosis 14.**
- Germ band elongation commences.
- Radial aim of mitoses in domains 8 & 9 internalizes gut precursors and brain neuroblasts (80–90 min).
- Domain A (amnioserosa) becomes squamous; dorsal folds disappear (95–105 min).
- Germ band elongation thrusts telson under the amnioserosa (95–105 min).
- Domain M cells change shape (100–105 min).
- Domain 8 invaginates (100–110 min).
- First burst of ventral neuroblast ingression.
- Cephalic fold opens ventrally.
- Antibodies reveal 15th *engrailed* stripe across the telson (105 min); four head patches appear sequentially (105–140 min).
- **First cells enter mitosis 15.**
- The clypeolabral lobe begins to form.
- **First cells enter mitosis 16.**
- The clypeolabral lobe is well formed; gnathal lobes start to form.
- Domain 14 cells (mesectoderm) ingress after 170 min.

**Figure A2.6.** Time line of the first 5 hours of development. (AED) After egg deposition. (Redrawn, with permission, from Foe et al. 1993.)

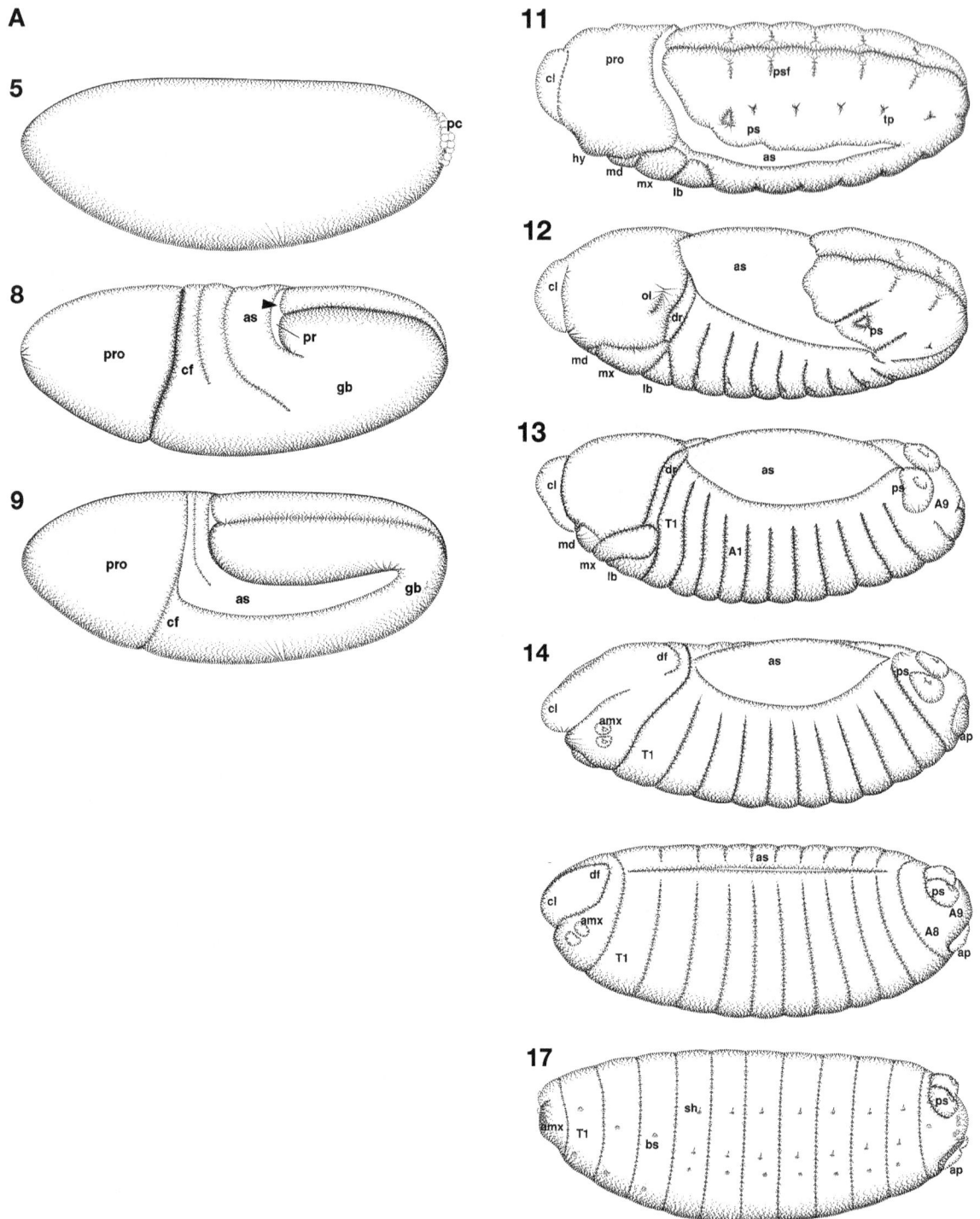

**Figure A2.7.** (*A*) Surface views of of key developmental events during *Drosophila* embryogenesis according to the stages described by Bownes (1975). (Stage 5) Cellular blastoderm; (8) gastrulation; (9) germ-band elongation; (11) segment formation; (12) germ-band shortening; (13) dorsal closure; (14) head involution; (15) completion of dorsal closure and head structure are displaced interiorly; (17) cuticle thickens and ventral denticles become invisible. (Reprinted, with permission, from Hartenstein 1993.) (*See facing page for part B and legend.*)

**B**

## Stages of Embryonic Development

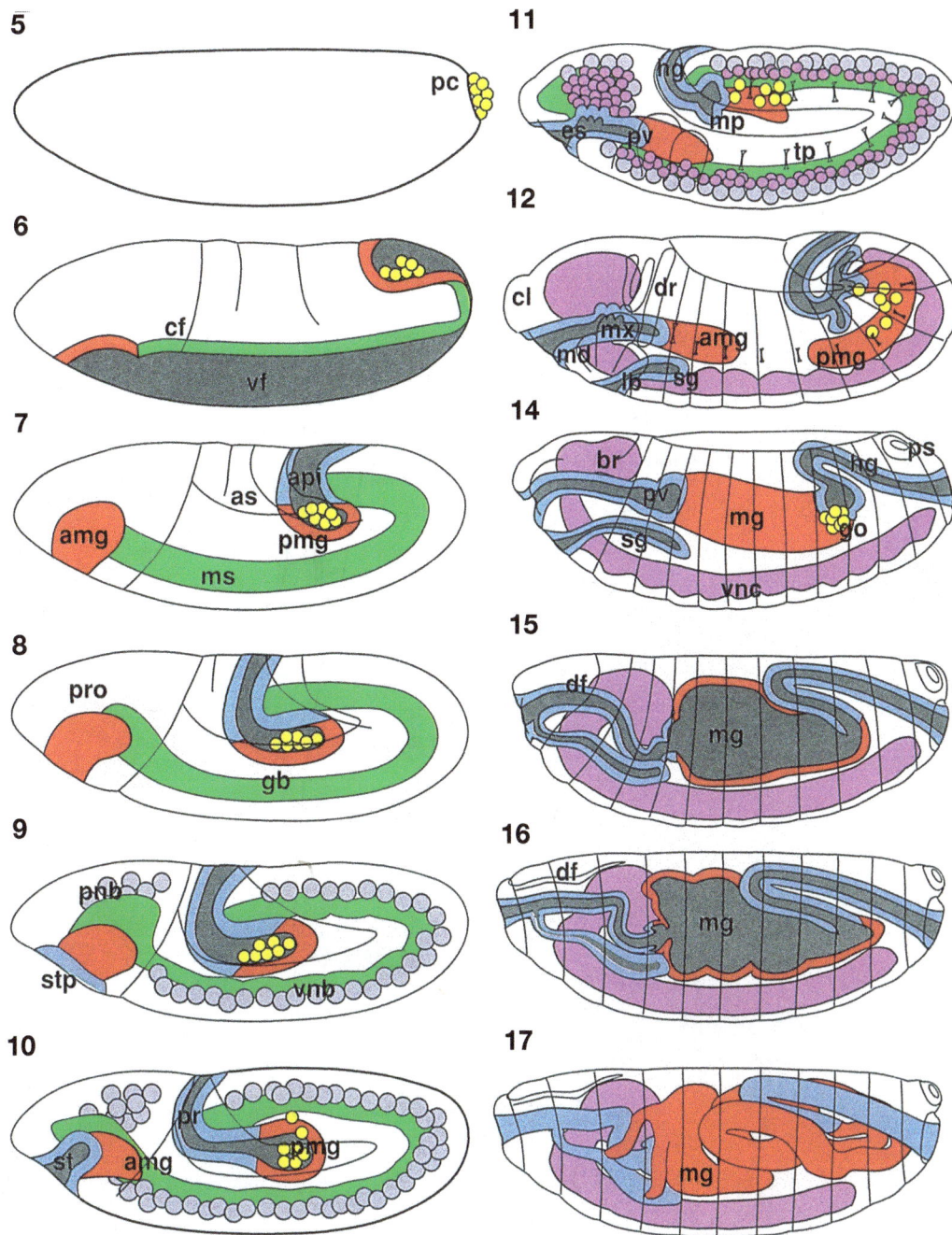

**Figure A2.7.** (*B*) Lateral views of the above developmental stages. (Reprinted, with permission, from Hartenstein 1993.) (A1,A8,A9) First, eighth, and ninth abdominal segment, respectively; (amx) antennomaxillary complex; (ap) anal plate; (as) amnioserosa; (bs) basiconical sensilla; (cf) cephalic furrow; (cl) clypeolabrum; (df) dorsal fold; (dr) dorsal ridge; (gb) germ band; (hy) hypopharyngeal lobe; (lb) labium; (md) mandible; (mx) maxilla; (ol) optic lobe; (pc) pole cells; (pr) amnioproctodeal invagination; (pro) procephalon; (ps) posterior spiracle; (psf) parasegmental furrows; (sh) trichoid sensilla; (T1) first thoracic segment; (tp) tracheal pits.

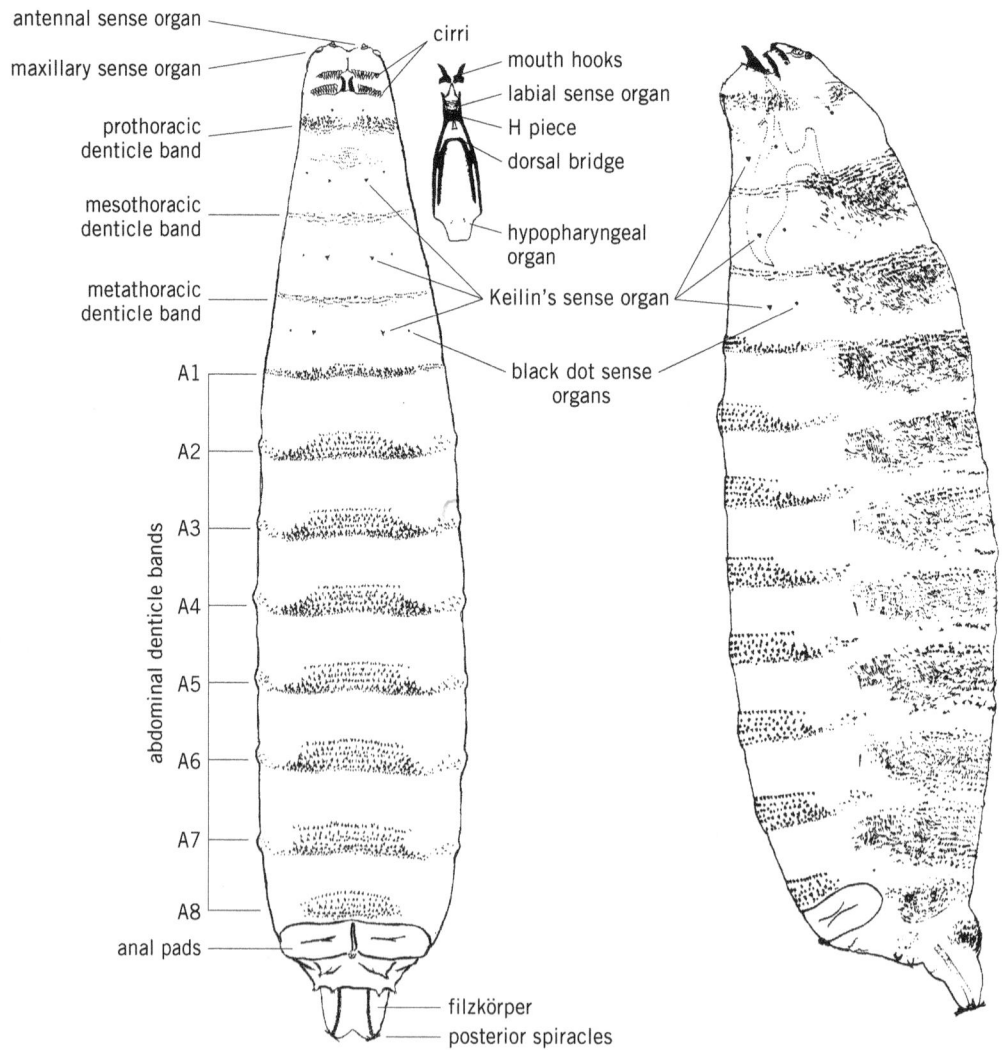

**Figure A2.8.** Ventral and lateral views of the larval cuticle. (Redrawn, with permission of Oxford University Press, from Wieschaus and Nüsslein-Volhard 1986).

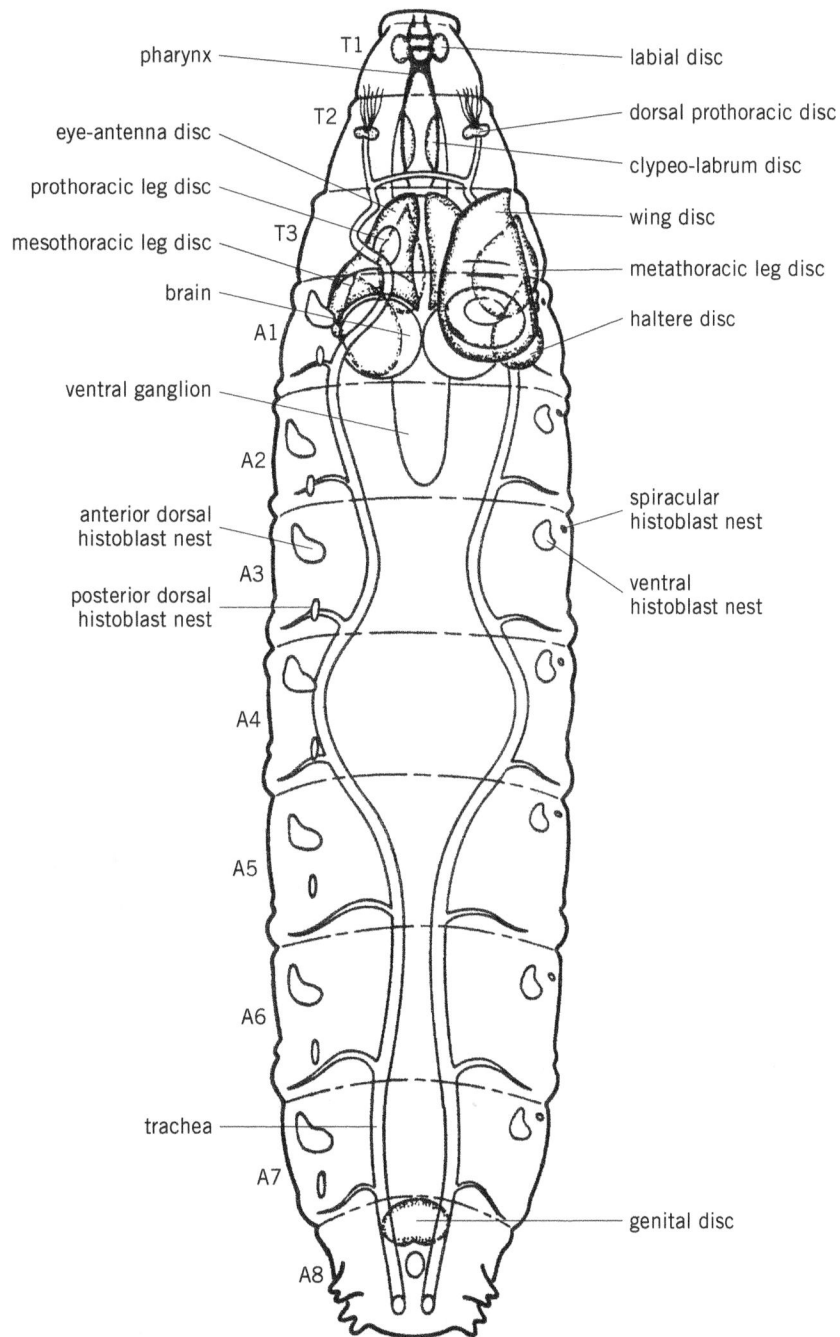

**Figure A2.9.** Position of the trachea, imaginal discs, and abdominal histoblasts in the larva. (Redrawn, with permission, from Poodry 1980.)

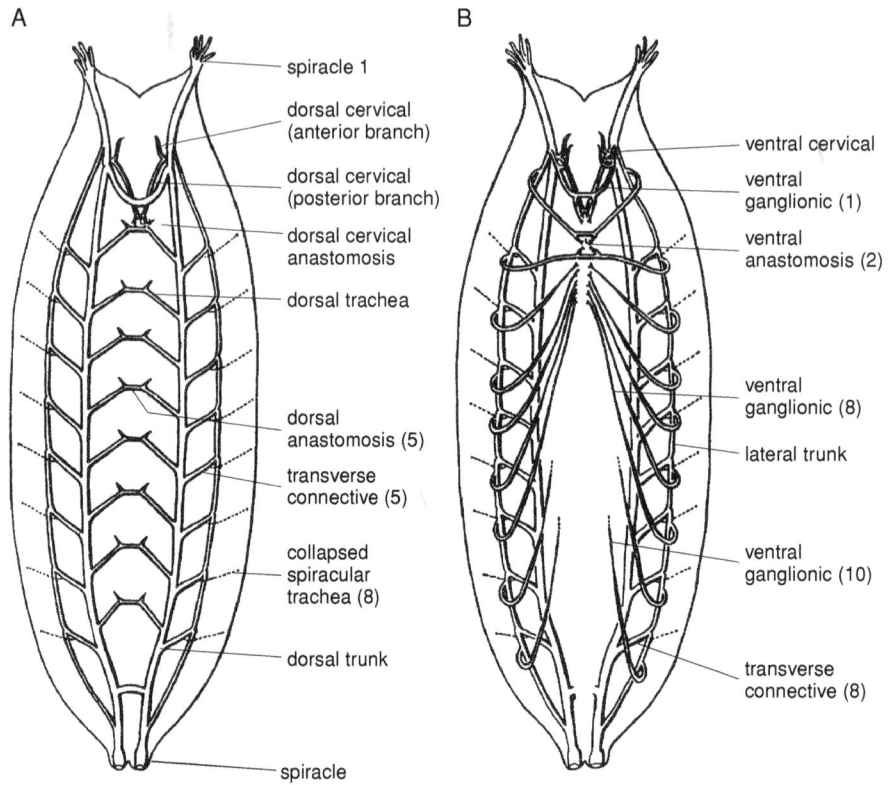

A

- spiracle 1
- dorsal cervical (anterior branch)
- dorsal cervical (posterior branch)
- dorsal cervical anastomosis
- dorsal trachea
- dorsal anastomosis (5)
- transverse connective (5)
- collapsed spiracular trachea (8)
- dorsal trunk
- spiracle

B

- ventral cervical
- ventral ganglionic (1)
- ventral anastomosis (2)
- ventral ganglionic (8)
- lateral trunk
- ventral ganglionic (10)
- transverse connective (8)

**Figure A2.10.** Dorsal and ventral views of the tracheal system in third-instar larvae. (Redrawn, with permission, from Whitten 1957 [©Company of Biologists, Ltd.].)

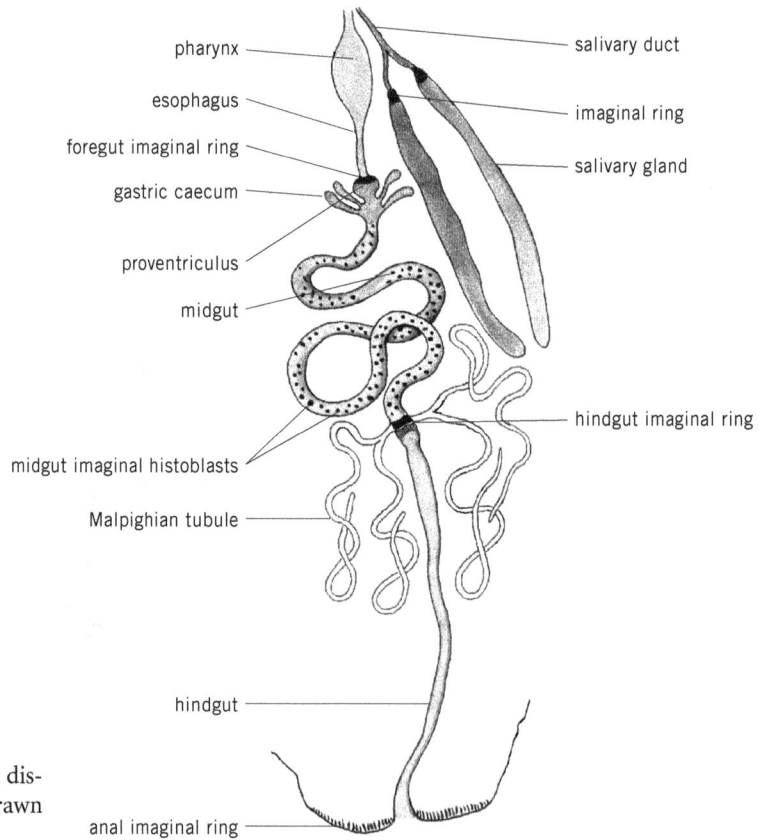

- pharynx
- esophagus
- foregut imaginal ring
- gastric caecum
- proventriculus
- midgut
- salivary duct
- imaginal ring
- salivary gland
- hindgut imaginal ring
- midgut imaginal histoblasts
- Malpighian tubule
- hindgut
- anal imaginal ring

**Figure A2.11.** View of a dissected larval gut. (Redrawn from Kowalevsky 1887.)

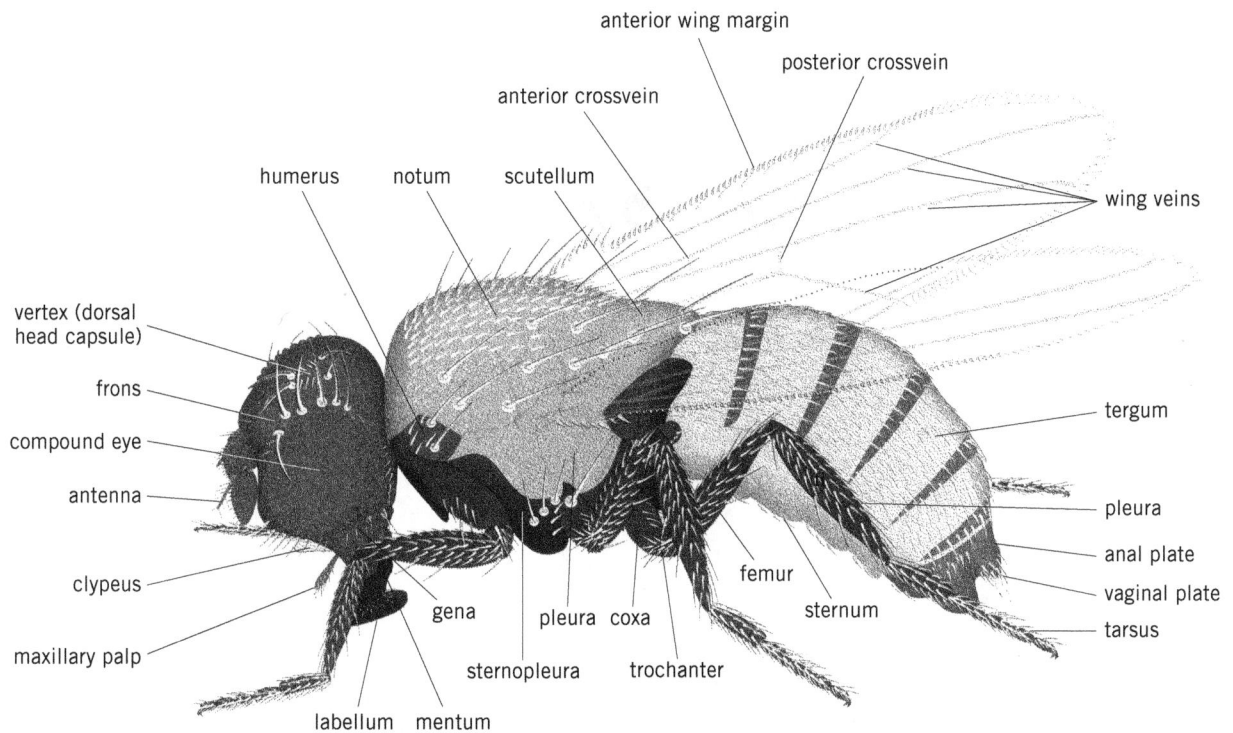

**Figure A2.12.** The epidermis of an adult female. (Reprinted, with permission, from Hartenstein 1993.)

A

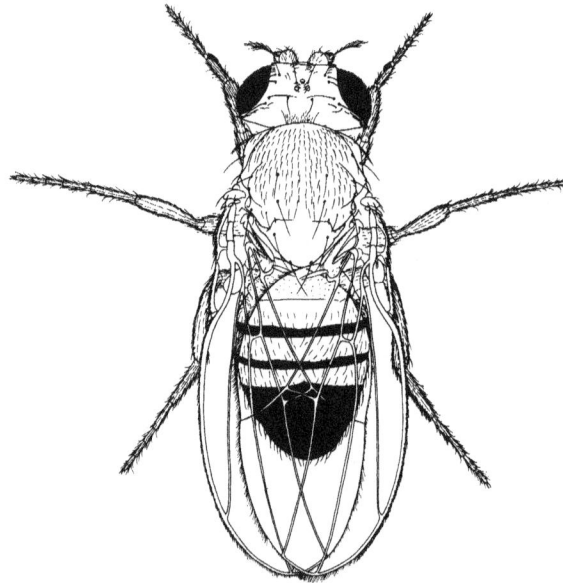

B

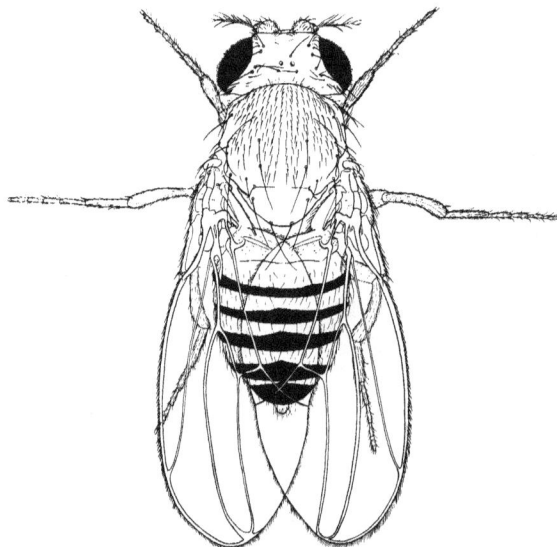

**Figure A2.13.** Dorsal view of an adult male (*A*) and female (*B*). (*A* Redrawn, with permission, from Bridges and Brehm 1944.) (*B* Redrawn, with permission, from Lindsley and Grell 1972.)

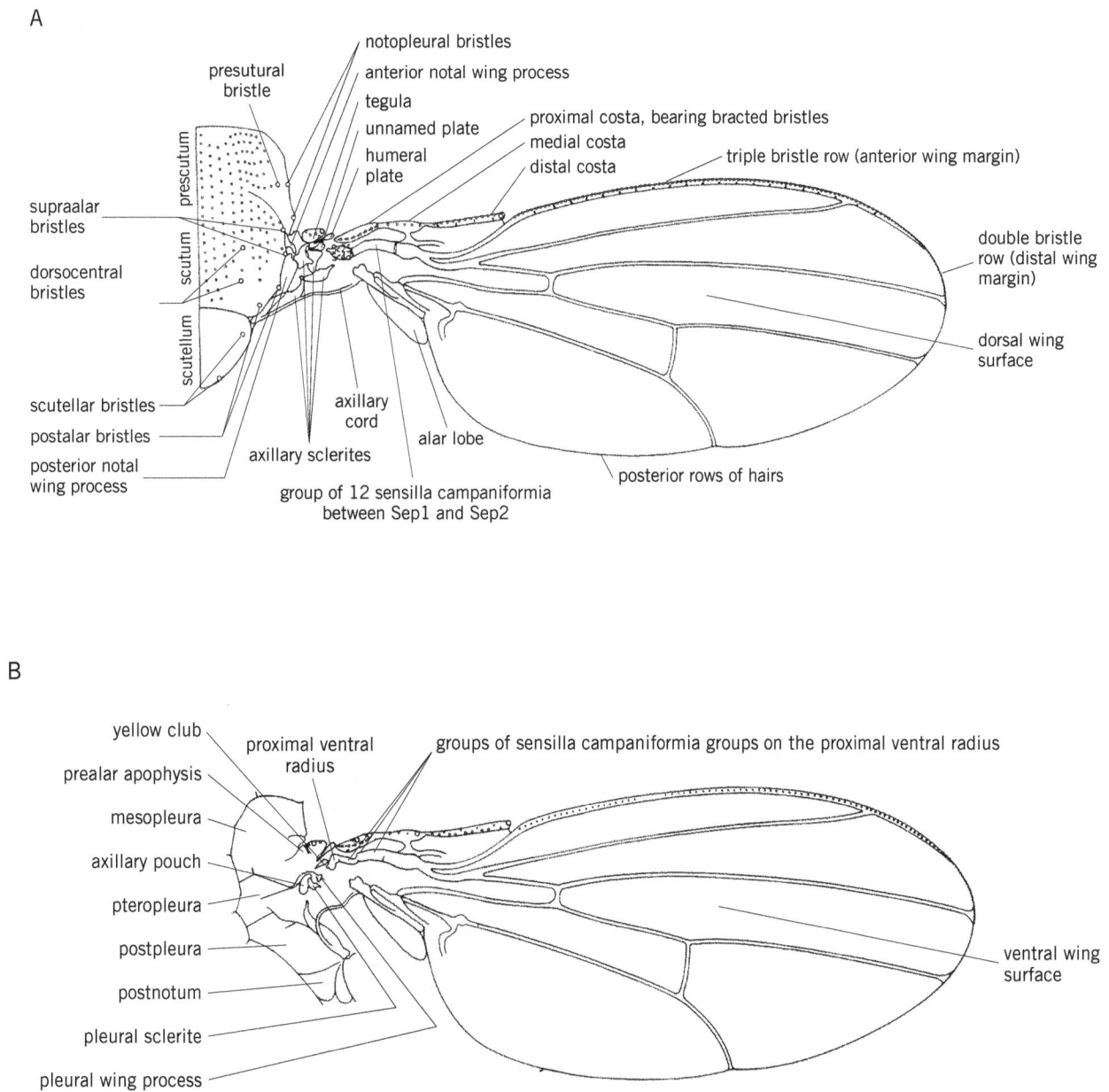

**Figure A2.14.** Dorsal (*A*) and ventral (*B*) views of the adult wing. (Redrawn, with permission from Bryant 1975 [©John Wiley & Sons, Inc.].)

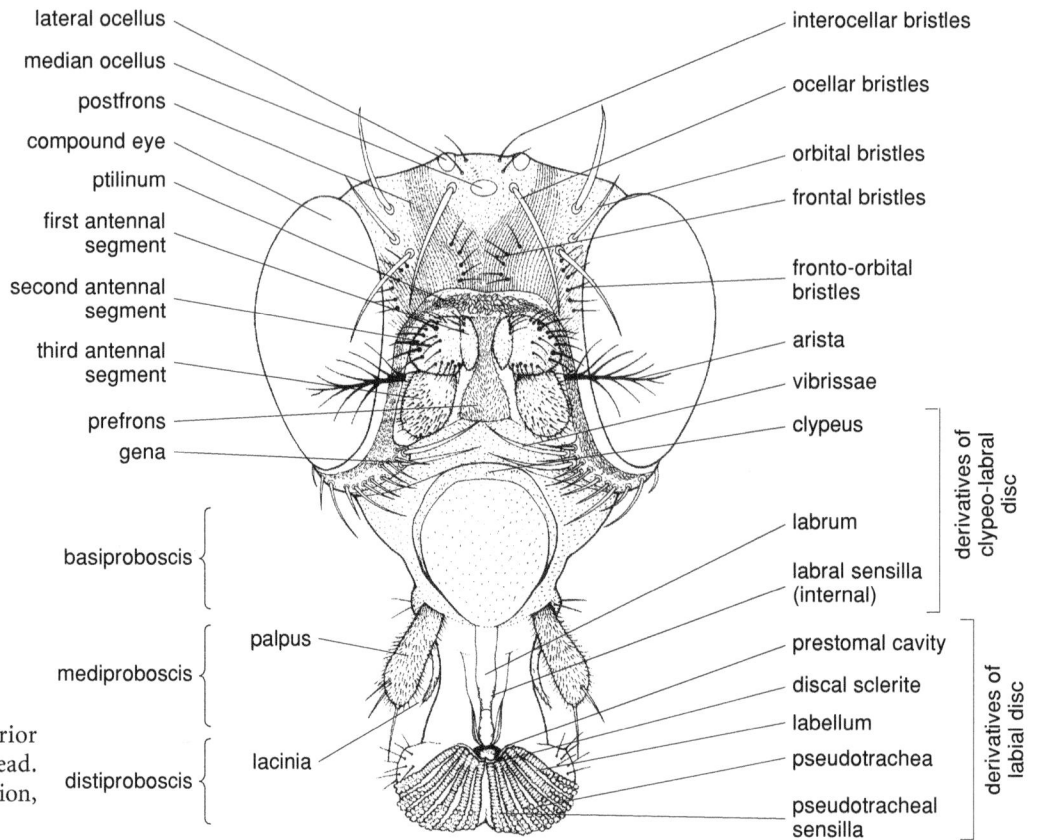

Figure A2.15. Anterior view of the adult head. (Redrawn, with permission, from Bryant 1978.)

Labels for Figure A2.15:
lateral ocellus
median ocellus
postfrons
compound eye
ptilinum
first antennal segment
second antennal segment
third antennal segment
prefrons
gena
basiproboscis
mediproboscis
palpus
distiproboscis
lacinia

interocellar bristles
ocellar bristles
orbital bristles
frontal bristles
fronto-orbital bristles
arista
vibrissae
clypeus
labrum
labral sensilla (internal)
derivatives of clypeo-labral disc
prestomal cavity
discal sclerite
labellum
pseudotrachea
pseudotracheal sensilla
derivatives of labial disc

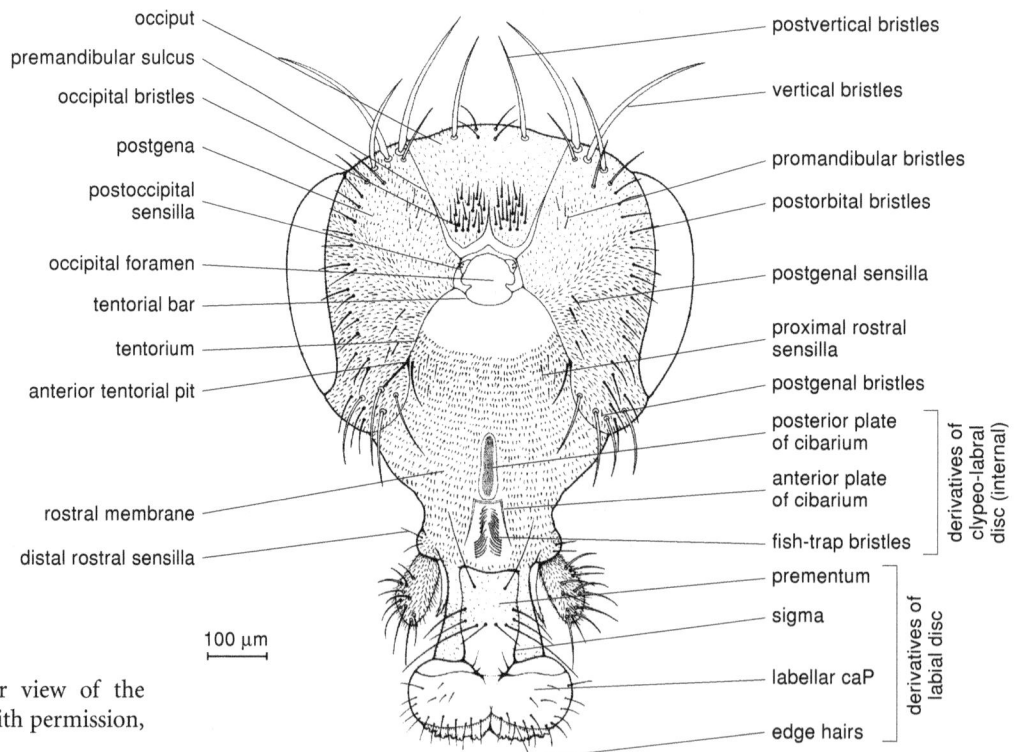

Figure A2.16. Posterior view of the adult head. (Redrawn, with permission, from Bryant 1978.)

Labels for Figure A2.16:
occiput
premandibular sulcus
occipital bristles
postgena
postoccipital sensilla
occipital foramen
tentorial bar
tentorium
anterior tentorial pit
rostral membrane
distal rostral sensilla

postvertical bristles
vertical bristles
promandibular bristles
postorbital bristles
postgenal sensilla
proximal rostral sensilla
postgenal bristles
posterior plate of cibarium
anterior plate of cibarium
fish-trap bristles
derivatives of clypeo-labral disc (internal)
prementum
sigma
labellar caP
edge hairs
derivatives of labial disc

100 μm

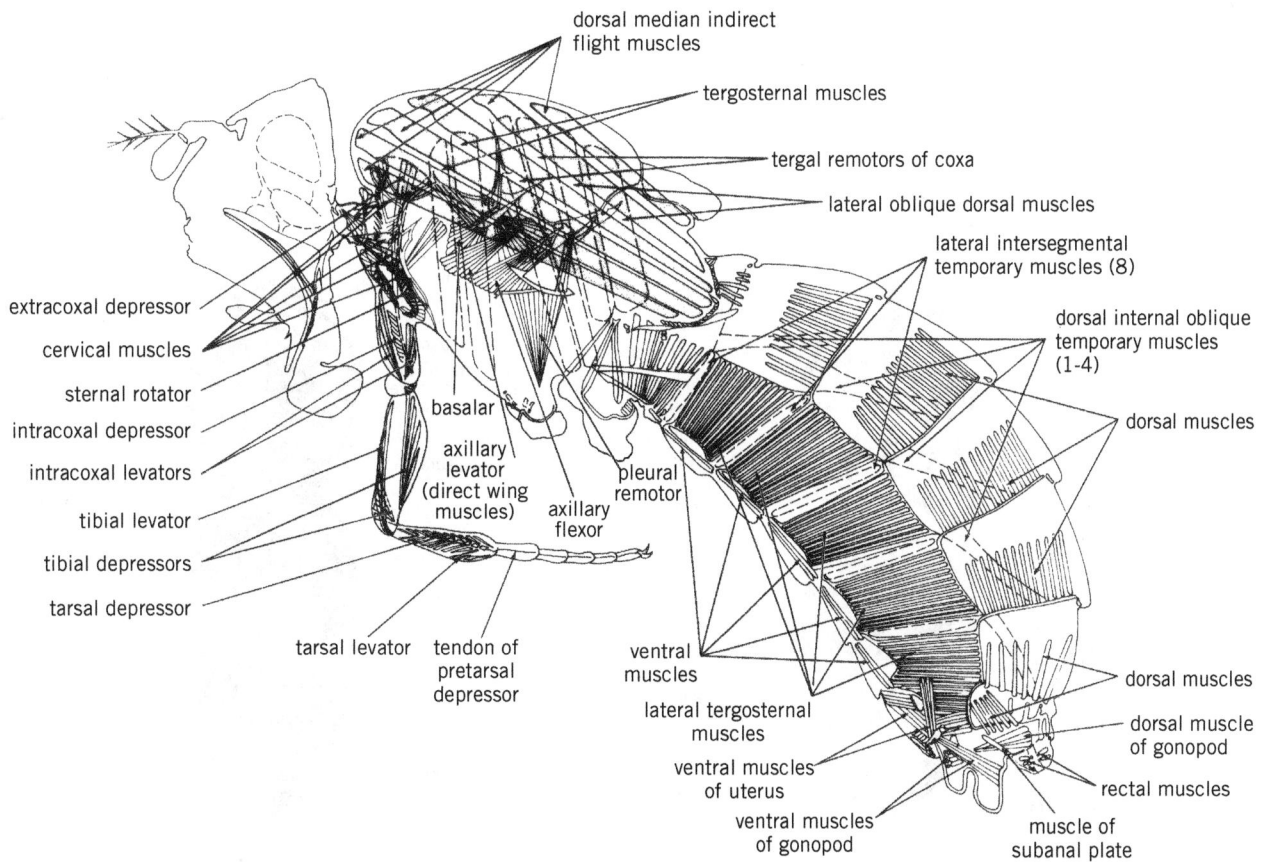

**Figure A2.17.** Lateral view of the major muscle groups in the adult. (Redrawn, with permission, from Miller 1950.)

A

medioscutal sac
lateroscutal sac
pleural sac
notopleural sac
anteroscutal sac
parenteric sac
anterior (meso-)
thoracic spiracle
propleural sac

postocular sac
dorsal sac of head
brain
antennal sac

oesophagus
frontal sac
postgenal sac
maxillary palpus
labial trachea
leg tracheae

postnotal sac
scutellar
sac

posterior (meta-)
thoracic spiracle

metathoracic
sac

metathoracic,
dorsal commissures

abdominal
first
air sac
abdominal
spiracle

dorsal tracheal
trunk

dorsal superficial
tracheae of third and
fourth abdominal
segments

dorsal segmental
trachea of fourth
abdominal segment

visceral trachea

sixth, abdominal dorsal
commissures

seventh abdominal
tergite

seventh abdominal
spiracle

ventral superficial
tracheae of fourth and
fifth abdominal trachea

sterno-
pleura
sac

hypopleural
sac

mediosternal
trachea

lateral
tracheal
trunk

pro-, meso-, and metathoracic
ventral commissures

first
abdominal
spiracle

B

compound eye
frontal sac

anteroscutal sac

parenteric sac

posterior (meta-)
thoracic spiracle
scutellar sac
abdominal
air sac

dorsal segmental
trachea of fourth
abdominal segment

dorsal superficial tracheae
of third and fourth
abdominal segments

visceral trachea

sixth, abdominal
dorsal commissures

antennal sac
frontal sac
dorsal sac of head
brain

postocular sec

cervical trachae
prothoracic dorsal
commissure

anterior (meso-)
thoracic spiracle

notopleural sac
lateroscutal sac
medioscutal sac
pleural sac
postnotal sac
metathoracic sac
first abdominal spiracle

metathoracic dorsal
commissures

dorsal tracheal trunk
anterior intestine
lateral tracheal trunk
spiracular trachea
of fifth abdominal
segment
atrium of spiracle
seventh, abdominal
spiracle
rectum

C

frontal sac
postgenal sac
prothoracic dorsal
commissure
anterior (meso-)
thoracic spiracle
leg tracheae
pro-, meso-, and
metathoracic ventral
commissures
mediosternal trachea
sternopleural sac
pro-, meso-, and
metathoracic ventral
commissures
hypopleural sac
leg tracheae

lateral tracheal
trunk

ventral superficial
tracheae of fourth
and fifth abdominal

cervical trachea
propleural sac

parenteric sac

posterior (meta-)
thoracic spriacle

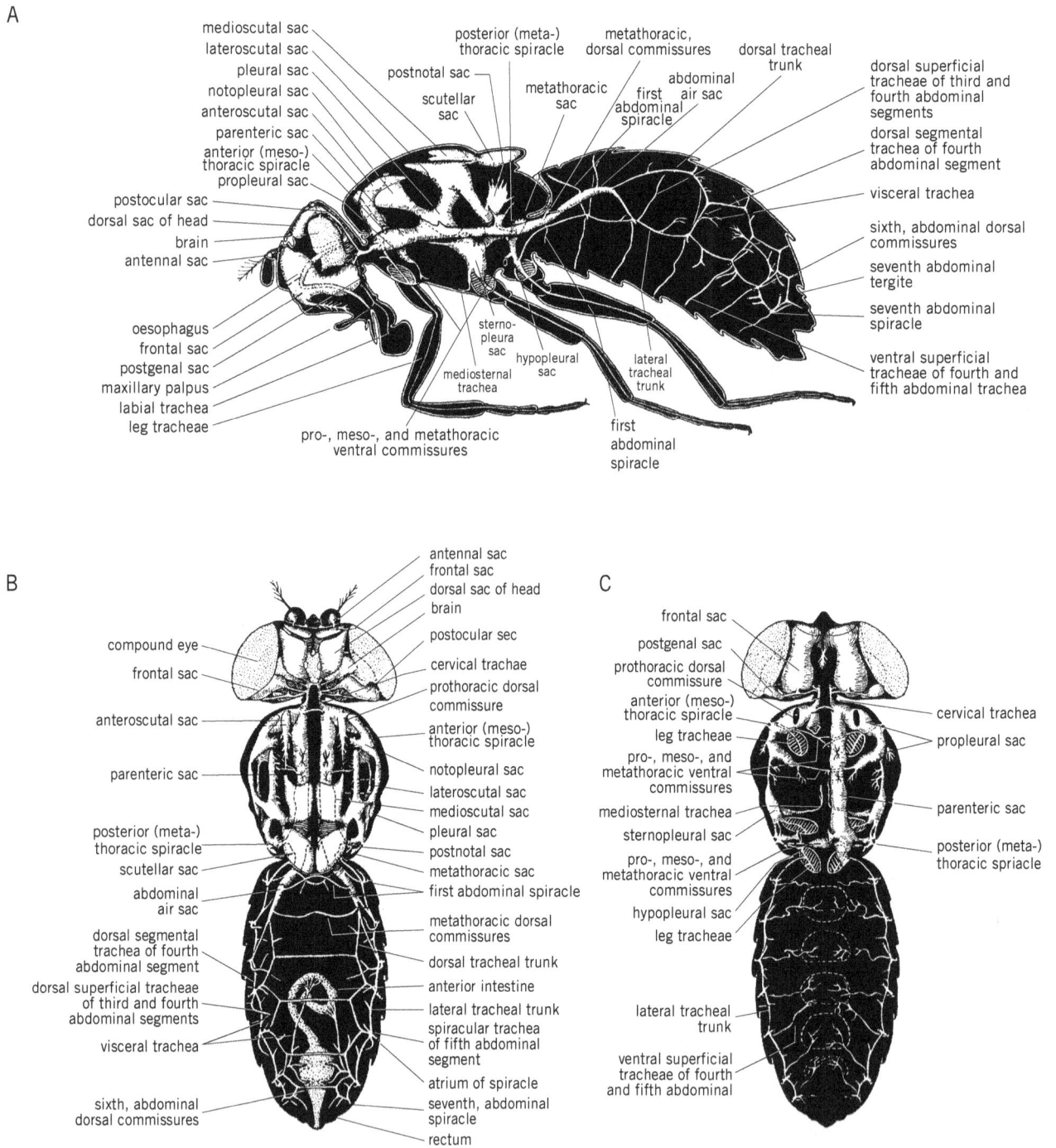

**Figure A2.18.** The adult respiratory system. (*A*) Lateral view. Dorsal view of the dorsal air sacs and tracheae (*B*) and ventral air sacs and tracheae (*C*). (Redrawn, with permission, from Miller 1950.)

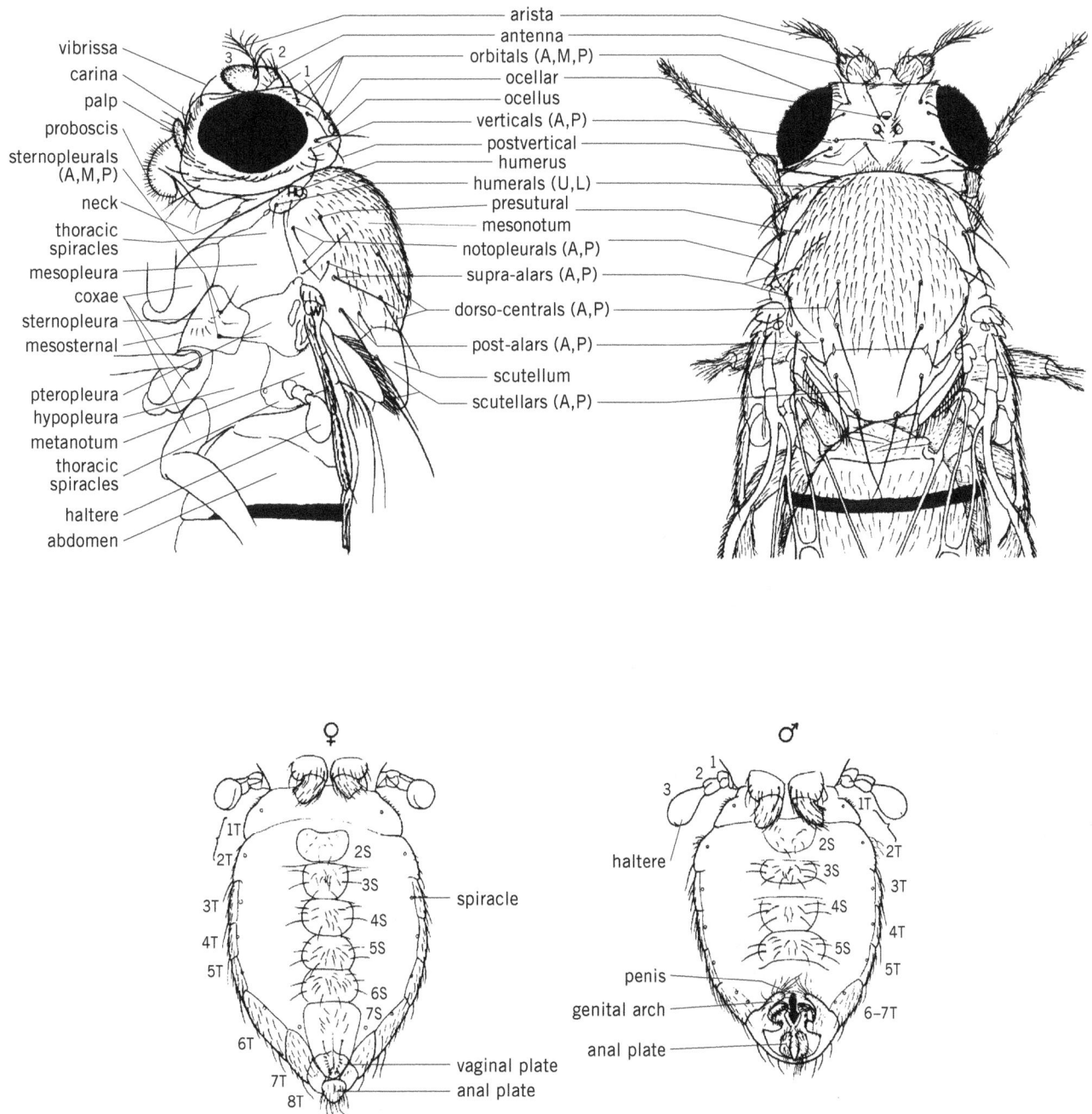

**Figure A2.19.** (*Top*) Lateral (*left*) and dorsal (*right*) views of adult thorax and head. (*Bottom*) Ventral views of female and male abdomen. (A) Anterior; (L) lower; (M) middle; (P) posterior; (T) tergite; (U) upper; (W) wing. (Redrawn with permission, from Bridges 1938.)

## REFERENCES

Bownes M. 1975. A photographic study of development in the living embryo of *Drosophila melanogaster*. *J. Embryol. Exp. Morphol.* **33:** 789–801.

Bridges C.B. 1938. *Drosophila melanogaster* mutants and linkage maps. *D.I.S.* **9:** 1–128.

Bridges C.B. and Brehme K.F. 1944. *The mutants of* Drosophila melanogaster. Carnegie Institution of Washington, Publication No. 552. Washington, D.C.

Bryant P.J. 1975. Pattern formation in the imaginal wing disc of *Drosophila melanogaster. J. Exp. Zool.* **193:** 49–77.

———. 1978. Pattern formation in imaginal discs. In *The genetics and biology of* Drosophila (ed. M. Ashburner and T.R.F. Wright), vol. 2c, pp. 229–335. Academic Press, London.

Foe V.E. and Alberts B.M. 1983. Studies of nuclear and cytoplasmic behaviour during the five mitotic cycles that precede gastrulation in *Drosophila* embryogenesis. *J. Cell Sci.* **61:** 31–70.

Foe V.E., Odell G.M., and Edgar B.A. 1993. Mitosis and morphogenesis in the *Drosophila* embryo. In *The development of* Drosophila melanogaster (ed. M. Bate and A. Martinez Arias), vol. 1, pp. 149–300. Cold Spring Harbor Laboratory Press, Cold Spring Harbor, New York.

Hartenstein V. 1993. *Atlas of* Drosophila *development.* Cold Spring Harbor Laboratory Press, Cold Spring Harbor, New York.

Kowalevsky A. 1887. Beiträge zur Kenntnis der nachembryonalen Entwicklung ser Musciden. I. *Z. Wiss. Zool.* **45:** 542–594.

Lawrence P.A. 1992. *The making of a fly.* Blackwell Science, Oxford, United Kingdom.

Lindsley D.L. and Grell E.H. 1972. *Genetic variations of* Drosophila melanogaster. Carnegie Institution of Washington Publication No. 627. Washington, D.C.

Lindsley D.L. and Tokuyasu K.T. 1980. Spermatogenesis. In *The genetics and biology of* Drosophila (ed. M. Ashburner and T.R.F. Wright), vol. 2d, pp. 225–294. Academic Press, London.

Miller A. 1950. The internal anatomy and histology of the imago of *Drosophila melanogaster.* In *Biology of* Drosophila (ed. M. Demerec), pp. 424–442. John Wiley & Sons, New York.

Parks S. and Spradling A.C. 1987. Spatially regulated expression of chorion genes during *Drosophila* oogenesis. *Genes Dev.* **1:** 497–509.

Poodry C.A. 1980. Imaginal discs: Morphology and development. In *The genetics and biology of* Drosophila (ed. M. Ashburner and T.R.F. Wright), vol. 2d, pp. 407–441. Academic Press, London.

Robinson D.N., Cant K., and Cooley L. 1994. Morphogenesis of *Drosophila* ovarian ring canals. *Development* **120:** 2015–2025.

Whitten J.M. 1957. The postembryonic development of the tracheal system in *Drosophila melanogaster. Q. J. Microsc. Sci.* **98:** 123–150.

Wieschaus E and Nüsslein-Volhard C. 1986. Looking at embryos. In Drosophila: *A practical approach* (ed. D.B. Roberts), pp. 199–227. IRL Press, Oxford.

# Appendix 3

## Solutions and Reagents

ALL CHEMICALS MUST BE REAGENT GRADE OR MOLECULAR BIOLOGY GRADE, and the water used in the preparation of all solutions must be of the highest quality available. Use sterile glass-distilled, deionized $H_2O$ (purified through a Milli-Q filter or similar type of system) whenever possible. Unless otherwise stated, most solutions require sterilization either by filtration through a 0.22-$\mu$m filter or by autoclaving at 15 psi on liquid cycle at 121ºC for 20–30 minutes. Use autoclaved $H_2O$ and sterile measuring devices for the preparation and use of solutions from sterile stock solutions and reagents. Most solutions can be stored at room temperature for at least 6 months, unless otherwise specified. For precautionary notes for reagents used in this appendix, see Appendix 4.

## 1× Phosphate-buffered Saline (PBS)

| Component and final concentration | Amount to add to make 1 liter |
|---|---|
| 137 mM NaCl | 8 g |
| 2.68 mM **KCl** | 0.2 g |
| 10.14 mM **Na$_2$HPO$_4$** | 1.44 g |
| 1.76 mM **KH$_2$PO$_4$** | 0.24 g |

Combine all components in 800 ml of H$_2$O and stir to dissolve. Adjust the pH to 7.2. Adjust the volume to 1 liter with H$_2$O. Dispense in convenient volumes and sterilize by autoclaving. Store at room temperature.

**CAUTION: KCl, KH$_2$PO$_4$, Na$_2$HPO$_4$** (see Appendix 4)

## 1× Tris-buffered Saline (TBS)

| Component and final concentration | Amount to add to make 1 liter |
|---|---|
| 137 mM NaCl | 8 g |
| 2.68 mM **KCl** | 0.2 g |
| 25 mM Tris-base | 3 g |

Combine all components in 800 ml of H$_2$O and stir to dissolve. Adjust the pH to 8.0 with 1 M HCl. Adjust the volume to 1 liter with H$_2$O. Dispense in convenient volumes and sterilize by autoclaving. Store at room temperature.

**CAUTION: HCl, KCl** (see Appendix 4)

## TE

*pH 7.4*
   10 mM Tris-Cl (pH 7.4)
   1 mM EDTA (pH 8.0)

*pH 7.6*
   10 mM Tris-Cl (pH 7.6)
   1 mM EDTA (pH 8.0)

*pH 8.0*
   10 mM Tris-Cl (pH 8.0)
   1 mM EDTA (pH 8.0)

## 20× SSC

| Component and final concentration | Amount to add to make 1 liter |
|---|---|
| 3 M NaCl | 175.3 g |
| 0.3 M Sodium citrate | 88.2 g |

Combine components in 800 ml of $H_2O$ and stir to dissolve. Adjust the pH to 7.0 with a few drops of 10 N **NaOH**. Adjust the volume to 1 liter with $H_2O$. Dispense into aliquots. Sterilize by autoclaving. Store at room temperature.

**CAUTION: NaOH** (see Appendix 4)

## Glycerol-based Mounting Medium

For the stock, prepare 10 mg/ml 1,4-**phenylenediamine** (Aldrich P2,396-2) in 10× PBS. Combine 10 ml of the stock with 90 ml of glycerol. Store in 1-ml aliquots at –20ºC.

**CAUTION: phenylenediamine** (see Appendix 4)

## 1× Embryo Wash Solution

| Component and final concentration | Amount to add to make 1 liter |
|---|---|
| 0.7% NaCl | 7 g |
| 0.05% Triton X-100 | 0.5 ml |

Combine components in sufficient $H_2O$ and stir to dissolve. Adjust the volume to 1 liter with $H_2O$.

## Robb's Minimal Saline (Robb 1969; Ashburner 1989)

### SOLUTION A

| Component | Amount to add to make 1 liter (g/liter) | Final concentration (in 1:1 mixture of Solutions A and B) |
|---|---|---|
| NaCl | 3.04 g | 2.6 mM |
| **KCl** | 2.98 g | 2.0 mM |
| glucose | 1.80 g | 0.5 mM |
| sucrose | 34.23 g | 5.0 mM |
| **MgSO₄·7H₂O** | 0.28 g | 0.06 mM |
| **MgCl₂·6H₂O** | 0.245 g | 0.06 mM |
| **CaCl₂·2H₂O** | 0.147 g | 0.05 mM |
| $H_2O$ | to 1 liter | |

**SOLUTION B**

| Component | Amount to add to make 1 liter (g/liter) | Final concentration (in 1:1 mixture of Solutions A and B) |
|---|---|---|
| $Na_2HPO_4 \cdot 2H_2O$ | 0.356 g | 0.1 mM |
| $KH_2PO_4$ | 0.05 g | 0.018 mM |
| $H_2O$ | to 1 liter | |

Adjust to pH 6.75 with 1 N HCl. Autoclave Solutions A and B separately; when sterile and cool, mix in 1:1 ratio.

**CAUTION: $CaCl_2$, KCl, $MgCl_2$, $MgSO_4$, $Na_2HPO_4$, $NaH_2PO_4$** (see Appendix 4)

## Grape–Apple Juice Agar Plates

700 ml of $H_2O$
25–30 g of agar
300 ml of juice concentrate (grape, apple, or other)
0.5 g of methyl paraben (*p*-hydroxymethylbenzoate)
20 ml of 95% ethanol

1. Autoclave the agar with $H_2O$ for 40 minutes. *This is Solution A.*

2. Add ethanol to a vial with methyl paraben in it. Add the solution of methyl paraben to the apple/grape concentrate. *This is Solution B.*

3. Add Solution B to Solution A and begin pouring; do this quickly so that the mixture will not harden in the pouring container.

This 1-liter mixture is enough to pour 20 or more trays of collecting plates. To prepare food plates, 1800 ml of the above fly food and 20 trays of collecting plates are required. Boil the food in 10-minute increments until all the food has melted. This will be approximately 45 minutes for the above volume of food. Pour as described above.

## Cornmeal Molasses-based food

**MIXTURE A**
10,000 ml of $H_2O$
1000 ml of molasses
118.4 g of agar

**MIXTURE B**
1000 ml of cornmeal
412 g of yeast
4000 ml of $H_2O$

**MIXTURE C**
tegosept (10% w/v methyl paraben in 95% ethanol)
to 22.5 g of methyl paraben, add 95% ethanol to a final
   volume of 225 ml
80 ml of propionic acid

1. Prepare Mixture A in a large steam kettle and bring to a boil. Let boil for 10 minutes.

2. While Mixture A is boiling, prepare Mixture B. To prepare Mixture B, mix the yeast and cornmeal together first before adding the $H_2O$. The mixture must be as smooth as possible. Clumps of yeast are undesirable.

3. Add Mixture B when Mixture A is ready and add 2 liters of $H_2O$ to adjust for evaporation. Bring to another boil, and let boil for 10 minutes.

4. Turn off heat and allow food to stop bubbling. Add tegosept (Mixture C) and propionic acid to inhibit mold growth. Be careful when adding the two mixtures because the vapors are strong enough to singe nose hairs! Quickly begin dispensing. Do not take too long because the food will harden.

## *Drosophila* Ringer's Solution (Tübingen and Düsseldorf)

| Component and final concentration | Amount to add to make 1 liter |
|---|---|
| 182 mM **KCl** | 13.6 g |
| 46 mM NaCl | 2.7 g |
| 3 mM **CaCl$_2$** | 0.33 g of $CaCl_2 \cdot 2H_2O$ |
| 10 mM Tris-Cl | 1.21 g of Tris-base |
| $H_2O$ | to make 1 liter |

Combine all components in 800 ml of $H_2O$ and stir to dissolve. Adjust to pH 7.2 with 1 N HCl. Adjust volume to 1 liter with $H_2O$. Pass through a 0.47-μm filter and autoclave.

**CAUTION: CaCl$_2$, HCl, KCl** (see Appendix 4)

## 1x Tris-Borate EDTA (TBE)

| Component and final concentration | Amount to add to make 1 liter |
|---|---|
| 89 mM Tris-base | 10.8 g |
| 89 mM Boric acid | 5.5 g |
| 2 mM EDTA (pH 8.0) | 4 ml of 0.5 M EDTA (pH 8.0) |

Combine all components in sufficient $H_2O$ and stir to dissolve. Adjust volume to 1 liter with $H_2O$. Store at room temperature.

## REFERENCES

Ashburner M. 1989. Drosophila. *A laboratory manual.* Cold Spring Harbor Laboratory Press, Cold Spring Harbor, New York.

Robb J.A. 1969. Maintenance of imaginal discs of *Drosophila melanogaster* in chemically defined media. *J. Cell Biol.* **41:** 876–885.

Sambrook J., Fritsch E.F., and Maniatis T. 1989. *Molecular cloning: A laboratory manual,* 2nd edition, Cold Spring Harbor Laboratory Press, Cold Spring Harbor, New York.

# Appendix 4

## Cautions

The following general cautions should always be observed.

- Become **completely familiar** with the properties of all substances used **before** beginning the procedure.
- **The absence of a warning** does not necessarily mean that the material is safe, since information may not always be complete or available.
- If **exposed** to toxic substances, contact the local safety office immediately for instructions.
- **Use proper disposal procedures** for all chemical, biological, and radioactive waste.
- For specific guidelines on **appropriate gloves**, consult the local safety office.
- Handle **concentrated acids and bases** with great care. Wear goggles and appropriate gloves, as well as a face shield if handling large quantities.
    Do not mix strong acids with organic solvents as they may react. Sulfuric acid and nitric acid especially may react highly exothermically and cause fires and explosions.
    Do not mix strong bases with halogenated solvent as they may form reactive carbenes which can lead to explosions.
- Never **pipette** solutions using mouth suction. This method is not sterile and can be dangerous. Always use a pipette aid or bulb.
- Keep **halogenated and nonhalogenated** solvents separately (e.g., mixing chloroform and acetone can cause unexpected reactions in the presence of bases). Halogenated solvents are organic solvents such as chloroform, dichloromethane, trichlorotrifluoroethane, and dichloroethane. Some nonhalogenated solvents are pentane, heptane, ethanol, methanol, benzene, toluene, *N,N*-dimethylformamide (DMF), dimethyl sulfoxide (DMSO), and acetonitrile.
- **Laser radiation**, visible or invisible, can cause severe damage to the eyes and skin. Take proper precautions to prevent exposure to direct and reflected beams. Always follow manufacturers safety guidelines and consult the local safety office. For more detailed information, see caution below.
- **Flash lamps**, due to their light intensity, can be harmful to the eyes and may explode on occasion. Wear appropriate eye protection and follow the manufacturer's guidelines.
- **Photographic fixatives and developers** contain harmful chemicals. Handle them with care and follow manufacturer's directions.
- **Power supplies and electrophoresis equipment** pose serious fire hazard and electrical shock hazards if not used properly.
- **Microwave ovens and autoclaves** in the lab require certain precautions. Accidents have occurred involving their use (e.g., to melt agar or bactoagar stored in bottles

661

or to sterilize). If the screw top is not completely removed and there is not enough space for the steam to vent, the bottles can explode when the containers are removed from the microwave or autoclave. Always completely remove bottle caps before microwaving or autoclaving.

- Use extreme caution when handling **cutting devices** such as microtome blades scalpels, razor blades, or needles. Microtome blades are extremely sharp! If unfamiliar with their use, have someone demonstrate proper procedures. For proper disposal, use the "sharps" disposal container in the lab. Discard used needles *unshielded*, with the syringe still attached. This prevents injuries (and possible infections; see Biological Safety) while manipulating used needles since many accidents occur while trying to replace the needle shield. Injuries may also be caused by broken pasteur pipettes, coverslips, or slides.

## GENERAL PROPERTIES OF COMMON CHEMICALS

The hazardous materials list can be summarized in the following categories:

- Inorganic acids, such as hydrochloric, sulfuric, nitric, or phosphoric, are colorless liquids with stinging vapors. Avoid spills on skin or clothing. Dilute spills with large amounts of water. The concentrated forms of these acids can destroy paper, textiles, and skin as well as cause serious injury to the eyes.
- Salts of heavy metals are usually colored powered solids that dissolve in water. Many of them are potent enzyme inhibitors and therefore toxic to humans and to the environment (e.g., fish and algae).
- Most organic solvents are flammable volatile liquids. Breathing their vapors can cause nausea or dizziness. Also avoid skin contact.
- Other organic compounds, including organosulphur compounds such as mercaptoethanol or organic amines, have very unpleasant odors. Others are highly reactive and must be handled with appropriate care.
- If improperly handled, dyes and their solutions can stain not only the sample, but also skin and clothing. Some of them are also mutagenic (e.g., ethidium bromide), carcinogenic, and toxic.
- Nearly all names ending with "ase" (e.g., catalase, β-glucuronidase, or zymolase) refer to enzymes. There are also other enzymes with nonsystematic names like pepsin. Many of them are provided by manufacturers in preparations containing buffering substances, etc. Be aware of the individual properties of materials contained in these substances.
- Toxic compounds often used to manipulate cells can be dangerous and should be handled appropriately.

## HAZARDOUS MATERIALS

**Acetic acid (concentrated)** must be handled with great care. It may be harmful by inhalation, ingestion, or skin absorption. Wear appropriate gloves and goggles and use in a chemical fume hood.

**Acetic anhydride** is extremely destructive to the skin, eyes, mucous membranes, and upper respiratory tract. It may be harmful by inhalation, ingestion, or skin absorption. Wear appropriate gloves and safety glasses and use in a chemical fume hood.

**Acetone** causes eye and skin irritation and is irritating to mucous membranes and upper respiratory tract. Do not breathe the vapors. It is also extremely flammable. Wear appropriate gloves and safety glasses.

**Acridine orange** is a mutagen and may be harmful by inhalation, ingestion, or skin absorption. Wear appropriate gloves and safety glasses. Do not breathe the dust.

**Acrolein** is extremely toxic and volatile. It may be harmful by inhalation, ingestion, or skin absorption. Wear appropriate gloves and use in a chemical fume hood.

**Acrylamide** (unpolymerized) is a potent neurotoxin and is absorbed through the skin (its effects are cumulative). Avoid breathing the dust. Wear appropriate gloves and a face mask when weighing powdered acrylamide and methylene-bisacrylamide and use in a chemical fume hood. Polyacrylamide is considered to be nontoxic, but it should be handled with care because it might contain small quantities of unpolymerized acrylamide.

**Actinomycin D** is a teratogen and a carcinogen. It is highly toxic and may be fatal if inhaled, ingested, or absorbed through the skin. It may also cause irritation. Avoid breathing the dusts. Wear appropriate gloves, safety glasses, and always use in a chemical fume hood. Solutions of actinomycin D are light-sensitive.

**α-Amanitin** is highly toxic and may be fatal by inhalation, ingestion, or skin absorption. Symptoms may be delayed for as long as 6–24 hours. Wear appropriate gloves and safety glasses and always work in a chemical fume hood.

**Ammonium acetate, $H_3CCOONH_4$,** may be harmful by inhalation, ingestion, or skin absorption. Wear appropriate gloves and safety glasses and use in a chemical fume hood.

**Ammonium persulfate, $(NH_4)_2S_2O_8$,** is extremely destructive to tissue of the mucous membranes and upper respiratory tract, eyes, and skin. Inhalation may be fatal. Wear appropriate gloves, safety glasses, and protective clothing and always use in a chemical fume hood. Wash thoroughly after handling.

**Ammonium sulfate, $(NH_4)_2SO_4$,** may be harmful by inhalation, ingestion, or skin absorption. Wear appropriate gloves and safety glasses.

**Ammonium sulfide** causes severe irritation to the respiratory tract and may be harmful by inhalation, ingestion, or skin absorption. Wear appropriate gloves and safety glasses.

**Ampicillin** may be harmful by inhalation, ingestion, or skin absorption. Wear appropriate gloves and safety glasses and use in a chemical fume hood.

**Aprotinin** may be harmful by inhalation, ingestion, or skin absorption. It may also cause allergic reactions. Exposure may cause gastrointestinal effects, muscle pain , blood pressure changes, or bronchospasm. Wear appropriate gloves and safety glasses and use only in a chemical fume hood. Do not breathe the dust.

**BCIG,** *see* **Bromochloroindolyl-β-D-galactopyranoside**

**BCIP,** *see* **5-Bromo-4-chloro-3-indolyl-phosphate**

**Benzoic acid** is an irritant and may be harmful by inhalation, ingestion, or skin absorption. Wear appropriate gloves and safety glasses. Do not breathe the dust.

**Bisacrylamide** is a potent neurotoxin and is absorbed through the skin (its effects are cumulative). Avoid breathing the dust. Wear appropriate gloves and a face mask when weighing powdered acrylamide and methylene-bisacrylamide.

**Bisbenzamide,** *see* **Hoechst 33258**

**Bleach (Sodium hypochlorite), NaOCl,** is poisonous, can be explosive, and may react with organic solvents. It may be fatal by inhalation and is also harmful by ingestion and destructive to the skin. Wear appropriate gloves and safety glasses. If possible, use in a chemical fume hood to minimize exposure and odor.

**Bradford dye** contains phosphoric acid and methanol. It is corrosive and toxic. Wear appropriate gloves and safety glasses.

**BrdU,** *see* **5-Bromo-2′-deoxyuridine**

**5-Bromo-4-chloro-3-indolyl-phosphate (BCIP)** is toxic and may be harmful by inhalation, ingestion, or skin absorption. Wear appropriate gloves and safety glasses. Do not breathe the dust.

**5-Bromo-2′-deoxyuridine (BrdU)** is a mutagen. It may be harmful by inhalation, ingestion, or skin absorption. It may cause irritation. Avoid breathing the dust. Wear appropriate gloves and safety glasses and always use in a chemical fume hood.

**Bromophenol blue** may be harmful by inhalation, ingestion, or skin absorption. Wear appropriate gloves and safety glasses and use in a chemical fume hood.

**CaCl₂,** *see* **Calcium chloride**

**Cacodylate** contains arsenic, is highly toxic, and may be fatal if inhaled, ingested, or absorbed through the skin. Wear appropriate gloves and safety glasses and use in a chemical fume hood. *See also* **Potassium cacodylate; Sodium cacodylate.**

**Cacodylic acid** is toxic and a possible carcinogen. It may be mutagenic and is harmful by inhalation, ingestion, or skin absorption. Wear appropriate gloves and safety glasses and use only in a chemical fume hood. Do not breathe the dust.

**Calcium chloride, $CaCl_2$,** may be harmful by inhalation, ingestion, or skin absorption. Wear appropriate gloves and safety glasses and use in a chemical fume hood.

**Carbon dioxide, $CO_2$,** in all forms may be fatal by inhalation, ingestion, or skin absorption. In high concentrations, it can paralyze the respiratory center and cause suffocation. Use only in well-ventilated areas. In the form of dry ice, contact with carbon dioxide can also cause frostbite. Wear appropriate gloves and safety goggles.

**$CHCl_3$,** *see* **Chloroform**

**$CH_3CH_2OH$,** *see* **Ethanol**

**Chloral hydrate** is extremely destructive to tissues of the mucus membranes and upper respiratory tract. It may be harmful by inhalation, ingestion, or skin absorption. Wear appropriate gloves and safety glasses and use in a chemical fume hood.

**Chloramphenicol** may be harmful by inhalation, ingestion, or skin absorption and is a carcinogen. Wear appropriate gloves and safety glasses and use in a chemical fume hood.

**Chloroform, $CHCl_3$,** is irritating to the skin, eyes, mucous membranes, and respiratory tract. It is a carcinogen and may damage the liver and kidneys. It is also volatile. Avoid breathing the vapors. Wear appropriate gloves and safety glasses and always use in a chemical fume hood.

**Chromic acid** is corrosive and extremely dangerous by inhalation, ingestion, or skin absorption. Wear appropriate gloves and safety goggles. Keep away from heat, sparks, and open flame.

**$CO_2$,** *see* **Carbon dioxide**

**Cobalt chloride, $CoCl_2$,** may be harmful by inhalation, ingestion, or skin absorption. Wear appropriate gloves and safety glasses.

**Cobalt nitrate, $Co(NO_3)_2 \cdot 6H_2O$,** is a strong oxidizer and may be harmful by inhalation, ingestion, or skin absorption. Wear appropriate gloves and safety goggles.

**Cocaine** is a controlled substance available to researchers who have a valid DEA license and an approved protocol. It is highly toxic and affects the central nervous system. It is harmful be inhalation, ingestion, or skin absorption. Wear appropriate gloves and safety glasses. Do not breathe the dust.

**$CoCl_2$,** *see* **Cobalt chloride**

**Colchicine** is highly toxic, may be fatal, and may cause cancer and be mutagenic. It may be harmful by inhalation, ingestion, or skin absorption. Wear appropriate gloves and safety glasses and use only in a chemical fume hood. Do not breathe the dust.

**$Co(NO_3)_2 \cdot 6H_2O$,** *see* **Cobalt nitrate**

**DAB,** *see* **3,3′-Diaminobenzidine tetrahydrochloride**

**DABCO,** *see* **1,4-Diazabicyclo-[2,2,2]-octane**

**DAPI,** *see* **4′,6-Diamidine-2′phenylindole dihydrochloride**

**Deoxycholate (DOC)** may be harmful by inhalation, ingestion, or skin absorption. Do not breath the dust. Wear appropriate gloves and safety glasses.

**4′,6-Diamidine-2-phenylindole dihydrochloride (DAPI)** is a possible carcinogen. It may be harmful by inhalation, ingestion, or skin absorption. It may also cause irritation. Avoid breathing the dust and vapors. Wear appropriate gloves and safety glasses and use in a chemical fume hood.

**3,3′-Diaminobenzidine tetrahydrochloride (DAB)** is a carcinogen. Handle with extreme care. Avoid breathing vapors. Wear appropriate gloves and safety glasses and use in a chemical fume hood.

**1,4-Diazabicyclo-[2,2,2]-octane (DABCO)** may be harmful by inhalation, ingestion, or skin absorption. Wear appropriate gloves and safety glasses and use in a chemical fume hood.

**Diethyl ether, $Et_2O$ or $(C_2H_5)_2O$,** is extremely volatile and flammable. It is irritating to the eyes, mucous membranes, and skin. It is also a CNS depressant with anesthetic effects. It may be harmful by inhalation, ingestion, or skin absorption. Avoid breathing the vapors. Wear appropriate gloves and safety glasses and always use in a chemical fume hood. Explosive peroxides can form during storage or on exposure to air or direct sunlight. Keep away from heat, sparks, and open flame.

**Digoxigenin** may be fatal if inhaled, ingested, or absorbed through the skin. Wear appropriate gloves and safety glasses and use in a chemical fume hood.

**$N,N$-Dimethylformamide (DMF), $HCON(CH_3)_2$,** is irritating to the eyes, skin, and mucous membranes. It can exert its toxic effects through inhalation, ingestion, or skin absorption. Chronic inhalation can cause liver and kidney damage. Wear appropriate gloves and safety glasses and use in a chemical fume hood.

**Dimethyl sulfoxide (DMSO)** may be harmful by inhalation or skin absorption. Wear appropriate gloves and safety glasses. Use in a chemical fume hood. DMSO is also combustible. Store in a tightly closed container. Keep away from heat, sparks, and open flame. *DMSO easily penetrates the skin and is added as a transport mediator in some creams. This property makes it a carrier for other substances thus potentiating their risks.*

**1,4-Dioxane** is highly flammable in both liquid and vapor form. It may be carcinogenic and is highly toxic by inhalation, ingestion, or skin absorption. Do not breathe the vapor. Wear appropriate gloves and safety glasses. Keep away from heat, sparks, and open flame.

**Dithiothreitol (DTT)** is a strong reducing agent that emits a foul odor. It may be harmful by inhalation, ingestion, or skin absorption. When working with the solid form or highly concentrated stocks, wear appropriate gloves and safety glasses and use in a chemical fume hood.

**DMF,** *see* **N,N-dimethylformamide**

**DMSO,** *see* **Dimethyl sulfoxide**

**DOC,** *see* **Deoxycholate**

**Dopamine** is an irritant and may be harmful by inhalation, ingestion, or skin absorption. Wear appropriate gloves and safety glasses. Do not breathe the dust.

**DPX** is composed of Distyrene, a plasticizer, and xylene and is commercially available. Follow the manufacturer's guidelines for handling DPX.

**Dry ice,** *see* **Carbon dioxide**

**DTT,** *see* **Dithiothreitol**

**EMS,** *see* **Ethyl methane sulfonate**

**ENU,** *see* **Ethylnitrosourea**

**Epoxy and acrylic resins,** *see* **Resins**

**Ethanol (EtOH), $CH_3CH_2OH$,** may be harmful by inhalation, ingestion, or skin absorption. Wear appropriate gloves and safety glasses.

**Ether,** *see* **Diethyl ether**

**Ethidium bromide** is a powerful mutagen and is moderately toxic. Consult the local institutional safety officer for specific handling and disposal procedures. Avoid breathing the dusts. Wear appropriate gloves when working with solutions that contain this dye.

**Ethyl methane sulfonate (EMS)** is a volatile organic solvent that is a mutagen and carcinogen. It is harmful if inhaled, ingested or absorbed through the skin. Discard supernatants and washes containing EMS in a beaker containing 50% sodium thiosulfate. Decontaminate all material that has come in contact with EMS by treatment in a large volume of 10% (w/v) sodium thiosulfate. Use extreme caution when handling. When using undiluted EMS, wear protective appropriate gloves and use in a chemical fume hood. Store EMS in the cold. DO NOT mouth-pipette EMS. Pipettes used with undiluted EMS should not be too warm; chill them in the refrigerator before use to minimize the volatility of EMS. All glassware coming in contact with EMS should be immersed in a large beaker of 1 N NaOH or laboratory bleach before recycling or disposal.

**Ethylnitrosourea (ENU),** *see* **N-Nitroso-N-ethylurea**

**EtOH,** *see* **Ethanol**

**FITC,** *see* **Fluorescein isothiocyanate**

**Fluoromount-G** contains sodium azide which is very toxic if ingested or inhaled. It is highly poisonous and blocks the cytochrome electron transport system. Solutions containing sodium azide should be clearly marked. Wear appropriate gloves and safety goggles and handle sodium azide with great care.

**Fluorescein isothiocyanate** may be harmful by inhalation, ingestion, or skin absorption. Wear appropriate gloves and safety glasses.

**Formaldehyde, HCOH,** is highly toxic and volatile. It is also a carcinogen. It is readily absorbed through the skin and is irritating or destructive to the skin, eyes, mucous membranes, and upper respiratory tract. Avoid breathing the vapors. Wear appropriate gloves and safety glasses and always use in a chemical fume hood. Keep away from heat, sparks, and open flame.

**Formamide** is teratogenic. The vapor is irritating to the eyes, skin, mucous membranes, and upper respiratory tract. It may be harmful by inhalation, ingestion, or skin absorption. Wear appropriate gloves and safety glasses and always use in a chemical fume hood when working with concentrated solutions of formamide. Keep working solutions covered as much as possible.

**Gamma rays,** *see* **Radioactive substances**

**Giemsa** may be fatal or cause blindness by ingestion and is toxic by inhalation and skin absorption. There is a possible risk of irreversible effects. Wear appropriate gloves and safety goggles and use only in a chemical fume hood. Do not breathe the dust.

**Glacial acetic acid,** *see* **Acetic acid**

**Glutaraldehyde** is toxic. It is readily absorbed through the skin and is irritating or destructive to the skin, eyes, mucous membranes, and upper respiratory tract. Wear appropriate gloves and safety glasses and always use in a chemical fume hood.

**Glycine** may be harmful by inhalation, ingestion, or skin absorption. Wear gloves and safety glasses. Avoid breathing the dust.

**Gum arabic** causes severe eye irritation and may cause allergic or respiratory reaction. It may form combustible dust concentrations in air. It may be harmful by inhalation, ingestion, or skin absorption. Wear appropriate gloves and safety goggles. Do not breathe the dust.

**HCl,** *see* **Hydrochloric acid**

**HCOH,** *see* **Formaldehyde**

**Heptane** may be harmful by inhalation, ingestion, or skin absorption. Wear appropriate gloves and safety glasses. It is extremely flammable. Keep away from heat, sparks, and open flame.

**$HNO_3$,** *see* **Nitric acid**

**$H_2O_2$,** *see* **Hydrogen peroxide**

**Hoechst 33258, bisbenzimide** may be harmful by inhalation, ingestion, or skin absorption. Wear appropriate gloves and safety glasses and use in a chemical fume hood. Do not breathe the dust.

**$H_2SO_4$,** *see* **Sulfuric acid**

**Hydrochloric acid (HCl)** is volatile and may be fatal if inhaled, ingested, or absorbed through the skin. It is extremely destructive to mucous membranes, upper respiratory tract, eyes, and skin. Wear appropriate gloves and safety glasses and use with great care in a chemical fume hood. Wear goggles when handling large quantities.

**Hydrogen peroxide, $H_2O_2$,** is corrosive, toxic, and extremely damaging to the skin. It may be harmful by inhalation, ingestion, and skin absorption. Wear appropriate gloves and safety glasses and use only in a chemical fume hood.

**Hydroxyurea** is toxic and may be mutagenic and teratogenic. It may be harmful by inhalation, ingestion, or skin absorption. Wear appropriate gloves and safety glasses and use only in a chemical fume hood. Do not breathe the dust.

**IPTG,** *see* **Isopropyl-β-D-thiogalactopyranoside**

**Isoamyl alcohol** may be harmful by inhalation, ingestion, or skin absorption and presents a risk of serious damage to the eyes. Wear appropriate gloves and safety goggles. Keep away from heat, sparks, and open flame.

**Isopropanol** is irritating and may be harmful by inhalation, ingestion, or skin absorption. Wear appropriate gloves and safety glasses. Do not breathe the vapor. Keep away from heat, sparks, and open flame.

**Isopropyl-β-D-thiogalactopyranoside (IPTG)** may be harmful by inhalation, ingestion, or skin absorption. Wear appropriate gloves and safety glasses.

**KCl,** *see* **Potassium chloride**

**$K_3Fe^{III}(CN)_6$,** *see* **Potassium ferricyanide**

**$K_4Fe^{II}(CN)_6 \cdot 3H_2O$,** *see* **Potassium ferrocyanide**

$KH_2PO_4/K_2HPO_4/K_3PO_4$, *see* **Potassium phosphate**

**KI,** *see* **Potassium iodide**

$KMnO_4$, **Potassium permanganate**

**KOH,** *see* **Potassium hydroxide**

**Lactic acid** is corrosive and causes severe irritation and burns to any area of contact. It may be harmful by inhalation, ingestion, or skin absorption. Wear appropriate gloves and safety goggles. Do not breathe vapor or mist.

**Laser radiation**, both visible and invisible, can be seriously harmful to the eyes and skin and may generate airborne contaminants, depending on the class of laser used. High-power lasers produce permanent eye damage, can burn exposed skin, ignite flammable materials, and activate toxic chemicals that release hazardous by-products. Avoid eye or skin exposure to direct or scattered radiation. Do not stare at the laser and do not point the laser at someone else. Wear appropriate eye protection and use suitable shields that are designed to offer protection for the specific type of wavelength, mode of operation (continuous wave or pulsed), and power output (watts) of the laser being used. Avoid wearing jewelry or other objects that may reflect or scatter the beam. Some nonbeam hazards include electrocution, fire, and asphyxiation. Entry to the area in which the laser is being used must be controlled and posted with warning signs that indicate when the laser is in use. Always follow suggested safety guidelines that accompany the equipment and contact the local safety office for further information.

**Ion lasers:** Present a hazard due to high-voltage high-current power supplies. Always follow manufacturer's suggested safety guidelines.

**Ultraviolet lasers:** Present a hazard due to invisible beam, high-energy-radiation. Always use beam traps, scattered light shields, and fluorescent beam-finder cards.

**Blue-green lasers:** Present a hazard due to photothermal coagulation. Blue and green wavelengths are readily absorbed by blood hemoglobin.

**L-DOPA** is irritating to the eyes, skin, and respiratory system. It may be harmful by inhalation, ingestion, or skin absorption. Wear appropriate gloves and safety glasses.

**Lead, Pb**, is a toxic metal. It presents a long-term danger as lead accumulates in the liver and interferes with its function. Avoid contact with skin and wear appropriate gloves when handling.

**Leupeptin** (or its **hemisulfate**) may be harmful by inhalation, ingestion, or skin absorption. Wear appropriate gloves and safety glasses and use in a chemical fume hood.

**Liquid nitrogen** can cause severe damage due to extreme temperature. Handle frozen samples with extreme caution. Do not breathe the vapors. Seepage of liquid nitrogen into frozen vials can result in an exploding tube upon removal from liquid nitrogen. Use vials with O-rings when possible. Wear cryo-mitts and a face mask.

**LiCl,** *see* **Lithium chloride**

**Lithium chloride, LiCl,** is an irritant to the eyes, skin, mucous membranes, and upper respiratory tract. It may be harmful by inhalation, ingestion, or skin absorption. Wear appropriate gloves, safety goggles, and use in a chemical fume hood. Do not breathe the dust.

**Magnesium chloride, $MgCl_2$,** may be harmful by inhalation, ingestion, or skin absorption. Wear appropriate gloves and safety glasses and use in a chemical fume hood.

**Magnesium sulfate, $MgSO_4$,** may be harmful by inhalation, ingestion, or skin absorption. Wear appropriate gloves and safety glasses.

**Manganese chloride, $MnCl_2$,** may be harmful by inhalation, ingestion, or skin absorption. Wear appropriate gloves and safety glasses and use in a chemical fume hood.

**MeOH or $H_3COH$,** *see* **Methanol**

**β-Mercaptoethanol (2-Mercaptoethanol), $HOCH_2CH_2SH$,** may be fatal if inhaled or absorbed through the skin and is harmful if ingested. High concentrations are extremely destructive to the mucous membranes, upper respiratory tract, skin, and eyes. β-Mercaptoethanol has a very foul odor. Wear appropriate gloves and safety glasses and always use in a chemical fume hood.

**Methanol, MeOH or $H_3COH$,** is poisonous and can cause blindness. It may be harmful by inhalation, ingestion, or skin absorption. Adequate ventilation is necessary to limit exposure to vapors. Avoid inhaling these vapors. Wear appropriate gloves and goggles and use only in a chemical fume hood.

**Methotrexate** is a carcinogen and a teratogen. It may be harmful by inhalation, ingestion, or skin absorption. Exposure may cause gastrointestinal effects, bone marrow suppression, liver or kidney damage. It may also cause irritation. Avoid breathing the vapors. Wear appropriate gloves and safety glasses and always use in a chemical fume hood.

***N*-Methyl-2-pyrrolidinone (2-methyl-2-pyrrolidinone)** may be harmful by inhalation, ingestion, or skin absorption. Wear appropriate gloves and safety glasses and use in a chemical fume good. Do not breathe the vapors. Keep away from heat, sparks, and open flame.

**Methyl salicylate** is volatile and may be harmful by inhalation, ingestion, or skin absorption. Do not breathe the dust. Wear appropriate gloves and safety glasses and use in a chemical fume hood.

**$MgCl_2$,** *see* **Magnesium chloride**

**$MgSO_4$,** *see* **Magnesium sulfate**

**$MnCl_2$,** *see* **Manganese chloride**

**$Na_2CO_3$,** *see* **Sodium carbonate**

**$NaH_2PO_4/Na_2HPO_4/Na_3PO_4$,** *see* **Sodium phosphate**

$NaIO_4$, *see* **Sodium periodate**

$NaN_3$, *see* **Sodium azide**

**NaOCl,** *see* **Bleach**

**NaOH,** *see* **Sodium hydroxide**

$NaNO_2$, *see* **Sodium nitrite**

$Na_3VO_4$, *see* **Sodium orthovanadate**

**NBT,** *see* **4-Nitro blue tetrazolium chloride**

**Nickel chloride, $NiCl_2$,** is toxic and may be harmful by inhalation, ingestion, or skin absorption. Do not breathe the dust. Wear appropriate gloves and safety glasses.

$NiCl_2$, *see* **Nickel chloride**

**Nitric acid, $HNO_3$,** is volatile and must be handled with great care. It is toxic by inhalation, ingestion, and skin absorption. Wear appropriate gloves and safety goggles and use in a chemical fume hood. Do not breathe the vapors. Keep away from heat, sparks, and open flame.

**4-Nitro blue tetrazolium chloride (NBT)** is hazardous. Handle with care.

**6-Nitroveratryl chloroformate, NVOC-Cl,** is corrosive and causes burns to areas of contact. It may be harmful by inhalation, ingestion, or skin absorption. Do not breathe the dust. Wear appropriate gloves and safety goggles and use in a chemical fume hood.

**NVOC-Cl,** *see* **6-Nitroveratryl chloroformate**

**OCT** is composed of polyvinyl alcohol, polyethylene glycol, and dimethyl benzyl ammonium chloride. Follow the manufacturer's guidelines for handling OCT.

**Octane** is highly flammable in both liquid and vapor forms. Keep away from heat, sparks, and open flame. It may be harmful by inhalation, ingestion, or skin absorption. Wear appropriate gloves and safety glasses.

**Octopamine** may be harmful by inhalation, ingestion, or skin absorption. Exposure may cause an increase in blood pressure. Wear appropriate gloves and safety glasses. Do not breathe the dust.

$OsO_4$, *see* **Osmium tetroxide**

**Osmium tetroxide, $OsO_4$,** (osmic acid) is highly toxic if inhaled, ingested, or absorbed through the skin. Vapors can react with corneal tissues and cause blindness. There is a possible risk of irreversible effects. Wear appropriate gloves and safety goggles and always use in a chemical fume hood. Do not breathe the vapors.

**Paraformaldehyde** is highly toxic. It is readily absorbed through the skin and is extremely destructive to the skin, eyes, mucous membranes, and upper respiratory tract. Avoid breathing the dust. Wear appropriate gloves and safety glasses and use in a chemical fume hood. Paraformaldehyde is the undissolved form of formaldehyde.

**Pb,** *see* **Lead**

**PEG,** *see* **Polyethyleneglycol**

**Pepstatin A** may be harmful by inhalation, ingestion, or skin absorption. Wear appropriate gloves and safety glasses and use in a chemical fume hood.

**Phalloidin and Phalloidin CPITC-labeled, FITC-labeled, and TRITC-labeled** are extremely toxic and may be fatal by inhalation, ingestion, or skin absorption. Great care must be taken when using these compounds. Wear appropriate gloves and safety glasses and use in a chemical fume hood. Do not breathe the dust.

**Phenol** is extremely toxic, highly corrosive, and can cause severe burns. It may be harmful by inhalation, ingestion, or skin absorption. Wear appropriate gloves, goggles, and protective clothing and always use in a chemical fume hood. Rinse any areas of skin that come in contact with phenol with a large volume of water and wash with soap and water; do not use ethanol!

**Phenylenediamine** may be harmful by inhalation, ingestion, or skin absorption. Wear appropriate gloves and safety glasses and use in a chemical fume hood.

**Phenyl-methyl-sulfonyl fluoride (PMSF), $C_7H_7FO_2S$ or $C_6H_5CH_2SO_2F$,** is a highly toxic cholinesterase inhibitor. It is extremely destructive to the mucous membranes of the respiratory tract, eyes, and skin. It may be fatal by inhalation, ingestion, or skin absorption. Wear appropriate gloves and safety glasses and always use in a chemical fume hood. In case of contact, immediately flush eyes or skin with copious amounts of water and discard contaminated clothing.

**PMSF,** *see* **Phenyl-methyl-sulfonyl fluoride**

**Polyacrylamide** is considered to be nontoxic, but it should be treated with care because it may contain small quantities of unpolymerized material (see **Acrylamide**).

**Polyethyleneglycol (PEG)** may be harmful by inhalation, ingestion, or skin absorption. Avoid inhalation of powder. Wear appropriate gloves and safety glasses.

**Polyvinyl alcohol** may be harmful by inhalation, ingestion, or skin absorption. Wear appropriate gloves and safety glasses.

**Potassium chloride, KCl,** may be harmful by inhalation, ingestion, or skin absorption. Wear appropriate gloves and safety glasses.

**Potassium cacodylate,** *see* **Cacodylate**

**Potassium ferricyanide, K$_3$Fe(CN)$_6$,** may be harmful by inhalation, ingestion, or skin absorption. Wear appropriate gloves and safety glasses and use in a chemical fume hood. Keep away from strong acids.

**Potassium hydroxide, KOH and KOH/methanol,** can be highly toxic. It may be harmful by inhalation, ingestion, or skin absorption. Solutions are caustic and should be handled with great care. Wear appropriate gloves.

**Potassium iodide, KI,** may be harmful by inhalation, ingestion, or skin absorption. Wear appropriate gloves and safety glasses and use in a chemical fume hood.

**Potassium permanganate, KMnO$_4$,** is an irritant and a strong oxidant. It may form explosive mixtures when mixed with organics. Use all solutions in a chemical fume hood. Do not mix with hydrochloric acid.

**Potassium phosphate, KH$_2$PO$_4$/K$_2$HPO$_4$/K$_3$PO$_4$,** may be harmful by inhalation, ingestion, or skin absorption. Wear appropriate gloves and safety glasses. Do not breathe the dust.

**2-Propanol,** *see* **Isopropanol**

**Propylene oxide** is highly flammable, toxic, and may be carcinogenic. High concentrations are extremely destructive to the mucus membranes and upper respiratory tract. It may be harmful by inhalation, ingestion, or skin absorption. Wear appropriate gloves and safety glasses and use only in a chemical fume hood. Keep away from heat, sparks, and open flame.

***n*-Propyl gallate (NPG)** may be harmful by inhalation, ingestion, or skin absorption. It causes eye and skin irritation and is irritating to mucous membranes and upper respiratory tract. Wear appropriate gloves and safety glasses and use in a chemical fume hood.

**Quinacrine** may be fatal by inhalation, ingestion, or skin absorption. Wear appropriate gloves and safety glasses and use in a chemical fume hood.

**Radioactive substances:** While planning an experiment that involves the use of radioactivity, consider the physicochemical properties of the isotope (half-life, emission type, and energy), the chemical form of the radioactivity, its radioactive concentration (specific activity), total amount, and its chemical concentration. Order and use only as much as really needed. Always wear appropriate gloves, lab coat, and safety goggles when handling radioactive material. **X-rays** and **gamma rays** are electromagnetic waves of very short wavelengths either generated by technical devices or emitted by radioactive materials. They might be emitted isotropic from the source or may be focused into a beam. Their potential dangers depend on the time period of exposure, the intensity experienced, and the wavelengths used. Be aware that appropriate shielding is usually of lead or other similar material. The thickness of the shielding is determined by the energy(s) of the X-rays or gamma rays. Consult the local safety office for further guidance in the appropriate use and disposal of radioactive materials. Always monitor thoroughly after using radioisotopes. A convenient calculator to perform routine radioactivity calculations can be found at:

http://www.graphpad.com/www/radcalch.htm

**Reserpine** is toxic and may be carcinogenic and mutagenic. It may be harmful by inhalation, ingestion, or skin absorption. Wear appropriate gloves and safety glasses. Do not breathe the dust.

**Resins (epoxy and acrylic)** are suspected carcinogens. The unpolymerized components and dusts may cause toxic reactions including contact allergies with long-term exposure. Avoid breathing the vapors and dusts. Wear appropriate gloves and safety glasses and always use in a chemical fume hood. Sensitivity to these chemicals may develop with repeated contact.

**Saponin** is an irritant and may be harmful by inhalation, ingestion, or skin absorption. Do not breathe the dust. Wear appropriate gloves and safety glasses and use in a chemical fume hood.

**SDS,** *see* **Sodium dodecyl sulfate**

**Serotonin** may be harmful by inhalation, ingestion, or skin absorption. Wear appropriate gloves and safety glasses. Do not breathe the dust. Overexposure may cause reproductive disorders.

**Silane** may be harmful by inhalation, ingestion, or skin absorption. It is extremely flammable. Keep away from heat, sparks, and open flame. The vapor is irritating to the eyes, skin, mucous membranes, and upper respiratory tract. Wear appropriate gloves and safety glasses and always use in a chemical fume hood.

**Sodium azide, $NaN_3$,** is highly poisonous. It blocks the cytochrome electron transport system. Solutions containing sodium azide should be clearly marked. It may be harmful by inhalation, ingestion, or skin absorption. Wear appropriate gloves and safety goggles and handle it with great care.

**Sodium cacodylate** may be carcinogenic and contains arsenic. It is highly toxic and may be fatal by inhalation, ingestion, or skin absorption. It also may be teratogenic. Effects of contact or inhalation may be delayed. Do not breathe the dust. Wear appropriate gloves and safety goggles and use only in a chemical fume hood. *See also* **Cacodylate.**

**Sodium carbonate, $Na_2CO_3$,** may be harmful by inhalation, ingestion, or skin absorption. Wear appropriate gloves and safety glasses and use in a chemical fume hood.

**Sodium dodecyl sulfate (SDS)** is toxic, an irritant, and poses a risk of severe damage to the eyes. It may be harmful by inhalation, ingestion, or skin absorption. Wear appropriate gloves and safety goggles. Do not breathe the dust.

**Sodium hydroxide, NaOH, and solutions containing NaOH** are highly toxic and caustic and should be handled with great care. Wear appropriate gloves and a face mask. All other concentrated bases should be handled in a similar manner.

**Sodium hypochlorite, NaOCl,** *see* **Bleach**

**Sodium metabisulfite** may be harmful by inhalation, ingestion, or skin absorption. Wear appropriate gloves and safety glasses and use in a chemical fume hood.

**Sodium nitrite, NaNO$_2$,** is irritating to the eyes, mucous membranes, upper respiratory tract, and skin. It may be harmful by inhalation, ingestion, or skin absorption. Wear appropriate gloves and safety glasses and always use in a chemical fume hood. Keep away from acids.

**Sodium orthovanadate, Na$_3$VO$_4$,** may be harmful by inhalation, ingestion, or skin absorption. Wear appropriate gloves and safety glasses and use in a chemical fume hood.

**Sodium periodate, NaIO$_4$, and Sodium periodate meta** are strong oxidizers and may be harmful by inhalation, ingestion, and skin absorption. It causes eye, skin, mucous membrane, and respiratory tract irritation. Wear appropriate gloves and safety glasses and use in a chemical fume hood.

**Sodium phosphate, NaH$_2$PO$_4$/Na$_2$HPO$_4$/Na$_3$PO$_4$,** is an irritant to the eyes and skin. It may be harmful by inhalation, ingestion, or skin absorption. Wear appropriate gloves and safety goggles. Do not breathe the dust.

**Succinic anhydride** is a possible mutagen and is a severe eye irritant. It may be harmful by inhalation, ingestion, or skin absorption. Wear appropriate gloves and safety glasses and use only in a chemical fume hood. Do not breathe the dust.

**Sulfuric acid, H$_2$SO$_4$,** is highly toxic and extremely destructive to tissue of the mucous membranes and upper respiratory tract, eyes, and skin. It causes burns and contact with other materials (e.g., paper) may cause fire. Wear appropriate gloves, safety glasses, and lab coat and use in a chemical fume hood.

**Tannic acid** is an irritant and large amounts may cause liver and kidney damage. It may be harmful by inhalation, ingestion, or skin absorption. Wear appropriate gloves and safety glasses. Do not breathe the dust.

**TEMED,** *see N,N,N´,N´-***Tetramethylethylenediamine**

*N,N,N´,N´-***Tetramethylethylenediamine (TEMED)** is extremely destructive to tissue of the mucous membranes and upper respiratory tract, eyes, and skin. Inhalation may be fatal. Prolonged contact can cause severe irritation or burns. Wear appropriate gloves, safety glasses, and other protective clothing and use only in a chemical fume hood. Wash thoroughly after handling. Flammable: Vapor may travel a considerable distance to source of ignition and flash back. Keep away from heat, sparks, and open flame.

**TOTO-3 iodide** contains **Dimethylsulfoxide (DMSO),** *see* **Dimethylsulfoxide**

**Trichlorofluoroethane,** *see* **Trichlorotrifluoroethane**

**Trichlorotrifluoroethane** may be harmful by inhalation, ingestion, or skin absorption. Wear appropriate gloves and safety goggles and use in a chemical fume hood. Keep away from heat, sparks, and open flame.

**Triethanolamine** may be harmful by inhalation, ingestion, or skin absorption. Wear appropriate gloves and safety glasses and use only in a chemical fume hood.

**UV radiation** is dangerous, particularly to the eyes. To minimize exposure, make sure that the UV light source is adequately shielded. Wear protective goggles or a full safety mask that efficiently blocks UV light. Wear protective gloves when holding materials under the UV light source. UV radiation is also mutagenic and carcinogenic.

**Uranyl acetate** is toxic if inhaled, ingested, or absorbed through the skin. Wear appropriate gloves and safety glasses and use in a chemical fume hood.

**Urea** may be harmful by inhalation, ingestion, or skin absorption. Wear appropriate gloves and safety glasses.

**Vinblastine sulfate** is an irritant and may be harmful by inhalation, ingestion, or skin absorption. Wear appropriate gloves and safety glasses. Do not breathe the dust.

**X-gal,** *see* **5-Bromo-4-chloro-3-indolyl-β-D-galactopyranoside (BCIG)**

**X-rays,** *see* **Radioactive substances**

**Xylene** is flammable and may be narcotic at high concentrations. It may be harmful by inhalation, ingestion, or skin absorption. Wear appropriate gloves and safety glasses and use only in a chemical fume hood. Keep away from heat, sparks, and open flame.

**Xylene cyanol,** *see* Xylene

# Appendix 5

## Suppliers

$\mathrm{W}$ ITH THE EXCEPTION OF THOSE SUPPLIERS LISTED IN THE TEXT with their addresses, all suppliers mentioned in this manual can be found in the BioSupplyNet Source Book and on the Web site at:

http://www.biosupplynet.com

If a copy of BioSupplyNet Source Book was not included with this manual, a free copy can be ordered by using any of the following methods:

- Complete the Free Source Book Request Form at the Web Site at:

  http://www.biosupplynet.com

- E-mail a request to:

  info@biosupplynet.com

- Fax a request to (1) 516-349-5598

# Appendix 6

## Trademarks

The following trademarks and registered trademarks are accurate to the best of our knowledge at the time of printing. Further information can be obtained from individual manufacturers or other sources.

| TRADEMARK | COMPANY |
|---|---|
| ABI PRISM | The Perkin Elmer Corp. |
| Adobe Photoshop | Adobe Systems Inc. |
| BEEM | Better Equipment for Electron Microscopy Inc. |
| Big Dye Terminators | The Perkin Elmer Corp. |
| Cascade Blue | Molecular Probes Inc. |
| Centricon | Millipore Corp. |
| Centriprep | Millipore Corp. |
| Clorox | Clorox Co. |
| Cream-of-Wheat | Nabisco |
| Cy | Amersham Pharmacia Biotech |
| Drierite | W.A. Hammond Drierite |
| Dumont | Manufactures D'Outils Dumont S.A. |
| Durcupan | Fluka Chemie AG |
| ECL | Amersham Pharmacia Biotech |
| Elutip-d | Schleicher & Schuell Inc. |
| Falcon | Becton Dickinson and Co. |
| FileMaker Pro | Filemaker Inc. |
| Fluoromount-G | Southern Biotechnology Associates Inc. |
| FM | Molecular Probes Inc. |
| FuGene | Fugent LLC |
| Gastight | Hamilton Co. |
| GenePix | Axon Instruments Inc. |
| Genius | Boehringer Mannheim Corp. |
| Haemo-sol | Haemo-sol Inc. |
| Histoacryl | B. Braun Melsungen Aktiengesellschaft |
| Hybond | Amersham Pharmacia Biotech |
| InCert | FMC Corp. |
| Internet Explorer | SyNet Inc. |
| Isopac | Sigma Chemical Co. |
| Java | Sun Microsystems Inc. |

| | |
|---|---|
| Kimwipes | Kimberly-Clark Corp. |
| Kodak | Eastman Kodak Co. |
| Lowicryl | Chemische Werke Lowi Beteiligungs GmbH |
| Macintosh | Apple Computer Inc. |
| MEGAscript | Ambion Inc. |
| MetaMorph | Universal Imaging Corp. |
| MicroAmp | The Perkin-Elmer Corp. |
| Microcon | W.R. Grace & Co. |
| Micro-FLUOR | Dynatech Laboratories Inc. |
| Microsoft Excel | Microsoft Corp. |
| Microtiter | Cooke Engineering Co. |
| Microvap | Eaton-Williams Group |
| Milli-Q | Millipore Corp. |
| Miracloth | Calbiochem |
| Monoject | Sherwood Medical Co. |
| Mowiol | Hoechst Aktiengesellschaft |
| NALGENE | Nalge Co. |
| Nanogold | Nanoprobes Inc. |
| Netscape Communicator | Netscape Communications Corp. |
| NeutraLite | Belovo Chemicals |
| Nichrome | Driver-Harris Wire Co. |
| Nipagin | Nipa Laboratories Inc. |
| Nitex | Tobler, Ernst & Traber Inc. (Tetko Inc.) |
| Nonidet | Shell International Petroleum Co. Ltd UK |
| NucTrap | Stratagene |
| OliGreen | Molecular Probes Inc. |
| Parafilm | American National Can Co. |
| pBluescript | Stratagene |
| PC | International Business Machines Corp. |
| pClamp | California Institute of Technology |
| Pellet Pestle | Kontes Glass Co. |
| Permount | Fisher Scientific Co. |
| Perspex | ICI Chemicals and Polymers Ltd |
| PhosphorImager | Molecular Dynamics |
| PicoGreen | Molecular Probes Inc. |
| Picospritzer | General Valve Corp. |
| Pipetman | Rainin Instrument Co., Inc. |
| Pluronic | BASF Corp. |
| PTC-100 | MJ Research |
| PubMed | The National Library of Medicine |
| Pyrex | Corning Inc. |
| QIAquick | Qiagen Inc. |
| QuikHyb | Stratagene |
| RAIN-X | UNELKO Corp. |
| Rubbermaid | Rubbermaid Inc. |
| ScanArray | General Scanning Inc. |
| Scotch Tape | 3M Company |
| Seal-A-Meal | Dazey Corp. |
| Sephadex | Amersham Pharmacia Biotech |

| | |
|---|---|
| Sepharose | Amersham Pharmacia Biotech |
| Sigmacote | Sigma Chemical Co. |
| Speci-Mix | Thermolyne Corp. |
| SpeedVac | Savant Instruments Inc. |
| Stratalinker | Stratagene |
| Strip-Ease | Robbins Scientific Corp. |
| Styrofoam | Dow Chemical Co. |
| Superfrost | Sybron Corp. |
| Surfasil | Siltech Inc. |
| SYBR Green | Molecular Probes Inc. |
| Sylgard | Dow Corning Corp. |
| Teflon | E.I. DuPont deNemours and Co. |
| Texas Red | Molecular Probes Inc. |
| Tissue-Tek | Sakura Finetek |
| TopCount | Packard Instrument Co. |
| TOTO | Molecular Probes Inc. |
| Touch-o-matic | Teledyne Hanau |
| Triton | Rohm & Haas Co. |
| TSA (Tyramide signal amplification system) | NEN |
| Tupperware | Dart Industries Inc. |
| Tween | ICI Americas Inc. |
| Tygon | Norton Co. |
| UltraFree | Millipore Corp. |
| UNIX | Unix System Laboratories Inc. |
| Vaseline | Conopco Inc. DBA Chesebrough-Pond's USA Co. |
| Vectashield | Vector Laboratories Inc. |
| Vectastain | Vector Laboratories Inc. |
| Vyon | Porvair Sciences Ltd |
| Whatman | Whatman International Ltd |
| Windows | Microsoft Corp. |
| YOYO | Molecular Probes Inc. |

# Index

www.ingramcontent.com/pod-product-compliance
Lightning Source LLC
Chambersburg PA
CBHW080341220326
41598CB00030B/4571